DIGITAL CIRCUITS AND SYSTEMS

Douglas V. Hall

McGRAW-HILL PUBLISHING COMPANY
New York Atlanta Dallas St. Louis San Francisco
Auckland Bogotá Caracas Hamburg Lisbon London
Madrid Mexico Milan Montreal New Delhi Paris
San Juan São Paulo Singapore Sydney Tokyo Toronto

Sponsoring Editor: Brian Mackin
Editing Supervisor: Paul Farrell
Design and Art Supervisor: Janice Noto
Production Supervisor: Catherine Bokman

Cover design: Michael Jung
Cover photograph: The Image Bank

Library of Congress Cataloging-in-Publication Data

Hall, Douglas V.
    Digital circuits and systems.

    Bibliography: p.
    Includes index.
    1. Digital electronics. I. Title
TK7868.D5H297 1989   621.3815   88-27177
ISBN 0-07-025537-7

The manuscript for this book was prepared electronically.

Digital Circuits and Systems

1 2 3 4 5 6 7 8 9 0 DOWDOW 8 9 6 5 4 3 2 1 0 9

ISBN 0-07-025537-7

CONTENTS

CHAPTER 9

Synchronous Devices, Circuits, and Systems 234

CHAPTER 10

Digital Signal Generation, Transmission, and Interfacing 277

CHAPTER 11

A/D and D/A Converter Circuits and Applications 318

CHAPTER 12

Programmable Logic Devices and State Machines 360

CHAPTER 13

Digital Arithmetic Devices and Microprocessors 411

CHAPTER 14

Memory Devices and Systems 449

CHAPTER 15

Computer-Aided Engineering of Digital Systems 489

PREFACE

The world of digital electronics has undergone several major changes in the last twenty years.

One of the most important changes has been in the tools and techniques we use to design, prototype, test, and debug digital systems. Up until just a few years ago, we used a pencil and eraser to draw circuit diagrams. We built the prototype for a circuit by soldering or wire-wrapping components on a printed circuit board. We tested and debugged digital circuits with a simple oscilloscope, logic probe, and voltmeter.

Now we use a computer and a "mouse" to draw the circuit diagram for a digital system, and we print the diagram on a printer or plotter. We prototype a circuit by "simulating" its operation with a computer program. When a design is complete, we use a computer program to lay out a printed circuit board for the circuit and another program to compile a parts list for the circuit. The complexity of today's systems requires that we use a computer to test them and multichannel test instruments such as logic analyzers to troubleshoot them.

The point of all this is that an electronic engineer or electronic engineering technician in the 1990's must be skilled with computer-aided engineering tools and techniques as well as with basic digital theory and devices. The purpose of this book is to teach both the traditional and the new methods of analyzing, testing, troubleshooting, and designing digital systems. No previous electronics courses or training is required. Throughout the book we use a practical, hands-on approach so that you can develop some skill with real parts, circuits, and systems as you go. To give you an overview of the path we take in the book, here is a condensed list of chapter titles:

1. Atoms, Electrons, and Electricity
2. Binary Numbers and Logic Gates
3. Digital Devices and Data Sheets
4. Oscilloscopes, Pulse Generators, and Pulse Characteristics of Logic Gates
5. Analyzing, Testing, Troubleshooting, and Drawing Logic Circuits
6. Designing and Implementing Combinational Logic Circuits
7. MSI and LSI Combinational Logic Devices and Applications
8. Latches, Flip-Flops, Counters, and Logic Analyzers
9. Synchronous Devices, Circuits, and Systems
10. Digital Signal Generation, Transmission, and Interfacing
11. A/D and D/A Converter Circuits and Applications
12. Programmable Logic Devices and State Machines
13. Digital Arithmetic Devices and Microprocessors
14. Memory Devices and Systems
15. Computer-Aided Engineering of Digital Systems

Besides the topics suggested by this list of chapters, we have built the following strengths into the book:

Latest devices and conventions. Devices from the 74LS, 74HC, and 74ALS families are used for most circuit examples. Mixed logic conventions and IEEE/IEC symbols are introduced early and continued throughout the book, and Appendix C describes the IEEE/IEC symbol standards in detail. Also, in Chapters 13 and 14 we introduce you to the operation and memory interface of a 16-bit microprocessor.

Many examples of interfacing logic devices to each other and to a variety of analog and digital devices. Once an individual device is introduced and explained, it is used in conjunction with other devices in increasingly sophisticated systems.

Extensive emphasis on use of test equipment and systematic troubleshooting techniques. We show you how to use a logic analyzer on a wide variety of circuits and how to use a digital storage oscilloscope. Also, we introduce concepts of automatic testing of digital circuits.

A system approach throughout, rather than just individual devices, small pieces of systems, or generic block diagrams. As much as possible, we have tried to use example systems which you are likely to encounter in your daily life. Some of these examples include a frequency counter, a digital voltmeter, and a compact audio disc player. Also, throughout the book we show circuit diagrams for many complete digital systems that you can build and experiment with.

Detailed discussions of the actual use of computer-aided engineering tools. We show you how to use schematic capture programs in Chapter 5, and we show you how to use simulation programs and printed circuit board layout programs in Chapter 15. Incidentally, we have put most of the discussion of CAE tools in Chapter 15 so you can see how they are used together, but you may find it interesting to read the section of Chapter 15 on simulation after you read Chapter 9, so you can see how a circuit such as the frequency counter in Chapter 9 is "software breadboarded."

Implementation of combinational and sequential logic circuits in programmable logic devices, such as PALs, using commonly available computer programs. Also, in a section of Chapter 15 we discuss the computer-based tools used to design custom application-specific integrated circuits, or ASICs, with logic-cell-array, gate-array, standard-cell, and full-custom methods.

Continued repetition and reinforcement of basic concepts throughout. Karnaugh maps, for example, are used in several chapters throughout the book. The logic analyzer is introduced in Chapter 8, then used in Chapters 9, 12, 13, and 14. Computer simulation is introduced in Chapter 12, then expanded in Chapter 15.

In-depth coverage of topics. This book is large because we teach not only basic concepts, but a wide variety of circuits and applications. You can choose those materials which are appropriate for the time and equipment you have available.

As we said above, this book contains many circuits with part numbers and pin numbers so that you can build them and experiment with them. Working with test equipment and experimenting with circuits is an important step in becoming fluent with digital technology. For more practice, we highly recommend the exercises in the accompanying laboratory manual.

For further information about some of the key topics in this book, consult the materials referenced in the Bibliography. Also, if you want to continue your study of microprocessors beyond the hardware introduction given in Chapters 13 and 14 here, consult the sequel to this book, *Microprocessors and Interfacing, Programming and Hardware.*

At this point I wish to express my sincere thanks to the

people who helped make this book a reality. First and foremost my thanks to Pat Hunter, who acted as a sounding board for many of my ideas, worked out the answers to the end-of-chapter problems to make sure the intentions were clear, proofread the manuscript and prepared it for publication, and made many other contributions too numerous to mention. Thanks to Dennis Berkheiser of Electronics Diversified Inc. for his meticulous scrutiny of the manuscript. Thanks to my students, whose candid feedback helped me fill in many holes and smooth some rough spots. Thanks to Robert Silva of Middlesex Community College and Benjamin Suntag of Plessey Electronic Systems Division, who reviewed the manuscript and contributed several valuable suggestions. Finally, thanks to my wife, Rosemary, and my children, Linda, Brad, Mark, Lee, and Katie, for their patience and encouragement during the long effort of writing this book.

Just as the electronics industry continues to evolve, this book must continue to evolve and improve if it is to remain useful. If you have any questions about topics in the book or suggestions for improving the book, please feel free to contact me or the publisher.

DOUGLAS V. HALL

1 Atoms, Electrons, and Electricity

This book is written so that it can be used by someone with no previous knowledge of electricity or electronics. In this chapter we introduce you to the major concepts of dc electricity, ac electricity, and electrical safety that you need to understand and work with the circuitry discussed in the rest of the book. We have tried not to burden you with unnecessary details at this point. If you want more information about the topics in this chapter, consult an ac/dc circuits book such as the latest edition of Bernard Grob's *Basic Electronics*.

OBJECTIVES

At the conclusion of this chapter you should be able to:

1. Draw a diagram for the Bohr model of an atom.

2. Describe several ways in which electricity is produced.

3. Define the terms *static electricity, direct current,* and *alternating current.*

4. Draw a diagram showing how a power supply or battery is connected to a circuit.

5. Define the terms *voltage, current,* and *resistance.*

6. Define the terms *volt, millivolt, amp, milliamp, microamp, ohm, kohm,* and *Mohm.*

7. Draw diagrams showing how you can connect: *(a)* an ammeter to measure the current through a part of an electric circuit; *(b)* a voltmeter to measure the voltage across a circuit; and *(c)* an ohmmeter to measure the resistance of a part of an electric circuit.

8. Given a color-coded resistor, determine its value in ohms.

9. Given the value of a resistor in ohms and the voltage across the resistor, calculate the current through the resistor.

10. Given the voltage across a resistor and the current through the resistor, calculate the power dissipated by the resistor.

11. Distinguish between a series circuit and a parallel circuit.

12. Calculate the voltage across each resistor in a circuit with two resistors connected in series.

13. Draw a graph showing one cycle of an ac voltage or current waveform.

14. Describe the function of fuses and circuit breakers in electric circuitry.

15. Describe the safety precautions required when working with ac and dc electric circuits.

BASIC ATOMIC STRUCTURE, ELEMENTS, AND COMPOUNDS

The Bohr Model of the Atom

As you probably learned in general science class, most liquids, solids, and gases are made up of *atoms.* Atoms are unbelievably small, so the objects you are used to working with contain very large numbers of atoms. For example, a piece of aluminum weighing only 26.98 grams (a little less than an ounce) contains approximately 602,000,000,000,000,000,000,000, or 6.02×10^{23}, aluminum atoms.

No one knows what an atom really looks like, so we use various "models" to explain their observed properties. The model or mental picture used depends on the properties we are trying to explain. For this book the simple Bohr model is sufficient, so we will use it.

Atoms are made up of three major types of particles, called *protons, neutrons,* and *electrons.* Protons and, for most kinds of atoms, one or more neutrons are contained in the center part of the atom, called the *nucleus.* Electrons are thought of as circling around the nucleus in orbits, similar to the way planets circle around the sun. This simple model of an atom, with protons and neutrons in the nucleus and electrons circling around nucleus in orbits, is called the *Bohr model.* Figure 1-1a shows the Bohr model for a hydrogen atom.

The number of protons in the nucleus of an atom determines what type of atom it is. An atom with 13 protons in its nucleus, for example, is an aluminum atom.

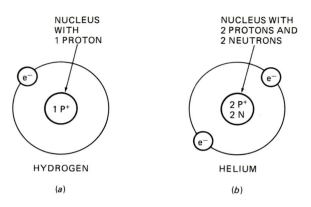

NUCLEUS
WITH
1 PROTON

1 P⁺

HYDROGEN

(a)

NUCLEUS WITH
2 PROTONS AND
2 NEUTRONS

2 P⁺
2 N

HELIUM

(b)

FIGURE 1-1 Bohr model of hydrogen and helium atoms. (a) Hydrogen atom. (b) Helium atom.

An atom with 29 protons in its nucleus is a copper atom. The number of protons in the nucleus of an atom is referred to as the *atomic number* for that atom. Substances that contain only one type of atom are called *elements*. Table 1-1 shows the atomic number and some other useful data for common elements.

You can determine the approximate number of neutrons in the nucleus of a particular type of atom by subtracting the number of protons from the atomic weight shown for that element in Table 1-1.

In any element the atoms are normally neutral. This means that the number of electrons circling around the nucleus is equal to the number of protons in the nucleus. As shown in Figure 1-1a, hydrogen atoms have one proton in their nucleus and one electron circling around the nucleus. Helium atoms, as shown in Figure 1-1b, have two protons in their nucleus and two electrons circling around the nucleus.

Now, as you look at these atom models, the question may occur to you, "Why don't the electrons simply fly off in a straight line?" A first thought might be that gravity holds the electrons in their orbits, just as gravity pulls the planets into orbits around our sun. The gravitational force between electrons and the nucleus, however, is too small to hold the electrons in their orbits. Electrons are held in their orbits by an *electric* force. Here's how this force works.

Electrons have a negative electric charge, and protons have a positive electric charge. Electric charge is a basic characteristic which can really only be described in terms of how particles with charge affect each other, or how moving charged particles are affected by the magnetic field near a magnet. All you need to know about electric charge at this point is:

Particles with the same electric charges *repel*, or push away from, each other.

Particles with different electric charges *attract* each other.

Since electrons each have a negative charge and the protons in the nucleus each have a positive charge, the electrical attraction between them holds the electrons in orbits. Because the electrons are moving, the electric

force of attraction pulls them into orbits, rather than pulling them directly into the nucleus. The situation is similar to that with the planets, which are pulled into orbits rather than pulled into the sun (we hope).

Incidentally, the neutrons, which are in the nucleus, have no charge, but they do help supply the "glue" which holds the positively charged protons together in the nucleus.

Now that you have an overview model of an atom we can take a closer look at how the electrons are distributed in orbits, or *shells*, around the nuclei of different types of atoms. This distribution determines the chemical and electrical properties of a particular type of material.

ELECTRON SHELLS AND CHEMICAL COMPOUNDS

Table 1-1 shows the distribution of the electrons in shells for the first 50 types of atoms, or *elements*. As you can see in the table, the major shells are identified with the letters K, L, M, N, and O. The K shell, closest to the nucleus, can hold a maximum of two electrons, and the L shell can hold a maximum of eight. For the elements shown in Table 1-1, the M shell and the N shell each hold a maximum of 18 electrons. As you go up in atomic number, the shells usually fill from the innermost shell outward, just like filling a hotel from the ground floor as more guests sign in. Using Table 1-1, you should be able to determine that oxygen atoms, for example, have eight protons in their nucleus, two electrons in their K shell, and six electrons in their L shell.

Atoms which have the same number of electrons and protons are electrically neutral, but some types of atoms aren't very "happy," or stable, that way. To make a long story short, most atoms like to have a full outer shell of electrons. To get in that state, they will donate electrons to some other type of atom, share electrons with some other type of atom, or accept donated electrons from some other type of atom—whichever they can most easily do. Here's an example of how this works.

According to Table 1-1, a chlorine atom has seven electrons in its outer (M) shell. To complete this outer shell, it would like to have eight. Again, according to Table 1-1, a sodium atom has one electron in its outer (M) shell. If it donates this one electron to some other atom, its (L) shell will then be the outer shell. And, since the L shell has its full number of eight electrons, the sodium atom will then be "happier." Now, when a sodium atom collides with a chlorine atom, the sodium atom donates an electron to the chlorine atom, as shown in Figure 1-2. The donated electron completes the outer shell of the chlorine atom, and the sodium atom is left with a complete outer shell. The resulting combination of sodium and chlorine is referred to as a chemical *compound*. It is given the name *sodium chloride* and represented with the formula $NaCl$. Sodium chloride, incidentally, is the chemical name for common table salt.

Another example of how chemical compounds are formed is the reaction that takes place between hydrogen atoms and oxygen atoms. According to Table 1-1, an oxygen atom contains six electrons in its outer shell, so it would like to get two more to complete the shell.

TABLE 1-1
ATOMIC NUMBER AND SHELL ARRANGEMENT OF ELECTRONS
FOR FIRST FIFTY ELEMENTS

Atomic Number	Element and Symbol		Shells					Atomic Weight
			K	L	M	N	O	
1	Hydrogen	H	1					1.00
2	Helium	He	2					4.00
3	Lithium	Li	2	1				6.94
4	Beryllium	Be	2	2				9.01
5	Boron	B	2	3				10.81
6	Carbon	C	2	4				12.01
7	Nitrogen	N	2	5				14.01
8	Oxygen	O	2	6				16.00
9	Fluorine	F	2	7				19.00
10	Neon	Ne	2	8				20.18
11	Sodium	Na	2	8	1			22.99
12	Magnesium	Mg	2	8	2			24.31
13	Aluminum	Al	2	8	3			26.98
14	Silicon	Si	2	8	4			28.09
15	Phosphorus	P	2	8	5			30.97
16	Sulfur	S	2	8	6			32.06
17	Chlorine	Cl	2	8	7			35.45
18	Argon	Ar	2	8	8			39.95
19	Potassium	K	2	8	8	1		39.10
20	Calcium	Ca	2	8	8	2		40.08
21	Scandium	Sc	2	8	9	2		44.96
22	Titanium	Ti	2	8	10	2		47.90
23	Vanadium	V	2	8	11	2		50.94
24	Chromium	Cr	2	8	13	1		52.00
25	Manganese	Mn	2	8	13	2		54.94
26	Iron	Fe	2	8	14	2		55.85
27	Cobalt	Co	2	8	15	2		58.93
28	Nickel	Ni	2	8	16	2		58.71
29	Copper	Cu	2	8	18	1		63.54
30	Zinc	Zn	2	8	18	2		65.37
31	Gallium	Ga	2	8	18	3		69.72
32	Germanium	Ge	2	8	18	4		72.59
33	Arsenic	As	2	8	18	5		74.92
34	Selenium	Se	2	8	18	6		78.96
35	Bromine	Br	2	8	18	7		79.91
36	Krypton	Kr	2	8	18	8		83.80
37	Rubidium	Rb	2	8	18	8	1	85.47
38	Strontium	Sr	2	8	18	8	2	87.62
39	Yttrium	Y	2	8	18	9	2	88.91
40	Zirconium	Zr	2	8	18	10	2	91.22
41	Niobium	Nb	2	8	18	12	1	92.91
42	Molybdenum	Mo	2	8	18	13	1	95.94
43	Technetium	Tc	2	8	18	14	1	99.00
44	Ruthenium	Ru	2	8	18	15	1	101.07
45	Rhodium	Rh	2	8	18	16	1	102.91
46	Palladium	Pd	2	8	18	18		106.40
47	Silver	Ag	2	8	18	18	1	107.87
48	Cadmium	Cd	2	8	18	18	2	112.40
49	Indium	In	2	8	18	18	3	114.82
50	Tin	Sn	2	8	18	18	4	118.69

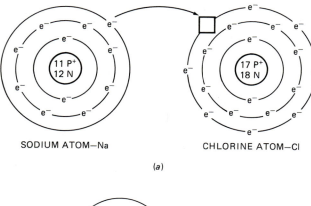

SODIUM ATOM—Na

CHLORINE ATOM—Cl

(a)

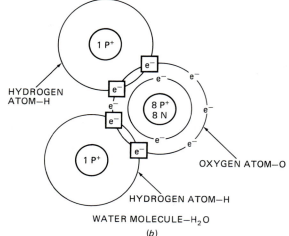

HYDROGEN ATOM—H

OXYGEN ATOM—O

HYDROGEN ATOM—H

WATER MOLECULE—H_2O

(b)

FIGURE 1-2   Formation of chemical compounds. (a) Sodium chloride. (b) Water.

Hydrogen has one electron in its outer shell, so it would like to get one more electron to complete its outer shell. To combine with an oxygen atom, a hydrogen atom has to share a pair of electrons, as shown in Figure 1-2b. Since oxygen needs two additional electrons, it effectively fills its outer shell by sharing a pair of electrons with each of two hydrogen atoms. The resulting compound then contains two hydrogen atoms for each oxygen atom. We might call this compound dihydrogen oxide, but usually we just call it *water*, or H_2O. To see if you understand how this works, try figuring out the compound that will be formed from aluminum and chlorine. Chlorine atoms need one additional electron to complete their outer shell. Aluminum atoms have three electrons in their outer shell that they can share. Therefore, aluminum will form the compound aluminum chloride ($AlCl_3$), with three chlorine atoms for each aluminum atom. Incidentally, elements such as helium, neon, argon, and krypton, which already have complete outer shells, are referred to as *inert* because they do not usually form compounds with other elements.

About now you may be wondering if a chemistry book, instead of an electronics book, was accidentally put in these covers. Don't worry, the brief introduction to atoms in this section will help you understand, in later sections, how electricity is produced and how it is conducted from one place to another.

GENERATING ELECTRICITY

Work and Gravitational Potential Energy

Before discussing electricity directly, we need to review how energy can be stored by bodies that are attracted to each other by gravity.

Now, suppose that you want to lift a rock off the ground. The rock is held on the ground by gravitational force. In order to lift the rock, you have to exert a force which overcomes the gravitational force. The amount of work you have to do to lift the rock depends on the weight of the rock and the height that you lift it. If you hold the rock up off the ground, the work you did in lifting the rock remains stored in the rock. When you allow the rock to fall back to the ground, the work you put into lifting the rock will be released as energy. The energy may go into producing sound, making a dent in the ground, or some other form. The work stored in the rock while you are holding it up is called *potential energy* because the rock has the potential to do work as it falls back to the ground.

A more useful example of storing gravitational potential energy might be raising the weights in a "grandmother" clock. The work or energy that you put into raising the weights runs the clock for some period of time.

The main concepts for you to get out of this are:

It takes work or energy to separate two bodies which attract each other.

The work put into separating the bodies will remain stored in the bodies as potential energy until they are allowed to come together again.

When the bodies are allowed to come together again, the potential energy will be released in some form, such as heat, light, sound, mechanical motion, etc.

Now let's see how all this relates to electricity.

Electric Potential Energy

In an earlier section we discussed how electrons are held in orbits around the nucleus by the electric force of attraction between the negatively charged electrons and the positively charged protons in the nucleus. Now, suppose that we want to move an electron away from its atom. Because of the attractive force, we will have to do work on the electron in order to do this, just as we have to do work overcoming gravity to lift a rock off the ground. The energy stored in the separated charges is referred to as *electric potential energy.* If we allow the electron to fall back into its old orbit, the work we put into separating the electron will be released in some form. The energy may, for example, be given off in the form of light or heat.

Separating electrons from their atomic nuclei in some way is referred to as *generating electricity.* Separating one electron from its nucleus does not store very much energy, so in most cases we separate billions and billions of outer-shell electrons from their nuclei when we generate electricity. As the example we gave before showed, 26.98 grams of aluminum contain 6.02×10^{23} atoms, so in a small amount of material there are a great many electrons available to be separated, even though only the outer-shell electrons are usually involved in this process. Later we will relate the number of electrons separated to common electrical units.

Methods of Generating Electricity

As you are probably aware, there are several ways to generate electricity. Here a few common ways are described.

RUBBING DIFFERENT MATERIALS TOGETHER

Rubbing certain materials together actually causes electrons to be removed from one material and deposited on the other. For example, if you walk across a rug in dry weather, your shoes may rub electrons off the rug and allow them to accumulate on your body. This type of electricity is often referred to as *static electricity* because the accumulated electrons just stay there until you give the electrons a path to "get home." The spark you feel and perhaps see when you touch another body is caused by the electrons leaving your body in an attempt to get back to their atoms again. Static electricity may also be produced by running a comb through your hair or pulling off a wool sweater. Thunder and lightning are caused by the release of static electricity built up on clouds by air currents.

CHEMICALLY

Batteries such as the ones shown in Figure 1-3 generate electricity by a chemical reaction. As a result of the reaction of two or more chemicals in the battery, some of the electrons are removed from the material connected to one terminal of the battery and deposited on the material connected to the other terminal of the battery. The energy required to separate the electrons from their shell comes from the chemical reaction.

The terminal with excess electrons is referred to as the *negative terminal*. The terminal which is deficient in electrons is referred to as the *positive terminal*. The chemical reaction prevents the electrons from returning to the nuclei through the battery. Therefore, the electric potential energy stored in a battery remains there until you supply an external path for the electrons to get back to the atoms from which they were separated. We will discuss this much more fully a little later.

Batteries produce what is referred to as *direct current*, or *dc*, electricity. The name comes from the fact that when you make an external connection to allow the electrons to "get back home" again, the electrons always flow in one direction through the connection. This will make more sense to you when you contrast it to the behavior of alternating current electricity (discussed later). Incidentally, some batteries, such as the battery in your car, can be "discharged" to supply the energy needed to start the car, then "recharged" by an electric generator called an *alternator*, which is turned by the engine as it runs.

(a)

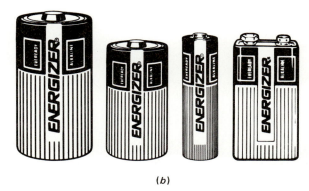

(b)

FIGURE 1-3   Photographs of common types of batteries. *(a)* Car battery. *(Courtesy of Pacific Chloride.)* *(b)* Different flashlight batteries. *(Courtesy of Eveready Battery Company.)*

LIGHT

In some substances, such as the element cesium, the outer electrons are so loosely attached to the atoms that a beam of light can knock the electrons off the atoms. These electrons can be collected and used. The production of electricity by light is often called the *photoelectric effect*.

Another way that electricity can be produced directly from light is with *solar cells*. These are made from two types of specially prepared silicon and produce a dc voltage when light is shined on them. Large panels of solar cells are used on space satellites to produce the electric energy needed to run pumps, radios, and other equipment.

ALTERNATORS AND GENERATORS

A magnet produces what is called a *magnetic field* in the space around it. It is this magnetic field which causes a magnet to be attracted to or repelled by another magnet without actually touching the other magnet.

If a piece of wire is moved through a magnetic field, some of the electrons in the wire will be separated from their atoms and moved to one end of the wire. If a coil of wire is rotated in a magnetic field as shown in Figure 1-4, then electrons will alternately be "piled up," first at one end of the coil and then at the other. In other words, during one half of each rotation, electrons will flow toward one end of the coil and make that end negative. During the second half of each rotation, the electrons

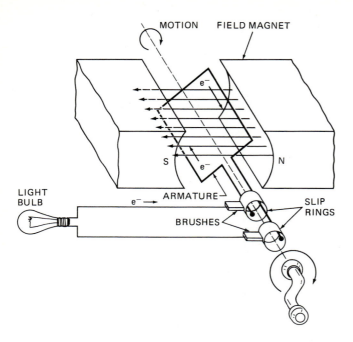

FIGURE 1-4   A simple electric generator.

an external connection is made between the negative terminal and the positive terminal, giving the electrons a path "home."

As the electrons return to the positive terminal, they release their stored electric potential energy in some form. If we use a light bulb as the path for the electrons to get back to the positive terminal, then the electric potential energy they release as they pass through the bulb will be given off as heat and light. If we use a motor as the path for the electrons to get back to the positive terminal, then the electric potential energy released by the returning electrons will go into turning the motor.

The connection of a source of electricity and some path for the electrons to pass through is called an *electric circuit.* Figure 1-5*a* shows a simple electric circuit consist-

will flow toward the other end of the coil and make *that* end negative. In either case, the end of the wire that has had some electrons removed will be positive. If an external conductor such as a light bulb is connected between the ends of a rotating coil, as shown in Figure 1-4, the electrons will alternately flow first in one direction through the bulb and then in the reverse direction. The electricity produced by a generator of this type is referred to as *alternating current,* or *ac,* electricity.

The electric generators used in coal, oil, nuclear, or water power plants effectively rotate large coils of wire in magnetic fields to produce the ac electricity you use in your home. In your car, an ac generator called an *alternator* produces the electricity needed to run the car and recharge the battery. The rotating coil of wire in a magnetic field in the alternator produces ac electricity, and devices called *diodes* convert the ac to the dc actually needed to charge the battery, etc.

Now that you have an overview of some of the ways that electricity is produced, we will show you how we measure and work with electricity.

ELECTRIC CIRCUITS AND UNITS
Electric Circuits and Circuit Diagrams

For this section we will work with direct current, or dc, electricity, such as that produced by a battery. As described in a previous section, a battery uses a chemical reaction to separate electrons from the material connected to one of its terminals and piles them up on the material connected to the other terminal. The terminal with an excess of electrons is called the *negative terminal* and the other terminal is called the *positive terminal.* Except for some small internal leakage, the electric potential energy stored in a battery remains there until

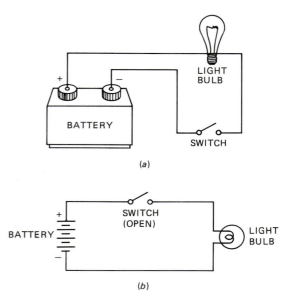

FIGURE 1-5   Simple electric circuit with battery, switch, and light bulb. *(a)* Circuit. *(b)* Schematic.

ing of a battery, a switch, and a light bulb. Closing the switch in this circuit completes a path for electrons from the negative terminal of the battery to the positive terminal through the light bulb.

Now, when you want to represent an electric circuit on paper, you don't have to draw pictures of the battery and the electrical components connected to the battery. You use symbols to represent each part of the circuit. Figure 1-5*b* shows a diagram for a simple electric circuit with a battery, a switch, and a light bulb. Circuit diagrams such as this are often called *schematics.* The schematic in Figure 1-5*b* represents a circuit similar to the one for the circuit used in a common flashlight.

In Figure 1-5*b* the short line on the symbol used for a battery represents the negative terminal. The long line on the symbol represents the positive terminal. If the switch is closed, electrons will flow from the negative terminal to the positive terminal through the light bulb, and the light will be on. If the switch is open, there will be no path for the electrons through the light bulb, so the light will be off. Now that you have an intuitive feel for what is going on in an electric circuit, it is time to introduce you to some of the basic measurements we make and the units we use in electric circuits.

Electric Charge and Current

We said previously that an electron has an electric charge. The basic unit of electric charge is the *coulomb*. One coulomb of charge is equal to 6.25×10^{18}, or 6.25 billions of billions of electrons. As you can see, a coulomb of charge passing through a light bulb represents a large number of electrons. In most electric circuits we don't work directly with the total number of coulombs that pass through the circuit, but with the rate (in coulombs per second) at which electrons are passing through the circuit.

The rate at which electrons flow through a light bulb or some other device is referred to as *electric current*. The basic unit of electric current is the *ampere*, or *amp* (*A*), which is defined simply as 1 coulomb of charge per second. The instrument we use to measure the current in a circuit is called an *ammeter*. Figure 1-6 shows how

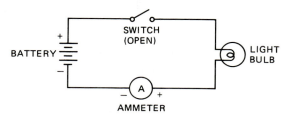

FIGURE 1-6 Connecting an ammeter in a circuit to measure current.

an ammeter can be inserted in our simple circuit to measure the electric current through the light bulb. Note that the ammeter is connected in the circuit so that the electrons must pass through it on their way back to the positive terminal of the battery. This is commonly referred to as connecting the ammeter *in series* with the circuit. To help you remember how to connect an ammeter in a circuit, think of an ammeter as functioning like a turnstile used to count the number of people who enter a stadium.

Also note in Figure 1-6 that the ammeter is connected in the circuit so that its positive terminal is nearest the positive terminal of the battery. This is referred to as connecting the meter with the correct *polarity*. Older types of ammeters may be destroyed if they are connected with the wrong polarity.

For most of the circuits that you will work with in this book, the currents will be considerably smaller than 1 A. Therefore, you will often use the standard prefixes shown in Table 1-2 to represent the size of currents. The prefix *milli*, for example, means $\frac{1}{1000}$, so instead of specifying a current as 0.020 A or $\frac{20}{1000}$ A, you can specify it as 20 milliamps, or 20 mA. An easy way to convert a current expressed in amps to its equivalent value in milliamps is to simply move the decimal point three places to the right and add *milli* or *m* before amps. The prefix micro

means $\frac{1}{1,000,000}$, or 10^{-6}, so you can represent a current of 0.000400 A as 400 microamps, or 400 μA. As you can see in Table 1-2, the prefix nano means 10^{-9}, or one billionth, and the prefix pico means 10^{-12}, or one trillionth. Believe it or not, we often refer to currents in the nanoamp and picoamp ranges.

The main points of this section are:

Current is the rate at which electrons flow through a circuit.

Current is measured by an ammeter, which is connected in the circuit in series so that the electrons must pass through it on their way back to the positive terminal.

Current is measured in units of amps (A), milliamps (mA), or microamps (μA).

By now the question that may have occurred to you is, "What determines how much current flows through a circuit?" In the next two sections we discuss the factors which determine the amount of current in a circuit.

Electric Voltage

One of the factors which determines how much current that a source of electricity causes to flow in a circuit is the *voltage*. The voltage of a battery or some other source of electricity represents the amount of electric potential energy that the electrons from that source have. A simple analogy may help you understand how the amount of potential energy affects the amount of current through a circuit.

Electric current through a circuit is often compared to the flow of water through a pipe. One factor that determines the rate at which water flows through a pipe is the pressure exerted on the water. The greater the pressure, the more water that will flow. In a gravitational water system the water pressure is determined, for the most part, by the height of the reservoir in which the water has been stored. In other words, the amount of water that will flow through the pipe is related to how much potential energy the water has. Likewise, the more potential energy the electrons have, the more electrons that will flow through a given circuit.

The unit of measurement for voltage is the *volt* (*V*), and 1 V is defined as 1 joule per coulomb. A joule is a unit of energy, so if a 1-V source of electricity is applied to a circuit, then each coulomb of electrons that flows through the circuit will release 1 joule of energy in some form. If a 6-V source of electricity is connected to the same circuit, then each coulomb of charge that flows through the circuit from that source will release 6 joules of energy in some form. Also, since the electrons from the 6-V source have six times as much energy, six times as many electrons will flow through the circuit in a given time. Thinking back to the water analogy may help you understand this.

To measure the electric voltage in a circuit, we use a *voltmeter*. Figure 1-7 shows how to connect a voltmeter in a circuit to measure the voltage of the source. Note

TABLE 1-2
PREFIXES COMMONLY USED WITH ELECTRICAL UNITS

Prefix	Symbol	Multiplier		Example	
Tera	T	1,000,000,000,000 or	10^{12}	4 Tohm	= 4,000,000 Mohm
Mega/Meg	M	1,000,000 or	10^6	2.2 Mohm	= 2,200,000 ohms
kilo	k	1,000 or	10^3	34.5 KV	= 34,500 volts
milli	m	0.001 or	10^{-3}	0.098 V	= 98 mV
micro	μ	0.000001 or	10^{-6}	20 μamps	= 0.000020 amps
nano	n	0.000000001 or	10^{-9}	100 ns	= 0.0000001 second
pico	p	0.000000000001 or	10^{-12}	100 ps	= 0.1 nsecond
femto	f		10^{-15}	1 μamp	= 10^9 femtoamps

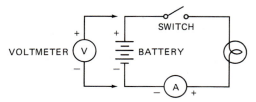

NOTE: + OF METER TOWARD + OF BATTERY
       − OF METER TOWARD − OF BATTERY

FIGURE 1-7   Connecting a voltmeter in a circuit to measure voltage.

carefully that a voltmeter is connected *across*, or in *parallel with*, the circuit. To help you remember how a voltmeter is connected, think of measuring voltage as measuring how high the reservoir is above the valley. For contrast, look at Figure 1-6 to refresh your memory as to how an ammeter is connected in a circuit. An ammeter is connected in series so that it can determine how many electrons pass through. A voltmeter is connected in parallel to determine how much potential energy the electrons have.

Voltage sources such as batteries can be connected in series, or "stacked up," to increase the voltage available for a circuit. Figure 1-8 shows how two 1.5-V batteries

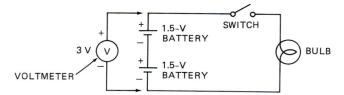

FIGURE 1-8   Circuit showing how batteries can be connected in series (stacked) to give greater voltage.

can be connected to supply 3 V to a circuit. This is the actual circuit for a flashlight containing two C-size or two D-size batteries. Note that the negative terminal of the top battery is connected to the positive terminal of the lower battery so that the potential energy of the top battery adds onto the potential of the bottom battery.

This is just like piling up bricks to make a wall higher. When voltage sources are connected in series in this way, the total voltage is simply equal to the sum of the individual voltages, just as the total height of a brick wall is equal to the sum of the heights of the individual courses.

In the laboratory or in electronic instruments we usually use a *power supply* to produce the dc voltages needed for our electronic circuits. A power supply is a device which converts the alternating current electricity available from wall outlets to direct current and maintains the dc-output voltage at the desired value. Many of the circuits in this book, for example, require a dc voltage source of 5 V. You will probably use a power supply such as the one shown in Figure 1-9 to produce this voltage for your circuits from the 120-V ac wall outlets. Laboratory power supplies such as the one in Figure 1-9 have large screw-type *terminals* which allow you to easily connect and disconnect the wires running to your circuits. Most laboratory power supplies produce several commonly needed voltages, such as 5 V, 12 V, and 15 V. The power supply shown in Figure 1-9 also has front panel knobs which allow you to adjust some of the voltage outputs precisely to any needed voltage.

The prefixes in Table 1-2 are also used with volts to simplify the specification of very large or very small voltages. For example, the large electric power lines which you see crossing the countryside may have a voltage of 500,000 V. For simplicity this voltage may be expressed as 500 kilovolts, or 500 kV. For voltages less than 1 volt, the *milli*, *micro*, and *nano* prefixes are used. A voltage of 0.015, or $^{15}/_{1000}$ V, for example, may be expressed as 15 mV.

To summarize this section, then:

The voltage of a source of electricity is a measure of the amount of electrical potential energy that the electrons from that source have. Voltage is like "electrical pressure," so the voltage of a source determines how much current it will cause to flow in a given circuit. If a voltage source with a larger voltage is connected to a given circuit, more electrons will flow through the circuit. Also, a higher voltage means the electrons have more energy, so the electrons flowing through the circuit will release more energy in some form.

FIGURE 1-9 Photograph of a common laboratory power supply. *(Courtesy of Tektronix Inc.)*

Voltage is measured with a voltmeter, which is connected in parallel with the voltage source.

Voltage is expressed in units of volts (V), kilovolts (kV), millivolts (mV), or microvolts (μV).

This section has shown you that one major factor which determines how much current flows in a circuit is the applied voltage. In the next section we discuss the other major factor which determines the amount of current which will flow in a given circuit.

Conductors, Insulators, and Semiconductors

In a previous section we told you that the electric potential energy stored in a voltage source such as a battery will remain stored there until some connection is made between the negative and positive terminals. Since there is air between the terminals, you might wonder why current doesn't just flow through the air from one terminal to the other. The answer to this is that air is a poor conductor of electricity. A material which allows an electric current to pass through it easily is referred to as a *conductor.* A material which does not allow an electric current to pass through it easily is called an *insulator.*

The Bohr model of the atom, which we discussed earlier, can help explain why some substances allow electrons to pass through easily and others don't. The key to all of this is that in order to conduct electricity, a substance must have electrons, or in some cases charged atoms, which are free to move easily through the substance. To start, let's take a look at the structure of aluminum atoms.

According to Table 1-1, aluminum atoms have three electrons in their outer shell. In aluminum these outer electrons are loosely held to the nuclei, so in a piece of aluminum these outer electrons tend to form an *electron cloud* which spreads evenly throughout the object, as shown in Figure 1-10a. The overall charge on the piece

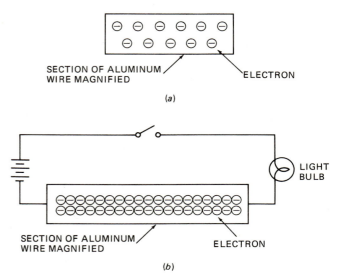

FIGURE 1-10 Electrons in a magnified section of aluminum wire. *(a)* Electron cloud shared by all atoms in wire, total number of electrons in wire is equal to total number of protons. *(b)* Connecting wire to negative terminal of battery causes resident electron cloud to be repelled toward right end of wire and forces more electrons into the wire, giving it a negative charge.

of aluminum is still neutral because the total number of electrons is still equal to the total number of protons. The outer electrons, however, are part of a shared cloud of electrons instead of being attached to a particular atom. The significance of this is that if one end of an aluminum wire is connected to the negative terminal of a battery, the electrons piled up at the negative terminal of the battery will repel the electron cloud in the aluminum toward the other end of the wire. Remember that like charges repel, or push away from, each other. Some electrons from the negative terminal of the battery will also flow into the aluminum wire, as shown in Figure 1-10b. The aluminum wire then contains more electrons than protons, so it has a negative charge.

Now, if the open end of the wire is connected to the positive terminal of the battery, some of the electrons from the electron cloud in the aluminum will flow into the positive terminal of the battery. As electrons flow out of the aluminum to the positive terminal, they will be replaced by electrons from the negative terminal of the battery. The result is effectively a flow of electrons from the negative terminal of the battery to the positive terminal of the battery through the aluminum wire. Because of its loosely held electrons, then, aluminum is a good conductor of electricity.

> NOTE: You should *never* connect a piece of wire directly across a battery in this way because a very large current will flow. This will quickly discharge the battery and may melt or burn up the wire.

Other than hydrogen and helium, atoms which have one, two, or three electrons in their outer shells are classified as *metals*. All metals conduct electricity because they have loosely held electrons in their outer shells. Silver is the best metallic conductor of electricity. Other metals that conduct electricity well are copper, gold, and aluminum.

Elements with five, six, seven, or eight electrons in their outer shells are insulators because their electrons are tightly held by the nucleus and are not free to move through the material, as they are in metals. Sulfur is an example of an element which is an insulator.

Elements such as carbon, silicon, and germanium, which have four electrons in their outer shells, conduct electricity, but very poorly as compared to metals. Therefore, they are referred to as *semiconductors*. The amount of current that a semiconductor material will conduct can be controlled by adding selected impurities to it. Most common electronic devices, such as diodes, transistors, microprocessors, and other integrated circuits that we discuss in this book, are made from silicon or some other semiconductor material.

When elements combine to form a compound, the amount of electricity that the compound will conduct depends on whether one atom donates electrons outright to another atom or they share pairs of electrons. For example, when hydrogen combines with oxygen to form water, as shown in Figure 1-2b, the oxygen atom shares a pair of electrons with each of two hydrogen atoms. The electrons are tightly held between the hydrogen and oxygen atoms. Therefore, as you may be surprised to learn, pure water is an electric insulator. Likewise, compounds of carbon and hydrogen, such as oil, waxes, and plastics, are insulators because the electrons are all held tightly between the hydrogen and carbon atoms.

When a compound such as sodium chloride is formed, one atom donates an electron to another atom to give them each complete outer shells, as we described earlier. As a solid crystal, sodium chloride (common table salt) does not easily conduct electricity because all the atoms and electrons are tightly held in place in the crystal structure. However, if table salt is dissolved in water, the chlorine atoms separate from the sodium atoms. Each chlorine atom has an extra electron that it accepted from the sodium atom, so it has a negative charge. Each sodium atom donated one of its electrons, so it is left with a positive charge. Charged atoms such as these are called *ions*.

If a battery is connected across a salt-and-water solution as shown in Figure 1-11, the positive sodium ions will migrate toward the negative terminal of the battery, and the negative chlorine atoms will migrate toward the positive terminal of the battery. A positive sodium ion can accept an electron from the negative terminal, and

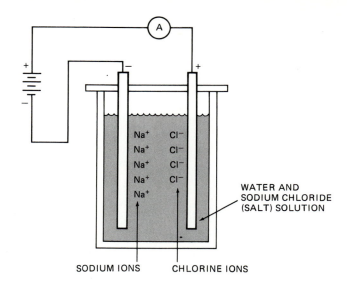

FIGURE 1-11 Separation of sodium (Na^+) and chlorine (Cl^-) ions when sodium chloride and water solution is connected to a battery.

the negative chlorine ion can donate an electron to the positive terminal. As the chlorine atom and the sodium atom drift around in the solution, the chlorine atom can again take an electron from the sodium atom to regenerate a chlorine ion and a sodium ion. To make a long story short, the ions provide a mechanism for electrons to get from the negative terminal of the battery to the positive terminal of the battery. Therefore, although pure water is an insulator, dissolving some salt in it converts water to a conductor. We discuss this more later, in a section on electrical safety.

Now that you are familiar with the terms *conductor*, *semiconductor*, and *insulator*, we want to tell you that there are no sharp dividing lines between these categories. There is really a continuum between good conductors such as silver and poor conductors such as plastics. Furthermore, the situation is complicated by the fact that some substances act as insulators if a low voltage is applied to them but as conductors if the applied voltage is high enough to pull the electrons of that substance free from their atoms. Lightning, for example, occurs when the voltage between two clouds or between a cloud and the earth is high enough to force the air to conduct electricity. The main point here is that the conductivity of the connecting path between the negative terminal and the positive terminal is the second factor which determines how much current flows through the circuit. The better conductor a substance is, the more electrons it will allow to pass through. Now let's look at how we specify the amount that a substance resists the flow of electric current through it.

Resistance and Resistors

RESISTANCE

Very simply, the amount that a conductor resists the flow of an electric current through it is called its *resistance*. The basic unit used to specify resistance is the *ohm.* A conductor is said to have a resistance of 1 ohm if a voltage of 1 V causes a current of 1 A to flow through it. The Greek letter Ω is often used in place of the word *ohm* in representing a resistance value. A resistance might, for example, be specified as 220 ohms or 220 Ω.

To provide the desired amount of resistance in a circuit, a device called a *resistor* is often used. Figure 1-12 shows a few common types of resistors you may see in electronic equipment. Some types of resistors are fixed in value, others are made so that the resistance value can be adjusted by turning or sliding a knob. The volume control on your stereo is an example of a variable resistor. Most of the circuits in this book use small carbon resistors such as those in Figure 1-12a or variable resistors such as those in Figures 1-12d and 1-12e. Fig-

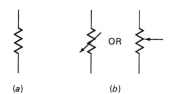

(a)   (b)

FIGURE 1-13 Symbols used to represent resistors in circuit diagrams. *(a)* Fixed-value resistor. *(b)* Variable-value resistors.

ure 1-13 shows the symbols used to represent resistors in electrical schematics.

Resistors with values from small fractions of an ohm to millions of ohms are commonly available. In order to make it easier to refer to the value of a resistor, prefixes are used with the basic ohms unit. Table 1-2 shows the prefixes commonly used in specifying resistance values. The prefixes we use most often are kilo, or k, for 1000 and Meg, or M, for 1,000,000. A resistor of 4700 Ω, for example, can be specified as 4.7 kohms or 4.7 kΩ, and a resistor of 3,300,000 ohms can be specified more easily as 3.3 megohms or 3.3 MΩ.

Some types of resistors have the value of the resistor in ohms or kohms printed directly on the resistor as a number. Most of the resistors that you will work with in this book and the accompanying laboratory manual have their resistance value represented by colored bands around the resistor. Once you catch on to how this color-coding scheme works, you should find it quite handy to be able to identify the value of a resistor at a glance. Here's how the scheme works.

RESISTOR COLOR CODES

Figures 1-14a and 1-14b show the two most commonly used color-coding systems. The one in Figure 1-14a is used for resistors with 5-, 10-, or 20-percent tolerance, and the one in Figure 1-14b shows the system used for resistors with a 1- or 2-percent tolerance. The basic principle of these systems is that the two or three bands nearest one end of the resistor tell you the value of the digits and the third or fourth band tells you the number of 0's to put after these digits.

As a first example, take a look at the resistor shown in Figure 1-15. Since this resistor has only four bands, it uses the coding scheme shown in Figure 1-14a. Note that the colored bands are closer to one end than they are to the other. You always start by reading the color band nearest one end of the resistor. This first band and the one next to it represent the two most significant digits of the resistor value. For our example resistor, the first band is red, which according to Figure 1-14c represents a 2. The second band is violet, which according to Figure 1-14c represents a 7. Therefore, the two most significant digits for the resistor value are 27.

The third colored band on the resistor tells you the multiplier—in other words, how many 0's to put after these two digits. Colors here represent the same numbers that they do for digit values. For our example resistor the third band is orange, which according to

(a)

(b)

(c)

(d)   (e)

(f)

FIGURE 1-12 Common resistor types. *(a)* Fixed carbon. *(b)* Surface mount. *(c)* High-power. *(d)* Variable. *(e)* Variable. *(f)* Variable.

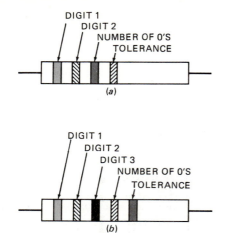

(a)

(b)

BAND COLOR	DIGIT VALUE	NUMBER OF 0'S	TOLERANCE %
BLACK	0	0	
BROWN	1	1	1
RED	2	2	2
ORANGE	3	3	
YELLOW	4	4	
GREEN	5	5	0.5
BLUE	6	6	0.25
VIOLET	7	7	0.1
GRAY	8	8	0.05
WHITE	9	9	
GOLD		DIGIT × 0.1	5
SILVER		DIGIT × 0.01	10
NO COLOR			20

(c)

FIGURE 1-14 Color-coding systems for resistors. (a) Bands for 5-, 10-, and 20-percent resistors. (b) Bands for 1- and 2-percent resistors. (c) Values for bands.

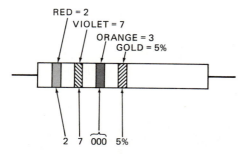

FIGURE 1-15 Example of 27-kΩ carbon resistor.

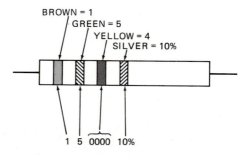

FIGURE 1-16 Second example of a carbon resistor.

Figure 1-14c represents 3. Therefore, the value of this resistor is 27 followed by three 0's, or 27,000 Ω. As we said previously, we would probably refer to the value of this resistor as 27 kΩ.

The fourth colored band on the resistor indicates its *tolerance.* Tolerance, in this case, is another name for the accuracy of the resistor. Common tolerances are 1, 2, 5, 10, and 20 percent. If a manufacturer identifies a resistor as having a tolerance of 10 percent, for example, this means that the actual value of the resistor will be in the range of 10 percent below the specified value to 10 percent above the specified value. For a 4.7-kΩ resistor with 10 percent tolerance, the actual value can be anywhere between $(4700 - 470)$ and $(4700 + 470)$ ohms, or between 4230 Ω and 5170 Ω. A gold fourth band indicates a tolerance of 5 percent, a silver fourth band indicates 10 percent, and no fourth band indicates 20 percent. The gold fourth band on our example resistor in Figure 1-15 then indicates that the resistor has a 5-percent tolerance.

On some 5- or 10-percent carbon resistors a fifth band may be present after the tolerance band to indicate the expected reliability of the resistor. For the work in this book, you don't have to worry about this fifth band if one is present.

To give you more practice with this color coding, take a look at the resistor in Figure 1-16. Use the color values in Figure 1-14c to determine the values for the first two digits. Brown represents 1 and green represents 5, so

the two most significant digits of the resistor value are 15. The third band is yellow, which represents 4, so the 15 is followed by four 0's. The value of the resistor then is 150,000 Ω, or 150 kΩ. Since the fourth band is silver, the tolerance of the resistor is 10 percent.

As a final example of this type of resistor, take a look at the resistor in Figure 1-17 and see if you can determine its

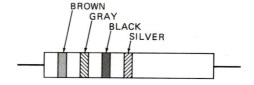

FIGURE 1-17 Third example of a carbon resistor.

value. The brown and gray bands tell you that the first two digits of the resistor value are 18. Since the color black represents 0, the black third band on this resistor indicates that there are no 0's after the 18. The value of the resistor then is simply 18 Ω. As indicated by the fourth silver band, the tolerance of the resistor is 10 percent.

Carbon resistors are typically available in certain standard values. Table 1-3 shows the standard values for the first two digits of 5-, 10-, and 20-percent resistors. Referring to this table may prevent you from trying to find a 3700-Ω resistor in your resistor collection when you build a circuit.

TABLE 1-3
STANDARD VALUES FOR RESISTORS

TOLERANCE		
20%	10%	5%
10	10	10
		11
	12	12
		13
15	15	15
		16
	18	18
		20
22	22	22
		24
	27	27
		30
33	33	33
		36
	39	39
		43
47	47	47
		51
	56	56
		62
68	68	68
		75
	82	82
		91
100	100	100

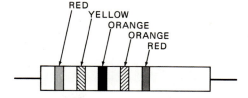

FIGURE 1-18 Example of a 2-percent-tolerance carbon resistor.

orange band indicates that these three digits are followed by three 0's. The final red band indicates that the tolerance of the resistor is 2 percent. Additional examples in the problems at the end of the chapter should give you some more practice with resistor color codes.

MEASURING RESISTANCE

In a previous section we described how you can connect an ammeter in series with a circuit to measure how much current is passing through the circuit. In another section we described how you can connect a voltmeter in parallel with a circuit to measure the voltage across the circuit. To directly measure the resistance of a circuit or device, you use an *ohmmeter*. An ohmmeter is usually part of a multipurpose test instrument called a *multimeter*. Figure 1-19 shows pictures of two common types of multimeters.

These instruments can be used to measure current, voltage, or resistance, depending on how you set the switches on the front of the meter. Test leads are used to connect the meter to the circuit or part that is being measured. On an analog meter the value of the measured voltage, current, or resistance is indicated by the position of a pointer needle on a numbered scale. On a digital meter the value of the measured quantity is given directly on a numerical display.

Before using an ohmmeter to measure the resistance in a circuit, *you must first disconnect all voltage sources* from the circuit so that the voltage source does not damage the ohmmeter. After disconnecting the voltage source, you connect the ohmmeter across the circuit or device as shown in Figure 1-20. The ohmmeter will then display the resistance of the circuit in ohms, kohms, or Mohms, depending on the meter scale you choose. Not only can you use an ohmmeter to measure the resistance of a resistor, you can also use it to determine if a light bulb or some other circuit has a conducting path through it.

If you touch the leads of an ohmmeter directly together, you should read 0 or near 0 Ω because the connecting leads have very low resistance. With the ohmmeter leads not connected to anything or connected to a circuit which has no conducting path through it, you should read infinite or a very large number of ohms, depending on how the particular meter indicates maximum value. Knowing these facts you can, for example, use an ohmmeter to determine if the filament in a light bulb still conducts or is burned out. If the bulb is good, the ohmmeter should show a low value when one lead of the ohmmeter is connected to each of the two metal contacts on the bulb. If the bulb is burned out, the ohmmeter will show

Resistors with 1 percent or better tolerance often have the resistance and tolerance printed in numbers directly on the resistor. For 1-percent and 2- percent resistors which are color-coded, the first three bands are used to represent the three most significant digits of the resistance value, as shown in Figure 1-14b. The fourth band on these resistors represents the multiplier, or number of 0's to be added after these three digits. The fifth band on these resistors represents the tolerance. A brown tolerance band is used to indicate a 1-percent-tolerance part, and a red band is used to represent a 2-percent-tolerance part.

Figure 1-18 shows an example of a resistor with this coding. See if you can determine its value. The result you should get is 243 kΩ, because for the three digits, the red band represents a 2, the yellow band represents a 4, and the orange band represents a 3. The second

(a)

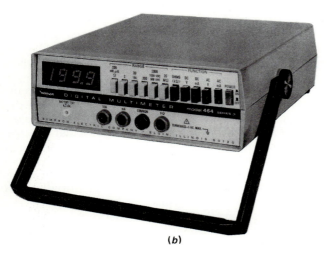

(b)

FIGURE 1-19 Photograph of multimeters. *(Courtesy of Simpson Electric Company.)* *(a)* Analog. *(b)* Digital.

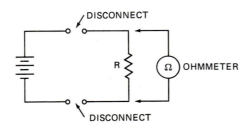

FIGURE 1-20 Measuring resistance in a circuit with an ohmmeter. NOTE: *Voltage source must be disconnected!*

infinite or very high resistance when connected across the bulb. In a similar manner you can use an ohmmeter to determine if the wire on a lamp or other appliance is broken.

Now that you know how to read resistor values and

measure resistance, we will show you how easy it is to calculate the current that will flow through simple circuits.

Ohm's Law

CALCULATING THE CURRENT IN A CIRCUIT

Earlier we said that if a circuit with 1 Ω of resistance has a voltage of 1 V applied to it, 1 A of current will flow through the circuit. Mathematically this specific case can be stated as:

$$1 \text{ amp} = \frac{1 \text{ volt}}{1 \text{ ohm}}$$

In more general terms the relationship between current, voltage, and resistance can be expressed as:

$$\text{Current} = \frac{\text{voltage}}{\text{resistance}}$$

This expression, called *Ohm's law*, is one of the most important laws you will learn in electronics because almost everything else depends on it. Let's see if we can give you a more intuitive feel for it.

Earlier in the discussion of voltage, we showed that if the voltage applied to a given circuit is increased, more current will flow through the circuit because the electrons have more energy. In the Ohm's law expression, the voltage is in the top part of the fraction, so increasing its value will increase the current through a circuit with a given resistance.

Likewise, if the resistance in a circuit with a given voltage is decreased, more current will flow, just as using a larger water pipe (less resistance) will allow more water to flow through the pipe. Since resistance is on the bottom of the fraction in the Ohm's law expression, you should be able to predict from Ohm's law that decreasing the resistance value will increase the current which will flow with a given voltage.

Now, as you may have already discovered, some people like to represent quantities such as current, voltage, and resistance with single letters instead of whole words. Ohm's law is stated in letters as follows:

$$I = \frac{V}{R}$$

The letter I was originally chosen to represent the *intensity* of the current through the circuit. We now just say that I represents the *current* through the circuit. The letter V represents the *voltage* applied to the circuit, and the letter R represents the *resistance* in the circuit. Some books use the letter E to represent voltage in Ohm's law because a less commonly used name for voltage is electomotive force, or EMF.

Some examples will help you see how to use Ohm's law. To start, Figure 1-21 shows a circuit with a 12-V battery and a 4-Ω resistor. You can calculate the current that will flow in this circuit by substituting the voltage and the resistance into Ohm's law as follows:

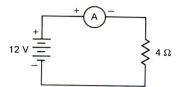

FIGURE 1-21   Ohm's law example circuit.

$$I = \frac{V}{R} = \frac{12\ V}{4\ \Omega} = 3\ A$$

Now, as another example, suppose that you want to calculate the resistor value that you can put in place of the 4-Ω resistor so that 6 A instead of 3 A will flow through the circuit. You can easily do this by first solving Ohm's law for R and then substituting in the known values of V and I. To solve the equation for R, multiply both sides of Ohm's law by R as follows:

$$I = \frac{V}{R}$$

$$I \times R = \frac{V \times R}{R}$$

Dividing both sides of the result by I gives:

$$\frac{I}{I} \times R = \frac{V}{I}$$

$$R = \frac{V}{I}$$

As you can see, the result of this is R = V/I. You can substitute the values of 12 V and 6 A in this equation and compute the result as follows:

$$R = \frac{V}{I} = \frac{12\ V}{6\ A} = 2\ \Omega$$

A resistance value of 2 Ω then will allow the desired current of 6 A to flow in the circuit.

As a third example of the use of Ohm's law we will show you how you can solve it to tell you the voltage that must be applied to a given resistor to produce a desired current through the resistor. The basic form of Ohm's law states:

$$I = \frac{V}{R}$$

Multiplying both sides by R gives:

$$I \times R = V \times \frac{R}{R}$$

Since R/R = 1, this reduces to just:

$$V = I \times R$$

Using this equation you can determine the voltage required to produce a current of 2 A in a 12-Ω resistor as follows:

$$V = I \times R = 2\ A \times 12\ \Omega = 24\ V$$

For more practice with Ohm's law and units of resistance, let's look at another circuit.

Figure 1-22 shows a circuit with a 2.2-kΩ resistor and a 5-V power supply. As discussed in a previous section, the prefix k in front of ohms tells you to multiply the

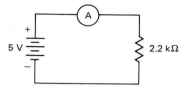

FIGURE 1-22   Another Ohm's law example circuit.

given digits by 1000. Therefore, the value of the resistor is 2200 Ω. To find the current through the resistor, you use:

$$I = \frac{V}{R} = \frac{5\ V}{2200\ \Omega} = 0.0023\ A$$

The value of the current expressed in amps is correct, but it is not in a very convenient form. As described in a previous section on current, we often use units of milliamps (mA) or microamps (μA) to represent small currents such as this. A milliamp is 0.001 A, so to convert a current from amps to the equivalent value in milliamps, just move the decimal point three places to the right and add the prefix *milli* to amps. A current of 0.0023 A, then, is equivalent to 2.3 mA.

Since many of the currents that you will work with in this book are in the range of a few milliamps, it is very useful to observe that if a voltage expressed in volts is divided by a resistance in kilohms, the resultant current value will automatically be in units of milliamps. For example, given a voltage of 20 V and a resistance of 10 kΩ, the current will be:

$$I = \frac{V}{R} = \frac{20\ V}{10\ k\Omega} = 2\ mA$$

From the preceding examples you can see that Ohm's law can be expressed in three equivalent forms:

$$I = \frac{V}{R} \qquad R = \frac{V}{I} \qquad V = I \times R = IR$$

Summary Example 1-1, on the next page, should help you decide which form to use to most easily find a desired current, voltage, or resistance.

Electric Power

Throughout this chapter we have explained that when you separate electrons from their atoms, the work that you put into separating them is converted to electric potential energy. When the electrons are allowed to return to their atomic orbits, the electric potential energy is given off in some form. If the electrons pass through a

SUMMARY EXAMPLE 1-1
OHM's LAW

Find the <u>current</u> I through a 2-kΩ resistor with an applied voltage of 5 V.

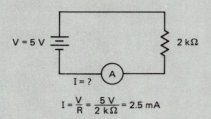

$$I = \frac{V}{R} = \frac{5\,V}{2\,k\Omega} = 2.5\,mA$$

Find the <u>voltage</u> required to give a current of 4 mA through a 2-kΩ resistor.

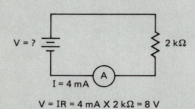

$$V = IR = 4\,mA \times 2\,k\Omega = 8\,V$$

Find the <u>resistance</u> which allows a current of 4 mA to flow in a 5-V circuit.

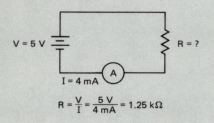

$$R = \frac{V}{I} = \frac{5\,V}{4\,mA} = 1.25\,k\Omega$$

light bulb on their way home, then the energy will be given off as light and heat. If the electrons pass through an electric heater or some other type of resistor on their way home, then most of the released potential energy will be given off as heat.

Now, if the amount of current being passed through a resistor is increased, then more energy will be given off in a given time. For example, if you increase the current through an electric heater, heat will be given off at a greater rate. The rate at which energy is given off or work is being done is called *power*. The unit we usually use to measure electric power is the *watt*, or *W*. You may have noticed that light bulbs are rated with values such as 40, 60, 75, or 100 W to give you an idea of the rate at which the bulb will give off light energy.

The resistors you use in electric circuits are also given wattage ratings to tell you how much power the resistor can dissipate without burning up. Resistors are available with power ratings of ⅛, ¼, ½, 1, 2 W, etc., up to several hundred watts. For your reference, Figure 1-23 shows the outlines of carbon resistors up to 2 W, which

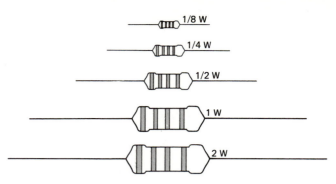

FIGURE 1-23   Outlines of common carbon resistors.

are the sizes you will usually encounter. Most of the circuits in this book and the accompanying lab manual use ¼-W resistors.

The electric power dissipated by a light bulb or some other type of resistor can be very easily calculated. The electric power in watts is simply equal to the voltage applied to the resistor multiplied by the current through the resistor. In letters this is represented as:

$$P = V \times I \quad \text{or} \quad P = VI$$

The power dissipated by an automobile headlight bulb which allows 6 A to pass through when 12 V is applied will then be:

$$P = VI = 12\,V \times 6\,A$$
$$= 72\,watts = 72\,W$$

For a toaster which allows 10 A to flow through when it is connected to a 120-V line, the power dissipated will be:

$$P = VI = 120\,V \times 10\,A$$
$$= 1200\,W = 1.2\,kW$$

In some cases you may know the voltage applied to a circuit and the resistance of the circuit, but not the current through the circuit. For these cases you can first use Ohm's law to find the current and then calculate the power using P = VI, as we showed in the preceding paragraph. An easier way is to make a substitution directly in the power equation. Ohm's law says that I = V/R, so V/R can be substituted for I in the power equation as follows:

$$P = V \times I$$
$$= V \times \frac{V}{R} = \frac{V^2}{R}$$

As an example of using this form of the power law, suppose that a 20-V supply is connected across a 10-kΩ resistor. The power dissipated by the resistor will be:

$$P = \frac{V^2}{R} = \frac{20\,V \times 20\,V}{10\,k\Omega}$$
$$= 0.040\,W = 40\,mW$$

The reason you need to do a calculation of this sort is to find out what power rating a resistor needs to have so that it operates safely in a circuit you are building. The 10-kΩ resistor in this example only dissipates 40 mW, so a resistor with a ⅛-W (125-mW) power rating would be large enough. Actually, however, you would probably use a ¼-W resistor because ¼-W resistors are more commonly available and more mechanically rugged.

In a case where you know the resistance in a circuit and the current through the circuit, you can solve the power law so you can calculate the power directly from these values. Since from Ohm's law V = IR, this can be substituted in the power law to give:

$$P = V \times I = I \times R \times I = I^2R$$

An old nursery rhyme, "Twinkle, twinkle little star, power equals I squared R," may help you remember this form of the power law.

To summarize all this, then, the power dissipated by an electric resistor can be simply calculated using one of the three forms of the power law:

$$P = VI \qquad P = \frac{V^2}{R} \qquad P = I^2R$$

The power dissipated by a resistor or some other device is expressed in watts (W) or milliwatts (mW). To give you an idea of how electric power is related to motors and mechanical power, 1 horsepower is equal to 746 watts (1 hp = 746 W).

Series Circuits and Parallel Circuits

Up to this point we have been discussing circuits which have only one resistor. For a variety of reasons many real circuits contain two or more resistors. Multiple resistors can be connected in *series*, in *parallel*, or in a combination of series and parallel. Figure 1-24 shows a simple series circuit and a simple parallel circuit so that you can easily compare the two basic configurations. In this section we show you how to work with these two basic circuit types.

SERIES CIRCUITS

Figure 1-24a shows a series circuit built with a voltage source and two resistors. The name *series* comes from the fact that the electrons have to flow through the complete series of resistors to get back to the positive terminal. In a series circuit such as this there is only one current path, so *the current is the same at any point in the circuit,* just as the amount of water flowing through a pipe is the same at any point in the pipe.

The total resistance, R_T, in a series circuit is just the sum of the individual resistors. In letters this is represented as:

$$R_T = R_1 + R_2 + \cdots + R_n$$

For the example circuit in Figure 1-24a, the total resistance, R_T, is 1 kΩ + 4 kΩ, or 5 kΩ. In other words, the current that will flow though this circuit is the same as

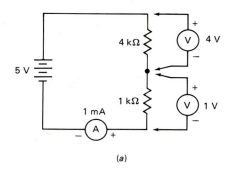

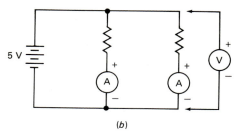

FIGURE 1-24 Series circuit and parallel circuit examples. (a) Series circuit. (b) Parallel circuit.

it would be if the 1-kΩ resistor and the 4-kΩ resistor were replaced with a single 5-kΩ resistor. One practical consequence of this is that you can make up any resistor value you need for a circuit by connecting smaller-value resistors in series. To calculate the current through the example circuit in Figure 1-24a, you can use Ohm's law as follows:

$$I = \frac{V}{R} = \frac{V}{R_T} = \frac{5\text{ V}}{5\text{ k}\Omega} = 1\text{ mA}$$

Now, as the electrons flow from the negative terminal to the positive terminal, they will lose some of their potential energy as they pass through the 1-kΩ resistor and the rest of their potential energy as they pass through the 4-kΩ resistor. Since *voltage* is the term we use to describe how much potential energy electrons have, another way of saying this is that the electrons will lose some of their voltage as they pass through the 1-kΩ resistor and the rest of their voltage as they pass through the 4-kΩ resistor. You can use Ohm's law to calculate the *voltage drop* across each resistor.

As we showed above, 1 mA of current is flowing through both the 1-kΩ resistor and the 4-kΩ resistor. The voltage drop across the 1-kΩ resistor then is:

$$V = IR = 1\text{ mA} \times 1\text{ k}\Omega = 1\text{ V}$$

The voltage drop across the 4-kΩ resistor then is

$$V = IR = 1\text{ mA} \times 4\text{ k}\Omega = 4\text{ V}$$

If you add these two voltage drops together, you will see that the sum of the voltage drops across the individual resistors is equal to 5 V, the voltage of the source. This will always be true in a resistive series circuit. This circuit is often used as a *voltage divider* because it divides the supply voltage into two or more parts. As you can see from the above calculations, the voltage across a resistor in a series circuit will be proportional to the relative size of the resistor. The larger resistor will have the larger voltage across it.

Here is a summary of the characteristics of a series circuit:

The total resistance, R_T, is equal to the sum of the individual resistors:

$$R_T = R_1 + R_2 + R_3 + \cdots + R_n$$

The same current flows through all of the resistors in the circuit. This current will be equal to the applied voltage, V_A, divided by the total resistance, R_T:

$$I = \frac{V_A}{R_T}$$

The voltage across any resistor, V_{R_n} in the circuit will be equal to the total current, I_T, times the resistance of that resistor, R_n:

$$V_{R_n} = I_T \times R_n$$

To further help you fix these concepts in your mind, work your way through Summary Example 1-2.

PARALLEL CIRCUITS

Figure 1-24b shows an example of a circuit with two resistors connected in parallel. The name comes from the fact that there are two or more parallel paths for electrons to flow from the negative to the positive terminal of the supply. The way the circuit is drawn should help you to see that in this circuit the *voltage* applied to each resistor is the same. For the example circuit in Figure 1-25, then, each resistor will have 5 V applied to it. Since

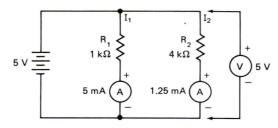

FIGURE 1-25   Parallel circuit example.

Ohm's law is true for a complete circuit or for any part of a circuit, you can use it to calculate the current through each resistor.

The current I_1 through the 4-kΩ resistor R_1 is:

$$I_1 = \frac{V}{R_1} = \frac{5\ V}{1\ k\Omega} = 5\ mA$$

and the current I_2 through the 4-kΩ resistor R_2 is:

$$I_2 = \frac{V}{R_2} = \frac{5\ V}{4\ k\Omega} = 1.25\ mA$$

SUMMARY EXAMPLE 1-2
RESISTANCES IN SERIES

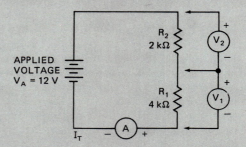

Find the <u>total resistance</u> R_T in the circuit.

$$R_T = R_1 + R_2$$
$$= 2\ k\Omega + 4\ k\Omega = 6\ k\Omega$$

Find the <u>total current</u> I_T in the circuit.

$$I_T = \frac{V_A}{R_T} = \frac{12\ V}{6\ k\Omega} = 2\ mA$$

Find the <u>voltage across</u> R_1.

$$V_1 = I_T \times R_1 = 2\ mA \times 4\ k\Omega$$
$$= 8\ V$$

Find the <u>voltage across</u> R_2.

$$V_2 = I_T \times R_2 = 2\ mA \times 2\ k\Omega$$
$$= 4\ V$$

<u>Sum of voltage drops</u> through resistors = V_A.

$$V_A = V_1 + V_2$$
$$= 8\ V + 4\ V = 12\ V$$

The total current, I_T, flowing from the negative terminal of the battery to the positive terminal is the sum of these two currents

$$I_T = I_1 + I_2$$
$$= 5\ mA + 1.25\ mA = 6.25\ mA$$

The digital devices we will be teaching you about in this book are connected to the power supply in parallel, so the total current required by a system is the sum of the currents for each of the individual devices. Also, the appliances in your house are connected in parallel with each other across the 120-V line, so if your toaster requires 12 A and your coffee maker takes 4 A, the total current required by the two is 16 A.

There are two ways of calculating the *equivalent resistance*, or the single resistance needed to cause the same current to flow from the supply as flowed when there were two resistances in parallel. One way is to determine the total current by adding the individual currents and dividing the applied voltage by this current. For the example circuit in Figure 1-25, the applied voltage is 5 V and the total current, which we calculated before, is 6.25 mA. The equivalent resistance, R_{eq}, of the circuit then is:

$$R_{eq} = \frac{V}{I_T} = \frac{5 \text{ V}}{6.25 \text{ mA}} = 0.8 \text{ k}\Omega$$

The second way of calculating the equivalent resistance, R_{eq}, of two resistors in parallel is with the formula:

$$R_{eq} = \frac{R_1 \times R_2}{(R_1 + R_2)}$$

For the circuit we are using here:

$$R_{eq} = \frac{1 \text{ k}\Omega \times 4 \text{ k}\Omega}{(1 \text{ k}\Omega + 4 \text{ k}\Omega)} = 0.8 \text{ k}\Omega$$

Note that the equivalent resistance for resistors connected in parallel is always less than the smallest resistor.

Here is a summary of the characteristics of a parallel circuit.

The voltage across each resistor or other device in a parallel circuit is the same and is equal to the applied voltage.

The current through each resistor, I_n, in a parallel circuit is equal to the applied voltage, V_A, divided by the resistance of that resistor, R_n:

$$I_n = \frac{V_A}{R_n}$$

The total current I_T for a parallel circuit is equal to the sum of the individual currents for each resistor:

$$I_T = I_1 + I_2 + I_3 + \cdots + I_n$$

The value of the single resistance R_{eq}, which is equivalent to the resistors in parallel, can be calculated by dividing the applied voltage, V_A, by the total current, I_T:

$$R_{eq} = \frac{V_A}{I_T}$$

To further help fix these concepts in your mind, work your way through Summary Example 1-3.

ALTERNATING CURRENT ELECTRICITY

So far in this chapter we have been talking mostly about *direct current*, or dc, electricity, such as that produced by a battery. When a resistor is connected across a source of direct current electricity, electrons always flow in one direction through the resistor, from the negative terminal to the positive terminal.

The alternator in your car and the generator which supplies electric power to your house produce what is called *alternating current*, or ac, electricity. This name

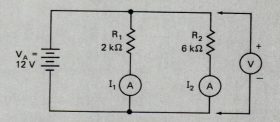

SUMMARY EXAMPLE 1-3
RESISTANCES IN PARALLEL

Find the current I_1 through resistor R_1.

$$I_1 = \frac{V_A}{R_1} = \frac{12 \text{ V}}{2 \text{ k}\Omega} = 6 \text{ mA}$$

Find the current I_2 through resistor R_2.

$$I_2 = \frac{V_A}{R_2} = \frac{12 \text{ V}}{6 \text{ k}\Omega} = 2 \text{ mA}$$

Find the total current I_T flowing through the circuit.

$$I_T = I_1 + I_2$$
$$= 6 \text{ mA} + 2 \text{ mA} = 8 \text{ mA}$$

Find the equivalent resistance, R_{eq}.

$$R_{eq} = \frac{V_A}{I_T} = \frac{12 \text{ V}}{8 \text{ mA}} = 1.5 \text{ k}\Omega$$

or

$$R_{eq} = \frac{R_1 R_2}{R_1 + R_2} = \frac{2 \text{ k}\Omega \times 6 \text{ k}\Omega}{8 \text{ k}\Omega} = 1.5 \text{ k}\Omega$$

comes from the fact that in an ac circuit the current alternately flows first in one direction, then in the other direction through the circuit. An ac current is generated by rotating a coil of wire in the magnetic field present between a magnetic south pole and a magnetic north pole. Figure 1-4 shows a very simple example of an ac generator. As the coil of wire is rotated, electrons in the coil are alternately pushed first toward one end of the coil, then toward the other end. The end of the coil that electrons have been moved away from will be positive, and the end that electrons have been piled up at will be negative. Since the electrons are being pushed first toward one end of the coil and then toward the other, the polarity of the voltage across the coil will change as the coil rotates. Figure 1-26a shows an end view of a coil of wire rotating in the field between two magnets. Figure 1-26b shows the voltage that will be produced on the leads of the coil as it rotates through one turn. Here's how this voltage waveform is produced.

When side X of the coil is moving parallel to the magnetic field lines, as it is at point A, no voltage (0 V) will be produced in the coil. As the coil rotates to point B, the X side of the coil passes down through magnetic flux lines, so a voltage is induced in the coil. One way of thinking of what happens here is that moving the wire through

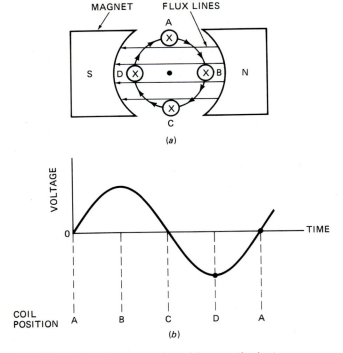

(a)

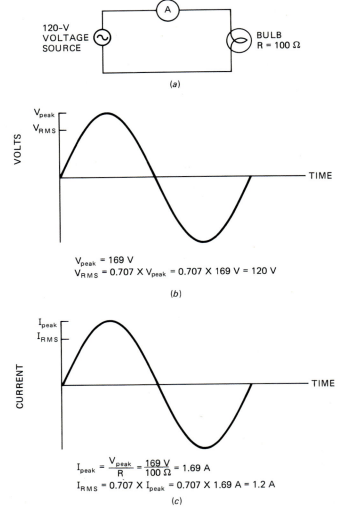

(a)

$$V_{peak} = 169 \text{ V}$$
$$V_{RMS} = 0.707 \times V_{peak} = 0.707 \times 169 \text{ V} = 120 \text{ V}$$

(b)

$$I_{peak} = \frac{V_{peak}}{R} = \frac{169 \text{ V}}{100 \text{ } \Omega} = 1.69 \text{ A}$$
$$I_{RMS} = 0.707 \times I_{peak} = 0.707 \times 1.69 \text{ A} = 1.2 \text{ A}$$

(c)

FIGURE 1-27   Voltage and current in an ac circuit. (a) Example circuit. (b) Voltage-versus-time graph for the circuit. (c) Current-versus-time graph for the circuit.

current will alternately flow first in one direction through the resistor, then in the other direction. According to Ohm's law, the current in the circuit at each point in time is proportional to the voltage at that point in time. The point when the voltage is 0 V, the current is 0 A. When the voltage is a maximum, the current is a maximum, etc. Therefore, a graph of current and time will have the same shape as the graph of voltage and time. Figure 1-27 shows an alternating current circuit, the

FIGURE 1-26   Voltage produced by a coil of wire rotating in a magnetic field.

the magnetic flux lines causes some of the electron cloud in the wire to be moved toward one end of the wire, making that end negative and leaving the other end of the wire positive. As shown in Figure 1-26b, the induced voltage is a maximum in one direction when the X side of the coil is at point B. As the X side of the coil rotates to point C, the induced voltage drops to 0 V again.

As the X side of the coil rotates toward point D, the electron cloud will be moved toward the other end of the wire because the X side of the coil is now moving up through the magnetic field lines instead of down. (There are rules for predicting all of this, but you don't need to know them for now.) The induced voltage will be a maximum in the reverse direction as the X side of the coil passes point D. As the X side of the coil comes back to point A, the induced voltage again drops to 0 V.

The graph of voltage as a function of time, shown in Figure 1-26b, is a very important one for you to understand. It shows that as the coil rotates, the induced voltage goes from 0 V to a maximum in one direction, then back to 0 V. It then goes to a maximum in the other direction, and finally back to 0 V as it returns to the starting position. Figure 1-26b shows what is called *1 cycle* of this alternating voltage waveform. The cycle will repeat over and over as the generator turns. For the ac electricity supplied to your house by the electric company, the cycle repeats 50 or 60 times each second, depending on what country you live in. Incidentally, the waveform in Figure 1-26b is often referred to as a *sine wave* because it has the same shape as a graph of the mathematical *sine function*.

If a resistor is connected across an ac voltage source,

voltage-versus-time graph for the circuit, and the current-versus-time graph for the circuit. Note that although they use different units, the voltage graph and the current graph have basically the same shape.

At this point the question that may occur to you is, "If the voltage and current are constantly changing in an alternating current circuit, how can I make any meaningful calculations for the circuit?" The answer is that for ac circuits we use special average values of current and voltage called the *RMS* values. The RMS value of an ac voltage with a sine waveform is equal to the peak value of the voltage multiplied by 0.707. The RMS value of a voltage is a measure of the effective value of the voltage in, for example, heating a resistor. A source of alternating voltage with an RMS value of 120 V, for example, will cause your toaster to give off the same amount of heat a 120-V dc source of electricity would. The RMS value of the current in a sine-wave ac circuit is equal to the peak value of the *current* multiplied by 0.707 as well.

The point of all this is that Ohm's law and the power law work just as well for resistive ac circuits as they do for dc circuits if you simply use the RMS values of voltage and current. Since most ac voltages are specified by their RMS values, you usually don't have to calculate them. Your house wiring, for example, may be specified as 120 V. This is the RMS value for the voltage. If you connect a light bulb with 100-Ω resistance across the 120-V_{RMS} ac line, as shown in Figure 1-27a, the RMS current is:

$$I = \frac{V}{R} = \frac{120 \text{ V}}{100 \text{ }\Omega} = 1.2 \text{ A}$$

The power dissipated by the light bulb is:

$$P = VI = 120 \text{ V} \times 1.2 \text{ A} = 144 \text{ W}$$

For the most part, the circuits in this book and the accompanying laboratory manual use dc voltage sources rather than ac. However, you need some awareness of ac because most test equipment plugs into ac outlets. Also, some knowledge of ac will help you understand the electrical safety precautions discussed in the next section.

ELECTRICAL SAFETY

As we are sure you already know, electricity can kill you if it is not used correctly. However, it is not something to fear unreasonably. The main point to remember when working with any electric or electronic circuit is:

Don't let electrons use your body as a path to get back home!

For the most part this means keep your fingers out of electric circuitry when it is connected to a source of voltage (the power is on). In order to give you a better understanding of electric safety, we first need to show you how some electric circuits are connected and introduce you to the concept of electric *ground*.

AC Power Wiring and Electric Ground

Figure 1-28 shows a circuit diagram for the ac power connections from a transformer at the end of your street to an electric drill and clothes dryer in your house. Here's how all of this works.

In order to reduce the amount of power lost in the connecting wires, ac power is sent long distances at voltages of 30,000 to 750,000 V. Large electrical devices called *transformers* are used to step these high voltages down to 240 V and 120 V, which are much safer to use in your house or laboratory. Three wires usually come from the transformer into your house. The voltage between the center tap and either outside wire is 120 V RMS. The voltage between the two outer wires is 240 V RMS. As shown by the ground connection symbol in Figure 1-28, the center tap of the transformer is connected to a metal rod or metal water pipe which goes directly into the earth. This connection is referred to as a *ground* connection.

The earth is a relatively good conductor of electricity, so this connection allows the earth to act as a reference, or zero point, for various voltages. Each of the outer wires from the transformer is then said to have a voltage of 120 V *with respect to ground*. The fact that one of the wires from a voltage source is connected to the earth has some important safety considerations, which we will discuss a little later.

Soon after the ac wires enter your house, they pass through a fuse or circuit-breaker box, as shown in Figure 1-28. The purpose of fuses or circuit breakers is to open a circuit if too much current begins to flow through it. A main circuit breaker will open if the total current used in the house is too large. The individual power lines which run around the house to lights, outlets, and appliances also have individual circuit breakers which will

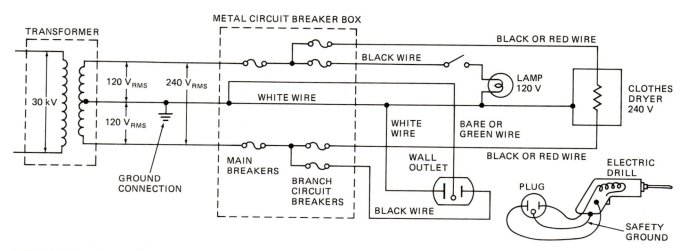

FIGURE 1-28   Circuit diagram for 120-V ac home wiring.

open if the current in that particular circuit exceeds some value, such as 20 A. One reason for a fuse or circuit breaker to "blow," or open, is a condition called a *short circuit*, or just a *short*. This condition occurs when a good conductor, such as a piece of wire or a screwdriver, is placed directly across the wires. The low resistance of the short will allow a very large current to flow through the wires. Without a fuse or circuit breaker to open and shut off the current, the wires would soon get hot enough to set the house on fire.

Large appliances such as clothes dryers, which require 240 V for their operation, are connected between the outer wires from the transformer, as shown on the right-hand side of Figure 1-28. Lights and outlets which require 120 V for their operation are connected in parallel between the center wire and one of the outer, or "hot," wires (also shown in Figure 1-28). Let's look more closely at how a standard 120-V ac outlet is connected to the power and ground wires.

The small slot in a standard electric outlet is connected to the "hot" lead from the transformer. The large slot in the outlet is connected to the center tap, or ground, from the transformer. When you plug in a light or an electric drill, it is connected across these two wires. Note that a third connection is made to the round hole in the outlet through a green wire, or in some systems, through the metal conduit that the wires are encased in. This connection also goes to the center tap (earth ground) of the transformer. This connection is called the *safety ground*, and normally no current flows through it. When you plug in your drill or test instrument, the metal case of the instrument is connected directly to ground through this pin. The safety ground may seem unnecessary, but it is not. Here's why.

Suppose that as your drill gets older, the insulation on the hot lead inside the drill wears through and allows the wire to touch (short to) the metal case. With the safety ground present, a large current will flow from the hot lead through the safety ground. This will blow the fuse or circuit breaker and open the circuit, protecting you from an electric shock. If you have removed the round safety ground pin from your drill cord, the safety ground is not connected, and when the hot lead shorts to the drill case, the *case* will be at a voltage of 120 V with respect to ground. If you hang onto the drill with one hand and a water pipe or something else connected to ground with the other hand, then current will flow through your body to ground. This could be fatal! The point here is that you want any leakage current to flow through the safety ground wire instead of through your body! For even more protection, the ac outlets in bathrooms and kitchens often have a ground fault indicator circuit, which will light if even a small current flows through the safety ground wire.

Here are some simple rules to help you prevent a potentially fatal shock when working with 120 V or 240 V ac:

1. Make sure the safety ground (the round prong), if present, is connected on any test instruments or tools you use. Note that some newer tools are completely enclosed in plastic, so that the safety ground connection is not required.

2. Always turn the power to any electric or electronic circuit off before you work on it, unless you absolutely have to work on it with the power on.

3. If you have to work on a "hot" circuit, make sure you do not in any way make a connection between the "hot" lead and the other lead or ground. Because ground can supply a return path to the center tap of the transformer, make sure that no part of your body is touching any grounded objects, such as water pipes, metal benches, or metal instrument cases. Do not touch any electric appliances while standing in your bathtub because the bathtub is grounded. Also remember that salty water conducts electricity.

4. Use the *one-hand survival rule*, which says, "Keep one hand behind your back, so that if you do accidentally touch a hot wire, there is no chance of a current path from one arm to the other through your heart."

DC Circuit Connections and Grounds

To supply electric power to the circuits in this book and the accompanying laboratory manual, you will be using dc voltage sources with voltages such as 5 V and 12 V. You could use batteries to produce these voltages, but it is easier to use a dc power supply such as the one shown in Figure 1-9. Remember that a power supply is a device which contains a transformer to step 120-V ac down to lower voltages, and circuitry to convert these voltages from ac to the desired dc voltages. The obvious advantage of a power supply over batteries is that as long as you pay your electric bill, the voltage remains constant and doesn't run down with age, the way the voltage from a battery voltage does.

One terminal of a dc voltage source is often tied to ground just as is done with the ac voltage sources we described in the previous section. Figure 1-29a shows how the negative terminal of a 12-V battery can be connected to ground. The voltage on the positive terminal of the battery is then referred to as +12 V *with respect to ground*. For simplicity, the supply is usually just referred to as a +12-V supply, and the reference to ground implied.

To make electrical schematics less cluttered, the wire connecting the grounded side of the power source and circuit components is often represented as shown in Figure 1-29b. By convention it is understood that all circuit points which have the ground connection symbol are actually connected together by a wire or other conductor, so the connection does not have to be drawn in the schematic. The ground shown in a dc schematic such as this is referred to as a *circuit ground*. This ground may or may not be connected to an earth ground such as the outer case of an electronic instrument. Incidentally, most automobile electrical systems connect one side of the battery to the metal frame of the car and use the frame as a "ground" conductor to one side of all of the lights, etc.

If the positive terminal of a battery or dc power supply is connected to ground as shown in Figure 1-30, the volt-

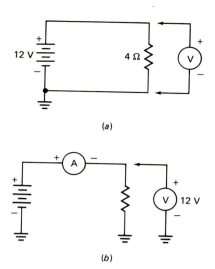

(a)

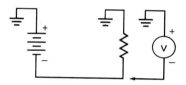

(b)

FIGURE 1-29   Ground connections in dc circuits.
(a) Battery or dc power-supply circuit with negative terminal grounded (positive supply with respect to ground). *(b)* Positive voltage supply circuit showing how connections to common ground conductor can be indicated on a schematic.

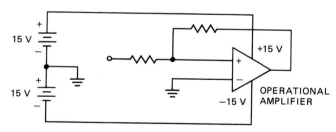

FIGURE 1-30   Battery or dc power-supply circuit with positive terminal grounded (negative voltage supply with respect to ground).

FIGURE 1-31   Circuit showing how a positive voltage supply and a negative voltage supply can be produced at the same time using two batteries or two dc power supplies.

age of the supply will be referred to as −15 V with respect to ground.

Some of the circuits that you will see later in this book require both a supply which is positive with respect to ground and a supply which is negative with respect to ground. Figure 1-31 shows how two batteries or dc power supplies can be connected up to do this. Note that the positive terminal of one supply and the negative terminal of the other supply are connected to ground.

Final Word on Electrical Safety

The shock hazard of any voltage supply is a function of how much current the voltage source can make flow through your body. If a very small current flows through your body, you probably won't notice it. With a current above a few microamps you will feel a shock. A current of a milliamp or so may interfere with your nervous system so that you cannot let go of a "hot" wire. A current of a few milliamps through your chest can kill you because it interferes with the electrical signals which control the beating of your heart.

Low ac and dc voltages, such as the 5 V dc used to power most of the circuitry described in this book, will not give you a shock, so you don't have to worry about getting a shock as you test and troubleshoot this type of circuitry. However, in any ac or dc circuitry where there is a chance of voltages higher than about 25 V, strictly obey the rules we gave you for working with 240-V and 120-V ac circuits! Specifically:

- Make sure the case of electrical equipment is grounded so that if an internal "hot" wire touches the case, the fuse will blow and prevent you from getting a shock.

- Turn off the power before working on a circuit unless absolutely necessary to work on it with the power on.

- If you have to work on a live circuit, don't touch any grounded objects with your body.

- When working on a live circuit, keep one hand behind your back so there is no chance of an electric current path from one hand to the other through your heart.

CHECKLIST OF IMPORTANT TERMS AND CONCEPTS IN THIS CHAPTER

If you do not remember any of the terms or concepts in this list, use the index to find them in the chapter.

Bohr model of the atom
   Proton, neutron, electron
   Atomic number
   Electron shells
Element, compound
Electrical potential energy
Generating static electricity
Batteries
   Direct current
Alternators and generators
   Alternating current, ac
   Magnetic field
Electrical circuit, schematic
Electrical current
   Unit of charge, coulomb
   Ampere, amp (A), mA, μA, nA, pA
   Ammeter
Polarity
Voltage
   Volt (V), kV, mV, μV

Voltmeter
Power supply
Series circuit
Parallel circuit
Conductor, insulator, semiconductor
Metals
Ions, charged atoms
Resistance
 Ohmmeter
 Multimeter
 Resistor
 Ohms (Ω), kΩ, MΩ
 Resistor color code
 Tolerance
Resistances in series

Summing resistance in series
 Voltage drop
Resistances in parallel
 Summing resistances in parallel
 Voltage divider
Ohm's Law
Power law
Power, watt (W), mW
RMS voltage
RMS current
Ground
 Electrical, circuit, chassis
Short circuit
Electric shock
One-hand survival rule

REVIEW QUESTIONS AND PROBLEMS

1. List the three major types of particles found in atoms and indicate for each type whether it has a positive charge, negative charge, or no charge.

2. What type of force holds electrons in orbits around the nucleus of an atom?

3. Draw a Bohr model picture of a magnesium atom, showing the number of electrons in each shell.

4. a. Why is work required to separate an electron from an atom?
   b. If an electron is separated from an atom and held separated, what happens to the work that was done to separate the electron?
   c. What happens to this work or energy if the electron is allowed to "fall" back to its orbit?

5. List and briefly describe three methods of generating electricity.

6. Figure 1-5a shows a diagram for an electric circuit containing a battery, a switch, and a light bulb.
   a. Redraw the circuit from part a, showing how you would connect an ammeter to measure the current in the circuit. Label the + and − terminals of the ammeter.
   b. Show how you would connect a voltmeter to the circuit from part a in order to measure the voltage of the battery. Label the + and − terminals of the voltmeter.

7. a. The unit used to measure electric current is the _____ .
   b. The unit used to measure voltage is the _____ .
   c. The unit used to measure resistance is the _____ .

8. a. A current of 0.045 A can also be represented as _____ mA.
   b. A current of 0.000126 A can also be represented as _____ µA.

9. If the voltage applied to a circuit is increased, the current flowing through the circuit will _____ (increase, decrease, remain the same).

10. Draw a diagram showing how two 6-V batteries can be connected to produce a 12-V supply.

11. Define the terms *conductor*, *insulator*, and *semiconductor*.

12. a. Use Table 1-1 to determine how many electrons magnesium atoms have in their outer shells.
    b. Based on the number of electrons in the outer shell of magnesium atoms, predict whether magnesium is a conductor or an insulator. Explain your prediction.
    c. Pure water is an insulator, but a solution of water and magnesium chloride is a good conductor. Explain why.

13. Determine the value and tolerance of carbon resistors with the following color bands:
    a. brown, red, orange, silver
    b. yellow, violet, red
    c. brown, green, brown, gold
    d. red, red, yellow, silver
    e. blue, gray, black
    f. brown, black red, gold
    g. red, violet, green, silver

14. a. Draw a diagram and describe how you would use an ohmmeter to determine if the bulb in the circuit in Figure 1-5 is burned out.
    b. Describe how you could use an ohmmeter to determine if the wire in an extension cord is broken.

15. Figure 1-32 shows a voltage source of 120 V connected across a heater which has 8 Ω of resistance. Calculate the current which will flow through the heater.

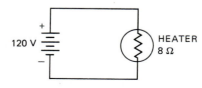

FIGURE 1-32 Circuit for Problem 15.

16. A 150-Ω resistor has a voltage of 3 V across it. Calculate the current through the resistor. Express your answer in milliamps.

17. Calculate the voltage required to cause a current of 2 A to flow through a 25-Ω resistor.

18. Calculate the voltage required to cause a current of 5 mA to flow through a 1-kΩ resistor.

19. Calculate the value of resistance which will allow 5 A to flow through it from a 12-V supply.

20. Calculate the power dissipated in the resistors in:
    a. Problem 15
    b. Problem 16
    c. Problem 17

21. Calculate the power dissipated in the resistors in:
    a. Problem 18
    b. Problem 19

22. What power rating resistor would you need to use for a 470-Ω resistor which will have 20 mA flowing through it?

23. For the series circuit in Figure 1-33, answer the following:

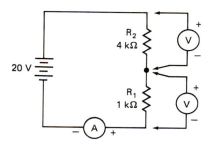

FIGURE 1-33   Series circuit for Problem 23.

a. The total resistance is _____ .
b. The current through the circuit is _____ .
c. The voltage across the 1-kΩ resistor is _____ _____ .
d. The voltage across the 4-kΩ resistor is _____ _____ .
e. The sum of the voltages across the two resistors is _____ .
f. Compare the sum of the voltages across the two resistors with the voltage of the source.
g. The power dissipated in the 1-kΩ resistor is _____ .
h. The power dissipated in the 4-kΩ resistor is _____ .
i. If the 4-kΩ resistor burns out (opens), the current through the circuit will be _____ .
j. With the 4-kΩ resistor burned out, the voltage across the 1 kΩ resistor then will be _____ . *Hint:* Use Ohm's law and the current you determined in part *i*.

24. Figure 1-34 shows a circuit commonly used to drive a device called a light-emitting diode, which gives off light when a few milliamps of current are passed

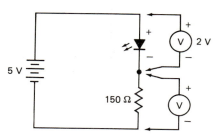

FIGURE 1-34   Circuit for Problem 24.

through it. Light-emitting diodes, or LEDs, may be red, yellow, or green. When an LED has current passing through it, it has a fixed voltage drop across it, as shown in Figure 1-34. (The actual voltage depends on the color of the LED.) Knowing this voltage drop, you can calculate the voltages and current in the circuit without having to know a "resistance" value for the LED. For the series circuit in Figure 1-34, answer the following:
a. The voltage across the 150-Ω resistor in the figure is _____ .
b. According to Ohm's law, the current through this resistor is _____ .
c. The current through the LED then must be _____ .
d. The power dissipated by the LED is _____ .

25. For the parallel circuit in Figure 1-35, answer the following:

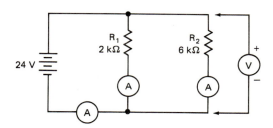

FIGURE 1-35   Parallel circuit for Problem 25.

a. The voltage read by the voltmeter is _____ .
b. The current through the 2-kΩ resistor is _____ .
c. The current through the 6-kΩ resistor is _____ .
d. The total current in the circuit is _____ .
e. The single resistor which will allow the same total current to flow is _____ .
f. The power dissipated by the 2-kΩ resistor is _____ .
g. If the 2-kΩ resistor burns out (opens), the voltage read by the voltmeter will be _____ .

26. Assume three kitchen appliances are connected to one 120-V circuit, as shown in Figure 1-36. The current required for each appliance as determined from its electrical label is shown next to each appliance.
    a. What voltage will the voltmeter show?
    b. Will the fuse blow if all three appliances are turned on at the same time?

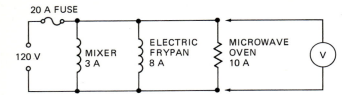

FIGURE 1-36   Circuit for Problem 26.

c.   How much power does the microwave oven use?

27.   Figure 1-37 shows the circuit for a string of Christmas tree lights.

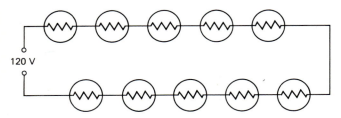

FIGURE 1-37   Circuit for Problem 27.

a.   Is this a series-connected string or a parallel string?
b.   What is the voltage across each bulb when the string is lit?
c.   Draw a diagram showing how the bulbs are connected for the other type of string.
d.   What is the voltage across each bulb connected in this way?

28.   a.   Draw a graph which shows one cycle of the voltage produced by an ac generator.
b.   If the positive half of the waveform you drew in part *a* causes current to flow in one direction through a circuit, what effect will the voltages represented by the negative half of the waveform have?
c.   A 120-V RMS source is connected across a 24-Ω resistor. Calculate the RMS current that will flow through the resistor.
d.   Calculate the power that will be dissipated by the 24-Ω resistor in part *c*.

29.   a.   Why are circuit breakers or fuses put in the ac power wires in your house?
b.   Describe the function of the round prong on a standard 120-V power plug and explain why you should never remove this prong.
c.   Explain why you can get an electrical shock by touching the "hot" ac power wire and a grounded object, such as a water faucet.

30.   List four important safety rules to remember when working with electrical circuits.

31.   a.   Draw a circuit showing how you would connect a 5-V power supply to give a voltage of +5 V with respect to ground.
b.   Draw a circuit showing how you would connect two 12-V power supplies to give both a voltage of +12 V with respect to ground and a voltage of −12 V with respect to ground.

32.   Describe the "one hand survival rule" and give the circuit voltage above which you should be especially careful to follow this rule.

2 Binary Numbers and Logic Gates

This chapter begins our actual trek into the world of digital electronics. The term *digital* is used to represent quantities or systems that can have only certain fixed values. A simple lamp, such as the one shown in Figure 2-1*a*, is an example of a digital system because the lamp

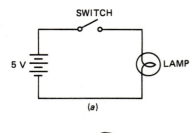

SWITCH

5 V

LAMP

(a)

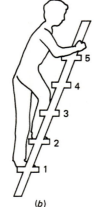

(b)

FIGURE 2-1 Digital systems. *(a)* On-off switch circuit. *(b)* Ladder.

can only have two states, on or off. A ladder, such as that shown in Figure 2-1*b*, is another common example of a digital system. For a ladder the only stable places to stand are on the steps, so the fixed values for your position on a ladder are step 1, step 2, step 3, etc.

By contrast, quantities or systems which can have any value within a given range are referred to as *analog*. The voltage between the center tap of the variable resistor and ground in Figure 2-2*a* is an example of an analog quantity because it can be adjusted to any value between

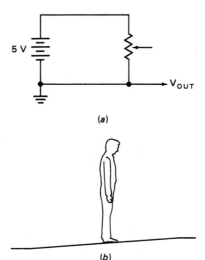

5 V

V_{OUT}

(a)

(b)

FIGURE 2-2 Analog systems. *(a)* Voltage from a variable resistor. *(b)* Sidewalk ramp.

0 V and 5 V. A sidewalk ramp such as that shown in Figure 2-2*b* is another example of an analog system because you can stand at any one of an infinite number of positions on the ramp.

Digital computers and other electronic digital systems work with electrical signals that can have only two possible values, such as 0 V and +5 V. Our familiar decimal number system is not directly suitable for representing numbers in a digital system because each decimal digit has 10 possible values, 0 through 9. Therefore, in digital systems we use the *binary* system to represent numbers because in a binary, or *base-2*, number system each digit can only be a 0 or 1. In this chapter we teach you how to work with the binary number system and how devices called *logic gates* make decisions based on the binary states on their inputs.

OBJECTIVES

At the conclusion of this chapter you should be able to:

1. Define the term *digital* and give an example of a digital quantity or system.

2. Define the term *analog* and give an example of an analog quantity or system.

3. Count in the binary number system.

4. Convert a decimal number to its binary equivalent.

5. Convert a binary number to its decimal equivalent.

6. Identify and draw the symbols for common logic gates.

7. Write the truth table for AND, OR, NAND, NOR, Exclusive OR, and Exclusive NOR gates, and for an inverter.

8. Write the Boolean algebra expressions for individual logic gates and for simple combinations of logic gates.

9. Use signal assertion levels and the shapes of logic symbols to describe the operation of a logic circuit directly from a schematic.

THE BINARY NUMBER SYSTEM

Review of the Decimal Number System

To help you better understand the binary number system, we will first review the familiar *decimal* number system. The decimal number system is often referred to as the *base-10* number system because each digit represents 10 raised to some power. Shown here, for example, is the decimal number 5346.72 with the weight of each placeholder, or digit, expressed as a power of 10:

	5	3	4	6	.	7	2
10^4	10^3	10^2	10^1	10^0	.	10^{-1}	10^{-2}	10^{-3}
10,000	1000	100	10	1	.	$\frac{1}{10}$	$\frac{1}{100}$	$\frac{1}{1000}$
						0.1	0.01	0.001

The first digit to the left of the decimal point represents 10 to the 0 power (10^0) and has a weight of 1. The next digit to the left represents 10^1 and has a weight of 10. The next digit to the left has a weight of 10^2, or 100. The digit to the left of the 10^2 digit has a weight of 10^3, or 1000. The power of 10 increases by 1 for each digit you move to the left. In other words, the weight of each digit is 10 times the digit to its right.

Digits to the right of the decimal point represent decimal fractions. The digits in the decimal number 5346.72 then tell you that the total value of the number is equal to five 1000's + three 100's + four 10's + six 1's + seven ¹⁄₁₀'s + two ¹⁄₁₀₀'s.

In the decimal number system there are 10 possible symbols, 0 through 9, for each digit. The odometer in your car provides a good example of what happens to each digit as you count up in decimal. When the count in any digit position reaches 9, the next count will cause that digit to roll around to 0, and a "carry" of 1 to be added to the next higher digit.

A number system can be built using powers of any number or base, but some numbers are much more useful as a base than others. It is difficult to build electronic devices which can store and manipulate ten different voltage levels but relatively easy to build devices which can work with two voltage levels. A base-2, or binary, number system has only two possible values for each digit, so it is a natural choice for representing numbers in digital systems.

Binary Numbers and Counting in Binary

In the binary, or base-2, number system the weight of each digit is equal to some power of 2. The weights of some binary digits as powers of 2 are shown here:

2^8	2^7	2^6	2^5	2^4	2^3	2^2	2^1	2^0.	2^{-1}	2^{-2}	2^{-3}
256	128	64	32	16	8	4	2	1 .	$\frac{1}{2}$	$\frac{1}{4}$	$\frac{1}{8}$

The first digit to the left of the binary point represents 2 to the 0 power (2^0), which has a weight of 1. The next digit to the left of the binary point represents 2^1, which has a weight of 2. The next binary digit represents 2^2, or 4. Since each digit is just 2 times the digit to its right, you can easily extend the example to as many digits as you want. For your reference, the example shows the values of powers of 2 from 2^{-3} to 2^8. Now let's see how you count in binary.

The number of symbols needed to represent digit values in any number system is equal to the base used. The decimal, or base-10, number system requires ten symbols, 0 through 9. The binary, or base-2, number system uses only two symbols, 0 and 1. Therefore, when a binary digit that is a 1 is incremented, that digit will roll around to 0 and a carry will be passed on to the next higher digit. Figure 2-3 shows how you count up from 0 in binary. If you work your way through this illustration, you should quickly see the pattern.

DECIMAL NUMBER	BINARY EQUIVALENT NUMBER								
	2^8	2^7	2^6	2^5	2^4	2^3	2^2	2^1	2^0
0									0
1									1
2								1	0
3								1	1
4							1	0	0
5							1	0	1
6							1	1	0
7							1	1	1
8						1	0	0	0
9						1	0	0	1
10						1	0	1	0
11						1	0	1	1
12						1	1	0	0
13						1	1	0	1
14						1	1	1	0
15						1	1	1	1
16					1	0	0	0	0
17					1	0	0	0	1
18					1	0	0	1	0
19					1	0	0	1	1
20					1	0	1	0	0

FIGURE 2-3 Binary numbers equivalent to the decimal numbers 0 through 20.

As you go from 0 to 1, the rightmost or *least significant digit* (LSD) increments from 0 to 1, as you would expect. When the count goes from 1 to 2, however, the 1 in the least significant digit of the binary count rolls around to 0 and a carry of 1 goes over to the next higher digit. This is logical because, if you look at the digit weights at the top of Figure 2-3, you can see that the value of this next digit is 2^1, or 2. The binary number 10 then represents $(1 \times 2^1) + (0 \times 2^0)$. A count of 3 is represented in binary as just 11, which is equal to $2^1 + 2^0$, or $2 + 1$. Incrementing the count to 4 rolls both of the rightmost digits around to 0 and sends a 1 to the third digit. The binary representation of 4, then, is 100, which is equal to:

$$(1 \times 2^2) + (0 \times 2^1) + (0 \times 2^0) = (4 + 0 + 0) = 4$$

It may help you to visualize the count sequence if you think of how a binary odometer would function.

One more important point that we need to tell you before going on is that the term *binary digit* is often shortened to simply *bit* for ease in speaking and writing. The binary number 10001110, for example, is referred to as an *8-bit* number. The leftmost bit of the number is referred to as the *most significant bit*, or *MSB*, and the rightmost bit of the number is referred to as the *least significant bit*, or *LSB*.

Converting Binary Numbers to their Decimal Equivalents

When you convert a number from one number system to another, you are not changing the value of the number. You are just changing how the value is represented.

The principle of converting a number expressed in binary to its equivalent decimal form is shown here:

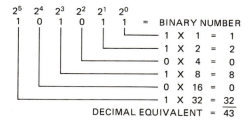

To do the conversion you add up the equivalent decimal values of all the digits which are 1 in the binary number. The 2^0 digit of the example binary number is a 1, so you add a 1 for the value of this digit. The 2^1 digit is also a 1. Since this digit has a decimal value of 2, you add 2 to the total for it. The 2^2 digit in the binary number is a 0, so 0 (nothing) is added to the total for it. The 2^3 digit in the binary number is a 1. Since the decimal equivalent value of this digit is 8, you add 8 to the total for it. The 2^4 digit in the binary number is a 0, so it contributes nothing to the total. Finally, the 2^5 digit in the binary number is a 1, so you add 32, the decimal equivalent value for this digit, to the total. The total of these values is 43 decimal, or 43_{10}, which is equivalent

to 101011 binary (101011_2). For larger binary numbers you continue the process until you have added the decimal value of all the binary digits that are 1 to the total.

We won't be using any fractional binary numbers in this book, but we want to show you that binary numbers with fractional parts can be converted to their decimal equivalents in the same way that whole numbers can. For numbers other than whole numbers, the decimal equivalent of each fractional part that is a 1 is added to the total along with the equivalents for the nonfractional digits, as shown in Summary Example 2-1.

SUMMARY EXAMPLE 2-1

1. Determine value of each digit that is a 1.

2. Add all of these values together.

2^3	2^2	2^1	2^0	.	2^{-1}	2^{-2}		
1	0	0	1	.	1	1	=	BINARY NUMBER

$1 \times 0.25 = 0.25$
$1 \times 0.5 = 0.50$
$1 \times 1 = 1$
$0 \times 2 = 0$
$0 \times 4 = 0$
$1 \times 8 = 8.00$
DECIMAL EQUIVALENT $= 9.75$

Converting Decimal Numbers to their Binary Equivalents

There are two common methods of converting decimal numbers to binary. For the first method you keep subtracting powers of two from the number and put a 1 in the binary number for each power that you subtract. This method is simply a reverse of the binary-to-decimal conversion discussed above. The decimal number 22, for example, is converted with this method as follows.

You start by looking at the number to find the largest power of 2 that is less than or equal to the number.

Decimal number $= 22$
$\underline{\quad - 16} \quad (2^4 \text{ bit } = 1)$
remainder $\quad 6$
$\underline{\quad - 4} \quad (2^2 \text{ bit } = 1)$
remainder $\quad 2$
$\underline{\quad - 2} \quad (2^1 \text{ bit } = 1)$
remainder $\quad 0$

64	32	16	8	4	2	1	Decimal weights
2^6	2^5	2^4	2^3	2^2	2^1	2^0	Binary weights
0	0	1	0	1	1	0	Binary digits

$$22_{10} = 10110_2$$

For our example number of 22, 2^6, or 64, is too large, and 2^5, or 32, is too large, so you put 0's in these digit positions in your binary number or leave these positions blank.

Since 2^4, or 16, will fit in 22, you subtract 16 from the decimal number you are converting and put a 1 in the 2^4 digit position of the binary number you are building. You then continue the process, using the remainder of 6 from the first subtraction as shown.

Since 2^3, or 8, is too large to fit in the remainder of 6, you put a 0 in the 2^3 digit position of the binary number you are building. Because 2^2, or 4, will fit in the remainder of 6, you subtract 4 from the 6 and put a 1 in the 2^2 position of the binary number. The remainder left from this subtraction is 2, which is exactly equal to 2^1. Therefore, you put a 1 in the 2^1 digit position of the binary number. Subtracting 2 from the previous remainder of 2 leaves no remainder, so you put a 0 in the 2^0 digit position of the binary number. The binary equivalent of 22 decimal, then, is 10110.

To check your answer, you can convert 10110 to its decimal equivalent with the binary-to-decimal technique we showed you in the previous section. For review, here's the conversion for 10110:

```
2⁴  2³  2²  2¹  2⁰
 1   0   1   1   0  = BINARY NUMBER
                └── 0 X 1  = 0
             └───── 1 X 2  = 2
          └──────── 1 X 4  = 4
       └─────────── 0 X 8  = 0
    └────────────── 1 X 16 = 16
                    CHECK = 22 DECIMAL
```

SUMMARY EXAMPLE 2-2

Decimal number = 46
                 −32 (2^5 Bit = 1)
      Remainder   14
                 − 8 (2^3 Bit = 1)
      Remainder   6
                 − 4 (2^2 Bit = 1)
      Remainder   2
                 − 2 (2^1 Bit = 1)
      Remainder   0

```
64   32   16   8    4    2    1
2⁶   2⁵   2⁴   2³   2²   2¹   2⁰
 0    1    0    1    1    1    0
        46₁₀ = 101110₂
```

$46_{10} = 101110_2$

CHECK by converting 101110 to its decimal equivalent.

```
2⁶  2⁵  2⁴  2³  2²  2¹  2⁰
 0   1   0   1   1   1   0  = BINARY NUMBER
                       └── 0 X 1  = 0
                    └───── 1 X 2  = 2
                 └──────── 1 X 4  = 4
              └─────────── 1 X 8  = 8
           └────────────── 0 X 16 = 0
        └───────────────── 1 X 32 = 32
                           CHECK  46 DECIMAL
```

This first conversion method is messy to describe but easy to do. Try using the method to convert the decimal number 46 to its binary equivalent. As shown in Summary Example 2-2, you should get 101110 as your result.

The second method of converting decimal numbers to their binary equivalent uses a sequence of divisions by 2. Here's how you use this method to build the binary equivalent of the decimal number 227:

					Binary Result		Check	
$2\overline{)227}$ =	113	remainder	1 (LSB)	1 ×	1	=	1
$2\overline{)113}$ =	56	remainder	1	1 ×	2	=	2
$2\overline{)56}$ =	28	remainder	0	0 ×	4	=	0
$2\overline{)28}$ =	14	remainder	0	0 ×	8	=	0
$2\overline{)14}$ =	7	remainder	0	0 ×	16	=	0
$2\overline{)7}$ =	3	remainder	1	1 ×	32	=	32
$2\overline{)3}$ =	1	remainder	1	1 ×	64	=	64
$2\overline{)1}$ =	0	remainder	1 (MSB)	1 × 128	=	128
					Total	=	227

Therefore:

$$227_{10} = 11100011_2$$

Here are the steps in using a sequence of divisions by 2. First divide the initial decimal number by 2 and write down the quotient and the remainder as shown. The remainder, which can only be a 0 or a 1, is the least significant bit (LSB) of the binary number you are trying to find. The quotient is used for the next division. Divide the quotient of 113 by 2 and write down the new quotient and the new remainder. The remainder produced by this division is the next bit in your binary number. The quotient produced here will be used in the next division. You continue the division process until the quotient produced by a division is 0. The column of remainders is the binary equivalent of the given digital number. Note that the MSB (most significant bit) of the binary number is at the bottom of the column and the LSB (least significant bit) is at the top of the column. The right-hand column (*Check*) of this conversion example shows how you can convert the binary result back to decimal to make sure that it is correct.

The advantage of this successive division conversion method is that it requires only simple division by 2. Also, as we show later, the same general method is useful for converting decimal numbers to their equivalents in other number systems, such as base 8 or base 16.

Fractional parts of decimal numbers can be converted to their binary equivalents by a repetitive multiplication technique. You will seldom have to do this, but for reference, here's how you convert 0.625 decimal to binary:

	Binary		Check	
2×0.625	= 1.25 (MSB)	1×0.5	=	0.5
2×0.25	= 0.5	0×0.25	=	0
2×0.5	= 1.0 (LSB)	1×0.125	=	0.125
				0.625

$$0.625_{10} = 0.101_2$$

To start, you multiply the fractional decimal number by 2. After the multiplication, the number to the left of the decimal point is the value of the MSB of the binary equivalent fraction you are trying to find. For the example shown, $0.625 \times 2 = 1.25$. Since the number to the left of the decimal point is a 1, the MSB of the desired binary fraction is a 1. The next step is to multiply the *decimal* part of the first result by 2. The number to the left of the decimal point in the new result is the next binary digit that you want. For our example here, $0.25 \times 2 = 0.50$. Since the digit to the left of the decimal point is a 0, the next binary digit is a 0. You continue this multiplication process until a multiplication produces a result of 0 on the right of the decimal point. For our example, the third multiplication produces a result of 1.00. The 1 to the left of the decimal point in this result indicates that the next binary digit is a 1. The 0 to the right of the decimal point indicates that we don't need to go any further with the process.

As you can see from the *Check* column, on the right side of this example, 0.101 is the binary equivalent of 0.625 decimal. This decimal number converts exactly to a binary equivalent fraction, so the number to the right of the decimal point became 0 after only a few operations. For decimal numbers that do not convert (the quantity to the right of the decimal never becomes 0), you can continue the multiplication process until you get the number of binary digits you need.

Comparing Decimal Numbers with Binary Numbers

Now that you have had some practice converting decimal numbers to binary and binary numbers to decimal, it is interesting to compare the number of digits required to express a given number in binary with the number of digits required to express the same number in decimal. In decimal a single digit can represent any one of 10 values, 0 through 9; two digits can represent 100 values, 0 through 99; and three digits can represent 1000 values, 0 through 999. In binary a similar pattern exists. In binary a single digit can represent two values, 0 through 1; two digits can represent four values, 00 through 11; and three digits can represent eight values, 000 through 111. The pattern here is that N decimal digits can represent 10^N values, and N binary digits can represent 2^N values. A 4-bit binary number, for example, can represent 2^4 values, 0000 through 1111 binary (0 through 15 decimal), as shown in Figure 2-3. As another example, an 8-bit binary number can represent 2^8 values, 00000000 through 11111111 (0 through 255 decimal).

In later sections of this book we show you how to work with a base-8 number system, a base-16 number system, and a variety of binary-based codes commonly used in digital systems. Next we want to show you how devices called *logic gates* work with binary data.

BASIC LOGIC FUNCTIONS, LOGIC GATES, AND BOOLEAN EXPRESSIONS

The Logical AND Function and Truth Tables

Previously we told you that digital circuitry represents numbers and other data in binary-coded form because binary codes require only two values, such as 0 and 1. Several different terms commonly used to refer to these two values, or *logic states* as they are called, are:

HIGH	LOW
1	0
TRUE	FALSE
+ 5 V	0 V
ON	OFF

The operation of some circuits is best described in terms of 1 and 0, and the operation of other circuits is best described in terms of high and low. Still others are best described in terms of on and off. We will use a simple example to show you how logical decisions can be made using devices which have only two states.

Figure 2-4 shows a simple circuit with a lamp, a battery, and two switches. The switches are digital devices

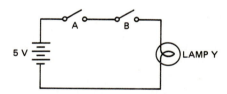

FIGURE 2-4 A simple two-switch and lamp circuit which demonstrates the AND function.

because they have only two possible states, on or off. What we want you to do here is mentally determine if the lamp will be lit or dark for each of the four possible on-and-off combinations of the two switches. To keep track of what state the lamp will be in for each switch combination, we use a table format, shown here, called a *truth table*:

A	B	LAMP

The first two columns of the truth table are used to show all of the possible on and off combinations for the A and B switches. The third column in the table is used to write in the state of the lamp for each of the switch combinations.

In order for the lamp to be on, there must be a conducting path from the negative terminal of the battery to the positive terminal of the battery through the lamp and the switches. If switch A and switch B are both off (open), there is obviously no continuous path for the electrons to get back home through the lamp, so the lamp is off. To indicate this, write OFF in the third column next to this switch combination. If the B switch is on (closed) but the A switch is still off, there will still be no conducting path, so the lamp will be off. Write OFF in the table next to this combination. Likewise, the lamp will be off if the A switch is on but the B switch is off, so write OFF in the table next to this combination. The final switch combination in the truth table is both switches on. With both switches on, the circuit is complete and the lamp will light. Write ON in the truth table next to this switch combination. Here's how your completed truth table for this circuit should look:

Switches		Lamp
A	B	Y
OFF	OFF	OFF
OFF	ON	OFF
ON	OFF	OFF
ON	ON	ON

From the truth table or from your own logic you can see that the lamp will be on only if the A switch is on AND the B switch is on. This relationship is referred to as a *logical AND function.* In addition to the truth table form we have just shown, the AND function and other logic functions can be represented with algebra-like expressions or with graphic symbols. The following sections show you how.

Boolean Algebra Logic Expressions

A truth table is a valid way to describe the operation of a digital circuit, but it takes up quite a bit of space. A more compact way of describing a digital circuit or a logic function is with a *Boolean algebra expression.* Boolean algebra expressions are similar to but not quite the same as those for ordinary algebra. For example, the Boolean algebra expression for the AND function in Figure 2-4 is:

$$LAMP = A \cdot B$$

The quantities A and B in the equation are referred to as *input variables,* and the quantity LAMP is referred to as an *output variable.* The dot ($\cdot$) between the A and the B indicates the AND function. The expression then tells you that the LAMP will be in its ON state if switch A is in its ON state AND switch B is in its ON state. Implied in the expression is the fact that the LAMP will be off for all the other combinations of switches A and B.

Textbooks and other references often use a single letter such as Y to represent the output variable when writing the Boolean expression for simple logic functions. Therefore, you will commonly see the generic expression for an AND function simply written as:

$$Y = A \cdot B$$

In real circuits, however, you will usually give input and output variables meaningful names, such as LAMP-ON.

Also, the dot is often left out of the Boolean expression for the AND function, and the relationship simply written in the form:

$$LAMP = Y = AB$$

The AND Logic Function Symbol

In schematics for digital systems we use graphic symbols to represent logic functions. Figure 2-5 shows the

FIGURE 2-5   Graphic symbol used to represent a 2-input AND logic function.

graphic symbol we most commonly use to represent a 2-input AND logic function. The lines labeled A and B on the symbol represent the inputs. The line labeled Y on the AND symbol represents the output. By definition this symbol tells you that if the A input is ON or HIGH AND the B input is ON or HIGH, the Y output will be ON or HIGH. Logic symbols are almost always drawn on schematics so that their inputs are to the left and their outputs are to the right. Although the symbol in Figure 2-5 shows only two inputs, an AND symbol can have any number of inputs.

Another Example of the AND Function

As another example of the logical AND function, take a look at the three-switch circuit in Figure 2-6. As in the previous example, we want you to use a truth table to determine what switch conditions will cause the lamp to be lit. Whenever you write a truth table for a digital circuit, it is important that you take into account all of the possible input combinations. Since there are three switches in this circuit, there are 2^3, or 8, possible switch

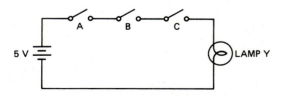

FIGURE 2-6   Another example of the AND logic function: a three-switch and lamp circuit.

combinations. You could make a truth table for this circuit using the terms ON and OFF as we did for the two-switch circuit example, but it is much easier to use 1's and 0's as shown here:

Switches			Lamp
A	B	C	Y
0	0	0	0
0	0	1	0
0	1	0	0
0	1	1	0
1	0	0	0
1	0	1	0
1	1	0	0
1	1	1	1

A 1 in the A, B, or C column represents a switch on, and a 0 represents a switch off. Also, in the output (LAMP) column of the truth table, a 1 is used to represent the on state of the lamp and a 0 is used to represent the off state of the lamp. If you use 1's and 0's for the states of the input variables you can make sure you include all of the possible input combinations by simply writing them down in binary counting sequence as shown. For this example, the eight possible states are represented by the binary counts 000 through 111.

Now, from the truth table you should see that the lamp will be on only if all three switches are on. See if you can write the Boolean expression for this circuit, the correct expression for which is:

$$\text{LAMP} = Y = A \cdot B \cdot C = ABC$$

Next draw the logic symbol that you think represents this logic function. Compare your drawing with Figure 2-7 to see if you drew it correctly.

FIGURE 2-7   Logic symbol for a 3-input AND gate.

Now that you have become familiar with the AND function, we will introduce you to another important logic function, the OR function.

The OR Logic Function

Figure 2-8 shows another simple switch and lamp circuit for you to analyze. On a separate piece of paper, write a truth table for this circuit using 1's and 0's for the switch states as we did in the preceding example. Then determine whether the lamp will be on or off for

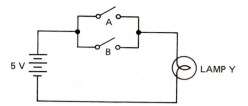

FIGURE 2-8   A simple two-switch and lamp circuit which demonstrates the OR logic function.

each switch combination and write your conclusions in the truth table. Compare your truth table with the correct truth table shown here, where 0 = OFF for a switch or the lamp and 1 = ON for a switch or the lamp:

A	B	Y
0	0	0
0	1	1
1	0	1
1	1	1

From the truth table you should see that the lamp will be ON if either the A switch is on OR the B switch is on. This relationship is called an *Inclusive OR function*, or more commonly just a *logical OR function*. The term *inclusive* here refers to the fact that the case of both switches on is included in the conditions that will light the lamp. In other words, the result (output) of this type of OR will be a 1 if any one of the inputs is a 1 OR if all of the inputs are 1's. Later you will meet an Exclusive OR function, which does not include this latter condition.

A + is used to represent the OR function in Boolean algebra, so the Boolean expression for the example in Figure 2-8 is written:

$$\text{LAMP} = Y = A + B$$

This expression tells you that the LAMP will be ON if the A switch is on OR the B switch is on. It is important for you to remember that in Boolean algebra the + represents the logical OR operation, not addition.

Figure 2-9 shows the graphic symbol we most commonly use to represent the OR logic function. Again, the lines labeled A and B represent the inputs, and the line labeled Y represents the output. Although only two inputs are shown in this example, an OR symbol can have more than two inputs.

FIGURE 2-9   Logic symbol for a 2-input OR gate.

To make sure you understand the operation of the OR function, let's see if you can write the truth table for the 3-input OR symbol in Figure 2-10. To start, draw on a piece

FIGURE 2-10   Logic symbol for a 3-input OR gate.

of paper the basic outline for the truth table as shown here:

A	B	C	Y

Under the A, B, and C headings use 1's and 0's to list all of the possible combinations of the three inputs, as we did for the 3-input AND symbol in a previous section. Remember that putting the input combinations down in binary counting sequence helps you make sure you include all eight possible combinations. Now, for each input combination think about whether the output will be a 1 or a 0 for that combination and write your conclusion in the output (Y) column of the truth table. When you have finished, your truth table should be the same as the one shown here:

A	B	C	Y
0	0	0	0
0	0	1	1
0	1	0	1
0	1	1	1
1	0	0	1
1	0	1	1
1	1	0	1
1	1	1	1

The output will be a 1 if any one of the inputs is a 1 or if more than one of the inputs are 1's. The Boolean expression for this gate then is:

$$Y = A + B + C$$

INTRODUCTION TO ACTUAL LOGIC GATES AND SIGNAL VOLTAGES

An AND Gate Application Example

To help make this discussion of logic functions more real for you and to give you an idea of how logic func-

tions are used, we will show you how a simple digital furnace controller might be built.

The physical circuits used to actually implement simple logic functions are usually called *logic gates*. A circuit which implements the AND function, for example, is referred to as an AND gate, and a circuit which implements the OR logic function is called an OR gate. Usually several gates are contained in a single physical package.

For most of the logic gates and other devices we use in this book, a signal voltage of near 0 V represents a logic low or 0, and a signal voltage of about +3 V to +5 V represents a logic high or 1. In other words, if a voltage of near 0 V is applied to the input of a gate, the gate will respond as if that input is a logic 0 or LOW. If +3 V to +5 V is applied to the input of a gate, then the gate will respond as if that input is a logic 1 or HIGH. Likewise, the output of the gate will be very near 0 V for a logic 0, and at +3 V to +5 V for a logic 1. In the next chapter we discuss these input and output voltage levels more. For now all you need is the concept that for logic gates one voltage level represents a logic LOW or 0 and a distinctly different voltage level represents a logic HIGH or 1. Now let's see how logic gates can be used to build a gas furnace controller.

Gas home furnaces typically operate as follows. When the thermostat indicates that the house is colder than the desired temperature, the furnace controller activates a circuit which ignites a pilot light. If the controller detects that the pilot light lit properly, it turns on the main gas valve. The pilot light ignites the flow of gas from this valve. The burning of this gas heats the house until the thermostat indicates that the house is at the desired temperature.

> NOTE: If the gas were turned on by the thermostat alone, without checking that the pilot light is lit, the gas might be turned on without a pilot light to ignite the gas. In this case the house might fill up with gas, which would explode if ignited by a spark!

The part of the controller we want to implement as an example here is the logic to produce a signal which turns on the gas if the thermostat is ON (house too cold) AND the pilot light is ON (lit). Here is the truth table we want for the controller:

PILOT	THERMOSTAT	GAS
OFF	OFF	OFF
OFF	ON	OFF
ON	OFF	OFF
ON	ON	ON

The variables in the truth table are defined as follows:

PILOT—If this signal is on, the pilot light is burning and will properly ignite the gas when it is turned on.

THERMOSTAT—This signal will be on if the room is too cold.

GAS—As shown in the truth table, we want this signal to be on if the pilot signal and the thermostat signal are both on.

We can write the Boolean expression for this function as:

$$GAS = PILOT \cdot THERMOSTAT$$

or, in shortened form, as $G = PT$. This is an AND function, so we will use an AND gate device to implement the function. The next thing we need to do is look at how the pilot and thermostat switches might be connected to produce the required input signals.

Mounted over the pilot light is a heat-activated switch which closes if the pilot light is lit (on). Figure 2-11 shows

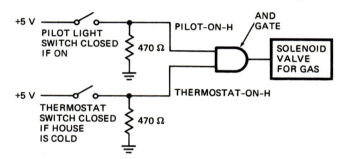

FIGURE 2-11   Circuit diagram for a simple digital furnace controller showing pilot-light switch, thermostat switch, and AND gate.

one way that this switch can be connected to produce a logic signal which is either 0 V or +5 V on the input of an AND gate. If the switch in the circuit shown in Figure 2-11 is closed, the gate input will be connected directly to +5 V. This represents a logic high or 1. If the switch is open, the gate input will be pulled to near 0 V by the resistor connected to ground. This voltage represents a logic low or 0 for the circuit. The lower part of Figure 2-11 shows an identical circuit for the thermostat switch. This circuit will produce +5 V for a logic 1 on the gate input when the switch is closed and 0 V for a logic 0 when the switch is open. If both switches are on, then both inputs of the AND gate will be at logic 1 levels. According to the truth table for a 2-input AND gate shown below, these two 1's will cause the output of the AND gate to be at a logic 1 level:

P	T	G
0	0	0
0	1	0
1	0	0
1	1	1

As we will show in more detail later in this book, the output of the gate can be connected to an electrically

controlled (solenoid) valve in such a way that this logic 1 turns on the main gas valve. The flow of gas will be ignited by the pilot light and heat the house until the thermostat switch opens, indicating the house is now warm.

When the temperature in the house reaches the desired temperature, the thermostat switch will open. The 470-Ω resistor will cause the thermostat input of the AND gate to be pulled to a voltage near 0 V, which is a logic 0. According to the truth table for an AND gate, if one input of an AND gate is made low (0), the output will go low (0). For the circuit here, a low on the output of the AND gate will turn off the gas valve. Likewise, if the pilot light blows out or is not lit, the pilot-light switch will open and cause that input of the AND gate to be pulled to a low (0). A low on the PILOT input will cause the output of the AND gate to be low regardless of the state on the THERMOSTAT input. This means that if the pilot is not on, the gas will not be turned on even if the thermostat is on. Take another look at the AND gate truth table if you need to convince yourself of this.

Hopefully, this example has given you an idea how logic gates can be used for controlling everyday systems and how switches can be used to produce logic level signals. Next we show you how logic devices called *inverters* are used to change the assertion levels of logic signals.

INVERTERS

The thermostat switch in Figure 2-11 produces a logic 0 voltage when the switch is open and a logic 1 voltage when the switch is closed. We said initially that the switch would close when the house got cold, so a logic 1 would then be applied to the input of the AND gate. Now, suppose that the thermostat switch is constructed such that it opens instead of closes when the house gets cold. With the circuit shown in Figure 2-11, this thermostat would produce a logic 0 signal when the house is cold and a logic 1 signal when the house is warm. If we connected this type of thermostat switch directly to the input of the AND gate as shown in Figure 2-11, the system will not work correctly. If you work through the logic of the circuit, you will see that the gas will be turned on when the house is warm.

The problem here is that the logic signals from the thermostat switch are reversed from the way we want them to be. The signal is a 0 when we want it to be a 1, and it is a 1 when we want it to be a 0. There are several ways to solve this problem. One way is to use a device called an *inverter*. Figure 2-12 shows two equivalent graphic symbols we use to represent an inverter. For each of the inverter symbols, the input is on the left side of the symbol and the output is on the right. As we show you later, you use the symbol which makes a particular circuit easiest to analyze. The point for you to remember for now is that the two symbols are equivalent.

The truth table for an inverter is as follows:

A	Y
0	1
1	0

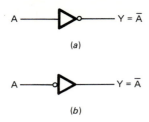

(a)

(b)

FIGURE 2-12 Inverter symbols. (a) Inverter symbol with bubble on output. (b) Inverter symbol with bubble on input.

As you can see from the truth table, the logic state on the output of an inverter is simply the opposite of the state on the input. The opposite of a state is often referred to as its *complement,* so another way of describing the operation is to say that an inverter produces the complement of the variable applied to its input.

The Boolean expression for the inverters shown in Figure 2-12 is:

$$Y = \overline{A}$$

The bar over the A indicates inversion, and the expression is read as, "Y equals A not," "Y equals A bar," or "Y equals not A." Figure 2-13 shows how an inverter can be used to solve our problem of a thermostat switch which is off instead of on when the house is cold.

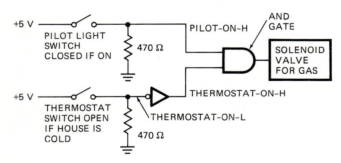

FIGURE 2-13 Inverter used to complement logic signal from thermostat switch in furnace controller.

The inverter is connected between the thermostat switch and the input of the AND gate. Now, if the switch is open (house cold), the 0 produced by the switch will be applied to the input of the inverter. The 0 on the input of the inverter will cause a 1 to be present on its output. The output of the inverter is connected to the input of the AND gate, so the AND gate will have a 1 on its input, as desired, when the house is cold. When the house is warm, the switch will be on and the input of the inverter will have a 1 on it. A 1 on the input of inverter will produce a 0 on the thermostat input of the AND gate. A low on this input will cause the output of the AND gate to go low and turn off the gas. To summarize, then, an inverter is simply used to complement a logic signal.

Most logic designers do not consider an inverter a logic gate in the same way that AND and OR circuits are gates because the output of an inverter is not the result of any logical operation on input variables. Inverters simply complement, or "change the assertion level of," logic signals. In the next short section we explain what we mean by "change the assertion level."

LOGIC SIGNAL ASSERTION LEVELS, NAMING, AND BUBBLES

Logic Signal Assertion Levels and Signal Naming

Digital signal sources such as gate outputs or switches have two logic states, but in most systems one level of a signal more obviously causes some action to occur. The signal level which causes some action to occur is referred to as the *asserted* level or the *active* level for that particular signal.

As an example of this, think about the THERMOSTAT signal in the furnace controller in Figure 2-11. If the thermostat is on, a logic high (1) will be applied to the input of the AND gate. A high on this input will enable the AND gate to turn on the gas when the pilot light turns on and puts a high on the other input of the AND gate. Since a logic high on the THERMOSTAT signal enables the AND gate, the THERMOSTAT signal is referred to as an *active high* signal. Another common way of describing this signal is to say "the asserted level for the THERMOSTAT signal is a high." The signal from the PILOT switch in Figure 2-11 is also an active high signal because it is the high level of this signal which enables the other input of the AND gate so that the gas is turned on in the furnace.

Here's another example. The pilot switch in Figure 2-11 produces a logic high (1) if the switch is closed (pilot on), and a logic low (0) if the switch is open (pilot off). The active state here is the high state because this is the state that will cause the gas to be turned on if the thermostat is on. Therefore, we refer to the PILOT signal from this circuit as an *active high* signal, or we say that the *asserted level* for the PILOT signal is a high.

To indicate on a schematic that a signal is active high, several different conventions are commonly used. The name of the signal may be followed by -H, .H, -1, or no additional symbol. To indicate that the THERMOSTAT signal is active high, for example, it might be labeled on a schematic as THERMOSTAT-H, THERMOSTAT.H, THERMOSTAT-1, or just THERMOSTAT.

For an example of an active low signal, we will look at the thermostat switch circuit in Figure 2-13. This circuit looks the same as the one we used in the previous example, but as we said earlier, this switch works in the opposite direction. In other words, this thermostat switch will be open if the house is too cold. The open condition of the switch will produce a voltage of about 0 V, or a logic low state. Since the switch condition (house cold) that we are most interested in produces a logic low, the signal from this switch is referred to as an active low signal. Active low signals are identified on schematics by adding -L, .L, -0, /, xi, or * after the signal name or by putting a bar over the signal name. The active low signal from the thermostat switch in Figure 2-13 might, for example, be identified as THERMOSTAT-L, THERMO-STAT.L, THERMOSTATxi, THERMOSTAT-0, THERMOSTAT/, THERMOSTAT*, or simply $\overline{\text{THERMOSTAT}}$.

The advantage of including the active or asserted level for a signal in the name is that you don't have to figure out the operation of, for example, a switch when you analyze a logic circuit such as the furnace controller in Figure 2-11. When you see a signal name such as THERMOSTAT-H you can immediately tell that when the thermostat indicates the house is cold, this signal will be asserted high on the input of the AND gate. The high here will cause the AND gate to turn on the gas if the pilot lights. In the same manner, the signal from the pilot-light switch is labeled with a name such as PILOT-H so you can see at a glance that if the pilot is on, this signal will be asserted high.

Here's the point of all of this. You know from the basic definition of an AND gate that the output will be asserted high if all of the inputs are asserted high. Therefore, if the inputs are labeled with THERMOSTAT-H and PILOT-H, as they are in Figure 2-11, you can see directly from the diagram that the output of the AND gate will be asserted high if the thermostat is on AND the pilot is on. Another example should help you further understand how these signal-naming conventions make it very easy to analyze digital circuits.

As we described in detail before, the asserted or active level for the THERMOSTAT signal from the thermostat switch in Figure 2-13 is a low because this signal will be low when the house is cold. Giving the signal from the switch a label such as THERMOSTAT-L or THERMO-STAT* helps you to see that this signal would not produce the desired result if it were connected directly to the input of the AND gate because the input of the AND gate requires a logic high to assert it. The active low signal, THERMOSTAT-L, then needs to be inverted so that it will assert the active high input of the AND gate. The inverter in Figure 2-13 changes the THERMOSTAT-L signal to THERMOSTAT-H as needed. Note in Figure 2-13 that we drew the inverter symbol with the bubble on its input. In the next section we discuss the meaning of bubbles on logic devices and show you how this makes the operation of the circuit clearer. Before going on, however, we want to make one more point about signal assertion levels and names.

Some digital signals cause one action to happen when they are high and a different action to happen when they are low. Signals such as this are given dual names to indicate the two actions caused. A signal input commonly found on memory devices, for example, is labeled READ/$\overline{\text{WRITE}}$, or just R/$\overline{\text{W}}$. The signal name tells you that if that input is high, the memory will be enabled for reading data from it. If the R/$\overline{\text{W}}$ input is low, the memory will be enabled for writing data to it.

Bubbles on Logic Gates

A small round circle, or "bubble," on the output of the symbol for a logic device indicates that the active or asserted level for that output is a logic low. Likewise, a bubble on the input of a logic device indicates that the input will be asserted by a logic low. No bubble on an input or output signal line indicates that the active level for that input or output is a high. The inverter symbol in Figure 2-12a, for example, has no bubble on its input and a bubble on its output. This tells you directly from the symbol that the input is active high and the output is active low. In other words, if the active high input of the inverter is asserted (high), the active low output of the inverter will be asserted (low). If the active high input is not asserted (low), the active low output will be not asserted (high).

The inverter symbol in Figure 2-12b has a bubble on its input and no bubble on its output. This symbol then describes an inverter as having an active low input and an active high output. In other words, this symbol tells you that if the input of the inverter is asserted low, the output will be asserted high. If you look at the truth table for an inverter, you should see that these two symbols are simply two different ways of representing the same truth table. In a particular schematic you use the symbol which makes the operation of the circuit more easily understood.

Here's an example. The thermostat circuit in Figure 2-13 shows an inverter symbol with a bubble on its input and no bubble on its output. This symbol tells you that if the input is asserted low, the output will be asserted high. The THERMOSTAT-L signal is active low by definition, so if it is asserted (low), the input of the inverter will be asserted (low) and the output of the inverter will be asserted (high). The thermostat input of the AND gate has no bubble, so it is active high. A high from the output of the inverter will assert it. Drawing the inverter symbol in this way shows you directly how the inverter converts the active low signal from the switch to the active high signal needed to assert the input of the AND gate. To carry this story a little further, the output of the AND gate has no bubble, so the AND gate output is active high. The AND symbol tells you then that the output will be asserted (high) if the thermostat input to the AND gate is asserted (high) and the pilot-light input is asserted (high).

To summarize this section, then, digital designers usually identify and label logic signals as active high or active low. An inverter is used to change the assertion level of a logic signal as needed to make a logic circuit work correctly. A bubble on an input or output of a digital device indicates that the input or output is active low. If an input or output has no bubble, then that input or output is active high.

The point of all this is that if a digital circuit is drawn and labeled using these signal conventions, it is very easy to analyze the operation of the circuit directly from the drawing when you are trying to find out why the circuit is not working correctly. In the next sections we introduce you to some other common digital devices and give you some more practice analyzing digital circuits.

NAND GATES

All possible logic circuits can be produced by combining basic AND gates, OR gates, and inverters. However, since certain logic functions made from these basic circuits are needed very often, manufacturers produce digital devices which implement them directly. One commonly available device, called a *NAND* gate, for example, functions as an AND gate with an inverter connected on its output. Figure 2-14*a* shows the equivalent circuit for a

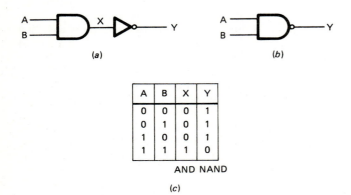

A	B	X	Y
0	0	0	1
0	1	0	1
1	0	0	1
1	1	1	0

AND NAND

(c)

FIGURE 2-14   NAND logic circuits. *(a)* Equivalent circuit. *(b)* Schematic symbol. *(c)* Truth table.

NAND gate, and Figure 2-14*b* shows the schematic symbol most commonly used to represent a 2-input NAND gate. From the equivalent circuit you might expect that the output states for a NAND gate are simply the complements of the corresponding states for an AND gate. The truth table shown in Figure 2-14*c* for the NAND gate equivalent circuit shows that this is indeed the case.

The X column in the truth table represents the values at the output of the AND part of the equivalent circuit, and the Y column in the truth table represents the output of the complete equivalent circuit.

The Boolean expression for the X point in the equivalent circuit is:

$$X = A \cdot B$$

One way of writing a Boolean expression for the Y output of the equivalent circuit and for a 2-input NAND gate in general is:

$$Y = \overline{AB}$$

The fact that the overbar extends over both A and B indicates that A and B are first ANDed together, and the result complemented, just as shown in the equivalent circuit diagram and the truth table. The term NAND is in fact just a simplified way of representing Not-AND.

We have used a 2-input NAND gate for our initial example here, but NAND gates are commonly available with 3, 4, 8, and 13 inputs. Figure 2-15, for example, shows the schematic symbol for an 8-input NAND gate. A Boolean expression for this gate is $Y = \overline{ABCDEFGH}$.

Alternative Ways of Describing NAND Gate Operation

There are several different ways of showing and describing the operation of a NAND gate. We will show you these

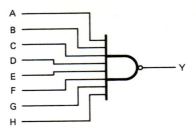

FIGURE 2-15   Logic symbol for 8-input NAND gate.

different ways and in the process give you more practice working with signal assertion levels and bubbles.

CLASSIC NAND SYMBOL AND EXPRESSION

In the preceding section we showed you the classic symbol for a NAND gate and said that one way of writing the expression for this gate is $Y = \overline{AB}$. This expression tells you that the Y output will be high if the A input AND the B input are not both high. In other words, the output will be high if A is low, B is low, or both A and B are low. This expression then describes the operation of a NAND gate in terms of the first three lines in the truth table. Implied but not stated in this expression is the fact that the output will be low if both A AND B are high.

The $Y = \overline{AB}$ expression correctly describes the operation of the gate, but as we will show you later, the overbar in the expression often makes it somewhat difficult to analyze circuits that contain several gates. Here's an easier way of describing the operation of a NAND gate.

ACTIVE LOW OUTPUT AND DESCRIPTION OF A NAND GATE

Remember from a previous discussion that a bubble on an input or output of a logic gate indicates that the input or output is active low. An input or output with no bubble is active high. The inputs on the NAND symbol in Figure 2-16 have no bubbles, so they are active high. In other words, logic highs will assert these inputs. The output of this NAND symbol shows a bubble, so according to the meaning we have given to bubbles, the asserted state for the output is a low.

If you temporarily forget about the bubble on the output and look just at the basic shape of the symbol in Figure 2-16, you can see that the logic function being

SYMBOL SHAPE INDICATES **AND**
FUNCTION IS BEING PERFORMED

NO BUBBLES, SO ACTIVE HIGH INPUTS

BUBBLE, SO OUTPUT IS ACTIVE LOW

FIGURE 2-16   NAND logic symbol showing active levels.

performed is the AND function. The bubble on the output of this NAND symbol can be thought of as simply indicating that the output is active low. In other words, the output will be asserted (low) when the input AND conditions are met. More specifically, if the A input is asserted (high) AND the B input is asserted (high), the output will be asserted (low). Implied in this is the understanding that any of the other input combinations will cause the output to be unasserted (high).

If you look again at the truth table for a NAND gate shown in Figure 2-14c, you will see that this bubbled output AND description simply describes the operation of a NAND gate in terms of the A = 1 and B = 1 line in the truth table. As we discussed before, the $Y = \overline{AB}$ expression describes the operation of a NAND gate in terms of the top three lines in the truth table.

To indicate that the output of this symbol can be thought of as active low, we give it a name such as Y* or Y/. The expression for this output signal then is Y* = AB, as shown in Figure 2-16.

The method we have just shown you for analyzing a logic circuit is called the *functional logic approach* or the *mixed logic approach*. The term *mixed logic* comes from the fact that active low and active high signals are represented on the same circuit drawing. The term *functional logic* comes from the fact that when analyzing the circuit you look at the shape of the symbol to determine the logic function that is being performed, and you look for bubbles to determine if the inputs and output are active low or active high. We use the functional logic approach for representing and analyzing digital circuits throughout the rest of this book.

The advantage of this way of analyzing, for example, the operation of a NAND gate is that you don't have to try to remember the truth table. You just remember that the basic symbol shape represents the AND logic function, then determine the assertion levels for the signals around the symbol by checking whether the signal lines have bubbles or not. The next example will give you more practice with this.

BUBBLED INPUT OR REPRESENTATION OF A NAND GATE

Another way that a NAND gate is commonly represented on electronic schematics is with the symbol shown in Figure 2-17. The obvious question that may come to your mind when you see this symbol is, "How can this OR-shaped symbol represent a NAND gate?" To answer this question, see if you can use the shape of the symbol and the signal assertion levels as indicated by bubbles to predict how the device represented by this symbol operates.

SYMBOL SHAPE INDICATES **OR** FUNCTION IS BEING PERFORMED

BUBBLE INDICATES ACTIVE LOW INPUTS

$Y = \overline{A} + \overline{B}$

NO BUBBLE, SO OUTPUT IS ACTIVE HIGH

FIGURE 2-17   Alternative symbol for NAND gate.

The basic symbol represents the OR function, so the input signals are being ORed. Since the inputs of the symbol have bubbles, they are active low. The fact that the output has no bubble indicates that it is active high. The functional logic description of the device using the shape of the symbol and the assertion levels for the signals is as follows:

If the A input is asserted (low) OR the B input is asserted (low), then the output will be asserted (high).

Another way of saying this is that if any one of the inputs is low or, for that matter, if all of the inputs are low, the output will be high. The symbol tells you directly that this device functions as an OR gate with active low inputs. Implied in this description is the fact that if both inputs are not-asserted (high), the output will be not-asserted (low). The expression for this symbol then can be written as:

$$Y = \overline{A} + \overline{B}$$

This Boolean expression corresponds exactly to our verbal description. In Chapter 6 we will prove that this expression for a NAND gate and the $Y = \overline{AB}$ expression we gave you above are equivalent, but for now we will use a truth table to show you that the bubbled output AND symbol and the bubbled input OR symbol equally describe the operation of a NAND gate.

To refresh your memory, Figure 2-18 shows, with a few additions, the truth table for a NAND gate. As you can see from the figure, the OR symbol with bubbles on its inputs describes the operation of a NAND gate based on the first three combinations in the truth table. With this symbol the operation of the device for the A = 1 and B = 1 input conditions is implied.

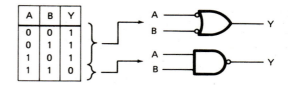

A	B	Y
0	0	1
0	1	1
1	0	1
1	1	0

FIGURE 2-18   Truth table for NAND gate showing relationship of alternative symbols.

Figure 2-18 also shows that the bubbled output AND symbol describes the operation of the device based on the A = 1 and B = 1 input condition as we described earlier. With this symbol the operation of the device for the other 3-input combinations is implied.

The main point for you to understand here is that the NAND symbol in Figure 2-16 and the inverted input OR symbol in Figure 2-17 are completely equivalent. When drawing schematics, a digital designer will use the symbol which best describes the logic function being performed in a particular part of the circuit. If a NAND gate is being used to AND active high signals together to give an active low output, then a symbol like that in Figure 2-16 will be used. If a NAND gate is being used to OR two active low signals together to give an active high output, then a symbol like that in Figure 2-17 will be used. In Chapter 6 we will teach you more about this, but it is very important that you start learning now how to describe the operation of logic gates using their symbol shapes and signal assertion levels.

NOR GATES

A NOR gate is another type of circuit which is so commonly needed that it is manufactured as a single circuit in logic devices. Figure 2-19a shows the equivalent cir-

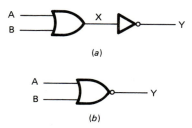

(a)

(b)

FIGURE 2-19   NOR logic circuits. (a) Equivalent circuit. (b) Traditional schematic symbol for 2-input NOR gate.

cuit for a NOR gate, and Figure 2-19b shows the most commonly used symbol for a 2-input NOR gate. The truth table for this equivalent circuit and for a NOR gate is as follows:

A	B	X	Y
0	0	0	1
0	1	1	0
1	0	1	0
1	1	1	0
		OR	NOR

The X column of the truth table represents the simple OR function. The states shown in the Y column are, as you can see, just complemented, or "notted," from those in the X column. The name *NOR* is a contraction of Not-OR. One way of writing the Boolean expression for a 2-input NOR circuit is:

$$X = \overline{A + B}$$

The fact that the overbar extends over both the A and the B tells you that A and B are first ORed, and then the result is inverted. NOR devices are available with more than two inputs. Figure 2-20, for example, shows the symbol used to represent a 3-input NOR gate. One way to write the Boolean expression for this 3-input NOR gate is $Y = \overline{A + B + C}$.

FIGURE 2-20   Schematic symbol for 3-input NOR gate.

Alternative Ways of Describing NOR Gate Operation

In a previous section we showed you that the operation of a NAND gate can be described in several equivalent

but different ways. The operation of a NOR gate can likewise be described in several different ways. Here are three.

TRADITIONAL NOR SYMBOL AND EXPRESSION

As shown in Figure 2-19b, the traditional symbol for a NOR gate is an OR symbol with a bubble on its output. The traditional expression for this symbol is $Y = \overline{A + B}$. The expression tells you that the output will be high if A OR B is not high. This expression accurately describes the operation of the gate, based on the A = 0, B = 0 line in the truth table, but the overbar in the expression makes it somewhat difficult to use in analyzing circuits which contain several gates. As you will see in the next section, taking a functional logic approach to the device is easier.

OR GATE WITH AN ACTIVE LOW OUTPUT

One symbol commonly used for a NOR gate is an OR symbol with a bubble on its output, as shown in Figure 2-21a. The inputs of this classic NOR symbol have no bubbles, so they are active high. Since the output has a bubble, it can be thought of as active low.

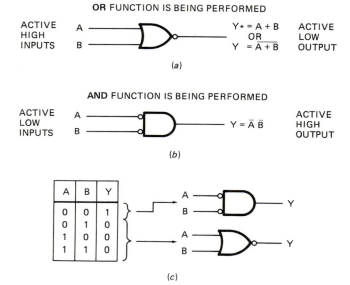

FIGURE 2-21   (a) A traditional NOR symbol. (b) Bubbled input AND representation for NOR gate. (c) Truth table for NOR gate showing relationship of two equivalent symbols.

The characteristic OR shape of the symbol tells you that the OR function is being performed, so the operation of the gate can be described directly from the symbol as follows:

If the A input is asserted (high) OR the B input is asserted (high), the output will be asserted (low).

In other words, if any input or all of the inputs of a NOR gate are high, the output will be low. This description covers the input conditions represented by the last three entries in the truth table in Figure 2-21c. Implied in this description is the fact that the output will be high if both inputs are low. A NOR gate then can be thought of as simply an OR circuit with active high inputs and an active low output. To indicate that the output is active low, it is given a name such as Y*. The logic expression for this bubbled output OR symbol can then be written as Y* = A + B, which corresponds directly to the verbal description we derived from the symbol.

BUBBLED INPUT AND REPRESENTATION OF A NOR GATE

The second way of representing a NOR gate is an AND symbol with bubbles on its inputs, as shown in Figure 2-21b. The bubbles on the inputs indicate that they are active low. Since the output has no bubble, it is active high. The characteristic shape of the symbol tells you that the input signals are being ANDed. The operation of a NOR gate then can be described directly from this symbol as follows:

If the A input is asserted (low) AND the B input is asserted (low), the output will be asserted (high).

This representation of a NOR gate then tells you that if all of the inputs of a NOR gate are asserted low, the output will be high. As shown in Figure 2-21c, this description corresponds to the truth table entry for A = 0 and B = 0. Implied in this description is the fact that any other input signal conditions will cause the output to be unasserted, which for this symbol is a low. The expression for the symbol can be written as Y = $\overline{A}$ AND B or Y = $\overline{A}\,\overline{B}$. Hopefully you can see that this expression comes directly from the shape of the symbol and the indicated assertion levels for the inputs and the output.

The point here is that on a particular schematic, a digital designer will use the NOR gate symbol which makes the operation of the circuit more easily understood. If a NOR gate is being used to OR active high signals together to give an active low output, then the bubbled output OR symbol in Figure 2-21a will be used. If a NOR gate is being used to AND active low signals together to give an active high output, then the bubbled input AND symbol, such as the one in Figure 2-21b, will be used.

EXCLUSIVE OR AND EXCLUSIVE NOR GATES

Exclusive OR Gates

Two other common logic functions which manufacturers produce as single gates are *Exclusive OR* gates and *Exclusive NOR* gates. Figure 2-22a shows the symbol used for an Exclusive OR gate, and Figure 2-22b shows the truth table for it.

The additional curved line at the front of the symbol tells you that this gate is different from the "inclusive" OR gate, which we described in the previous sections. A look at the truth table will quickly show you the difference.

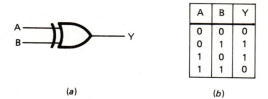

A	B	Y
0	0	0
0	1	1
1	0	1
1	1	0

FIGURE 2-22 Exclusive OR gate. (a) Schematic symbol. (b) Truth table.

The output of an Exclusive OR will be high if the A input is high OR the B input is high but, unlike the regular OR gate, the output will be low if *both* inputs are high. The name of the gate comes from the fact that the output will be high if one input exclusively is high OR the other input exclusively is high. You might think of an Exclusive OR gate as a "disagreeable" gate because its output will be high only if its two inputs have different logic states, or "disagree" with each other. If the two inputs have the same state, both low or both high, the output of the Exclusive OR will be low.

There are two common ways of writing the Boolean algebra expression for an Exclusive OR gate. The first form simply summarizes the input conditions which will cause the output to be high. This expression is:

$$Y = A\overline{B} + \overline{A}B$$

The expression tells you that the output will be high if A is high and B is low OR if A is low and B is high.

The second way of writing the expression for an Exclusive OR function is:

$$Y = A \oplus B$$

The $\oplus$ between the A and B is a shorthand way of indicating that A and B are Exclusive ORed. You should use the first form to represent the Exclusive OR function until the operation of the gate is firmly fixed in your mind. Likewise, the name Exclusive OR is often abbreviated as simply *XOR*, but you should use the full name until you become familiar with the operation of the gate.

Exclusive OR gates typically have only two inputs, but as we show in later chapters, several gates can be connected together to produce the Exclusive OR function for as many inputs as needed.

Exclusive NOR Gates

Figure 2-23a shows the symbol used to represent an Exclusive NOR gate. To make it easier to compare the two, Figure 2-23b shows the truth table for both an Exclusive OR and an Exclusive NOR.

If you compare the Exclusive NOR column (XNOR) in the truth table with the Exclusive OR (XOR) column in the truth table you should see that the output states for this gate are just the inverse of the corresponding states for the Exclusive OR. The output of an Exclusive NOR gate will be high if the two inputs are both high or if they are both low. You might think of an Exclusive NOR as an "agreeable" gate because its output will be high if

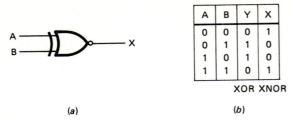

A	B	Y	X
0	0	0	1
0	1	1	0
1	0	1	0
1	1	0	1

XOR XNOR

(a)   (b)

FIGURE 2-23   Exclusive NOR gate. *(a)* Schematic symbol. *(b)* Truth table.

the logic states on the two inputs are the same (agree). As you will see later, this gate is often used as an "equality detector" to determine if two binary numbers are equal to each other.

One way to write the Boolean expression for an Exclusive NOR gate is to simply summarize the input conditions which cause the output to be high. This gives:

$$Y = AB + \overline{A}\,\overline{B}$$

Another, more compact way of representing the Exclusive NOR function is the expression:

$$Y = \overline{A \oplus B}$$

An Exclusive NOR gate is often referred to simply as an *XNOR gate.*

COMBINING LOGIC GATES—INTRODUCTION

The preceding sections of this chapter have introduced you to the individual operations of logic gates. Chapter 6 and following chapters describe in detail how logic gates can be combined to produce the logic function needed for a given digital system. However, to give you an introduction to how this is done, we will show you a couple of small examples.

The whimsical example in Figure 2-24 shows with a

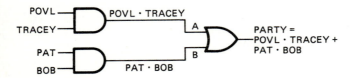

FIGURE 2-24   Example showing use of AND gates with OR gate.

logic diagram a couple of the conditions that cause a party to occur at our house. The party output will be asserted if the A input or the B input of the OR gate is asserted. The A input of the OR gate will be asserted if the Povl and Tracey inputs of the upper AND gate are asserted. The logic expression for this is A = Povl AND Tracey. The B input of the OR gate will be asserted if the Pat and Bob inputs on the lower AND gate are asserted. The logic expression for this is B = Pat AND Bob. The

logic expression for the entire circuit is generated by combining the expressions for the two inputs of the OR gate as follows:

PARTY = (POVL AND TRACEY) + (PAT AND BOB)

Maybe you can use expressions like this to make your life more logical.

Figure 2-25 shows a second, more traditional, example of combining logic gates. Here the outputs from two OR gates are connected to the inputs of a NAND gate.

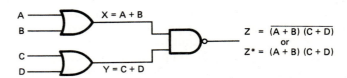

FIGURE 2-25   Example showing use of OR gates with NAND gate.

The output expression for the top OR gate is X = A + B, and the output expression for the bottom OR gate is Y = C + D. These two signals are combined by the output NAND gate. The expression for the final output can be written in two different ways. Based on the traditional overbarred expression for a NAND gate, the output expression is:

$$Z = \overline{(A + B)(C + D)}$$

The second way of writing the final output expression for this circuit is based on our previous discussion of signal assertion levels, bubbles, and the logic function indicated by the shape of the symbol.

The bubble on the output of the NAND gate indicates that it is active low, so the Z signal can be labeled as Z/ or Z*. The inputs and outputs of the OR gates and the inputs of the NAND gate are all active high, so the expression for the complete circuit can then be written as:

$$Z* = (A + B)(C + D)$$

What this expression says is that Z* will be asserted (low) if (A is high OR B is high) AND (C is high OR D is high). Since this expression does not have the overbar that the traditional expression does, it is much more useful for testing and troubleshooting the circuit. This expression makes it easier to see, for example, that if A is high, B is low, C is low, and D is high, the Z* output will be asserted (low). All you have to do to evaluate these conditions is to write a 1 over the A, a 0 over the B, a 0 over the C, and a 1 over the D. Since each pair of parentheses contains a 1, the AND condition is met and Z* will be asserted (low).

In Chapter 6 we show you in detail how to work with multiple gate logic circuits and logic expressions, but this preview should give you some feeling for how logic expressions relate to circuit diagrams.

ANSI/IEEE STANDARD 91-1984 LOGIC SYMBOLS

The familiar curved logic symbols that we have been using for AND, OR, NAND, NOR, Exclusive OR, and Exclusive NOR gates and inverters are defined in Military Standard 806 (MIL STD 806). These symbols have the advantage that you can immediately determine the logic function being performed just by the shape of the symbol. However, the symbols specified for medium-scale integration (MSI) and large-scale integration (LSI) devices in MIL STD 806 do not clearly indicate the function being performed by the devices and they do not indicate the functions of input and output signals on the devices. For 15 years or so various committees have been working on a symbol standard which solves these problems. At this point the latest set of graphic standards for logic functions is standard 91-1984, produced by the American National Standards Institute (ANSI) and the Institute of Electrical and Electronics Engineers (IEEE). This standard uses a scheme called *dependency notation* to represent the relationship of input and output signals on the symbols for MSI and LSI devices. For simple logic gates the 91-1984 standard allows you to use either the traditional MIL STD 806 curved symbols or equivalent rectangular symbols. In this book we will use mostly the traditional curved symbols because they are so easily recognized at a glance. However, since many data books now show both the curved and the rectangular symbols, we want you to become familiar with them. Also, the method used to identify the logic function being performed by each of these devices is used in the MSI and LSI dependency notation symbols, so learning these now gives you a head start on understanding the more complex symbols.

Figure 2-26 shows the ANSI/IEEE standard 91-1984 symbols for the logic gates we have discussed in this chapter. A 1 in a rectangle indicates that the device is an inverter, a & in a rectangle indicates that the device performs the AND function, a ≤ 1 indicates the device performs the OR function, and an $= 1$ indicates that the device performs the Exclusive OR function. For future work it is important that you learn these identifying symbols.

Active low inputs or active low outputs are indicated on these symbols either with round bubbles, as you have seen previously, or with small triangles, as shown in Figure 2-26.

SUMMARY OF BASIC LOGIC GATES

In the next two chapters we show you how to measure the input and output characteristics of actual logic gate devices and how to build circuits with logic gates. In Chapter 5 we show you more about analyzing logic circuits, and in Chapter 6 we will show you how to design and implement logic circuits which solve specified problems. To conclude this chapter, Figure 2-27 shows the logic symbols, truth tables, and logic expressions for all the devices we have discussed in this chapter so you can directly compare the operations of the different devices.

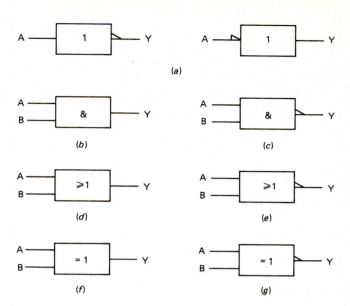

FIGURE 2-26   ANSI/IEEE standard 91-1984 alternative symbols for common logic gates. *(a)* Inverter. *(b)* AND. *(c)* NAND. *(d)* OR. *(e)* NOR. *(f)* Exclusive OR. *(g)* Exclusive NOR.

CHECKLIST OF IMPORTANT TERMS AND CONCEPTS IN THIS CHAPTER

Digital

Analog

Decimal, or base-10, number system

Binary, or base-2, number system

Least significant digit, LSD

Most significant digit, MSD

Binary digit, bit

Least significant bit, LSB

Most significant bit, MSB

Converting binary numbers to their decimal equivalents

Converting decimal numbers to their binary equivalents

Logic state

Truth table

Logical AND function

Boolean algebra expressions

Input and output variables

Logical OR, Inclusive OR, and Exclusive OR functions

Logic gate

Inverter

Inverse, or complement

Signal assertion level
   Active high, asserted high, -H, .H, -1
   Active low, asserted low, -L, .L, xi, -1, /, *

Bubbled inputs and bubbled outputs

Functional and mixed logic

Dependency notation

AND, OR, NAND, NOR, XOR, and XNOR gates

ANSI/IEEE standard logic symbols

REVIEW QUESTIONS AND PROBLEMS

1. State whether each of the following quantities is analog or digital:
   a. the speed of a car
   b. the energy levels of electrons around atoms
   c. time
   d. the number of water molecules in the ocean
   e. a socket wrench ratchet.

2. Give the decimal equivalents for the powers of 2 from 2^0 to 2^8.

3. Write the sequence of binary numbers from 0 to 17.

4. Convert the following binary numbers to their decimal equivalents:
   a. 110
   b. 1010101
   c. 1111
   d. 101101110
   e. 0.1101
   f. 0.011

5. Convert the following decimal numbers to their binary equivalents:
   a. 12
   b. 73
   c. 207
   d. 4096
   e. 0.35

6. Why do digital computers and other digital electronic circuits use only two signal levels?

7. Write the truth table and the Boolean algebra expression for each of the logic symbols shown in Figure 2-27.

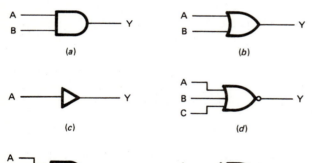

FIGURE 2-27 Figure for Problem 7.

8. Predict the output voltage level for each of the circuits shown in Figure 2-28. Assume that +5 V represents a logic high and 0 V represents a logic low.

9. a. Write the traditional Boolean expression for each gate in Figure 2-29.
   b. Using the shape of the logic symbol and the signal assertion levels indicated on each input and

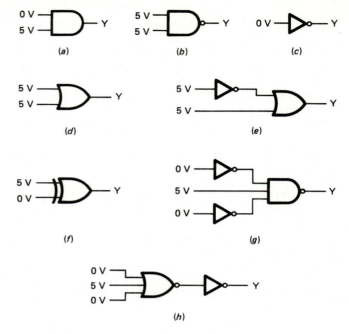

FIGURE 2-28 Figure for Problem 8.

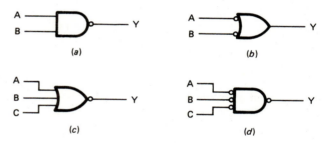

FIGURE 2-29 Figure for Problem 9.

output, describe in words the operation of each gate in Figure 2-29
   c. Write a logic expression which uses the symbol shape and the signal assertion levels to describe the operation of each gate in Figure 2-29.

10. Draw a logic symbol which represents each of the following verbal descriptions:
   a. If the A input is high, the B input is high, and the C input is high, the output is low.
   b. If the A input is high or the B input is high, the output is low.
   c. If the A input is high or the B input is high, the output is high; but when both inputs are high, the output is low.
   d. If the A input is low, the B input is low, and the C input is low, the output is low.

11. Given the two-switch signal sources shown in Figure 2-30:
   a. Draw a gate circuit which will produce a high output only when both switches are closed.

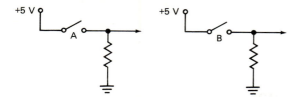

FIGURE 2-30 Figure for Problem 11.

b. Draw a gate circuit which will produce a logic high only if switch A is open and switch B is open.

c. Draw a logic circuit which will produce a low output only when both switches are open.

12. Given the switch signal sources shown in Figure 2-31:

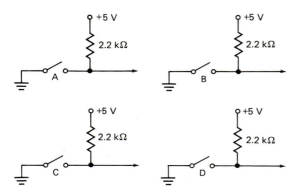

FIGURE 2-31 Figure for Problem 12.

a. Draw a gate circuit which will produce a high output if A is closed or if B is closed, but not if both are closed or open.

b. Draw a gate circuit which will produce a low output only if A is closed and B is closed.

c. Draw a logic circuit which will produce a high output only if A is open, B is closed, and C is open.

d. Draw a logic circuit which will produce a high output only if A and B are closed or C and D are closed.

13. Identify the logic function represented by each of the ANSI/IEEE standard 91-1984 logic symbols shown in Figure 2-32.

14. Draw the ANSI/IEEE standard 91-1984 logic symbols for each of the logic gates in Figure 2-27.

15. a. Write a truth table for the simple lamp circuit in Figure 2-33. Assume SWITCH ON is represented by a 1 and LAMP ON is represented by a 1 in the truth table.

b. Use the truth table to help you think about how you might verbally describe the switch conditions which will cause the lamp to be lit, then use your verbal description to write the Boolean expression for the circuit.

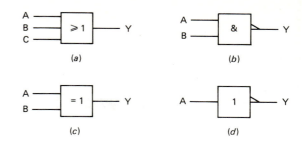

FIGURE 2-32 Figure for Problem 13.

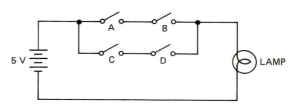

FIGURE 2-33 Figure for Problem 15.

c. From your truth table or Boolean expression, determine the combination of logic gates which can be used to implement this logic function.

16. Figure 2-34 shows the schematic diagram for a household light circuit commonly called a *three-way switch circuit*. You might use a circuit such as this to control the light over a stairway. One switch is located at the top of the stairs and the other is located at the bottom of the stairs. This circuit allows you to turn the light on at the bottom of the stairs and then turn it off when you reach the top of the stairs. It also allows you to turn the light on at the top of the stairs and off at the bottom.

a. Write a truth table for this circuit. Assume that SWITCH UP represents a logic high and SWITCH DOWN represents a logic low. Also assume that LIGHT LIT is represented by a logic high.

b. Look at the truth table in part *a* and think about how you might verbally describe the switch conditions which will cause the light to be lit. Use this verbal description to help you write the Boolean expression for the circuit.

c. Draw the symbol for the logic gate which performs the same logic function as this circuit.

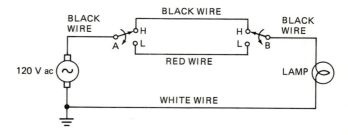

FIGURE 2-34 Figure for Problem 16.

3 Digital Devices and Data Sheets

In the last chapter we introduced you to the graphic symbols, the truth tables, and the Boolean expressions for common logic gates. The major purposes of this chapter are to show you how to build circuits with actual logic gate devices, how to measure some basic input and output characteristics of logic gates, and how to find the information you need about a logic gate in the manufacturer's data sheet. We also show you in this chapter how to predict and observe the signal waveforms that will appear on the outputs of logic gates when their input signals are pulsed high and low rather than being held at fixed levels.

OBJECTIVES

At the conclusion of this chapter you should be able to:

1. Identify the package type for a given logic gate device.

2. Describe how a circuit is built in a systematic manner using either solderless protoboard, soldered connections, or wire wrap.

3. Identify the manufacturer's part number on a given device and use this part number to determine the function and the characteristics of the device from the manufacturer's data book.

4. Use the data sheet for a given logic gate device to determine the pin connections for it.

5. Use a data sheet for a given logic gate device to determine basic input and output characteristics, such as minimum input high voltage, maximum input high voltage, and maximum output low voltage.

6. Define the terms *risetime, falltime,* and *propagation delay time* and use a data sheet to find the values of these parameters for a logic gate device.

7. List the four most common logic device families and describe the major characteristics of each.

8. Describe the three TTL output structures and explain why totem-pole outputs should never be connected together on a common line.

9. Describe the operation of a CMOS transmission gate.

10. Describe the precautions required to prevent static electricity from damaging MOS and other types of ICs.

INTEGRATED CIRCUITS, PACKAGES, AND NUMBERING

In the last chapter we showed you how an AND gate circuit could be used to turn on the gas in a furnace when the pilot light was on and the thermostat indicated that the room was cold. That example was just intended to give you an introduction to how logic gates are used in the real world. Therefore, for that example all we told you about the actual device was that a voltage of near 0 V represented a logic low, and a voltage of +3 V to +5 V represented a logic high. Now we will introduce you to some actual logic gate devices and show you how to build and test circuits with them.

Digital Integrated Circuits

In the early days of digital electronics, logic gates were constructed from large, individual components. Now, all the electronic circuitry required to produce one or more logic gates is built on a single, small piece of specially processed silicon or some other semiconductor material. The basic silicon chip containing the circuitry is referred to as a *die.* Figure 3-1a shows an example of a die for a digital logic device. Depending on the amount of circuitry contained on it, a die may be anywhere from 0.1 to 0.4 inch square.

In order to protect the die from damage and allow connections from it to external circuitry, the die is mounted in a package such as that shown in Figure 3-1b. Figure 3-1b shows the package with its cover removed so you can see how a die is mounted. If you look closely at the figure, you should be able to see that tiny wires connect pads on the die to the larger pins around the edge of the package. Connections can then be easily made from these larger pins to other devices as needed.

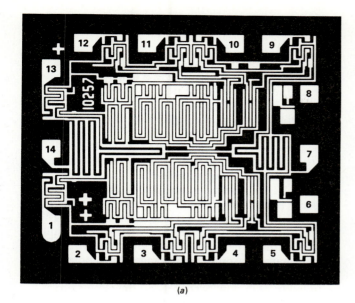

(a)

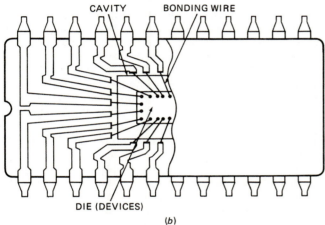

CAVITY · BONDING WIRE

DIE (DEVICES)

(b)

FIGURE 3-1 Die for a digital logic device. *(a)* Die for CD4012BH chip. *(Courtesy RCA Corporation.)* *(b)* Die mounted in package with cover removed.

Logic gates and other electronic devices which contain many electronic components in a single package are commonly called *integrated circuits,* or just *ICs.* Digital ICs are categorized according to the number of basic gates or gate-equivalent circuits that they contain. The commonly used categories and the range of gates for each category are as follows:

Number of Gates	Integrated Circuit Category Name	Abbreviation
1–12	Small-scale integration	SSI
13–99	Medium-scale integration	MSI
100–9999	Large-scale integration	LSI
10,000–99,999	Very large-scale integration	VLSI
Over 100,000	Ultra high-scale integration	UHSI

Simple logic gate devices such as the ones we discuss in this chapter are examples of SSI devices. The counters and code converters we introduce you to in Chapters 7 and 8 are examples of MSI devices. Microprocessor ICs, used as the "brains" of personal computers, are examples of LSI or VLSI devices, depending on their complexity. Some gate array devices, such as those we introduce you to in Chapter 15, are large enough to be classified as UHSI devices. Now let's look at some of the physical packages that ICs come in so you will recognize them when you see them in your computer or some other electronic instrument.

IC PACKAGE TYPES

Most of the devices that you will work with in this book are in dual in-line packages, or *DIPs.* These packages are intended to be soldered into holes, plugged into sockets, or surface-mounted in a circuit board such as the one in Figure 3-2*a.* This type of board is often called a *printed-circuit* board because connections between devices are made with thin metal traces which are produced on the board by a method similar to making photo-

(a)

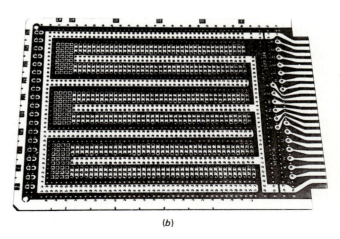

(b)

FIGURE 3-2 Printed-circuit board. *(a)* Quadram's QuadEGA ProSync graphics board. *(b)* A universal PC board. *(Courtesy of Vector Electronics Company Inc.)*

graphic prints. Printed-circuit boards are often referred to as *PC boards*, or just *boards*.

Integrated circuits come in a wide variety of package types. The factors which determine the type of package used are the amount of circuitry contained on the silicon chip and the number of external connections that need to be made to it, the temperature and humidity environment that the IC is intended to operate in, and the method that is to be used to mount the device on a circuit board with other devices. Figure 3-3 shows some of the IC package types that you may see in various pieces of electronic equipment.

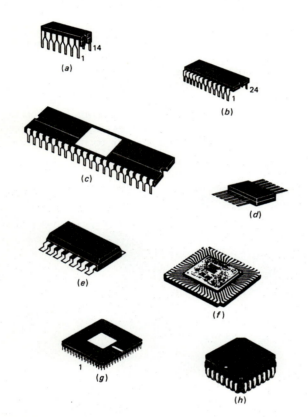

FIGURE 3-3   Common package types used for digital ICs. *(a)* 14-pin dual in-line, or DIP. *(b)* 24-pin DIP. *(c)* 40-pin DIP. *(d)* Ceramic flat-pack. *(e)* Surface mount DIP. *(f)* Ceramic chip-carrier package. *(g)* Pin-grid array. *(h)* J-lead surface mount.

The types shown in Figures 3-3a to 3-3c are called *dual in-line packages*, or *DIPs*. The name comes from the fact that the external connecting pins are in parallel rows along the two long edges of the package. DIPs range in size from 4 to 64 pins, depending on the number of connections that need to be made to the internal circuitry. DIPs usually have a notch in one end of the package, as shown in Figure 3-3a. If you face the DIP so the notch is on the top at the end farthest away from you, then pin number 1 for the device is to the left of the notch. Pin numbers increase as you go down the left side of the device, across the bottom, and back up the right side of the device. In other words, you count pin numbers

around the DIP from pin 1 in a counterclockwise direction.

> NOTE: Some DIPs have a notch in both ends of the package. On these devices a small painted dot or an indentation is put next to pin 1.

For applications where high operating temperatures and/or high humidity conditions are not expected, inexpensive epoxy plastic packages are used. For high temperatures or for devices that dissipate large amounts of power, ceramic packages are used.

Figure 3-3d shows a ceramic flat package that is used for some ICs. These packages are hermetically sealed, which means that they are totally immune to the effects of humidity. These packages are often used in military equipment that must be able to stand harsh environments. Pins are counted around the package from a notch or dot, just as they are for DIPs. These packages are usually mounted in high-quality sockets on the circuit board.

Figure 3-3e shows an IC package called a *surface mount* package. This popular package is very similar to the standard DIP except that for a given part type it is smaller and, as the name implies, its pins are constructed so that it can be soldered directly to metal pads on the surface of a circuit board. The pins on surface mount devices do not need to be as large as those on standard DIPs because they do not need to stand the strain of being pushed into sockets. Therefore, the package can be smaller than the equivalent DIP. Since surface mount packages are soldered on one surface of the circuit board, holes do not have to be drilled in the circuit board for the IC pins or socket pins, as they do for standard DIPs. Surface mount devices have the further advantage that they are more easily handled by equipment which automatically mounts components in the correct position on circuit boards during manufacturing.

Figures 3-3f and 3-3g show IC packages that are used for digital circuits such as microprocessors, which dissipate large amounts of power and require many circuit connections. The first type shown in Figure 3-3f is a thin ceramic package with gold-plated pads as pins. These packages are commonly referred to as *chip-carrier* packages. They are intended to be clamped into a socket so that the pads press against contacts which are connected to circuit board signal lines. Pin 1 on this package is to the right of the notched corner. As you look at the top of the IC, pins are counted off counterclockwise around the device. Note that some chip carrier packages have pin 1 in the center of one edge of the package, so you should check the data sheet if in doubt.

Figure 3-3g shows another type IC package which is used for VLSI digital circuits such as the Intel 80386 microprocessor. This type of package is called a *pin-grid array*. The number of pins in the array depends on the particular device. Some devices use two or three rows of pins around the outside of the array, and some devices use all of the pins in the array. The four corner pin positions are usually left without pins. Common array sizes are 10×10, 13×13, and 14×14. Large ICs such as these are put in sockets so they can easily be replaced if the device fails.

If you look carefully at the IC packages in Figure 3-2a you should notice that some of them have sequences of numbers and letters. These numbers and letters are used to identify the manufacturer of the device, the family that the device is a member of, the functions performed by the part, and, in some cases, the date the device was manufactured. In the next section we show you how to interpret the numbers and letters on common digital logic ICs.

Introduction to Logic Families

Digital ICs such as logic gates are used in a wide variety of electronic instruments and systems. Some systems require logic gates that change their outputs very quickly after a change is made on their inputs. The time it takes for a change on the input of a logic gate to cause a change on the output of the gate is called the *propagation delay*, or t_{pd}, of the gate. Another way of saying that a device must switch quickly, then, is to say that it must have a short propagation delay. Propagation delay of gates is usually specified in nanoseconds (ns). (A nanosecond, remember, is 10^{-9}, or 1 billionth of a second.) In order to produce a device with a short propagation delay, the currents in the internal circuit must be relatively large. This means that a logic device that switches quickly will probably dissipate more power than a slower device.

In other electronic systems, such as battery-powered medical equipment, the logic gates may not need a short propagation delay but do need to use an absolute minimum amount of power, so they don't drain the power-supply battery too quickly.

In order to accommodate this wide diversity of logic gate needs and to take advantage of advances in technology, manufacturers of digital ICs have produced several different families of digital parts. Later in this chapter we discuss in detail the characteristics of some of these different logic families. At this point, however, we will give you a brief overview of the commonly available families so you can identify a variety of logic devices. This will make it possible for you to start building and testing some actual logic gate circuits.

TTL LOGIC FAMILIES

Throughout the late 1960s and the 1970s, the most commonly used logic family was *standard transistor-transistor logic,* which is usually referred to as *TTL,* or "T squared L." The term *transistor* in the name refers to the fact that the circuitry in TTL ICs is built with devices called *bipolar transistors.* Most standard TTL ICs require a power-supply voltage between +4.75 V and +5.25 V to operate properly.

Devices in the standard TTL family are identified by numbers that start with 74 or, for military specification devices, 54. Two or three digits after the 74 are used to identify the logic functions performed by the device. A standard TTL device that contains four 2-input NAND gates, for example, is identified by the number 7400 printed on the package. A standard TTL device that contains four 2-input NOR gates has the number 7402 printed on the package. Table 3-1 shows the identifying numbers and the logic functions performed by several

TABLE 3-1
DEVICE NUMBERS OF COMMONLY AVAILABLE TTL GATE FUNCTIONS

Device Number	Description of Device	Number of Gates
7400	Quadruple 2-input NAND gates	4
7402	Quadruple 2-input NOR gates	4
7404	Hex inverters	6
7408	Quadruple 2-input AND gates	4
7410	Triple 3-input NAND gates	3
7420	Dual 4-input NAND gates	2
7430	8-input NAND gate	1
7432	Quadruple 2-input OR gates	4
7486	Quadruple 2-input XOR gates	4
7493	4-bit binary counter	1

commonly available TTL devices. Take a look at these to get an idea of some of the devices you can choose from when you build simple logic circuits.

TTL devices such as those in Table 3-1 are usually produced in dual in-line packages, or DIPs. Figure 3-4 shows the pin connection diagrams for four common TTL devices to give you an idea of how gate inputs and outputs are connected to the pins of these packages. Consult Appendix A of this book or a TTL data book to find the pin connection diagrams for other TTL devices.

TTL LOGIC SUBFAMILIES

Soon after the introduction of the standard TTL family of devices, IC manufacturers started producing other families of logic devices which have the same power supply and signal voltages as standard TTLs but are designed to minimize propagation delay or to minimize power dissipation. Numbers given to parts in these TTL *subfamilies* start with 74, as they do for standard TTLs. The logic functions performed by the device are indicated by an additional two- or three-digit number. For the most part, devices in these TTL subfamilies have the same logic functions and pin diagrams as the standard TTL parts with the same identifying numbers. A 74LS00, for example, is a quadruple (or *quad*) 2-input NAND gate with the same pin diagram as a standard 7400 TTL device.

The subfamily that a device belongs to is indicated in a device number by one to three letters after the 74. Here is a list of the major TTL subfamilies with the identifying letters and a brief description of the features of each.

7400—*Standard TTL.* The best seller for many years, seldom used in new designs because newer families are faster and use less power.

74H00—*High-Speed TTL.* Faster than standard TTL but uses too much power, so now obsolete. Replaced by Schottky TTL.

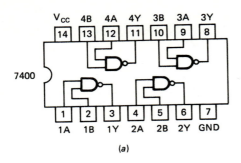

(a)

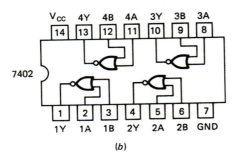

(b)

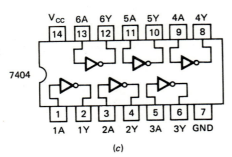

(c)

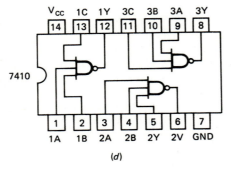

(d)

FIGURE 3-4 Pin diagrams showing how gates are connected to DIP pins. (a) 7400. (b) 7402. (c) 7404. (d) 7410. (Courtesy of Texas Instruments Incorporated.)

74L00—*Low-Power TTL.* Uses less power than standard TTL but is also much slower. Obsolete. Replaced by newer families.

74S00—*Schottky TTL.* Uses Schottky transistors, which switch faster than standard bipolar transistors, so gates have short propagation delay time, t_{pd}.

74AS00—*Advanced Schottky TTL.* Improved version of Schottky TTL.

74LS00—*Low-Power Schottky TTL.* Good compromise of speed and power dissipation. Slightly faster than standard TTL but only about 1/5 the power dissipation per gate.

74ALS00—*Advanced Low-Power Schottky TTL.* Improved version of low-power Schottky TTL.

CMOS LOGIC SUBFAMILIES

All of the TTL logic devices use bipolar transistors in their internal circuitry. *Metal-oxide semiconductor*, or *MOS*, transistors can also be used to make logic gates in integrated circuits. The advantage of logic gates made with MOS transistors is that they dissipate very little power when operated at low frequencies. This makes them ideal for use in many battery-operated digital instruments and computers.

Early MOS logic devices did not use the same supply and signal voltage levels as TTL. Also, the device numbers and pin diagrams were very different from those of TTL, so many logic designers resisted using the devices. Eventually, National Semiconductor came out with a CMOS logic family which operated with a +5-V power supply and had the same numbers, logic functions, and pin diagrams as standard TTL devices. The C in the name CMOS stands for *complementary*, which simply means that the circuits in these devices use two different types of MOS transistors so the circuit will operate with only a +5-V power supply. We discuss all of the MOS logic families in detail later in the chapter, but here is a list of the TTL-compatible subfamilies to help you identify the devices you find on a circuit board:

74C00—*First TTL-Compatible CMOS Family.* Microwatt power dissipation at low frequency, but very slow compared to TTL.

74HC00—*High-Speed CMOS.* Low power dissipation at low-frequency operation, but almost as fast as standard TTL.

74HCT00—Another *High-Speed CMOS.* Has input and output voltage levels that are more compatible with TTL subfamily devices than 74HC00 family devices are.

More about Digital IC Numbering Systems

As we described above, numbers and letters on IC packages identify the logic family and the logic function of a device. In addition to these numbers and letters, an IC may have numbers and letters which indicate which company made the device, the factory where the device was manufactured, the year and month the device was manufactured, the package type, and a code which indicates how thoroughly the device was tested.

Different manufacturers use somewhat different formats, so at first glance the identification may seem to be a confusing array of letters and numbers. Fortunately, most of the time all you need are the numbers and letters which identify the logic family and the function performed by the device. To give you a general idea, here is how Signetics organizes the information for its 74HCT00N device:

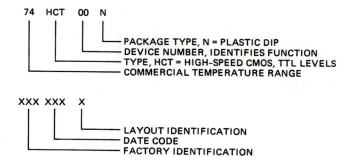

PACKAGE TYPE, N = PLASTIC DIP
DEVICE NUMBER, IDENTIFIES FUNCTION
TYPE, HCT = HIGH-SPEED CMOS, TTL LEVELS
COMMERCIAL TEMPERATURE RANGE

LAYOUT IDENTIFICATION
DATE CODE
FACTORY IDENTIFICATION

The first step in finding the actual device number is to look for the one- or two-letter code that each IC manufacturer puts in front of the device number to indicate the company that manufactured the part. A 74LS10 TTL device manufactured by National Semiconductor, for example, will be marked DM74LS10. An identical device manufactured by Texas Instruments will be marked SN74LS10. Here are some company codes for the most common IC manufacturers:

Code	Manufacturer
AM	Advanced Micro Devices
CD	GE/RCA
DM	National Semiconductor
F	Fairchild
GD	Goldstar
H	Harris
HD	Hitachi
IM	Intersil
KS	Samsung
LR	Sharp
M	SGS
MC	Motorola
MM	Monolithic Memories
MN	Panasonic
N	Signetics
P	Intel
SN	Texas Instruments
SP	SPI
US	Sprague
TC	Toshiba

After you work with ICs for a while, many of these codes will stick in your mind, and you will automatically use them to center in on the device number when you are looking at an IC.

Most manufacturers also add a letter on the end of a device number to indicate the type of package. A 74LS10 device manufactured in a plastic dual in-line package, or DIP, by Texas Instruments, for example, will be labeled SN74LS10N. The same device in a ceramic DIP will be labeled SN74LS10J. Here are some letter codes for common package types:

N	Plastic DIP
J	Ceramic DIP
D	Glass/metal DIP
W	Flat pack

To summarize, then, how you identify an IC: First find the one- or two-letter code which identifies the manufacturer of the device. The numbers following these letters identify the logic family and the function performed by the part. You can use these numbers to look up the characteristics of the device in a manufacturer's data book, if you need to. Finally, a letter after the device number indicates the package type. As another example of all this, a device labeled MC74LS02J is a low-power Schottky TTL device manufactured by Motorola in a ceramic DIP. As shown in Table 3-1, the 02 in the number indicates that the device contains four 2-input NOR gates.

There is a very important point for you to remember about these IC numbers when you are repairing electronic equipment. Suppose that, using techniques we show you later in this book, you have determined that an IC labeled DM74LS08J is defective and needs to be replaced. Your local electronics store doesn't have a DM74LS08J but does have an SN74LS08J. The obvious question is, "Can I use the SN part?" The answer is yes, because the parts are identical except that the DM part was made by National Semiconductor and the SN part was made by Texas Instruments. In other words, quality considerations aside, you care only about the function of a device, not the manufacturer.

Now, further suppose that the part your local electronics store has is an SN74LS08N instead of an 74LS08J, as we originally assumed. The question now is, "Can I use an N package device in place of a J package device?" A J package is made of a ceramic material and is intended for use in higher-temperature and higher-humidity applications than the plastic N package is intended for. Therefore, if the instrument originally has a J package device, it is best to replace it with a J package device, especially if the instrument is going to operate in high-temperature conditions.

Now that you have an overview of some common logic devices and are familiar with how ICs are identified, the next step is to show you how to build circuits containing these devices.

BUILDING AND TESTING DIGITAL CIRCUITS

At some time in your work in electronics, you will most likely have to build some digital circuits. This may be for testing a new circuit (*prototyping*) or for adding a modification to an existing circuit. The success of the circuit will depend not only on its design but on the care you put into laying out the circuit and building it. Quickly putting a pile of parts together in a "bird's nest" configuration usually produces a circuit that is unlikely to work, difficult to troubleshoot, and undependable—if you do finally get it working. The purpose of this section is to give you some specific directions on how to build digital circuits that work. We start by discussing some of the methods used to mount and connect ICs.

Device Mounting and Wiring Methods

There are several different methods of mechanically mounting ICs for a circuit and connecting their inputs and outputs as needed. The method chosen depends mostly on how long the circuit is going to be used and how many copies of the circuit are to be made.

SOLDERLESS PROTOBOARDS

For laboratory use and other applications where a permanent circuit is not needed and the signals used have frequencies which are less than 10,000,000 cycles per second (10 MHz), solderless protoboard such as that shown in Figure 3-5a is appropriate. Individual sections such as those shown in Figure 3-5b snap together to form as large an area as needed for a particular circuit. The square dots in the board in Figure 3-5b represent

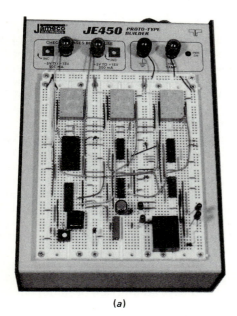

(a)

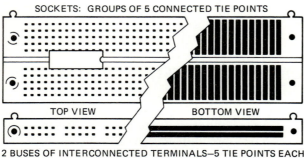

SOCKETS: GROUPS OF 5 CONNECTED TIE POINTS

TOP VIEW   BOTTOM VIEW

2 BUSES OF INTERCONNECTED TERMINALS—5 TIE POINTS EACH

(b)

FIGURE 3-5   Solderless protoboard. (a) Jameco Electronics JE450 Solderless Proto-Type Builder. (b) Top and bottom view of socket strip and bus strip section showing IC plugged in. (Courtesy of Global Specialties Corporation.)

holes that ICs, resistors, other circuit components, and connecting wires can be plugged into. To make circuit building easier, holes are connected together by metal strips inside the board. The bottom view in Figure 3-5b shows the metal strips inside the board so you can see which holes are connected to one another.

In socket sections, holes are connected in lines of five on each side of a central trough. If an IC in a DIP is plugged into the board as shown, each pin on the IC will contact one of the lines of holes. The other four holes in each line can then be used to plug in wires and the leads of other components that you want to connect to that IC pin. To insert connecting wires, you first strip off about ¼ inch of insulation on each end of the wire.

In bus-type protoboard sections, two metal strips run the full length of the section. These bus-type sections are useful for running the power-supply voltage (V_{CC}) and ground to the ICs plugged into an adjacent socket section. V_{CC} and ground leads from the power supply are inserted into the holes at one end of the strip. Short wires can then be used to connect the V_{CC} and ground pins of the ICs to these bus lines. This produces much neater wiring than would be produced by running V_{CC} and ground wires from the power supply to each IC. Multiple bus strips can be used for circuits that require additional voltages, such as +12 V and −12 V.

If you are putting together more than a couple of solderless protoboard sections, you should screw the sections on a piece of metal, plywood, or plastic so they don't keep separating from each other. When you are making connections on a solderless protoboard, use only 22-gage or 24-gage wire. Smaller-diameter wire will not make dependable connections, and larger-diameter wire will stretch out the metal contacts in the protoboard so that they will not make dependable connections in future circuits.

> NOTE: The leads on ½-W resistors and many other parts are too large in diameter to use directly in this type of protoboard. Solder a short piece of thinner wire on these leads to plug into the protoboard. When you strip the insulation off the ends of connecting wires, make sure you don't strip more than about ¼ inch, or the bare wires in adjacent lines may accidentally touch each other.

The major advantage of solderless protoboard is that components can easily be inserted and removed without damaging the board. Therefore, the board can be used many times. The disadvantage of this type of board is that wires and components tend to get pulled out if the board is moved around very much.

UNIVERSAL PRINTED-CIRCUIT BOARDS, SOLDERING, AND WIRE WRAPPING

For more permanent circuits, universal printed-circuit boards such as that in Figure 3-2b are used. ICs, sockets for ICs, connecting wires, and other components can be inserted in the holes in this type of board and soldered in place. The main points to keep in mind when soldering components into a board such as this are:

1. Make sure you heat the connection enough so that solder flows smoothly and evenly around the connection, but not so much that you lift the metal pads off the board.

2. Make sure you use enough solder to make a solid connection, but not so much that adjacent pads are shorted together.

Figure 3-6 shows an example of what a good solder connection on an IC socket pin should look like.

FIGURE 3-6   Good solder connection on a circuit board. *(Courtesy of Graymark International.)*

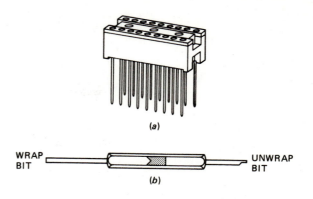

(a)

WRAP BIT

UNWRAP BIT

(b)

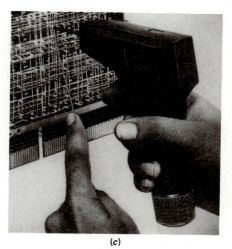

(c)

FIGURE 3-7   *(a)* Wire-wrap socket. *(b)* Hand wire-wrap tool. *(c)* Wire-wrap gun. *(Courtesy OK Machine and Tool Corporation.)*

Note that the board in Figure 3-2b has two metal bus strips running around the board. These can be used to supply V_{CC} (+5 V) and ground to the ICs on the board. The metal strips along the edge of the board allow the board to be plugged into a standard edge connector socket if desired. Using an edge connector socket to make connections to the board makes it easier to remove the board from a system to work on.

Soldered connections have the advantage that they are quite durable, but they have the disadvantage that they are somewhat difficult to remove if you want to change the circuit. Also, the heat required to remove components may loosen the metal pads on the board. Therefore, we now often use a technique called *wire wrapping* to make circuit connections on a universal printed-circuit board such as that in Figure 3-2b.

Wire-wrap connections are made by tightly wrapping the stripped end of a connecting wire many times around square pins on special IC sockets and other devices. Figure 3-7a shows a typical wire-wrap socket with the relatively long square pins that the wires are wrapped on. A small hand tool such as that shown in Figure 3-7b or a wire-wrap gun such as that in Figure 3-7c is used to wrap the wire around the pin. If properly done, these connections are as dependable as soldered connections because the square corners on the pins hold the wire tightly wrapped.

Figure 3-8 shows the steps you perform to make a wire-

wrap connection. You first strip about 1 inch of the insulation from the end of the wire with a small slotted stripping tool. The wire used for standard wire wrapping is usually 28 or 30 gage, so you may want to practice a little to get used to working with wire this thin. The stripped end of the wire is inserted in the small hole in the end of the wire-wrap tool tip, not the larger hole in the center. The wire is then bent back and the center hole in the tip placed over the wire-wrap pin. Don't press the tool too tightly against the circuit board. Then turn the hand tool or press the trigger on the wire-wrap gun. To get a smooth wrap, make sure you let the wire-wrap tool or gun lift up as the wire wraps up the pin. After you work your way through these steps a few times, you should have little trouble making neat wire-wrap connections.

Circuit Construction Hints

1. Plan the layout of ICs on the board so that, as much as possible, signals go from left to right as they do on schematics. This helps signal tracing during troubleshooting.

2. Point all of the ICs in the same direction to reduce the chance of one being put in backwards. Make sure to leave room between ICs for connecting wires and other components where needed.

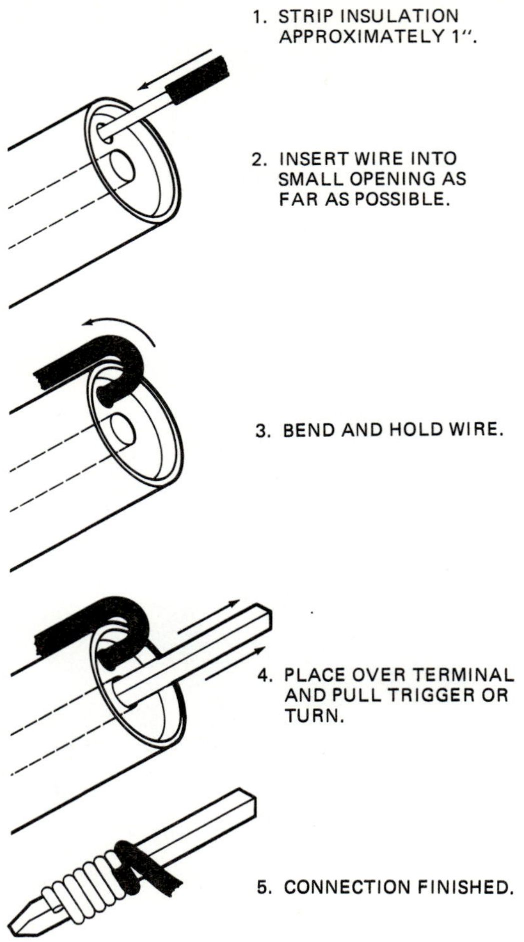

1. STRIP INSULATION
   APPROXIMATELY 1".

2. INSERT WIRE INTO
   SMALL OPENING AS
   FAR AS POSSIBLE.

3. BEND AND HOLD WIRE.

4. PLACE OVER TERMINAL
   AND PULL TRIGGER OR
   TURN.

5. CONNECTION FINISHED.

FIGURE 3-8   Wire-wrapping steps. *(Courtesy of OK Machine and Tool Corporation.)*

3. Use color-coded wiring. You might, for example, use red wires for V_{CC} connections and black wires for ground connections. This improves your chances of getting connections to the correct pins and makes signal tracing easier. For solderless protoboard construction, the multiwire cable used by telephone companies can be stripped of its outer jacket to yield a large variety of color combinations. (Make sure the wire is 22 or 24 gage. Some phone wire is 26 gage, and this size does not always make dependable connections.)

4. On solderless protoboard, keep connecting wires short so the circuit does not look like a "bird's nest." Route wires around ICs so that if an IC proves defective, it can be removed without having to tear the circuit apart.

5. If possible, build and test one small section of a circuit at a time. When you have one small section working correctly, add another small section to it. Then test and, if necessary, troubleshoot the combination. Experience has shown that this is the fastest way to get a large circuit built and working.

6. As you wire each section of a circuit, trace the corresponding lines of the circuit on the schematic with a yellow or other transparent-color marking pen to keep track of what has been completed. This shows you where to start when you come back to work after a hard weekend.

7. When you have completed wiring a section, check each IC pin from pin 1 to the highest number, then from the highest-number pin back to pin 1, to make sure each pin is connected correctly. This may seem laborious, but a wiring error is the most common reason for a prototype circuit to initially not work. The time spent in careful checking is repaid by having more circuits work correctly the first time—and by having fewer blown ICs!

8. Connect a 100-μF electrolytic capacitor from V_{CC} to ground and a 0.1-μF capacitor from V_{CC} to ground where the V_{CC} and ground wires connect to the protoboard. These capacitors filter out electrical noise produced on the V_{CC} line by gate outputs changing from high to low or low to high. The larger capacitor filters out low-frequency noise and the smaller capacitor filters out high-frequency noise. We explain more about this at the end of the next chapter.

Initial Circuit Testing

After carefully wiring and checking your circuit, turn on the power but keep your hand on the power switch so you can quickly turn off the power if necessary. Listen and look for violent signs of an unhappy circuit, such as smoke, pops, fizzes, or the smell of hot components. If you see, hear, or smell any of these, quickly turn off the power. Then, if the problem device is not obvious, check for excessively hot components by touching them *lightly* with your finger. We emphasize the *lightly* because, when they are inserted backwards, many ICs and electrolytic capacitors quickly get hot enough to give you a nasty burn if you hold your finger on them too long.

If you find a hot component, check to see that it is not inserted in the circuit backwards and check that it is wired correctly. Also look for incorrect resistor values and for IC pins that are accidentally connected together by excess solder or connecting wires that have had too much insulation stripped off.

If visual inspection does not find any apparent problems, you can disconnect the power supply and use an ohmmeter to see if some component is connecting V_{CC} directly to ground. You can also use an ohmmeter to check if the output of a gate has accidentally been connected directly to V_{CC} or ground. Remember to turn off the power before using an ohmmeter in a circuit!

NOTE: You will probably have to temporarily remove some connecting wires in order to isolate the defective component.

If you find a capacitor or an IC that has been inserted backwards in the circuit, you should replace it with a new one. When you are removing ICs from sockets or solderless protoboards, use a small screwdriver (or IC remover) to lift first one end of the IC and then the other until the IC pops out. If you try to pull out the IC with your fingers, you will most likely end up with a perforated thumb and bent IC leads!

After you have replaced the component causing a problem, turn on the power. Again watch for signs of discontent from the circuit and quickly turn off the power if you detect any. If a capacitor is still getting hot, check its polarity markings carefully. The polarity markings on some types of electrolytic capacitors are hard to read. If an IC is still getting hot, there may still be a wiring error on the board or there may be an error in the schematic. Get a data sheet for the IC and check that the pin numbers on the schematic are correct. When you work with TTL ICs, it is very easy to get used to connecting V_{CC} to pin 14 or 16 and ground to pin 7 or 8 because this is where these connections are for many devices. However, a common TTL latch, the 74LS75, has its V_{CC} connection on pin 5 and its ground connection on pin 12. The "normal" supply pins, 8 and 16, are both outputs for this device. Watch for this kind of error.

Functional Testing and Troubleshooting

CHOOSING INPUT SIGNALS AND PREDICTING OUTPUTS

After your circuit passes the smoke and heat test, the next step is to determine if the circuit is functioning correctly. It is very important that you test and troubleshoot the circuit in a systematic manner rather than wasting a great deal of time just poking, probing, and "witch hunting" in the hope that the circuit will magically start working correctly. We will use an example to show you how to systematically approach testing and troubleshooting a digital circuit.

Let's assume that you have built the three-gate circuit in Figure 3-9a and need to see if it functions correctly. To start a systematic approach, some of the points you need to think about are:

1. What test signals do I need to apply to the inputs?

2. What results will these signals produce on the outputs?

3. What instruments will I need to measure these input and output signals?

To answer the first two questions you need to construct a truth table for the circuit. The circuit has three inputs, A, B, and C, so the truth table requires eight entries to represent all of the possible combinations. If you remember that these gates are NAND gates and that if the inputs of a NAND gate are high, the output is low, you can fill in the truth table for the circuit as shown in Figure 3-9b. From the truth table you can see that the X output of the circuit will be high if A and B are high or C is high. The expression for the circuit then is X = AB + C.

Now that you know the output that is expected for each of the possible input combinations, the next step is to decide how you are going to produce the input signals. If you only need to determine that the circuit works with static input signals, you can simply use a piece of wire to con-

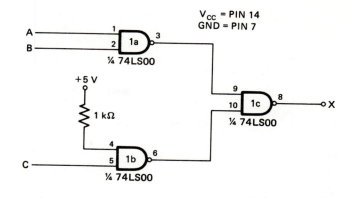

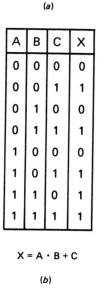

(a)

A	B	C	X
0	0	0	0
0	0	1	1
0	1	0	0
0	1	1	1
1	0	0	0
1	0	1	1
1	1	0	1
1	1	1	1

$$X = A \cdot B + C$$

(b)

FIGURE 3-9 (a) Three-gate circuit. (b) Truth table.

nect an input to ground for a logic low or to V_{CC} for a logic high. If you need to test the circuit with pulsed input signals, you will use a pulse generator or a word generator to produce the required signals. In a later section of this chapter we describe the use of a pulse generator.

TEST INSTRUMENTS FOR NONPULSED DIGITAL SIGNALS

To measure the output signal there are several different instruments you can use. The first instrument is a digital multimeter such as the one we described in Chapter 1. A digital voltmeter will very accurately tell you the voltage on an input or output, and you can mentally convert this value to a high or low. However, many digital voltmeter leads are not physically designed for probing on IC pins. While you are looking at the readout to determine the value, the probe may, if you are not careful, slip and accidentally short adjacent pins together. Shorting pins in this way may destroy all or part of a device. Also, the continual looking back and forth between the circuit and the readout is tiring. Manufacturers of digital meters have attempted to solve these problems by making small, hand-held meters which can be placed closer to the circuit, and by making meters which verbally tell you the voltage being read.

FIGURE 3-10 Hewlett-Packard 545A logic probe.

A second instrument for determining logic levels is a logic probe such as the HP545A, shown in Figure 3-10. These probes have one light-emitting diode (LED) that lights if the tip of the probe is connected to a logic low signal and another LED that lights if the tip is connected to a logic high signal. Voltage comparison levels in the probe can be adjusted to correspond to the threshold voltages of common logic families. Most logic probes also have a way of telling you if a pulsed signal is present on a signal line.

Figure 3-11 shows the circuit for a "homemade" logic probe you can build in one corner of your protoboard and use for testing logic levels. The green LED will light if the input is connected to a logic low, and the red LED will light if the input is connected to a logic high. This is easy to remember because the red light is above the green light in most traffic signals. If the input of the circuit in

Figure 3-11 is connected to a "floating" output or is left open, both LEDs will light. Here's an overview of how this circuit works.

An LED will give off light if a current, typically 10 to 20 mA, is passed through it. Most logic gate outputs will not supply the current needed to drive an LED directly, or will not supply that much current without changing the output logic levels. The circuit in Figure 3-11 solves this problem by using transistors Q1 and Q2 as logic-level-controlled switches. The basic principle of a transistor is that a small current applied at its base terminal enables a much larger current to flow between its emitter and its collector. A base current of 0.2 mA, for example, may produce a collector-emitter current of 100 times that much, or 20 mA. The Greek letter β is often used to represent the ratio of collector-emitter current to base current for a transistor. The point here is that only a small amount of current from the output of the logic gate is required to turn on one of the transistors in Figure 3-11. Once it is turned on, the transistor supplies a current path for the 20 mA or so needed to light the LED.

The transistor symbols used in Figure 3-11 represent bipolar transistors, such as those used in TTL logic devices. As you can see by the symbols in Figure 3-11, there are two different types of transistors. Q1 is an example of a *PNP-type* transistor and Q2 is an example of an *NPN* transistor. The name PNP comes from the fact that a PNP-type transistor consists of a crystal which has a region of P-type silicon, a region of N-type silicon, and another region of P-type silicon. In the same manner, an NPN device consists of a crystal which has a region of N-type silicon, a region of P-type silicon, and another region of N-type silicon. The main skills we want you to take with you from this section are identifying the symbol for each type of transistor and predicting the logic level which will turn on each type transistor so that current flows between its collector and emitter.

In addition to telling you the types of silicon that a transistor consists of, the letters in its name tell you about how the device is connected in a circuit by associating the first letter with the emitter and the second letter with the base of a device. Q1, for example, is a PNP device. The P in the name tells you that the emitter of the device is connected to the more positive (P) terminal of the supply voltage. In the circuit in Figure 3-11 note that the emitter, represented in the symbol by an arrow, is connected to the +5-V supply. In a similar manner, Q2 is an NPN transistor, so its emitter is connected to the more negative (N) terminal of the supply voltage. The emitter of Q2 is connected to ground in Figure 3-11.

The second letter in the name of a transistor tells you what you have to do to the base of that type transistor to turn it on. The N in the middle of PNP, for example, tells you that you turn on a PNP transistor by supplying a current path from its base to a voltage more negative (N) than its emitter. Connecting the open end of the 10-kΩ resistor (R1) on the base of Q1 in Figure 3-11 to ground or to a logic low will supply a current path to a voltage which is more negative than the +5 V connected to the emitter. This will turn on Q1 and supply current to light the green LED. If the open end of the 10-kΩ resistor (R1) is connected to a logic high or to +5 V, the necessary

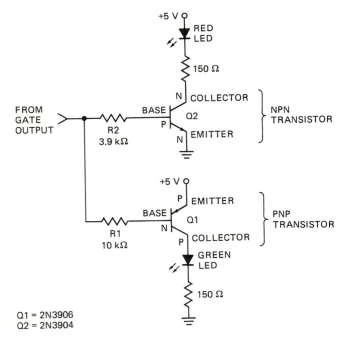

Q1 = 2N3906
Q2 = 2N3904

FIGURE 3-11 "Homemade" logic probe using LEDs.

current path from the base to a more negative voltage will not be present, and Q1 will be off.

In a similar manner, the P in NPN tells you that you turn on an NPN transistor by supplying a current path from the base to a voltage more positive than the voltage on the emitter. Connecting the open end of the 3.9-kΩ resistor (R2) on the base of Q2 to a logic high or to +5 V, for example, will supply the necessary current path, and Q2 will be turned on. Q2 will then supply the current required to light the red LED. If the open end of this resistor (R2) is connected to ground or to a logic low, the necessary base current path will not be present, and Q2 will be off.

In Chapter 10 we show you how to interface a wide variety of devices to logic gate inputs and outputs.

Two other test instruments commonly used to observe and measure digital signals are an oscilloscope and a logic analyzer. These instruments have the advantage that they can be used to observe pulsed signals as well as static dc logic levels. In Chapter 4 we show you how to use an oscilloscope and in Chapter 8 we show you how to use a logic analyzer. For our example circuit in Figure 3-9a, a voltmeter or a logic probe is sufficient to check the output logic level.

TESTING AND TROUBLESHOOTING PROCEDURE

To test the example circuit in Figure 3-9a, start by jumpering the input pins with the first input signal combination in the truth table, 000. If the output signal is correct for this input combination, change the jumpers on the inputs to 001, the next input combination from the truth table. Then check the output logic level to see if it is correct. You should check the output for all of the possible input combinations because a circuit may work for some input combinations but not for others.

If the output logic level is not correct for an input signal combination, check the intermediate logic levels directly on the input pins of the final gate. For example, if the output of the circuit in Figure 3-9a is not correct, check pin 9 to see if the correct signal from gate 1a is getting to that point. If you find the correct signal on pin 9, then check pin 10 to see if the correct signal is getting to this point from gate 1b. Using this approach quickly determines which branch of the circuit contains the problem. You can then work your way back through the branch containing the problem instead of having to face the whole circuit.

To carry this example further, suppose you find that pin 9 of gate 1c is high when it should be low for an input combination of 111. Check the signal directly on pin 3 of gate 1a to see if this gate is producing the correct signal. Gate 1a may be producing the correct signal, but it may not be getting to pin 9 because of a broken wire, a loose wire, a miswired circuit, or a broken printed-circuit board trace. If the output signal on pin 3 is not correct, check on pins 1 and 2 to see if the desired input signals are actually present on the input pins. If the inputs are correct but the output is not correct, then either the output of the device is shorted to V_{CC} or the device is defective. To see if the output is shorted to V_{CC}, turn off the power and use an ohmmeter to check if the line between pin 3 and pin 9 is shorted to V_{CC}. If this is

not the problem, then the IC is most likely defective and should be replaced. The point here is to work your way systematically back through the circuit until you find the problem, and then fix it.

In this section of the chapter we have shown you how to build and test simple digital circuits. In the next section we show you how to read and interpret the manufacturer's data sheets for logic gates. These data sheets contain more detailed information about the required voltages and currents for the devices. You need this information to determine, for example, if the voltage you measure on the input of a logic gate is a legal logic low or a legal logic high for that device. You also need this detailed information to determine factors such as how many gate inputs can be connected to one gate output.

INTERPRETING LOGIC DEVICE DATA SHEETS

IC manufacturers produce large data books containing data sheets for each of the devices they produce. Learning to interpret these data sheets is an important part of understanding digital electronics. If you are going to be working with a particular family of logic devices, you should obtain a copy of the data book for that family. The *TTL Data Book, Volume 2*, published by Texas Instruments Incorporated, for example, contains data sheets for a great many devices in the 7400, 74L00, 74H00, 74S00, and 74LS00 families. We will use selected portions of the 54LS00/74LS00 data sheet from this book for the discussion in this section.

To start, take a quick look at Figures 3-12 to 3-15 to get an overview of the information typically contained in the data sheet for a logic gate device. Some of the information in the data sheet should already be familiar to you from our previous discussions.

Logic Functions and Pin Diagrams

The first part of the data sheet shown in Figure 3-12 tells you that devices with 00 as their identifying number contain four independent 2-input NAND gates. The data sheet also indicates that the devices are available in a 74LS00 series of devices and in a 54LS00 series of devices. Parts given a 54LS00 number are functionally the same as those given 74LS00 numbers, but 54LS00 parts are guaranteed to operate properly over a wider range of temperatures and power-supply voltages. Parts with 54LS00 numbers are often specified for military equipment.

The data sheet also shows the truth table, the logic symbol, and two Boolean expressions for a NAND gate. The truth table shown here is a little bit different from the one we showed you in the last chapter, but it gives the same information. The center line in this truth table shows an L for the A input, an X for the B input, and an H for the output. The X in the B column means that the logic state on the B input doesn't matter. In other words, if the A input is low, the output will be high whether the B input is high or low. If you look at the truth table we gave you for a NAND gate in Figure 2-27, you should see that this is indeed the case. The bottom line in the truth

- **Package Options Include Both Plastic and Ceramic Chip Carriers in Addition to Plastic and Ceramic DIPs**

- **Dependable Texas Instruments Quality and Reliability**

description

These devices contain four independent 2-input NAND gates.

The SN5400, SN54H00, SN54L00, and SN54LS00, and SN54S00 are characterized for operation over the full military temperature range of −55°C to 125°C. The SN7400, SN74H00, SN74LS00, and SN74S00 are characterized for operation from 0°C to 70°C.

FUNCTION TABLE (each gate)

INPUTS		OUTPUT
A	B	Y
H	H	L
L	X	H
X	L	H

logic diagram (each gate)

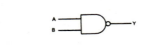

positive logic

$$Y = \overline{A \cdot B} \quad \text{or} \quad Y = \overline{A} + \overline{B}$$

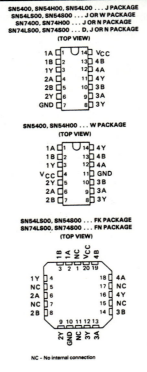

SN5400, SN54H00, SN54L00 . . . J PACKAGE
SN54LS00, SN54S00 . . . J OR W PACKAGE
SN7400, SN74H00 . . . J OR N PACKAGE
SN74LS00, SN74S00 . . . D, J OR N PACKAGE
(TOP VIEW)

SN5400, SN54H00 . . . W PACKAGE
(TOP VIEW)

SN54LS00, SN54S00 . . . FK PACKAGE
SN74LS00, SN74S00 . . . FN PACKAGE
(TOP VIEW)

NC - No internal connection

TEXAS INSTRUMENTS
POST OFFICE BOX 225012 • DALLAS, TEXAS 75265

FIGURE 3-12   Page 1 of the Texas Instruments 74LS00 data sheet.

table on the data sheet uses an X in the A column to indicate that if the B input of the NAND gate is low, the A input is a "don't care" because the output will be high regardless of the state on the A input.

Another point to notice on the data sheet is that two positive-logic Boolean expressions are given for the NAND gate. These are called *positive-logic* expressions because the asserted level for all signals is assumed to be a logic high. The first expression:

$$Y = \overline{AB}$$

indicates that Y will be high for the cases where A and B are not both asserted high. This is the traditional expression for a NAND gate, but it is not as easy to understand as the second expression. The second form:

$$Y = \overline{A} + \overline{B}$$

tells you that the Y signal will be asserted (high) if the

A signal is NOT asserted (high) OR the B signal is NOT asserted (high). This second expression corresponds to the bubbled input OR symbol representation of a NAND gate that we showed you in Figure 2-27.

As you can see from Figure 3-12, the data sheet also shows you the packages that the devices are available in and the pin numbering for those packages. You will need these pin diagrams when you draw schematics for circuits you are going to build. The numbers inside the chip outline represent the pin numbers. Note that for the J and N packages, which you will be mostly using, pin 1 is to the left of a notch in one end of the package, and the numbers increase around the IC in a counterclockwise direction when you look at the top of the IC. The letters on the outside of the pins in Figure 3-12 indicate the internal circuit point that the pin connects to. The pin labeled 1A, for example, is connected to one input of a NAND gate in the device, and the pin labeled 1B is connected to the other input of that NAND gate. The pin labeled 1Y is connected to the output of that

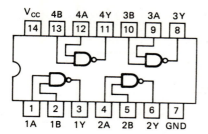

FIGURE 3-13   Internal gate connections for an
SN74LS00.

NAND gate. To help you visualize this, Figure 3-13 shows graphically the internal gate connections for a 74LS00 in a J or N package.

The pin labeled GND in Figures 3-12 and 3-13 will be connected to the ground terminal of the power supply when you build a circuit with this device. The pin labeled V_{CC} will be connected to the +5-V terminal of a power supply. The V_{CC} and GND connections are separate from the signal inputs. These pins allow current to flow through the device from the power supply as needed to operate the internal logic gate circuitry.

Schematics and Maximum Ratings

SCHEMATICS FOR INTERNAL CIRCUITRY

The next part of the data sheet to look at contains the schematics of the circuits used for this basic NAND gate in the different subfamilies. This is shown in Figure 3-14. As we said before, you do not need to know how these circuits operate, but it is interesting to see that the circuits have only a few types of devices in them. You should recognize the resistor symbols in the drawings. The small triangular symbols with a line across one point represent devices called *diodes*. A diode allows an electric current to flow in one direction through it but not in the other. The other symbol in these schematics is an NPN bipolar transistor, such as we introduced you to in the discussion of the "homemade" logic probe in a previous section. Diode and transistor symbols which have hooked lines on them represent Schottky devices. It is these devices which give names to the Schottky and the low-power Schottky TTL subfamilies. If you are interested in how circuits such as the ones in Figure 3-14 work, consult Appendix B.

MAXIMUM RATINGS

The maximum ratings section at the bottom of the data sheet shown in Figure 3-14 is quite straightforward, but there is one point we need to make sure is clear in your mind. The point is that, when discussing any logic family, there are two different types of voltage ratings to be concerned with. The first is the power-supply voltage. This is the voltage that must be connected between the V_{CC} and GND pins of the device to supply current for the internal circuitry. The second voltage rating to be concerned with is the signal input voltage. This is the logic signal voltage that comes to a gate input from some external signal source, such as a switch, or from the output of another gate.

As you can see in the data sheet in Figure 3-14, most TTL devices have a maximum supply voltage of +7 V. This is the maximum voltage that can be applied between the V_{CC} and GND pins of the devices without damaging them, but as you will see a little later, the devices will not operate properly with a supply voltage this high.

Looking again at Figure 3-14, you should see that the maximum input signal voltage for most TTL subfamily devices is 5.5 V. If a logic signal with a voltage greater than 5.5 V is applied to a gate input, it may destroy the gate, even if the power-supply voltage is at the proper value of +5 V. This means you should always measure the voltage from a logic signal source to make sure it is correct *before you connect it to the input of a logic gate*.

The operating temperature ranges shown in the data sheet tell you that parts whose numbers start with 54 are guaranteed to operate correctly anywhere in the temperature range of −55°C to +125°C. This is sometimes referred to as the *military temperature range* because it is the range required for most military electronic equipment. Parts whose numbers start with a 74 are guaranteed to operate correctly over a temperature range of 0°C to 70°C. This range is often referred to as the *commercial temperature range* for electrical parts. Note that devices can be safely stored in a wider temperature range than they can be operated in.

Operating Conditions and Characteristics

INTRODUCTION TO PARAMETER NAMES AND SUBSCRIPTS

Figure 3-15 shows the recommended operating conditions and the operating characteristics for the standard 74LS00 and 54LS00 NAND gates. Let's start by looking at the recommended operating conditions and showing you how to read "subscriptese."

Each voltage, current, or temperature parameter on a data sheet is identified by a letter which tells what kind of parameter it represents. For example, V is used to represent a voltage parameter, I is used to represent a current parameter, t is used to represent a time parameter, and T is used to represent a temperature parameter. Subscripts after each letter are used to identify specific parameters of each type. The power-supply voltage required by the device, for example, is represented by V_{CC}.

Most of the subscripts used are the initials of the name for the specified parameter. As shown in Figure 3-15, for example, V_{IH} represents an *Input High* logic signal voltage, V_{IL} represents an *Input Low* logic signal voltage, I_{OH} represents an *Output High* logic signal current, and I_{OL} represents an *Output Low* logic signal current. Don't worry if the meaning of these parameters is not clear to you at this point. We will soon discuss them in greater detail.

T_A on the data sheet represents *Ambient temperature*, which is another name for the air temperature around the system containing the device.

Next take a look at the actual recommended values given for V_{CC} in the data sheet. The NOM (nominal, or recommended) power-supply voltage for TTL devices is +5 V, but the data sheet indicates that a 74LS00 is guaranteed to operate properly with a power-supply voltage

schematics (each gate)

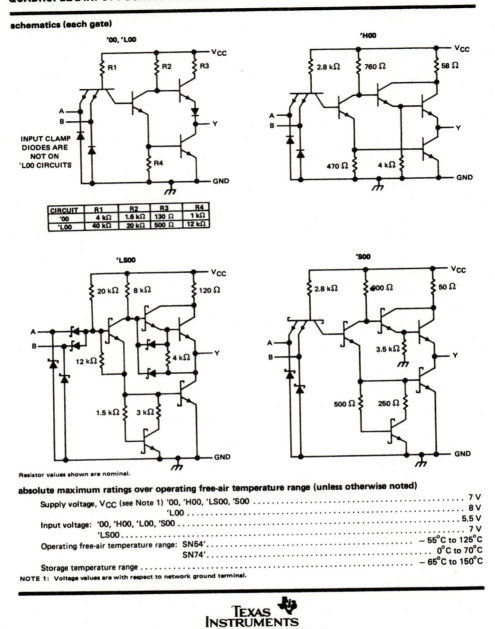

CIRCUIT	R1	R2	R3	R4
'00	4 kΩ	1.6 kΩ	130 Ω	1 kΩ
'L00	40 kΩ	20 kΩ	500 Ω	12 kΩ

Resistor values shown are nominal.

absolute maximum ratings over operating free-air temperature range (unless otherwise noted)

Supply voltage, V$_{CC}$ (see Note 1) '00, 'H00, 'LS00, 'S00 .... 7 V
                               'L00 .... 8 V
Input voltage: '00, 'H00, 'L00, 'S00 .... 5.5 V
              'LS00 .... 7 V
Operating free-air temperature range: SN54' .... − 55°C to 125°C
                         SN74' .... 0°C to 70°C
Storage temperature range .... − 65°C to 150°C

NOTE 1: Voltage values are with respect to network ground terminal.

FIGURE 3-14   Page 2 of the Texas Instruments 74LS00 data sheet.

anywhere between a minimum (MIN) of 4.75 V and a maximum (MAX) of 5.25 V. This voltage range is important to remember because if the power-supply voltage in a digital system drops below the required minimum voltage, the logic gates and other ICs may no longer work correctly. Note that for devices labeled with a 54LS00 number, the minimum supply voltage is 4.5 V and the maximum is 5.5 V. Also note that, as we mentioned before, the 54LS00 parts are guaranteed to operate cor-

rectly over a wider temperature range than are parts with a 74LS00 number.

INPUT AND OUTPUT VOLTAGE PARAMETERS

In order to discuss the current and voltage parameters shown for the 74LS00 in the data sheet, we are going to treat it as a "black box," with only an input, an output, and power-supply connections. The term "black box" refers to the fact that you can understand the voltage and

recommended operating conditions

		SN54LS00			SN74LS00			UNIT
		MIN	NOM	MAX	MIN	NOM	MAX	
V_{CC}	Supply voltage	4.5	5	5.5	4.75	5	5.25	V
V_{IH}	High-level input voltage	2			2			V
V_{IL}	Low-level input voltage			0.7			0.8	V
I_{OH}	High-level output current			−0.4			−0.4	mA
I_{OL}	Low-level output current			4			8	mA
T_A	Operating free-air temperature	−55		125	0		70	°C

electrical characteristics over recommended operating free-air temperature range (unless otherwise noted)

PARAMETER	TEST CONDITIONS †			SN54LS00			SN74LS00			UNIT
				MIN	TYP‡	MAX	MIN	TYP‡	MAX	
V_{IK}	V_{CC} = MIN,	I_I = −18 mA				−1.5			−1.5	V
V_{OH}	V_{CC} = MIN,	V_{IL} = MAX,	I_{OH} = −0.4 mA	2.5	3.4		2.7	3.4		V
V_{OL}	V_{CC} = MIN,	V_{IH} = 2 V,	I_{OL} = 4 mA		0.25	0.4		0.25	0.4	V
	V_{CC} = MIN,	V_{IH} = 2 V,	I_{OL} = 8 mA					0.35	0.5	
I_I	V_{CC} = MAX,	V_I = 7 V				0.1			0.1	mA
I_{IH}	V_{CC} = MAX,	V_I = 2.7 V				20			20	µA
I_{IL}	V_{CC} = MAX,	V_I = 0.4 V				−0.4			−0.4	mA
I_{OS} §	V_{CC} = MAX			−20		−100	−20		−100	mA
I_{CCH}	V_{CC} = MAX,	V_I = 0 V			0.8	1.6		0.8	1.6	mA
I_{CCL}	V_{CC} = MAX,	V_I = 4.5 V			2.4	4.4		2.4	4.4	mA

† For conditions shown as MIN or MAX, use the appropriate value specified under recommended operating conditions.
‡ All typical values are at V_{CC} = 5 V, T_A = 25°C.
§ Not more than one output should be shorted at a time, and the duration of the short-circuit should not exceed one second.

switching characteristics, V_{CC} = 5 V, T_A = 25°C (see note 2)

PARAMETER	FROM (INPUT)	TO (OUTPUT)	TEST CONDITIONS		MIN	TYP	MAX	UNIT
t_{PLH}	A or B	Y	R_L = 2 kΩ,	C_L = 15 pF		9	15	ns
t_{PHL}						10	15	ns

NOTE 2: See General Information Section for load circuits and voltage waveforms.

TEXAS
INSTRUMENTS
POST OFFICE BOX 225012 • DALLAS, TEXAS 75265

FIGURE 3-15   Page 3 of the Texas Instruments 74LS00 data sheet.

current parameters without having to understand the operation of the circuitry inside the "box." Figure 3-16 shows the basic gate model that we will use for the following explanations.

In our digital furnace controller example in Chapter 2, we told you that the logic gate we used there would treat an input signal voltage near 0 V as a logic low and a voltage near +5 V as a logic high. The term *near* is not very precise, so a question that may occur to you is, "How close does the input signal voltage have to be to 0 V in order for the gate to treat it as a logic low, and how close to +5 V does an input signal voltage have to be in order for the gate to accept the signal as a logic high?" Another question that may occur to you is, "What happens if an input signal halfway between 0 V and +5 V is applied to a gate input?" The answers to these questions can be determined from values given on the data sheet.

According to the data sheet for the 74LS00 in Figure 3-15, the maximum value for a low-level input voltage, V_{IL}, is 0.8 V. This means that the gate is guaranteed to treat any input voltage below 0.8 V as a logic low. Again, according to the data sheet, the minimum value for a high-level input voltage, V_{IH}, is 2.0 V. This means that the gate is guaranteed to treat any input signal of 2.0 V or greater as a logic high. The manufacturer makes no promises about the behavior of the gate for input signal voltages between 0.8 V and 2.0 V, so input signal voltages in this range should be considered illegal.

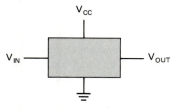

FIGURE 3-16   "Black box" model of a logic gate.

In most cases the input of a gate is connected to the output of another gate, so the voltages output by this gate will determine if the input signals meet these requirements. Let's look at the data sheet to see what the manufacturer specifies for output voltage levels.

The electrical characteristics section of the 74LS00 data sheet in Figure 3-15 shows that the maximum output low voltage, V_{OL}, for a gate in this device is 0.4 V with an output low current, I_{OL}, of 4 mA. This means that if a gate output is properly connected to the following gate inputs, the output voltage will never be greater than 0.4 V. Note that this maximum output low voltage of 0.4 V is 0.4 V *less* than the maximum voltage (0.8 V) that an input will accept as a logic low signal (V_{IL}). This guaranteed 0.4-V (400-mV) difference between the worst-case output voltage of 0.4 V and the maximum allowable input voltage of 0.8 V is called a *noise margin*. Figure 3-17 illustrates this lower noise margin. Noise margins

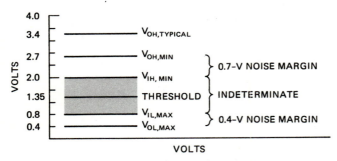

FIGURE 3-17   TTL voltage levels and noise margins.

are designed into logic gates and other digital devices so that a small noise voltage induced on a signal line cannot cause the input voltage to go into the indeterminate range of 0.8 V to 2.0 V.

For a logic high signal on the output of a gate, the data sheet shows that the minimum output high voltage, V_{OH} minimum, is equal to 2.7 V. This means that the output high signal voltage from the gate will always be 2.7 V or greater. As shown in Figure 3-15, this voltage is 0.7 V greater than the minimum voltage, V_{IH}, that an input will accept as a high. The difference of 700 mV between these levels then provides a noise margin of 0.7 V for logic high signals.

OUTPUT VOLTAGE VERSUS INPUT VOLTAGE TRANSFER CURVE

Figure 3-18 shows how the output voltage from an inverting gate such as a NAND gate changes as the input voltage is changed from 0 to +5 V. This graph is known as a *transfer curve*. From this graph you can perhaps see why an input signal voltage between 0.8 V and 2.0 V is considered illegal.

As the input voltage is increased above the $V_{IL,MAX}$ of 0.8 V, the output voltage of the gate drops below the guaranteed output high, $V_{OH,MIN}$, of 2.7 V. As the input voltage is increased further, the output voltage eventually drops to a legal output low level. Likewise, as the input voltage is decreased below a legal input high voltage of 2.0 V, the output voltage rises above the specified $V_{OL,MAX}$

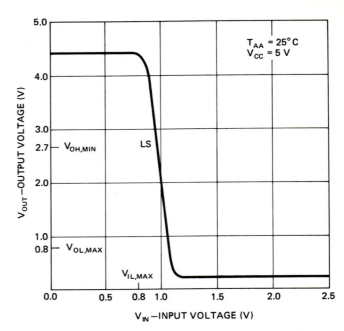

FIGURE 3-18   Transfer curve for a low-power Schottky TTL inverter.

of 0.8 V. The main point here is that you must make sure that an input signal to a gate does not stay in the range of 0.8 V to 2.0 V, or the output voltage may not be at a legal level!

Incidentally, the input voltage where the V_{OUT} is changing most rapidly is called the *threshold voltage*. For standard TTL, this threshold voltage is about 1.35 V. For a low-power Schottky TTL device the threshold voltage is about 1 V if the device has diode inputs and approximately 1.35 V if the device has transistor inputs. Other logic family devices have different threshold voltages.

INPUT CURRENT PARAMETERS

Before we can meaningfully discuss the current parameters of a gate we need to introduce you to the terms *source current* and *sink current*.

As you know from the discussion in Chapter 1, most electric currents consist of a flow of electrons from the negative terminal of a battery or power supply to the positive terminal. However, for historical reasons, current flow in diodes, transistors, and logic gates is usually defined in terms of an equivalent positive or *conventional* current flowing from the positive terminal to the negative terminal. It really doesn't matter which convention you use to describe the operation of a particular circuit, as long as you don't change conventions half way through. To be consistent with the data sheet definitions, we will use the positive current convention to describe the operation of logic gates.

Figure 3-19 shows a TTL output in a logic low state connected to the input of another TTL gate. The resistors and diodes shown in the gates represent the equivalent circuitry for an input and for an output of a typical TTL gate. You don't have to know how all these function, we just want you to see the current path from V_{CC} to ground through the devices.

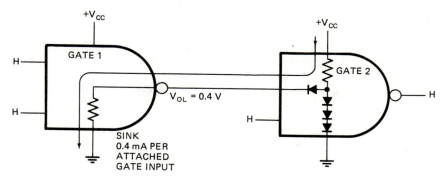

FIGURE 3-19   Output of a TTL gate in the low state sinks current from the input of a following gate.

If the output of gate 1 in Figure 3-19 is low, then a positive current will flow from the +5-V V_{CC} supply through the input circuitry of gate 2 into the output of gate 1, and through the output circuitry of gate 1 to ground, as shown by the arrow. An output is said to *sink current* if it creates a path for a positive current to get to ground or, for some logic families, to a negative voltage supply. The output of gate 1 in Figure 3-19 then sinks current when it pulls the input of gate 2 low. The data sheet for a 74LS00 gate in Figure 3-15 indicates that the maximum input low sink current, $I_{IL,MAX}$, is −0.4 mA. This means that for a 74LS TTL input to be pulled to the low state, the output connected to it may have to sink as much as 0.4 mA. A minus sign in front of a current specification indicates that a positive current is flowing out of the specified pin on the device. The $I_{IL,MAX}$ specification, for example, has a minus sign because, for a logic low, a positive current is flowing out of the input pin of the gate. A plus sign or no sign on a current specification indicates that a positive current is flowing into the device.

Figure 3-20 shows the equivalent circuit for a 74LS TTL gate with an output high connected to the input of a following gate. With a high on its output, conventional (positive) current will flow from the +5-V V_{CC} supply through the output circuitry of gate 1 into the input circuitry of gate 2, and through the input circuitry of gate 2 to ground. An output is said to *source current* if it creates a path from the positive supply to the input of the following gate. When the output of gate 1 is in the high state, it must source the current required to pull the input of the following gate to a logic high state. According to the data sheet in Figure 3-15, the maximum input high current, $I_{IH,MAX}$, for a 74LS00 TTL gate is 20 μA.

To summarize, then, an output may have to sink as much as 0.4 mA to pull a 74LS TTL gate input low, and it may have to source as much as 20 μA to pull a gate input high.

NOTE: A TTL input which is left unconnected or open will actually act as if it has a logic high on it. However, this high is not very dependable, so you should always connect an unused TTL input to the output of another gate, to V_{CC}, or to ground—whichever supplies the desired logic level to that input. Unused inputs on CMOS devices must never be left open!

OUTPUT CURRENT PARAMETERS

Now let's look at the data sheet to determine how much current a TTL output can sink in the low state and how much current it can source in the high state. The data sheet shows this information in two places.

The first place to look is in the test conditions column of the electrical characteristics section of the data sheet, next to the V_{OL} and V_{OH} parameters. As we said previously, the $V_{OH,MIN}$ specification of 2.7 V indicates that the voltage for an output high will always be higher than 2.7 V, as long as the gate is operated within certain specified conditions. The test conditions next to the V_{OL} parameter indicate that the device will meet its V_{OL} and V_{OH} specifications with a minimum supply voltage, V_{CC}, of 4.75 V, a worst-case input high, V_{IH}, of only 2.0 V, and an output sink current, I_{OL}, of 4 mA. What this means is that if the supply voltage is ever dropped below 4.75 V, or the input high signal voltage is dropped below 2.0 V, or the output is forced to sink more than 4 mA,

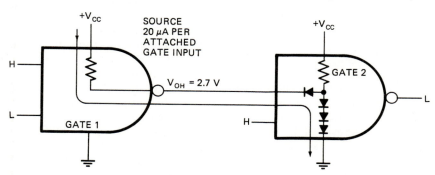

FIGURE 3-20   Output of a TTL gate in the high state sources current for the input of a following gate.

the manufacturer will not guarantee that the output low voltage will be less than 0.4 V. If the circuit can tolerate a $V_{OL.MAX}$ of 0.5 V instead of 0.4 V, the second line under V_{OL} on the data sheet tells you that the output can sink as much as 8 mA. This value of 8 mA is also listed as the maximum sink current, $I_{OL.MAX}$, in the recommended operating conditions section of the data sheet in Figure 3-15.

The maximum current that a 74LS00 TTL output can source in the high state can be determined from the data sheet in a similar manner. The test conditions next to the V_{OH} parameter in the data sheet in Figure 3-15 indicate that the manufacturer guarantees a $V_{OH.MIN}$ of 2.7 V with a source current of −0.4 mA, or −400 μA.

Using the data sheet values for input and output currents, you can determine how many gate inputs can be connected to one 74LS00 TTL gate output without overloading it. This number is usually referred to as *fanout*. The number of inputs that can be connected on an output is calculated as follows:

$$\text{Fanout} = \frac{\text{output current}}{\text{required current per input}}$$

This calculation must be done for both the high level and the low level because in some logic families the results are different. For a low-power Schottky TTL, the calculation for the output low state is:

$$\text{Fanout} = \frac{8\ \text{mA}}{0.4\ \text{mA}} = 20$$

For a 74LS00 TTL, the calculation for the output high state is:

$$\text{Fanout} = \frac{400\ \mu\text{A}}{20\ \mu\text{A}} = 20$$

Since the result of each calculation is 20, this means that you can connect as many as twenty 74LS TTL inputs to one 74LS TTL output without causing the low output voltage to exceed $V_{OL.MAX}$ or the output high voltage to be less than $V_{OH.MIN}$.

One more output current that we should briefly explain is I_{OS}. This is the current that an output in the high state will source if it is connected directly to ground. This rating is included to show you the amount of current the output can source if you are driving some device other than the input of another gate.

POWER-SUPPLY CURRENT RATINGS

The data sheet in Figure 3-15 shows two values for the power-supply current required to operate a 74LS00 device. The first current, I_{CCH}, is the supply current that the device will require when the outputs of all four NAND gates in the package are high. As shown in the test conditions column of the data sheet, this is accomplished by making all of the inputs of the NAND gates low (0 V). I_{CCH} has a maximum value of 1.6 mA. The second current parameter, identified in the data sheet as I_{CCL}, is

the amount of current that will be required by the device if all of the outputs are in the low state. As indicated in the test conditions column, all the inputs are made high (4.5 V) to cause all the outputs to go low. The maximum value for I_{CCL} is 4.4 mA.

Now, suppose that you are totaling up the current requirements for all of the devices on a circuit board to see how large the power supply needs to be. The obvious question is, "Which value do I use?" The answer to this is you use the worst-case value of 4.4 mA. We actually like to round values such as this to perhaps 5 mA. This gives a margin of safety and makes power available if we later have to connect additional devices.

PROPAGATION DELAY TIME AND PULSED SIGNAL CHARACTERISTICS

The final data sheet parameters to discuss here for a TTL gate are the propagation delay times, t_{PLH} and t_{PHL}. Propagation delay is the time needed for a change on the input of a device to cause a change on the output. To measure this parameter, a *pulsed* logic signal such as that shown in the top half of Figure 3-21 is used. The term *pulsed* means that the signal changes from one logic state to the other for a while, and then changes back to its previous state. Before we discuss how propagation delay is measured, here are a few basic pulse parameters from Figure 3-21 that you should start getting familiar with.

Pulse Width (t_{PW}) is defined simply as the time between the 50-percent point on the leading edge of the pulse and the 50-percent point on the trailing edge of the pulse.

Risetime (t_R) is usually defined as the time it takes a pulse to go from 10 percent of its amplitude to 90 percent of its amplitude.

Falltime (t_F) is usually defined as the time it takes a pulse to go from 90 percent of its amplitude to 10 percent of its amplitude.

> NOTE: Risetime and falltime may also be referred to in data books as *transition time low to high*, t_{TLH}, and *transition time high to low*, t_{THL}.

To observe and make measurements on pulsed signals we use a test instrument called an *oscilloscope*. In Chapter 4 we show you how to use an oscilloscope, but for now let's just see how propagation delay times are defined for a logic gate.

If one input of a 74LS00 gate is made high by connecting it to +5 V, the gate will function as an inverter. When the other input is made high, the output will go low after a propagation delay time, as shown in the lower half of Figure 3-21. Likewise, if the input is made low again, the output will go high. Propagation delay time is measured from the 50-percent point of a transition (change) on the input signal to the 50-percent point of the corresponding change on the output signal. If the *output* is making a transition from a low to a high, the propagation delay time is identified as t_{PLH}. If the *output* transition is from a high state to a low state, then the

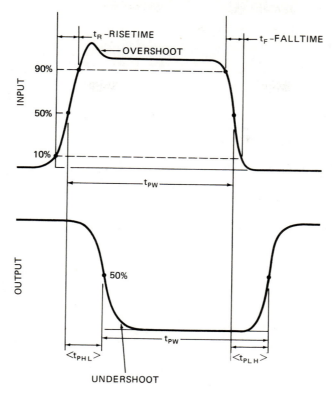

FIGURE 3-21   Pulse parameters.

propagation delay time is identified as t_{PHL}. Note on the data sheet in Figure 3-15 that the worst-case value for both t_{PLH} and t_{PHL} is 15 ns. As we will show you in later chapters, propagation delay time is an important consideration in digital systems.

Summary of 74LS TTL Data Sheet Parameters

A 74LS series TTL gate device requires a power-supply voltage, V_{CC}, of between 4.75 V and 5.25 V and a power-supply current of a few milliamperes.

To pull a 74LS TTL input to a logic low state, you must apply a voltage of between 0.0 V and +0.8 V to it. The device or circuit used to pull an input low must be able to *sink* up to a maximum of 0.4 mA of conventional (positive) current to ground.

To pull a 74LS TTL input to a logic high state, you must apply a voltage of between 2.0 V and 5.0 V to it. The device or circuit used to pull an input high must be able to *source* up to a maximum of 20 μA of conventional current to the input from the V_{CC} supply.

The output of a 74LS TTL gate can sink enough current in the low state or source enough current in the high state to drive as many as 20 74LS TTL inputs. To indicate this ability, the gate is said to have a *fanout* of 20.

The guaranteed output high voltage, $V_{OH,MIN}$, of 2.7 V for a 74LS TTL gate is 0.7 V greater than the minimum required input high voltage, $V_{IH,MIN}$, of 2.0 V. For the logic high state this gives a 700-mV *noise margin* between the voltage produced by an output and the voltage required by the input of the gate it is driving. A noise margin of 0.4 V exists between the maximum output

low voltage, $V_{OL,MAX}$, of 0.4 V and the maximum input low voltage, $V_{IL,MAX}$, of 0.8 V.

It takes about 15 nanoseconds (ns) for a signal change on the input of a 74LS00 TTL gate to cause a change in the output state of the gate. This time is called the *propagation delay time* for the gate. The time between the 50-percent point of a change on the input and the 50-percent point on a low-to-high output transition is called t_{PLH}. The time between the 50-percent point of a change on the input and the 50-percent point on a high-to-low output transition is called t_{PHL}.

COMPARISON OF COMMON LOGIC FAMILIES

In the last section we showed you how to read the data sheet for a device in a common logic family, the 74LS TTL family. Now that you know some of the parameters used to describe the characteristics of a logic device, we can describe and compare the features of common logic families more thoroughly than we did in the brief overview earlier in the chapter.

Table 3-2 shows a list of common logic families with some typical operating characteristics for devices in each family. Don't be overwhelmed by the amount of information in the table. This material is presented as a reference, not as something you should expect to memorize. At this point we just want to use the table to give you an overview of the device families you are likely to encounter. Later you will use some of the information in this table to help you interface a device from one logic family to a device from another family.

TABLE 3-2
COMPARISON OF LOGIC FAMILY CHARACTERISTICS.

Logic Family Name	Family	Specific Device	Operating Supply Voltage V_{CC} MIN (V)	Operating Supply Voltage V_{CC} MAX (V)	Minimum Logic 1 Input Voltage $V_{IH,MIN}$ (V)	Maximum Logic 0 Input Voltage $V_{IL,MAX}$ (V)	Minimum Logic 1 Output Voltage $V_{OH,MIN}$ (V)	Maximum Logic 0 Output Voltage $V_{OL,MAX}$ (V)	Maximum Logic 0 Input Current $I_{IL,MAX}$ (mA)	Maximum Logic 1 Input Current $I_{IH,MAX}$ (mA)	Minimum Logic 0 Output Current $I_{OL,MIN}$ (mA)	Minimum Logic 1 Output Current $I_{OH,MIN}$ (mA)	Prop. Delay t_{PHL} (ns)	t_{PLH} (ns)	C_L (pF)	Typical Power Dissipation per gate (mW)
TTL	7400	SN7400	4.75	5.25	2.0	0.8	2.4	0.4	−1.6	0.04	16	−0.4	7	11	15	10
	5400	SN5400	4.5	5.5	2.0	0.8	2.4	0.4	−1.6	0.04	16	−0.4	7	11	15	10
	74H	SN74H00	4.75	5.25	2.0	0.8	2.4	0.4	−2.0	0.05	20	−0.5	6	6	25	22
	54L	SN54L00	4.5	5.5	2.0	0.7	2.4	0.3	−0.18	0.01	2	−0.1	31	35	50	1
	74S	SN74S00	4.75	5.25	2.0	0.8	2.7	0.5	−2.0	0.05	20	−1.0	5	5	50	19
	74AS	SN74AS00	4.5	5.5	2.0	0.8	$V_{CC}-2$	0.5	−0.5	0.02	20	−2.0	4	4.5	50	8
	74LS	SN74LS00	4.75	5.25	2.0	0.8	2.7	0.5	−0.4	0.02	8	−0.4	10	9	15	2
	74ALS	SN74ALS00	4.5	5.5	2.0	0.8	$V_{CC}-2$	0.4	−0.1	0.02	8	−0.4	8	11	50	1.2
ECL	ECL10KH	MC10H101 @−5.2V	V_{EE} −8	0	−1.13	−1.48	−0.98	−1.95	0.0005	0.265	−50	−50	1.5	1.5	3	25
	ECL100K	F100101@ −4.5V	V_{EE} −7	0	−1.16	−1.47	−1.03	−1.62	0.0005	0.35	−55	−55	1.1	1.1	3	40
MOS	Low-thresh-old PMOS	1702A ROM	V_{CC}=+5 V_{DD}=−9 V_{GG}=V_{DD}		3.0	0.65	3.5	V_{CC}−6V MOS or 0.45	0.001	0.001	1.6 @ 0.45 V	0.2 @ 3.5 V	—	—	—	—
	+12, +5,−5 NMOS	8080A	V_{CC}=+5 V_{DD}=+12 V_{BB}=−5		3.3	0.8	3.7	0.45	±0.001	±0.01	1.9	0.15	—	—	—	—
	+5 only NMOS	2102 RAM	4.5	5.5	2.0	0.8	2.4	0.4	0.01	0.01	2.1	0.1	—	—	—	—
CMOS	+5V 4000A	CD4011	3	15	$\frac{2}{3}V_{CC}$	$\frac{1}{3}V_{CC}$	V_{CC}−.01	0.01	10 pA	10 pA	0.12 @ 0.5 V	0.12 @ 0.5 V	180	125	50	0.001 @ 1KHz
	+5V 4000B	CD4081B	3	18	$\frac{2}{3}V_{CC}$	$\frac{1}{3}V_{CC}$	V_{CC}−0.1	±0.01	±0.001	0.001	0.4 @ 0.4 V	1.6 @ 2.5 V	160–320	210–420	50	1 @ 1 MHz
	+15V 4000B	CD4081B	3	18	$\frac{2}{3}V_{CC}$	$\frac{1}{3}V_{CC}$	V_{CC}−0.1	±0.01	±0.001	0.001	3 @ 1.5 V	3 @ 13.5 V	50	65	50	10 @ 1 MHz .01 @ 1 kHz
	74C	74C00 @+5 V	3	15	3.5	1.5	4.5	0.5	0.001	−0.001	1.75	−1.75	90	90	50	2.5 nW
	74HC	74HC00 @ 4.5 V	2	6	3.15	0.9	4.4	0.1	−0.001	0.001	4	4	23	23	50	0.5 @ 1 MHz
	74HCT	74HCT00	4.5	5.5	2	0.8	4	0.33	0.0001	0.001	4	4	20	20	50	
GaAs	10G000A		V_{DD}=0 V_{SS}=−3.5 V_{EE}=−5.5	0 −3.3 −5.1	−0.7	−1.8	−0.7	−1.8	0.4	0.4		−60	0.29	0.15		220

To start, look at the column labeled "Logic Family Name" in Table 3-2 and note that there are five major families of logic gates. Each of these families uses a different technique to produce logic gates on the tiny chip in the IC. Also note in Table 3-2 that within each family there are subfamilies which all use basically the same technology, but have slightly different operating characteristics.

For now you don't need to know much about the actual circuitry and technology used for the logic families we discuss in the following sections. If you are interested in this information, consult Appendix B of this book.

General Comments about TTL Logic Devices

We introduced you to the TTL subfamilies and some of the MOS subfamilies in an earlier section of this chapter so that you could build some circuits with devices from one of these subfamilies. To help refresh your mem-

ory of these, here's some brief notes about each TTL subfamily in addition to the characteristics shown in Table 3-2.

7400—Standard TTL. Best seller for many years, but seldom used in new designs because 74LS is faster and uses less power.

74H00—High-Speed TTL. Faster than standard but uses too much power, so now obsolete.

74L00—Low-Power TTL. Uses less power than standard TTL, but is also much slower. Obsolete, replaced by newer families.

74S00—Schottky TTL. Uses Schottky transistors, which switch faster than standard bipolar transistors, so gates have short t_{PD}.

74LS00—Low-Power Schottky TTL. Good compromise of speed and power dissipation. Slightly faster than standard TTL, but only about ⅕ the power dissipation.

Fanout of 20 instead of 10, and high-level noise margin of 0.7 V instead of 0.4 V.

74AS00—*Advanced Schottky TTL.* Improved version of Schottky TTL.

74ALS00—*Advanced Low-Power Schottky TTL.* Improved version of Low-Power Schottky TTL. Note that the propagation delay is 8 ns instead of 10 ns, and power dissipation is 1 mW per gate instead of the 2 mW per gate of 74LS devices.

TTL Output Types

TOTEM-POLE OUTPUTS

TTL devices commonly have one of three different types of outputs. The NAND gates in the 74LS00 that we discussed in a preceding section have what are called *totem-pole* outputs. Figure 3-22 shows again the circuit used for one of the gates in a 74LS00. The transistors labeled Q1, Q2, and Q3 in this circuit make up the totem-pole output. When the output is in a logic high state, transistors Q1 and Q2 are turned on and Q3 is turned off. If you think of an on transistor as a closed switch, then you can see that Q1 and Q2 source current from V_CC to the inputs of the following gates. When the output is in a logic low state, Q1 and Q2 are turned off, and Q3 is turned on. Q3 then sinks current from the inputs of the following gates to ground.

The advantage of a totem-pole output is that it has a transistor driving the output in both the high state and the low state, so t_{PLH} and t_{PHL} are nearly equal. A major disadvantage of totem-pole outputs is that two or more totem-pole outputs cannot be connected together on a common signal line. Figure 3-22b shows you why they can't. If the gate output on the left is high, then Q1 and Q2 will be supplying a low-resistance current path from V_CC to the common point. If the output on the right side of Figure 3-22b is low, then Q4 in this device will be supplying a low-resistance path from the connection point to ground. The result then is that in this condition there is a low-resistance current path from V_CC to ground through the two outputs. This will cause a large current to flow and possibly destroy the outputs of the gates. The rule that comes from this is:

> *Never connect a totem-pole output to another totem-pole output, to ground, or to V_CC.*

OPEN-COLLECTOR OUTPUTS

In some applications we want to have any one of several logic gates be able to affect the logic level on a single signal line. One way of doing this is to use devices which have *open-collector* outputs. Figure 3-23 shows the internal circuitry for one of the NAND gates in a 74LS01 quad 2-input NAND gate device. The circuitry for this device is similar to that for the 74LS00 in Figure 3-22a except that the transistors which supply current from V_CC to the output are missing. For an open-collector output these transistors are replaced by an external resistor of perhaps 2.2 kΩ connected from V_CC to the output as shown.

When the output is in the logic low state, Q3 is turned on. Q3 then will sink the required input low current from any inputs connected to it. When the output is in the logic high state, Q3 will be off. The 2.2-kΩ resistor will then source the input high current required by any inputs connected to the output.

If two open-collector outputs are connected together, as shown in Figure 3-23b, either output can go low and pull the common point low. However, regardless of the combination of high and low states on open-collector outputs connected together, the current from V_CC to ground will be limited to a small value by the 2.2-kΩ resistor.

Figure 3-23c shows the logic function that is accomplished by tying the outputs of two open-collector NAND gates together. The output will be pulled low if A is high and B is high OR if C is high and D is high. A Boolean expression for the circuit is then $Y* = AB + CD$. Because of the logic function that is performed, this connection is often called a *wired OR* or a *dotted OR* connection. Note how the wired OR connection is indicated on the circuit drawing.

Figure 3-23d shows the ANSI/IEEE symbol used to represent a NAND gate with an open-collector output. The small diamond shape after the & at the top of the symbol indicates that the output is open collector and must have a pull-up resistor to operate properly.

THREE-STATE OUTPUTS

Open-collector devices are useful in applications where we want any one of several gates to pull a signal line to a low level. In other applications we want any one of several devices connected on a line to be able to pull the line low, pull the line high, or not affect the logic level on the line. In these applications we use devices which have *three-state* outputs. The three states for the outputs of these devices are logic low, logic high, and high impedance, or "floating." To see what determines the output state, let's take a look at one of the three-state output buffers of a noninverting buffer such as that found in a 74LS367 device, shown in Figure 3-24.

If the output enable input, $\overline{E}$, is asserted low, the output buffer will be *enabled*. The logic level present on the input of this buffer will also be present on the output. If the data input is low, then the output will be at a logic low, and if the input is at a logic high, the output will be at a logic high. If the $\overline{E}$ input is high, the output buffer will be *disabled*. In this condition both the upper and lower transistors in the output buffer are turned off, so the buffer output has no effect on the voltage present on the line to which it is connected. With the output in this high-impedance state, the line will "float," or be pulled to whatever voltage level is applied to it by some other device. When several three-state devices are connected to a line, care must be taken that only one output is enabled at a time. If two three-state outputs on a line are enabled at the same time, a disaster similar to that which happens when totem-pole outputs are connected together may occur. In later chapters of this book you will see many applications of three-state outputs.

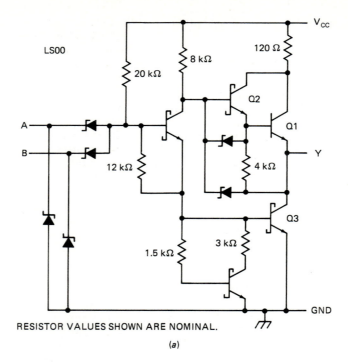

RESISTOR VALUES SHOWN ARE NOMINAL.

(a)

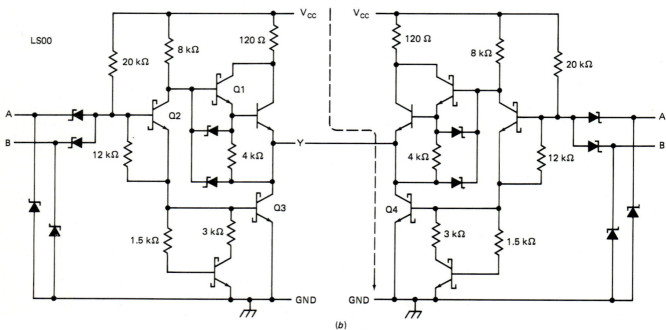

(b)

FIGURE 3-22 (a) Circuit used for gate in 74LS00. *(Courtesy Texas Instruments Incorporated.)* (b) Connecting totem-pole outputs.

SUMMARY OF OUTPUT TYPES

Totem-pole outputs should never be connected together because if one is high and the other is low, one or both may be destroyed. Open-collector outputs can be connected together so that any one of several device outputs can pull the line low. The outputs of devices with three-state outputs can be connected to a common signal line, but only one output can be enabled at a time. When a three-state output is enabled, the output logic level is a logic high or a logic low. When the output is disabled, the output "floats," or is pulled to whatever voltage is applied to it by another device connected to the line.

ECL Logic Devices

The next major family of logic devices shown in Table 3-2 is *emitter-coupled logic,* or *ECL.* Devices in this family also use bipolar transistors, but they use a different

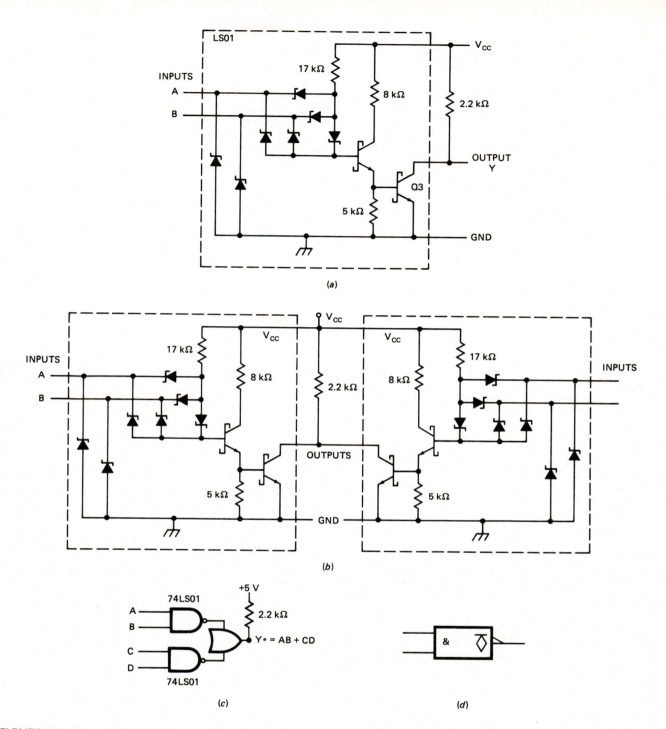

FIGURE 3-23 Internal circuit for a NAND gate with open-collector output. *(a)* Circuit. *(Courtesy Texas Instruments Incorporated.)* *(b)* Two open-collector outputs connected together. *(c)* Logic function of two open-collector NAND gates tied together. *(d)* IEEE symbol used to represent NAND gate with an open-collector output.

internal circuit so that the gates have a very short propagation delay. ECL gates are used in systems which must operate at very high frequencies. Note in Table 3-2 that ECL logic subfamilies are usually operated with power-supply voltages which are negative with respect to ground. Also note that the voltage level of a logic high for devices in these families is about -1 V, and the voltage level of a logic low is about -1.5 V.

The disadvantages of ECL gates are:

1. The gates dissipate relatively large amounts of power.

2. Each signal line connecting the output of one gate to the input of another must be "terminated" with resistors. In Chapter 10 we show you how to terminate signal lines when needed.

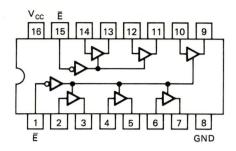

TRUTH TABLE

INPUTS		OUTPUT
$\overline{E}$	D	
L	L	L
L	H	H
H	X	(Z)

FIGURE 3-24   Three-state output buffers of a Motorola 74LS367 device.

3. The negative power-supply voltage and logic levels make ECL gates difficult to interface to devices from other logic families.

The second two points mean that many extra parts besides the gates are usually needed on a circuit board with ECL devices.

Gallium Arsenide Devices

All of the other logic families we discuss here use transistors made from silicon in their internal circuitry. A new logic family which is still in a relatively early stage of development uses transistors made from a combination of the elements gallium (Ga) and arsenic (As). The advantage of gallium arsenide (Ga As) transistors is that they can be turned on and off very quickly. This means that logic gates made from these transistors have very short risetimes, falltimes, and propagation delay times. As you can see in Table 3-2, current devices, such as the 10G000A quad 2-input NOR gate device made by GigaBit Logic, have risetimes and falltimes of about 150 *picoseconds* (ps) and propagation delay times of about 290 ps. (To perhaps give you a feeling for these times, a beam of light only travels a little over an inch in 100 ps.) Devices under development have propagation delays under 100 ps. Currently the parts available in GaAs devices include basic gates, flipflops, counters, RAM, and arithmetic circuits. Many of the currently available devices have ECL-compatible inputs and outputs. As with ECL logic devices, the signal lines of GaAs devices must be terminated with resistors to prevent signal reflections on the lines. Due to the extremely high frequencies at which these devices can operate, many other devices are being developed.

MOS Logic Devices

The number of gates that can be put on a single silicon chip using the bipolar circuits we have discussed so far

is limited by two major factors. First, each gate dissipates a relatively large amount of power. Therefore, if too many bipolar gates are put on a single chip, the chip will overheat and no longer function. The second factor is that each gate circuit takes up a sizable amount of space, or "real estate," on the chip when it is made with bipolar transistors. If a chip is made large enough to fit many bipolar circuits, there is a high probability that an imperfection in the silicon will occur in that chip and cause it to malfunction. With small chips it is possible to fit many chips between the imperfections and, therefore, get a greater yield of good devices.

To avoid some of these problems and others, ICs were developed that use *metal-oxide semiconductor,* or *MOS,* transistors instead of the bipolar transistors used in the TTL devices. For reference, a section of Appendix B shows the schematic symbols used for MOS transistors and describes their operation. The main points for you to know about MOS transistors for now are:

1. When operated at low frequencies they require much less power than the corresponding bipolar device.

2. They take up less space on an IC than the comparable bipolar circuit, so many more devices can be put on a single chip.

3. They can be used to build circuits, such as dynamic memories, that are impossible to build with bipolar devices.

4. They require very low input current for both an input low and an input high logic state.

5. Their input and output structures can easily be destroyed by static electricity.

MOS devices contain circuitry to help protect them from being destroyed by static electricity. However, the static voltage generated by walking across a carpet on a low-humidity day may be ten times as much as the protection circuit is capable of draining off. The protection circuit cannot be improved without degrading the performance of the gate, so care in handling is required. In a later section of this chapter we tell you in detail how to prevent static electricity from destroying MOS ICs and other ICs.

MOS SUBFAMILIES

MOS technology allows a very large number of circuits to be built in a single IC. It is this technology which has made possible the microprocessors, memories, and other LSI devices which are used to build microcomputers. For the most part this technology is used for these LSI devices rather than for simple logic gates such as the TTL devices we discussed earlier.

The first MOS ICs used what are called *P-channel transistors,* so these devices were referred to as *PMOS.* PMOS devices had the disadvantages that they were slow and required several supply voltages. Table 3-2 shows some typical values for PMOS devices.

The next step in the evolution of MOS was to use N-channel MOS transistors in the internal circuits. *NMOS* devices are about three times as fast as the comparable

PMOS devices. As shown in Table 3-2, the first NMOS devices also required several different power-supply voltages. Within a few years, however, manufacturers learned how to make NMOS devices which require only a single 5-V power supply. The memories and microprocessors which we discuss in later chapters use this 5-V-only NMOS technology.

NOTE: A sometimes confusing array of subscripts are used to identify the supply voltages on MOS devices. For MOS and CMOS devices which require only +5 V and ground, the positive supply is now almost universally referred to as V_{CC}, and ground is referred to as V_{SS} or simply not mentioned. V_{DD} is used to refer to the most negative supply voltage for PMOS devices and to refer to the most positive supply voltage for NMOS devices operating with a supply voltage greater than +5 V. The names V_{BB} and V_{GG} are used to identify additional bias voltages required by some devices.

CMOS Logic Devices

For a variety of reasons, neither pure PMOS circuitry nor pure NMOS circuitry is suitable for making simple logic gate devices such as the TTL devices listed in Table 3-1. However, by building the circuits on the IC with a combination of N-channel and P-channel transistors, it is possible to produce logic gate devices which have the desired characteristics. Since N-channel and P-channel transistors are in some ways complementary to each other, the resultant IC devices were referred to as *complementary metal-oxide semiconductors*, or just *CMOS*. Here are some of the major characteristics that you should know about CMOS for now:

1. CMOS gates are easily destroyed by static electricity, so you must follow the handling precautions required for all MOS devices.

2. The major advantage of CMOS devices is that when they are operated with dc or low-frequency logic signals, they use only a few microwatts of power per gate. When operating with a +5-V supply, a typical CMOS gate dissipates only about 10 μW of power with a 1-kHz input signal. Note in Table 3-2 that the typical power dissipation for TTL is a few milliwatts per gate. This is about a thousand times greater than the dissipation for a CMOS gate. The very low power dissipation of CMOS gates makes them ideal for use in portable computers and other battery-powered equipment.

3. The power dissipation of CMOS gates increases as the frequency of the signals applied to them increases. With a 1-MHz input signal the power dissipation of a CMOS gate operating with a +5-V supply is about the same as the power dissipation of a low-power Schottky TTL gate.

4. MOS transistors are voltage-controlled devices, so CMOS gates require very little input current.

5. CMOS devices can operate with a wider range of supply voltages than TTL devices. As shown in Table 3-2, the devices in all but one of the CMOS subfamilies can operate properly with a power-supply voltage of +3 V to +15 V.

6. The high-level noise margin and the low-level noise margin are both about ⅓ of the applied power-supply voltage.

7. Signals applied to CMOS inputs must not go more negative than ground or more positive than the V_{CC} being used. If an input signal exceeds these limits, the CMOS device may enter a *latchup* condition. In this condition the device will no longer respond to changes of the input signal. Once a CMOS device is latched up in this way, the only way to get it out of the latchup condition is to turn off the signal, turn off the power, turn the power on again, and reduce the amplitude of the applied signal. Latchup destroys some CMOS devices, so you should always try to avoid it.

8. Most CMOS devices have outputs similar to the totem-pole outputs of TTL devices. Therefore, you should never connect two or more CMOS outputs on a common line unless they are open-drain or three-state outputs. As with TTL, only one TTL three-state output should be enabled at a time on a common line.

9. CMOS technology can be used to produce unique devices called *transmission gates*. The following section shows you how transmission gates work and introduces you to how they are used.

CMOS TRANSMISSION GATES

Figure 3-25 shows the schematic symbol commonly used for a CMOS transmission gate. A transmission gate is essentially a logic-controlled switch. If the enable input is high, there is essentially no connection between one side of the device and the other. If the enable input is low, the two sides of the device are connected by a resistance of a few hundred ohms. The output voltage will therefore be almost the same as the voltage applied to the input, depending on the current through the gate.

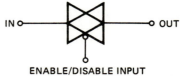

FIGURE 3-25 Schematic symbol for a transmission gate.

Figure 3-26 shows how a transmission gate can be connected to the output of a CMOS gate to give a *three-state* output. If the transmission gate in Figure 3-26 is enabled by making its enable input low, the logic level on the output of the NAND gate will be transmitted to the output of the transmission gate. The output of the transmission gate will then be a logic low or a logic high, the same as the logic level on the output of the NAND gate.

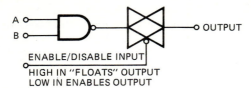

ENABLE/DISABLE INPUT
HIGH IN "FLOATS" OUTPUT
LOW IN ENABLES OUTPUT

FIGURE 3-26   Transmission gate connected to the output of a CMOS gate to give a three-state output.

However, if the transmission gate in Figure 3-26 is not enabled, it acts like an open switch. The output of the transmission gate will therefore "float," or be pulled to whatever voltage is applied to the signal line by some other device, just as we described for TTL three-state outputs in a previous section. This floating state is sometimes referred to as the *high-Z state*.

CMOS SUBFAMILIES

The first CMOS logic subfamily to be developed was the 4000A series. The family includes logic functions such as those shown for TTL in Table 3-2, but the devices are numbered differently. For example, the device numbered 4001A contains four 2-input NOR gates, and the device numbered 4012A has two 4-input NAND gates.

The 4000A series devices have very little output drive capability, so the 4000B series was developed. Devices in the 4000B subfamily have the same functions and numbers as those in the 4000A series, but the numbers are followed by a B instead of an A. Note the much longer propagation delay of these devices as compared to TTL devices.

By the time the 4000 series of CMOS was developed, most logic designers had become very familiar with the logic functions, part numbers, and pin connections of the devices in the standard TTL subfamily. These designers were for the most part unhappy about learning a whole different set of device numbers and pin numbers for the 4000 series devices, and they therefore tended to avoid using CMOS unless it was absolutely necessary. To solve this problem manufacturers came out with the 74C series of CMOS parts. Parts in this series have the same logic functions and pin connections as the TTL part with the same basic number. A 74C00, for example, contains four 2-input NAND gates, just as a 7400 device in any of the TTL subfamilies does. The outputs of the 74C00 parts are also designed so that they are more easily interfaced to TTL devices than are those of the 4000 series devices.

Another CMOS subfamily is the 74HC *high-speed CMOS* series. Parts in this subfamily also have logic functions and pin numbers which correspond to the similarly numbered TTL part. As shown in Table 3-2, a 74HC series device has a shorter propagation delay than the corresponding 74C part. Also note in Table 3-2 that the output low current of 4 mA and the output high current of 4 mA are large enough to drive one or more TTL inputs directly.

A still newer CMOS family is the 74HCT family. Devices in this family have input low and high voltage re-

quirements which are more compatible with TTL output voltage levels than are 74HC inputs.

Summary of the Common Logic Families

We have worked our way through a considerable list in the preceding sections in order to give you an overview of most of the available logic families. However, the major subfamilies you will encounter in this book and in common electronic equipment are the 7400, 74LS00, and 74ALS00 series of TTL devices; NMOS memories and microprocessors; and the 74C and 74HC series of CMOS devices. All of these subfamilies have the advantage that they can operate with only a +5-V power supply. TTL subfamily devices have the advantage that they have relatively short propagation delays, but they have the disadvantage that they use more power than devices made with MOS technology. NMOS technology is used to make memories, microprocessors, and other LSI devices which have a large amount of circuitry on a single IC. CMOS technology is used to make logic gates and other devices which dissipate very little power when operated with dc or low-frequency input signals.

Handling Precautions for MOS and Other Digital Devices

Walking across a carpet, sitting on a foam cushion, or picking up a standard plastic bag can can cause an electrostatic voltage of as much as 35,000 volts to accumulate on your body. When you touch some conductor, the accumulated charge leaves your body. Even though the voltage is very high, the amount of current involved is low, so you may at most feel a slight shock. This discharge of static electricity from your body, however, is more than enough to destroy many electronic devices.

It has been estimated that *electrostatic discharge*, or ESD, is responsible for as much as $10 billion per year in damage to ICs and electronic equipment. Due to this large sum, manufacturers have spent a considerable amount of time and money finding the sources of the problem. Their research has shown that most of this damage can be prevented by some simple handling precautions. It was originally thought that these handling precautions were only necessary for MOS devices. However, further research has shown that resistor packs, chip capacitors, and even bipolar devices such as 74LS TTL gates and operational amplifiers are also damaged or destroyed by ESD if they are not handled properly. This is especially true as components and internal circuits are made smaller and smaller.

The point we want to make very strongly here is that the following precautions are *not* optional. They are required in engineering, production, and field service departments by most companies:

1. Connect your body to an earth ground with a wrist strap such as that shown in Figure 3-27 whenever you are working with ICs or printed-circuit boards. This will allow any charge that

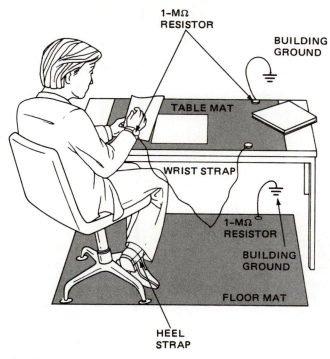

1-MΩ
RESISTOR

BUILDING
GROUND

TABLE MAT

WRIST STRAP

1-MΩ
RESISTOR

BUILDING
GROUND

FLOOR MAT

HEEL
STRAP

FIGURE 3-27 Antistatic precautions.

accumulates on your body to drain off to ground. If you do not have a wrist strap as shown, connect your wrist to ground with a length of wire and a 1-MΩ series resistor. The resistor prevents you from getting a fatal shock if you touch a voltage source but allows static electricity to drain off.

2. Work only on an antistatic-treated workbench or use an antistatic mat such as that shown in Figure 3-27.

3. Do not wear clothes which generate static electricity every time you move. Usually you can tell if your clothes are doing this by the snap and crackle you hear and little shocks you feel when you touch some grounded object.

4. Store MOS integrated circuits in *conductive foam* (usually dull black or pink) or in aluminum foil. Never store any ICs in white styrofoam, which is an excellent static electricity generator.

5. Keep spare PC boards containing ICs in conductive plastic or metallic envelopes. The best of these are gray in color.

6. Connect to an earth ground (the round prong of all properly wired 120-V ac outlets) your soldering-iron tip, your metal bench, and the chassis of any test equipment used.

7. Be sure to turn *off* the power before removing ICs from their sockets or inserting them in their sockets.

8. Make sure the power is *on* at the same time or before applying any signal to a device input. With no V_{CC} present, a low-output-impedance signal generator may force enough current through an input to destroy it.

9. Connect all unused input pins on ICs to V_{CC} or V_{DD}, whichever provides the desired logic state for the input. Never leave inputs open because open inputs provide a path for static voltages to get into the IC.

10. Put resistors in series with any MOS inputs that receive signals from another circuit board.

Interfacing Logic Gate Devices

In later chapters we show you in detail how to interface logic gates to a wide variety of other devices. For now we want to briefly describe the major factors you have to take into account when you connect gate outputs to the inputs of other gates in the same or in different logic families.

DEVICES WITHIN THE SAME SUBFAMILY

When you are connecting devices from the same subfamily, the main point to make sure of is that you do not exceed the specified fanout for that subfamily. You can use Table 3-2 to determine the fanout of a particular subfamily. For example, the fanout for standard TTL is 10, so if you are connecting standard TTL gate inputs to the output of a standard TTL gate, you want to make sure that you do not connect more than 10 inputs to one output. Now, here are some points about connecting logic devices from different families to each other.

CONNECTING DIFFERENT TTL SUBFAMILIES TO EACH OTHER

Connecting different TTL subfamily devices to each other is relatively easy because the input and output voltage levels are all compatible. The main point that you have to make sure of is that a gate output can sink enough current in the low state to pull the inputs connected to it low. When you have a load made up of gate inputs from different subfamilies, use Table 3-2 to determine the maximum input low current for each gate input that you are going to connect. Add these currents and make sure the total does not exceed the output low current for the driving gate. The following example shows how to determine if a 74LS00 output can drive one 7400 TTL input and five 74LS00 TTL inputs. First use Table 3-2 to find the maximum logic 0 input currents and add them up:

$$7400 \text{ Max } I_{IL} = 1.6 \text{ mA}$$

$$5 \times 74LS00 \text{ Max } I_{IL} = 5 \times 0.40 = 2.0 \text{ mA}$$

$$\text{Total required } I_{IL} = 1.6 + 2.0 = 3.6 \text{ mA}$$

$$74LS00 \text{ } I_{OL} @ V_{OL} \text{ of } 0.4 \text{ V} = 4.0 \text{ mA}$$

The total required I_{IL}, then, is less than the 4.0 mA that the 74LS00 output is capable of sinking in the low state, so the 74LS00 can drive this mixture of gate inputs.

TTL TO MOS TO TTL

Early MOS devices required special level translator ICs to interface to TTL parts. Current MOS devices for the most part have TTL-compatible inputs and outputs. The main point that you have to consider when connecting TTL inputs on a MOS output is that the MOS output can sink enough current in the low state to pull the TTL inputs low. You determine this in the same way we showed in the preceding section for TTL-to-TTL connections.

TTL TO CMOS TO TTL

When connecting a TTL output to the input of a 4000A or 4000B series CMOS device operating with a V_{CC} of +5 V, connect a 10-kΩ resistor from the TTL output to V_{CC}. This makes sure the CMOS input will be pulled to a legal high level because the worst-case output high of 2.4 V for a TTL device is not quite high enough to be a legal high for CMOS devices in these series.

When connecting the output of a CMOS device to one or more TTL inputs, the main point to consider is whether the CMOS output can sink enough current in the low state to pull the TTL inputs low. You calculate this in the same way that we showed for TTL-to-TTL connections in a previous section.

If you need more current to drive following gates, you can use a *buffer*. Two commonly available CMOS buffers are the CD4050 noninverting buffer and the CD4049 inverting buffer. Each of these devices has six buffers which can drive as many as two standard TTL inputs. The outputs of the 4000B, 74C00, and 4500 series CMOS devices are designed to drive the input of one 74LS00 family device, so you can also use a device in this family as a buffer from CMOS to TTL.

Note in Table 3-2 that CMOS devices in the 74HC00 series are capable of sinking up to 4 mA in the low state, so these devices can directly drive several TTL inputs.

TTL TO ECL TO TTL

Since the power-supply voltages and the logic-level voltages for TTL and ECL are so different from each other, special level translator/buffers are required to interface them. The MC10124, for example, is a quad TTL-to-ECL translator. The MC10125 is a quad ECL-to-TTL translator.

CHECKLIST OF IMPORTANT TERMS AND CONCEPTS IN THIS CHAPTER

Die
Integrated circuit, IC
  Small-scale integration, SSI
  Medium-scale integration, MSI
  Large-scale integration, LSI
  Very large-scale integration, VLSI
  Ultra high-scale integration, UHSI
Printed-circuit board
Packages
  Dual in-line (DIP), ceramic flat pack, ceramic-chip carrier, pin-grid array
Propagation delay, t_{PD}
Bipolar transistor
TTL, transistor-transistor logic
  Standard, high-power, low-power, Schottky, low-power Schottky, advanced low-power Schottky
Digital IC package numbering systems
Prototyping
Solderless protoboard
Universal printed-circuit board
Wire wrapping
Diode
Data sheet
  Pin numbering
  Internal gate connections
  Maximum ratings
  Operating characteristics, V_{CC}, V_{IH}, V_{IL}, I_{OH}, I_{OL}, T_A
Noise margin
Transfer curve
Threshold voltage
Source and sink current
Conventional current
Fanout
Propagation delay, t_{PLH}, t_{PHL}
Pulse width, t_{PW}
Risetime, t_R or t_{TLH}
Falltime, t_F or t_{THL}
TTL output types
  Totem-pole, open-collector, wired OR, dotted OR, three-state
Enable, disable
Emitter-coupled logic, ECL
Gallium arsenide, GaAs, devices
Metal-oxide semiconductor, MOS, NMOS, PMOS
Complementary metal-oxide semiconductor, CMOS
Transmission gate
  High, low, and "floating" (high-Z) states
Electrostatic discharge, ESD
Antistatic precautions
Buffer

REVIEW QUESTIONS AND PROBLEMS

1. Draw a top view of a typical 16-pin dual in-line package IC. Identify pin 1 on the package and write the appropriate number next to each pin.

2. What two numbers are used at the start of the identifying numbers for TTL devices?

3. What are the main differences between the 7400 series of TTL devices and the 5400 series of TTL devices?

4. What are the two main advantages of 74LS subfamily devices over standard TTL devices?

5. How is the manufacturer of an IC identified on the package?

6. a. What are the main advantages of the solderless protoboard?
   b. What are bus strips on a solderless protoboard used for?
   c. What size and type of wire should be used with solderless protoboard?
   d. About how long a section of wire should be stripped of insulation to be inserted in solderless protoboard?
   e. Why should wiring be done neatly on a solderless protoboard?

7. a. Give an advantage of wire-wrap circuit construction over solderless protoboard construction.
   b. Describe the steps in making a wire-wrap connection.

8. Why should you connect a 100-μF capacitor and a 0.1-μF capacitor in parallel between V_{CC} and ground where V_{CC} connects to a prototype circuit?

9. What is one handy way to keep track of which parts of a circuit you have completed when you are building a circuit?

10. Describe the initial testing procedure for a newly built circuit.

11. What are the most likely causes of a hot IC in a circuit?

12. What is the advantage of a logic probe over most digital voltmeters for checking logic levels?

13. List the five major types of logic families.

14. What is the major advantage of CMOS logic devices?

15. Describe the handling precautions you must observe when working with MOS or CMOS logic devices to avoid damage caused by static electricity.

16. Use the data sheet and Table 3-2 to determine for a low-power Schottky gate:
    a. The maximum V_{CC} voltage.
    b. The maximum allowable input signal voltage with V_{CC} at 5 V.
    c. The lowest (worst-case) voltage a device is guaranteed to accept as a logic high.
    d. The highest voltage that the device is guaranteed to accept as a logic low.
    e. The lowest voltage that the output will ever be at for a logic high.
    f. The highest voltage the output will ever be at for a logic low.

17. a. Define the term *noise margin*.
    b. Use Table 3-2 to determine the high-level noise margin and the low-level noise margin for a 74LS00 device.

18. For the circuit in Figure 3-28:
    a. Predict the logic level that should be present on the output with the signals shown on the A, B, and C inputs.

b. What logic level does the voltage shown on the output represent for a 74LS family device?
c. Identify the probable source of this problem and describe how you would go about verifying it.

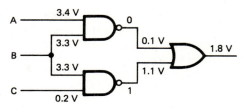

FIGURE 3-28   Circuit diagram for Problem 18.

19. a. According to Table 3-2, what is the maximum current that may be required to pull a 74LS00 device input to a logic low level?
    b. What is the maximum current that may be required to pull the input of a 74LS device to a logic high state?

20. a. Define the term *fanout*.
    b. Use Table 3-2 to determine the fanouts of a standard TTL gate and a Schottky TTL gate.

21. a. How much power-supply current is required to operate a 74LS00 device with all of its outputs high?
    b. How much power-supply current is required to operate a 74LS00 device with all of its outputs low?
    c. Which of these ratings should you use when you are computing the amount of current that will be required by the ICs on a board you are building? Why?

22. a. Define the general term *propagation delay*.
    b. Use a drawing to help you explain the meaning of the parameters t_{PHL} and t_{PLH}.
    c. Between what two points on a pulse are the risetime and falltime of a pulse usually measured?

23. For the circuit in Figure 3-29:
    a. Predict the change that will occur on the X output when the A input signal changes from a low to a high as shown.
    b. What name is given to the time between this change on the input and the corresponding change on the output?
    c. Use Table 3-2 to determine the value of this parameter for a 74HC00 gate.
    d. How long will it take for the change on the A input to produce a change on the Y output?

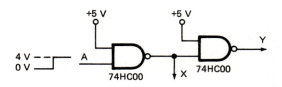

FIGURE 3-29   Circuit diagram for Problem 23.

e. Draw a timing diagram showing the relationship between the change on the A input and the corresponding change on the Y output.

24. Explain why you should never connect TTL totempole outputs together, but you can safely connect open-collector outputs together.

25. Figure 3-30 shows the circuit diagram of a 74LS366 hex inverting buffer with three-state outputs.
    a. What is meant by the term *three-state output?*
    b. In what applications are three-state output devices such as this used?
    c. Describe the signal conditions necessary to enable the three-state outputs of the buffers in the 74LS366.

26. What is the main advantage of gallium arsenide logic devices?

27. a. Describe the major advantage of CMOS logic families.
    b. Why must CMOS and other MOS devices be kept in conductive material when they are not in a circuit board?
    c. Draw the schematic symbol used to represent a CMOS transmission gate.
    d. How is a CMOS transmission gate enabled?

28. Why is it important to connect a wrist strap to ground whenever you work with ICs and circuit boards?

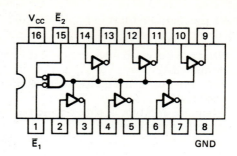

TRUTH TABLE

INPUTS			OUTPUT
$\bar{E}_1$	$\bar{E}_2$	D	
L	L	L	H
L	L	H	L
H	X	X	(Z)
X	H	X	(Z)

FIGURE 3-30   Circuit diagram for Problem 25.

29. How many Schottky TTL inputs can be connected to one 74LS TTL output without exceeding the output current capability of the 74LS? (Use Table 3-2 to help you.)

4 Oscilloscopes, Pulse Generators, and Pulse Characteristics of Logic Gates

Voltmeters and logic probes can be used to determine static logic levels, but they are not useful for making measurements on pulsed logic signals. To observe and measure pulsed signals, we use an instrument called an *oscilloscope* which draws pictures of signal waveforms on a screen similar to that of a television set. In the first section of this chapter we show you how to use an oscilloscope. In later sections we discuss *signal generators*, which are used to produce pulsed signals for testing digital circuits, and we show you how to predict the signal that will be present on the output of a logic gate which has pulsed input signals.

OBJECTIVES

At the conclusion of this chapter you should be able to:

1. Describe the functions of the major controls on an oscilloscope

2. Describe how an oscilloscope can be used to observe the signals present on the inputs and the output of a logic gate

3. Describe the function of the major controls on a pulse generator and on a function generator

4. Predict and draw the waveform that will be present on the output of a given logic gate circuit which has pulsed signals applied to its inputs

5. Describe how an oscilloscope can be used to measure risetime, falltime, and propagation delay time for logic gates

6. Explain why bypass capacitors must be connected between V_{CC} and ground in logic circuits which have pulsed signals.

OSCILLOSCOPES—OBSERVING PULSED SIGNALS

Basic Theory of Operation

As we said previously, an oscilloscope is an instrument which draws a picture of input waveforms on a screen

similar to that of a television set. The screen on which signal waveforms are displayed is the front part of a bottle-shaped vacuum tube called a *cathode-ray tube*, or *CRT*. Figure 4-1 shows a diagram of a typical CRT.

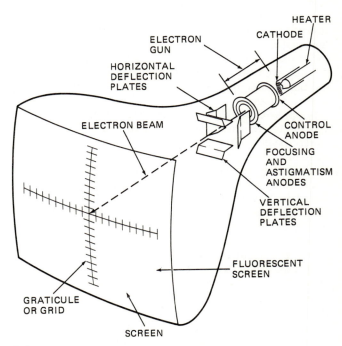

FIGURE 4-1 Diagram of a cathode-ray tube.

The inside of the screen portion of the CRT is coated with a phosphor substance which gives off visible light when it is hit by electrons. An electron gun at the back of the CRT emits free electrons. The negatively charged electrons are accelerated toward the phosphor-coated screen on the front of the tube by a positive voltage of a few thousand volts. Some electrodes in the tube focus the accelerated electrons into a narrow beam. When the beam of electrons hits the phosphor, it causes the phosphor to give off a bright dot of visible light. Pictures of signal waveforms are drawn out on the screen by moving the bright dot around on the screen.

The electron beam is moved around on the screen by two sets of *deflection plates* in the CRT. Figure 4-1 shows the location of these plates. The vertical deflection plates are used to move the beam up or down on the screen, and the horizontal deflection plates are used to move the beam back and forth across the screen. Here's how this works.

If no voltage is applied to the plates, the beam of electrons will pass straight through the plates and produce a bright spot in the center of the screen. Now, suppose that the top vertical deflection plate is made positive and the bottom vertical deflection plate is made negative. As the negatively charged electrons pass between the plates, they will be repelled by the negatively charged bottom plate and attracted by the positively charged upper plate. This attraction-repulsion will cause the beam of electrons to be deflected upward so that it hits above center on the screen. If the voltage applied to the plates is reversed, the beam will be deflected downward so that it makes a bright spot below center on the screen.

In the same manner, a voltage applied to the horizontal deflection plates will cause the beam to be deflected to the left or to the right on the screen, depending on the polarity of the applied voltage. If an alternating voltage is applied to the horizontal deflection plates, the beam will sweep back and forth across the screen. This will cause a bright horizontal line to be displayed on the screen. The pattern produced by the beam sweeping across the screen is often referred to as a *trace*.

Suppose that at the same time an ac voltage is being applied to the horizontal plates to cause the beam to sweep back and forth across the screen, a pulsed waveform such as that shown in Figure 4-2a is applied to the vertical deflection plates. For now assume that the bottom plate is at ground and the upper plate is alternating between 0 and 5 V as shown in Figure 4-2b. During the time that the pulsed signal is at 0 V, there is no voltage between the vertical plates, so the beam sweeps across the center of the screen. When the pulsed signal waveform changes to +5 V, the upper plate will now have a positive voltage with respect to the bottom plate. This positive voltage will cause the electron beam to be deflected upward on the screen. During the time the signal is high, then, the beam will be sweeping at a higher position on the screen. When the pulsed signal returns to 0 V, the electron beam will return to the center of the screen. During the time the pulsed signal is low, the beam will sweep along the centerline of the screen. The point of all this is that if the rate at which the beam sweeps across the screen horizontally is synchronized in some way with the rate at which the pulsed signal is repeating, a section of the pulsed signal waveform will be displayed on the screen. Now that you have an overview of how an oscilloscope works, we will describe how you set the basic controls on an oscilloscope to obtain stable displays of simple pulsed signals. Later in the book we discuss, as needed, how to use some of the more advanced features of oscilloscopes and how to use another instrument called a *logic analyzer*, which allows you to observe many digital signals at the same time.

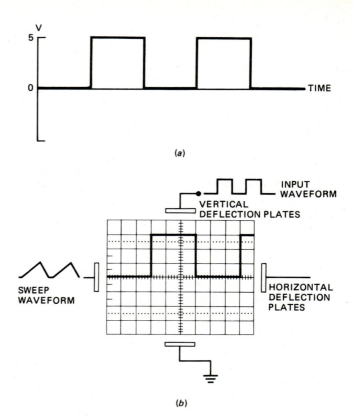

FIGURE 4-2 *(a)* Pulsed waveform. *(b)* Trace of pulsed waveform.

Oscilloscope Controls and Use

Figure 4-3 shows a photo of a medium-complexity laboratory oscilloscope which we will use as an example in this section. Our discussion is general enough that if you have a different type of oscilloscope, you should have little trouble finding the equivalent controls on your oscilloscope. Incidentally, in practice an oscilloscope is almost always referred to as simply a *scope*. For the most part we will refer to it in this way throughout the rest of the book.

At first look, a scope may seem to be a confusing array of knobs and switches, but in reality it is much easier to operate a scope than to drive an automobile. You don't

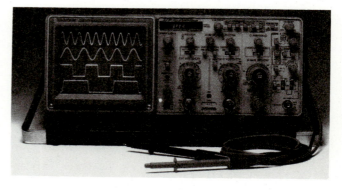

FIGURE 4-3 Photograph of a Tektronix 2236 oscilloscope with multimeter input probes.

have to learn all of the functions of a scope at once. The major scope controls and concepts you need to know about for the work in the next few chapters are:

Vertical input connector(s)

Probes

Vertical volts/division control

Vertical position control

Input coupling

Horizontal time/division control

Trigger slope control

Trigger level control

Trigger stability control

Chop/alternate control

Intensity or brilliance control

Focus control

We will work our way through these in the following sections. In later chapters we will explain how to use other controls on the scope, such as the B-timebase controls and the controls along the top of the scope, which allow you to measure the voltage between two points on a signal waveform or the time between two points on a signal waveform. If possible, sit in front of a scope as you read these sections so that you can find each part as you read about it.

INPUT CONNECTORS AND PROBES

Figure 4-4 shows a diagram of the vertical section controls of the Tektronix 2236 scope shown in Figure 4-3. On your scope or in Figure 4-4 find the front panel connector(s) labeled CH1 or Vertical Input. This input is connected through an amplifier to the vertical deflection plates. A signal applied to this input will cause the beam to move up or down as it sweeps across the screen. Most modern scopes have two or more vertical inputs so that you can look at and compare two or more signals. Later we will explain to you how a single electron beam can be used to draw waveforms for both inputs at the same time on the screen.

You connect the signal you want to look at to the vertical input connector with a coaxial cable or with a special lead called a *probe*. Figure 4-5a shows a common type of scope probe. The main advantage of probes is that they have internal circuitry, which allows them to take a small sample of a signal without "loading down" or otherwise affecting the operation of the circuit being measured.

A disadvantage of probes is that the signal output from the probe to the input of the oscilloscope is much lower in voltage than the signal applied to the input of the probe. If you look at the circuit for a ×10 probe in Figure 4-5b, perhaps you can see why this is so. The 9-MΩ series resistor in the probe and the 1-MΩ resistor in the input circuitry of the scope form a 9:1 voltage divider. The voltage that actually gets to the input of the scope,

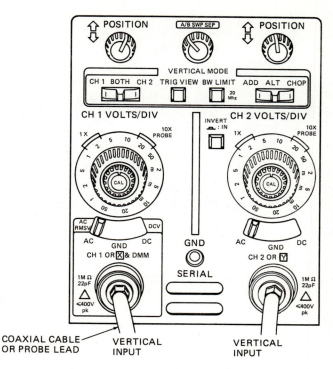

FIGURE 4-4 Vertical section controls of Tektronix 2236 oscilloscope.

then, is only one-tenth of the voltage applied to the tip of the probe. The probe is said to *attenuate* the input signal by a factor of 10. This attenuation is not a problem, however, because a vertical amplifier inside the scope can be set to amplify or, in some cases, further attenuate the signal so that it has the correct levels needed for deflecting the electron beam. The VOLTS/DIV control, which we discuss in the next section, determines the amount that the signal applied to the vertical input is amplified or attenuated.

From Figure 4-5b perhaps you can also see why connecting a scope probe to a circuit affects the circuit less than connecting the scope input directly. Connecting the scope probe to a circuit is equivalent to connecting a 10-MΩ resistor (9 MΩ + 1 MΩ) from the circuit test point to ground. Connecting the scope input itself to a circuit is equivalent to connecting a 1-MΩ resistor from the circuit test point to ground. The higher resistance of the probe will not cause as much additional current to flow into or out of the circuit being tested as would connecting the scope input directly to a circuit. In other words, connecting a ×10 probe will "load down" the circuit being tested less than connecting a cable directly to the scope input.

Test probes are commonly specified as ×1, ×10, or ×100. These numbers mean that the signal at the output of a probe is reduced, or *attenuated*, by the specified number. To get the true value of a signal from the output of the probe, you multiply it by 1, 10, or 100, respectively. After you read the next section, you will have a better idea of how you use this attenuation factor in determining the amplitude of a signal. Later, in the scope start-up checklist, we also tell you how to adjust, or *compensate*, a probe for the best signal response.

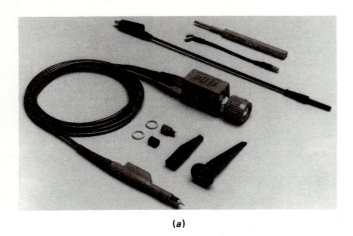

(a)

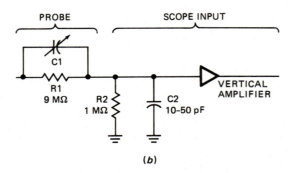

PROBE   SCOPE INPUT

C1

R1
9 MΩ   R2   C2
         1 MΩ   10–50 pF

VERTICAL
AMPLIFIER

(b)

FIGURE 4-5   (a) Typical scope probe. *(Courtesy Tektronix Inc.)* (b) Equivalent circuit for ×10 probe and oscilloscope input.

VERTICAL VOLTS/DIVISION CONTROLS

Note in the photograph of an oscilloscope in Figure 4-3 that the screen has a grid of horizontal and vertical lines over it. The spaces between the lines are commonly referred to as *divisions*. Note also in Figure 4-3 that within each division there are small dashes to help you estimate fractional parts of divisions. As we will show you in a later section, the vertical lines on the screen are most often used to represent units of *time*. These lines help you to determine the time between, for example, two points on a signal waveform. The horizontal lines across the screen of the scope are used to represent units of *voltage*. These lines, together with the setting of the VOLTS/DIV control and the attenuation factor of the probe, allow you to determine the size, or amplitude, of a signal waveform. Here's how you do this.

Each input channel on a scope has a VOLTS/DIV control to set the amount a given input voltage will deflect the trace vertically on the CRT screen. The VOLTS/DIV control has settings such as 1, 2, 5, and 10 V/division. A division in this case refers to the distance between two solid horizontal lines on the CRT. In other words, if the VOLTS/DIV control is set for 1 V/division, each division in the vertical direction represents 1 V. The amplitude of a signal applied to the scope input then can be read directly from the display on the CRT. For example, if your scope is set for 2 VOLTS/DIV and a voltage applied directly to the input of the scope causes the trace to move

up two divisions on the screen, then the voltage is equal to 2 divisions × 2 V/division, or 4 V. Another example with the same scope setting is shown in Figure 4-6. The

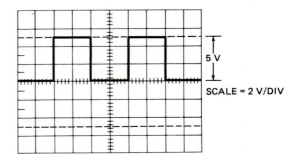

5 V

SCALE = 2 V/DIV

FIGURE 4-6   Example of measuring signal voltages with a scope.

low level of this signal is at the center, or zero, line on the display, and the high level of the signal is 2½ lines up from this. Since each division represents 2 V, the high level of the signal is 2.5 × 2 V, or 5 V, above the low level. The pulse is said to have an *amplitude* of 5 V.

NOTE: In addition to the main VOLTS/DIV control, most scopes also have a *variable volts/division* control (often in the center of the volts/division knob), which allows you to adjust the volts/division to values other than those available with the main volts/division switch. If you want to use the volts/division scale readings as we have described, make sure this variable volts/division control is in its *calibrated position*. On the Tektronix 2236 and many other scopes, this is done by turning the red CAL knob in the center of the VOLTS/DIV control clockwise until it clicks.

If you are using a probe on a scope input, you have to take into account the attenuation factor of the probe. To do this you simply multiply the VOLTS/DIV setting by the probe attenuation factor. For example, if the VOLTS/DIV control is set for 1 V/division and you are using a ×10 probe, each division on the CRT actually represents 10 × 1 V, or 10 V. With this combination, then, a display such as the one in Figure 4-6 would represent a signal with an amplitude of:

2.5 divisions × 2 V/division
          × attenuation factor of 10 = 50 V

Here's a summary of how you measure voltages with a scope:

1. Start by setting the volts/division to a scale somewhat higher than any voltages that you expect to encounter.

2. When you get a display on the screen, turn the volts/division knob to successively lower voltage scales until the waveform fills a sizable portion of the screen.

3. Count the number of large squares, and estimate any fractional parts that the trace moves from the zero position or the number of squares between the bottom of the waveform and the top of the waveform.

4. Multiply the number of squares, including fractional parts, by the volts/division setting, and then multiply the result by the attenuation factor of the probe.

As another example of this computation, suppose that your scope is set for 0.2 V/division, you are using a × 10 probe, and you touch the probe to a dc voltage, which causes the trace to move down 2.6 divisions. The dc voltage (V) then is:

$$V = -2.6 \text{ divisions} \times 0.2 \text{ V/division}$$
$$\times \text{ attenuation of } 10 = -5.2 \text{ V}$$

INPUT COUPLING AND VERTICAL POSITION CONTROLS

Figure 4-4 shows a three-position switch with the labels AC, GND, and DC just over each vertical input connector. The switch determines how an input signal is coupled between the input connector and the input of the vertical amplifier.

The ground (GND) setting of this switch is used for setting the 0-V position of the trace on the screen. If the input coupling switch is in the ground position, the input of the vertical amplifier is connected to ground, or 0 V. With no voltage on the vertical amplifier, the trace should be a straight line across the center of the screen. The vertical POSITION control of the scope, shown at the top of Figure 4-4, can be used to adjust the trace to the center of the screen or to adjust the 0-V position of the trace to some other convenient point on the screen.

> NOTE: The coupling switch in the ground position connects the input of the vertical amplifier to ground but does not connect the input signal from a probe or circuit to ground. Connecting the input signal to ground would "short out" the circuit being tested.

To explain ac and dc coupling, we will use the waveform shown in Figure 4-7a. This waveform represents the signal that might be seen on the output of a 10-V power supply. As you can see in the figure, the voltage is not steady at exactly 10 V; it alternates slightly back and forth across the 10-V line. One way of describing this waveform is to say that it has a dc component of 10 V and an ac component of a few millivolts.

Now, if you want to see mostly the dc component displayed on your scope, you set the input coupling switch for DC. For the waveform in Figure 4-7a, you might set the VOLTS/DIV control for 5 V/division. When the signal from the power supply is connected to the input of the scope, the trace will move up 2 divisions on the screen (10 V = 5 V/division × 2 divisions). The trace on the scope will look very much like Figure 4-7b. With these settings, then, you will be able

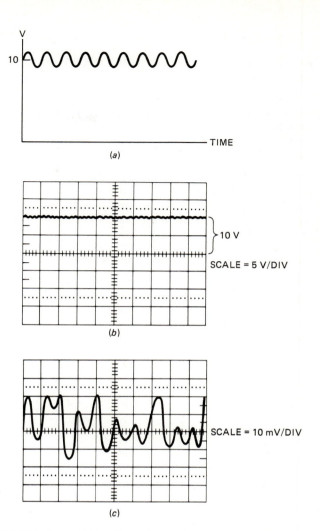

FIGURE 4-7 (a) Waveform from 10-V power supply. (b) Trace of waveform with dc coupling. (c) Trace of waveform with ac coupling.

to see that the 10 V is present on the input, but the ac component on top of the dc will be so small that you won't be able to see it in much detail. If you turn the VOLTS/DIV control to a setting low enough that you can see the ac component clearly, the dc component of the signal will push the trace so far off the top of the screen that you will not be able to get it back on again with the vertical POSITION control.

To see just the ac component of the signal, you use the AC position of the input coupling switch. With the switch in this position, a device called a *capacitor* is inserted between the input connector and the input of the vertical amplifier. This capacitor blocks the dc component and allows only the ac, or changing, portion of the signal to get to the amplifier and be displayed. The display of the ac component will then be centered around the 0-V position of the trace. With the dc component blocked out, you can turn the VOLTS/DIV control to successively lower settings until the ac component fills a large enough part of the screen that you see the details in it. The trace on your scope might then be similar to the one shown in Figure 4-7c.

To summarize the use of the input coupling switch:

The GND position is used to adjust the 0-V position of the trace to the desired point on the screen.

The DC position of the switch is used to determine the dc levels of a signal. With your scope set for DC coupling, for example, you can use it to determine if V_{CC} is present on an IC or to see if a digital signal is reaching legal low and legal high levels.

The AC position of the switch is used for looking at the ac, or changing, component of a signal. You should not use AC input coupling for looking at logic signals because this mode will not show you if the signal is reaching legal low and high voltage levels.

HORIZONTAL TIME/DIVISION CONTROL

As we mentioned previously, an ac voltage is used to sweep the electron beam across the screen over and over to make a trace. Figure 4-8 shows the ac waveform usually used for this.

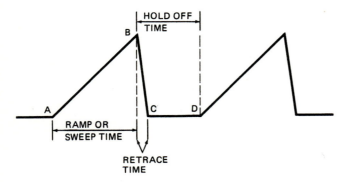

FIGURE 4-8   Waveform of signal used to sweep electron beam across the oscilloscope screen.

The portion of the waveform labeled AB causes the beam to sweep across the screen from left to right. The time required to do this is called the *ramp time*, or *sweep time.* It is during this time that a section of the signal waveform will be displayed. The section of the waveform labeled BC in Figure 4-8 causes the electron beam to quickly return to the left side of the screen to get ready for the next sweep. During this time, known as the *retrace time*, the beam is turned off so that it does not produce an extra line back through the display. The portion of the waveform labeled BD in Figure 4-8 represents the *holdoff time*, which is the time that the beam is made to wait before it starts its next sweep across the screen. Later we will talk more about this holdoff time and how you control it.

The ramp section of the sweep waveform in Figure 4-8 is a straight line, so the beam is swept across the screen at a constant rate. In other words, equal distances across the screen represent equal increments of time. The vertical lines on the scope screen are usually used to mark off units of time, just as the horizontal lines are used to mark off units of voltage.

The *horizontal time/division* control on the scope is used to set the amount of time represented by each horizontal division on the screen. Figure 4-9 shows the time/division (SEC/DIV) control of the Tektronix 2236 scope, shown in Figure 4-3. The settings for this SEC/DIV control are labeled with values such as these: 1 ms/division, 0.5 μs/division, 0.2 μs/division, 0.1 μs/division, etc.

Here's how you use this control and the vertical lines to measure the time between two points on a waveform:

1. After you get a stable trace of the waveform on the screen, turn the SEC/DIV knob to successively lower scales until only one or two cycles of the waveform are visible on the screen.

2. Count the number of divisions and fractions of divisions between, for example, the rising edges of two pulses on a waveform.

NOTE: You can use the horizontal POSITION control shown at the top of Figure 4-9 to move the display horizontally so that one of the points is on a vertical line. This makes it easier to determine the number of divisions.

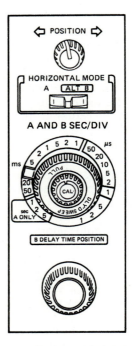

FIGURE 4-9   Horizontal system controls of oscilloscope.

3. Multiply the number of divisions between the two points on the waveform by the setting of the SEC/DIV knob. For example, suppose that your scope is set for 2 ms/division and you want to determine the time between point A and point B on the waveform in Figure 4-10. Since there are 4.2 squares between points A and B, the time between them is 4.2 divisions × 2 ms/division, or 8.4 ms.

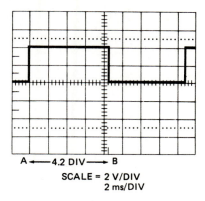

A ← 4.2 DIV → B

SCALE = 2 V/DIV
2 ms/DIV

FIGURE 4-10   Example of measuring time between signal points.

NOTE: Most scopes have a variable time/division control (usually in the center of the time/division knob) which allows you to adjust the time to any desired amount per division. Make sure this control is in its calibrated position if you want to make a measurement using the fixed values of the time/division control. On the Tektronix 2236 and many other scopes, this is done by turning the red CAL knob in the center of the SEC/DIV control clockwise until it clicks.

TRIGGER CONTROLS OVERVIEW

Before discussing the trigger controls, let's review what we have told you about the operation of a scope up to this point.

A trace is produced on the CRT by sweeping a beam of electrons across the screen over and over. As the beam sweeps across the screen, a signal applied to the vertical input is amplified by the vertical amplifier. This amplified signal causes the beam to move up or down as it sweeps across the screen. The sweeping beam then draws a representation of the signal waveform on the screen. The block diagram of a scope in Figure 4-11 should help you see how the major sections and controls are related.

In the upper left corner of Figure 4-11, first find the CH1 input and follow the signal path from there through the input coupling switch, the VOLTS/DIV control, and the vertical preamplifier to the vertical deflection plates. This circuitry controls the vertical motion of the electron beam.

Then, in the bottom right corner of Figure 4-11, find the horizontal sweep generator and the horizontal amplifier. As we described before, this circuitry controls the horizontal motion of the beam. Note that there is a switch before the horizontal amplifier that allows you to select either the output of the sweep generator or the output of the CH2 vertical preamplifier as input to the horizontal

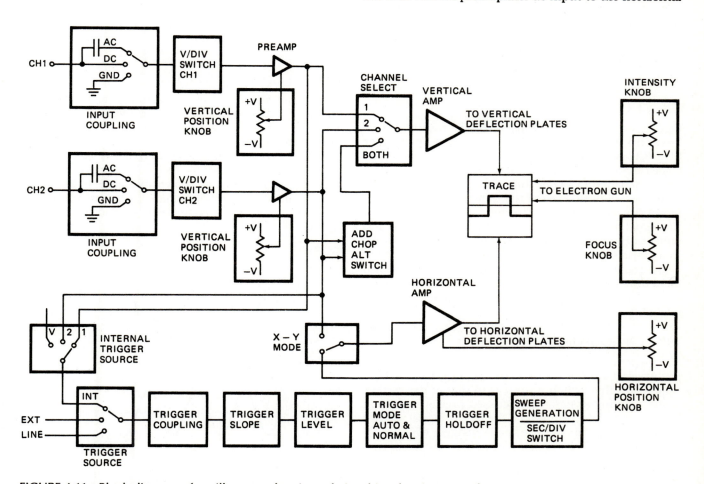

FIGURE 4-11   Block diagram of oscilloscope showing relationship of major controls.

amplifier. If the signal from the CH2 vertical preamplifier is used, then the trace on the screen will be a graph of CH1 voltage versus CH2 voltage, instead of the usual voltage versus time. This voltage-versus-voltage mode is called the *X-Y mode*. The X-Y mode is useful for drawing voltage transfer curves such as the one shown for a TTL gate in Figure 3-18.

Now, in order to produce a stable display on a standard scope, the signal being looked at must be repetitive. In other words, the waveform must be made up of repeated, identical sections. Since each sweep of the beam across the screen traces out a later section of the signal waveform, the displayed sections must be identical so that each sweep traces out the same path across the screen. If the waveform is not repetitive, each sweep will trace out a different path across the screen and all you will see on the screen is a bright band as wide as the amplitude of the signal.

An additional requirement for a stable display is that the time at which the beam starts sweeping across the screen must be synchronized with the waveform being looked at. To do this we *trigger* the horizontal sweep generator so that each sweep of the beam across the screen draws out an identical section of the waveform in the same position on the screen. The *trigger* controls on the scope are used to synchronize the internal sweep generator with the signal being displayed.

Read the names of the trigger controls connected to the input of the sweep generator in Figure 4-11. Then find these controls on your scope, or on the scope photograph in Figure 4-3, so you have an idea where they are located. In the following paragraphs we will explain the function and operation of these trigger controls.

TRIGGER SOURCE CONTROL

Figure 4-12 shows the *trigger source* control of the Tektronix 2236 scope shown in Figure 4-3. The trigger SOURCE switch determines whether the trigger signal comes from some external source related to the signal being displayed (EXT) or internally (INT) from the displayed signal itself. Most of the time you will use the internal trigger source, INT. In this mode a small part of the signal to be displayed is "picked off" the vertical amplifier output and shaped to produce the trigger signal that starts the beam sweeping across the screen.

The trigger source switch on some scope models also has a LINE position. You select this LINE position if want to get a stable trace of a signal with a frequency which is related to the ac power-line frequency.

TRIGGER COUPLING CONTROL

The *trigger coupling* switch shown in Figure 4-12 allows you to select ac or dc coupling for the trigger signal. The function of this switch is similar to that of the vertical input coupling switch we discussed previously. If you want to trigger the beam when the trigger signal reaches a specified dc level, you use the DC trigger coupling. If you want to block the dc component of a signal and trigger on just the ac component of, for example, a power supply noise signal, you use the AC trigger coupling. For digital signals you will usually use DC trigger coupling.

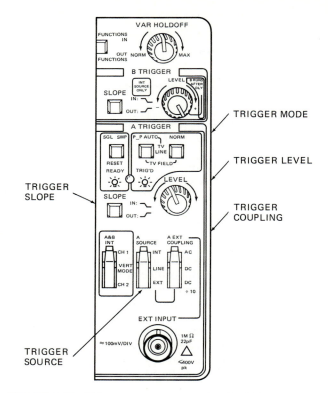

FIGURE 4-12   Trigger source controls of a Tektronix 2236 oscilloscope.

TRIGGER SLOPE CONTROL

The *trigger slope* control shown in Figure 4-12 is used to specify whether the sweep starts when a rising (+) edge of the trigger signal occurs, or when a falling (−) edge of the trigger signal occurs. If, for example, the trigger SOURCE control is set for internal (INT) and the trigger SLOPE control is set to positive, the display of a sine-wave signal will start as shown in Figure 4-13a. If the trigger SLOPE control is switched to negative, the display will start with a falling edge as shown in Figure 4-13b.

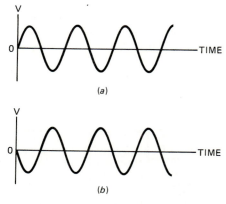

FIGURE 4-13   Display of sine wave. *(a)* Positive trigger slope setting. *(b)* Negative trigger slope setting.

TRIGGER LEVEL CONTROL

The *trigger level* control shown in Figure 4-12 is used to set the voltage at which you want triggering to occur. If, for example, the trigger LEVEL is set for 0 V, the sweep will be triggered when the trigger signal crosses the 0-V line in the direction selected by the trigger SLOPE control. Figure 4-14a shows a trace produced with an internal trigger source (INT), a rising trigger slope, and a 0-V trigger level. Figure 4-14b shows a trace of the same

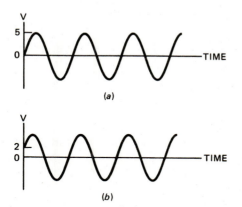

FIGURE 4-14 Display of sine wave. (a) Trigger level set for 0 V. (b) Trigger level set for +2 V.

signal, but with a trigger level setting of 2 V. As you can see, the trace now starts at the point where the input signal reaches 2 V. In practice you usually adjust the trigger LEVEL until you get a stable trace on the screen. When looking at a waveform such as that in Figure 4-15, for example, you will more easily get a stable dis-

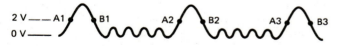

FIGURE 4-15 Complex waveform showing advantage of adjustable trigger level.

play if you raise the trigger level above the level of the small bumps, so the scope only triggers at a point such as A1, A2, or A3.

Incidentally, the trigger SLOPE and trigger LEVEL controls are independent, so you can trigger on a negative slope and a positive voltage, as represented by point B1 in Figure 4-15.

TRIGGER MODE CONTROL

As shown under the label A TRIGGER in Figure 4-12, the Tektronix 2236 has two main trigger modes, AUTO and NORM(AL).

In the *automatic triggering mode* the sweep generator "free-runs" so that a trace is always visible on the screen even with no signal applied to the vertical input of the scope. When a signal is applied to the vertical input, this input signal, together with the settings of the SLOPE and LEVEL controls, determines when triggering occurs.

Automatic triggering mode is useful for checking dc

signal levels such as power-supply voltages and logic levels. With no signal applied, the trace will be a horizontal line at the 0-V position on the screen. When you connect a dc signal to the vertical input, the line will move up or down on the screen, depending on the polarity of the applied voltage and the setting of the VOLTS/DIV control. The dc voltage can then be determined by multiplying the number of divisions the trace moves by the setting of the volts/division control and the probe attenuation factor, as we described earlier. The automatic triggering mode also works well for observing simple digital signals.

When the triggering mode is set for NORM, the sweep will not start until the input signal reaches the voltage set by the trigger LEVEL control. This mode is used for triggering on more complex signals, such as that in Figure 4-15, where you don't want the trace to start until the signal reaches, for example, point A1. The difficulty with the normal mode is that no trace will be visible on the screen unless the scope is triggered, so you have to adjust the trigger LEVEL control until you get a trace. We often start with this control in the AUTO position and then switch to the NORMAL position if we cannot get a stable trace in the automatic mode.

TRIGGER HOLDOFF CONTROL

As we have said before, in order to get a stable display of a waveform on the screen, each sweep of the beam across the screen must trace out an identical section of the waveform. At the end of each sweep across the screen, the beam is retraced back to the left side of the screen. The beam waits here until the sweep generator receives a trigger signal which tells the sweep generator to start another sweep across the screen. The sweep will be triggered the next time the input signal crosses the voltage level set by the trigger level control. For signals such as a sine wave where all of the pulses are identical, this produces a stable display. For signals such as the one at the top of Figure 4-16, however, this may not produce a

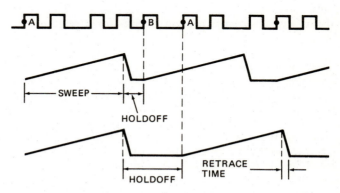

FIGURE 4-16 Complex waveform showing the use of trigger holdoff.

stable display. In this signal a pattern of pulses repeats, but the individual pulses are not identical. To help you visualize the problem this can cause, take a look at the sweep waveform shown directly under the signal waveform in Figure 4-16. The start of the first sweep corresponds to point A in the signal waveform. When this sweep is done, the beam is retraced to the left side of the screen to wait for the next trigger. The horizontal section of the sweep waveform represents the time the sweep generator is waiting for a trigger signal. The signal transition at point B will cause the beam to sweep again. This second sweep of the beam traces out a different section of the waveform, so what you will see on the screen is a mixture of the two traces. The result is usually not understandable.

The *trigger holdoff* control solves this problem by allowing you to adjust the amount of time the sweep generator waits before it accepts a new trigger signal. The bottom sweep signal shown in Figure 4-16 shows the trigger holdoff time adjusted so that the second sweep starts at the desired point on the signal waveform. In practice you first adjust the trigger LEVEL control to attempt to stabilize a trace, and if this does not work, you try to adjust the trigger HOLDOFF control.

THE CHANNEL SELECTOR SWITCH

As shown in Figure 4-3 or Figure 4-4, dual trace scopes have a CH1/BOTH/CH2 switch which allows you to display only the signal on channel 1, only the signal on channel 2, or both signals. If the switch is set for the BOTH position, then the ADD/CHOP/ALT switch determines the mode that is used to display the two signals. In the following sections we describe how these functions work.

THE CHOP/ALTERNATE CONTROL

Earlier in this section of the chapter we told you that most modern scopes are capable of drawing two different signal waveforms on the screen at the same time with a single electron beam. There are two ways this can be done.

One way is the *alternate mode*. In this mode the beam draws out a section of the signal on the CH1 input during one sweep across the screen, and then draws out a section of the signal on the CH2 input during the next sweep across the screen. During the following sweep the beam again draws a section of the signal on the CH1 input. At higher sweep rates the alternation is fast enough that your eye does not detect it, so you simply see a stable display for each of the two signals. The vertical POSITION controls can be used to position one display on the top half of the screen and the second display on the bottom half of the screen.

The second way of drawing two signal waveforms on the screen at the same time is with the *chopped mode*. In this mode the input to the vertical amplifier is rapidly switched between the CH1 signal and the CH2 signal. As the beam sweeps across then, it first draws a tiny section of the CH1 signal and then draws a tiny section of the CH2 signal. It then draws another tiny section of the CH1 waveform and another tiny section of the CH2 waveform. The process continues until the beam reaches the right edge of the screen. One sweep of the beam across the screen draws both waveforms. At low sweep rates the chopping between channels is fast enough that your eye does not detect it. Therefore, you simply see the two waveforms displayed on the screen. Again, the vertical POSITION controls on each channel can be used to separate the two displays on the screen.

In summary, then, you use the chopped mode (CHOP) when you have the SEC/DIV control set for relatively long times such as 1 ms/div because the display would appear to blink in the alternate mode. At high sweep rates you use the alternate mode (ALT) because the chop mode will not accurately show the time at which an input signal changes from, for example, low to high if the change occurs while the beam is drawing a section of the signal on the other channel.

Getting a Display on a Dual-Trace, Triggered Oscilloscope

When you first sit down to use a scope to look at a signal waveform, the array of switches and knobs may seem overwhelming and you may not have an idea of just where to start. To help you, we have made a checklist of how to set the controls on a dual-trace, triggered scope to get a basic signal display. Once you get a basic display, you can "fine-tune" the settings of the controls to get the best display. You might want to make a copy of Figure 4-11 and a few copies of this checklist to use until you become thoroughly familiar with the controls on your scope.

For this start-up procedure you will be using the signal from the PROBE ADJUST or calibration output on your scope.

SCOPE START-UP CHECKLIST

- Turn on the power to the scope.
- Adjust the INTENSITY control to about 75 percent of maximum.
- Connect probes to CH1 and CH2 vertical input connectors.
- Set the CH1 VOLTS/DIV control to 1 V/division.
- Set the CH2 VOLTS/DIV control to 1 V/division.
- Turn the VARIABLE VOLTS/DIV controls in the centers of the VOLTS/DIV controls to their calibrated positions.
- Set the Vertical Mode channel selector to BOTH.
- Set channel 1 and channel 2 vertical POSITION controls to near the centers of their ranges.
- Set the CH1 and CH2 input coupling switches to their DC positions.
- Set the ALT/CHOP switch to CHOP.
- Set the SEC/DIV control to 1 ms/division.
- Turn the VARIABLE TIME/DIV control in the center of the SEC/DIV control to its calibrated position.
- Set the HORIZONTAL MODE control to the A timebase position.
- Set the horizontal POSITION control to the center of its range.

- Set the trigger SOURCE switch for internal (INT).

- Set the A&B INT switch to pick the trigger signal from CH1.

- Set the trigger MODE control to AUTO.

- Set the trigger SLOPE control to +.

- Set the trigger LEVEL control to just above its center, or 0-V, position.

- Connect the probe from the A channel vertical input to the PROBE ADJUST or calibrator output of your scope.

- Adjust the CH1 VOLTS/DIV control so that the waveform fills about 2½ vertical divisions on the screen.

- Adjust the SEC/DIV control so that about two repetitions of the signal waveform are shown on the screen.

- If necessary compensate the probe. This is done by connecting the probe to the calibrator output on the scope and using a small screwdriver to adjust the probe so that the corner of the waveform at the top of the rising edge is as square as possible. Note that when you compensate a probe you are essentially matching the probe to the individual characteristics of the input it is connected to. Therefore, this adjustment usually has to be made only when you first connect a probe to an input.

- To prolong the life of your CRT, adjust the intensity of the trace so that it is not too bright but is still clearly visible.

PULSES AND PULSE GENERATORS

Most electronic lab benches have a function generator and/or a pulse generator to produce the signals needed for testing digital circuits. Before we can tell you how to set the controls on a function generator or pulse generator, we need to show you more about the parameters used to describe pulsed signals.

Characteristics of Pulses

AMPLITUDE

Figure 4-17 shows a pulsed waveform with some of the important characteristics labeled. The first characteristic to note is the amplitude, which is simply the difference in voltage between the low level of the waveform and the high level of the waveform. The amplitude is specified in volts. A waveform with a low level of -10 V and a high level of $+10$ V, for example, will be specified as having an amplitude of 20 V. A pulsed waveform, or *pulse train* as it is sometimes called, that has a low level of 0 V and a high level of $+4$ V will be referred to as having an *amplitude* of 4 V.

OFFSET

Another pulse characteristic you very much need to know about when working with digital circuitry is *voltage offset*. A waveform that is centered about 0 V, such as the 3-V signal shown in Figure 4-18a, is said to have *zero*

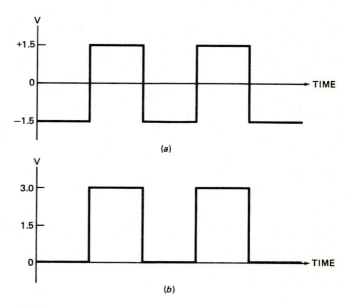

FIGURE 4-18 Three-volt square-wave signal. *(a)* Offset of 0 V. *(b)* Offset of 1.5 V.

offset. This signal has the correct amplitude for a TTL signal, but not the correct voltage levels. The low level of the signal is -1.5 V, which is not a legal value for a TTL input low, and the high level of this signal is $+1.5$ V, which is not a legal value for a TTL input high. In order to be used as an input signal for a TTL gate, the signal in Figure 4-18a must be *offset* so that the low level of

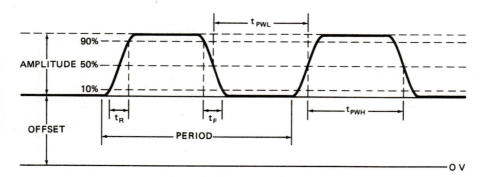

FIGURE 4-17 Pulsed waveform showing important characteristics.

the signal is near 0 V, as shown in Figure 4-18b. Now both the low and high levels of the signal are clearly legal TTL levels.

RISETIME AND FALLTIME

The next parameters to look at in Figure 4-17 are risetime and falltime. As we showed in Chapter 3, *risetime*, or t_R, is the time it takes for a pulse to go up from 10 percent of its amplitude to 90 percent of its amplitude. *Falltime*, or t_F, is the time it takes for a pulse to go from 90 percent of its amplitude down to 10 percent of its amplitude. The t_R and t_F times are usually specified in microseconds (μs), nanoseconds (ns), or picoseconds (ps).

PULSE WIDTHS

The *time-high pulse width*, t_{PWH}, for a pulse is measured from the 50 percent point on a rising edge to the 50 percent point on the next falling edge, as shown in Figure 4-17. As also shown in Figure 4-17, the *time-low pulse width* for a pulsed signal, t_{PWL}, is measured from the 50 percent point on a falling edge to the 50 percent point on the next rising edge. A pulsed waveform in which the time low is equal to the time high is referred to as a *square wave*.

PERIOD AND FREQUENCY

For a pulsed waveform which repeats over and over again, the *period* of a signal is the time it takes the signal to complete one high-low cycle of its waveform. Since the period is a measure of time, it will be expressed in units of time such as milliseconds, nanoseconds, microseconds, or picoseconds.

The rate at which a pulsed signal repeats is called its *frequency*, or its *pulse repetition rate*. The frequency of a signal can be calculated directly from the period:

$$\text{Frequency} = \frac{1}{\text{period}}$$

The units used for frequency are cycles per second, or *hertz*. Hertz is usually abbreviated Hz.

As an example of this calculation, suppose you look at a signal with an oscilloscope and find that it has a period of 1 ms. The frequency then is:

$$\text{Frequency} = \frac{1}{1 \text{ ms}} = \frac{1}{0.001 \text{ s}}$$
$$= 1000 \text{ cycles/s}$$
$$= 1000 \text{ Hz} = 1 \text{ kHz}$$

As another example, suppose that the period of a signal is 0.25 μs. The frequency of the signal then is:

$$\text{Frequency} = \frac{1}{0.25 \text{ }\mu\text{s}} = 4,000,000 \text{ Hz} = 4 \text{ MHz}$$

DUTY CYCLE

The final pulse characteristic you need to know about for now is called *duty cycle*. Duty cycle of a pulse waveform is usually expressed as a percent. To calculate

duty cycle, you divide the time the pulse is high in a cycle by the total time for one cycle and multiply the result by 100 percent. In equation form this is:

$$\text{Duty cycle} = \frac{t_{PWH} \times 100 \text{ percent}}{\text{period}}$$

As an example computation, suppose that a pulsed signal is high for 4 ms during each cycle, and that the period, or total time, for one cycle is 10 ms. The duty cycle then is:

$$\text{Duty cycle} = \frac{4 \text{ ms} \times 100 \text{ percent}}{10 \text{ ms}} = 40 \text{ percent}$$

Pulse Generator and Function Generator Controls

Figure 4-19 shows a photograph of a typical function generator, and Figure 4-20 shows some of the wide variety of signal waveforms that most function generators can be set to produce. Signal generators described as pulse generators usually only produce pulsed waveforms such as those shown in Figure 4-20d, 4-20e, and 4-20f, but the pulse characteristics such as risetime, falltime,

FIGURE 4-19   Photograph of a Tektronix FG501A function generator.

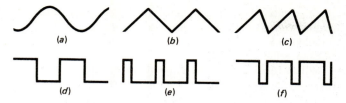

FIGURE 4-20 Function generator signal waveforms. (a) Sine. (b) Triangle. (c) Sawtooth. (d) Square. (e) Pulse with low duty cycle. (f) Pulse with high duty cycle.

and pulse width are more adjustable than they are on a function generator. Since most lab benches have a function generator, in this section we explain the use of the basic controls on a function generator such as the one in Figure 4-19.

FUNCTION—Selects the desired output waveform from choices such as the sine, triangle, sawtooth, square, and pulse waveforms shown in Figure 4-20.

FREQUENCY—Used to set the leading digit value for the output frequency in the same way that the first two or three colored bands on a resistor specify the leading digit values of the resistance. The selected digit value is multiplied by the setting of the MULTIPLIER control to determine the actual output frequency.

FREQUENCY MULTIPLIER—Sets the power of 10 that the digit value selected by the FREQUENCY control is multiplied by to determine the actual frequency of the output signal. For example, if the FREQUENCY control is set for 2.5 Hz, and the MULTIPLIER is set for 10^3, then the output frequency can be calculated as follows:

$$\text{Output frequency} = 2.5 \text{ Hz} \times 1000$$

$$= 2500 \text{ Hz}$$

$$= 2.5 \text{ kHz}$$

NOTE: On some function generators such as the one in Figure 4-19, the output frequency calculation is done as above for sine-, triangle-, and square-wave signals, but for pulse and sawtooth waveforms, the output frequency is about 1.6 times the product of the dial settings.

AMPLITUDE—Sets the difference in voltage between the lowest point on the signal waveform and the highest point on the signal waveform. This difference in voltage is referred to as the *peak-to-peak* amplitude of the signal. Typically, the amplitude of the output signal can be adjusted from a few millivolts to perhaps 20 V. Both the amplitude and the offset of a signal should be checked with a scope before the signal is connected to a circuit.

OFFSET—Used to adjust the voltage level of the output signal with respect to ground. The offset of a signal can be adjusted without affecting its amplitude. Figure 4-18a, for example, shows a square wave signal centered around 0 V. The *amplitude* of this signal is appropriate for a CMOS gate input, but the low level of −1.5 V is not

a legal low and the high level voltage of +1.5 V is not a legal high level for a CMOS or TTL input. By adjusting the OFFSET control on the function generator, you can raise the signal shown in Figure 4-18a upward so that the low level of the signal is at 0 V and the high level is about 3 V, as shown in Figure 4-18b. The signal now has both the correct amplitude and voltage levels for a CMOS or TTL input. The important point to remember here is:

Always use a scope to help you adjust the amplitude and the offset of a signal waveform before you connect the signal to a logic gate or other circuit. Remember that you use the controls on the function generator, not the controls on the scope, to actually adjust the amplitude and the offset of the signal.

OTHER CONTROLS—In addition to the basic controls described above, a function generator or pulse generator may have a variety of other controls. Here's a list of a few other controls you may have and a few hints on how to set these controls.

PULSE WIDTH—This adjusts the amount of time that a pulse is high during each cycle. On some generators this control is also used to adjust the slope of sawtooth waveforms. As a starting point, adjust this control so that the pulse width is about equal to half the period of the desired output signal.

RISETIME and FALLTIME—This adjusts the 10 percent to 90 percent and the 90 percent to 10 percent transition times for pulse and square-wave signals. For use with TTL and CMOS gates, set these controls for about 5 ns.

GATED, TRIGGERED, AND FREE-RUN MODES—In *gated* mode a generator will only output pulses when a signal applied to the GATE INPUT connector is at a logic high. If a low-frequency square wave is applied to the gate input, the generator will output bursts of pulses, which are useful for testing some types of circuits.

In *triggered* mode a generator will output a pulse each time it receives a trigger pulse on its TRIGGER INPUT connector. This mode allows you to generate pulses which have the same frequency as, but different characteristics from, those of the trigger pulse.

In *free-run* mode the generator outputs a continuous series of cycles or pulses. For most applications you will want to set your signal generator in FREE-RUN mode.

Summary of Using a Function Generator

With a little practice and the aid of your scope, you should soon become familiar with setting the controls on your function generator or pulse generator to produce an output waveform with the desired parameters. The main point to keep in mind when you are working with a signal generator is:

Always use a scope to check the amplitude and offset of a signal before you connect the signal to a circuit.

USING A FREQUENCY COUNTER

As we explained previously, you can use a scope to measure the period of a pulse or sine-wave signal and to calculate the frequency of the signal with the simple relationship, frequency = 1/period. A more accurate method of determining the frequency, or period, of a signal is with a common lab instrument called a *frequency counter*. Figure 4-21 shows a picture of a common frequency counter. Since frequency counters have few controls, they are usually quite simple to use.

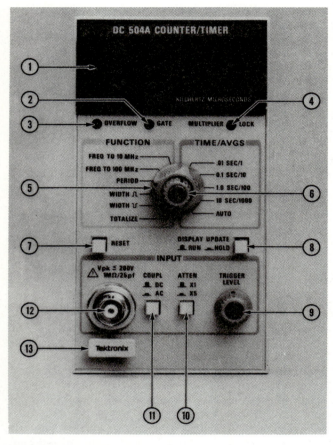

FIGURE 4-21 Tektronix DC 504A counter/timer.

Signals are connected to the input connector (no. 12 in the figure) with a coaxial cable or a probe such as that used for a scope.

A master selector control (no. 5 in the figure) is used to choose the desired function. The Tektronix DC 504A counter/timer shown in Figure 4-21 allows you to measure frequency, period, pulse-high time, pulse-low time, or total number of counts.

The TIME/AVERAGES control in the center of the function knob (no. 6) has two different roles, depending on the measurement being made. For frequency measurements this control determines the amount of time that input cycles are counted for each measurement. For example, if the control is set for 1 s, the counter will count the number of input cycles that occur in 1 s and display this number on the seven-segment displays (no. 1). The

time during which a measurement is being taken is called the *gate time*, or the *count window*. [The LED labeled GATE (no. 2) will blink on and off as measurements are being taken.]

When measuring a high-frequency signal, the number of cycles that occur in 1 s may overflow the six digits that can be displayed on the LEDs. To avoid this, you can set the counter for a shorter gate time such as 0.1 s or 0.01 s. The circuitry in the frequency counter will automatically compensate for this setting and display the correct frequency reading.

In a similar manner, if you are measuring a low-frequency signal such as 20 Hz, not enough cycles will occur during a 1-s window to accurately determine the frequency. Since the counter cannot measure fractions of cycles, a 1-s window only allows you to measure the frequency to the nearest 1 cycle/s. An error of ±1 Hz for a frequency of 20 Hz is an error of ±5 percent, which is about the same as you can get by measuring the frequency with a scope. One way to improve the accuracy of low-frequency measurements is to use a longer count window such as 10 s. In 10 s the counter will count 200 cycles of a 20-Hz signal and automatically adjust the display to compensate for the increased count. An error of ±1 count out of 200 is only an error of 0.5 percent, which is much better. The difficulty with using a long count window is that you have to wait a relatively long time for a new reading.

If the TIME/AVGS control is set for AUTO, the counter will automatically select a gate time which produces a display of 6 digits. High-frequency signals will be internally divided by 10 if necessary to fit. Low-frequency signals in the range of 10 Hz to 25 kHz will be internally multiplied by 100 with a circuit called a *phase-locked loop*. Multiplying the frequency allows more cycles to be counted in a given gate time and therefore gives a more accurate measurement without having to use a long count window. (In Chapter 10 we show you how a phase-locked loop circuit works.)

For period and pulse width measurements, the TIME/AVGS control specifies how many measurements are taken and averaged before being displayed. For very low frequency signals, you can improve the accuracy by measuring the period of the input signal and calculating the frequency of the signal with the relationship, frequency = 1/period. The period will be displayed in units of seconds, milliseconds, or microseconds.

Another frequency counter control you need to know about is the *DISPLAY UPDATE* (no. 8). If this control is in the HOLD position, the latest reading of frequency or period will be held on the display. No new readings will be taken. If the control is in the RUN position, the counter will keep taking new measurements and displaying them.

The final frequency counter control you need to know about is the *trigger* control (no. 9), which functions similarly to the trigger level control on a scope. With this control you set a voltage level that an input signal must cross over in order to be counted by the counter. You adjust the trigger control so that the counter counts the desired pulses but does not count noise spikes on the signal.

OPERATION OF LOGIC GATES WITH PULSED INPUT SIGNALS

Now that you know how to use a scope to look at pulsed logic signals and how to adjust a function generator to produce pulsed logic signals, we will show you how to predict the output of a logic gate which has pulsed signals on its inputs.

AND Gate with Pulsed Input Signals

Figure 4-22*a* shows some pulsed signals being applied to the inputs of an AND gate and the pulsed signal these inputs will produce on the output of the AND gate. Showing the signals in this way is accurate, but in this form it is difficult to see the relationship between the output and input signals. Figure 4-22*b* shows the form usually used to show the relationship between pulsed input signals and the result they produce on the output of a digital device. This graph is referred to as a *timing diagram* because it shows how the signals change as time increases from some specified starting time.

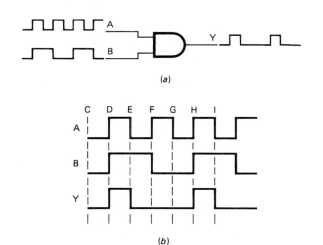

(a)

(b)

FIGURE 4-22 *(a)* Pulsed signal applied to input of AND gate. *(b)* Timing diagram.

The most important point to keep in mind when you look at waveforms such as the ones in Figure 4-22*b* is:

At any instant in time, the output of a gate can be predicted from its input signals using the basic truth table for the gate.

This means that you can analyze the operation of the gate for one section of the input waveforms at a time. We will work through the timing diagram in Figure 4-22*b* to show you how to do this.

To start, notice in Figure 4-22*b* that we drew a dashed vertical line at each point where one of the input signals changes. Between each pair of vertical lines, the inputs do not change. Therefore, the output for that section of the timing diagram can be predicted directly from the truth table for an AND gate. For the time interval labeled CD in Figure 4-22*b*, the A input is low and the B input

is low. Therefore, as predicted by the truth table for an AND gate, the output Y is low.

For the section of the timing diagram labeled DE in Figure 4-22*b*, the A input is now high and the B input is now high. As predicted by the truth table, the output is high for this time interval. For the next time interval, labeled EF, the A signal is low and the B signal is high, so the output is low. For time interval FG, the A signal is high and the B signal is low, so the output is low. Finally, for the GH section of the waveforms, the A signal is low and the B signal is low, so the output is low.

To summarize this method of analyzing the response of a gate with pulsed input signals, then:

1. Draw a vertical line at each point where one of the input signals changes.

2. Use the truth table for the gate, or the symbol shape and the assertion levels of inputs and the output to predict the output logic level during each of the time intervals you have marked off.

3. Draw the predicted output level for each time interval.

As another example of this process, let's predict the signal that will be produced on the output of another common gate when pulsed signals are applied to its inputs.

NOR Gate with Pulsed Input Signals

Figure 4-23*a* shows some pulsed waveforms applied to the inputs of a NOR gate. Figure 4-23*b* shows the timing diagram for these input signals and for the resultant output signal. You can use the OR symbol and the

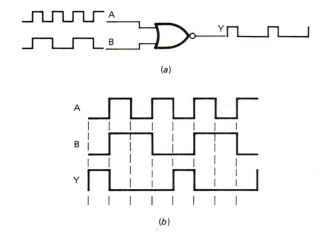

(a)

(b)

FIGURE 4-23 *(a)* Pulsed waveform applied to input of NOR gate. *(b)* Timing diagram.

signal assertion levels of the input and output signals as indicated by bubbles to help you predict each of the output levels.

The inputs on the standard NOR symbol are active high because they have no bubbles. The output can be thought of as active low because it has a bubble. With this in mind, the operation of the gate can be described as follows:

> If the A input is asserted (high) or the B input is asserted (high), the output Y will be asserted (low).

Using this description, you can simply draw a low on the output waveform for each section where one or both of the inputs is high. For the remaining sections, where the A input and the B input are both low, the output will be a high. Work your way across the waveforms in Figure 4-23b using this technique.

Enabling and Disabling Logic Gates

Figure 4-24a shows another pulsed input circuit for you to analyze. In this circuit one input of the AND gate is always high, and a pulsed signal is applied to the other input. When the pulsed signal is high, both inputs of the AND gate will be high, so the output will be high. When the pulsed signal is low, one input of the gate will be low and the other will be high, so the output will be low. With one input of the AND gate tied high, the pulsed signal applied to the other input simply passes through the gate, as shown by the timing diagram in Figure 4-24b. Making one input of an AND gate high in this way is said to *enable* the gate because it allows a signal applied to the other input to pass through.

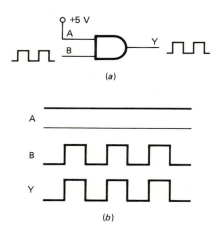

FIGURE 4-24   Enabling an AND gate. *(a)* Circuit. *(b)* Timing diagram.

For contrast, let's analyze the circuit in Figure 4-25a. In this circuit one input of the AND gate is tied low, and a pulsed signal is applied to the other input. When the pulsed signal is low, both inputs of the AND gate are low, so the output is low. When the pulsed signal is high, one input of the AND gate is still low, so the output is low. For this circuit, then, the output is always low, as shown by the timing diagram in Figure 4-25b. If one input of an AND gate is made low, the gate is said to be *disabled,* because a pulsed signal applied to the other input will not cause any change on the output.

To see if you understand this, try predicting the out-

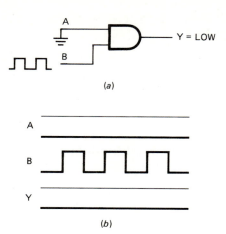

FIGURE 4-25   Disabling an AND gate. *(a)* Circuit. *(b)* Timing diagram.

put waveform for a 2-input NAND gate with one input tied high and a pulsed signal applied to the other input. When the pulsed signal is high, both inputs of the gate are high, so the output is low. If you continue with this analysis, you should find that with the gate enabled in this way, the output signal is simply the inverse of the input. A 2-input NAND gate with one input tied high, then, simply acts as an inverter. You should find that if one input of a NAND gate is tied low, the gate is disabled and the pulsed signal on the other input will have no effect on the output.

To summarize, then, an AND gate or a NAND gate will be *enabled* by making all its inputs but one high. An AND gate or a NAND gate will be *disabled* by making any one of its inputs low.

For further practice let's use the circuit in Figure 4-26a to see how a NOR gate is enabled and disabled. If

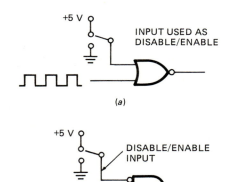

FIGURE 4-26   Enabling and disabling a NOR gate. *(a)* NOR gate. *(b)* Alternative NOR gate symbol.

the switch on the disable/enable input of the gate is in the high position, the truth table for a NOR gate tells you that the output will be low, regardless of the logic level on the other input. Therefore, a high on this input *disables* the NOR gate. If the disable/enable switch is in the low position, the truth table for a NOR gate tells you that the output will be the inverse of the input. In other words, if a NOR gate is enabled by making one input low, it simply functions as an inverter. To more clearly show that a low on an input enables a NOR gate, we can use the alternative NOR gate symbol shown in Figure 4-26b. The bubble on the disable/enable input tells you that this input is active low. The symbol further tells you that if the gate is enabled by making this input low and the pulsed signal is low, the output will be high. The pulsed signal is then inverted as it passes through an enabled NOR gate.

To summarize then, when AND gates and NAND gates are used as logic-controlled switches, they can be thought of as having active high enable inputs. When OR and NOR gates are used as logic-controlled switches, they can be thought of as having active low enable inputs.

Introduction to Logic Gates as State Decoders

Another way of looking at the operation of the NOR gate in Figure 4-23a is to say that the gate detects the time when both inputs are low and outputs a high pulse during this time. Detecting a particular binary pattern, or *state*, on logic signal lines is often referred to as *decoding* that state. The NOR gate in Figure 4-23a then can be described as decoding the 00 state for the input waveforms because the output of the NOR gate is only high when the inputs are both lows or zeros. Likewise, the AND gate in Figure 4-22a can be described as decoding the 11 state on its input signal lines because the AND gate will only go high when both inputs are high. The two examples we chose here both output highs when the decoded state is on the signal lines; other gates will output a low when the decoded state is on the inputs. A NAND gate, for example, will output a low when both inputs are high. An OR gate will output a low when both inputs are low. We will return to the concept of decoding many times throughout the rest of the book.

Using a Scope to Observe and Draw Timing Diagrams

Suppose that you need to observe the input and output signals of a logic circuit and draw a timing diagram such as the one in Figure 4-22b or the one in Figure 4-23b. Even though it has only two input channels, you can use a dual-trace, triggered scope such as we described in previous sections of this chapter to do this. Here's how you do it.:

1. First connect the A input signal to channel 1 of your scope and adjust the scope controls to obtain a stable trace of this signal.

2. Connect the B input signal to the channel 2 input of your scope and adjust the scope controls so that you have a stable display of both the A and B signals. *Hints:* If the A and B signals have different frequencies, trigger on the signal which has the lower frequency. For low-frequency signals, use the chopped mode; for high-frequency signals, use the alternate mode.

3. Carefully draw these waveforms on graph paper.

4. Disconnect the scope probe from the input signal that you are not using as the trigger source and connect this input to the output signal from the gate.

5. Adjust the scope controls, if necessary, to get a stable trace of these two signals.

6. Using the one input signal still displayed as a reference, draw the output waveform under the two input waveforms in the proper position.

You can use this method for observing and drawing the timing diagrams for any number of pulsed signals. An exercise in the accompanying laboratory manual gives you some practice predicting and observing pulsed input and output waveforms.

Using a Scope to Measure Risetime, Falltime, and Propagation Delay Time for Logic Gates

As shown in Table 3-2, the risetimes, falltimes, and propagation delay times of common logic family devices are on the order of a few nanoseconds. In order to measure times in this range, you need a scope with a bandwidth rating of at least 100 MHz. (The actual relationship for the minimum required bandwidth is: Bandwidth = 0.35/risetime in seconds.) The scope should also have a delay line in the vertical signal path so that the signal edge which causes the scope to trigger will be displayed on the screen.

To produce a test signal for these measurements, you need a signal generator capable of outputting a 1-MHz square-wave signal with the correct logic levels for the device you are testing. This high-frequency test signal is necessary so that you can turn the SEC/DIV control on the scope to a sweep speed of 10 or 20 ns/division and still have the trace bright enough to see.

If your scope does not have the bandwidth and delay line needed to measure the delay times for faster gates such as the 74LS00 devices, you can learn how to make these measurements by simply using a slower device such as a 74C00 gate. The 100-ns risetime and 100-ns propagation delay times of this gate can be measured with more moderate equipment as follows:

1. Set your signal generator for a 1-MHz square-wave output signal. Use your scope to display the signal and help you adjust the amplitude and offset of the signal so they are correct for a 5-V CMOS input.

2. Wire up a 74C00 device as shown in Figure 4-27a. Note that all unused inputs are connected to ground to avoid damage by static electricity.

3. Turn on the power to the CMOS test gate.

4. Connect the 1-MHz square-wave signal to the test input of the CMOS gate.

5. Connect the CH1 scope probe to the signal from the signal generator and connect the CH2 probe to the output of the gate.

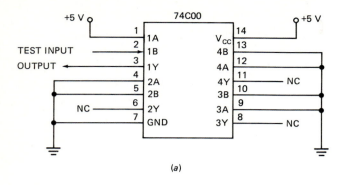

(a)

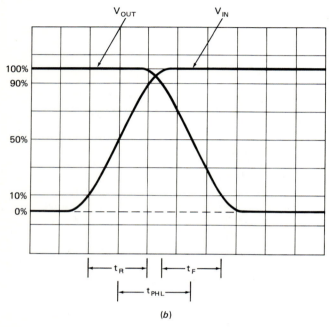

(b)

FIGURE 4-27   CMOS test gate. (a) Circuit.
(b) Oscilloscope trace of input and output signals.

6. Adjust the scope controls to get a stable display of both signals; then turn the SEC/DIV control to successively lower scales until only one cycle or less of each waveform is shown on the screen.

7. Use the vertical position control to move the output trace so that it is centered on the screen and move the input signal trace out of the way for now. Adjust the VOLTS/DIV control and the variable VOLTS/DIV control until the displayed waveform exactly fills the space between two dotted horizontal lines on the screen. The 10 percent and the 90 percent points for this waveform display are then specified along the left-hand side of the display as shown in Figure 4-27b.

NOTE: On some scopes you adjust the size of the display until it fills the entire screen, and dotted horizontal lines indicate the 10 percent and the 90 percent points.

8. Once you find the 10 percent point on a rising edge and the 90 percent point on that edge, determine

the number of horizontal divisions between these two points. Multiply the number of divisions by the setting of the SEC/DIV control to determine the risetime. For the example in Figure 4-27b, this time is 100 ns (2 divisions × 50 ns/division).

9. To determine the falltime, count the number of horizontal divisions between the 90 percent point and the 10 percent point on the falling edge of the signal. [If necessary, you can switch the trigger slope control to " − " (minus) to get the falling edge of the signal displayed.]

10. To measure the propagation delay times for the gate, move the trace of the input signal to the center of the screen. Then adjust the channel 1 controls so that this trace is the same size as the trace of the output signal.

11. As we told you earlier, propagation delay times are measured from the 50 percent point on the input waveform to the 50 percent point of the corresponding edge on the output signal. Since the 50 percent points are all on the centerline, you should be easily able to count the number of divisions between the input going high and the output going low, t_{PHL}, and the number of divisions between the input going low and the output going high, t_{PLH}. Multiply the number of divisions by the SEC/DIV control setting to convert these values to times. For the example in Figure 4-27b, the t_{PHL} is equal to 125 ns (2.5 divisions × 50 ns/division).

POWER-SUPPLY TRANSIENTS AND BYPASS CAPACITORS

In Chapter 3 we told you that when you build a digital circuit, you should connect *bypass capacitors* between V_{CC} and ground at several points in the circuit. Now that you know how to use a scope to look at pulsed waveforms, we can show you the function these capacitors perform.

For a start, suppose that we connect V_{CC} and ground to an HCMOS gate as shown in Figure 4-28a and then apply a 100-kHz square-wave signal to its input. If you connect one scope probe to the input signal and another scope probe to the output signal, you will see a display such as that in the top two lines in Figure 4-28b. The waveform produced on the output of the inverter is simply the complement of the 100-kHz signal on the input, but it is not a very "clean" signal. If you move the scope probe from the gate output to the V_{CC} line, you will see a trace such as that on the third line from the top in Figure 4-28b.

Now, V_{CC} is the dc voltage supply for the device. The trace for it should be a perfectly flat line, not the ragged waveform shown on the third line of Figure 4-28b. The unwanted pulses present on the V_{CC} waveform are often referred to as *transient voltages*, or just *transients*, because they are only present for short time intervals. These transients are caused by *changes* in the current flowing through the V_{CC} and ground wires.

In addition to their dc resistance, the V_{CC} and ground wires connecting the IC to the power supply have a prop-

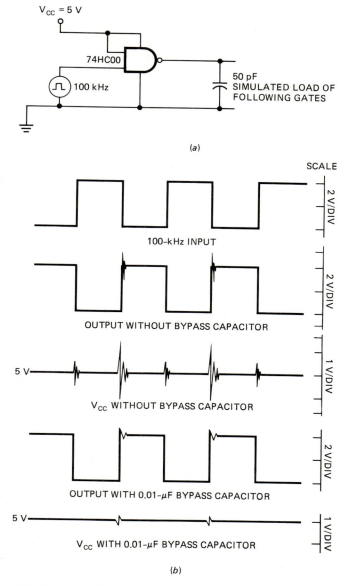

(a)

(b)

FIGURE 4-28   Effect of bypass capacitor. (a) Circuit. (b) Waveforms.

erty called *inductance.* The inductance of a circuit or device is a measure of how much the circuit opposes or resists a *change* in the current through it. Inductance in an electric circuit is analogous to inertia in a mechanical system. *Inertia,* you may remember from general science, is the resistance of a body to a change in its velocity. The greater the inertia of a body, the more work you have to do to get it moving, and once you get it moving, the more work you have to do to change its velocity or stop it. If, for example, you try to push a car by hand, it is hard to get the car moving. Once you get the car moving, however, inertia keeps the car moving at the same rate, so it is difficult to slow down or stop the car. Likewise, in a circuit with inductance, it is harder to start a current flowing though the circuit. Once a current is flowing in the circuit, the inductance will try to keep the current flowing at a constant rate. If we try to change the current through the circuit, the inductance

will oppose this attempted change and try to keep the current constant. Before we discuss how lead inductance affects the logic circuit in Figure 4-28a, we will take a brief look at how inductance affects the operation of a simple switch and resistor circuit.

Figure 4-29a shows a simple circuit containing a resistor, an inductor, and a 5-V power supply. Note the schematic symbol used to represent an inductor. This symbol comes from the fact that an inductor usually consists of a coil of wire. To increase the amount of inductance an inductor has, the coil of wire is often wound around a core of iron or some other magnetic material. The parallel lines over the spiral in the symbol indicate that the coil of wire is wound around a core of iron or some other magnetic material. To help you recognize inductors when you see them on a circuit board, Figure 4-30 shows two common types.

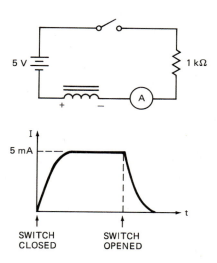

FIGURE 4-29   (a) Circuit with inductor and resistor. (b) Current-versus-time waveform as switch is closed, then opened.

For purposes of this discussion, assume the inductor in Figure 4-29a has no dc resistance. When the switch is closed, you would expect the current through the circuit to immediately go to a value of 5 mA (5 V/1 kΩ). The inductor, however, limits the rate at which the current can change, so the current goes to 5 mA, as shown in the first section of the current-versus-time graph in Figure 4-29b. During the time that the current is increasing to its final value, work or energy is going into building up a magnetic field in the inductor. When the switch in Figure 4-29a is opened, the current will not immediately drop to zero. The collapsing magnetic field of the inductor will induce a voltage across the inductor which tries to keep the current flowing in the circuit. The + and − next to the inductor in Figure 4-29a show the polarity of this induced voltage. Because of the inductance in the circuit, the current will slowly decrease to 0, as shown by the trailing edge of the graph in Figure 4-29b.

The question that may occur to you at this point is, "Once the switch is opened, where does the current that

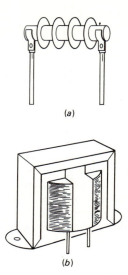

FIGURE 4-30   Inductors. (a) Air-core coil. (b) Iron-core coil.

keeps flowing go to, since there is now no conducting path back to the positive terminal of the supply?" The answer to this question is that the electrons which the inductor forces to keep flowing will "pile up" on the right-hand side of the inductor and on the parts connected to it. Piling up electrons on this part of the circuit will give it a negative voltage with respect to the other end of the inductor and with respect to the positive terminal of the supply. As you will soon see, this mechanism is what causes the transients on the V_{CC} line of a pulsed logic circuit.

The voltage produced across an inductor when the current through it is changed can be predicted by the relationship:

$$V = -L \frac{dI}{dt}$$

where L is the inductance of the inductor in units called *henrys* and dI/dt is the rate at which the current is changed. We only show you this relationship so that you can see that the voltage induced by changing the current through an inductor is a function of the amount of inductance and the rate at which the current through the inductor is changed. Now let's look at our pulsed logic circuit again.

Even a single, straight piece of wire has a small amount of inductance, so the leads connecting the gate in Figure 4-28a to the power supply have inductance. This means that if the current through the leads is changed, a voltage will be induced across each of the leads. A small but rapid change in current causes induced voltages of several volts. These induced voltages will add to, or subtract from, the supply voltage, depending on whether the current in the circuit is increased or decreased. The result of this is that the actual voltage between the V_{CC} and ground pins of the IC will change as the current through the supply wires changes.

As we showed you for the 74LS00 in Figure 3-15, the amount of supply current that most logic gates require

for a high output is different from that required for a low output. The 74HC00 in Figure 4-28a requires a different amount of supply current for an output low than for an output high, so the amount of current through the supply lines will change as the gate input is pulsed high and low. The trace labeled "V_{CC} voltage without bypass capacitor" in Figure 4-28b shows you the effect that this will have on the V_{CC} voltage actually present on the IC. As you can see in this waveform, the voltage transients produced on the V_{CC} line go from about 3.7 V to about 6.3 V. Note that the start of each transient section corresponds to an output transition of the gate.

This trace shows the effect that switching just one gate will have on V_{CC} if no bypass capacitors are used. Can you imagine what the V_{CC} line would look like if we tried to operate a board full of pulsed ICs with no bypass capacitors on V_{CC}? The transients on V_{CC} could easily be large enough to cause the gates to malfunction.

The bottom two traces in Figure 4-28b show the effect that connecting a 0.1-μF capacitor between V_{CC} and ground next to the 74HC00 has on the V_{CC} and gate output waveforms. As you can see, the transients have for the most part been removed or "filtered out." In order to explain how a capacitor does this, we will first discuss how a capacitor is constructed and, in general terms, how it functions.

As shown in Figure 4-31, a capacitor effectively con-

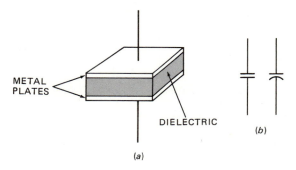

FIGURE 4-31   Capacitor. (a) Construction. (b) Schematic symbols.

sists of two metal plates separated by an insulator. The insulator is often referred to as a *dielectric*. Because of the insulator between the plates, no dc current can flow through a capacitor.

Now, suppose we connect a capacitor in a circuit such as that in Figure 4-32a and close switch SW1. Because no current can actually flow through the capacitor, you might expect that the voltage across the capacitor would immediately go to 5 V, the supply voltage. As shown by the first part of the voltage-versus-time graph in Figure 4-32b, however, the voltage across the capacitor takes some time to reach 5 V. When you first close switch SW1, electrons flow from the negative terminal of the supply into the lower plate of the capacitor in Figure 4-32a. These electrons cannot flow through the dielectric, but they can repel electrons from the upper plate of the capacitor. The electrons repelled from the upper plate flow through the resistor to the positive terminal of the sup-

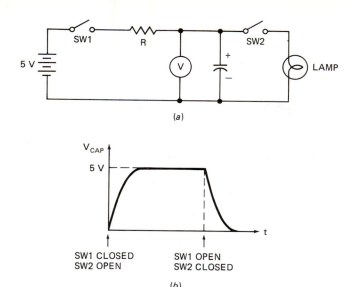

(a)

(b)

FIGURE 4-32 (a) Switch and capacitor circuit. (b) Waveforms produced when charging and discharging capacitor.

ply. The number of electrons repelled from the upper plate is equal to the number that flows into the lower plate. Electrons continue to flow into the lower plate and to be repelled from the upper plate until the capacitor charges up to a voltage which is equal to the supply voltage. The rate at which the capacitor charges is determined by the size of the capacitor *and* the size of the resistor.

The capacitor now has a charge stored in it. If switch SW1 is opened, the voltmeter will show that the capacitor has a voltage of 5 V. Electrons have flowed into the lower plate, so it has a negative charge. Electrons have been repelled from the upper plate, so it will have a positive charge. If switch SW2 is closed, the electrons stored on the lower plate of the capacitor will flow through the lamp to the upper plate of the capacitor until the capacitor is discharged. The second half of the voltage-versus-time waveform in Figure 4-32*b* shows how the voltage across the capacitor goes down as it discharges through the lamp.

Perhaps you can see from the voltage-versus-time waveforms in Figure 4-32*b* that *a capacitor resists, or opposes, a change in the voltage across it.* This feature is exactly what we need to eliminate the transients caused by lead inductance trying to keep the current in a circuit constant. To see this, let's take a look at the circuit in Figure 4-33.

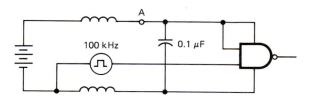

FIGURE 4-33 Gate circuit showing power supply.

When the power in the circuit in Figure 4-33 is first turned on, the capacitor charges up to +5 V. Assume the gate output is at first in the state where it uses its lower amount of current. When the gate is switched to its higher current state, the lead inductance will initially prevent the increased current from flowing to the gate. If the bypass capacitor were not present, this would cause the V_{CC} voltage on the gate to temporarily decrease. (The voltage produced by the inductance to oppose the increase in current will subtract from the V_{CC} voltage available for the gate.) As soon as the voltage at point A in the circuit drops below the voltage stored on the capacitor, however, some of the charge stored on the capacitor will flow out into the circuit and, for a short time, supply the additional current needed by the gate. Once the inductance of the leads has been overcome, the power supply will charge the capacitor up again and supply the additional current needed by the gate. All of this is just another way of saying that a capacitor connected in parallel with a circuit tries to keep the voltage across the circuit constant.

As another example of this, think about what happens when the gate is switched from its high-current state to its low-current state. In this case the inductor will try to keep the higher current flowing toward point A in the circuit. If it were not for the bypass capacitor, this would cause the voltage at point A to increase as in the waveforms in Figure 4-28*b*. With the bypass capacitor present, however, the excess current simply charges up the capacitor a little bit more. Once the inductive effect of the leads is overcome, the power supply will provide the lower value of current for the gate and allow the capacitor to discharge back to its steady-state value. The bottom waveform in Figure 4-28*b* shows you that the bypass capacitor holds the voltage on the gate nearly constant as it is switched back and forth between its low-current state and its high-current state. A common way of referring to the action of a bypass capacitor is to say that it "filters out transients or noise" on the V_{CC} line, or that it "bypasses noise to ground."

From the above examples, perhaps you can see that bypass capacitors should be as close to the ICs as possible so that they help hold the voltage actually present on the ICs constant. The final points we need to tell you about bypass capacitors are how you read the values on various capacitors, and how you decide what value(s) of capacitor you should use for bypassing in a particular circuit.

Capacitors have many uses in digital circuitry, so to help you recognize one when you see it on a circuit board, Figure 4-34 shows some examples of common capacitor types. The three factors you need to be able to determine from the package markings for any capacitor are the capacitance, the maximum working voltage, and the polarity, if any.

The basic unit of capacitance is the *farad*, which is abbreviated as simply F. Capacitors of 1 F or more are available, but a farad represents a very large amount of capacitance. We usually work with capacitors in the range of 1 *picofarad* (pF) to 10,000 *microfarads* (μF). A picofarad is 10^{-12} F, and a microfarad is 10^{-6} F. A microfarad is 10^6 pF.

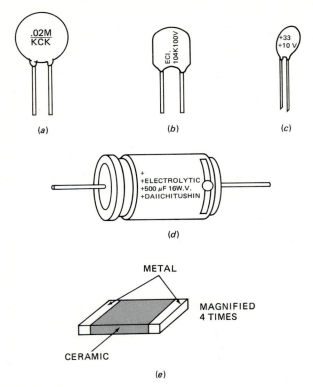

FIGURE 4-34 Common capacitor types. *(a)* Ceramic disk. *(b)* Mylar. *(c)* Tantalum electrolytic. *(d)* Aluminum electrolytic. *(e)* Surface-mount ceramic chip.

For capacitors of 1 μF or greater, the capacitance is usually written directly on the package as, for example, 100 μF. Because of the way the dielectric is constructed in these larger capacitors, they are often referred to as *electrolytics*. Most electrolytic capacitors are *polarized*, which means that they have a positive terminal and a negative terminal. When used in a circuit, the positive terminal must be connected to the positive of the supply and the negative must be connected to the negative of the supply. Large electrolytics may explode if they are connected backward in a circuit. Manufacturers indicate the polarity on electrolytics by marking the terminals with + and − signs, or with a circled − sign and an arrow pointing to the negative terminal. Markings on the capacitor also indicate the dc working voltage, or the maximum voltage that can be continuously applied to the capacitor without breaking down the dielectric.

For capacitors with capacitance less than 1000 pF, the capacitance and working voltage are usually written directly on the package with numbers. A small capacitor might, for example, have the markings 75J 100V on it. The small size of the capacitor and the 75 tell you that the capacitance is 75 pF, the J is an indication of how much the capacitance changes with temperature, and the 100V tells you that the dc working voltage for the capacitor is 100 V. Small capacitors such as this are seldom, if ever, polarized.

The capacitance of capacitors between 1000 pF and 1 μF may be indicated on the package in one of two ways. One way is to express the capacitance as a decimal fraction in microfarads. Examples are 0.01, 0.0027, and

0.22. The other way of representing the value of the capacitor is with two digit values and a multiplier, similar to the way resistor values are indicated. A capacitor using this scheme might have numbers such as 104K100V on it. The first three numbers tell you that the value of the capacitance in picofarads is 10 with 4 zeros after it, or 100,000 pF. Since $1 \mu F = 10^6$ pF, this is equal to 0.1 μF. The 100V in the label tells you that the working voltage for the device is 100 V. As another example of this, a capacitor might have two red bands and an orange band on it. These bands tell you that the value of the capacitor in picofarads is 22 followed by 3 zeros, which is 22,000 pF, or 0.022 μF. Capacitors with values below 1 μF are usually not polarized.

Now that you know how to read the value of capacitors, the next step is to tell you how to determine what size capacitors to use for bypassing in a particular application. Your first thought might be that bigger is better, so you should use the largest capacitor you can find for bypassing. In reality this doesn't work. For reasons that we don't have space to explain here, the lead inductance of large capacitors prevents their use for bypassing V_{CC} noise signals with frequencies above a few kilohertz. Large electrolytic capacitors, then, are useful for filtering out low-frequency noise on the power-supply lines, but are useless for filtering out noise such as that produced by switching a gate high and low at a 100-kHz rate. For applications in this frequency range, a bypass capacitor of about 0.1 μF is a good choice. To provide effective bypassing in circuits operating with signals of 1 MHz or so, 0.01 μF is an appropriate value. For higher frequency circuits, we use capacitors of a few hundred picofarads to achieve effective bypassing. Ceramic chip capacitors such as the one shown in Figure 4-34 have lower lead inductance than wire lead types, so they will effectively bypass higher frequency signals than the same value in a wire lead capacitor. For any bypass capacitor, it is important to keep the leads short so that the capacitance in it can more effectively do its job.

On most circuit boards we use several different values of bypass capacitors to make sure that all frequencies are effectively bypassed. A 25- to 100-μF capacitor and a 0.1-μF capacitor are usually connected in parallel between V_{CC} and ground where the leads first enter the board from the power supply. Then 0.1- and 0.01-μF capacitors are connected between V_{CC} and ground at various points around the circuit board. It is common to put one or more bypass capacitors next to each LSI and VLSI device on a board. It is much better to have too many bypass capacitors than to have not enough.

CHECKLIST OF IMPORTANT TERMS AND CONCEPTS IN THIS CHAPTER

Oscilloscope
    Cathode-ray tube, CRT
    Trace
    Horizontal and vertical deflection plates
    Vertical input connector(s)
    Input coupling
    Probes, attenuation
    Vertical volts/division control

Horizontal time/division control
Ramp time, sweep time, retrace time, holdoff time
X-Y mode
Trigger coupling, slope, level, stability controls
Automatic triggering mode
Trigger holdoff
Chop/alternate control
Intensity or brilliance control
Focus control
Probe compensation
Pulse train
Pulse or function generator
Pulse characteristics
    Voltage offset, risetime (t_R), falltime (t_F), period,
       frequency, duty cycle, pulse width (t_{PWH}, t_{PWL})

Sine, triangle, sawtooth, square waves
   Peak-to-peak amplitude
Pulse/function generator controls
Frequency counter
Timing diagram
Enabling and disabling gates
State decoding
Transient voltage, transients
Inductor, inductance, henry (H)
Capacitor, capacitance
   Farad (F), microfarad (μF), picofarad (pF)
Electrolytic
Polarized
Dielectric
Bypass capacitor

REVIEW QUESTIONS AND PROBLEMS

1. Describe how a beam of electrons is used to draw a trace on the screen of an oscilloscope.

2. Besides being convenient to handle, describe the main advantage of using a ×10 probe to connect a scope input to a test point in a circuit.

3. Briefly describe the function of each of the following controls on a scope:

   a. VOLTS/DIV
   b. VERTICAL POSITION
   c. SEC/DIV
   d. TRIGGER SLOPE
   e. TRIGGER LEVEL
   f. CHOP/ALTERNATE
   g. INTENSITY

4. A scope's VOLTS/DIV control is set for 0.2 V/division and you are using a ×10 probe. When you connect the probe to a dc voltage point in a circuit, the trace moves up 3 divisions. Calculate the circuit voltage that this display represents.

5. A scope's VOLTS/DIV control is set for 0.05 V/division and you are using a ×10 probe. The low level of a pulsed signal is 3 divisions below the centerline, and the high level for the signal is 3 divisions above the center-line. Calculate the amplitude of the signal represented by this display.

6. The SEC/DIV control on a scope is set for 2 ms/division. One cycle of a signal displayed on the screen occupies 5 horizontal divisions. For the signal calculate:
   a. The period.
   b. The frequency.

7. *a.* Why must a signal be repetitive in order to be displayed on a standard oscilloscope?
   b. What is the purpose of a trigger signal in a triggered scope?
   c. Describe the difference between the trigger slope control and the trigger level control.
   d. Briefly describe the function of the TRIGGER HOLDOFF control.

8. Describe the process used to "compensate" a scope probe for best frequency response.

9. On the pulse waveform in Figure 4-35, label the risetime, the falltime, and the pulse high time.

FIGURE 4-35   Waveform for Problem 9.

10. The period of a pulsed signal waveform is 25 μs. The signal is high for 5 μs during each cycle. Calculate the duty cycle for this waveform.

11. Briefly describe the function of each of the following controls on a function generator:
   a. FUNCTION
   b. FREQUENCY
   c. FREQUENCY MULTIPLIER
   d. AMPLITUDE
   e. OFFSET

12. Why must you always use a scope to check the amplitude and offset of a signal before you connect the signal to a circuit?

13. *a.* When using a frequency counter for measuring high frequencies, why do we use a 0.1- or 0.01-s gate time instead of a 1-s window?
   b. For what type of signals can you more accurately determine the frequency by measuring the period and computing the frequency?
   c. What is the purpose of the trigger control on a frequency counter?
   d. What symptom will you observe if the DISPLAY UPDATE control of a frequency counter accidentally gets left in the HOLD position.

14. On the waveforms in Figure 4-36 identify t_{PLH} and t_{PHL}.

15. The pulsed signals shown in Figure 4-37 are applied to the inputs of a 2-input OR gate. Redraw these waveforms on a separate paper and below

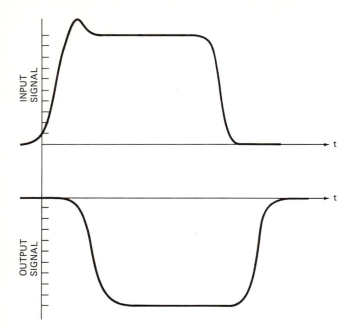

FIGURE 4-36   Waveforms for Problem 14.

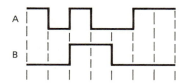

FIGURE 4-37   Pulsed signals for Problem 15.

them draw the waveform that will be produced on the output of the OR gate by these input signals.

16. *a.* The pulsed signals shown in Figure 4-38*a* are applied to the inputs of the gate shown in Figure 4-38*b*. Redraw these waveforms on a separate paper, and below them draw the waveform that will be produced on the output of the gate.

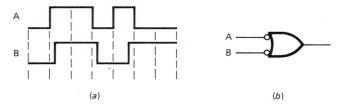

FIGURE 4-38   Pulsed signals and gate for Problem 16. (a) Pulsed signals. (b) Gate.

   b. What binary state on the input waveforms is being decoded by this gate?

17. The pulsed signals shown in Figure 4-39 are applied to the inputs of a 3-input NOR gate.
   a. Redraw these waveforms and below them draw the waveform that will be produced on the output.
   b. What binary state of the input signals is being decoded by this gate?

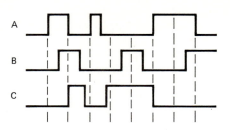

FIGURE 4-39   Waveforms for Problem 17.

18. *a.* Which position of the switch in Figure 4-40*a* enables the gate?
   b. Is this an active low or an active high enable input?
   c. Given the input waveforms shown in Figure 4-40*b*, show the waveform that will appear on the output if the gate is enabled.

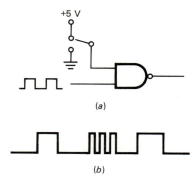

FIGURE 4-40   Circuit and waveforms for Problem 18.

19. What must be done with the unused inputs of the gate in Figure 4-41 in order to enable the gate so that pulsed input signal will appear on the output?

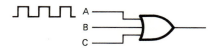

FIGURE 4-41   Diagram for Problem 19.

20. Describe how you would use a dual-trace, triggered scope to observe and draw the input signals shown in Figure 4-39 and the signal these produce on the output of a 3-input NOR gate.

21. When the current required by an IC changes from a low value to a high value, why does the V_{CC} voltage on the IC often initially change much more than would be predicted by the dc resistance of the supply leads?

22. *a.* Draw a graph which shows how the current thorough the circuit in Figure 4-42 changes when the switch is closed.
   b. Explain why the current requires some time to reach its final value.

23. Describe how a capacitor is constructed.

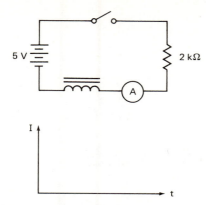

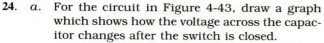

FIGURE 4-42   Circuit for Problem 22.

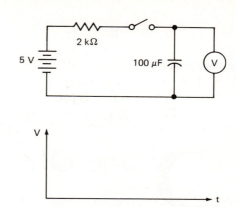

FIGURE 4-43   Circuit for Problem 24.

24. *a.* For the circuit in Figure 4-43, draw a graph which shows how the voltage across the capacitor changes after the switch is closed.
    b. If the switch has been closed for a long time, what voltage will the voltmeter show after the switch is opened?

25. Explain the meaning of the following capacitor markings:
    a. + 100μF 25 VDC –
    b. 150K 100V 10%
    c. 103 50V
    d. .015 1000V

26. *a.* Show how you would connect a bypass capacitor in the circuit in Figure 4-27a.
    b. What would be an appropriate value for this bypass capacitor if the circuit is going to operate at a frequency of 2 MHz?
    c. Why should the leads on the bypass capacitor be kept as short as possible?
    d. Why is a 0.1-μF capacitor often connected in parallel with a 25-μF capacitor across the supply leads where they first enter a printed-circuit board?

5

Analyzing, Testing, Troubleshooting, and Drawing Logic Circuits

In Chapter 2 we introduced you to the truth tables, schematic symbols, and Boolean expressions for simple logic gates. Chapters 3 and 4 showed you how to build circuits with real logic devices and how to make measurements on logic circuits. In the first part of this chapter we show you how to analyze multiple-gate circuits. In the second part of the chapter we show you how to test/ troubleshoot multiple-gate circuits that once worked. Finally, in the third part of this chapter we show you how to use computer-based tools to draw schematics for digital circuits. You can then use these tools to draw your schematics as you work your way through this book.

In the past, digital systems often contained printed-circuit (PC) boards with hundreds of discrete gate ICs such as 7400s, 7404s, and 7410s. In present digital systems, a few large-scale integration (LSI) devices usually perform those functions that once required several PC boards full of ICs. Discrete logic gates, however, are still often used as the "glue" to interface medium-scale integration (MSI) and LSI devices with each other, so you need to be able to analyze, test, troubleshoot, design, and implement simple circuits of this type.

If a required glue circuit is complex, we usually use a computer to design the logic for it and use a single, programmable logic device such as a PAL to implement the circuit. In a later chapter we show you how to work with programmable logic devices. Here we show you the basics of working with logic gate devices and the concepts on which programmable devices are based.

OBJECTIVES

At the conclusion of this chapter you should be able to:

1. Determine the expected logic levels at intermediate points in a multiple-gate logic circuit.

2. Predict the output state of a multiple-gate circuit for specified input logic levels.

3. Write the truth table for a multiple-gate logic circuit.

4. Given the logic levels at intermediate points in a multiple-gate circuit, determine the probable cause of a circuit malfunction.

5. Given waveforms at intermediate points in a multiple-gate logic circuit, identify the gate(s) most likely to be causing an incorrect final output.

6. Test the operation of a combinational logic circuit using a binary counter to produce a sequence of all the possible input combinations, and a scope to determine if the output state is correct for each input combination.

7. Use a Schematic Capture Program such as Schema II, OrCAD/SDT III, or EE Designer to produce a schematic for a digital system.

ANALYZING LOGIC CIRCUITS

The three most common ways of representing the function of a logic gate circuit are with a *circuit diagram,* a *truth table,* or a *Boolean algebra expression.* It is important that you learn to work with each one of these and to convert each representation to either of the other two as needed. Figure 5-1 shows the possible conversions and summarizes how each is done. We spend most of this chapter showing you how to work your way around this triangle so you can analyze and troubleshoot circuits. We start by showing you how to analyze a multiple-gate circuit.

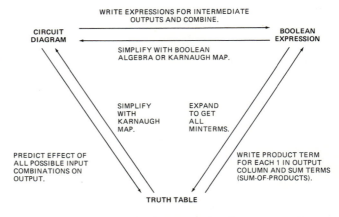

FIGURE 5-1 Converting between the three ways of representing the operation of a logic gate circuit.

When testing a logic gate circuit the first step is to make a complete truth table for the circuit. This is important so that you can determine if the output state is correct for each input combination. Also, if the circuit does not work correctly, the input combinations that produce incorrect outputs can help you troubleshoot the circuit by pointing you toward the part of the circuit which has a problem. The following sections show two methods of producing a circuit's truth table.

Constructing a Truth Table Directly from the Circuit

The first step is to set up the basic truth table with a column for each input variable and a column for the output. For the example circuit in Figure 5-2a, you need

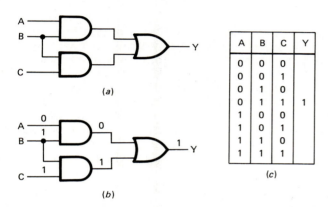

FIGURE 5-2 Writing a truth table for a combinational logic circuit. (a) Circuit diagram. (b) Outputs produced by A = 0, B = 1, and C = 1. (c) Partially completed truth table.

columns for A, B, C, and Y. Next, you write in all of the possible combinations for the input variables. Remember that since there are three input variables, the truth table must contain 2^3, or 8, input combinations. As we said in Chapter 2, it is best to write these down in binary count sequence so you don't forget any.

The second step is to trace the effect of each input combination through the circuit to see the effect that each combination has on the output. As shown in Figure 5-2b, for example, the input combination A = 0, B = 1, and C = 1 produces a 0 on the output of the upper AND gate and a 1 on the output of the lower AND gate. The 0 from the top AND gate and the 1 from the lower AND gate are ORed together to produce a 1 on the Y output.

Work your way through the other seven input combinations in this way and complete the truth table in Figure 5-2c. The output column should contain a 1 for the three input combinations A = 0, B = 1, C = 1 (011); A = 1, B = 1, C = 0 (110); A = 1, B = 1, C = 1 (111).

For more practice with this method, work out a truth table for the circuit in Figure 5-3. This circuit has 4 variables, so the truth table will have 16 possible input combinations, 0000 to 1111. For the first combination, 0000, the output of the upper AND gate will be a 0 because

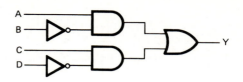

FIGURE 5-3 Four-input combinational logic circuit.

A = 0. The output of the lower AND gate will also be a 0 because C = 0. The 0 from the upper AND gate ORed with the 0 from the lower AND gate produces a 0 on the output. You can trace the effect of the other 15 input combinations through the circuit in the same manner. Figure 5-4 shows the complete truth table for the circuit in Figure 5-3.

A	B	C	D	Y
0	0	0	0	0
0	0	0	1	0
0	0	1	0	1
0	0	1	1	0
0	1	0	0	0
0	1	0	1	0
0	1	1	0	1
0	1	1	1	0
1	0	0	0	1
1	0	0	1	1
1	0	1	0	1
1	0	1	1	1
1	1	0	0	0
1	1	0	1	0
1	1	1	0	1
1	1	1	1	0

FIGURE 5-4 Truth table for circuit in Figure 5-3.

As this last example probably showed you, developing a truth table one line at a time directly from the circuit is not difficult, but it is somewhat tedious. In the next section we show you a method which usually makes the process much easier.

Using a Boolean Expression to Help Construct a Truth Table

A THREE-VARIABLE EXAMPLE

The second method of constructing the truth table for a circuit is to write the Boolean expression for the circuit and then use the Boolean expression to help you determine the entries for the truth table.

As a first example of how to do this, we will use the circuit in Figure 5-5a. The Boolean expression can be derived directly from the schematic drawing as follows. The output gate performs the OR function, so the overall expression for the output will be the OR, or sum, of the signals applied to the two inputs of this gate. To represent this, first write an expression in the form:

$$Y = \underline{\hspace{2cm}} + \underline{\hspace{2cm}}$$

Then, when you find the expressions for each of the terms that are ORed together, you can simply write the expressions for these terms in the spaces you left.

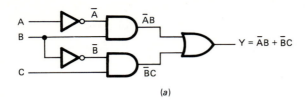

(a)

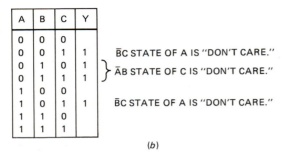

A	B	C	Y
0	0	0	
0	0	1	1
0	1	0	1
0	1	1	1
1	0	0	
1	0	1	1
1	1	0	
1	1	1	

$\overline{B}C$ STATE OF A IS "DON'T CARE."

$\overline{A}B$ STATE OF C IS "DON'T CARE."

$\overline{B}C$ STATE OF A IS "DON'T CARE."

(b)

A	B	$\overline{A}B$
0	0	0
0	1	1
1	0	0
1	1	0

(c)

B	C	$\overline{B}C$
0	0	0
0	1	1
1	0	0
1	1	0

(d)

FIGURE 5-5   Three-variable example. *(a)* Circuit with inputs A, B, and C. *(b)* Partially completed truth table for the circuit. *(c)* Truth table showing output of upper AND gate. *(d)* Truth table showing output of lower AND gate.

The signal into the OR gate from the upper AND gate can be represented by the expression $\overline{A}B$. The signal from the lower AND gate can be represented by the expression $\overline{B}C$. The result of ORing these together then is $Y = \overline{A}B + \overline{B}C$, as shown next to the circuit in Figure 5-5a. Here's how you use this Boolean expression to help you fill in the truth table.

The first term in the Boolean expression tells you that the output will be a 1 if A = 0 AND B = 1. In the truth table find all the input combinations where A = 0 and B = 1; then write a 1 in the output column for these combinations. For the example here, the combinations which satisfy this are A = 0, B = 1, C = 0 (010), and A = 0, B = 1, C = 1 (011). Note if A = 0 and B = 1, the output will be a 1, whether C = 1 or C = 0, because C is not part of the $\overline{A}B$ term in the expression.

The second term in the Boolean expression tells you that the output will be a 1 if B = 0 AND C = 1. In the truth table find all the input combinations where B = 0 and C = 1, then write a 1 in the output column for these combinations. For the example in Figure 5-5a, the two combinations which satisfy this are A = 0, B = 0, and C = 1 (001), and A = 1, B = 0, and C = 1 (101). Since A is not present in the $\overline{B}C$ term of the expression, the output will be a 1, whether A = 0 or A = 1, as long as B = 0 and C = 1.

Figure 5-5b shows a three-variable truth table with 1's in the output column for the four combinations represented by $\overline{A}B + \overline{B}C$. Since none of the other four input

combinations will cause the output to be a 1, you can just write a 0 in the output column for each of them.

Writing the Boolean expressions for the different parts of the circuit, combining them, and using them to help write the truth table is usually much faster than working each of the possible input combinations through the circuit. An additional advantage of this method is that it allows you to easily write the truth table for an intermediate point in the circuit if you need it for troubleshooting the circuit. The expression for the output of the upper AND gate in Figure 5-5a, for example, is $\overline{A}B$. Knowing this, you can write the truth table for this point in the circuit as shown in Figure 5-5c. The expression for the output of the lower AND gate in Figure 5-5a is $\overline{B}C$, so you can write the truth table for this point in the circuit as shown in Figure 5-5d. Once you have the truth tables for each point in a circuit, you can easily see what logic level should be on each point in the circuit for a given input combination. This is a great help in troubleshooting, because, when you find the final output is incorrect for a given input combination, you can compare the measured level at each point in the circuit with the level predicted by the truth table for that point. We will discuss troubleshooting more later. For now we want to work through some more examples of using Boolean expressions to write the truth table for logic circuits.

AN EXAMPLE WITH BUBBLED INPUTS AND OUTPUTS

We will use the simple circuit in Figure 5-6 to show you more about how to take bubbles into account when you write the Boolean expressions for a circuit. First of all, in analyzing the circuit in Figure 5-6a, you can assume the A, B, and C signals are active high because they have

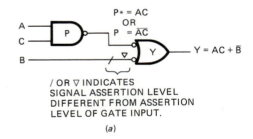

/ OR ▽ INDICATES
SIGNAL ASSERTION LEVEL
DIFFERENT FROM ASSERTION
LEVEL OF GATE INPUT.

(a)

A	C	P
0	0	1
0	1	1
1	0	1
1	1	0

(b)

A	B	C	Y
0	0	0	1
0	0	1	1
0	1	0	0
0	1	1	0
1	0	0	1
1	0	1	1
1	1	0	0
1	1	1	1

(c)

FIGURE 5-6   Analyzing a circuit with bubbled inputs and outputs. *(a)* Circuit diagram. *(b)* Truth table for output of gate P. *(c)* Truth table for complete circuit.

no .L, −1, *, or / after them. The output expression for gate P in the circuit, then, can be written in two different ways.

In a data book the symbol used for gate P in Figure 5-6a is described as a "positive logic NAND gate." The *positive logic* part of this name refers to the fact that the operation of the gate is described only in terms of active high input and output signals. The positive logic expression for this NAND gate, $P = \overline{AC}$, tells you that P will be asserted high if A AND C are NOT both asserted high. This expression is accurate, but it is not the easiest one to use in tracing circuit operation.

The second way of writing the expression for gate P in the circuit is based on the basic shape of the symbol and the assertion levels of the inputs and the output. As we showed you in Chapter 2, a *bubble* indicates an active low input or output, and no bubble indicates an active high input or output. The inputs of gate P are active high because they have no bubbles. The bubble on the output of gate P indicates that it can be thought of as active low. To identify the signal at this point as an active low signal, we give it a name such as P*. Directly from the symbol, then, you can see that the gate is performing the AND function with active high inputs and an active low output. If the A signal is asserted (high), it will assert the A input, and if the C signal is asserted (high), it will assert the C input of gate P. If the A signal is asserted AND the C signal is asserted, the P* output will be asserted. Then, the Boolean expression for gate P can be written simply as P* = AC. Now let's look at how gate Y in Figure 5-6a functions.

The shape of the symbol for gate Y tells you that the gate is performing the OR function on its two input signals. Therefore, you can write a template in the form Y = _____ + _____ for the output expression and determine what to write in each blank in the expression. Since there are bubbles on the inputs of the OR symbol, the output of gate Y will be asserted (high) if the upper input is asserted low OR the lower input is asserted low. As indicated by the expression P* = AC, the upper input of the Y gate will be asserted (low) if the A signal AND the C signal are asserted. If the active low P* signal is asserted then, it will assert the active low input of the bubbled input OR symbol. Therefore, you can write just AC as the first term of the output expression. Another way of looking at this is to think of the bubble on the output of the AND symbol and the bubble on the input of the OR symbol as "canceling" each other. With the bubbles out of the way, you can perhaps more easily see that the output term for this input is just AC. Now, let's see how the B signal appears in the second term of the Y output expression.

The B signal is active high, and it is connected to an active low input on the Y gate. This means that the active low input of the Y gate will be asserted if the B signal is NOT asserted. To indicate this in the expression for the Y output, you write the B term with an overbar. The complete expression for the circuit in Figure 5-6a, then, is Y = AC + $\overline{B}$.

The term *functional logic*, or *mixed logic*, is often given to the method of analyzing logic circuits that we have just shown you. The "big" rules for you to remember

when you are using functional logic to analyze circuits are as follows:

1. Use the shape of the symbol to tell you the logic function being performed.

2. Use bubbles or the absence of bubbles to tell you the assertion levels of input and output signals.

3. If the assertion level of a signal is different from that of the gate input that it is connected to, the term for that input signal will be complemented in the Boolean expression for the gate.

To indicate that the assertion level of a signal is different from the assertion level of an input, some references put a slash on the signal line or a small triangle over the signal line. The B input line in Figure 5-6a shows examples of these two ways of "flagging" a signal that will be complemented in the output expression. Although it is not done this way in industry, you may find it useful to mark signal lines in this way until you become more practiced at writing the expressions for functional logic circuits.

Once you get the Boolean expressions for the circuit in Figure 5-6a, you can use them to develop truth tables. The output of gate P will be asserted low if A = 1 and C = 1, so the truth table for this point is the familiar one shown in Figure 5-6b. The truth table for the entire circuit will contain eight input combinations because there are three input variables. As predicted by the output expression, the output will be a 1 for any input combination where A and C are both 1. Input combinations 101 and 111 satisfy these conditions, so write a 1 in the output column for each of these. The output expression also indicates that the output will be a 1 if the B signal is not asserted, or a 0. Input combinations 000, 001, 100, and 101 satisfy this condition, so write a 1 in the output column of the truth table for each of these combinations. Write a 0 in the output column of the truth table for each of the three remaining input combinations. Figure 5-6c shows the completed truth table for the circuit in Figure 5-6a.

Using Alternative Gate Symbols to Simplify an Output Expression

Figure 5-7a shows a four-variable circuit for you to analyze and develop the truth table for. To start, see if you can write the Boolean expression for each intermediate point in the circuit and for the final output; then check your results with those here.

The expression for the output of the P NAND gate is P* = $\overline{A}BC$, the expression for the Q NAND gate is Q* = C$\overline{D}$, and the expression for the R NAND gate is R* = A$\overline{BC}$. The three active low signals, P*, Q*, and R*, are all connected to active high inputs on the output NAND gate, S. Therefore, each of these three terms will be complemented in the output expression. The output of gate S has a bubble, so the final output expression is:

$$S* = \overline{(\overline{A}BC)(C\overline{D})(A\overline{B}\,\overline{C})}$$

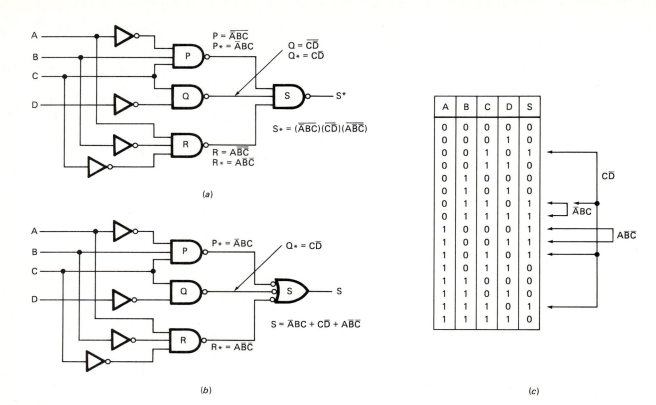

FIGURE 5-7 Use of alternative gate symbols to simplify circuit analysis. *(a)* Circuit with only positive logic gate symbols. *(b)* Same circuit using bubbled input OR symbol to represent output NAND gate. *(c)* Truth table for the circuit.

It may also be expressed as follows:

$$S = \overline{(\overline{ABC})(\overline{CD})(\overline{AB\,C})}$$

Both of these expressions are correct for the circuit in Figure 5-7a, but the many overbars make it difficult to relate these expressions directly to a truth table. In Chapter 6 we show you how to use Boolean algebra to simplify expressions such as this, but a simpler method to use when you are analyzing a circuit is to replace some of the gates in the circuit diagram with their alternative symbols. Here's how you do this.

Remember from previous work that a positive logic NAND gate can be represented by an OR symbol with bubbles on its inputs, and a NOR symbol can be replaced by an AND symbol with bubbles on its inputs. The question then is, "When do you replace a symbol with its alternative symbol?" The rule we use is:

Replace a symbol which causes complex intermediate expressions to be complemented in the output expression.

Gate S in the circuit in Figure 5-7a is a good candidate for replacement with its alternative symbol because it has three active low gate outputs connected to its active high inputs. As we said before, this means that the terms for these three inputs will be complemented (overbarred) in the final output expression. Figure 5-7b shows the circuit from Figure 5-7a drawn with a bubbled input OR symbol in place of the final output NAND

symbol. Let's see how this affects the analysis of the circuit.

The outputs of the three intermediate gates are still $P* = \overline{A}BC$, $Q* = C\overline{D}$, and $R* = A\overline{B}\overline{C}$. However, since these active low signals are connected to active low inputs on the output OR symbol, the expressions for these gate outputs will not be complemented in the output expression. The expression for the final output, then, is:

$$S = \overline{A}BC + C\overline{D} + A\overline{B}\,\overline{C}$$

Since this expression does not contain as many overbars as the expression derived from the three NAND symbols, you can more easily use it to help you write a truth table. For practice, use this expression to write a truth table for the circuit in Figure 5-7b, then read through the following discussion of how to do it. The circuit has four variables, so the truth table has 16 entries. The first term in the expression is $\overline{A}BC$, so find all the input combinations where A = 0, B = 1, and C = 1. You should find two combinations that satisfy these conditions, 0110 and 0111. Mark a 1 next to each of these in the output column of the truth table. The second term in the expression is $C\overline{D}$, so find all of the input combinations where C = 1 and D = 0. You should find four input combinations which satisfy these conditions. Mark a 1 in the output column next to each of these combinations. Finally, the last term of the expression is $A\overline{B}\overline{C}$, so find all the input combinations where A = 1, B = 0, and C = 0. Mark a 1 next to each of these in the output column of the truth table. Figure 5-7c shows the completed truth table for the circuit.

An Example with Active High and Active Low Input Signals

So far in this book we have, for simplicity, used mostly active high signals as inputs to our logic circuit examples. Figure 5-8a shows an example of a circuit with a

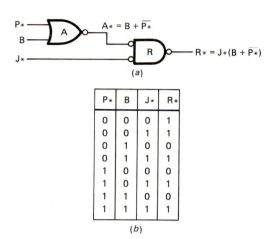

(a)

P*	B	J*	R*
0	0	0	1
0	0	1	1
0	1	0	0
0	1	1	1
1	0	0	0
1	0	1	1
1	1	0	0
1	1	1	1

(b)

FIGURE 5-8   Analyzing a circuit with a mixture of active high and active low input signals. (a) Circuit diagram and Boolean expressions. (b) Truth table.

mixture of active high and active low input signals. Here's how you analyze a circuit such as this.

The active low P* signal is connected to an active high input on gate A, so the term for this signal will be complemented in the output expression for gate A. The B signal is active high, so it will not be complemented. The output expression for gate A, then, is $A* = B + \overline{P*}$, as shown in the figure. If the A* signal is asserted low, it will assert the active low input of gate R. Therefore, the $B + \overline{P*}$ term will appear in the final output expression without an overbar. Likewise, the active low J* signal is connected directly to an active low input on gate R, so it will appear uncomplemented in the final output expression. The output of gate R is active low, so the expression for its output is $R* = J*(B + \overline{P*})$. This means that the output of gate R will be asserted (low) if the J* input is asserted (low) AND the B input is asserted (high) OR the P* input is NOT asserted. Since the P* signal is defined as active low, its NOT asserted state is a high.

Given this expression, which you can read directly from the circuit, it is quite easy to make a truth table for the circuit. The expression tells you that the output will be a low or zero for those input combinations where J* is low AND B is high OR P* is high. You can write a 0 in the output column of the truth table for each of the three input combinations which satisfy these conditions, and a 1 for the remaining five input combinations, as shown in Figure 5-8b.

Another Mixed Logic Example Circuit Analysis

Since most logic diagrams now contain both active low and active high signals, it is very important for you to become comfortable analyzing this type of circuit. Therefore, here's one more example before we move on to the next topic. Look carefully at the circuit in Figure 5-9a.

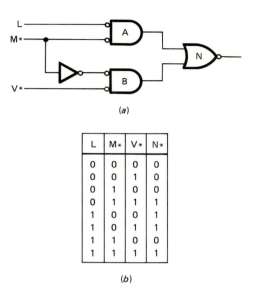

(a)

L	M*	V*	N*
0	0	0	0
0	0	1	0
0	1	0	0
0	1	1	1
1	0	0	1
1	0	1	1
1	1	0	0
1	1	1	1

(b)

FIGURE 5-9   Another example of a circuit with a mixture of active high and active low input signals. (a) Circuit diagram. (b) Truth table.

See if you can write the expression for each intermediate point in the circuit and the expression for the output before you read further.

The expression for gate A is $\overline{L}M*$, the expression for the B gate is $\overline{M}*V*$, and the expression for gate N is just the logical sum of these, $N* = \overline{L}M* + \overline{M}*V*$. This expression tells you that the N output will be asserted (low) if the L input is not asserted (low) AND the M* input is asserted (low) OR the M* input is not asserted (high) AND the V* input is asserted (low). To make a truth table for this circuit, then, just write a 0 in the output column for each input combination where L is low and M* is low and each input combination where M* is high and V* is low. Write a 1 for each of the remaining input combinations. Figure 5-9b shows the completed truth table for the circuit.

Exclusive OR and Exclusive NOR Circuits

Up to this point we have analyzed circuits built with just AND, NAND, OR, and NOR gates and inverters. Any combinational logic circuit can be built with a combination of these types, but as we told you in Chapter 2, some combinational circuits are so commonly needed that they are produced ready-made. Examples of this are *Exclusive OR* gates and *Exclusive NOR* gates.

For review, Figure 5-10 shows the commonly used symbols and the truth tables for each of these. The output of the Exclusive OR gate will be high if the two inputs "dis-

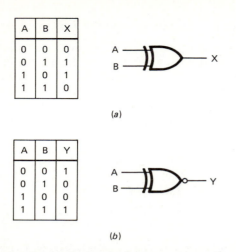

A	B	X
0	0	0
0	1	1
1	0	1
1	1	0

(a)

A	B	Y
0	0	1
0	1	0
1	0	0
1	1	1

(b)

FIGURE 5-10   Truth table and symbols for: *(a)* Exclusive OR gate. *(b)* Exclusive NOR gate.

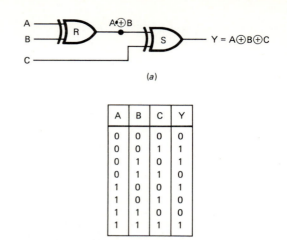

(a)

A	B	C	Y
0	0	0	0
0	0	1	1
0	1	0	1
0	1	1	0
1	0	0	1
1	0	1	0
1	1	0	0
1	1	1	1

(b)

FIGURE 5-11   Circuit to determine the parity of a binary number. *(a)* Circuit. *(b)* Truth table for circuit.

agree" or, in other words, are at different logic levels from each other. Another way of looking at an Exclusive OR is to say that the output will be high if the number of highs on the inputs is odd. The number 1 is an odd number, and as you can see in the truth table in Figure 5-10a, the output of an Exclusive OR gate will be high if one of the inputs is high. The numbers 0 and 2 are considered even numbers, and as you can see in the truth table, the output will be low for an even number of ones on its inputs.

A binary word which contains an odd number of 1's is said to have *odd parity.* The binary number 01011, for example, has an odd number of 1's (three), so it is said to have odd parity. A binary number which has an even number of 1's is said to have *even parity.* The binary number 011011, for example, has even parity. Later in the book we will show you the different uses of parity. A circuit made up of Exclusive OR gates is often used to determine if a binary number has odd parity or even parity. Figure 5-11a shows how two Exclusive OR gates can be connected to determine if a 3-bit binary number contains an odd or even number of 1's. You can construct a truth table for this circuit in one of two ways.

The first way is to work your way through the circuit for each of the eight input combinations and write the output logic level produced by each as shown in Figure 5-11b. The second way of producing the truth table for the circuit is to first recognize that the output will be high if the number of 1's on the input signal lines is odd. By counting 1's in the input combinations, you can then immediately see that the number of 1's is odd for states 1, 2, 4, and 7. All you have to do is write a 1 in the output column next to each of these combinations as shown. For the rest of the input combinations, the output will be low, so you can put 0's in the output column for these combinations.

To summarize then, the output of this circuit will be asserted high if the input word has odd parity. If the input word has even parity, the output of the circuit will be asserted low. The Boolean expression for the intermediate point in the circuit is A XOR B. This signal is Exclusive ORed with the C signal by the output gate, so

the expression for the complete circuit is Y = A XOR B XOR C. Whenever you see an expression such as this, think of it as telling you that the output will be high if the number of 1's on the input signal lines is odd, regardless of the number of inputs. Now let's look at how an Exclusive NOR device is often used in circuits.

The key point in analyzing a circuit containing an Exclusive NOR gate is to remember that the output of an Exclusive NOR gate will be high if the signals on the two inputs are the same. Another way of looking at this is to say the device "compares" the two signals on its inputs and outputs a high if the two signals are equal. Figure 5-12 shows an Exclusive NOR circuit which can be used to compare two 2-bit binary numbers. If the least significant bits of the two numbers, A_0 and B_0, are equal, the output of the upper XNOR gate will be high. If the most significant bits of the two numbers are equal, the output of the lower XNOR gate will be high. If the outputs of both the XNOR gates are high, then the output of the AND gate will be high, indicating that the two numbers are equal.

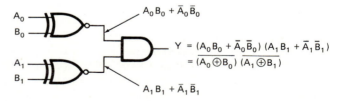

FIGURE 5-12   Exclusive NOR circuit used to compare two 2-bit binary numbers.

The Boolean expression for the circuit is:

$$Y = (A_0B_0 + \overline{A}_0\overline{B}_0)(A_1B_1 + \overline{A}_1\overline{B}_1)$$

The basic circuit in Figure 5-12 is often expanded with more XNOR gates and a larger AND gate to compare binary numbers with more bits. The 74LS688, for example, contains circuitry to compare two 8-bit numbers.

Open-Collector Circuits

In Chapter 3 we introduced you to the open-collector output structure found on some TTL devices such as the 74ALS01 and the 74AS09. The main advantage of open-collector outputs is that, unlike standard totem-pole outputs, several outputs can be connected on a common line. Figure 5-13a shows how the outputs of two open-

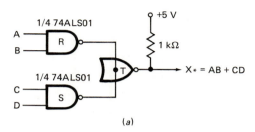

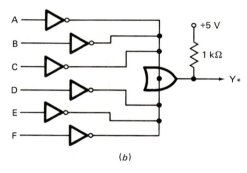

FIGURE 5-13 Wired OR circuits. (a) Connecting outputs of two open-collector NAND gates. (b) Another example of a wired OR circuit.

collector NAND gates can be connected in this way. Because open-collector outputs have no internal circuitry to pull the output line to a high state, an external pull-up resistor is connected between the common output and V_{CC} as shown. This resistor pulls the output to a logic high level when the outputs of both NAND gates are in the high state. Here's how you analyze the logic function performed by this circuit.

The common output line will be asserted low if the output of gate R is low, the output of gate S is low, or both outputs are low. This connection then ORs the outputs of gate R and gate S. The output of gate R will be low if A is high and B is high. The output of gate S will be low if C is high and D is high. Therefore, the expression for the whole circuit can be written as $X* = AB + CD$. As shown in Figure 5-13a, we use an OR symbol with a dotted connection in the center to indicate that the connection performs the OR function. This connection is often referred to as a *dotted OR*, or *wired OR*, connection.

Figure 5-13b shows another wired OR circuit for you to analyze. See if you can write the Boolean expression for it. The result you should get is:

$$Y* = A + B + C + D + E + F$$

Now that you have some practice analyzing combina-

tional logic circuits, the next step is to learn how to test and troubleshoot multiple-gate circuits.

TESTING AND TROUBLESHOOTING COMBINATIONAL LOGIC CIRCUITS

The most important tool for testing and/or troubleshooting a digital circuit is a *systematic approach*. Random poking, probing, and hoping is seldom productive. The first step in this systematic approach is to develop a truth table(s) for the circuit as we showed you in the last section of this chapter. With a truth table you can determine if the output is correct for each combination of the input variables. If the output is not correct for one or more input combinations, you may need the truth tables for intermediate points in the circuit to help you find the source of the problem. The main point here is that you should write down truth tables instead of trying to keep all the logic levels for a circuit in your head.

In the following sections we first show you how to test a logic circuit manually, then semiautomatically.

Manual Testing and Troubleshooting Digital Circuits

One way to test a digital circuit is to apply combinations of highs and lows to the circuit inputs by manually connecting a jumper wire from each input to ground or V_{CC}. The basic testing process involves setting the jumpers for one input combination from the truth table, measuring the logic level produced on the output by this input combination, and then changing the jumpers to produce the next input combination in the table, etc. Since both the input signals and the output signal are just dc voltage levels, you can measure them with either a DVM, a logic probe, or a scope.

As an example of this process, we will use the simple circuit in Figure 5-14a. The output expression for gate R in the circuit is $R* = AB$, the expression for the S gate is $S* = \overline{B}C$, and the expression for the entire circuit is $T = AB + \overline{B}C$ as shown on the drawing. The truth table for the circuit is shown in Figure 5-14b.

To test this circuit with the static method, you start by connecting the three inputs to ground with jumper wires to simulate the first input combination. Then you power up the circuit and determine the output logic level with a DVM, logic probe, or scope. For the input combination of 000, the output should be a 0. Let's assume you find the output is low as predicted and go on to the next combination.

For the 001 combination, you lift the C input jumper from ground and connect it to V_{CC}. The A and B inputs are left tied to ground. According to the truth table, the output should be a 1 for this input combination. In a similar manner you work your way through all of the input combinations and determine if the output logic level is correct for each.

Now, suppose that as you are testing the circuit, you find that for the ABC input condition of 001, the output is a 0 instead of a 1. Look at the circuit diagram and truth table in Figure 5-14 to see if you can think of some

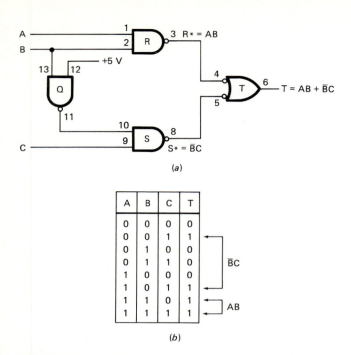

(a)

A	B	C	T
0	0	0	0
0	0	1	1
0	1	0	0
0	1	1	0
1	0	0	0
1	0	1	1
1	1	0	1
1	1	1	1

(b)

FIGURE 5-14 *(a)* Circuit to be used for static testing method. *(b)* Truth table for the circuit.

reasons why gate T does not have a logic high on its output as predicted. Compare your thoughts with the list of possibilities below:

- If the circuit is a prototype, it may not be wired correctly.
- V_{CC} is not getting to the device.
- One of the input signals to gate T is incorrect.
- Gate T is not responding to the signals on its inputs.
- The output of gate T is shorted to ground.

Once you get a list of the possible causes of the problem, the next step is to check out each of these possibilities, one at a time. Here's how you might go about finding the problem in the circuit in Figure 5-14a:

1. Visually check the circuit to see if it is wired correctly.

2. Check with a DVM, logic probe, or scope to see if V_{CC} is getting to the correct pins on the ICs.

3. Predict the level that should be present on each input of gate T for ABC = 001. The expression for gate R is R* = AB, so for the input combination 001, the output of gate R and pin 4 of gate T should be high. The expression for gate S is S* = $\overline{B}$C, so for the input combination 001, the output of gate S and pin 5 of gate T should be low.

4. Check the signal levels on the inputs of gate T. If pin 5 of gate T is low and pin 4 is high as predicted, then the problem is probably caused by a malfunction in gate T or by a short between the output of gate T and ground.

5. If the IC containing gate T is in a socket, you can determine which of these is the case by removing the IC from its socket, gently bending pin 6 up, putting the IC back in the socket, and testing the circuit again. If the output directly on pin 6 is now correct, turn off the power. Then check visually and with an ohmmeter to see if the output lead that was connected to pin 6 is shorted. If the output is still incorrect, try replacing the IC containing gate T with an IC of the same type that is known to be good.

6. If signal on pin 5 of gate T is high instead of low as predicted for the ABC input combination 001, then:
   a. The input signals to gate S may not be correct.
   b. Gate S may be defective.
   c. The output of gate S may be shorted to V_{CC} or to another signal line.

7. To check the first of these possibilities, predict the level that should be on pins 10 and 9 of gate S for ABC input conditions 001.
   Gate Q is simply acting as an inverter, so if B = 0 as specified, then pin 10 should be high. Pin 9 is jumpered to V_{CC}, so it should also be high. If both inputs of gate S are high as predicted and the output is not low, then the trouble is probably caused by possibility 6b or 6c.

8. To find out if the gate is defective, remove the IC from its socket, carefully bend up pin 8, re-insert the IC, and test the circuit again. If the output of gate S is still not correct, try replacing the IC with an IC of the same type that you know is good. If the output of gate S is now correct, turn off the power, then check visually and with an ohmmeter to see if the connecting wire between gates S and T is shorted.

We have found it much more effective to troubleshoot circuits backward from the output to the inputs in this way, rather than from the inputs to the output. We will use the circuit in Figure 5-14a to show you why.

Suppose you find that the output is incorrect with the ABC input combination 001, as we assumed for the preceding discussion. Upon further checking, you find that the logic level on pin 4 of the output gate is correct for these input conditions, but the logic level on pin 5 of the output gate is incorrect. This discovery automatically eliminates gate R as the cause of the problem in the circuit. The result is that you have narrowed the problem down to about half of the total circuit. If the output gate had been a 3- or 4-input gate, finding an incorrect signal on one of the inputs would have narrowed the problem down to an even smaller part of the total circuit. In a similar manner, finding an incorrect signal on an input of gate S would further narrow down your search for the part of the circuit causing the problem. This approach is another example of the old "divide and conquer" technique.

The manual method we have described for testing a digital circuit works, but it is time-consuming, especially for circuits with four or more input variables. In the next section we show you a faster method.

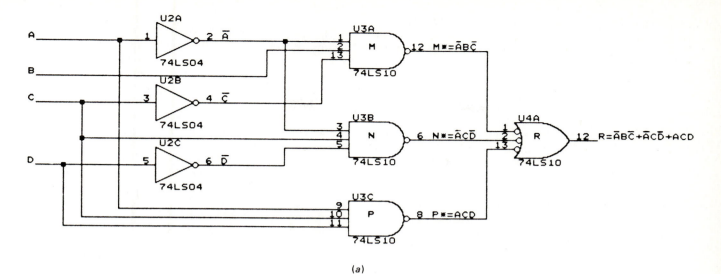

$$(a)$$

State	QD A	QC B	QB C	QA D	R
0	0	0	0	0	0
1	0	0	0	1	0
2	0	0	1	0	1
3	0	0	1	1	0
4	0	1	0	0	1
5	0	1	0	1	1
6	0	1	1	0	1
7	0	1	1	1	0
8	1	0	0	0	0
9	1	0	0	1	0
10	1	0	1	0	0
11	1	0	1	1	1
12	1	1	0	0	0
13	1	1	0	1	0
14	1	1	1	0	0
15	1	1	1	1	1

$\overline{A}B\overline{C}$ $\overline{A}C\overline{D}$

ACD

$$(b)$$

FIGURE 5-15   Combinational test circuit used to demonstrate circuit testing. *(a)* Circuit. *(b)* Truth table.

A "Semiautomatic" Method of Testing Combinational Logic Circuits

After they are first built, digital circuit boards are usually tested by computer-controlled test machines which are commonly called *automatic test equipment,* or ATE.

The basic principle of ATE is that the computer outputs a sequence of binary test patterns to the inputs of a board or circuit. For each input combination the computer reads in the logic level(s) produced on the output(s) of the circuit, and uses a table of expected values in its memory to determine if the output is correct for that input combination. If the output is incorrect for any input test pattern, the computer displays an error message. In this section we introduce you to some of the concepts of ATE. We also show you a way of "automatically" producing the sequence of input combinations needed to test a digital circuit and a way of using a scope

to display the sequence of output states produced by this input sequence.

To start, suppose that you want to test the simple combinational circuit in Figure 5-15a. The circuit has four input variables, so the truth table for the circuit will have 16 entries, as shown in Figure 5-15b. As you well know, the input combinations are listed in the truth table in binary count sequence, both for consistency and so that you don't miss any combinations when you write the truth table. When you test a combinational logic circuit such as this, you usually test the circuit in the same sequence. Instead of producing the sequence of input combinations with jumper wires as we described in the previous section, you can use a common digital device called a *binary counter* to do the job for you.

Figure 5-16a shows an inexpensive 74LS93 binary counter circuit that we will use for our discussion. Figure 5-16b shows the count sequence that this circuit will output on its QA, QB, QC, and QD outputs if a TTL-compatible square-wave signal is applied to its CLKA in-

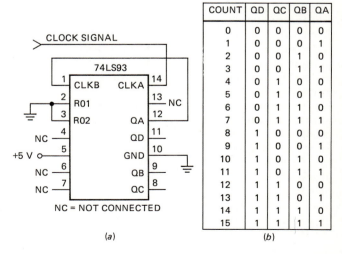

COUNT	QD	QC	QB	QA
0	0	0	0	0
1	0	0	0	1
2	0	0	1	0
3	0	0	1	1
4	0	1	0	0
5	0	1	0	1
6	0	1	1	0
7	0	1	1	1
8	1	0	0	0
9	1	0	0	1
10	1	0	1	0
11	1	0	1	1
12	1	1	0	0
13	1	1	0	1
14	1	1	1	0
15	1	1	1	1

FIGURE 5-16   Using a 74LS93 binary counter to produce a binary count sequence for testing logic circuits. *(a)* Circuit. *(b)* Count sequence.

put. In a later chapter we will discuss in detail how this counter works. For now, all you need to know is that each pulse applied to the CLKA input of the counter will cause the binary count on the Q outputs to increment to the next count in the sequence.

Note that the count sequence for the 74LS93 circuit is exactly the same as the sequence of input signals you need to test the circuit in Figure 5-15a. The only difference between the two is the way the columns are labeled. In a truth table with four variables labeled A, B, C, and D, variable A is usually the most significant digit. For the 74LS93 counter, the QA output is the least significant bit (LSB) of the count, and the QD output is the most significant bit (MSB) of the count. This minor problem is solved by simply connecting the QA output of the counter to the D input of the circuit to be tested, the QB output to the C input of the circuit, the QC output to the B input, and the QD output to the A input of the circuit, as shown in Figure 5-17. A 10-kHz TTL level sig-

NOTE: When you are using a scope to look at signals of two different frequencies, and you are using an internal trigger source as you are in this case, it is easier to obtain a stable display of both signals if you trigger on the *lower frequency* signal.

Adjust the other scope controls to obtain a stable trace showing eight cycles or more of the QA waveform. Draw these two waveforms.

Next, move the channel 1 scope probe to the QC output of the counter circuit and adjust the scope controls as needed to see a stable display of these two waveforms. Draw the QC waveform directly below the QB and QA waveforms you drew before.

Then move the channel 2 scope probe to the QD output of the counter circuit. Again adjust the scope controls as needed to obtain a stable display of both signals. Draw the observed QD waveform directly below the QA,

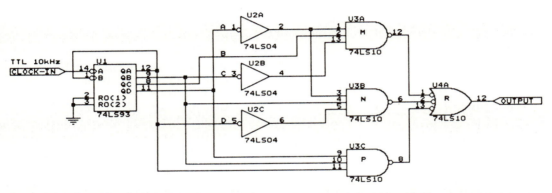

FIGURE 5-17 Connecting 74LS93 counter circuit to the test circuit shown in Figure 5-15a.

nal applied to the CLKA input of the 74LS93 counter will cause the counter to cycle through its output count sequence over and over. Therefore, the inputs of the circuit being tested will also be cycled through all of their possible input combinations over and over.

The term *word generator*, or *pattern generator*, is often used to describe the instrument or circuit used to produce a sequence of binary words for testing a circuit. Using the counter circuit as a word generator for our circuit eliminates the need for moving jumper wires around to step through the input combinations. However, you now have the problem of how to determine what output level is present for each input combination, or *state*. Since the signals are all rapidly changing, a DVM or logic probe won't do you much good. A scope or logic analyzer is the obvious choice for this job. For now we will show you how to use a basic dual-trace scope to do the job. Later, when you really need one, we will show you how to use a logic analyzer. To start, here's how you use a dual trace scope to make a timing diagram for the four counter outputs.

First, connect the channel 1 input of the scope to the QA output of the counter and the channel 2 input of the scope to the QB output of the counter. Set the trigger slope control for "−" and the trigger source switches for INTERNAL, CHANNEL 2.

QB, and QC waveforms you drew before. Figure 5-18 shows how your waveform drawings should look. There are two ways to describe the relationship between these signal waveforms.

The first way of describing the waveforms relates to the frequencies of the signals. Although it doesn't show in these waveforms, the frequency of the QA signal is half the frequency of the signal applied to the CLKA input of the counter. The frequency of the QB signal is half the frequency of the QA signal, the frequency of the QC signal is half that of the QB signal, and the frequency

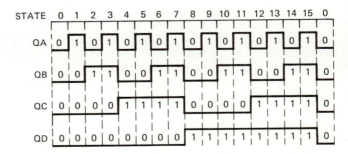

FIGURE 5-18 Timing diagram for the four 74LS93 counter outputs.

of the QD signal is half the frequency of the QC signal. In a later chapter we will discuss in detail why the frequencies are related to each other in this way, but for now, the second way of describing these waveforms is more useful.

The second way of describing the 74LS93 waveforms is in terms of the *sequence of states* that they step through. To see this sequence of states, draw vertical lines through the waveforms at each transition of the QA signal as shown in Figure 5-18. The waveform section between each pair of vertical lines represents one state of the counter outputs. To identify these states, mark 1's and 0's on the waveforms as also shown in Figure 5-18. Now, if you rotate the waveforms clockwise one-quarter turn, you should see that, starting from the top, the first output state is 0000, the second output state is 0001, the third output state is 0010, etc. Looking at the waveforms in this way shows you that the sequence of states that the waveforms step through is indeed the same as the output count sequence in Figure 5-16b. Here's how you use this sequential state way of looking at the waveforms to help test a circuit.

Each section of the QA waveform corresponds to one state of the counter output waveforms. Therefore, if you connect one probe of your scope to the QA signal, each section of the displayed QA waveform will identify the time when a different state is present on the inputs of the circuit being tested. If you connect the second probe of your scope to the output of the circuit being tested, then the display of this signal will show the sequence of logic levels produced on the output by the sequence of states applied to the input. The QA display then marks off states, and the second display shows you what is present on the output during each state. There is, however, still one minor problem.

The problem is that the QA waveform marks off states, but it doesn't tell you which state is on the inputs of the test circuit during each time interval. The solution to this problem is to make the scope display always start with state 0000 at the left side of the screen. If you know that the display starts with state 0000, you can then use sections of the QA waveform to count off states just as we showed in Figure 5-18. To get a display that you know starts with state 0000, you need some way to trigger the scope when the counter outputs roll from 1111 to 0000. There are several ways to do this.

One way is to use the QD output signal from the counter. If you take a look at the counter output waveforms in Figure 5-18, you will see that when the QD output goes from high to low, the other three outputs also go low. The falling edge of the QD signal then indicates the start of state 0000 on the counter outputs. Therefore, if the QD output of the counter is connected to the EXTERNAL TRIGGER input of the scope, and the TRIGGER SLOPE control of the scope is set for "−," the waveform displays on the scope will start with 0000.

Another way to produce the required trigger signal is with a circuit called a *state decoder*, or *word recognizer*. Figure 5-19 shows a circuit for a simple 4-input word recognizer. If you work through the logic of the circuit with the switches in the positions shown, you should see that the TRIGGER output will go low when state 0000

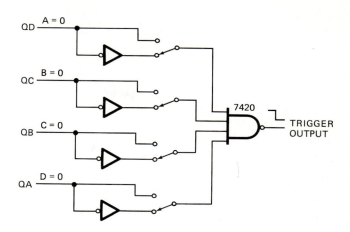

FIGURE 5-19 Circuit for a simple 4-input word recognizer.

appears on the A, B, C, and D inputs. The inputs of this circuit are connected to the outputs of the counter, and the output of the circuit is connected to the EXTERNAL TRIGGER input of the scope. If the TRIGGER SLOPE control of the scope is set for "−," then the signal from the word recognizer will trigger the scope display to start with state 0000 as desired. The advantage of using this state decoder or word recognizer circuit instead of simply using the falling edge of the QD signal is that this circuit is programmable! By changing the position of the switches, you can program the word recognizer so that the display starts with any desired state. For example, if you set the A switch in the down position, the B switch in the up position, the C switch in the up position, and the D switch in the up position, the output of the circuit will go low and trigger the scope when state 0111 or 7 is on its inputs. Logic analyzers and some scopes have a programmable word recognizer such as this built in, so that you can trigger on any desired pattern of 1's and 0's. This feature is needed because many circuits for which you want to display a sequence of signals do not have a unique signal edge such as the falling edge of the QD signal on the simple counter.

Now that you know how to trigger the scope so that the display starts with state 0000, let's look at the display produced for our example circuit in Figure 5-15a. Figure 5-20 shows the waveform for the QA output of the counter and the waveform for the R output of the circuit under test, as seen on a dual-trace scope.

As we said previously, each section of the QA waveform represents a different input combination or state applied to the inputs of the circuit under test. Therefore, you can use the sections of this waveform to count off states on the output waveform. Since you know the sequence of output states on the display starts with that for input state 0000, you can compare the sequence of 1's and 0's on the output waveform directly with the R column of the truth table in Figure 5-15b. If you do this, you will see that the sequences are identical. This tells you the circuit is producing the correct output for all of the input combinations.

To summarize then, the principles involved in this method of testing a circuit are as follows:

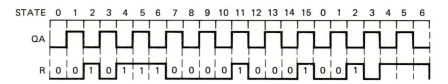

FIGURE 5-20 Waveform for the QA counter output and the R output of the test circuit shown in Figure 5-15a.

1. Use a word generator or pattern generator to produce a known sequence of signals as inputs to the circuit being tested.

2. Use one scope trace to display the most rapidly changing signal produced by the word generator. This signal marks off individual states.

3. Use the other scope trace to display the output signal of the circuit being tested.

4. Use a word recognizer to trigger the scope on a known state, so you can identify each state on the display of the output signal.

5. Compare the output for each state with the corresponding truth table entry to see if it is correct.

A major advantage of this testing method is that the circuit is stepped through all the possible input combinations and the entire output column of the truth table is displayed on one scope trace. This is obviously much easier than jumpering the inputs one at a time with wires. If you want to test a combinational circuit with more than four input variables, the QD output of one 74LS93 counter circuit can be connected to the CLKA input of another 74LS93 to produce the required sequence of binary combinations for up to eight inputs.

Now that you know how to test a circuit using this semiautomatic method, the next step is to show you how to use the displayed waveforms to find the source of a problem in the circuit if it doesn't work correctly.

For example, suppose that you were testing the circuit in Figure 5-15a as we described above and you observed the waveforms shown in Figure 5-21. The

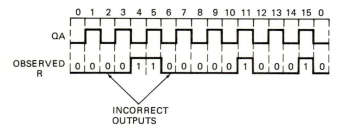

FIGURE 5-21 Waveform used to describe troubleshooting method.

sequence of states in the output waveform is 000011000001000010, instead of the sequence 00101110000100010 predicted by the R column of the truth table in Figure 5-15b. By comparing these two you can see that the output is incorrect for input states $ABCD = 0010$ (state 2) and $ABCD = 0110$ (state 6). The

expression for 0010 is $\overline{A}\overline{B}C\overline{D}$, and the expression for the 0110 term is $\overline{A}BC\overline{D}$. Since the B term is a 1 in one of these and a 0 in the other, the two terms can be represented by the single expression $\overline{A}C\overline{D}$, as shown next to the truth table in Figure 5-15b. The Boolean expressions next to the gates in Figure 5-15a show that gate N is responsible for producing this term on the center input of gate R. This conclusion points you toward gate N or the connection between the output of gate N and the input of gate R as the source of the problem.

The term *node* is often used to refer to a circuit connection point such as the connection of the output of gate N to the input of gate R. A circuit node which does not change when a changing signal is applied to it is commonly referred to as a *stuck node*. We will use the term node from now on so you get used to how it is used.

To troubleshoot the circuit in Figure 5-15a, first move the channel 2 scope probe from the output of gate R to the node at the output of gate N. The output node of gate N should be asserted (low) for input states 0010 (state 2) and 0110 (state 6). You can use the channel 1 display of the QA signal to count off states on this waveform as you did before to see if the N output goes low for these states. If it doesn't, then you know that either the inputs of gate N are not correct, gate N is defective, or the output of gate N is shorted to something that holds it high.

To check if the inputs to gate N are correct, you can again use channel 2 of your scope. First connect the channel 2 probe to pin 3 of gate N. The signal at this node is $\overline{A}$, so it should be high for states 0000 to 0111 and low for states 1000 to 1111. If this signal is not correct, suspect the inverter on the A signal or the line connecting the output of the inverter with the input of gate N.

If the $\overline{A}$ signal is correct at pin 3 of gate N, move the channel 2 scope probe to pin 4 of gate N. The signal on this node is simply the C signal, so, as shown in the truth table in Figure 5-15b, this signal should be low for states 0, 1, 4, 5, 8, 9, 12, and 13, and high for states 2, 3, 6, 7, 10, 11, 14, and 15. The waveform this produces, of course, is a square wave of one-half the frequency of the QA reference signal. If this signal is not correct on pin 4 of gate N, then the most likely cause is a broken connecting lead.

Finally, move the channel 2 scope probe to pin 5 of gate N. The signal at this node should be $\overline{D}$. Since you are using the D or QA signal on channel 1 of your scope as a reference, the waveform you see for the $\overline{D}$ signal should be just the inverse of this channel 1 signal. If the signal at pin 5 of gate N is not correct, suspect the inverter on the D signal or the connecting wires.

Once you narrow the problem down to a single IC or a small section of the circuit, you can use techniques such as visual inspection, lifting pins, and replacing ICs to find the cause of a stuck node or incorrect signal level.

As we said at the beginning of this section, a computerized ATE system can be programmed to apply a sequence of test states to the inputs of a circuit and to check if the output is correct for each input state. Each input combination and the expected output level for that combination are usually stored together as a binary word in one of the computer's memory locations. A binary word containing both an input combination and the expected output for that input combination is referred to as a *test vector*. Each line of the truth table in Figure 5-15b represents a test vector for the circuit in Figure 5-15a.

An ATE computer can be further programmed so that if the output is not correct for any input state, the computer checks the level at intermediate nodes in the circuit and prints out a message that identifies the gate or circuit section which is most likely to be causing the problem. Later in this book, we will be showing you how ATE is used with a variety of devices and systems.

USING A COMPUTER TO DRAW DIGITAL SCHEMATICS

Introduction

Electronic circuit drawings, or schematics, are usually drawn in one of five standard sizes. These sizes are identified with the letters A, B, C, D, and E. For reference, the actual dimensions of these five are as follows:

Size	Height, in	Width, in
A	8.5	11
B	11	17
C	17	22
D	22	34
E	34	44

Up until a few years ago, digital designers drew schematics with mechanical pencils and plastic *logic templates* such as that shown in Figure 5-22a. Now we almost always use an *engineering workstation* type of microcomputer, such as the Apollo DN4500 shown in Figure 5-22b, to draw schematics. Using a computer to draw schematics has the same advantages over hand drawing that using a word processor has over using a standard mechanical typewriter. Since the schematic is drawn on the computer screen, you don't have to erase anything on paper. You can move symbols around on the screen, change connecting wires, add or delete symbols, and print out the result on a printer or plotter when it looks just the way you want it to. If you change your mind about some part of the schematic, you can just edit the drawing on the screen and do a new printout.

A further advantage of the computer-aided drafting approach is that you usually don't even have to draw the symbols! Most electronic drafting programs have large library files containing common logic device symbols, complete with pin numbers, on disk. All you have to do when you want to put a particular IC on a schematic, then, is to pull the symbol for it from a library file. Once you have the IC symbols for a circuit on the screen, you can draw the con-

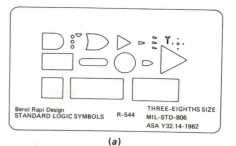

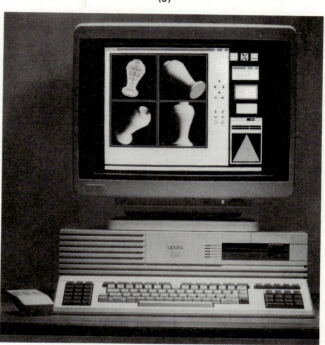

FIGURE 5-22 *(a)* Logic template. *(b)* Apollo DN4500 workstation. *(Courtesy Apollo Computers Inc.)*

necting wires between them, then add junctions, connectors, labels, etc., to complete the drawing.

Most computer drafting programs allow you to use the keys on the computer keyboard to draw schematics, but a faster and more convenient way is to use a device called a *mouse*. Figure 5-22b shows a mouse on the table next to the Apollo workstation.

There are two common mouse types. A mechanical type mouse has a large ball bearing on the bottom. As you move the mouse around on a table, the ball rotates and causes the cursor to move around on the screen of the computer. An optical mouse contains an LED and a phototransistor, which detect when the mouse is moved over grid lines on a special plate. As the mouse is moved around on the plate, the cursor moves around on the screen of the computer. Buttons on the top of each type of mouse are used to select items from a menu displayed on the screen or to enter commands.

Programs used to draw electronic schematics are commonly called *schematic capture* programs. Schematic capture programs are available for most computers. The Ideaware programs from Mentor Graphics run on Apollo engineering workstations such as the one shown in Figure 5-22b. Systems and software such as these are used for designing large, complex digital systems or ICs. For smaller projects we often use a microcomputer such as an IBM PC/AT or a Macintosh. Schematic capture programs for IBM-type PCs include OrCAD/SDT III's Draft from OrCAD Systems Corporation, Schema II from OMATION, and EE Designer III from Visionics. Schematic capture programs available for the Macintosh include Schematic from Douglas Electronics Inc. and LogicWorks from Capilano Computing Systems Ltd.

In the following sections we show you how you use one of these programs to draw simple schematics. The program that we use as an example here is the OrCAD/SDT III Draft program from the OrCAD Systems Corporation. Most of the schematic capture programs are quite similar in how they operate, so if you have some other one, you should have little difficulty adapting to it.

Overview of Menus

Schematic capture programs typically have a great many different commands you can use to help you draw schematics. To avoid your having to memorize all these commands, they are arranged in menus which you can display on the screen any time you need to select a command. If the commands were all listed in one menu, the menu would be so long that it would be tedious and time consuming to scroll through to the command you wanted. Therefore, the commands are grouped in a series of menus and submenus as shown in Figure 5-23. The major command types are listed in a menu called the *MAIN menu*. Each of the entries in the MAIN menu leads to a submenu which contains the specific commands of the type that you can choose. Each item in the submenu leads to a still lower level menu which gives you the options available for that command.

As a first example of how this works, suppose that you want to set the drawing size for a drawing you are going to make. You first click the left key on the mouse to get a display of the MAIN menu. You then use the mouse to move a highlighted box to the SET entry in the MAIN menu and click the left button on the mouse again. As soon as you click the mouse key, the SET menu will appear on the screen. You next move the mouse around to position a highlighted box over the WORKSHEET SIZE line in this menu and click the left key on the mouse again. Clicking the key this time will display a list of the choices you can make for the size of the worksheet. To select one of these, you use the mouse to move the highlighted box to the desired drawing size in the list and click the left key on the mouse again. Clicking the mouse key this time will set up the program for the desired sheet size and return you to the MAIN menu level.

The point of the preceding example was to give you an overview of how you work with the multilevel command menus which most systems now use. Basically, you move the mouse to choose a menu or a command, and then click the left button on the mouse to select the menu or execute the command. With a little practice you can quickly find and execute commands in this type of menu

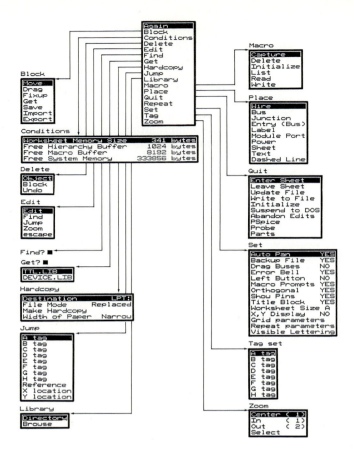

FIGURE 5-23   Menu system for OrCAD/SDT III Draft schematic capture program.

structure. To exit a menu, click the right button on the mouse.

In the following sections we show you how we used OrCAD/STD III's Draft program to draw the schematic in Figure 5-17 and introduce you to some of the terms commonly used in schematic capture programs. Exercises in the accompanying laboratory manual provide practice in using a schematic capture program.

A Schematic Capture Example

INITIAL SETUP

The following discussions assume you have an IBM PC type computer with a Hercules monochrome graphics adapter, a hard disk, and a mouse. Also assumed is that all of the programs and library files are installed on the hard disk in a directory named C:\ORCAD, as described in the OrCAD/SDT III manual. To start the Draft program running, you boot up the system, change to the ORCAD directory, and type:

draft fig5-17.sch

and press the carriage return (enter) key. After you press the carriage return twice more, the computer will clear the screen, show a cursor arrow on the screen, and display the message ——— New Worksheet \\\ at the top of the screen.

Before you can start drawing, you have to set some parameters to different values from their default values. The first of these is to tell Draft to display the X-Y coordinates of the cursor at the upper right corner of the screen. To do this, click the left key on the mouse to display the MAIN menu. Then use the mouse to move the highlighted box to the SET command in the menu and click the left button on the mouse again. As soon as you click the mouse key, the SET menu will appear on the screen. You next move the mouse around to position the highlighted box over the X,Y DISPLAY command in the SET menu and click the left mouse key again. A YES/NO menu will appear. The highlighted box is already on the YES command, which is the choice you want, so just click the left mouse key again to enter this choice. The 0,0 position for the X-Y coordinates is the upper left corner of the drawing sheet. X increases as you go to the right across the sheet, and Y increases as you go down the sheet. If you move the mouse around the screen, you will see a change in the coordinates displayed at the top right of the screen.

Clicking the left mouse key again will return you to the MAIN menu. Select AGAIN (a shortcut for repeating your last MAIN menu selection) to return to the SET menu so you can tell Draft to display a grid of dots on the screen. These dots aren't necessary, but we have found that they make it much easier to line devices up neatly on the drawing. Use the mouse to move the highlighted box to the GRID PARAMETERS line of the SET menu and click the left mouse key. The GRID PARAMETERS submenu will now be displayed. Use the mouse to move the highlighted box to the VISIBLE GRID DOTS command and click the left mouse key again. When the YES/NO menu appears, click the left mouse key twice to enter the YES command and return to the MAIN menu. The screen should now look like Figure 5-24. The dis-

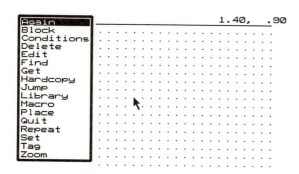

FIGURE 5-24 OrCAD's Draft screen showing grid dots and grid parameters.

tance between dots on the screen represents 0.1 in on the actual drawing. The next step is to start making the drawing.

GETTING PARTS FROM A LIBRARY

For most of the parts you will want to put on a schematic, you don't have to draw them out line by line, you can simply pull the symbol for the part from a library of devices stored on the hard disk. To get a part from a library, you use the mouse to move the highlighted box to the GET command in the MAIN menu and click the left mouse key. Draft will respond with GET? at the top of the screen.

If you know the exact number of the part you want the symbol for, you can use the keyboard to type in the identifying number for the part. In the schematic you want to draw here, for example, you need a 74LS93, so you could just type in 74LS93 followed by a carriage return.

If you do not know whether a device is available in a particular logic family, you can respond to the GET? prompt by clicking the left mouse key. Draft will respond with a list of available libraries. Since you want a TTL numbered device, move the highlighted box to the TTL.LIB and click the left mouse key again. Along the left side of the screen, you will now see a list of numbers starting with 00. These numbers represent the last two digits of the identifying numbers of the devices in the TTL library. The 00, for example, represents all the devices which are identified as 74XX00. Use the mouse to scroll down the list to 93 and click the left mouse key again. You will now see a list of the available 74XX93 type devices. Move the highlighted box to the 74LS93 line and click the left mouse key. The schematic symbol for a 74LS93 should now appear on the screen as shown in Figure 5-25. Note that the symbol shows both the IC pin numbers and the names of the pins. The next task is to place the symbol in a reasonable position on the drawing.

In its default mode the screen only shows a portion of the actual sheet. This often makes it somewhat difficult to position ICs on the sheet. To overcome this problem you can use the zoom feature. To see how this works, click the left mouse key to put the PLACE menu on the screen and use the mouse to move the highlighted box to the ZOOM line. When you click the left mouse key again, the ZOOM menu should appear. You want to zoom out to see more of the sheet, so move the highlighted box to the OUT command in this menu and click the left mouse key again. Most of the sheet should now be visible on the screen. If you want to zoom out further on this sheet or if you are working on larger sheets, you can repeat the zoom operation.

Now that you can see nearly the entire sheet, you can use the mouse to move the symbol around to where you want it on the sheet. The X,Y coordinates at the top of the screen tell you where the symbol is on the sheet at

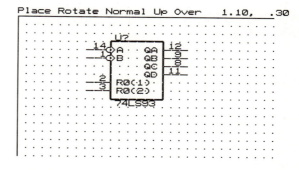

FIGURE 5-25 Draft screen showing schematic symbol for 74LS93.

any time. The displayed coordinates actually represent the upper left corner of the symbol. We initially chose to position the 74LS93 at coordinates 1.4, 3.00.

Once you decide where you want to put a symbol, you need to tell Draft to hold it in that position. You do this by clicking the left mouse key to display the PLACE menu, moving the highlighted box to the PLACE command, and clicking the left mouse key again.

The next step is to get the inverter symbols you need for the next part of the drawing. Click the right mouse key to leave the PLACE menu and then click the left mouse key to display the MAIN menu. Use the mouse to select the GET command of this menu and click the left mouse key. Draft will respond with the GET? prompt. In response to this prompt, type 74LS04 and press the carriage return key. When this command executes, a square box will appear on the screen. After a few seconds the box will change into the traditional inverter symbol. Use the mouse to position the inverter symbol where you want it on the sheet. Then use the PLACE command to hold the symbol in that position as you did with the 74LS93.

You can put the other two inverter symbols on the sheet without having to go to the library again. If you move the mouse after you place the first inverter, another inverter symbol will move around on the screen. Just move this symbol to where you want it and place it there. Move the mouse around again to position the third inverter symbol and place it. Click the right mouse key to leave the PLACE menu and click the left mouse key to return to the MAIN menu.

The next step is to get and place four 74LS10 symbols. You do this by using the GET command and the PLACE command as you did previously for the inverters. You now have all the basic logic symbols on the drawing as shown in Figure 5-26. In the next section we tell

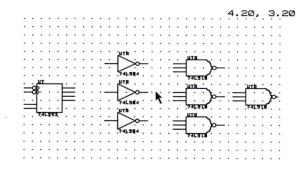

FIGURE 5-26 Zoom-out of Draft screen showing symbols for Figure 5-17.

you how to delete a symbol, move a symbol, or change a symbol to its alternative form.

DELETING SYMBOLS

Suppose that in placing the NAND gates in the previous step you accidentally put in an extra NAND gate. You can delete the extra gate as follows. Click the left mouse key to display the MAIN menu, select the DELETE line in the MAIN menu, and click the left mouse key again.

The DELETE menu gives you the choices OBJECT, BLOCK, and UNDO. Select OBJECT and use the mouse to move the on-screen arrow into the gate symbol you want to delete. Click the left mouse key twice to delete the device. If you want to delete another device, you can just move the arrow to that device and click the left mouse key twice again. To exit the DELETE menu, just click the right mouse key.

MOVING SYMBOLS

Now, suppose that you placed the last NAND gate and you decide that on second look that you don't like where you put it. To move a symbol, select the BLOCK line from the MAIN menu to display the BLOCK menu. The options in this menu allow you to MOVE a block, EXPORT a block to a file on a disk for future use, IMPORT a block from a disk file, etc. Select the MOVE command and click the left mouse key.

To select the block you want to move, position the on-screen arrow in the upper left-hand corner of an imaginary box containing the gate or parts you want to move. Click the left mouse key twice. If you now move the mouse down and to the right, you should see a dotted box appear. Move the mouse so that the box encloses the parts you want to move as shown in Figure 5-27. When you

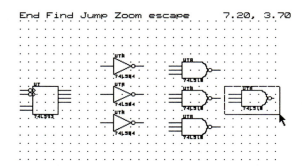

FIGURE 5-27 Defining a block on the Draft screen.

have the block outlined correctly, click the left mouse key twice again. You can now use the mouse to move the marked block around on the screen. When the block is in the correct position, click the left mouse key twice to place it there.

NOTE: The DELETE command allows you to select and delete a block in a similar manner.

CHANGING A SYMBOL TO ITS DE MORGAN EQUIVALENT FORM

To make the logic performed by the output NAND gate in Figure 5-27 clearer, you want to represent it with its *De Morgan equivalent*, an OR symbol with bubbled inputs. To do this, you select the EDIT line from the MAIN menu. Put the arrow in the symbol and click the left mouse key twice more to display the EDIT menu. Select the ORIENTATION line in this menu and click the left mouse key once more to get to the ORIENTATION menu. The commands available in this menu allow you to point a symbol UP, point a symbol DOWN, produce the MIR-ROR image of a symbol, or CONVERT a symbol to its De Morgan equivalent. For the task you want here, select CONVERT. At this point your drawing should look like the one in Figure 5-28.

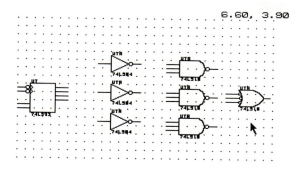

FIGURE 5-28 Symbol on schematic changed to De Morgan equivalent.

DRAWING SIGNAL LINES

Once you have placed all the gate symbols on the drawing, the next step is to place the connecting wires in the drawing. To do this, select the PLACE line in the MAIN menu. In the PLACE menu select WIRE. Move the on-screen arrow to a point where you want to start a wire, and click the left mouse key twice. Use the mouse to draw the wire to where you want it to turn or end.

If you want the wire to turn, click the left mouse key twice and draw the wire in the new direction with the mouse. You can keep repeating this sequence until you reach the point where you want the wire to end.

When you want to end a wire, you click the left mouse key once to display the WIRE menu. Select the END command from this menu if you don't need to draw any more lines. Select the NEW command from the menu if you still have lines to draw. Click the left mouse key to execute the command. To draw more lines, just repeat the procedure of moving the arrow around with the mouse and clicking the left key twice to change direction. When you have all the basic wires drawn, your screen should look something like Figure 5-29.

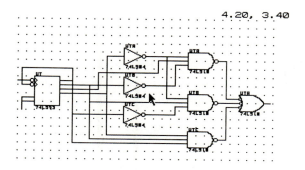

FIGURE 5-29 Schematic with connecting wires in place.

To remove an unwanted line, select the DELETE line in the MAIN menu, put the on-screen arrow at one end of the line you want to delete, and click the left mouse key twice. Click the right mouse key to leave the DELETE menu.

RUBBERBANDING—DRAGGING A BLOCK

Suppose that your drawing is to the point shown in Figure 5-29 and that you decide the output NAND gate is too close to the right-hand side of the sheet to leave space for the output identifier. You can use the BLOCK DRAG command to move the gate and the connecting wires all at once.

First you have to tell Draft to drag the buses. To do this, from the MAIN menu select the SET menu and then choose the DRAG BUSES command. To move the gate, you select BLOCK from the MAIN menu. In the BLOCK menu choose the DRAG option and position the on-screen arrow in the upper left corner of an imaginary box which includes the gate and parts of the wires you want to move. Click the left mouse key twice to set this beginning point; then move the mouse around to form a box around the gate as shown in Figure 5-30a. Click the left mouse key twice more to fix the block size. Then use the mouse to move the block to where you want it. When the block is where you want it, click the left mouse key twice to place it. Note that as shown by the slanted lines in Figure 5-30b, the wires are dragged along with the gate. Dragging wires along in this way is commonly called *rubberbanding*.

For the finished drawing we want all the wires to be at right angles, or *orthogonal*. You can make the connecting wires orthogonal with the BLOCK FIXUP command. To do this, select BLOCK from the MAIN menu. Select FIXUP from the BLOCK menu and use the mouse to move the arrow to the end of a wire where segments need to be added or removed to make the wire vertical or horizontal. Click the left mouse key twice more to select the PICK menu. In this menu select the DRAG ALL command. You will now be able to move the arrow and a dotted line connected to it back to a position where the wire is vertical as shown in Figure 5-30c.

To enter this correction, click the left mouse key to reach the menu which contains the DROP and END commands. Select the END command to clean up the drawing. Some dangling wires will remain; you can remove these with the DELETE command. The point of rubberbanding is that it lets you move a part without losing all the connections and having to figure them out again.

MAKING JUNCTIONS BETWEEN WIRES

As when you manually draw a schematic, lines which simply cross each other are not connected. The usual way of indicating a connection between two wires is with a dot. To make junctions on your drawing, select PLACE from the MAIN menu and then select JUNCTION from the PLACE menu. Move the on-screen arrow to a point where you want to make a junction. Click the left mouse key twice to place the junction, then move the arrow to the next point where you want a junction and repeat the process. Click the right mouse key to leave the menu.

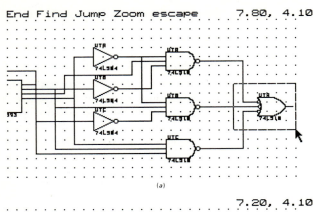

(a)

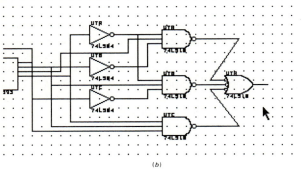

(b)

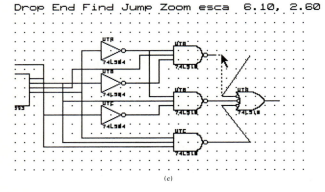

(c)

FIGURE 5-30 Dragging a block using rubberbanding. *(a)* Defining the block. *(b)* Block after being dragged. *(c)* Partial fixup of dragged block.

CONNECTING GATE INPUTS TO V$_{CC}$ OR GROUND

The V$_{CC}$ and ground power pins on all devices in the OrCAD libraries are automatically connected to V$_{CC}$ and ground when a program called NETLIST generates a wiring list, or *netlist,* as it is often called. However, the signal inputs R0(1) and R0(2) on 74LS93 must both be connected to ground for the device to work correctly.

You use the GET command to retrieve the GND POWER symbol from a library file called DEVICE.LIB in the same way you retrieved inverters from the library file TTL.LIB. Once the ground symbol is in position on the worksheet, you place it as you did other parts.

You can connect a gate input to V$_{CC}$ by selecting PLACE from the MAIN menu and selecting POWER from the PLACE menu. Move the on-screen arrow to the point you

want connected to V$_{CC}$ and click the left mouse key twice to place the connection.

EDITING THE LABELS ON A DRAWING

For positioning gates and wires on the sheet, you zoomed out so that you could see the entire sheet at once. To edit the labels on the drawing, you need to zoom in so that you can see the details of each symbol. To do this, select ZOOM in the MAIN menu and in the ZOOM menu select IN. Click the left mouse key twice to do the zoom and return to the MAIN menu.

In the MAIN menu select EDIT. Move the on-screen arrow to the first symbol on which you want to edit the reference label and pin numbers. Click the left mouse key twice for the EDIT PART menu. In this menu, select NAME and then select REFERENCE. A line at the top of the screen prompts you to type in the reference number you want associated with that gate or device. A logical designation for the 74LS93, for example, is U1, so use the keyboard to delete the ?, then type a 1 and press the carriage return. The reference name, U1, will then appear over the device.

As you may or may not remember from previous discussions, a 74LS04 contains six inverters. Now, if you look at your drawing in its current state, you will see that all of the inverters have the same pin numbers. For these devices you must edit the reference designation and the pin numbers.

Edit the reference designation as you did for the 74LS93 above. If you identify the top inverter as U2, Draft will automatically add an A to this so the part becomes identified as U2A. To get the correct pin numbers on an inverter, you select the EDIT PART menu, choose WHICH DEVICE, and a list containing numbers 1 through 6 will be displayed on the screen. For the top inverter, select 1 and click the left mouse key.

When you finish editing one part, click the right mouse key to leave that part. Move the arrow to the next part and click the left mouse key twice to reach the EDIT PART menu for that part. Repeat the process until all the devices have correct labels and pin numbers.

NOTE: The ORIENTATION option in the EDIT PART menu can be used to convert symbols to their alternative form. As you see in the finished drawing in Figure 5-17, we used this option to produce the alternative symbols for the inverters.

ADDING INPUT AND OUTPUT LABELS

The final item we want to tell you is how to add input and output labels to your drawing. These are referred to as *module ports.*

To add these to your drawing, select PLACE in the MAIN menu, and then select MODULE PORT. You will be prompted at the top of the screen to type in the name you want to give to that input or output. As an example of this, you might give the name CLOCK-IN to the signal that will be connected to pin 14 of the 74LS93. After you type in this name followed by a return, you will be prompted to indicate whether this is an input or an output. Select INPUT and click the left mouse key. On the screen you will then see an arrow-shaped box containing this label. You can use the mouse to position this box where you want it, and then use the PLACE command to put it there.

MAKING A PRINTOUT OF THE DRAWING

To make a printout, or hard copy, of your finished drawing, select the HARDCOPY option from the MAIN menu then choose the MAKE HARDCOPY command.

SAVING A DRAWING TO DISK AND LEAVING DRAFT

To save your drawing on the hard disk, select QUIT from the MAIN menu. In the QUIT menu select UPDATE FILE. If you want to return to editing after you make a backup of your drawing on disk, press the right mouse key. If you are all done editing, select the ABANDON EDIT command in the QUIT menu.

NETLISTS AND PRINTED-CIRCUIT BOARDS

The preceding sections have shown you how you can use a computer and a schematic capture program to produce a drawing of a digital circuit. In addition to being used to produce a hard copy of the drawing, the file you create with the schematic capture program can be used for many other purposes.

The first step is to process the drawing file with a program called CLEANUP which cleans up errors such as duplicate symbols, overlapped lines, etc. This step is analogous to checking the output from a word processor with a spelling checker program.

The second step is to process the drawing file with a program called Electrical Rules Check (ERC), which checks for wiring errors such as two outputs connected together, an output connected to V_{CC}, etc. This check is analogous to checking the output from a word processor with a syntax checker.

Once all the errors have been removed from the file, a program called NETLIST is used to produce a wiring list from it. The wiring list, commonly called a *netlist*, can be used with another program to develop a printed-circuit board for the circuit. A program called PARTLIST will make a list of all the parts in your circuit.

After processing by another program, the file produced by Draft can be used to *simulate* the operation of the circuit. The term simulate here means to run a program which predicts the levels that will be produced on the output(s) of a circuit when specified signals are applied to the input(s). Simulation is a way of testing the logic and timing of a circuit without actually building the circuit and testing it with signal generators and an oscilloscope. In Chapter 15 we show you how to use computer programs to simulate the operation of a device or circuit.

The file produced by Draft may also be used to develop a printed-circuit board for the circuit or a custom IC which implements the function performed by the circuit. In Chapter 15 we will also tell you more about how this is done.

CHECKLIST OF IMPORTANT TERMS AND CONCEPTS IN THIS CHAPTER

Positive logic
Functional or mixed logic
Combinational logic circuits
Boolean expression
Sum-of-products
Testing and troubleshooting circuits
Odd and even parity
Open-collector circuit
Automatic test equipment (ATE)
Binary counter
Word generator
State, state decoder
Word recognizer
Node
Stuck node
Test vector
Logic template
Engineering workstation
Mouse
Schematic capture
Library of parts
Rubberbanding
Orthogonal
Module ports
Netlist
Simulate

REVIEW QUESTIONS AND PROBLEMS

1. Construct the truth table and determine the output expression for the circuit in Figure 5-31.

2. For the circuit in Figure 5-32:
   a. Write a truth table for each intermediate point and one for the final output.
   b. Write a Boolean expression for each intermediate point and one for the final output.

3. For the circuits in Figure 5-33:
   a. Write the Boolean expression for the output.
   b. Use the Boolean expression to help you write the truth table for each circuit.

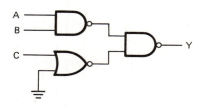

FIGURE 5-31   Circuit for Problem 1.

4. a. Redraw the logic circuits in Figure 5-34a and b using alternative logic symbols which make the

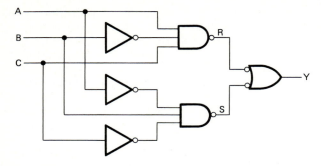

FIGURE 5-32   Circuit for Problem 2.

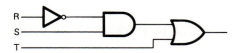

FIGURE 5-33   Circuit for Problem 3.

operation performed by the circuit more easily determined.

b.   Write the Boolean expression for each of the redrawn circuits.

c.   Use the Boolean expressions to help you write a truth table for each of the circuits.

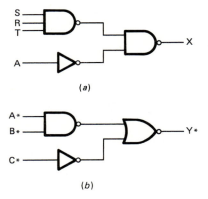

(a)

(b)

FIGURE 5-34   Circuits for Problem 4.

5.   Write the truth table and the Boolean expression for each logic circuit shown in Figure 5-35.

6.   Write the truth table and Boolean expression for each mixed logic circuit shown in Figure 5-36.

7.   Show the truth table and the output expression for the circuit in Figure 5-37.

8.   Write the Boolean expression and the truth table for the Exclusive OR circuit in Figure 5-38.

9.   For the Exclusive OR circuit in Figure 5-39:

a.   Write the Boolean expression for the output.

b.   Write a truth table.

c.   Describe what this circuit output tells you about the 4-bit binary word applied to its inputs.

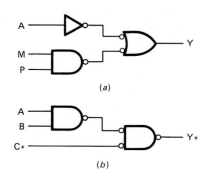

FIGURE 5-35   Circuits for Problem 5.

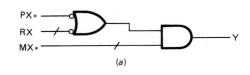

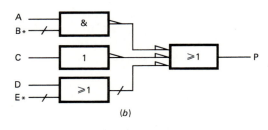

FIGURE 5-36   Circuits for Problem 6.

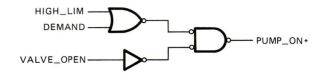

FIGURE 5-37   Circuit for Problem 7.

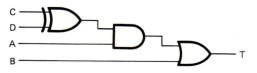

FIGURE 5-38   Circuit for Problem 8.

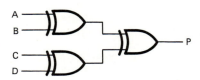

FIGURE 5-39   Circuits for Problem 9.

10. For the Exclusive NOR circuit in Figure 5-40:
   a. Write the Boolean expression for the output.
   b. How many entries would the truth table for this circuit have?
   c. What information does this circuit give you about the two 3-bit binary words, A2 A1 A0 and B2 B1 B0, applied to its inputs?

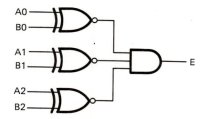

FIGURE 5-40   Circuit for Problem 10.

11. Write the truth table and Boolean expression for the open collector wired OR circuit in Figure 5-41.

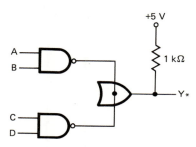

FIGURE 5-41   Circuit for Problem 11.

12. a. Write the output expression for the open-collector circuit in Figure 5-42.
   b. Draw the symbol for a single gate which performs the same logic function.

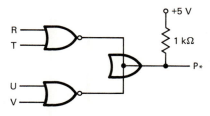

FIGURE 5-42   Circuit for Problem 12.

13. For the circuits in Figure 5-43a and b, show the output waveforms that will be produced by the given input waveforms shown in Figure 5-43c. Note that the output must obey the truth table for the circuit at each instant. Using this fact, you can construct the output waveforms a section at a time under the input waveforms.

14. Show the waveform that will be produced on the output of the circuit in Figure 5-44a when the

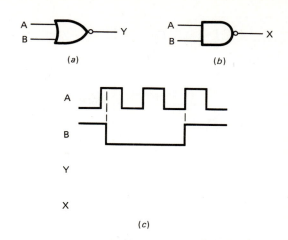

FIGURE 5-43   Circuits and waveforms for Problem 13.

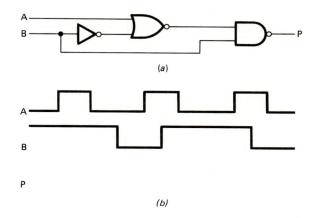

FIGURE 5-44   Circuit and waveforms for Problem 14.

waveforms shown in Figure 5-44b are applied to the inputs. Note that the output must obey the truth table for the circuit at each instant. Using this fact you can construct the output waveform under the input waveform a section at a time under the input waveforms.

15. The TTL logic block shown in Figure 5-45 is not working correctly. A technician measures the input and output voltages for each gate. The results are shown in Figure 5-45.
   a. Where is there a problem in the circuit?
   b. Give two possible causes for the problem.
   c. Describe how you would attempt to solve the problem.

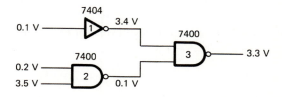

FIGURE 5-45   Circuit for Problem 15.

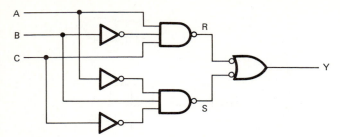

FIGURE 5-46   Circuit for Problem 16.

16. You find that node S in the circuit shown in Figure 5-46 is stuck high. Describe how you would find the cause of the problem.

17. a. Write a truth table for the circuit in Figure 5-47a.
    b. Draw the waveform that will be produced on the output of this circuit if the binary count sequence waveforms shown in Figure 5-47b are applied to the inputs of the circuit as shown. Assume a word recognizer such as that in Figure 5-19 is connected to the circuit and is set to trigger the scope when state 0000 is on the inputs of the circuit.
    c. Suppose that you observe the QA waveform and the output waveform with a dual-trace scope

and you find that the output 1 for state 0111 is missing. What part of the circuit would you first check out to find out why this 1 is missing?

18. What letters are used to represent the standard sheet sizes for schematic drawings?

19. What are the major advantages of using a computer and a schematic capture program to draw schematics?

20. With reference to a schematic capture program:
    a. What are the uses of a mouse?
    b. What is the purpose of the library files which are included with the basic schematic capture program?

21. In general terms:
    a. How would you get the symbol for a digital device such as a 74HC20 at coordinates 2.00, 4.50 on a drawing?
    b. How would you move a section of a circuit drawing to a different location on a sheet?
    c. How would you remove an unwanted section of a circuit?

22. Briefly explain what is meant by the term *simulation*.

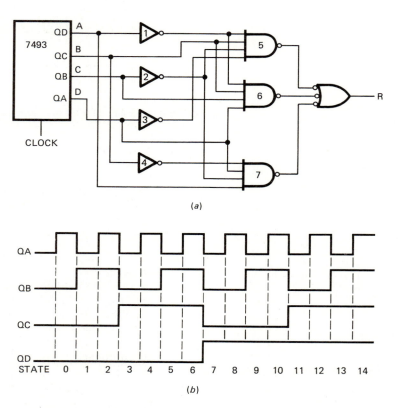

(a)

(b)

FIGURE 5-47   Circuit and waveforms for Problem 17.

6 Designing and Implementing Combinational Logic Circuits

In Chapter 5 we showed you how to analyze, trouble-shoot, and use a computer to draw the schematics for combinational logic circuits. Now we will show you how to design combinational logic circuits which solve some stated problems, and how to implement the designs with common logic gates. The major steps in this design and implementation process are as follows:

1. Carefully define the problem that the circuit is meant to solve, the available input signals, and the needed output signals.

2. Write a truth table which shows the desired relationship between the input variables (signals) and the output(s).

3. Use Boolean algebra or Karnaugh maps to produce a simplified Boolean expression for the circuit.

4. Choose the gate devices needed to implement the simplified expression.

5. Build the circuit and test it as described in Chapter 5.

OBJECTIVES

At the conclusion of this chapter you should be able to:

1. Given the truth table for a logic function, write the Boolean expression for the truth table in sum-of-products form.

2. Use Boolean algebra to simplify a logic expression.

3. Use a Karnaugh map to simplify a logic expression with up to four variables.

4. Given a written description of a logic problem, develop a truth table which represents the function.

5. Use commonly available gates to implement a specified logic function.

DEFINING THE PROBLEM AND WRITING A TRUTH TABLE

As we said above, the first step in designing a combinational logic circuit is to carefully define the problem you

want the circuit to solve. Included in this is determining the input signals available as well as the required output signals.

For our first design example, we will use an automobile seat belt warning system which functions as follows. We want a buzzer to sound for 15 s when there is weight on the driver's seat, the ignition key is in the ignition, and the seat belt is not fastened. Also, to get double use out of the buzzer and sensors, we want the buzzer to sound when there is no weight on the seat and the key is in the ignition switch. This second function warns the driver that he or she has left the car with the key still in the ignition switch.

The input variables for this system are W, weight in the seat; F, the seat belt fastened; and K, the key in the ignition. We will identify the output variable, buzzer on, with a B.

The next step in the design process is to make a truth table showing the relationships you want between the input signals and the output(s). Figure 6-1 shows a pos-

W	F	K	B	MINTERMS	MAXTERMS
0	0	0	0		W + F + K
0	0	1	1	$\overline{W}\,\overline{F}\,K$	
0	1	0	0		$W + \overline{F} + K$
0	1	1	1	$\overline{W}\,F\,K$	
1	0	0	0		$\overline{W} + F + K$
1	0	1	1	$W\,\overline{F}\,K$	
1	1	0	0		$\overline{W} + \overline{F} + K$
1	1	1	0		$\overline{W} + \overline{F} + \overline{K}$

FIGURE 6-1 Truth table and Boolean expressions for seat belt problem.

sible truth table for this seat belt warning system. The ON, or TRUE, condition for each variable is represented by a 1 in this truth table. Think your way through this truth table to see if you agree with the combinations for which we have decided to sound the buzzer.

Once you have a truth table for a circuit, the next step in the design process is to develop a Boolean expression which will require a minimum number of gates to implement. Assuming that the circuit is simple enough that you are not going to implement it directly in a programmable logic device such as a PAL, there are two major ways to develop a simplified expression from a truth table. One way is to write a Boolean expression from the truth table and use the rules of Boolean algebra to simplify the expression. The second way is to write the truth table in an alternative form called a *Karnaugh map* and extract a simplified expression directly from the map. In the following sections, we will show you how to use each of these techniques to produce a simplified expression for the seat belt warning system and for other combinational logic circuits.

Writing a Sum-of-Products Expression from a Truth Table

Boolean expressions derived from truth tables are usually represented in a *sum-of-products* form or in a *product-of-sums* form. In this section and the next we show how to derive these two standard forms. We will start with the sum-of-products form.

When writing an expression for a truth table, you can write the expression based on the input conditions that will cause the output to go high or an expression based on the input conditions that will cause the output to go low. For the seat belt warning system the output signal, B, is defined as active high, so we will write an expression for the conditions which cause the output to go high.

The first step in doing this is to write the term for each of the input combinations which produces a 1 in the output column, as shown in the column labeled MINTERMS in Figure 6-1. As you can see, the expressions for the three output 1's are $\overline{W}\overline{F}K$, $\overline{W}FK$, and $W\overline{F}K$. Each of these terms is commonly referred to as a *minterm*, or *product term*. By definition, a minterm contains each of the input variables in its true or complemented form, but not both forms.

The Boolean expression for the entire truth table can be made by logically summing (ORing) all of the product terms that produce a 1 on the output. For the truth table in Figure 6-1, then the complete Boolean expression is:

$$B = \overline{W}\,\overline{F}K + \overline{W}FK + W\overline{F}K$$

The resultant expression is referred to as a *sum-of-products*, or a *sum-of-minterms*, expression. This is the form in which you will usually derive Boolean expressions from a truth table.

A shorthand often used in engineering texts to represent an expression such as this is:

$$B = f(W, F, K) = \Sigma m(1, 3, 5)$$

What this expression tells you is that B is a *function* (*f*) of input variables W, F, and K, and that B is asserted for the *sum* (Σ) of *minterms* (m) 1, 3, and 5. The 1, 3, and 5 here represent the decimal equivalent numbers for the input combinations of W, F, and K that produce a 1 on the B output.

Writing a Product-of-Sums Expression from a Truth Table

In the preceding section we showed you how to write the minterm or product terms which produce a 1 in the output column of a truth table, and how to sum these terms to produce a sum-of-products expression for the truth table. An alternative approach is to write a product-of-sums expression for the truth table.

The first step in doing this is to identify each input combination that produces a 0 on the output. For each of these input combinations, you then write what is called a *sum term*, or *maxterm*. The maxterms for the seat belt warning system are shown in the far right column of Figure 6-1. You produce a maxterm by inverting the input variables for that combination and then ORing them together. In the truth table in Figure 6-1, for example, the input combination 0,0,0 for $\overline{W}$, $\overline{F}$, $\overline{K}$ produces a 0 on the output. Inverting each of these and ORing them together gives W + F + K as the maxterm for this input combination. As another example, the input combination 110 or $WF\overline{K}$ also produces a 0 on the output. Inverting each variable in this combination and ORing the result gives $\overline{W} + \overline{F} + K$ as the maxterm for this combination.

The maxterm Boolean expression for the complete truth table is formed by ANDing (multiplying) all the maxterms. For the example here, the result of this is:

$$B = (W + F + K)(W + \overline{F} + K)(\overline{W} + F + K)$$
$$(\overline{W} + \overline{F} + K)(\overline{W} + \overline{F} + \overline{K})$$

As implied by the method used to form it, a Boolean expression in this form is referred to as a product-of-sums form. A shorthand commonly used to represent product-of-sums expressions such as this is:

$$B = f(W, F, K) = \Pi M(0, 2, 4, 6, 7)$$

The *f*(W,F,K) in the shorthand form tells you that B is a function of input variables W, F, and K. The Π and the M in the right-hand expression tell you that B will be asserted for the product of maxterms (M) 0, 2, 4, 6, and 7.

Introduction to Implementing Boolean Expressions

Since the sum-of-products expression and the product-of-sums expression we derived above represent the same truth table, they are equivalent. Each can easily be implemented with common logic gates. In most cases, however, the sum-of-products expression is easier to understand and easier to implement. Therefore, we will concentrate mostly on it.

The sum-of-products expression $B = \overline{W}\overline{F}K + \overline{W}FK + W\overline{F}K$ consists of three terms ORed together, so the output gate needs to be a 3-input OR gate. The three terms that are being ORed together each consist of three variables being ANDed together, so a 3-input AND gate is required to produce each of these terms. The variables W and F need to be complemented for some of the inputs to the AND gates, so two inverters are also needed. Figure 6-2 shows how these gates can be connected in a circuit which will directly implement the sum-of-products expression for B.

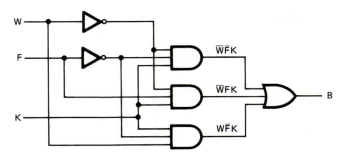

FIGURE 6-2   One solution to the seat belt problem.

The circuit in Figure 6-2 will perform the desired logic function, but it is not the minimum circuit which will do the job. In other words, it is possible to build a circuit which uses less gates and uses simpler gates to perform the same logic function. The two most common manual techniques used to determine the minimum solution for a given logic function are Boolean algebra and Karnaugh maps. In the following sections we show you how to work with each of these.

BOOLEAN ALGEBRA

Introduction

Throughout this book we have used Boolean expressions to represent the functions performed by logic gates. A set of rules referred to as *Boolean algebra* can be used to simplify and change the form of Boolean expressions. Figure 6-3 shows the major rules of Boolean algebra. The rules are valid for any number of variables, but for simplicity, they are shown here with the minimum number needed to demonstrate each rule. We will discuss each of the rules in Figure 6-3 using truth tables and logic diagrams, and then we will show you how they are used to simplify a variety of logic expressions. Carefully work your way through these Boolean algebra rules, because some of them are different from the rules for ordinary algebra.

Basic Boolean Identities

RULE 1

Rule 1, A AND $0 = A0 = 0$, can be thought of as telling you that if any input of an AND gate is low, the output will be low, regardless of the logic level on the other input(s). The truth table shown in Figure 6-4 shows you that this is indeed the case. Holding one input of an AND or NAND gate low in this manner is often referred to as *disabling* the gate, because any signal on the other input(s) will have no effect on the output.

RULE 2

Rule 2, A AND $1 = A1 = A$, can be explained with a gate diagram and truth table shown in Figure 6-5 as

MAJOR RULES OF BOOLEAN ALGEBRA

Basic Boolean Identities

*1. A AND $0 = A0 = 0$ *2. A AND $1 = A1 = A$

*3. A AND A = AA = A   *4. A AND $\overline{A} = A\overline{A} = 0$

*5. $A + 0 = A$ *6. $A + 1 = 1$

*7. $A + A = A$ *8. $A + \overline{A} = 1$

*9. $\overline{\overline{A}} = A$

Boolean Identities Pertaining to Order of Operations

10. $AB = BA$ 11. $A + B = B + A$

12. $(AB)C = A(BC)$ 13. $(A + B) + C = A + (B + C)$

Identities Showing Factoring of Boolean Expressions

*14. $AB + AC = A(B + C)$ 15. $(A + B)(A + C) = A + BC$

Identities Which Eliminate Unneeded Variables and Terms in Boolean Expressions

16. $A + AB = A$ *17. $A + \overline{A}B = A + B$

*18. $AB + \overline{A}B = B$ 19. $A(A + B) = A$

*20. $A(\overline{A} + B) = AB$

De Morgan's Theorem

*21. $\overline{AB} = \overline{A} + \overline{B}$ *22. $\overline{A + B} = \overline{A}\,\overline{B}$

*Rules used most often for simplifying Boolean expressions

FIGURE 6-3   Boolean algebra equivalents.

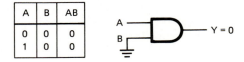

FIGURE 6-4   Truth table and logic diagram for rule 1.

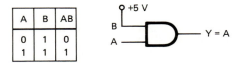

FIGURE 6-5   Truth table and logic diagram for rule 2.

follows. If one input of a 2-input AND gate, or all of the inputs but one of a multiple-input AND gate, are tied high, the output state will be the same as the state on the remaining input. For the 2-input example shown in Figure 6-5, the output will "follow" or, in other words, be the same as the signal on the A input. Making all but one input of an AND or NAND gate high, then, *enables* the gate to pass the signal applied to the remaining input.

RULE 3

Rule 3, AA = A, can be demonstrated by tying the inputs of an AND gate together as shown in Figure 6-6.

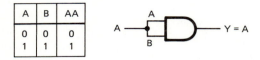

FIGURE 6-6   Truth table and logic diagram for rule 3.

Since both inputs of the AND gate then have to be the same, the output is a 0 for the case where A = 0 and B = 0. For the case where A = 1 and B = 1, the output is a 1. Therefore, the output is the same as, or "follows," the signal on the connected inputs.

RULE 4

Rule 4, $A\overline{A} = 0$, shows that if a variable is ANDed with its complement, the result is always a 0. From the logic gate diagram for this case in Figure 6-7, you can see that the inverter between the inputs will cause one input to always be low. If the input signal is a 0, then the top input of the AND gate will be low. If the input signal is a 1, the output of the inverter will apply a 0 to the lower input of the AND gate. According to the truth table for an AND gate in Figure 6-7, if one input is 0, the

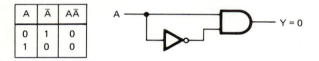

FIGURE 6-7   Truth table and logic diagram for rule 4.

output will always be 0. The circuit illustrating this rule is obviously not very practical, but the rule itself is useful for reducing Boolean expressions to simpler forms.

Rules 5, 6, 7, and 8 show the results that will be produced if a variable is ORed with a 0, with a 1, with itself, or with its complement. Compare each of these with the corresponding rule for the AND function which we discussed in the preceding paragraphs.

RULE 5

Rule 5, A + 0 = A. Figure 6-8 shows that if a variable is ORed with a 0, the result will always be the same as, or "follow," the variable. This rule tells you that a 2-input OR or NOR gate is *enabled* by applying a 0 to one input.

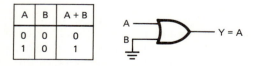

FIGURE 6-8   Truth table and logic diagram for rule 5.

RULE 6

Rule 6, A + 1 = 1. As shown by the truth table in Figure 6-9, this means that if one input of an OR gate is high,

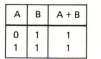 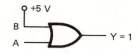

FIGURE 6-9   Truth table and logic diagram for rule 6.

the output will be a high regardless of the signal level on the other input(s). In practical terms then, rule 6 tells you that a 2-input OR or NOR gate can be *disabled* by applying a 1 to one input.

RULE 7

Rule 7, A + A = A, can be explained by the truth table and logic diagram shown in Figure 6-10. This figure shows that an OR or NOR gate can be used as a buffer by simply tying its inputs together.

FIGURE 6-10   Truth table and logic diagram for rule 7.

RULE 8

Rule 8, $A + \overline{A} = 1$. If a variable is ORed with its complement as shown in rule 8, the result is a 1 because one input is always high. Again, the circuit shown in Figure 6-11 to illustrate this rule is not very useful, but, as you will see, the rule itself is very helpful in reducing Boolean expressions.

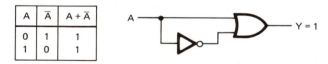

FIGURE 6-11   Truth table and logic diagram for rule 8.

RULE 9

Rule 9, $\overline{\overline{A}} = A$, shows that if a variable is complemented twice, the result is just the variable. In other words, as shown by the circuit diagram in Figure 6-12, two inver-

FIGURE 6-12   Truth table and logic diagram for rule 9.

sions of a logic signal "cancel." A zero on the input of the first inverter produces a 1 on its output. This 1 on the input of the second inverter produces the original 0 signal level on the final output.

Boolean Identities Pertaining to Order of Operations

RULES 10 AND 11

Rules 10 and 11, AB = BA and A + B = B + A, can be thought of as showing you that the inputs of a gate are equivalent and interchangeable.

RULES 12 AND 13

Rules 12 and 13, (AB)C = A(BC) and (A + B) + C = A + (B + C), show that the order in which variables are ANDed or ORed doesn't matter. For example, if two variables, A and B, are ORed together and the result ORed with variable C, the final result will be the same as ORing B with C and then ORing the result with A. These rules are illustrated in the logic diagrams shown in Figure 6-13.

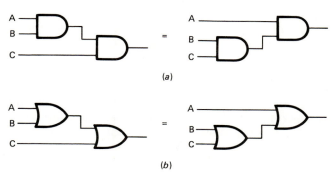

(a)

(b)

FIGURE 6-13   Logic diagrams for Boolean identities pertaining to the order of operations. *(a)* Rule 12. *(b)* Rule 13.

Identities Showing Factoring of Boolean Expressions

RULE 14

Rule 14, AB + AC = A(B + C), is equivalent to factoring in ordinary algebra. Since variable A is present in both terms on the left side of the equation, it can be separated out as shown on the right side of the equation. You can see from the logic diagrams for this rule in Figure 6-14 that the factored form of the expression requires fewer gates to implement than does the non-factored form. This is a simple example of how manipulating the Boolean expression for a desired logic function can simplify its implementation with gates. The truth tables for the two forms are, of course, identical because they are simply two equivalent ways of expressing the same logic function.

RULE 15

Rule 15, (A + B)(A + C) = A + BC, can be thought of as another type of factoring. This type, however, only works in Boolean algebra. Intuitively the rule works as follows. Since A is present in each term, the result will be a 1 if A = 1. The other condition that will cause the left side of the equation to be a 1 is for B to be a 1 and C to be a 1. Then, the two conditions that will cause

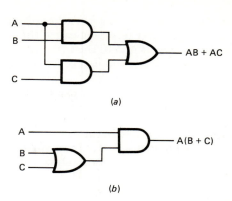

(a)

(b)

FIGURE 6-14   Logic diagram representation of rule 14.

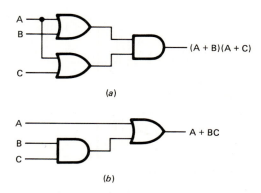

(a)

(b)

FIGURE 6-15   Logic diagram representation of rule 15.

the expression to equal 1 can be summarized simply as A + BC, which is the form shown in the right side of rule 15. From the logic diagrams for these two forms shown in Figure 6-15, you can see that the A + BC form is easier to implement with gates than the (A + B)(A + C) form. This is another example of how these rules can be used to simplify a Boolean expression.

Identities Which Eliminate Unneeded Variables and Terms in Boolean Expressions

RULE 16

Rule 16, A + AB = A, can be thought of intuitively as follows. If A = 1, then the left side of the expression will be a 1, regardless of the state of B. Since the state of B doesn't matter, the AB term can be dropped. You can prove this rule more formally by factoring A out of both terms on the left side of the expression to give A(1 + B) as shown in rule 14. According to rule 6, 1 + B = 1, so A(1 + B) simplifies to A AND 1. Rule 2 states that A1 = A, so the final result is A + AB = A.

RULE 17

Rule 17, A + $\overline{A}$B = A + B, can also be proved intuitively or formally. Intuitively, if A = 1, then the expression on the left will be equal to 1, regardless of the state of B. If B = 1, the state of A doesn't matter, because if A = 1, the first term will cause the expression to be equal to 1, and if A is 0, then the $\overline{A}B$ term will be equal to 1. More formally you can prove this rule as follows.

Use rule 15 in the reverse direction to expand A + $\overline{A}B$ to (A + $\overline{A}$)(A + B). Since, according to rule 8, A + $\overline{A}$ = 1, this expanded form reduces to 1(A + B). Rule 2 shows that a variable ANDed with a 1 is equal to just the variable, so 1(A + B) is equal to just A + B. Again, you can see from the logic diagrams for the two forms shown in Figure 6-16 that the reduced expression, A + B, is simpler to implement with gates than the original expression, A + $\overline{A}B$.

A	B	$\overline{A}$	$\overline{A}B$	A + $\overline{A}B$	A + B
0	0	1	0	0	0
0	1	1	1	1	1
1	0	0	0	1	1
1	1	0	0	1	1

(a)

(b)

FIGURE 6-16   Truth table and logic diagram for rule 17.

RULE 18

Rule 18, AB + $\overline{A}B$ = B, is one of the most often used Boolean identities because it shows how a *sum-of-products* expression such as you might construct from a truth table can be simplified. Since the left-hand side of the expression will be a 1 if either A = 1 and B = 1 or A = 0 and B = 1, then perhaps you can see that if B is a 1, the output will be a 1, regardless of the state of A. Therefore, the expression reduces to simply B.

More formally you can prove this identity by first factoring B out of both terms on the left side of the equation to give B(A + $\overline{A}$). As shown in rule 8, A + $\overline{A}$ = 1, so B(A + $\overline{A}$) simplifies to B AND 1. Rule 2 shows that a variable ANDed with a 1 is equal to just the variable, so B AND 1 simplifies further to just B as shown on the right side of rule 18.

RULE 19

Rule 19, A(A + B) = A, is a less used identity which can be intuitively thought of as follows. If A = 1, then both terms on the left side of the expression will be equal to 1, and the entire expression will be equal to 1. If A = 0, then the left side of the expression will be equal to 0, regardless of the state of B. Therefore, the state of B doesn't matter, and the expression reduces to just A.

Formally you can prove this identity by first "multiplying" each term in the parentheses by A to give AA + AB. According to rule 3, AA is equal to A, so this expression can be reduced to A + AB. If you factor A out

of both of these terms, you get A(1 + B). Rule 6 states that a variable ORed with a 1 is equal to 1, so the expression further reduces to A AND 1. As shown in rule 2, A1 reduces to just A.

RULE 20

Rule 20, A($\overline{A}$ + B) = AB, can be proved in a similar way to the method we used for rule 19. Multiplying each term in the parentheses by A gives A$\overline{A}$ + AB. According to rule 4, A$\overline{A}$ is equal to 0, so the expression reduces to 0 + AB. Rule 5 states that a variable ORed with 0 is equal to just the variable, so the expression reduces further to just AB, as shown in the right-hand side of the rule above.

De Morgan's Theorem

Rules 21 and 22 in Figure 6-3 are usually thought of as the two parts of De Morgan's theorem. As you will soon see, you have actually been using these rules since Chapter 2, but up until now we didn't specifically state them.

RULE 21

Rule 21 states that $\overline{AB}$ = $\overline{A}$ + $\overline{B}$. The easiest way to prove this rule is with a truth table. As you can see in Figure 6-17, the $\overline{AB}$ column of the truth table is identical to the $\overline{A}$ + $\overline{B}$ column of the truth table. Since the truth tables for the two expressions are identical, the expressions must be equivalent.

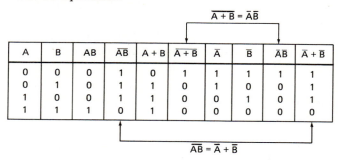

					$\overline{A + B} = \overline{A}\overline{B}$				
A	B	AB	$\overline{AB}$	A + B	$\overline{A + B}$	$\overline{A}$	$\overline{B}$	$\overline{A}\overline{B}$	$\overline{A}$ + $\overline{B}$
0	0	0	1	0	1	1	1	1	1
0	1	0	1	1	0	1	0	0	1
1	0	0	1	1	0	0	1	0	1
1	1	1	0	1	0	0	0	0	0

$\overline{AB} = \overline{A} + \overline{B}$

FIGURE 6-17   Truth table demonstrating De Morgan's theorem.

The practical consequences of rule 21 are shown by the logic diagram representations for it in Figure 6-18. The $\overline{AB}$ expression on the left side of rule 21 represents a positive logic NAND gate, as shown by the left drawing in Figure 6-18. The $\overline{A}$ + $\overline{B}$ expression on the right side of rule 21 represents an OR gate with active low input signals, as shown by the drawing in the center of Figure 6-18. De Morgan's theorem then tells you that a NAND gate can be represented either by an AND symbol with an active low output or by an OR symbol with active low inputs. This part of De Morgan's theorem, then, is just a formal statement that the two symbols you have been using to represent a NAND gate are indeed equivalent.

The rightmost drawing in Figure 6-18 shows that an OR gate with inverters on its inputs performs the same logic function as a NAND gate. This means that you can use an OR gate with inverters on its inputs to perform a NAND operation.

FIGURE 6-18   Logic diagram representation of rule 21.

FIGURE 6-19   Logic diagram representation of rule 22.

RULE 22

Rule 22, the second part of De Morgan's theorem, states that $\overline{A + B} = \overline{A}\overline{B}$. To prove this identity, compare the $\overline{A + B}$ column of the truth table in Figure 6-17 with the $\overline{A}\overline{B}$ column in the table. Since the truth table entries for the two expressions are identical, the two expressions must be equivalent.

One practical consequence of rule 22, shown by the logic diagrams for it in Figure 6-19, is that you can represent a NOR operation with either a bubbled output OR symbol or a bubbled input AND symbol. This part of De Morgan's theorem is a formal statement that the two symbols you have been using to represent a NOR gate are indeed equivalent.

Another practical consequence of rule 22, shown by the rightmost diagram in Figure 6-19, is that you can use an AND gate with inverters on its inputs to perform the NOR operation.

A useful way to think of De Morgan's theorem when you are using it to simplify Boolean expressions is as follows:

> Splitting or joining the bar over the terms of a Boolean expression changes the sign between the terms.

For example, splitting the bar over $\overline{AB}$ changes the sign between A and B to give $\overline{A} + \overline{B}$, as shown in rule 21. Joining the bars in the expression $\overline{A}\overline{B}$ changes the sign to give $\overline{A + B}$, which is an equivalent expression according to rule 22.

NOTE: We have shown you all of these Boolean identities for your reference, but you do not have to memorize them all. The rules that you will use most often in simplifying Boolean expressions you get from truth tables are rules 1 through 9, 14, 17, 18, 21, and 22, so study these carefully.

USING BOOLEAN IDENTITIES TO SIMPLIFY LOGIC EXPRESSIONS

Now that we have shown you the basic identities or rules of Boolean algebra, we will show you how they can be used to simplify logic expressions and thereby reduce the number and complexity of gates required to implement the logic function. We will start with some simpler examples and then show you how to simplify the expression for the seat belt warning system that we introduced at the start of this section of the chapter.

For a first example, suppose that you derive the Boolean expression $S = AB + \overline{A}B + A\overline{B}$ by summing the minterms from a truth table and you want to simplify the expression. Here's one way to do it.

The first two terms of the expression both contain B, so use rule 14 to factor B out of each to give:

$$S = B(A + \overline{A}) + A\overline{B}$$

Since $A + \overline{A} = 1$ (rule 8), this expression reduces to:

$$S = B(1) + A\overline{B}$$

According to rule 2, $B(1) = B$, so the expression further reduces to:

$$S = B + A\overline{B}$$

Rule 17 shows you that $\overline{B}$ is not needed in the expression, so the final result is:

$$S = B + A$$

If you like to have variables in alphabetical order, you can write the expression as $S = A + B$ (rule 11).

Now let's look at an example with three variables. Again, suppose that from a truth table you derive the sum-of-products expression:

$$R = \overline{A}BC + AB\overline{C} + ABC$$

The major technique in simplifying an expression such as this is to look for terms which have all but one variable the same, and factor the like variables out of each term. In the expression above, for example, the last two terms each contain AB, but one has C and the other has $\overline{C}$. You can then factor AB out of these terms to give:

$$R = \overline{A}BC + AB(\overline{C} + C)$$

Since $C + \overline{C} = 1$ (rule 8), and AB ANDed with 1 is just AB (rule 2), the expression reduces to:

$$R = \overline{A}BC + AB$$

The two remaining terms each contain B, so you can factor out B to give:

$$R = B(\overline{AC} + A)$$

According to rule 17, $\overline{A}$ is not needed in the expression, so the final result is:

$$R = B(C + A)$$

As our next example we will show you how to simplify the expression we derived from the truth table for our seat belt controller:

$$B = \overline{\overline{W}FK} + \overline{WF}K + W\overline{F}K$$

Again, look for terms which have all but one variable the same. In this expression the first two terms both have $\overline{W}$ and K, so you can factor these out to give:

$$B = \overline{W}K(\overline{F} + F) + W\overline{F}K$$

Since $\overline{F} + F = 1$ and $\overline{W}K(1) = \overline{W}K$, the expression reduces to:

$$B = \overline{W}K + W\overline{F}K$$

The two remaining terms each contain K, so you can factor out K to see if the remainder can be reduced further. The result of this factoring is:

$$B = K(\overline{W} + W\overline{F})$$

Rule 17 tells you that W is not needed in the $W\overline{F}$ term, so this expression reduces further to:

$$B = K(\overline{W} + \overline{F})$$

After we show you some more examples of using Boolean identities to simplify expressions and how to use Karnaugh maps to simplify expressions, we will show you several ways this simplified expression for the seat belt warning system can be implemented with gates. For now let's see how De Morgan's theorem can help you simplify logic expressions.

Remember that De Morgan's theorem (rules 21 and 22) shows you that splitting a bar over terms in a logic expression or joining the bars over terms changes the sign between the terms. These simple rules are often very helpful in simplifying Boolean expressions derived from logic diagrams. Here's an example.

Figure 6-20a shows a three-gate circuit and the Boolean expressions for each point in the circuit. Because of all the overbars in the output expression, it is somewhat difficult to determine the truth table for the circuit. As we showed you in an earlier section, one way to make a circuit such as this easier to analyze is to replace the output NAND gate symbol with its De Morgan equivalent symbol, an OR symbol with bubbles on its inputs.

For review, Figure 6-20b shows the circuit redrawn in this way. Since the bubbles on the inputs of the OR symbol effectively "cancel" the bubbles on the outputs of the NAND gates, you can read the output expression for this equivalent circuit directly off the diagram as $Y = AB + CD$.

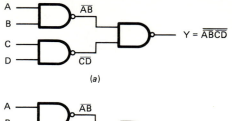

(a)

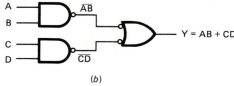

(b)

FIGURE 6-20   Logic circuit using positive NAND gates. (a) Standard all-NAND schematic. (b) Circuit using inverted input OR representation of output NAND.

Now that you know De Morgan's theorem, you can simplify the output expression for the circuit in Figure 6-20a without having to redraw the circuit. To start, split the top overbar on the expression and change the sign between the AB term and the CD term. This gives:

$$Y = \overline{\overline{AB}} + \overline{\overline{CD}}$$

As shown by rule 9 in Figure 6-3, double bars over a variable or term cancel, so this expression reduces to simply $Y = AB + CD$, which is the same result you got by replacing the output NAND gate in the drawing with its De Morgan equivalent symbol. It is certainly much easier to write a truth table for this expression than it is to write one for the initial overbarred expression.

As another example of how De Morgan's theorem can help you simplify the output expression for a circuit, take a look at the circuit and the Boolean expressions for it in Figure 5-7a. The output expression taken directly from the circuit is:

$$S = \overline{(\overline{ABC}(\overline{CD})(\overline{ABC})}$$

Splitting the top overbar into three sections and changing the sign between the three terms gives:

$$S = \overline{\overline{ABC}} + \overline{\overline{CD}} + \overline{\overline{ABC}}$$

Again, according to rule 9, double bars over a variable or a term cancel, so this expression reduces to just:

$$S = \overline{ABC} + C\overline{D} + \overline{ABC}$$

So far in this section on Boolean algebra we have, for simplicity, used expressions that contain only active high signals. Since most real systems have both active low and active high signals, here are some mixed logic examples.

As a first example, suppose that you need to simplify the expression:

$$Y* = A*BC + ABC + A*B*C$$

To simplify this expression, you use the same techniques we have described previously. Start by looking for common factors which can be factored out of two or more terms. The factor BC, for example, can be factored out of the expression above to give:

$$Y* = BC(A* + A) + A*B*C$$

Since A* is defined as active low, A is the complement of A*. Therefore, A* + A = 1, and the expression reduces to:

$$Y* = BC + A*B*C$$

The common factor C can be factored out of the remaining two terms to give:

$$Y* = C(B + A*B*)$$

According to rule 17, the B* in this expression is not needed, so the expression reduces to simply:

$$Y* = C(B + A*)$$

As a second example of simplifying mixed logic expressions, here's an example which contains both active low signals and overbars:

$$X = \overline{P*}RS* + P*RS*$$

The first two terms in the expression both contain R and S*. These can be factored out to give:

$$X = RS*(\overline{P*} + P*)$$

Again, according to rule 8, a variable ORed with its complement equals 1, so the expression reduces to just:

$$X = RS*$$

The purpose of these last two examples is to show you that the rules of Boolean algebra work with mixed logic expressions as well as with positive logic expressions. The main point to remember when working with mixed logic expressions is:

Handle signals (variables) with an * the same as you do signals (variables) without an *.

In this section we have shown you how you can use Boolean algebra rules to simplify a logic expression derived from a circuit or one derived from a truth table. For an expression with three or four variables and many terms, however, Boolean algebra manipulations can be somewhat like doing a proof in geometry. If you make one wrong move, you can wander around for a long time trying to simplify the expression. In the next section we show you a graphic technique called a *Karnaugh map* which you can use to simplify Boolean expressions. You may find these maps much easier to use than Boolean algebra.

KARNAUGH MAP SIMPLIFICATION

Developing Karnaugh Maps from Truth Tables and Extracting Simplified Boolean Expressions

A *Karnaugh map*, or *K-map*, is simply a different way of writing the truth table for a logic function. However, because of the way terms are arranged in a Karnaugh map, you can derive a simplified expression for the logic function directly from the map. This is usually easier than working through Boolean algebra manipulations to simplify the sum-of-products expression you get from a standard truth table. Karnaugh maps can be used for logic functions of any number of input variables, but we will only show you how to use them for functions of up to four input variables. For a logic function with more than four input variables, you will usually use one of several commonly available microcomputer programs to simplify the expression, and you will most likely use a single, programmable logic device such as a PAL, FPLA, or ROM to implement the circuit for the simplified expression. In a later chapter we show you how to do this. For now, here's how you use Karnaugh maps to simplify logic expressions.

Figure 6-21 shows how you draw Karnaugh maps for two-, three-, and four-variable logic functions. Each square in the map corresponds to one of the possible

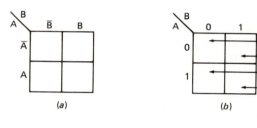

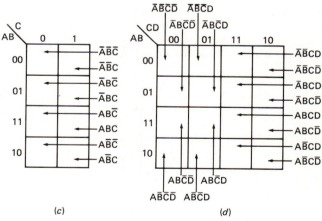

FIGURE 6-21   Karnaugh maps for logic functions with different numbers of variables. *(a)* Map for two variables. *(b)* Alternative map for two variables. *(c)* Map for three variables. *(d)* Map for four variables.

combinations of the input variables. The input combinations are written along the edges of the maps as shown. Some designers like to identify the possible input combinations on the map with $\overline{A}$ and A, $\overline{B}$ and B, etc., as shown in Figure 6-21a. Other designers like to identify the possible input combinations with 0's and 1's, as shown in Figure 6-21b. We will use this latter form because it allows you to more easily see the correspondence between a line in the standard truth table and a box in the Karnaugh map. Basically you can read an input combination as a series of 1's and 0's off the truth table, and find the same series of 1's and 0's on the map.

For each input combination that produces a 1 in the output column of a standard truth table, you write a 1 in the box that corresponds to that input combination on the map. In two-variable maps such as those shown in Figure 6-21a and b, for example, the combination A = 0 and B = 0 is represented by the boxes in the upper left corners of the maps. If A = 0 and B = 0 produce a 1 on the output of a logic circuit, then you write a 1 in this box. The boxes in the upper right corners of the maps in Figure 6-21a and b represent the input combination A = 0 and B = 1, the box in the lower left corner represents the input combination A = 1 and B = 0, and the box in the lower right corner of the map represents the input combination A = 1 and B = 1.

The truth table for a three-variable logic function has eight possible input combinations, so the map for it has eight boxes as shown in Figure 6-21c. Note carefully that the input combinations for A and B are written in the order 00, 01, 11, 10 down the left side of the map. The combinations are written in this order so that only one variable changes as you go from one row in the map to the next. You *must* write the combinations in this order so that you can easily read the simplified Boolean expression off the map. Note that for the four-variable map in Figure 6-21d, the combinations for C and D are written in this same order of 00, 01, 11, 10 so that only one variable at a time changes as you go from one column to the next. For future reference we have shown the product expression for each box in these Karnaugh maps. Now we will show you how to draw the Karnaugh map for a common gate and extract a Boolean expression from it.

Figure 6-22a shows the standard truth table for an

To extract the Boolean expression for the gate from the map, first note that, as shown by the bottom two boxes in the map, the output will be a 1 if A = 1 and B = 0, or if A = 1 and B = 1. In other words, if A = 1, then the output will be a 1, regardless of whether B = 0 or B = 1. Therefore, if you group these two 1's together and write an expression for the two 1's together, the expression does not need to contain B. The expression which represents these two 1's is simply Y = A, which tells you that the output will be a 1 if A = 1.

You can use Boolean algebra as follows to prove that this simplification is indeed valid. From the standard truth table the sum-of-products expression for these two 1's is $Y = A\overline{B} + AB$. Factoring A out of each of these terms gives $Y = A(\overline{B} + B)$. The expression $\overline{B} + B = 1$, so, as expected, the final result is just Y = A.

The advantage of a Karnaugh map over a standard truth table is that it allows you to easily see which adjacent 1's you can group together and write a simplified expression for. To keep track of which 1's you have grouped together and written an expression for, you circle them as shown in Figure 6-22c.

To produce the complete expression for a truth table, all of the 1's on the Karnaugh map must be included in at least one circled group. Above we showed how the two 1's in the bottom boxes of the Karnaugh map in Figure 6-22b could be grouped and represented with the expression Y = A. To get a complete expression for Y, we need to also take into account the 1 in the upper-right-hand corner of the map in Figure 6-22b.

The two 1's in the right-hand column of this Karnaugh map tell you that Y will be a 1 for the input combinations A = 0 and B = 1 or A = 1 and B = 1. In other words, if B = 1, the output will be a 1, regardless of whether A = 0 or A = 1. Therefore, we can group these two 1's together and represent them with the expression Y = B. Again, to indicate that you have grouped these two adjacent 1's together, you circle them as shown in the right column of the map in Figure 6-22c.

To produce the complete Boolean expression for the

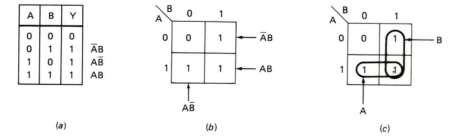

FIGURE 6-22 (a) OR gate truth table. (b) Karnaugh map for OR gate truth table. (c) Grouping 1's in Karnaugh map.

OR gate, and Figure 6-22b shows the truth table for the OR gate in Karnaugh map form. As shown in the standard truth table, the input combination A = 0 and B = 0 produces a 0 on the Y output, so you put a 0 in the 00 box in the map. All of the other input combinations produce a 1 on the Y output, so you put a 1 in each of the other three boxes in the map.

map, you sum (OR) the expressions you have written for circled groups of 1's. For the example in Figure 6-22c, then, the complete expression derived from the map is Y = A + B, which, of course, is the minimum expression for an OR gate.

To compare this method of simplification with the Boolean algebra method for the same truth table, first note that the sum-of-products expression taken directly from the standard truth table is:

$$Y = \overline{A}\overline{B} + A\overline{B} + AB$$

If you factor A out of the last two terms, the result is:

$$Y = \overline{A}B + A(\overline{B} + B)$$

which, as we showed in earlier Boolean examples, reduces to:

$$Y = \overline{A}B + A$$

Rule 17 from Figure 6-3 shows that this reduces to just:

$$Y = A + B$$

Here's another simple example to help you get an intuitive feeling for how the Karnaugh map technique works. Figure 6-23a shows the truth table for another common logic gate, and Figure 6-23b shows the Karnaugh map for it. As shown by the two 1's in the top row of the Karnaugh map, the output will be a 1 if A = 0 and B = 0 or if A = 0 and B = 1. In other words, if A = 0, then the output will be a 1, whether B = 0 or B = 1. The term which represents these two 1's, then, is just $\overline{A}$. To indicate that you have grouped these two 1's to derive the simplified expression from the map, you circle them as shown in Figure 6-23c.

The two 1's in the left column of the Karnaugh map in Figure 6-23b tell you that the output of this gate will be a 1 if B = 0 and A = 0 or if B = 0 and A = 1. The output then will be a 1 if B = 0, regardless of whether A = 0 or A = 1. The expression which represents these two 1's is just $\overline{B}$. To indicate that you have grouped these two 1's together, you circle them as shown in Figure 6-23c.

To write the expression for the complete circuit from the map, you OR the expressions you wrote for the two groups of circled 1's. The complete expression for this gate then is:

$$Y = \overline{A} + \overline{B}$$

Rule 21 in Figure 6-3, the first part of De Morgan's theorem, shows you that another way of writing this expression is:

$$Y = \overline{AB}$$

which is, of course, the positive logic expression for a NAND gate. For practice you can use the rules of Bool-

ean algebra to show that the sum-of-products expression in Figure 6-23a reduces to this same result.

The general principles of using a Karnaugh map are as follows:

1. Circle groups of adjacent 1's.

2. Write the expression for each circled group of 1's.

3. OR these expressions to produce the complete expression for the logic function.

Here's why grouping adjacent 1's produces a reduced expression. Because of the way that entries are arranged in a Karnaugh map, the expressions for adjacent 1's in a column or in the same row differ in only one variable. For example, the expression for the box in the top left corner of the map in Figure 6-23c is $\overline{A}\overline{B}$ and the expression for the box directly to the right of this one is $\overline{A}B$. A = 0 for both of these boxes, but B is a 0 for one box and a 1 for the other. As we discussed before, the fact that these two adjacent boxes contain 1's shows you that if A = 0, the output will be a 1, regardless of whether B = 0 or B = 1. For these two terms, then, B is a "don't care" and does not need to be included in the expression. All that is needed to represent these two 1's is the expression Y = $\overline{A}$. Writing a truth table in Karnaugh map form and circling groups of adjacent 1's allows you to quickly see which variables are "don't cares" and which variables must be included in the expression for those terms. In larger Karnaugh maps, such as those for three- and four-variable truth tables, you can circle groups of two, four, or eight adjacent 1's to produce simplified expressions.

The rules for circling groups of 1's in Karnaugh maps are as follows:

1. Only groups of one, two, four, eight, or sixteen 1's can be circled.

2. Circled groups must be square or rectangular in shape.

3. Ones which are only diagonally adjacent to each other cannot be circled because there is a change in more than one variable between these two squares.

4. Groups should be made as large as possible without violating rules 1 and 2.

5. Every 1 on the map must be included in at least one group.

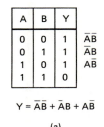

$$Y = \overline{A}\overline{B} + \overline{A}B + A\overline{B}$$

(a)

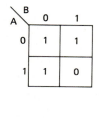

(b)

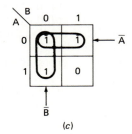

(c)

FIGURE 6-23 (a) Truth table for common logic gate. (b) Karnaugh map. (c) Karnaugh map with circled ones.

6. A 1 can be included in several groups in order to make them larger.

7. Loops may wrap around from one edge of a map to the opposite edge because boxes on opposite edges differ in only one variable.

To help you see how to use these rules, Figure 6-24 shows some examples of legal groups in three- and four-variable maps. Each group of letters in these examples assumes that you only have 1's in those locations, so you can't join 1's in these circled groups with 1's in other circled groups.

pression for the group. The resultant expression for this circled group is $N = \overline{A}B$.

For the two locations labeled M in Figure 6-24a, B = 0 and C = 0, but A = 0 for one location and A = 1 for the other location. Therefore, A can be left out and the expression for the circled group is $M = \overline{B}\overline{C}$. Note that this group illustrates rule 7, which states that 1's on opposite edges of a map can be grouped because they differ by only one variable. Here's some more grouping examples.

Figure 6-24b shows some possible groups on a four-variable map. The boxes containing X's in Figure 6-24b

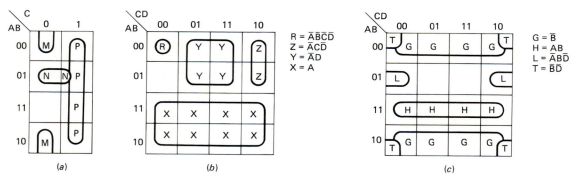

FIGURE 6-24   Examples of legal groups in three- and four-variable maps.

In the three-variable map in Figure 6-24a, for example, if the four boxes labeled with P's contain 1's, they can be circled as shown. If the two boxes labeled N contain 1's, they can be circled as shown. Likewise, if the two boxes labeled M in Figure 6-24a contain 1's, they can be circled as shown. Now, before we look at some more ways of circling 1's, we need to show you more about how to write the expression for a group of circled 1's.

The rule you use for writing the expression for a group of circled 1's on a map is:

Include in the expression only those variables which have the same state for all of the boxes in the circled group.

To illustrate how this works, look at the four circled P's in Figure 6-24a. For this circled group, A = 0 for two locations and A = 1 for the other two locations. This means that A can be left out of the expression for this group. Likewise, B = 0 for two of the locations in the circled group, and B = 1 for the other two locations. This means that B can also be left out of the expression for the circled group of 1's represented by P's. This leaves only C to consider. Since C = 1 for all four locations in the circled group, C is included in the expression for the group. The expression for the group identified with P's is just P = C.

For both map locations labeled with N's in Figure 6-24a, A = 0 and B = 1, so $\overline{A}$ and B must be included in the expression for the circled group. C, however, is a 0 for one location in the group and a 1 for the other location in the group. Therefore, C can be left out of the ex-

show how you can circle a group of eight adjacent 1's. A = 1 for all eight of the circled 1's, but all of the other three variables change as you go from box to box in the group. Therefore, the expression for this group is just X = A.

For the four locations identified with Y's in Figure 6-24b, A = 0 and D = 1. Therefore, the expression for this group is just $Y = \overline{A}D$ as shown. For the two locations labeled with Z's in Figure 6-24b, the variables which are the same are A = 0, C = 1, and D = 0. Therefore, the expression for the Z group is $Z = \overline{A}C\overline{D}$.

The R in the map in Figure 6-24b represents the case of a single, isolated 1 which cannot be grouped with any other 1's in the map. The expression for an isolated 1 must contain all of the variables which identify the position of the 1 on the map. For the location labeled R, then, the expression is $R = \overline{A}\overline{B}\overline{C}\overline{D}$.

Figure 6-24c shows still more examples of how you can circle adjacent 1's on a four-variable Karnaugh map. For the eight locations labeled with G's in the figure, the only variable that is the same is B = 0. Therefore, the expression for the G group is just $G = \overline{B}$. This is another example of how 1's on opposite edges of the map can be grouped together.

For the locations identified with H's in Figure 6-24c, A = 1 and B = 1 for all four, but C and D have different values. Therefore, the expression for the H group is H = AB.

The circled locations identified with L's in Figure 6-24c are another example of how 1's on opposite edges of a map can be circled. The variables which are the same for these two locations are A = 0, B = 1, and D = 0. The expression for the group, then, is $L = \overline{A}B\overline{D}$.

The final example in Figure 6-24c shows another interesting case of how the edges of a Karnaugh map are adjacent to each other. For the four corner squares of the map, B = 0 and D = 0, but A and C differ. Therefore, 1's in these four squares can be grouped together and represented by the expression T = $\overline{B}\overline{D}$.

As you can see from these examples, the main objective in circling 1's is to include as large a group as possible. The larger you can make the circled group, the fewer variables you have to include in the expression for the group. Now that we have shown you in general how to group 1's in Karnaugh maps and how to write the simplified expression for circled groups, we will show you a variety of specific examples.

Karnaugh Map Examples

As a first example here, Figure 6-25a shows the truth table and the unsimplified Boolean expression for a three-variable logic function. Figure 6-25b shows the truth table in Karnaugh map form with the 1's circled in

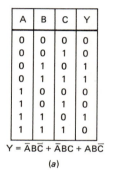

A	B	C	Y
0	0	0	0
0	0	1	0
0	1	0	1
0	1	1	1
1	0	0	0
1	0	1	0
1	1	0	1
1	1	1	0

$Y = \overline{A}B\overline{C} + \overline{A}BC + AB\overline{C}$

(a)

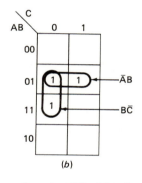

(b)

FIGURE 6-25 (a) Truth table and unsimplified Boolean logic function. (b) Karnaugh map.

appropriate groups. Note that, as specified in rule 6, a 1 can be included in more than one group. The expression for the circled horizontal group on the map is $\overline{A}B$, and the expression for the circled vertical group is $B\overline{C}$. Therefore, the complete expression for the function is the sum of these: $Y = \overline{A}B + B\overline{C}$. As you can see, using the map quickly reduces the expression for the function from three terms with three variables each to an equivalent expression which has only two terms with two variables per term. Also, the reduced expression, derived from the Karnaugh map, is in a sum-of-products form which is easily implemented with simple AND and OR gates.

As our next example, let's look at how a Karnaugh map can be used to produce a simplified expression for the seat belt warning system we described at the start of this section. To refresh your memory of this problem, Figure 6-26a shows the truth table and the unsimplified Boolean expression for it. Figure 6-26b shows the Karnaugh map for the truth table. Note that because of the "wrap-around" feature of Karnaugh maps and the fact that a 1 can be used in more than one group, the 1 in the lower right corner can be grouped with the 1 in the upper right

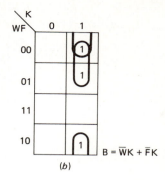

W	F	K	B
0	0	0	0
0	0	1	1
0	1	0	0
0	1	1	1
1	0	0	0
1	0	1	1
1	1	0	0
1	1	1	0

$B = \overline{W}\overline{F}K + \overline{W}FK + W\overline{F}K$

(a)

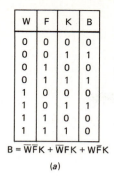

$B = \overline{W}K + \overline{F}K$

(b)

FIGURE 6-26 Seat belt warning system. (a) Truth table and unsimplified Boolean expression. (b) Karnaugh map.

corner. The expression for these two 1's is $\overline{F}K$, and the expression for the two 1's in the upper right corner of the map is $\overline{W}K$. The simplified expression for the seat belt warning system, then, is just the sum of these: $B = \overline{W}K + \overline{F}K$. Later we will show you several ways this simplified expression can be implemented with gates. First, however, we want to show you a few more examples of working with Karnaugh maps.

Figure 6-27a shows the truth table and the unsimplified Boolean expression for a four-variable logic function. This expression is somewhat messy to simplify

A	B	C	D	Y
0	0	0	0	0
0	0	0	1	0
0	0	1	0	1
0	0	1	1	0
0	1	0	0	1
0	1	0	1	0
0	1	1	0	1
0	1	1	1	0
1	0	0	0	0
1	0	0	1	1
1	0	1	0	1
1	0	1	1	1
1	1	0	0	1
1	1	0	1	0
1	1	1	0	1
1	1	1	1	0

$Y = \overline{A}\overline{B}C\overline{D} + \overline{A}B\overline{C}\overline{D} +$
$\quad \overline{A}BC\overline{D} + A\overline{B}\overline{C}D +$
$\quad A\overline{B}C\overline{D} + A\overline{B}CD +$
$\quad AB\overline{C}\overline{D} + ABC\overline{D}$

(a)

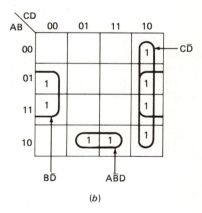

(b)

FIGURE 6-27 (a) Truth table and unsimplified four-variable logic function. (b) Karnaugh map.

with Boolean algebra, but, as shown in Figure 6-27b, a Karnaugh map does it quite easily. Because of the wrap-around rule and the multiple use rule, all eight 1's can be grouped in just three groups. The simplified expression is $Y = A\overline{B}D + B\overline{D} + C\overline{D}$.

As a final example in this section, take a look at the truth table in Figure 6-28a and the Karnaugh map for it in Figure 6-28b. Due to the wraparound rule, all six 1's

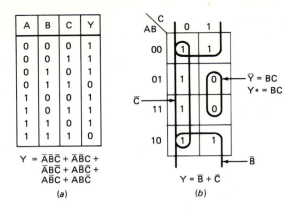

A	B	C	Y
0	0	0	1
0	0	1	1
0	1	0	1
0	1	1	0
1	0	0	1
1	0	1	1
1	1	0	1
1	1	1	0

$Y = \overline{A}\overline{B}\overline{C} + \overline{A}\overline{B}C +$
$\overline{A}B\overline{C} + A\overline{B}\overline{C} +$
$A\overline{B}C + AB\overline{C}$

(a)

$\overline{Y} = BC$
$Y* = BC$

$Y = \overline{B} + \overline{C}$

(b)

FIGURE 6-28 Example showing simplification of a function by circling 0's on the Karnaugh map.

in the map can be circled with just two groups. The expression for the vertical group is $\overline{C}$, and the expression for the wraparound group is $\overline{B}$. The complete expression, then, is $Y = \overline{B} + \overline{C}$. This can be implemented with two inverters and an OR gate, but there is an easier way. Here are two methods of determining the easier way.

The first method is to apply De Morgan's theorem to $Y = \overline{B} + \overline{C}$, the expression we derived from the Karnaugh map. Joining the bars over B and C changes the sign between them to give $Y = \overline{BC}$. Since this is the expression for a positive logic NAND gate, you can now see that the truth table in Figure 6-28a can be implemented with a single NAND gate.

The second method of determining this simpler implementation is to circle groups of adjacent 0's on the map and write the expression for the conditions which will cause the output to be low. The rationale here is that if you build a circuit which produces lows for these input combinations, the output will be high, as desired, for the other input conditions.

For the Karnaugh map in Figure 6-28b, the two circled 0's can be represented by the expression BC. Since these input conditions produce a low on the output, you write the complete expression as $Y* = BC$ or $\overline{Y} = BC$. The expression $Y* = BC$ tells you that you need a gate whose output will be low when its B input is high and its C input is high. This, of course, is one way of describing the operation of a NAND gate. Alternatively, you can move the overbar in the expression $\overline{Y} = BC$ to the right-hand side of the equation to give $Y = \overline{BC}$, which is the standard expression for a NAND gate.

In some cases circling adjacent 0's on a Karnaugh map and writing an expression for the conditions which will cause the output to be low, produces a simpler expression than circling 1's and writing an expression for the input conditions which will cause the output to go high. The expressions derived by the two methods must be equivalent because they represent the same truth table. Circling 0's and writing the low-output expression for the circled groups of 0's is really just a way of letting the map do a De Morgan's theorem operation on the circled 1's expression so that you can see how to implement a function easier. If the expression you derive from a Karnaugh map contains a lot of overbars, try "De

Morganizing" it or pulling the 0's expression off the map to see if this produces an expression which is easier to implement with gates.

Using Karnaugh Maps for Functions with Active Low and Active High Signals

For simplicity, the preceding Karnaugh map examples used only active high input signals. In real applications, however, the inputs are often a mixture of active high and active low signals. In a case such as this, you could assume that inverters will be used to change the assertion levels of all the active low signals to active high before they are applied to the logic circuit. With this assumption, you can then use a Karnaugh map to find the minimum logic expression, just as we have shown in the preceding section. However, as we show you now, this is not necessary. You can use a Karnaugh map to directly produce a simplified expression for a truth table which has both active high and active low signals.

Figure 6-29a shows a truth table with two active low

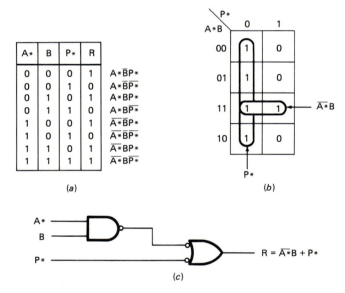

FIGURE 6-29 Example showing Karnaugh map use with mixed logic. (a) Truth table. (b) Karnaugh map. (c) Circuit.

signals, A* and P*, and two active high signals, B and R. The 0's and 1's in this truth table represent, as usual, the actual logic levels present on the inputs and output of the circuit. With LS TTL devices, for example, a 0 represents a voltage of about 0 V, and a 1 represents a voltage of 2.5 to 5 V. It is important to write the truth table in this form so that it tells you what voltages to expect when you test or troubleshoot the circuit.

Figure 6-29b also shows the Boolean expression for each input combination in the truth table. Note that A* is an active low signal, so if A* = 0 in an input combination, A* will appear in the Boolean expression with no overbar. If A* is a 1 in an input combination, then A* will appear in the output expression for that term with an overbar. The B signal in the truth table is active high, so if B = 1 for an input combination, it will appear in the expression without an overbar. The expression for the 000 input combination, for example, tells you that R will be a 1 if A* is asserted (0), B is not asserted (0), and P* is asserted (0).

Figure 6-29b shows how you draw a Karnaugh map for this truth table. Variables are identified with their name and their assertion levels. Input combinations are written along the edges of the map in normal order. As usual, you write a 1 in each box that represents an input combination which produces a 1 in the truth table. Then you circle adjacent 1's in groups of one, two, four, or eight, as we showed you in previous examples. The next step is to write the expression for each circled group.

The rule for writing expressions for groups on Karnaugh maps which have both active low and active high input signals can be summarized as follows:

> If the specified assertion level of a signal is different from the signal level which causes the output to be asserted, the signal name will be overbarred in the product term for any group containing it.

Here's an example of how this works.

For the horizontal group of two adjacent 1's in Figure 6-29b, A* = 1 and B = 1. Since A* is an active low signal, a 1 represents the unasserted state for the A* signal. Therefore, A* will have an overbar in the expression for the group. The B signal is active high, so B = 1 represents the asserted state for it. Therefore, B will not have an overbar in the expression for the group. The expression for this group, then, is just $\overline{A*}B$ as shown. Now let's look at the vertical group of circled 1's.

A* and B are "don't cares" for this group of four because both A* and B change as you go from one end of the group to the other. Therefore, all you have to consider is the P* signal, which is a 0 for all four of these circled 1's. Since 0 represents the asserted level for the P* signal, the expression for this group is just P* without an overbar. The complete Boolean expression is R = $\overline{A*}B$ + P*.

Figure 6-29c shows a circuit you might use to implement the simplified expression derived from the Karnaugh map. For practice, analyze the circuit to prove to yourself that it does implement the desired truth table.

Developing Karnaugh Maps from Boolean Expressions

So far in this section we have concentrated on showing you how to develop Karnaugh maps from truth tables because this is usually the route you will take when designing a simple combinational logic circuit to solve some problem. As a final point here we will briefly show you how to develop a Karnaugh map for a Boolean expression, so that you can see if the expression is in its simplest form. There are basically two simple ways of developing a Karnaugh map from a Boolean expression.

The first way is to expand the expression so that each term contains all variables in either their true or com-

plemented form and so that all represented minterms are present in the expression. Once you get the expression expanded to this maximum form, you can just write a 1 in the Karnaugh map box which corresponds to each minterm. As an example of this method, we will show you how to expand the expression Y = $\overline{A}B$ + BC + AC.

The variable C is not present in the $\overline{A}B$ term, so if A is low and B is high, the expression will be true whether C = 0 or C = 1. Therefore, the $\overline{A}B$ term can be expanded to a term with C and a term with $\overline{C}$ or $\overline{A}BC + \overline{A}B\overline{C}$. This process is just the reverse of the factor-and-cancel method you use to simplify two terms which are the same except that one contains a variable and the other contains its complement. In the same manner the BC term can be expanded to ABC + $\overline{A}BC$, and the AC term can be expanded to ABC + $AB\overline{C}$. The result of these expansions when summed gives:

$$Y = \overline{A}BC + \overline{A}B\overline{C} + ABC + \overline{A}BC + ABC + AB\overline{C}$$

The $\overline{A}BC$ term and the ABC terms each appear twice in the expression, so one of each can be dropped to give:

$$Y = \overline{A}BC + \overline{A}B\overline{C} + ABC + AB\overline{C}$$

To make a Karnaugh map for this expression, you just write a 1 in each of the Karnaugh map squares represented by these four terms as shown in Figure 6-30a.

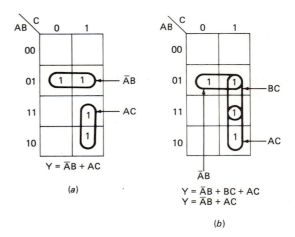

FIGURE 6-30 Writing a Karnaugh map for a Boolean expression. (a) First method. (b) Second method.

Once you have the Karnaugh map, you can circle adjacent 1's to find the minimum expression as we described in the preceding sections. For this map, only the two groups, $\overline{A}B$ and AC, are needed to contain the four 1's, so the minimum expression for the function is Y = $\overline{A}B$ + AC. This shows you that the BC term in the original expression, Y = $\overline{A}B$ + BC + AC, is not needed.

The second method of writing a Karnaugh map for a Boolean expression is much more preferable to us because you don't have to manipulate the Boolean expression. You simply write a 1 in the box or boxes represented by each term in the original expression. To help you use this method, remember that if a variable is missing in a term of an expression, it means that the missing variable can be a 0 or a 1 and the term will still be equal to 1. Therefore, the term really represents two squares in the Karnaugh map. If two variables are not present in a term, it means that each of those variables can be 0 or 1 and the term will still be equal to 1. Therefore, if two variables are not represented, it means that the term represents four squares in the Karnaugh map. As you will see, this second method is essentially just the reverse of the method you use for extracting the simplified expression from a map. To illustrate the method, we will use the expression $Y = \overline{A}B + BC + AC$, the same expression we used to show you the first method of getting a Karnaugh map from an expression.

The term $\overline{A}B$ represents two horizontally adjacent boxes in the Karnaugh map as shown in Figure 6-30b, so you write 1's in the two boxes for this term. The term BC represents two vertically adjacent boxes in the map, as also shown in Figure 6-30b. So you write a 1 in each of these boxes. Note that if you put in the 1's for the $\overline{A}B$ term, one of the boxes already has a 1 in it. Finally, the AC term also represents two vertically adjacent boxes as shown in Figure 6-30b, so you put 1's in the two boxes to represent this term. Now if you look carefully at the completed map, you should see that the four 1's can be contained in just two circled groups, the $\overline{A}B$ group and the AC group, because the two 1's in the BC group are each included in other groups. This means that the original expression, $Y = \overline{A}B + BC + AC$, can be simplified to just $Y = \overline{A}B + AC$, which is the same result produced by the Boolean expansion method.

The preceding sections have shown you how to use Karnaugh maps to produce a simplified logic expression for a given truth table or Boolean expression. In the next section we show you some techniques for implementing the simplified expressions with commonly available logic gates.

IMPLEMENTING LOGIC FUNCTIONS WITH COMMON GATES

Throughout the previous sections on producing simplified Boolean expressions for logic functions, we have, for the most part, derived expressions in the sum-of-products form. The reasons we did this are:

1. Expressions in this form more closely follow the way people think, so they are easier to understand than other forms.

2. Sum-of-product expressions are easily implemented with simple AND gates, OR gates, and inverters.

3. Expressions in this form can be directly implemented in programmable logic devices such as PALs, FPLAs, and ROMs.

As a first example of implementing logic functions, suppose that you want to implement $R = AB + CD$. The expression represents the ORing of two ANDed terms. Therefore, one way to implement this function is with a

2-input OR gate and two, 2-input AND gates as shown in Figure 6-31a. If you were working with devices in the 74HC family of CMOS logic gates, you might use an OR gate from a 74HC32 quad, 2-input OR device and two AND gates from a 74HC08 quad, 2-input AND device. The difficulty with implementing the logic function in this way is that two different IC packages are required. You may not need the other gates in these packages, so they are essentially wasted. Here's another, more efficient way to implement AND-OR logic functions.

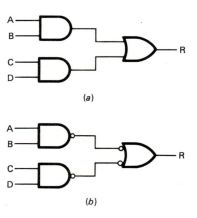

FIGURE 6-31   Implementing the logic function $R = AB + CD$. (a) Using two different IC packages. (b) Using one NAND gate package.

Implementing Logic Functions with only NAND Gates and Inverters

The classic symbol for a NAND gate tells you that, with active high input signals, the device functions as an AND gate with an active low output. As we have shown you many times before, a NAND gate can also be represented by an OR symbol with bubbles on its inputs. This symbol tells you that a NAND gate performs the OR function if it has active low input signals. In short, then, the function $R = AB + CD$ can be implemented with just three NAND gates as shown in Figure 6-31b. Since this way of implementing the function uses only NAND gates, a single 74HC00 quad, 2-input NAND gate IC is all that is needed to implement it.

As another example of how to implement logic functions with logic gates, let's look again at the seat belt warning system we have been designing throughout this section. As we showed in Figure 6-26b, the simplified expression for the system is $B = \overline{W}K + \overline{F}K$. Using the technique we described in the previous paragraph, this expression can be directly implemented with three NAND gates and two inverters as shown in Figure 6-32a. You might, for example, use a 74LS00 quad, 2-input NAND gate device and a 74LS04 inverter. This implementation is quite straightforward, but if you manipulate the expression slightly, you can implement the function with fewer gates. Here's how.

Each term in the expression $B = \overline{W}K + \overline{F}K$ contains the variable K. The K can thus be factored out to give

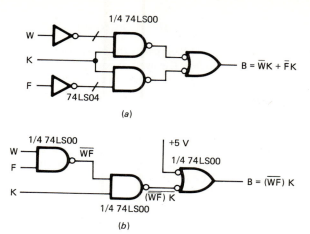

$$B = \overline{W}K + \overline{F}K$$

$$B = (\overline{WF})\,K$$

FIGURE 6-32 Seat belt example implementation
(a) Using B = WK + FK. (b) Using B = WFK.

$B = K(\overline{W} + \overline{F})$. According to rule 21 in Figure 6-3, the expression in the parentheses can be written as $\overline{WF}$, so another form for the expression then is $B = K\overline{WF}$. This expression says that the function can be implemented by ANDing K with the complemented AND (NAND) of W and F. Figure 6-32b shows how this function can be implemented using just three NAND gates. Since one input of the output NAND gate is tied high, this gate simply functions as an inverter. This implementation, then, only requires one IC package, a 74LS00 quad, 2-input NAND gate device, for example.

This last example brings to mind an important point we need to make at this time. In the old days, when digital systems consisted of large PC boards containing many discrete gate devices, we used to spend many hours working though logic functions to implement them with the absolute minimum number of gates. The development of MSI and LSI devices which perform most of the logic functions in a system has drastically changed our approach. In the first place, as we indicated previously, for logic functions of over four variables, we almost always manipulate the expressions with a computer program and implement the function with a single programmable IC. For functions with fewer variables, we first use a Karnaugh map to produce a simplified sum-of-products expression from the truth table. Then, if we are only going to produce one or a few copies of the circuit, we implement the function directly with NAND gates and inverters as shown in Figure 6-32a. The logic expression for the circuit is more directly visible with the circuit implemented in this form, and the chance of introducing an error during further manipulation is eliminated. Also, the cost of the added IC (about $0.25) is usually less than the cost of the time required to simplify the implementation further.

If we are designing a system that is going into mass production, then we manipulate the sum-of-products expression somewhat to see if we can implement the function with fewer gates as we showed you for the seat belt example in Figure 6-32b.

Implementing Logic Functions with Only NOR Gates and Inverters

When deriving a Boolean expression from a truth table or Karnaugh map, you usually end up with a sum-of-products expression. However, some problems may produce a product-of-sums expression. There are two ways you can implement a product-of-sums expression.

The first way is to use the rules of Boolean algebra to convert the expression to sum-of-products form and then implement the result with NAND gates and inverters as we described in the previous section. The Boolean algebra for this involves simply "multiplying" all the ANDed terms and eliminating duplicated terms in the result.

The second method is to implement the function directly with NOR gates and inverters in a manner similar to the way you implement a sum-of-products expression using only NAND gates and inverters. Figure 6-33, for example,

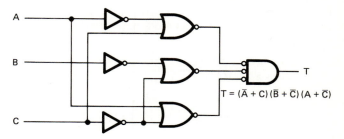

$$T = (\overline{A} + C)(\overline{B} + \overline{C})(A + \overline{C})$$

FIGURE 6-33 Implementing the expression $T = (\overline{A} + C)(\overline{B} + \overline{C})(A + \overline{C})$ with only NOR gates and inverters.

shows how the expression $T = (\overline{A} + C)(\overline{B} + \overline{C})(A + \overline{C})$ can be implemented with only NOR gates and inverters. Note that using the bubbled input AND symbol for the output NOR gate makes it easier to see the logic function performed by the circuit.

Using Exclusive OR and Exclusive NOR gates

As we explained in the previous sections, any combinational logic circuit can be implemented with NAND gates and inverters or with NOR gates and inverters. Some circuits, however, are more easily implemented with Exclusive OR or Exclusive NOR gates. Here's an example.

Suppose that you need a circuit that outputs a high if the number of 1's on its four inputs is odd. In other words, you need a circuit which detects odd parity for a 4-bit input word. Figure 6-34a shows a truth table for the circuit, and Figure 6-34b shows a Karnaugh map for it. None of the 1's in the Karnaugh map are adjacent to each other and there are no don't cares, so none of the 1's can be circled with any other. In other words, this function will not simplify! The sum-of-products expression for this map will contain eight terms with all four input variables in each one. Implementing this eight-term expression directly would require eight 4-input AND gates, an 8-input OR gate, and four inverters.

Figure 6-35 shows how this circuit can be implemented with just three 2-input Exclusive OR gates. The output expression for this circuit is $Y = A \oplus B \oplus C \oplus D$.

A	B	C	D	Y
0	0	0	0	0
0	0	0	1	1
0	0	1	0	1
0	0	1	1	0
0	1	0	0	1
0	1	0	1	0
0	1	1	0	0
0	1	1	1	1
1	0	0	0	1
1	0	0	1	0
1	0	1	0	0
1	0	1	1	1
1	1	0	0	0
1	1	0	1	1
1	1	1	0	1
1	1	1	1	0

(a)

AB \ CD	00	01	11	10
00		1		1
01	1		1	
11		1		1
10	1		1	

(b)

FIGURE 6-34   Truth table and Karnaugh map for a circuit that detects odd parity for a 4-bit input word.

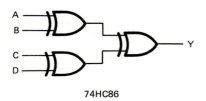

74HC86

FIGURE 6-35   Circuit that detects odd parity for a 4-bit word.

Now, the question that probably comes to your mind is, "How did we know to use Exclusive OR gates to implement this function?" The answer is that, whenever we see a checkerboard Karnaugh map such as this, we recognize that it can be implemented with Exclusive OR gates or with Exclusive NOR gates, depending on where the checkerboard is on the map. If the 0000 box on the map contains a 0, then we can implement the function with Exclusive OR gates. If the 0000 box in the Karnaugh map contains a 1, then we look to Exclusive NOR gates to implement the circuit.

For the example here, a quad 2-input Exclusive OR device such as a 74HC86 allows us to implement the entire function with a single package.

DESIGN EXAMPLES

So far in this chapter we have worked with logic circuits which produce only one output signal from a group of input signals. In many applications we need logic circuits which produce several different output signals from the same group of input signals. This is usually not difficult because you can use the methods we have shown you in the preceding sections to independently develop the circuit required for each output. Therefore, we will show you a couple of multiple output design examples to summarize the techniques we have shown you in the preceding sections.

A Solar Hot Water Heater Controller

Figure 6-36 shows a block diagram of a solar hot water heater system. Basically the system consists of two parts. The solar collection system consists of a collector, a circulation pump, and a large storage tank. The liquid used in this part of the system is usually a mixture of water and antifreeze, similar to the coolant used in an automobile cooling system. The sun shining on the collector heats the liquid in it. If the temperature of the liquid in the collector is above the temperature of the liquid in the storage tank, the solar collector pump is turned on to circulate liquid from the tank through the collector. The effect of this is to heat up the liquid in the storage tank. As the name implies, the storage tank acts like a reservoir of stored energy.

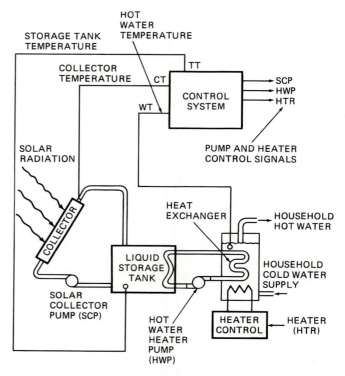

FIGURE 6-36   Block diagram of solar hot water heater system.

The second part of the system is a conventional gas or electric hot water heater, a second circulation pump, and two heat exchangers (one in the storage tank and one in the hot water tank). If the temperature of the water in the tank drops below some setpoint temperature, one of two actions will occur. If the temperature of the storage tank is above the temperature of the water, the controller will turn on the circulation pump to transfer heat from the storage tank to the water. If the storage tank is colder than the water, the controller will turn on the electric or gas heater in the water tank. The job we want to do here is to design a controller which uses signals from three sensors to produce the control signals for the two pumps and the water heater. In a later chapter we show you how the signals from sensors such as these are produced and how signals from logic gates are interfaced to devices such as motors and heaters. For now, here's a list of the signal names, descriptions, and abbreviations we will use for the design:

CT—*Collector temperature.* CT = 1 if temperature is above storage tank temperature.

TT—*Tank temperature (storage).* TT = 1 if temperature of storage tank is above temperature of hot water tank.

WT—*Water temperature.* WT = 1 if hot water temperature is above setpoint temperature.

SCP—*Solar collector pump.* SCP = 1 turns on pump to circulate heat transfer fluid.

HWP—*Hot water pump.* HWP = 1 turns on pump to circulate water through heat exchanger in storage tank.

HTR—*Heater.* HTR = 1 turns on electric or gas heater.

After defining the input signals and output signals, the next step in the design process is to make a truth table showing the desired relationships between these signals. Figure 6-37a shows a possible truth table for this design. First, note that we want the heater on for the case where the collector is cold, the storage tank is cold, and the hot water tank is cold and that we want the heater on for the case where the collector is hot, but the temperature of the storage tank is still below the temperature of the water tank. Second, notice that we turn the solar collector pump on for the 110 case so that heat transferred to the water is replaced by heat from the collector. Finally, note that we turn on the solar collector pump for the 111 case to store as much energy in the tank as possible and to prevent the collector from overheating.

The next step in the design process is to produce a simplified Boolean expression for each of the desired output signals. By inspection we see that the SCP column of the truth table is identical to the CT column, so the expression is just SCP = CT for this output.

The HWP output column has two 1 terms, so a Karnaugh map as shown in Figure 6-37b is an easy way to produce a simplified expression for it. The simplified expression from the map is:

$$HWP = TT \, \overline{WT}$$

Finally, the HTR column in the truth table also has two 1 terms which can be circled as shown in Figure 6-37c The expression for this group is:

$$HTR = \overline{TT} \, \overline{WT}$$

Once you have the simplified expression for each output, you then implement each expression with the appropriate gates. For this control system, you can do it as follows.

The SCP expression is just SCP = CT, so this function can be implemented with a piece of wire between the CT sensor and input of the electronically controlled switch that turns on the solar collector pump as shown in Figure 6-37d.

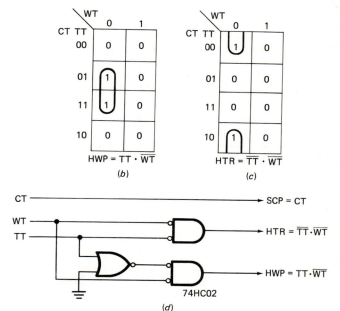

CT	TT	WT	SCP	HWP	HTR
0	0	0	0	0	1
0	0	1	0	0	0
0	1	0	0	1	0
0	1	1	0	0	0
1	0	0	1	0	1
1	0	1	1	0	0
1	1	0	1	1	0
1	1	1	1	0	0

(a)

$$HWP = TT \cdot \overline{WT}$$
(b)

$$HTR = \overline{TT} \cdot \overline{WT}$$
(c)

SCP = CT

$$HTR = \overline{TT} \cdot \overline{WT}$$

$$HWP = TT \cdot \overline{WT}$$

74HC02

(d)

FIGURE 6-37   Solar heater problem. *(a)* Truth table. *(b)* Karnaugh map for HWP. *(c)* Karnaugh map for HTR. *(d)* Controller circuit.

The HTR expression we derived from the truth table, HTR = $\overline{TT}\overline{WT}$, would require two inverters and a 2-input AND gate to implement directly in this form. Applying De Morgan's theorem to the expression will quickly put it in a form which can be implemented with a single gate. If you join the bars over the two terms and change the signs between them, the expression becomes:

$$HTR = \overline{TT + WT}$$

This expression can be implemented with a single 2-input NOR gate as shown in Figure 6-37d. Note that we used the alternate symbol for a NOR gate so that the logic being performed can be more easily read directly from the circuit diagram. Further note that we used a CMOS device to minimize power dissipation, and that we tied unused inputs of the 74HC02 device to ground to prevent them from picking up noise or static electricity.

The HWP expression, HWP = $TT\overline{WT}$, is similar in form to the expression for HTR, so it can also be converted to an OR expression. The bar over $\overline{WT}$ can be extended over the rest of the expression if you first put an overbar over

TT and then change the sign between TT and $\overline{WT}$. The result is:

$$HWP = \overline{\overline{TT} + WT}$$

This expression tells us that, if we invert the TT signal, the function can be implemented with a 2-input NOR gate as shown in Figure 6-37d. Again, we used the bubbled input AND representation for the NOR gate, so the original logic function can more easily be read from the circuit diagram. Note that we used one of the NOR gates in the 74HC02 as an inverter so that the entire controller could be implemented with a single IC package.

A Seven-Segment LED Driver Circuit

As a second complete design example we will show you how to make a circuit which converts a 4-bit binary number in the range 0 to 9 to the signals necessary to display the equivalent decimal numbers on a seven-segment LED display.

For review, Figure 6-38a shows how an LED can be connected to the output of a logic gate so that when the output of the gate is low, the LED will light. Remember

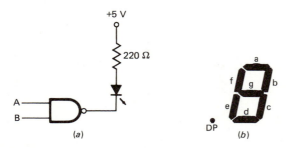

FIGURE 6-38 (a) LED connected to the output of a logic gate. (b) Seven-segment LED display showing segment labels.

that the 220-Ω resistor is necessary to limit the current through the LED to about 15 mA. To display decimal digits, seven LEDs are put in a single device as shown in Figure 6-38b. Each LED produces one segment of a number or letter when lit. Segments are usually identified with the letters shown next to each in Figure 6-38b.

Figure 6-39 shows the equivalent circuits of the two types of seven-segment LED displays. The connection shown in Figure 6-39a is referred to as a *common cathode* device because the negative ends, or cathodes, of all the LEDs are connected together as shown. For this type display, a logic high applied to the positive end, or *anode*, of a diode will light it. The connection shown in Figure 6-39b is referred to as a *common anode* device because the anodes of the LEDs are all connected together as shown. To light an LED in this type of display, you apply a logic low to its cathode as shown in Figure 6-39b. For the design example here, we will use a common anode display, so a logic low is required to light each segment.

Figure 6-40 shows in block diagram form the problem we want to solve with this example. The input to the

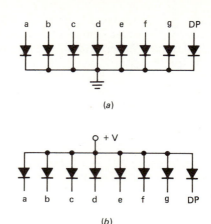

FIGURE 6-39 Seven-segment LED display. (a) Common cathode device. (b) Common anode device.

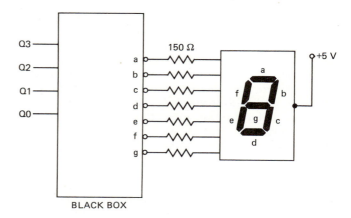

FIGURE 6-40 Block diagram for LED problem.

"black box" is a binary number between 0 and 9. We want the circuitry in the black box to output the segment code required to display the decimal digit which corresponds to the 4-bit binary code applied to the inputs.

As before, the first step in the design process is to make a truth table which shows the input signal combinations and the logic level we want on each output line for each input combination. For this design, we have seven output lines, so we need seven output columns in the truth table as shown in Figure 6-41. Study this truth table to see if you agree with the values we have put in the table. Remember that a low lights a segment for a common anode display. Incidentally, because of personal preference we lit the "a" segment for a 6, but this isn't necessary.

Once you have a complete truth table, the next step is to derive a simplified Boolean expression for each desired output. We will use this example to show you how to take advantage of "don't care" states in Karnaugh maps. We will also show you again how to circle 0's on a Karnaugh map to produce a sum-of-products expression for the complement of a function. First, let's see how you can use X's.

For numbers above 1001 in the input columns of the truth table, note that we have put X's in all the output

Q3	Q2	Q1	Q0	a	b	c	d	e	f	g	DIGIT DISPLAYED
0	0	0	0	0	0	0	0	0	0	1	0
0	0	0	1	1	0	0	1	1	1	1	1
0	0	1	0	0	0	1	0	0	1	0	2
0	0	1	1	0	0	0	0	1	1	0	3
0	1	0	0	1	0	0	1	1	0	0	4
0	1	0	1	0	1	0	0	1	0	0	5
0	1	1	0	0	1	0	0	0	0	0	6
0	1	1	1	0	0	0	1	1	1	1	7
1	0	0	0	0	0	0	0	0	0	0	8
1	0	0	1	0	0	0	1	1	0	0	9
1	0	1	0	X	X	X	X	X	X	X	X
1	0	1	1	X	X	X	X	X	X	X	X
1	1	0	0	X	X	X	X	X	X	X	X
1	1	0	1	X	X	X	X	X	X	X	X
1	1	1	0	X	X	X	X	X	X	X	X
1	1	1	1	X	X	X	X	X	X	X	X

FIGURE 6-41   Truth table for LED problem.

columns. An X in this case represents a "don't care" condition for that input combination. The definition of the problem states that the inputs can never get to states above 1001, so the value that would be produced on the outputs by states above 1001 doesn't matter. As you will see, writing X's in the Karnaugh map for these don't care states often leads to a simpler expression because you can include them with either 1's or 0's to make larger groups.

Figure 6-42a shows the Karnaugh map for the "a" segment output of the black box. Since, by definition, an X can be either a 1 or a 0 without affecting the output of the circuit, an X can be grouped with a 1 as shown by the circled group in the left column of the map. Without including the X, the expression for this 1 is $\overline{Q3}\,Q2\,\overline{Q1}\,\overline{Q0}$. The expression for the circled group which includes the X is $Q2\,\overline{Q1}\,\overline{Q0}$. Including the X makes Q3 unnecessary in the expression. No other 1's or X's can be grouped with the 1 in the top line of the map, so the complete expression derived from the map is:

$$a = \overline{Q3}\,Q2\,\overline{Q1}\,Q0 + Q2\,\overline{Q1}\,\overline{Q0}$$

Before we discuss how to implement this expression, let's try a small experiment. When working with Karnaugh maps that have many don't cares, it is a good idea to first circle the 1's and X's and write the expression which produces a high output. Then circle 0's and X's and write the expression which produces a low output. These expressions are equivalent because they represent the same truth table, but as we showed in a previous example, one form may make it easier to see how to implement the function than the other form.

Figure 6-42b shows the Karnaugh map for the "a" segment with 0's and X's circled in appropriate groups. The expression derived in this way is:

$$a* = Q1 + Q3 + \overline{Q2}\,\overline{Q0} + Q2\,Q0$$

With some Boolean algebra experimentation, you can show that this expression is equivalent to the expres-

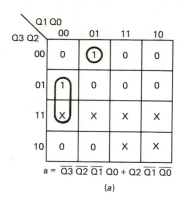

$$a = \overline{Q3}\,\overline{Q2}\,\overline{Q1}\,Q0 + Q2\,\overline{Q1}\,\overline{Q0}$$

(a)

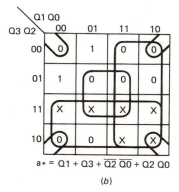

$$a* = Q1 + Q3 + \overline{Q2}\,\overline{Q0} + Q2\,Q0$$

(b)

FIGURE 6-42   Karnaugh maps for the "a" segment in the seven-segment LED problem. (a) Ones circled. (b) Zeros circled.

sion derived by circling 1's in the Karnaugh map. For now, however, let's see how each can be implemented with gates.

The expression derived by circling 1's can be implemented with a 4-input NAND gate, a 3-input NAND gate, a 2-input NAND gate, and four inverters as shown in Figure 6-43a. The expression derived by circling 0's and X's can be implemented with two 2-input AND gates, a 4-input NOR gate, and two inverters as shown in Figure 6-43b. You will usually choose the implementation which uses gates left over from other combinational logic circuits on the board or the implementation which leaves gates that can easily be utilized in other circuits on the board. We leave the derivation of the expressions for the remaining segment outputs and the circuits for these expressions for you as a practice problem at the end of the chapter.

CHECKLIST OF IMPORTANT TERMS AND CONCEPTS IN THIS CHAPTER

Minterm
Boolean algebra
Enabling and disabling gates
De Morgan's theorem
Karnaugh map
Common anode and common cathode LED displays
Don't care states (X) in Karnaugh map

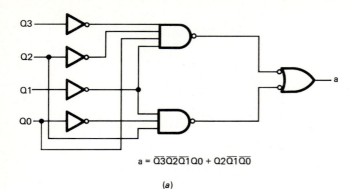

$$a = \overline{Q3}\,\overline{Q2}\,\overline{Q1}\,Q0 + Q2\,\overline{Q1}\,\overline{Q0}$$

(a)

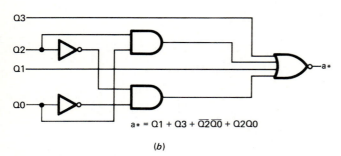

$$a* = Q1 + Q3 + \overline{Q2}\,\overline{Q0} + Q2\,Q0$$

(b)

FIGURE 6-43   Circuit for LED problem.

REVIEW QUESTIONS AND PROBLEMS

1. *a.* Write the minterm for each output 1 in the truth table in Figure 6-44.
    b. Write the sum-of-products Boolean expression for the truth table.

A	B	C	Y
0	0	0	1
0	0	1	1
0	1	0	0
0	1	1	1
1	0	0	0
1	0	1	0
1	1	0	0
1	1	1	0

FIGURE 6-44   Truth table for Problems 1, 2, and 10.

2. *a.* Write the maxterms for each of the output 0's in the truth table in Figure 6-44.
    b. Write the product-of-sums Boolean expression for the truth table in Figure 6-44.

3. Write the sum-of-products Boolean expression for the truth table in Figure 6-45.

4. *a.* Write a truth table for the expression: $R = f(A, B, C, D) = \Sigma (8, 10, 12, 14)$
    b. Write the sum-of-products Boolean expression for the truth table.

5. *a.* A 3-input NAND gate is available on a circuit board. Show how the gate can be connected so that it functions as an inverter.

A	B	C	Y
0	0	0	0
0	0	1	1
0	1	0	1
0	1	1	0
1	0	0	1
1	0	1	0
1	1	0	0
1	1	1	1

FIGURE 6-45   Truth table for Problem 3.

   b. Show how a 2-input NOR gate can be connected so that it functions as an inverter.
   c. Can a 2-input AND gate be connected so that it functions as an inverter?

6. Using Boolean algebra, simplify the following:
    a. $Y = (A + \overline{B})(A\overline{B} + C)C$
    b. $T = (A + \overline{B})C + (A\overline{B} + C)$

7. Using Boolean algebra, simplify the following:
    a. $S = AB + ABC + \overline{A}B + A\overline{C}$
    b. $T = \overline{\overline{AB} + \overline{CD}}$
    c. $X = \overline{A+B} + \overline{\overline{CD}}$

8. The expression $T = \overline{AR} + \overline{ST}$ requires two 2-input NAND gates and an OR gate to implement in this form. Use De Morgan's theorem to convert the expression to a form that can be implemented with a single gate.

9. Use Boolean algebra to simplify the following expressions:
    a. $Y = \overline{A}*\overline{B}*C + \overline{A}*B*C + \overline{A}*B*\overline{C}$
    b. $L* = R*\overline{S}*T + R*\overline{S}*\overline{T}$

10. *a.* Use a Karnaugh map to produce a simplified expression for the truth table in Figure 6-44.
     b. Show how you can implement the simplified expression with common gates.

11. Simplify the equation $X = AB + ABC + \overline{A}B + AC$ using:
     a. Boolean algebra.
     b. A Karnaugh map.
     c. Show how to implement the simplified expression with gates. Remember, to get a minimum expression, try to group as many boxes as possible in groups of 2, 4, 8, or 16.

12. Use a Karnaugh map to simplify:

$$X = AB\overline{C}\,\overline{D} + A\overline{B}CD + ABCD + ABC\overline{D} + \overline{A}BCD + \overline{A}BCD$$

and then show the circuit for the simplified expression.

13. *a.* Use a Karnaugh map to produce a simplified expression for:

$$R = f(A, B, C, D) = \Sigma m(8, 10, 12, 14)$$

b. Show the circuit for the simplified expression.

14. *a.* Use a Karnaugh map to simplify the expression:

$$Y = \overline{A*B*C} + \overline{A}*B*C + \overline{A}*B*\overline{C}$$

b. Show how the circuit can be implemented with gates.

15. *a.* Use a Karnaugh map to simplify the expression:

$$L* = R*\overline{S}*T + R*\overline{S}*\overline{T}$$

b. Show how the circuit can be implemented with gates.

16. Three secretaries and a boss work in an office. The radio will be on if the boss or at least two secretaries want it on.
   a. Write the truth table for the radio ON function. (Let a yes vote equal a 1 in the truth table and ON equal a 1.)
   b. Use a Karnaugh map to produce the simplified expression for the function.
   c. Show how the function can be implemented by using only NAND gates.
   d. Show how the function can be implemented by using only NOR gates.

17. Figure 6-46 shows the truth table for an auto system which sounds a warning buzzer if the key is in the ignition and the driver has not fastened his or her seat belt. The buzzer also sounds if the key is in the ignition and no one is sitting in the seat. To

W = WEIGHT IN SEAT
F = SEAT BELT FASTENED
K = KEY IN IGNITION
B = BUZZER ON

W	F	K	B	
0	0	0	0	
0	0	1	1	$\overline{W}\overline{F}K$
0	1	0	0	
0	1	1	1	$\overline{W}FK$
1	0	0	0	
1	0	1	1	$W\overline{F}K$
1	1	0	0	
1	1	1	0	

FIGURE 6-46   Truth table for Problem 17.

get further use out of the buzzer, suppose that you want the buzzer to sound if any door is open. Assume that the car has a door sensor which goes high when any one of the doors is open.
   a. Redo the truth table of Figure 6-46 to add this condition.
   b. Use Boolean algebra or a Karnaugh map to produce a simplified expression for the truth table.
   c. Show how the circuit can be implemented with 74HC00 gates.

18. *a.* Write a truth table for a circuit which outputs a 1 if the number of 1's in a 4-bit binary number applied to its inputs is odd.
   b. Use Boolean algebra or a Karnaugh map to produce a simplified expression for the truth table.
   c. Draw a circuit diagram showing how to implement the expression with common gates.

19. Design a circuit which outputs a high if two 4-bit binary numbers applied to its eight inputs are equal. *Note:* We did not show you how to simplify circuits with eight input variables, but if you use XNOR gates for part of the circuit, you don't need to deal directly with the eight input variables.

20. In this chapter we showed you how to develop a simplified expression for the "a" segment portion of a circuit which converts a BCD code to the signals necessary for driving a seven-segment display. A key part of this was using "don't care" states to help simplify the expression. Use the same technique to develop the simplified expressions for the b, c, d, e, f, and g segments.

21. The 3-bit binary counter in Figure 6-47 outputs a continuous binary count sequence from 000 to 111 on its Q outputs. QC is the most significant bit of the count. For this problem, we want you to design a logic circuit which functions as follows.

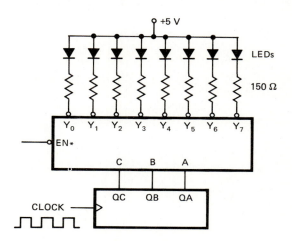

FIGURE 6-47   Circuit for Problem 21.

If the enable input, EN*, is asserted, the binary number applied to the inputs of the circuit should cause the Y output with that number to be asserted. Since the Y outputs are active low, the LED connected to that output will light when the output is asserted. In other words, if 000 is present on the C, B, and A inputs of the logic circuit, you want the Y_0* output to be asserted (low), so that LED will be lit. If 011 is present on the CBA inputs, you want the Y_3* output to be asserted, so that LED will be lit. As the counter steps through its count, the LEDs will light in sequence over and over.

a. Write a truth table for the logic circuit. (It is helpful to put the variables in the truth table in the order EN*, C, B, A.)

b. Write a Karnaugh map for the Y_0* output and derive a simplified expression for the Y_0* from it. Remember that since you want Y_0*, you should look at the 0's on the map.

c. Based on the expression you derived for the Y_0* output, write Boolean expressions for each of the other Y outputs.

d. Examine the output expressions to determine the basic type of gate which will most easily implement each expression and what is required to generate the input signals for each of these gates.

e. Draw a diagram for the complete logic circuit.

22. A technician designed the circuit in Figure 6-48a to implement the truth table in Figure 6-48b.

a. Use Boolean algebra or a Karnaugh map to determine why the circuit does not work correctly.

b. Draw a circuit which will correctly implement the truth table in Figure 6-48b.

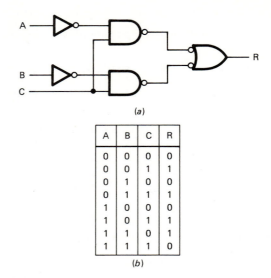

(a)

A	B	C	R
0	0	0	0
0	0	1	1
0	1	0	0
0	1	1	1
1	0	0	0
1	0	1	1
1	1	0	1
1	1	1	0

(b)

FIGURE 6-48   Truth table and circuit for Problem 22.

7 MSI and LSI Combinational Logic Devices and Applications

In the preceding chapters we showed you how to analyze, test, troubleshoot, and design combinational logic circuits with simple logic gates. Fortunately, many commonly needed combinational logic functions are now available in single MSI or LSI devices. Some of these devices have fixed functions. Others such as ROMs, PALs, and FPLAs can be programmed to perform a needed function. In this chapter we introduce you to some of these devices and show you how they are used in a variety of applications.

OBJECTIVES

At the conclusion of this chapter you should be able to:

1. Draw gate circuits which act as logic-controlled switches for digital signals.

2. Describe how a multiplexer and a demultiplexer can be used to transmit several digital signals on a single wire.

3. Use a multiplexer to implement a combinational logic function.

4. Describe how a parity generator/checker can be used to detect an error in transmitted digital data.

5. Predict the signal(s) that will be produced on the outputs of an encoder by given input signals.

6. Predict the signal(s) that will be produced on the outputs of a decoder by given input signals.

7. Given the IEEE/IEC (dependency notation) logic symbol for an MSI device, describe the function of the inputs and outputs of the device.

8. Describe the differences in internal structures of a ROM, PAL, and FPLA.

9. Show how a ROM, a PAL, or an FPLA can be used to implement any desired combinational logic function.

10. Show how a ROM, a PAL, and an FPLA can be used to convert one binary code to another.

DIGITAL SWITCHES AND MULTIPLEXERS

Logic-Controlled Switches

In a previous chapter we showed you how an AND gate can be used as a logic-controlled switch for digital signals. To refresh your memory, Figure 7-1a shows an ex-

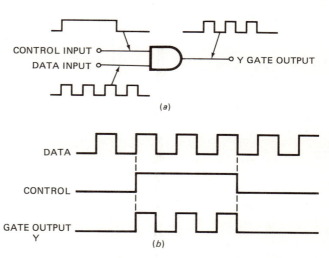

FIGURE 7-1   AND gate used a logic-controlled switch. *(a)* Schematic. *(b)* Timing diagram.

ample of this. If the AND gate input labeled CONTROL is low, the output of the gate will be low, regardless of the signal level on the other input of the gate. If the AND gate is enabled by making the CONTROL input high, then a signal applied to the DATA input will appear on the output after a few nanoseconds' delay. Figure 7-1b shows this in timing diagram format. During the time the CONTROL input is high, the pulsed input signal is passed through to the output.

You can use a NAND gate as a switch in the same way, but the output signal levels will be inverted from those of the AND gate switch. For example, when a NAND gate is disabled by applying a low to one input, the output will be high instead of low as it is for an AND gate. Likewise, if a NAND gate is enabled by making one input high, a signal applied to the other input will be inverted as it passes through the gate.

OR gates or NOR gates can also be used as simple logic-controlled switches. As indicated by their truth tables, however, these devices will be enabled by applying a low to the input you designate as the CONTROL input.

Figure 7-2a shows another type of logic-controlled switch you can build with simple gates, and Figure 7-2b shows the mechanical-switch equivalent circuit for it. For this switch circuit, an input signal will be routed to the X output or to the Y output, depending on the logic level on the CONTROL input. If the CONTROL input is high, the upper AND gate will be enabled, and the lower AND gate will be disabled. Therefore, the input signal will be passed through to the Y output. If the CONTROL input is low, the upper AND gate will be disabled and the lower AND gate will be enabled. As shown by the mechanical-switch equivalent circuit in Figure 7-2b, this circuit functions as a *single-pole, double-throw (SPDT)* switch. Unlike a mechanical switch, however, signals can only pass in one direction through the logic gate switch. The reason for this, of course, is that you can't send logic signals backward through standard logic gates.

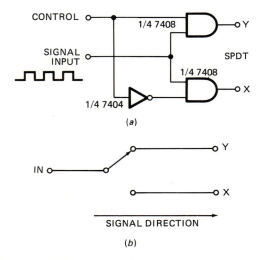

FIGURE 7-2   AND gates used as logic-controlled switches. *(a)* SPDT output schematic.
(b) Mechanical-switch equivalent for the schematic.

Figure 7-3a shows a circuit which functions as the reverse of the circuit in Figure 7-2a. This circuit will route one of two input signals to a single output, depending on the logic level applied to the CONTROL input. If the CONTROL input is high, then the upper NAND gate will be enabled and the signal applied to the A input will be passed through to the Y output. If the CONTROL input is low, then the lower NAND gate will be enabled and the signal applied to the B input will be passed through to the Y output.

For reasons we will explain a little later, the circuit in Figure 7-3a, is called a *2-input multiplexer.* If you need just one of these multiplexers in a system, you can build it with four NAND gates as shown in Figure 7-3a. If you need several of these in a system, you can use a 74LS157,

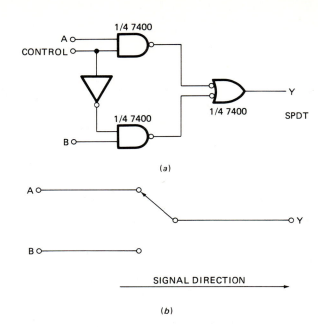

FIGURE 7-3   NAND gates used as logic-controlled switches. *(a)* SPDT input schematic. *(b)* Mechanical-switch equivalent for schematic.

quad 2-input multiplexer, which has four complete multiplexers in a single package. In the next section we teach you more about multiplexers and show you how they are used in systems.

MULTIPLEXERS AND DEMULTIPLEXERS

Introduction

As we described in the preceding section, the term *multiplexing* means transmitting several signals on a single wire by interleaving samples of each. The term *demultiplexing* means separating the output signal from a multiplexer into separate signals again. As a first example of how this works, suppose that a pulsed signal is applied to the CONTROL input of the circuit in Figure 7-4. When the CONTROL input is high, a section of the A signal will be passed through to the Y output. When the CONTROL input is low, a section of the B signal will be passed through to the Y output. The Y output line then will contain alternating sections of the A signal and the B signal.

This scheme may seem familiar to you because this is the same technique that is used to display two signals on a scope with just one electron beam. Remember from the discussion on scopes in Chapter 4 that for a scope, we either "chop" rapidly from input to input as the beam sweeps across the screen, or we "alternate" between doing a trace of one signal and a trace of the other.

The multiplexer on the left side of Figure 7-4 rapidly chops back and forth between the two input signals. If the chopping or multiplexing frequency is higher than the frequency of the data signals, then the original A and B data signals can be accurately reconstructed at the other end of the transmission line. The circuit or

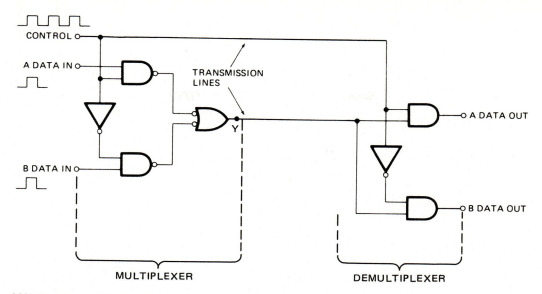

CONTROL

A DATA IN

B DATA IN

TRANSMISSION LINES

Y

A DATA OUT

B DATA OUT

MULTIPLEXER

DEMULTIPLEXER

FIGURE 7-4   NAND gate switch circuits used to transmit two data signals on one wire.

device used to separate the multiplexer output signal into separate signals again is called a *demultiplexer.* For our example here, the simple switch circuit on the right side of Figure 7-4 functions as a two-channel demultiplexer. Here's how it works.

When the CONTROL line is high, the signal on the A input will be routed on to the transmission line by the multiplexer on the left side of Figure 7-4. The high on the CONTROL line will also enable the upper AND gate in the demultiplexer part of the circuit on the right side of Figure 7-4. This will cause the signal on the transmission line to be routed to the A output. For the time that the CONTROL line is high, the circuit functions as a wire between the A input and the A output. When the CONTROL input is low the circuit functions as a wire between the B input and the B output.

About now the question that may have occurred to you is, "Why go to the trouble of chopping up the signals, sending them down a wire, and reconstructing them on the other end of the line?" The answer to this is that multiplexing-demultiplexing reduces the number of wires required to send data signals from one point to another. For the two-channel circuit in Figure 7-4, there is no reduction in the number of lines because you still need two signal wires, a data line, and a control line to transmit two channels of data. When a greater number of signals are multiplexed on a line, however, more lines are saved. After we introduce you to a commonly available multiplexer device, we will show you how this is done.

A Common Multiplexer, the 74ALS151

DATA SHEET ANALYSIS

There are several single-chip multiplexers commonly available. Examples of these are the 74ALS157 quad 2-input multiplexer, the 74HC153 dual 4-input multiplexer, the 74ALS151 8-input multiplexer, and the 74LS150 16-input multiplexer. Figure 7-5 shows the first two pages of the data book entry for a 74ALS151 8-input multiplexer. Let's see what you can learn from these.

First, look at the logic diagram to determine the inputs and outputs on the device. As expected, it has eight data inputs, D0 to D7. It also has a strobe (enable) input, $\overline{G}$, and three DATA SELECT inputs, A, B, and C. The device has an active high output, Y, and an active low output, W. Figure 7-6 shows a switch model for the device. This model and the function table (truth table) in the data sheet should help you see how the device functions.

From the function table you can see that if $\overline{G}$ is high, the Y output will be a low and the W output will be high, regardless of the logic levels present on the data and select inputs. Therefore, the $\overline{G}$ input must be low to enable the device. From the switch model you can see that if the device is enabled, one of the data inputs will be effectively connected to the Y output and its complement sent to the W output. The function table tells you that the 3-bit binary code present on the A, B, and C select inputs determines which data input is connected to the Y output. For example, if 000 is applied to the select inputs, the data present on the D0 input will be transferred to the Y output. If 011 is present on the select inputs, the data on the D3 input will be transferred to the Y output. As a review of basic gates, you can work your way through the logic diagram to see how a particular 3-bit word on the select inputs connects that data input to the Y output.

IEEE/IEC LOGIC SYMBOL FOR A MULTIPLEXER

Figure 7-7*a* shows the symbol that has traditionally been used to represent a 74151-type multiplexer on a schematic. This symbol identifies the pins on the device, but it does not give you any information about how the pin functions relate to each other. As we discussed in an earlier chapter, ANSI/IEEE standard 91-1984 defines symbol conventions which attempt to show the truth table information for a device directly in the symbol. The symbols used in this standard are called *IEEE/IEC logic symbols.* The symbols are also called *dependency notation symbols* because they show how the output sig-

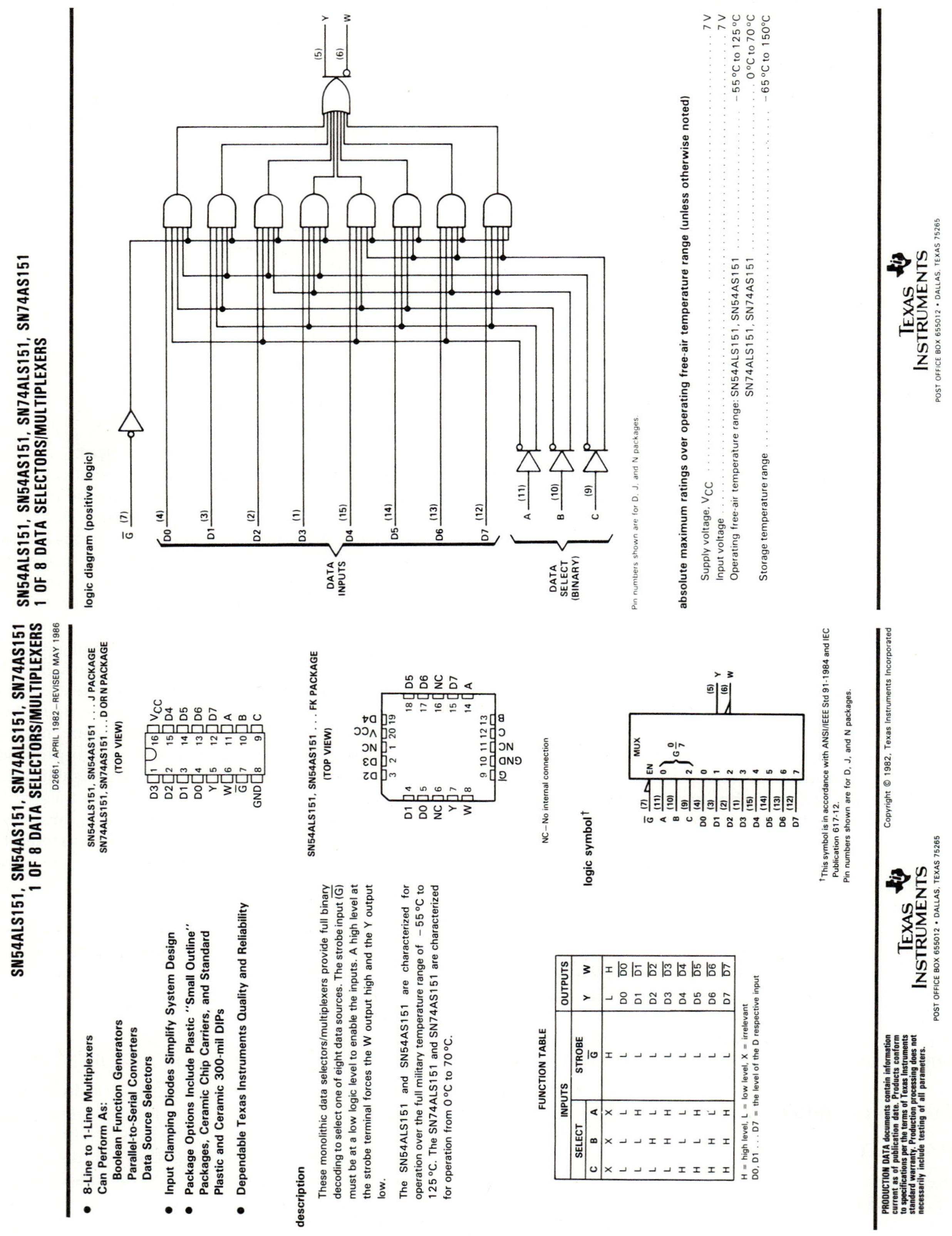

SN54ALS151, SN54AS151, SN74ALS151, SN74AS151
1 OF 8 DATA SELECTORS/MULTIPLEXERS

D2661, APRIL 1982—REVISED MAY 1986

- **8-Line to 1-Line Multiplexers
  Can Perform As:**
  Boolean Function Generators
  Parallel-to-Serial Converters
  Data Source Selectors
- **Input Clamping Diodes Simplify System Design**
- **Package Options Include Plastic "Small Outline" Packages, Ceramic Chip Carriers, and Standard Plastic and Ceramic 300-mil DIPs**
- **Dependable Texas Instruments Quality and Reliability**

description

These monolithic data selectors/multiplexers provide full binary decoding to select one of eight data sources. The strobe input (G) must be at a low logic level to enable the inputs. A high level at the strobe terminal forces the W output high and the Y output low.

The SN54ALS151 and SN54AS151 are characterized for operation over the full military temperature range of −55 °C to 125 °C. The SN74ALS151 and SN74AS151 are characterized for operation from 0 °C to 70 °C.

SN54ALS151, SN54AS151 . . . J PACKAGE
SN74ALS151, SN74AS151 . . . D OR N PACKAGE
(TOP VIEW)

D3	1	16	V_{CC}
D2	2	15	D4
D1	3	14	D5
D0	4	13	D6
Y	5	12	D7
W	6	11	A
G	7	10	B
GND	8	9	C

SN54ALS151, SN54AS151 . . . FK PACKAGE
(TOP VIEW)

NC—No internal connection

FUNCTION TABLE

INPUTS			OUTPUTS	
SELECT		STROBE	Y	W
C	B	A	G		
X	X	X	H	L	H
L	L	L	L	D0	D0
L	L	H	L	D1	D1
L	H	L	L	D2	D2
L	H	H	L	D3	D3
H	L	L	L	D4	D4
H	L	H	L	D5	D5
H	H	L	L	D6	D6
H	H	H	L	D7	D7

H = high level, L = low level, X = irrelevant
D0, D1 . . . D7 = the level of the D respective input

logic diagram (positive logic)

Pin numbers shown are for D, J, and N packages.

logic symbol†

†This symbol is in accordance with ANSI/IEEE Std 91-1984 and IEC Publication 617-12.
Pin numbers shown are for D, J, and N packages.

absolute maximum ratings over operating free-air temperature range (unless otherwise noted)

Supply voltage, V_{CC}	7 V
Input voltage	7 V
Operating free-air temperature range: SN54ALS151, SN54AS151	−55 °C to 125 °C
SN74ALS151, SN74AS151	0 °C to 70 °C
Storage temperature range	−65 °C to 150 °C

Pin numbers shown are for D, J, and N packages.

TEXAS INSTRUMENTS
POST OFFICE BOX 655012 • DALLAS, TEXAS 75265

FIGURE 7-5 First two pages of the data sheet for a 74ALS151 8-input multiplexer. *(Courtesy of Texas Instruments Inc.)*

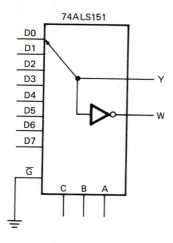

FIGURE 7-6   Switch model for the 74ALS151 device.

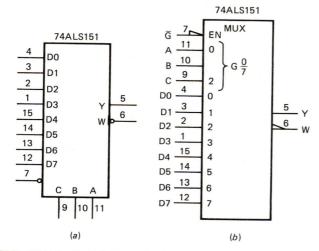

(a)

(b)

FIGURE 7-7   74ALS151 multiplexer. (a) Traditional symbol. (b) IEEE/IEC logic symbol.

nals depend on the input signals. The examples explained in this chapter should give you a good start at understanding this standard.

The IEEE/IEC logic symbol for a 74ALS151 multiplexer is shown again in Figure 7-7b so you can compare it with the traditional symbol. At the top of the symbol, the general function of the device is indicated by the label MUX, a common shorthand for multiplexer. Note that input signals are always drawn on the left side of the device and output signals are drawn on the right side. Since there are fewer of them, let's look at the outputs first.

The wedge on the W output shows that it is active low. As we have shown previously, a bubble may be used instead of the wedge to indicate an active low input or output. Since it has no wedge or bubble, the Y output is active high.

An enable input on an IEEE/IEC logic symbol is identified with the letters EN. The wedge on the EN input of the 74ALS151 indicates that this input must be low to enable the device. The data inputs on the multiplexer are each identified with a number. The number for each

is the decimal equivalent of the binary number that must be applied to the select inputs in order to connect that input to the Y output. The binary weights of the A, B, and C select inputs are indicated by numbers 0 to 2 next to them. From this you can tell that the least significant bit of the binary select code should be applied to the A select input, and the most significant bit of the binary select code should be applied to the C input.

The letter G in an IEEE/IEC logic symbol is used to represent AND dependency. For this device the characters $\}G_7^0$ tell you that the signal on the Y output depends on the 3-bit code on the select inputs AND the data on the input, 0 to 7, selected by that code.

At this point the function of the device may not be immediately obvious to you from the symbol. However, if you study the function table to see how the symbol relates to it, the next time you see the symbol it should trigger a memory of the truth table. The traditional symbol in Figure 7-7a is less likely to serve as a trigger in this way. In some cases, throughout the rest of this book, we will use the IEEE/IEC logic symbol and in other cases we will use the traditional symbol, whichever makes the operation of a particular circuit easier to see.

Now that you are familiar with the operation of a multiplexer, let's take a look at some multiplexer applications.

Multiplexer Applications

DIGITAL SIGNAL MULTIPLEXING

Figure 7-8 shows how a 74ALS151 can be used to multiplex eight signals onto one wire. The 3-bit code applied to the select inputs determines which of the eight signals is being sent on the transmission line at a particular time. This 3-bit code is often called the *address* of the input. You might, for example, say that a 3-bit address is required to select one of the eight inputs. The 3-bit binary counter connected to the select inputs determines the address present. As this counter is clocked, it continuously steps through a binary count sequence from 000 to 111 over and over. The output transmission line then will contain a sample of data from input 0 fol-

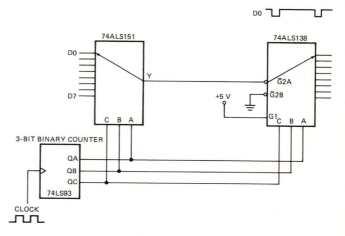

FIGURE 7-8   Multiplexing eight signals onto one wire using a 74ALS151.

lowed by a sample of data from input 1, followed by a sample of data from input 2, etc. When the counter cycles back to 000, another sample from input 0 will be sent down the wire. This result is equivalent to connecting a motor to the switch in Figure 7-6 so that the switch rotates from input 0 to input 7 and back to 0 over and over again.

The data samples sent down the transmission line are separated into eight individual signals again by an 8-output demultiplexer such as that represented by the mechanical-switch equivalent circuit on the right side of Figure 7-8. In a later section of this chapter, we show the timing waveforms for a system such as this and discuss in detail the 74ALS138 which can be used as the demultiplexer in the system.

Note in Figure 7-8 that the address lines are connected to the demultiplexer in the same order that they are connected to the multiplexer. This is done so that when the multiplexer reads in a data sample from an input, the data sample will be output on the corresponding output of the demultiplexer as we explained for the 2-input multiplexer in Figure 7-4.

The system in Figure 7-8 sends eight channels of data with only one data line and three address lines. This saves four wires. Multiplexers can be *cascaded* as shown in Figure 7-9 to save even more wires. The system in Figure 7-9, for example, multiplexes 64 signals with one transmission line and six address lines. This then makes a total of only 7 lines to send 64 data signals.

The three lower address lines, A0 to A2, are connected in parallel to all the input multiplexers. These lines select one of the input lines in each of the input multiplexers and route it to one of the inputs of the output multiplexer. Address lines A3 to A5, connected to the select inputs of the final multiplexer, determine which of these signals is routed to the final output. If address 111000 is applied to the system, for example, the signals on the D0 inputs of the 8-input multiplexers will be routed to the output multiplexer. Since the select inputs of this multiplexer have 111 or 7 applied to them, the signal from the D7 input of this multiplexer will be routed to the final output. If a 6-bit binary counter is connected to the inputs of this circuit, it will cycle through all 64 inputs over and over. Circuits such as the one in Figure 7-9 are sometimes called *multiplexer trees.*

AN OSCILLOSCOPE INPUT MULTIPLEXER

In Chapter 4 we described how a single beam oscilloscope traces out two waveforms on the CRT by either chopping rapidly back and forth between two input signals as the beam sweeps across the screen, or alternating between doing a trace of one input signal and doing a trace of the other input signal. Both the chop mode and the alternate mode are examples of multiplexing.

Figure 7-10 shows a circuit you can build that will display up to eight channels of digital data on a standard oscilloscope. Since this is the most complex schematic we have shown you so far, we will make a general comment about how you most effectively approach a new system before we discuss the operation of this one.

For most of us, the first tendency is to be overwhelmed when we are faced with analyzing a new system or sche-

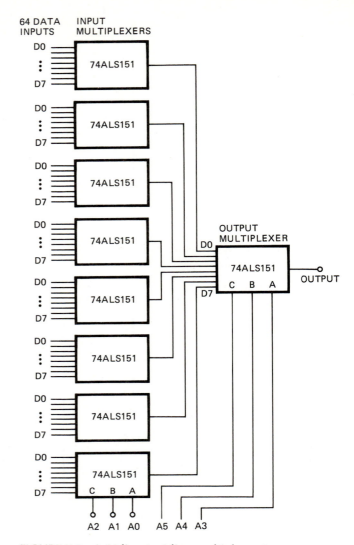

FIGURE 7-9 A 64-line to 1-line multiplexer tree.

matic. The method used to overcome this tendency we jokingly call the *Five-Minute Rule.* This rule states that:

> When faced with some seemingly overwhelming system or schematic, you get 5 minutes to "freak out" and think about changing vocations; then you have to dig in, find something you recognize, and work your way through the system from there.

For the scope multiplexer circuit in Figure 7-10 you might, after your 5 min, start with the 74ALS151 digital multiplexer because you remember it from the preceding discussions. The enable input, $\overline{EN}$, of the 74ALS151 is tied low, so the device is permanently enabled. Therefore, the signal on the Y output will be the same as that on the data input addressed by the 3-bit binary code on the select (A, B, C) inputs. The select inputs of the multiplexer are connected to the outputs of a 74LS93 binary counter, so the multiplexer will be cycled through all its inputs over and over, as we described in the preceding multiplexer application example.

A second point to keep firmly in mind when you are

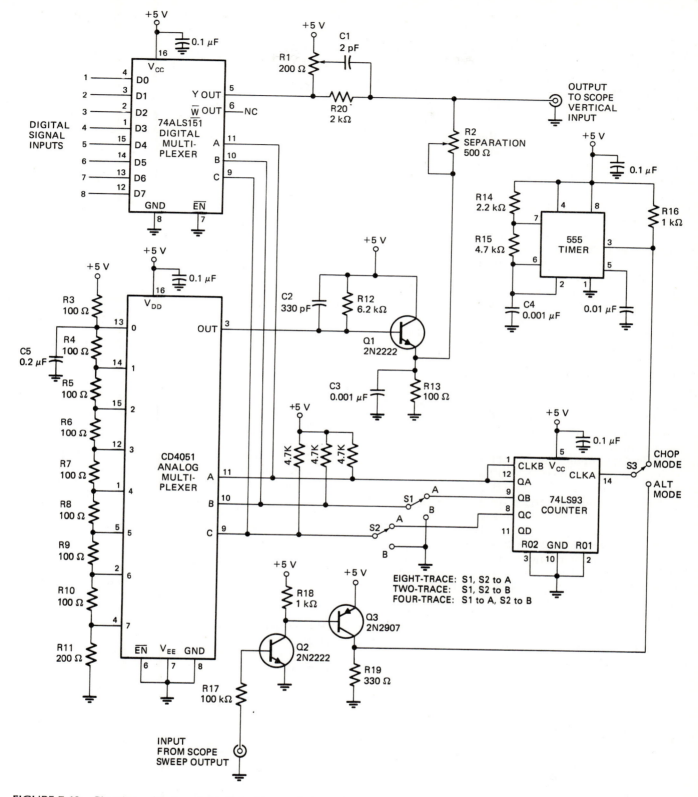

FIGURE 7-10 Circuit to display eight digital input signals on an oscilloscope. [*This circuit was modified from one found in the October 14, 1976, issue (vol. 49) of* Electronics *magazine, published by McGraw-Hill.*]

faced with learning a new system or analyzing a new schematic is that you usually don't have to understand the detailed operation of every part in the system in or-

der to build it, troubleshoot it, or use it. Often an understanding of the function of a block of circuitry is all that is needed.

In a later chapter, we discuss the internal operation of a 74LS93 counter, but for analyzing the circuit in Figure 7-10, all you need to know about the 74LS93 is that it outputs a binary count sequence on its QA, QB, and QC outputs when it receives a pulsed signal on its clock input.

The output of the 555 timer device in the upper right corner of Figure 7-10 is connected to the CLKA input of the 74LS93, so you can assume that this device generates the clock signal required to step the counter through its count. With the resistor and capacitor values shown, the 555 outputs a TTL level signal with a frequency of about 120 kHz.

The next major part of the circuit for you to determine the function of is the CD4051 analog multiplexer in the lower left corner of Figure 7-10. Figure 7-11 shows the internal structure of this device. Since the switches between the inputs and the output are CMOS transmission gates, the CD4051 can multiplex either analog or digital signals. Remember from Chapter 3 that when a transmission gate is enabled, it acts like a resistance of a few hundred ohms between its input and its output. Therefore, signals can be passed in either direction through transmission gates. As indicated by the IN/OUT labels on the data pins in Figure 7-11, this device can be used to route one of eight signals to a common output, or it can be used to route a single input signal to one of eight outputs. The 3-bit binary code applied to the A, B, and C select inputs determines which of the eight transmission gates in the CD4051 is turned on at any particular time.

To understand the function of the CD4051 in this particular circuit, you need to look back at the signal produced on the output of the 74ALS151 digital multiplexer. As its select inputs are cycled through the binary count 0 to 7, this device sends samples of the signals on its eight data inputs out on its Y output. If this signal were connected directly to the vertical input of a scope, the trace produced would simply show the sequence of eight samples along one line on the screen. What we really want here is to have the data for the eight inputs separated into separate traces as it is for the normal chop or alternate display of two input channels. Remember from the discussion of scope operation in Chapter 4 that we can move the electron beam up or down on the screen

by simply adding a dc voltage to the signal. This is how the vertical position control on a scope works. The point here is that if we add a different dc voltage to each of the samples from the digital multiplexer, the traces for each will be separated on the CRT as desired. Now look back to the CD4051 analog multiplexer section of Figure 7-10 to see how we do this.

The eight data inputs of the CD4051 are connected to the nodes of a voltage divider set up to produce voltages of 1.0, 1.5, 2.0, 2.5, 3.0, 3.5, 4.0, and 4.5 V. When one of the inputs is selected by applying the binary code for it to the A, B, and C inputs, the voltage on that input will pass through a transmission gate to the OUT pin of the device. The 2N2222 transistor circuit connected to this output buffers this voltage and adds it to the signal from the digital multiplexer. The outputs from the 74LS93 binary counter are connected to the select inputs of both the digital multiplexer and the analog multiplexer, so as each digital input is sampled, a different analog voltage is sampled by the CD4051 and added to the digital sample. This causes the digital samples each to be displayed at a different vertical position on the CRT as desired. Potentiometer R2 allows you to adjust the amount of separation between the traces.

If you do not need all eight input channels, you can display fewer channels by adjusting S1 and S2 as shown in Figure 7-10. If you display fewer channels, each channel will get sampled more often, so you can accurately display higher frequency signals.

To look at signals that are higher in frequency than you can accurately display in chopped mode, you can use alternate mode if your scope has a sweep output signal available. To do this, connect the sweep output signal from the scope to the input of the 2N2222 and 2N2907 buffer shown on the bottom of the circuit in Figure 7-10. Then move switch S3 to route the output of this buffer to the CLKA input of the 74LS93 counter. Each sweep of the scope beam will then step the multiplexers to the next input channel so it will be displayed on the next trace. Incidentally, to obtain a stable display in either the chop or alternate mode, you should connect the lowest frequency digital input signal to the EXTERNAL TRIGGER input of the scope and set the scope trigger controls as needed for this.

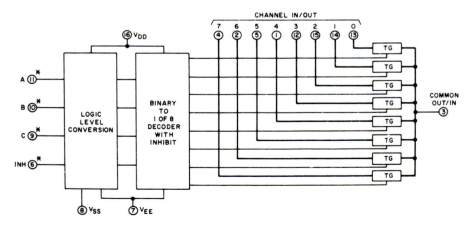

FIGURE 7-11 Functional diagram of CD4051B. *(Courtesy of RCA.)*

USING A MULTIPLEXER TO IMPLEMENT COMBINATIONAL LOGIC FUNCTIONS

Besides its use for multiplexing signals, a multiplexer can be used to implement logic functions such as those we showed you how to implement with discrete gates in Chapter 6. As a first example, suppose that you want to implement the logic function shown by the truth table in Figure 7-12a. Using the approach we showed you in

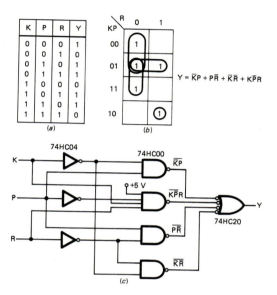

FIGURE 7-12 A logic expression implemented in a traditional way. (a) Truth table. (b) Karnaugh map. (c) Logic circuit.

Chapter 6, you would make a Karnaugh map for the truth table as shown in Figure 7-12b. As shown in the figure, the reduced expression derived from the Karnaugh map is $Y = \overline{K}P + P\overline{R} + \overline{K}\overline{R} + K\overline{P}R$. This expression can be implemented with NAND gates and inverters as shown in Figure 7-12c. As you can see in the figure, this implementation requires three IC packages, a 74HC20 triple 4-input NAND, a 74HC00 quad 2-input NAND, and a 74HC04 hex inverter. An alternative way to implement this circuit is with a single 74HC151 multiplexer as shown in Figure 7-13. Here's how this works.

The three input variables are connected to the select inputs of the multiplexer in the same order as they are in the truth table. Now, you know that when the K, P, and R inputs are 000, the logic level on the D0 input of the multiplexer will be sent out on the Y output of the multiplexer. According to the truth table, you want the output to be a 1 for this input combination. If you tie the D0 input of the multiplexer to a logic high (+5 V), then an input combination of 000 on the KPR lines will cause this high to be transferred to the Y output of the multiplexer.

The next entry in the truth table tells you that, for the input combination 001, you want the Y output to be a 0. For this input combination, the D1 input will be addressed, so to produce this output you just connect the D1 input to a logic low (ground). To implement the rest

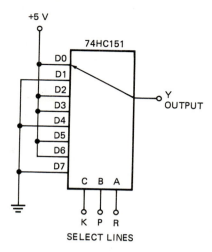

FIGURE 7-13 A 74HC151 multiplexer used to implement the logic circuit shown in Figure 7-12c.

of the truth table you just connect each data input to +5 V or to ground, whichever logic state you want the Y output to have for the corresponding input combination.

The advantage of this circuit is that you can implement any logic function of three input variables almost as fast as you can write the truth table. Also, if you want to change the output expression to, for example, a low output for the input combination 010, all you have to do is move the wire connected to the D2 input from +5 V to ground. With the previously described methods, you would have to do a new Karnaugh map, derive a new simplified expression, and then redesign and rebuild the circuit. For this example, the new circuit would probably require different ICs. Remember that the 74151-type devices have an active low W output, so you can easily implement functions which require an active low output.

To directly implement any logic function of four input variables, you can use this same multiplexer technique with a 16-input multiplexer such as the 74ALS150. The 74ALS150, however, comes in a larger, more expensive 24-pin package, and the device has only an active low output. A simple *folding trick* allows you to implement any logic function of four input variables with an 8-input multiplexer such as a 74ALS151 and perhaps one inverter. Here's an example.

Figure 7-14a shows a standard truth table for a four-variable logic function, and Figure 7-14c shows how this truth table can be implemented with a 74ALS151. As you can see in Figure 7-14c, the L, M, and N signals are simply connected to the C, B, and A inputs of the multiplexer in the same order they are written in the truth table. To see what to do with the P input variable, first draw a vertical line down through the truth table between the N column and the P column as shown in Figure 7-14b. Then draw horizontal lines after each group of two input combinations as also shown in Figure 7-14b.

Now, for both of the first two input combinations, L = 0, M = 0, and N = 0, so the Y output of the multiplexer will be connected to the D0 input. For the first of these combinations, P = 0 and we want the output to be a 0. For the second of these combinations, P = 1 and

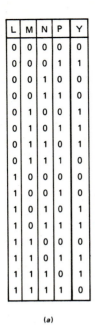

L	M	N	P	Y
0	0	0	0	0
0	0	0	1	1
0	0	1	0	0
0	0	1	1	0
0	1	0	0	1
0	1	0	1	1
0	1	1	0	1
0	1	1	1	0
1	0	0	0	0
1	0	0	1	0
1	0	1	0	1
1	0	1	1	1
1	1	0	0	0
1	1	0	1	1
1	1	1	0	1
1	1	1	1	0

(a)

L	M	N	P	Y	
0	0	0	0	0	} Y = P
0	0	0	1	1	
0	0	1	0	0	} Y = 0
0	0	1	1	0	
0	1	0	0	1	} Y = 1
0	1	0	1	1	
0	1	1	0	1	} Y = $\overline{P}$
0	1	1	1	0	
1	0	0	0	0	} Y = 0
1	0	0	1	0	
1	0	1	0	1	} Y = 1
1	0	1	1	1	
1	1	0	0	0	} Y = P
1	1	0	1	1	
1	1	1	0	1	} Y = $\overline{P}$
1	1	1	1	0	

(b)

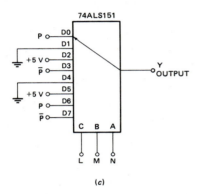

(c)

FIGURE 7-14   Implementation of a standard truth table for a four-variable logic function using a multiplexer. (a) Standard truth table. (b) Modified truth table. (c) Implementation using a 74ALS151.

we want Y to be a 1. For these combinations, then, the desired states on Y are the same as those on P. Therefore, the first two combinations in the truth table can be implemented by simply connecting the P signal to the D0 input of the multiplexer. Another example should help you see how easy this is.

Take a look at the P and Y values for L = 0, M = 0, and N = 1 combinations in Figure 7-14b. For the first of these combinations, P = 0 and Y = 0. For the second of these combinations, P = 1 and Y = 0. For these two combinations, then, we want the Y output to be a 0, regardless of the logic level on the P input. Since 001 on the multiplexer inputs addresses the D1 data input, we can implement this by simply connecting the D1 input to ground as shown in Figure 7-14c.

Likewise, for the two 010 input combinations of L, M, and N, we want Y to be a 1, regardless of the state on P. Therefore, all we have to do to implement this is to connect the D2 input of the multiplexer to +5 V as shown in Figure 7-14c. Finally, for the two 011 input combinations of L, M, and N, we want Y to be the complement of

P. This is easily accomplished by inverting the P signal and connecting the result to the D3 input of the multiplexer. Work your way through the rest of the modified truth table to make sure you see how this works. Note that the only possibilities for 74HC151 data inputs are ground, +5 V, P, or $\overline{P}$. If you need to make a change in the truth table, all you have to do is update the connections to the data inputs to match the new truth table using ground, +5 V, P, or $\overline{P}$.

With this folding trick, an 8-input multiplexer and an inverter can implement any logic function of four input variables. Likewise, a 16-input multiplexer such as the 74LS150 can be used with this folding trick to implement any function of up to five input variables.

A multiplexer used to implement logic functions in this way is a first example of a *programmable logic device* which can implement any desired function, depending on how you program it. You "program" a multiplexer for a desired logic function with the signals you connect to the data inputs. Later in this chapter we show you how to use more complex programmable logic devices such as *read-only memories* (ROMs), *programmable logic arrays* (PLAs), or *programmable array logic* devices (PALs). You usually use one of these devices for applications where you need a function of more than four or five variables, or for applications where you need several output functions from a set of input variables.

Demultiplexers

INTRODUCTION

In earlier sections of this chapter, we have shown you how a multiplexer can be used to transmit several digital signals on a single wire, and how a demultiplexer can be used to separate the multiplexed signals into individual signals again. In this section we discuss the operation of a commonly available demultiplexer and analyze the timing diagram for a multiplexer-demultiplexer data link.

Some commonly available demultiplexers are the 74ALS155 dual 4-output demultiplexer, the 74ALS138 8-output demultiplexer, and the 74ALS154 16-output demultiplexer. For our example here, we will look at the data sheet of the 74ALS138, which can be used to demultiplex data from the 74ALS151 multiplexer discussed in the preceding sections.

DEMULTIPLEXER DATA SHEET ANALYSIS

Figure 7-15 shows the logic symbols, logic diagram, and function table for a 74ALS138. Note that two different IEEE/IEC logic symbols are shown for it because the device can function as either a demultiplexer or as a 3-line to 1-of-8 line low decoder, depending on the application. In a later section of the chapter on code converters, we will describe how the 74138 functions as a decoder. For now we want to discuss how it functions as a demultiplexer.

The wedges on the Y0 to Y7 outputs of the logic symbol tell you that they are active low. As with a multiplexer, the select inputs C, B, and A are used to address one of the eight outputs. If the device is enabled, the 3-bit bi-

logic symbols (alternatives)

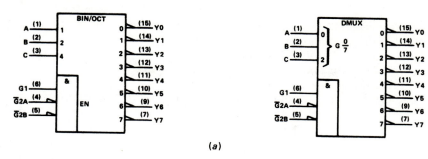

(a)

logic diagram (positive logic)

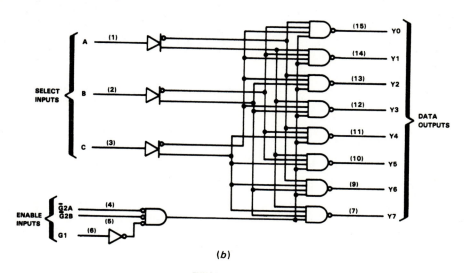

(b)

FUNCTION TABLE

ENABLE INPUTS			SELECT INPUTS			OUTPUTS							
G1	$\overline{G2A}$	$\overline{G2B}$	C	B	A	Y0	Y1	Y2	Y3	Y4	Y5	Y6	Y7
X	H	X	X	X	X	H	H	H	H	H	H	H	H
X	X	H	X	X	X	H	H	H	H	H	H	H	H
L	X	X	X	X	X	H	H	H	H	H	H	H	H
H	L	L	L	L	L	L	H	H	H	H	H	H	H
H	L	L	L	L	H	H	L	H	H	H	H	H	H
H	L	L	L	H	L	H	H	L	H	H	H	H	H
H	L	L	L	H	H	H	H	H	L	H	H	H	H
H	L	L	H	L	L	H	H	H	H	L	H	H	H
H	L	L	H	L	H	H	H	H	H	H	L	H	H
H	L	L	H	H	L	H	H	H	H	H	H	L	H
H	L	L	H	H	H	H	H	H	H	H	H	H	L

(c)

FIGURE 7-15 Extract from the 74ALS138 data sheet. *(a)* Logic symbols. *(b)* Logic diagram. *(c)* Function table. *(Courtesy of Texas Instruments Inc.)*

nary code applied to these inputs will determine which output will be asserted (low). To determine how the enable inputs on this device work, let's take a look at the logic diagram in Figure 7-15b. From the logic diagram you can see that the output of the bubbled 3-input AND gate in the lower left corner of the logic diagram must be high to enable the output NAND gates. If this point is low, the outputs of all eight NAND gates will be high, regardless of the 3-bit code on the select inputs.

In order for the output of the bubbled input AND to be high, the $\overline{G2A}$ input must be asserted (low), the $\overline{G2B}$

input must be asserted (low), and the G1 input must be asserted (high). On the logic symbol for the DMUX, these conditions are indicated by the rectangular box containing an "&" (ampersand) in the lower left corner of the symbol. The G in the logic symbol again represents AND dependency. For this DMUX symbol, the characters {G_7^0 mean that the output state depends on the 3-bit code on the select inputs AND the output of the box with the & in it. If the output of the & box is high, then the addressed output will be asserted (low). If the output of the & box is low, then the addressed output will be unasserted (high). Let's look at the function table in Figure 7-15c to verify this.

The first three lines in the function table show you that if any of the ENABLE inputs are not asserted, then the outputs will all be high, regardless of the logic levels on the select inputs. The rest of the function table shows you that if the ENABLE inputs are all asserted, the output addressed by the code on the select inputs will be asserted (low).

By now you may have a good feeling for how the device operates, but it is probably not clear to you how the device can be used as a demultiplexer. Figure 7-16 shows how you connect a 74ALS151 and a 74ALS138 in a multiplexer-demultiplexer data link.

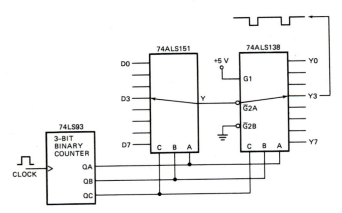

FIGURE 7-16   A 74ALS151 and a 74ALS138 multiplexer-demultiplexer data link.

First note in Figure 7-16 that the address lines are connected to the two devices in parallel so that when the multiplexer is pointing to a particular channel, the demultiplexer is pointing to the same channel. Next note in Figure 7-16 that we have connected the data from the multiplexer to the $\overline{G2A}$ input of the 74ALS138 and connected the other ENABLE inputs of the 74ALS138 so that they are permanently asserted. To see how this works, suppose that the data on the transmission line from the multiplexer is a low, and the address on the select lines is 011.

The low on the $\overline{G2A}$ input will enable the 74ALS138. According to the truth table for the device, the output addressed by the select lines, Y3, will be asserted (low), which corresponds to the data on the transmission line. Now, if the data on the transmission line is high, the $\overline{G2A}$ input of the 74ALS138 will not be asserted. As shown by the first line of the function table in Figure 7-15c, the 74ALS138 will now be disabled, and the Y3 output will be high, which again corresponds to the data on the transmission line. To summarize, then, a low data bit enables the 74ALS138 and causes the addressed output to be asserted low. A high data bit disables the device so that the output is high, regardless of the address on the select inputs.

For some applications the basic multiplexer-demultiplexer circuit in Figure 7-16 has a slight problem. The problem is caused by the fact that when the demultiplexer is disabled, all the output lines go high. In other words, when an output

of the demultiplexer is not selected, it will go high, regardless of the data that was on the output when it was selected. To see this problem more clearly, let's take a look at the timing diagrams for the system.

MULTIPLEXER-DEMULTIPLEXER SYSTEM TIMING DIAGRAM

Figure 7-17 shows a timing diagram for the various parts of the multiplexer-demultiplexer system in Figure 7-16. The clock signal applied to the 74LS93 binary counter causes it to output a binary count sequence of 000 to 111 over and over on its QA, QB, and QC outputs. The counter outputs are connected to the select (C, B, A) inputs of the 74ALS151 multiplexer, so when a particular state is present on the outputs of the counter, the data from the corresponding input of the multiplexer will be present on the Y output of the multiplexer. In the timing diagram in Figure 7-17, we show both a high and a low for each data bit on the output of the multiplexer because the bit may be high or it may be low, depending on the signal on the data input. You will see this convention used often in data sheets and throughout the rest of this book. The 50 percent point on the waveform is usually taken as the time when a valid high or low is on the specified signal line.

The address lines from the counter are connected in parallel to the multiplexer select inputs and to the demultiplexer select inputs, so when the multiplexer is pointing to a particular input channel, the demultiplexer is pointing to the same numbered channel. If, for example, the address lines contain 000, the data on the transmission line will be transferred to the Y0 output of the demultiplexer as shown by the Y0 waveform in Figure 7-17. Note that on this waveform, we again show both a high and a low for the data bit because the data may be high or low, depending on the signal input to the multiplexer at that time.

When the counter increments to 001, the Y0 output of the demultiplexer will always go high because that output is no longer selected. The 001 on the select inputs will select the Y1 output, so the data on the transmission line will be transferred to the Y1 output of the demultiplexer. As you can see in Figure 7-17 the data from a particular input of the multiplexer is only present on the corresponding output of the demultiplexer when the address for that channel is present on the select inputs. For some applications this is acceptable, but for other applications we would like the data on each output line to be held constant until it is updated by a new sample from the transmission line. Devices called *latches* or other devices called *flip-flops* can be used to hold data bits constant as needed here. In the next chapter we discuss the operation of these devices and show you how they can be used to produce demultiplexed signals which remain constant between updates.

DIGITAL SYSTEM CODES

Numerical Codes

One common type of MSI combinational logic device is the group of devices used to convert numbers or quan-

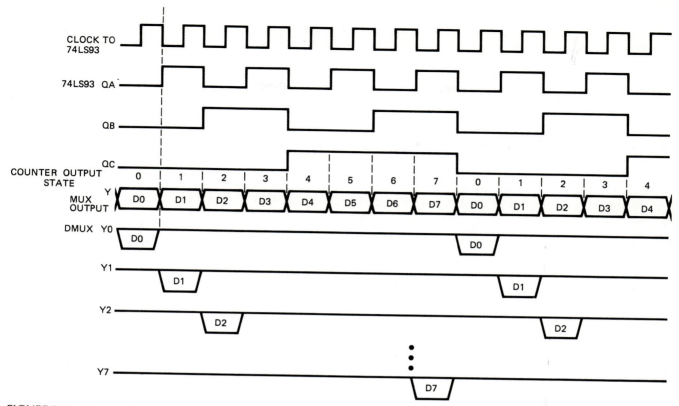

FIGURE 7-17 Timing diagram for the multiplexer-demultiplexer system in Figure 7-16.

tities in one digital code to an equivalent value in some other digital code. Before we introduce you to the devices, we need to discuss the structure of commonly used codes and how you manually convert a number from one code to another.

BINARY

In Chapter 2, we introduced you to the binary number system. We also showed you in Chapter 2 how to convert decimal numbers to their binary equivalents and how to convert binary numbers to their decimal equivalents. For review, here are the weights for a few binary digits to the left and a few digits to the right of the binary point:

$$2^8 \quad 2^7 \quad 2^6 \quad 2^5 \quad 2^4 \quad 2^3 \quad 2^2 \quad 2^1 \quad 2^0 \quad . \quad 2^{-1} \quad 2^{-2} \quad 2^{-3}$$

$$256 \; 128 \; 64 \quad 32 \quad 16 \quad 8 \quad 4 \quad 2 \quad 1 \quad . \quad \tfrac{1}{2} \quad \tfrac{1}{4} \quad \tfrac{1}{8}$$

For further review, we show at the right how decimal 134 can be converted to its binary equivalent of 10000110 by successive subtractions of powers of 2.

As shown on the right side of the calculation, the binary result can be checked by multiplying each binary bit times its equivalent weight, and adding up all the products. We don't usually bother to say it that way, but a number expressed in binary form can be thought of as being in *straight binary code.*

Since most computers and other digital systems can only work with two logic levels, straight binary or some other binary-based code is the natural choice to represent data in these systems. Most people, however, have trouble remembering long binary numbers. To solve this

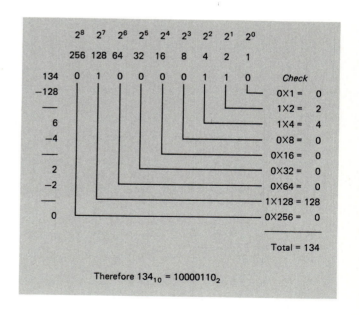

Therefore $134_{10} = 10000110_2$

problem, we often represent binary numbers in one of two other number systems, *octal* or *hexadecimal.*

OCTAL

As we told you in Chapter 2, you can build a number system using any number as a base, but some numbers are more useful to us than others. The octal number system uses 8 as a base, so each digit in an octal number represents a power of 8. Here are the decimal weights of a few octal digits to the left and a few to the right of the *radix* (octal) point. Note that as you move from right to left, each digit is 8 times the weight of the previous digit:

8^4	8^3	8^2	8^1	8^0	.	8^{-1}	8^{-2}	8^{-3}
4096	512	64	8	1	.	$\frac{1}{8}$	$\frac{1}{64}$	$\frac{1}{512}$

The number of different symbols required to express a digit value in any number system is equal to the base. The decimal system, for example, requires 10 symbols (0 to 9), the binary system requires 2 symbols (0 and 1), and the octal system requires 8 symbols (0 to 7). To count from 0 in octal, then, you count 0, 1, 2, 3, 4, 5, 6, 7, 10, 11, 12, 13, 14, 15, 16, 17, 20, 21, etc.

Each octal digit can represent eight different values, 0 to 7, or, in binary, 000 to 111. This means that one octal digit can be used to represent three binary digits. As an example of this, suppose that you want to represent the binary number 101011111 with its octal equivalent. Starting from the binary point and moving to the left, mark off groups of three binary digits. Then, below each group of three binary bits, write the equivalent octal digit. For the example shown below, the rightmost group of 111 is equal to 7, the center group of 011 is equal to 3, and the leftmost group of 101 is equal to 5. The binary number 101011111, then, is equal to 537 in octal:

Binary	101	011	111
Octal	5	3	7

To show that a number is in octal, a subscript 8 is usually written after the number, as, for example, 537_8. For most of us 537_8 is much easier to remember than 101011111.

An octal number can be converted to its binary equivalent by simply reversing the process we just used to get an octal equivalent number. Given the number 712_8, for example, you just write the 3-bit binary equivalent for each octal digit. The 7 is equal to 111, the 1 is equal to 001, and the 2 is equal to 010, so 712_8 is equal to 111001010 binary.

To convert a decimal number to its octal equivalent, you can convert it first to binary, and then to octal, or you can convert it directly to octal in one of two ways.

The first way to convert a decimal number directly to octal is with successive divisions by powers of 8. Here is an example of how to convert 249_{10} to octal:

Power of 8	8^4	8^3	8^2	8^1	8^0
Decimal weight	4096	512	64	8	1
	0	0	3	7	1

$$\begin{array}{r} 3 \rightarrow 3 \\ 64\overline{)249} \\ 192 \\ \hline 57 \end{array}$$

$$\begin{array}{r} 7 \rightarrow 7 \\ 8\overline{)57} \\ 56 \\ \hline 1 \end{array}$$

$$\begin{array}{r} 1 \rightarrow 1 \\ 1\overline{)1} \end{array}$$

$$249_{10} = 371_8$$

You first look for the highest power of 8 that is less than the number you are trying to convert. For the example of 249 decimal, 8^4, or 4096, is too large and 8^3, or 512, is too large, so 8^2, or 64, is the largest power of 8 that will fit in 249. As shown by the division, 8^2, or 64, fits 3 times into 249. Subtracting 3×64, or 192, from the original 249 leaves a remainder of 57. Therefore, you write a 3 in the 8^2 digit position of the octal number and continue the process with the remainder of 57. Since 57 divided by 8^1, or 8, is 7, you write a 7 in that position of the octal number. Subtracting 8×7, or 56, from the 57 leaves 1, which you simply write in as the least significant digit position of the octal number. As shown, 249 decimal is equal to 371_8.

The second method of converting a decimal number directly to octal involves successive division by 8 until the result of the division *(quotient)* is 0. Here's how to convert 327_{10} to octal using this method:

				Check			
$8\overline{)327}$	=	40 remainder	7 (LSD)	7	×	1	= 7
$8\overline{)40}$	=	5 remainder	0	0	×	8	= 0
$8\overline{)5}$	=	0 remainder	5 (MSD)	5	×	64	= $\underline{320}$
						Total	= 327

Therefore:

$$327_{10} = 507_8$$

The result of the first division gives a quotient of 40 and a remainder of 7. With this method, the 7 is the least significant digit of the octal result. Dividing the 40 by 8 gives a quotient of 5 and a remainder of 0, so the next most significant digit of the octal number is a 0. The next division gives a quotient of 0 and a remainder of 5, so the MSD of the octal number is a 5 as shown. Since the quotient from this division is 0, you don't do any further divisions. The octal result can be checked by multiplying each octal digit by its decimal weight and adding the results as shown on the right hand side of the example.

HEXADECIMAL

Several popular minicomputers of the 1970s worked with 12-bit binary words. Since an octal digit can represent three binary bits, four octal digits were an easy way to represent 12 binary bits so they could be more easily remembered. The microcomputers and minicomputers of the 1980s and 1990s, however, use binary words of 8, 16, and 32 bits. Since these numbers do not divide evenly into groups of three, octal is not a convenient code to help people remember binary numbers of these sizes. Therefore, we now, for the most part, use a base 16 number system called *hexadecimal*, or just *hex*, which allows us to represent each group of four bits with a single digit.

Each digit in the hexadecimal number system represents a power of 16. Here are the decimal weights of a few digits to the left and a few digits to the right of the radix point in the hexadecimal number system. Note that as you move from right to left, the weight of a digit is 16 times the weight of the preceding digit:

16^3	16^2	16^1	16^0	.	16^{-1}	16^{-2}	16^{-3}
4096	256	16	1	.	$\frac{1}{16}$	$\frac{1}{256}$	$\frac{1}{4096}$

Since hexadecimal is base 16, sixteen different symbols are required to represent the possible values for each digit. Here are the 16 symbols used:

Decimal	Hex	Decimal	Hex
0	0	8	8
1	1	9	9
2	2	10	A
3	3	11	B
4	4	12	C
5	5	13	D
6	6	14	E
7	7	15	F

The numbers 0 through 9 are used for the first 10 symbols and the letters A through F are used for values of 10 through 15. When you count up from 0 in hex, then, you count 0, 1, 2, 3, 4, 5, 6, 7, 8, 9, A, B, C, D, E, F, 10, 11, 12, 13, 14, 15, 16, 17, 18, 19, 1A, 1B, 1C, etc. As you can see from this series, after 15 decimal or F hexadecimal, the next number is 10 hexadecimal, which means $1 \times 16^1 + 0 \times 16^0$. To show that a number is hexadecimal, it may be followed by a subscript 16, as in $46A_{16}$, or it may be followed by an uppercase H, as in 46AH. The uppercase H is most often used because it is easier to display on a CRT screen or print out on paper.

A hexadecimal digit can represent values from 0 to 15 decimal or 0000 to 1111 binary. Therefore, you can use a hexadecimal digit to represent a group of four binary digits. To convert a binary number to its hexadecimal equivalent, then, all you have to do is divide the binary number into groups of 4 bits, starting from the radix point and moving to the left. To convert the binary number 11010110 to hexadecimal, for example, mark off the binary bits in groups of four as shown and write the hex equivalent of each group below it:

Binary	1101	0110
Hex	D	6

The 1101 group is equal to D in hex and the 0110 group is equal to 6 in hex, so 11010110 is equal to D6H. For most of us, D6H is certainly easier to remember than the binary equivalent 11010110.

One way to convert a decimal number to its hexadecimal equivalent is to first convert the number to binary with one of the previously described methods, and then mark off the binary number in groups of four as shown above.

Another way of converting a decimal number to hexadecimal is with the divide-and-subtract method we showed you in the discussion of octal. As an example of this method, suppose that you want to convert 227 dec-

imal to its hexadecimal equivalent. First, you look at the number to find the largest power of 16 that will fit in the number. For our example of 227, 16^2, or 256, is too large, so the largest power that will fit is 16^1, or 16. Next you divide this power of 16 into the number you want to convert (227) to find the value for the 16^1 digit. As shown below, 16 divides into 227 fourteen times and leaves a remainder of 3. Since decimal 14 is E in hexadecimal, the value of the 16^1 digit is E. The weight of the 16^0 digit is 1, so you just put a 3 in this digit position to represent the remainder of 3. The result of this process is that 227 decimal is equal to E3H:

Power of 16	16^2	16^1	16^0
Decimal weight	256	16	1
	0	E	3

$$16\overline{)227} \quad \begin{matrix} 14 \text{ R3} \\ \underline{} \\ 224 \end{matrix}$$

Therefore:

$$227_{10} = E3H$$

Still another way of converting a decimal number directly to its hexadecimal equivalent is to successively divide the decimal number by 16 until the result of the division (quotient) is 0. Here we show you how to use this method on our example number of 227 decimal:

				Check				
$16\overline{)227}$	=	14 remainder	3 (LSD)	3	×	1	=	3
$16\overline{)14}$	=	0 remainder	E (MSD)	E	×	16	=	224
						Total	=	227

Therefore:

$$227_{10} = E3H$$

The first division by 16 gives a quotient of 14 and a remainder of 3. The remainder of 3 is the least significant digit of the hex number you want. The next division by 16 gives a quotient of 0 and a remainder of 14. Since 14 decimal is equal to E in hex, you write E as the most significant digit of the desired hex result. Decimal 227 then is equal to E3H. To verify your result you can multiply each hex digit by its decimal weight and add the products as shown on the right side of the example.

If you ever need to convert an octal number to hex, the easiest way is to write the binary equivalent of the octal number, divide the binary number into groups of 4 bits, and write the hex digit for each group of 4 binary bits. Likewise, the easiest way to convert a hex number to its octal equivalent is to write the binary equivalent of the hex number, divide the binary number into groups of 3 binary bits, and then write the octal digit for each group of 3 bits.

To illustrate how hexadecimal numbers are used in practice, suppose that a service manual tells you that the data word on the outputs of a memory system at a particular time is 3A4CH. To determine the logic level you should find on each of the 16 outputs with your scope, just write the 4-bit binary equivalent of each hex digit. The result you should get is 0011 1010 0100 1100. The 3A4CH is easier to communicate and less prone to errors. After a little practice you should be able to convert numbers back and forth between binary and hex almost automatically.

BCD CODES

In digital applications such as frequency counters, digital voltmeters, or calculators, where the output is a decimal display, numbers are often represented in *binary-coded decimal*, or BCD, form. In BCD codes, 4 binary bits are used to represent the value of each digit in a decimal number. For your reference, Table 7-1 shows how the decimal numbers 0 to 15 are represented in two different BCD codes. The most commonly used of these is the 8421 code, so when we use the term BCD from now on, this is the code we will be referring to, unless we specify otherwise.

The whole point of BCD is that 4 binary bits are used to separately represent each decimal digit. As shown in Table 7-1, BCD uses the 4-bit binary codes 0000 to 1001 to represent the decimal values 0 to 9. To represent decimal 10 requires two decimal digits, so the BCD representation of decimal 10 contains two groups of 4 bits each. The leftmost group is 0001 to represent the value of the 10's digit and the rightmost digit is 0000 to represent the value of the 1's digit. As another example, suppose that you want to represent the decimal number 73

in BCD. The BCD code for 7 is 0111 and the BCD code for 3 is 0011, so the BCD code for decimal 73 is 0111 0011. Note that each group must contain 4 bits, so 0's are put in as placeholders for bits that are not 1's. Numbers expressed in BCD are often separated into groups of 4 bits as shown to distinguish them from straight binary numbers.

An important point to make here is that, to represent a decimal number in BCD, you simply replace each decimal digit with its 4-bit binary equivalent. The resulting binary number is usually not equal to the original decimal number. For example, the decimal number 73 is represented in BCD as 0111 0011. If you treat 01110011 as a binary number and convert it to decimal, you get a result of 64 + 32 + 16 + 2 + 1, or 115 decimal, which is certainly not equal to the original 73. It is then up to you to remember whether a group of binary bits in an application represents a standard binary number, a BCD number, or a number in some other binary based code.

GRAY CODE

Gray code is a binary code which is often used to indicate the position of a rotating shaft in computer-controlled machines such as lathes and milling machines. One mechanism commonly used to generate a gray code which corresponds to the position of a shaft is called an *optical encoder*. Figure 7-18 shows the parts of a simple gray code optical encoder. The first major part of the encoder is a plastic disk with transparent

TABLE 7-1
COMMON NUMBER CODES

Decimal	Binary	Octal	Hex	Binary-Coded Decimal		Reflected Gray Code	Seven-Segment Display (1 = on)	
				8421 BCD	EXCESS-3		a b c d e f g	Display
0	0000	0	0	0000	0011 0011	0000	1 1 1 1 1 1 0	0
1	0001	1	1	0001	0011 0100	0001	0 1 1 0 0 0 0	1
2	0010	2	2	0010	0011 0101	0011	1 1 0 1 1 0 1	2
3	0011	3	3	0011	0011 0110	0010	1 1 1 1 0 0 1	3
4	0100	4	4	0100	0011 0111	0110	0 1 1 0 0 1 1	4
5	0101	5	5	0101	0011 1000	0111	1 0 1 1 0 1 1	5
6	0110	6	6	0110	0011 1001	0101	1 0 1 1 1 1 1	6
7	0111	7	7	0111	0011 1010	0100	1 1 1 0 0 0 0	7
8	1000	10	8	1000	0011 1011	1100	1 1 1 1 1 1 1	8
9	1001	11	9	1001	0011 1100	1101	1 1 1 0 0 1 1	9
10	1010	12	A	0001 0000	0100 0011	1111	1 1 1 1 1 0 1	A
11	1011	13	B	0001 0001	0100 0100	1110	0 0 1 1 1 1 1	B
12	1100	14	C	0001 0010	0100 0101	1010	0 0 0 1 1 0 1	C
13	1101	15	D	0001 0011	0100 0110	1011	0 1 1 1 1 0 1	D
14	1110	16	E	0001 0100	0100 0111	1001	1 1 0 1 1 1 1	E
15	1111	17	F	0001 0101	0100 1000	1000	1 0 0 0 1 1 1	F

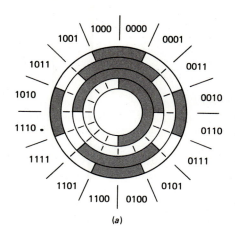

(a)

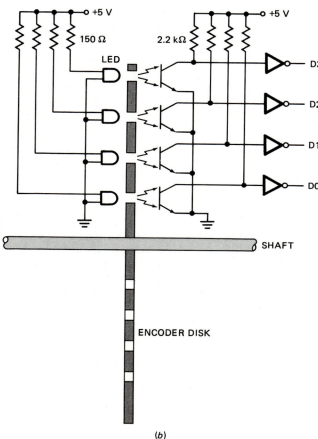

(b)

FIGURE 7-18 Parts of gray code optical encoder.
(a) Encoder disk. (b) Circuit to detect light and dark sections of encoder disk.

sections and dark sections as shown in Figure 7-18a. This disk is mounted on the shaft so that it rotates with the shaft. The disk has four circular tracks, so the encoder generates a 4-bit code.

To detect light and dark sections along a track, an LED is mounted on one side of each track and a phototransistor is mounted on the other side of the disk from each LED as shown in Figure 7-18b. When light from an LED passes through a transparent section of a track, it turns on the phototransistor. When a

phototransistor turns on, it effectively connects the bottom end of the 2.2-kΩ resistor on its collector to ground. This puts a low on the input of the inverter and produces a high on its output. The result of all this is that if a transparent section of a track is between an LED and a phototransistor, the output of the inverter will be high. If a dark section of a track is between an LED and a phototransistor, then the output of the inverter will be low.

As shown in Figure 7-18a, the light and dark sections of the tracks are arranged so that the disk is divided into 16 sectors. Each sector produces a different 4-bit code on the outputs of the inverters. Since this encoder disk contains 4 tracks, its resolution is 1 part in 2^4, or 16. The term *resolution* here refers to the size of the sectors, or in other words, how closely the encoder can tell you the position of the shaft. In degrees, the resolution of the system shown in Figure 7-18 is equal to 360°/16, or 22.5°. An encoder with 8 tracks would have a resolution of 1 part in 2^8, or 256, which is 1.4°.

Note in Figure 7-18a that, as you go around the encoder disk in a clockwise direction, the 4-bit gray codes do not follow the normal binary count sequence. If you carefully examine the sequence, you will note that only 1 bit in a word changes as you go from one sector to the next. To see why the codes are put in this sequence, let's take a look at the normal binary count sequence in Table 7-1 and predict how the encoder might act if the encoder disk were encoded in this way. Suppose, for example, that a disk encoded in standard binary was passing the boundary from 0111 to 1000. Further suppose that the LEDs and phototransistors picked up the lower 3 bits' change to 000, but did not quite pick up the most significant bit change to a 1. The outputs of the inverters would then be 0000 instead of the desired 1000. For a disk encoded with standard binary, this is an error of 1 MSB, or 180°.

To see how a gray code sequence reduces the maximum error, let's take a look now at the transition from 0100 to 1100 (gray code 7 to gray code 8). The lower 3 bits do not change between these two sectors, so these bits cannot cause any error for this transition. If the LED and phototransistor do not pick up the change of the most significant bit from a 0 to a 1, then the resultant code will be 0100 instead of the desired 1100. In other words, the encoder will output the code for 7 instead of the code for 8. This is a possible error of only 1 least significant bit. Since only 1 bit at a time changes as you go from one sector to the next, you can demonstrate that the maximum error for any transition is 1 least significant bit, instead of 1 most significant bit as it would be for an encoder designed to output a standard binary sequence.

We have used a 4-bit gray code example here, but a gray code number can contain any number of bits. If you need to construct a gray code table larger than that in Table 7-1, a handy method is to observe the pattern of 1's and 0's and just extend it. The least significant bit column starts with one 0 and then has alternating groups of two 1's and two 0's as you go down the column. The second most significant bit column starts with two 0's and then has alternating groups of four 1's and four 0's. The third column starts with four 0's and then has alternating groups of eight 1's and eight 0's. By now you should see the pattern. Try extending the pattern to see if you can predict the gray code for decimal 16. The result you should get is 11000. Incidentally, you may remember that the values you write along the edges of a Karnaugh map are in gray code sequence so that only one variable at a time changes as you go from one box to the next.

SEVEN-SEGMENT DISPLAY CODES

In Chapter 6 we introduced you to seven-segment LED displays. For review Figure 7-19 shows the letters used to identify the seven segments and the internal connections for the two types of displays. In a common-cathode display, the negative ends or cathodes of the LEDs are internally tied together as shown in Figure 7-19b. To light a segment in a common-cathode display, the positive end, or anode, of the LED is connected to a logic high or to a positive voltage through a current limiting resistor. The point here is that a logic high or 1 lights a segment in a common-cathode LED display.

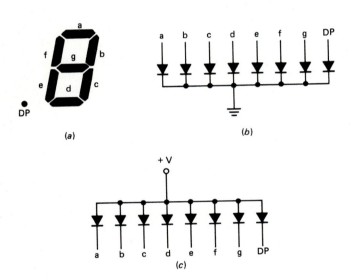

FIGURE 7-19   Seven-segment LED display. *(a)* Segment labels. *(b)* Schematic of common-cathode type. (c) Schematic of common-anode type.

In a common-anode display such as that shown in Figure 7-19c, the anodes of the LEDs are all internally tied together. The common anode is connected to a positive voltage. To light a segment in a common-anode display, the cathode of the LED for that segment is connected to a logic low or to ground through a current limiting resistor as shown in Figure 6-40.

With a little imagination a seven-segment LED can display the decimal digits 0 to 9 and recognizable symbols for the hexadecimal digits A to F. The rightmost section of Table 7-1 shows the codes which will display 0 to 9 and A to F on a common-cathode seven-segment display. Note that the "a" segment is turned on for a "6" to distinguish it from the symbol used to display a "b." Invert

each bit in Table 7-1 to get the codes required for a common-anode display.

Alphanumeric Codes

INTRODUCTION

To communicate between digital systems such as a smart cash register and a central computer or between a computer and a printer, we need a binary code that can represent the letters of the alphabet and numbers. Table 7-2 shows two commonly used *alphanumeric* codes which do this. In order to save space in Table 7-2, we have represented each of the binary codes with its hexadecimal equivalent. The binary code for the letter F in the ASCII code in Table 7-2, for example, is 01000110. In the table, we represent this with its hexadecimal equivalent, 46H.

PARITY BITS AND TRANSMISSION ERROR DETECTION

Before we discuss the individual codes in Table 7-2, we need to describe how an additional bit, called a *parity bit*, is often added to the basic data bits of these codes. If used, the parity bit is added to the left-hand side of the basic data bits so that it becomes the most significant bit of the total word. The purpose of the parity bit is to help a receiving system determine if the received code bits are correct. Here's how the parity-bit error detection scheme works.

Remember from Chapter 6 that the term *parity* is used to describe whether a digital word has an even number of 1's or an odd number of 1's. A binary number with an even number of 1's is said to have even parity and a binary number with an odd number of 1's is said to have odd parity. The binary number 1010100, for example, is an even number, but it has odd parity because it has an odd number of 1's (three).

As we also discussed in Chapter 6, an Exclusive OR circuit can be used to determine if the parity of a binary word is odd or even. To refresh your memory of this, make a truth table for the Exclusive OR "tree" in Figure 7-20. From this truth table you should see that the output of the circuit will be high if the binary word applied to its inputs has odd parity.

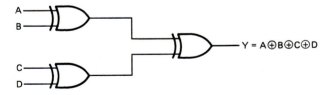

FIGURE 7-20   Exclusive OR tree circuit for four variables.

A common TTL device, the 74LS280 parity generator-checker, is essentially a 9-input Exclusive OR tree which can be used to check the parity of up to a 9-bit data word. If the number of 1's on the 9 inputs is even, the 74LS280 will output a high on its ΣEVEN output. If the

TABLE 7-2
COMMON ALPHANUMERIC CODES

ASCII Symbol	HEX Code for Seven-Bit ASCII	EBCDIC Symbol	HEX Code for EBCDIC	ASCII Symbol	HEX Code for Seven-Bit ASCII	EBCDIC Symbol	HEX Code for EBCDIC	ASCII Symbol	HEX Code for Seven-Bit ASCII	EBCDIC Symbol	HEX Code for EBCDIC
NUL	00	NUL	00	*	2A	*	5C	T	54	T	E3
SOH	01	SOH	01	+	2B	+	4E	U	55	U	E4
STX	02	STX	02	,	2C	,	6B	V	56	V	E5
ETX	03	ETX	03	-	2D	-	60	W	57	W	E6
EOT	04	EOT	37	.	2E	.	4B	X	58	X	E7
ENQ	05	ENQ	2D	/	2F	/	61	Y	59	Y	E8
ACK	06	ACK	2E	0	30	0	F0	Z	5A	Z	E9
BEL	07	BEL	2F	1	31	1	F1	[	5B	[	AD
BS	08	BS	16	2	32	2	F2	\	5C	NL	15
HT	09	HT	05	3	33	3	F3	]	5D	[	DD
LF	0A	LF	25	4	34	4	F4	^	5E	]	5F
VT	0B	VT	0B	5	35	5	F5	—	5F	—	6D
FF	0C	FF	0C	6	36	6	F6	`	60	RES	14
CR	0D	CR	0D	7	37	7	F7	a	61	a	81
S0	0E	S0	0E	8	38	8	F8	b	62	b	82
S1	0F	S1	0F	9	39	9	F9	c	63	c	83
DLE	10	DLE	10	:	3A	:	7A	d	64	d	84
DC1	11	DC1	11	;	3B	;	5E	e	65	e	85
DC2	12	DC2	12	—	3C	—	4C	f	66	f	86
DC3	13	DC3	13	=	3D	=	7E	g	67	g	87
DC4	14	DC4	35	\	3E	\	6E	h	68	h	88
NAK	15	NAK	3D	?	3F	?	6F	i	69	i	89
SYN	16	SYN	32	@	40	@	7C	j	6A	j	91
ETB	17	EOB	26	A	41	A	C1	k	6B	k	92
CAN	18	CAN	18	B	42	B	C2	l	6C	l	93
EM	19	EM	19	C	43	C	C3	m	6D	m	94
SUB	1A	SUB	3F	D	44	D	C4	n	6E	n	95
ESC	1B	BYP	24	E	45	E	C5	o	6F	o	96
FS	1C	FLS	1C	F	46	F	C6	p	70	p	97
GS	1D	GS	1D	G	47	G	C7	q	71	q	98
RS	1E	RDS	1E	H	48	H	C8	r	72	r	99
US	1F	US	1F	I	49	I	C9	s	73	s	A2
SP	20	SP	40	J	4A	J	D1	t	74	t	A3
!	21	!	5A	K	4B	K	D2	u	75	u	A4
"	22	"	7F	L	4C	L	D3	v	76	v	A5
#	23	#	7B	M	4D	M	D4	w	77	w	A6
$	24	$	5B	N	4E	N	D5	x	78	x	A7
%	25	%	6C	O	4F	O	D6	y	79	y	A8
&	26	&	50	P	50	P	D7	z	7A	z	A9
'	27	'	7D	Q	51	Q	D8	{	7B	{	8B
(	28	(	4D	R	52	R	D9	|	7C	|	4F
)	29	)	5D	S	53	S	E2	}	7D	}	9B
								~	7E	¢	4A
								DEL	7F	DEL	07

number of 1's on the 9 inputs is odd, the 74LS280 will output a high on its ΣODD output. The ΣODD signal or the ΣEVEN signal is transmitted along with the data bits as a parity bit.

Figure 7-21 shows how two 74LS280s can be connected to detect errors in transmitted parallel data. The system in Figure 7-21 is set up for even parity, which means that the parity of the eight data lines plus the parity-bit line will always be even. Here's how the system works.

The 74LS280 at the sending end of the data link checks the parity of the 8-bit word output by the sending system. For the example 8-bit data word shown on these lines in Figure 7-21, parity is odd. The fixed high connected to the I input of the 74LS280, however, makes the overall parity of the nine inputs of the 74LS280 even. This will cause the 74LS280 to output a high (1) on its ΣEVEN output (pin 5). As you can see in Figure 7-21, this ΣEVEN signal is transmitted as a parity-bit along

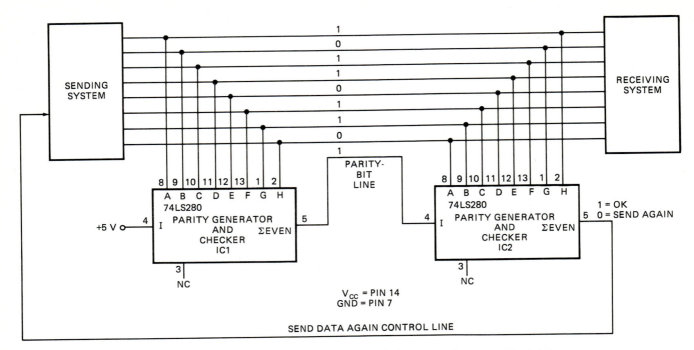

FIGURE 7-21 System using parity generator and checker to check for errors in transmitted data.

with the 8 data bits. The 1 on the parity-bit line makes the overall parity on the eight data lines and the parity-bit line even.

When the data bits and the parity bit reach the receiver end of the lines, the 74LS280 at that end checks the parity of the eight data lines plus the parity line. If the 74LS280 at the receiving end finds even parity on these nine lines as it should, it outputs a high on its ΣEVEN output. This high indicates that the data contained no parity error. If the 74LS280 at the receiving end of the system finds that the overall parity of the nine lines is odd, then it outputs a low on its ΣEVEN output. Odd parity on these lines indicates that electrical noise or some other problem has altered one or more of the data bits somewhere along the transmission lines. A low on this output can be used to tell the sending system that the data word should be sent again.

If the parity of the original 8-bit data word is even, the fixed 1 on the ninth (I) input of the 74LS280 at the sending end will make the number of 1's on the nine inputs odd. The ΣEVEN output will be a 0, so the parity on the eight data lines plus the parity lines is again even. The 74LS280 at the receiving end checks the parity of the eight data lines plus the parity-bit line as before and outputs the appropriate level on its ΣEVEN output.

To summarize this error-detecting method, then, the 74LS280 at the sending end checks the parity of the data word and generates a parity bit equal to 1 or 0 so that the overall parity of the nine lines is always even. The 74LS280 at the receiving end checks the overall parity of the nine lines and sends an error signal to the sending system if the overall parity is not even as it should be.

One difficulty with this parity-bit method of detecting errors is that two errors may effectively cancel each other. In other words, if 1 bit is accidentally changed from a low to a high and another bit is changed from a high to a low, the parity remains the same. In this case, the 74LS180 at the sending end will not detect the fact that the data is incorrect. A more complex method uses what are called *hamming codes* to detect and to some extent correct multiple-bit errors in transmitted data. If we are in a good mood when we get to Chapter 14, we will show you how this is done.

The main reason we included a discussion of parity at this point is so that you will not be confused when you are looking at alphanumeric data with a scope or logic analyzer. The data you see will depend on whether a parity bit has been added to the basic codes shown in Table 7-2. As an example of this, the 7-bit ASCII code for T with no parity bit is 1010100 binary, or 54H. If the code is being transmitted in a system that uses even parity, a 1 will be added as the MSB and you will see 11010100, or D4H, on the eight lines. If the data is being transmitted in a system which uses odd parity, you will see 01010100 binary, or 54H, on the eight lines.

Now that you know how and why a parity bit is often added to the basic data bits of an alphanumeric code, we will briefly discuss the ASCII and EBCDIC codes shown in Table 7-2.

ASCII

The acronym *ASCII* stands for the *American Standard Code for Information Interchange*. The ASCII symbols and codes are shown in the first two columns of Table 7-2. With no parity bit, basic ASCII is a 7-bit code. With 7 bits you can represent up to 128 characters, which is enough for full uppercase and lowercase alphabets plus numbers, punctuation marks, and control characters. The control codes are used for functions such as carriage return, line feed, backspace, and bell. For reference, Table 7-3 shows the full names of the ASCII control characters.

You may have seen a 1-in-wide metal or paper tape

TABLE 7-3
DEFINITIONS OF CONTROL CHARACTERS

NULL	Null	DLE	Data link escape
SOH	Start of heading	DC1	Direct control 1
STX	Start text	DC2	Direct control 2
ETX	End text	DC3	Direct control 3
EOT	End of transmission	DC4	Direct control 4
ENQ	Enquiry	NAK	Negative acknowledge
ACK	Acknowledge	SYN	Synchronous idle
BEL	Bell	ETB	End transmission block
BS	Backspace	CAN	Cancel
HT	Horizontal tab	EM	End of medium
LF	Line feed	SUB	Substitute
VT	Vertical tab	ESC	Escape
FF	Form feed	FS	Form separator
CR	Carriage return	GS	Group separator
SO	Shift out	RS	Record separator
SI	Shift in	US	Unit separator

with holes punched to represent ASCII codes as shown in Figure 7-22. Tapes such as this are often used to store a sequence of ASCII-coded instructions for *Computer Numerical Control* (CNC) machines such as lathes and milling machines.

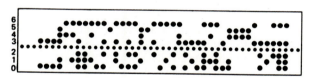

FIGURE 7-22   ASCII code on metal tape.

In a microcomputer such as a personal computer, a parity bit is not needed, so an eighth bit is often added to the basic ASCII codes to give additional control codes. The additional codes are used to tell the computer to perform functions such as moving the cursor around on the CRT screen, deleting a line of text on the screen, saving data on a disk, etc. These additional control codes are usually produced on a computer by holding down the CONTROL key on the keyboard and pressing a letter or number key at the same time. Different computers use different control codes, so you have to consult the manual for a specific machine to determine them.

EBCDIC

The acronym *EBCDIC* stands for *Extended Binary Coded Decimal Interchange Code*. This code, often used in IBM equipment, has 8 data bits, so it can directly represent a greater number of characters than basic ASCII. When it is needed, a parity bit is added as a ninth bit.

MSI CODE CONVERTERS
BCD-to-Seven-Segment Decoders

INTRODUCTION

In digital systems it is often necessary to convert a number in one code to the equivalent number in another code. Suppose, for example, that you have a counter that outputs a BCD code between 0000 and 1001, and you want to display the equivalent decimal number on a common-anode, seven-segment LED display as shown in Figure 7-23.

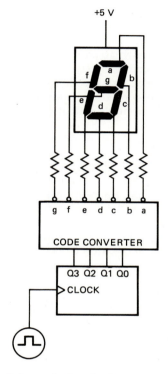

FIGURE 7-23   Schematic for displaying BCD code on seven-segment LED display.

Near the end of Chapter 6 we showed you the traditional method of solving this problem. The first step in the traditional method was to write a truth table such as that in Figure 6-41. The next step was to use a Karnaugh map to produce a simplified expression for each of the segment outputs. Figure 6-42 showed how to do this for the "a" segment. The final step was to implement the simplified expression for each segment with gates. Figure 6-43 showed two possible ways to implement the simplified expression for the "a" segment. As you know, if you worked out the solution to Problem 20 at the end of Chapter 6, implementing all the required segment functions with discrete logic gates would require a sizable number of devices.

One alternative approach to solving the problem might be to use multiplexers as we showed earlier in this chapter. Remember that with the "folding trick" we showed you, an eight-input multiplexer can be used to implement any function of four input variables. Since each of the seven segments requires a different function of the four input variables, seven multiplexers would be required to do this. This is obviously still a sizable number of devices. There must be a better way.

Many commonly needed functions such as code converters have been implemented in single MSI or LSI devices. A second, more feasible approach to solving a code conversion problem, such as the BCD-to-seven-segment code problem, is to check the functional index in a TTL or CMOS data book to see if the desired function is available in a single MSI or LSI device.

ENCODERS VERSUS DECODERS

Before you look in the functional index to find a solution to this problem, however, we need to clarify the terms *encoder* and *decoder* for you. The term *decoder* is used to describe a device which converts a compact code such as BCD to a less compact code such as seven-segment. By less compact we mean a code that uses more bits for each value. A device which converts BCD to seven-segment code, for example, will be called a BCD-to-seven-segment decoder.

The term *encoder* is used to describe a device which converts a low-density code, such as 16 sectors marked off on a disk, to a more dense code, such as a 4-bit gray code. Remember from the discussion of gray code in a previous section that the mechanism used to generate a gray code which represents the position of the shaft is called an *optical encoder*. Another example of an encoder is a 74LS148 eight-line to three-line encoder. In a later section we will discuss this device in detail. For now, let's see if we can find an appropriate decoder for our present problem.

TTL BCD-TO-SEVEN-SEGMENT DECODERS

Figure 7-24 shows a list of decoders from the functional index of a Texas Instruments TTL data book. As you can see, there are available several TTL and LS TTL devices listed as BCD-to-seven-segment display decoder/drivers. In this section we discuss the data sheets and the operation of some of these devices.

The rightmost column in Figure 7-24 tells you that the data sheets for these devices are in volume 2 of the

Texas Instruments (TI) *TTL Data Book* series. Figure 7-25 shows some major parts of the data sheets for the 7446, 7447, 7448, and 7449 types of BCD-to-seven-segment decoders. Let's use the table in Figure 7-25a to see which of these devices is appropriate for our example application.

As shown in Figure 7-23, our displays are common anode, so we need a decoder with active low outputs. This narrows our choices down to the 5446/7446 and 5447/7447 type devices. We don't need a military temperature range device, so our choices are further narrowed down to 7446 or 7447 type devices. To minimize system power dissipation of the circuit, we would like to use an 74LS family device if its outputs can sink enough current. According to the table in Figure 7-25a, a 74LS47 has active low outputs which can each sink up to 24 mA and the device has a typical power dissipation of only 35 mW.

A 74LS47 then seems to meet our needs. The next step is to study the logic symbol and the function table for the device in Figure 7-25b and c to determine the details of how the device operates.

The A, B, C, and D pins on the 74LS47 are the BCD inputs. The A pin is the LSB input and the D pin is the MSB input. As expected, the segment outputs are labeled a through g. The first 16 lines of the function table tell you that if the $\overline{LT}$ input is high, the $\overline{RBI}$ input is high, and if the $\overline{BI}/\overline{RBO}$ pin is high (open), then the BCD code applied to the ABCD inputs will determine the code present on the segment outputs. The character displayed for each segment code is shown over the function table in Figure 7-25c. Note that the first 10 of these are the decimal digits 0 to 9. The other symbols are sometimes used to show when a non-BCD code is on the inputs. To explain the function of the remaining three pins on the 74LS47, we need to take a close look at the function table.

The pin labeled $\overline{LT}$ is the *lamp test* input. According to the bottom line of the function table, if this input is asserted (low) and the $\overline{BI}/\overline{RBO}$ pin is high, the segment outputs will all be on, regardless of the code applied to the BCD inputs. As the name implies, asserting the $\overline{LT}$ input low is a way to test if all seven segments are working. During normal operation this pin will be tied high, so the segment outputs will be controlled by the other inputs.

The pin labeled $\overline{BI}/\overline{RBO}$ on the 74LS47 is used as a *blanking input* or as a *ripple blanking output*. The internal circuitry connected to this pin is unique in that a signal can be applied to it to force it to a low, or it can output a signal, depending on the application. As shown by the $\overline{BI}/\overline{RBO}$ line in the function table, if the $\overline{BI}/\overline{RBO}$ pin is made low, the segment outputs will all be off, regardless of the signals on the other inputs. Asserting this pin, then, unconditionally blanks the display.

Finally, the pin labeled $\overline{RBI}$ on the 74LS47 functions as a *ripple blanking input*. The $\overline{RBI}$ line in the function table shows that if the $\overline{RBI}$ input is made low AND the BCD code on the ABCD inputs is 0000, all the segment outputs will be off, and the $\overline{RBO}$ pin will be asserted low. In other words, if $\overline{RBI}$ is low, a 0 will not be displayed. For converting and displaying a single BCD digit, the $\overline{RBI}$ input is tied high as shown in Figure 7-26a so that

OPEN-COLLECTOR DISPLAY DECODERS /DRIVERS

DESCRIPTION	OFF-STATE OUTPUT VOLTAGE	TYPE	TECHNOLOGY					VOLUME
			STD TTL	ALS	AS	L	LS	
BCD-To-Decimal	30 V	'45	●					
	60 V	'141	●					
	15 V	'145	●				●	
	7 V	'445					●	
BCD-To-Seven-Segment	30 V	'46	A			●		
	15 V	'47	A			●	●	
	5.5 V	'48	●				●	
	5.5 V	'49	●				●	2
	30 V	'246	●					
	15 V	'247	●				●	
	7 V	'347					●	
	7 V	'447					●	
	5.5 V	'248	●				●	
	5.5 V	'249	●				●	

FIGURE 7-24   A list of decoders from the functional index of a Texas Instruments TTL data book.

TYPE	DRIVER OUTPUTS				TYPICAL POWER DISSIPATION	PACKAGES
	ACTIVE LEVEL	OUTPUT CONFIGURATION	SINK CURRENT	MAX VOLTAGE		
SN5446A	low	open-collector	40 mA	30 V	320 mW	J, W
SN5447A	low	open-collector	40 mA	15 V	320 mW	J, W
SN5448	high	2-kΩ pull-up	6.4 mA	5.5 V	265 mW	J, W
SN5449	high	open-collector	10 mA	5.5 V	165 mW	W
SN54L46	low	open-collector	20 mA	30 V	160 mW	J
SN54L47	low	open-collector	20 mA	15 V	160 mW	J
SN54LS47	low	open-collector	12 mA	15 V	35 mW	J, W
SN54LS48	high	2-kΩ pull-up	2 mA	5.5 V	125 mW	J, W
SN54LS49	high	open-collector	4 mA	5.5 V	40 mW	J, W
SN7446A	low	open-collector	40 mA	30 V	320 mW	J, N
SN7447A	low	open-collector	40 mA	15 V	320 mW	J, N
SN7448	high	2-kΩ pull-up	6.4 mA	5.5 V	265 mW	J, N
SN74LS47	low	open-collector	24 mA	15 V	35 mW	J, N
SN74LS48	high	2-kΩ pull-up	6 mA	5.5 V	125 mW	J, N
SN74LS49	high	open-collector	8 mA	5.5 V	40 mW	J, N

(a)

logic symbols

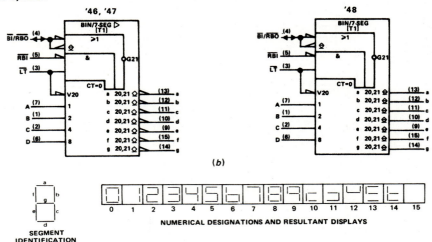

(b)

SEGMENT IDENTIFICATION

NUMERICAL DESIGNATIONS AND RESULTANT DISPLAYS

'46A, '47A, 'L46, 'L47, 'LS47 FUNCTION TABLE

DECIMAL OR FUNCTION	INPUTS						BI/RBO†	OUTPUTS							NOTE
	LT	RBI	D	C	B	A		a	b	c	d	e	f	g	
0	H	H	L	L	L	L	H	ON	ON	ON	ON	ON	ON	OFF	
1	H	X	L	L	L	H	H	OFF	ON	ON	OFF	OFF	OFF	OFF	
2	H	X	L	L	H	L	H	ON	ON	OFF	ON	ON	OFF	ON	
3	H	X	L	L	H	H	H	ON	ON	ON	ON	OFF	OFF	ON	
4	H	X	L	H	L	L	H	OFF	ON	ON	OFF	OFF	ON	ON	
5	H	X	L	H	L	H	H	ON	OFF	ON	ON	OFF	ON	ON	
6	H	X	L	H	H	L	H	OFF	OFF	ON	ON	ON	ON	ON	
7	H	X	L	H	H	H	H	ON	ON	ON	OFF	OFF	OFF	OFF	
8	H	X	H	L	L	L	H	ON	ON	ON	ON	ON	ON	ON	
9	H	X	H	L	L	H	H	ON	ON	ON	OFF	OFF	ON	ON	1
10	H	X	H	L	H	L	H	OFF	OFF	OFF	ON	ON	OFF	ON	
11	H	X	H	L	H	H	H	OFF	OFF	ON	ON	OFF	OFF	ON	
12	H	X	H	H	L	L	H	OFF	ON	OFF	OFF	OFF	ON	ON	
13	H	X	H	H	L	H	H	ON	OFF	OFF	ON	OFF	ON	ON	
14	H	X	H	H	H	L	H	OFF	OFF	OFF	ON	ON	ON	ON	
15	H	X	H	H	H	H	H	OFF	OFF	OFF	OFF	OFF	OFF	OFF	
BI	X	X	X	X	X	X	L	OFF	OFF	OFF	OFF	OFF	OFF	OFF	2
RBI	H	L	L	L	L	L	L	OFF	OFF	OFF	OFF	OFF	OFF	OFF	3
LT	L	X	X	X	X	X	H	ON	ON	ON	ON	ON	ON	ON	4

H = high level, L = low level, X = irrelevant

NOTES: 1. The blanking input (BI) must be open or held at a high logic level when output functions 0 through 15 are desired. The ripple blanking input (RBI) must be open or high if blanking of a decimal zero is not desired.
2. When a low logic level is applied directly to the blanking input (BI), all segment outputs are off regardless of the level of any other input.
3. When ripple-blanking input (RBI) and inputs A, B, C, and D are at a low level with the lamp test input high, all segment outputs go off and the ripple blanking output (RBO) goes to a low level (response condition).
4. When the blanking input/ripple blanking output (BI/RBO) is open or held high and a low is applied to the lamp-test input, all segment outputs are on.

† BI/RBO is wire AND logic serving as blanking input (BI) and/or ripple-blanking output (RBO).

(c)

FIGURE 7-25   Part of the data sheets for the 7446, 7447, 7448, and 7449 types of BCD-to-seven-segment decoders. *(Courtesy of Texas Instruments Inc.)*

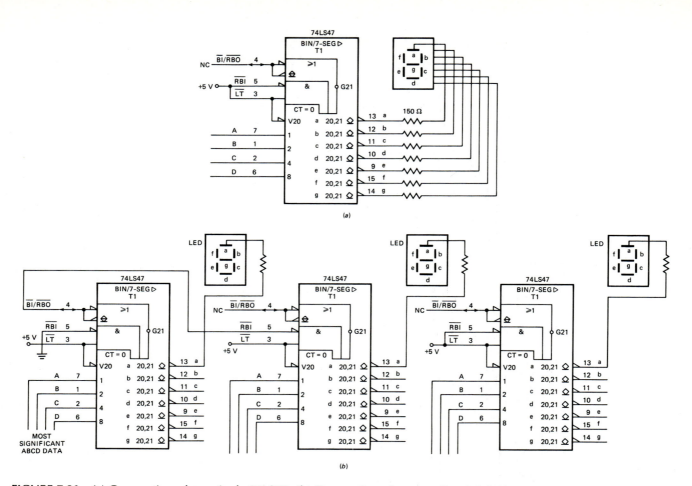

FIGURE 7-26 *(a)* Connections for a single 74LS47. *(b)* Connections for cascading 74LS47's.

all digits will be displayed. Note that, in this application, $\overline{LT}$ is also tied high and $\overline{BI}/\overline{RBO}$ is simply left open.

When two or more 74LS47's are cascaded to convert and display multiple BCD digits as shown in Figure 7-26b, the $\overline{RBI}$ pin of the MSD is tied low. The $\overline{BI}/\overline{RBO}$ of this 74LS47 is tied to the $\overline{RBI}$ of the next most significant 74LS47. Here's how the devices function with these connections.

If the BCD code applied to the most significant 74LS47 is 0000, then this zero will not be displayed because the $\overline{RBI}$ of this device is low. This is called *leading zero blanking*. Leading zeros are often blanked on instruments such as frequency counters and digital voltmeters so that they do not distract from the nonzero digits. For the system shown in Figure 7-26b, leading zero blanking is extended to the second digit by the connection from the $\overline{BI}/\overline{RBO}$ of most significant 74LS47 to the $\overline{RBI}$ of the next 74LS47. If the MSD is blanked because it is zero, the $\overline{BI}/\overline{RBO}$ pin will assert a low on the $\overline{RBI}$ of the second 74LS47. With its $\overline{RBI}$ input asserted low, a BCD code of 0000 applied to this device will cause it to turn off all its segment outputs. Because of the connection between the two devices, the second 74LS47 will only blank a zero if the MSD is also a zero. The scheme can be extended through as many digits as desired.

Now that you have an understanding of how a 74LS47 functions, let's see if we can explain how the IEEE/IEC

logic symbol for the device in Figure 7-25b summarizes its operation.

The BCD inputs are labeled with 1, 2, 4, and 8 to indicate their binary weights. The segment outputs have wedges on them to indicate that they are active low, and the small, diamond-shaped boxes next to each to show that they are NPN transistor, open-collector outputs.

In IEEE/IEC logic symbols, the letter V is used to indicate OR dependency. The V20 in the 74LS47 symbol, for example, tells you that the outputs on the pins labeled with 20 depends on the level on the $\overline{LT}$ input OR the code on the BCD data inputs.

To help you decipher the rest of the structure in the 74LS47 symbol, Figure 7-27 shows the internal logic blocks using conventional logic symbols. Remember from previous discussions that the letter G is used to indicate AND dependency. The bubble next to G21 in this symbol indicates that the outputs identified with 21 depend on the code on the BCD data inputs AND the output of the logic block being NOT asserted. From the logic diagram in Figure 7-27 you can see that there are two conditions that can cause the output of the block to be NOT asserted. One condition is if the $\overline{BI}$ input is asserted. This agrees with the function table in Figure 7-25c which tells you that the segment outputs will be off if the $\overline{BI}$ input is low. The other condition which can cause the output of the logic block in Figure 7-27 to be NOT as-

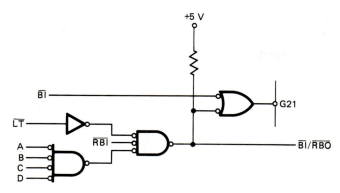

FIGURE 7-27   The internal logic blocks of the 74LS47.

serted is when the $\overline{LT}$ input is unasserted (high) AND the $\overline{RBI}$ input is asserted (low) AND the code on the BCD inputs is 0000. The RBI line in the function table shows that all the segment outputs will be off for these conditions.

Again, the IEEE/IEC logic symbol is intended to help you remember pin functions when you look at a schematic. It is not intended as a replacement for the function table.

One-of-N-Low Decoders

Another useful type of decoder consists of devices which assert one of some number of outputs when the code for that output is applied to the select inputs. Figure 7-28

DECODERS/DEMULTIPLEXERS

DESCRIPTION	TYPE OF OUTPUT	TYPE	STD TTL	ALS	AS	L	LS	S	VOLUME
4 To 16	3 State	154	●			●			
	OC	159	●						
4 To 10 BCD To Decimal	2 State	42	A			●	●		2
4 To 10 Excess 3 To Decimal	2 State	43	A			●			
4 To 10 Excess 3-Gray To Decimal	2 State	44	A			●			
3 To 9 with Address Latches	2 State	131		●	▲				3
		137		●	▲		●		2
3 To 9	2 State	138		●	▲				3
							●	●	2
	3 State	538		▲					3
Dual 2 To 4	2 State	139		▲	●				3
	2 State	155	●			A	●		2
	OC	156	●			A	●		
Dual 1 To 4 Decoders	3 State	539		▲					3

FIGURE 7-28   A list of available TTL decoders from the functional index of a Texas Instruments TTL data book.

shows some of the available TTL decoders of this type. A device described in the table as "4-to-16" (4-line to 16-line) uses 4 select lines to address one of 16 outputs. A 4-line to 10-line decoder uses 4 select inputs to address and assert a low on one of 10 outputs. A 3-line to 8-line decoder uses 3 select inputs to address and assert a low on one of 8 outputs. As an example of this group of devices, we will use the 74ALS138 3-line to 8-line decoder/demultiplexer.

Earlier in the chapter we discussed the data sheet for the 74ALS138 and showed you how it can be used as an eight-output demultiplexer. To refresh your memory of the device, take a look at the logic symbols and the function table for it in Figure 7-15a.

Previously we discussed the logic symbol labeled DMUX. Now we want to discuss the alternative symbol labeled BIN/OCT. The box in the lower left corner of the symbol tells you that the device will be enabled if the G1 input is asserted (high) AND the $\overline{G2A}$ input is asserted (low) AND the $\overline{G2B}$ input is asserted (low). The bottom eight lines of the function table in Figure 7-15c tell you that if the device is enabled, one of the outputs will be asserted low. The number of the output that will be low corresponds to the 3-bit binary number applied to the C, B, and A select inputs. If, for example, 011 is applied to the select inputs, the output labeled Y3 will be asserted low.

The outputs of devices such as the 74ALS138 are often connected to the $\overline{ENABLE}$ inputs of other devices so that one of several devices can be enabled, depending on the 3-bit code we apply to the select inputs. You will see many examples of this in later chapters. For now, Figure 7-29a shows a fun example you might like to build and experiment with.

NOTE: To reduce the cost of this circuit, we used a 74LS138 because you don't need the speed of the 74ALS138 in this application.

As the 3-bit binary counter (74LS93) steps through its count, the LEDs connected to the Y outputs of the 74LS138 will be lit one after the other in sequence. This is similar to the way lights rotate on a theater marquee or on the front of a car named Kit in a once popular TV series called "Knight Rider." Figure 7-29b shows how you can connect two 74LS138s so that 1 of 16 LEDs is lit at a time. Note that when the QD from the counter is low, only the left-hand 74LS138 is enabled, so the LED addressed by the lower 3 bits from the counter will be lit on this device. When the QD output from the counter is high, only the right-hand 74LS138 is enabled, so the LED addressed by the lower 3 bits from the counter will be lit on this device. For a problem at the end of the chapter, we give you a few hints how to expand the system to more LEDs.

Encoders

Previously we told you that the name *encoder* is usually given to a device which converts one code to another code that contains fewer bits. As an example of this type of device, we will discuss the 74HC147 and 74HC148 priority encoders.

PRIORITY ENCODERS

Figure 7-30a shows the function table and the logic symbol for a 74HC147 10-line to 4-line encoder. We'll explain later why there are only nine inputs. The wedges in the logic symbol tell you that the inputs and the outputs are all active low. According to the truth table, if one of the inputs is asserted, a 4-bit binary code corresponding to

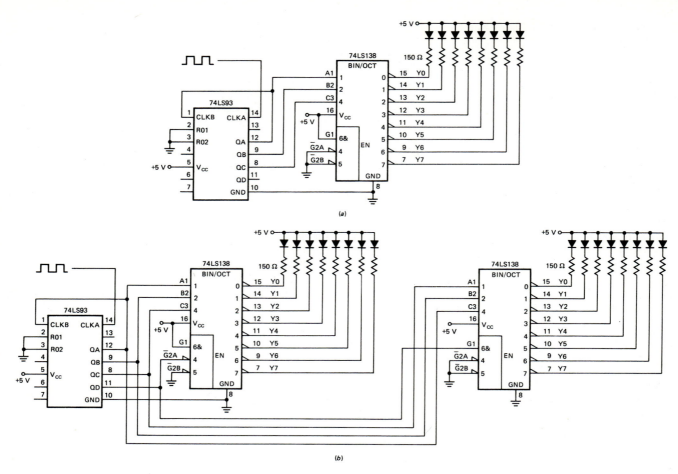

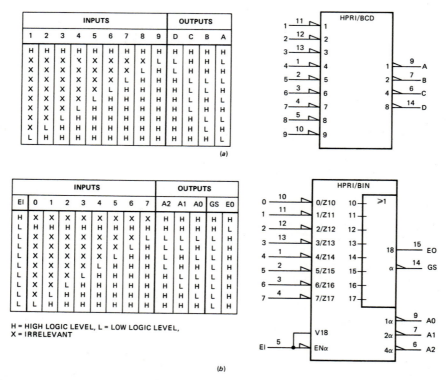

FIGURE 7-29 *(a)* Knight Rider example circuit. *(b)* Another example circuit.

FIGURE 7-30 74HC147 and 74HC148 function tables and logic symbols. *(a)* 74HC147. *(b)* 74HC148.

H = HIGH LOGIC LEVEL, L = LOW LOGIC LEVEL,
X = IRRELEVANT

the number of that input will be present on the outputs. Since the outputs are active low, the code will be the inverted BCD code for that input. For example, if the 3 input is asserted (low), the outputs will contain HHLL or 1100, which is the inverted BCD code for 3.

Now, suppose that the 3 input and the 5 input are asserted at the same time. The X's in the function table tell you that if a particular input is asserted, all the inputs with lower numbers are "don't cares." For the example here, 5 is a higher number than 3, so the outputs will contain the inverted BCD code for 5, HLHL. Here, the low on the 3 input has no effect on the output. The 5 input is said to have a *higher priority* than the 3 input. For the 74HC147 the 9 input has the highest priority and the 1 input has the lowest priority.

The 74HC147 is described as a 10-line to 4-line encoder but it has only nine inputs. As you can see from the first line in the function table, the tenth output state, HHHH, is produced when none of the nine inputs are asserted. The absence of an actual tenth input limits the usefulness of the device in some applications. Now let's look at the 74HC148.

Figure 7-30b shows the logic symbol and function table for a 74HC148 which is described as an 8-line to 3-line priority encoder. As shown in the logic symbol, it has eight active low data inputs and three active low data outputs. The device also has an enable input ($\overline{EI}$), an enable output (EO), and a group strobe output ($\overline{GS}$). The function table should help you see how these pins work.

The first line in the function table in Figure 7-30b tells you that $\overline{EI}$ must be asserted (low) to enable the device. If $\overline{EI}$ is not asserted, the data outputs will all be high, regardless of the signals on the data inputs. Also, if the device is not enabled by $\overline{EI}$ being low, the $\overline{GS}$ signal will not be asserted.

The bottom eight lines in the function table tell you that if the 74HC148 is enabled, the data outputs will contain the active low (inverted) binary code for the highest numbered data input that has a low on it. In other words, the outputs will contain the code for the highest priority asserted input. The bottom eight lines of the function table also show you that if any of the data inputs are asserted, the $\overline{GS}$ output and the EO output will both be asserted. The second line in the function table shows you that if none of the data inputs are asserted, these two outputs will not be asserted. To explain how these signals are used, we will show you a simple application of the device.

A common need in digital systems is to convert the signal from a key pressed on a keyboard to a binary-based code which represents that key. The device or circuit which does this is called a *keyboard encoder*. Figure 7-31 shows a system that uses two cascaded 74HC148s and some gates to generate a BCD code which corresponds to the number of a key on a 10-key keyboard. The BCD code from the encoder part of the circuit is applied to the inputs of a 74LS47 BCD-to-seven-segment decoder so the decimal digit for the pressed key will be displayed on the LED display. Here's how it works.

One side of each key switch is connected to ground, so pressing a key puts a low on an input of one of the 74HC148s. Since the $\overline{EI}$ input of the upper 74HC148 is grounded, this device is permanently enabled. If a key connected to this device is pressed, its EO will go high. Since the EO from the upper device is connected to the $\overline{EI}$ of the lower device, this high will disable the lower 74HC148. This connection then maintains the desired priority from 0 to 9. If no keys connected to the upper device are pressed, the EO signal from it will be low, which enables the lower device.

The NAND gates in the center of Figure 7-31 generate the BCD code from the 74HC148 data output signals. From the bubbled input OR symbol we used, you can see that the A bit of the BCD code will be a 1 if the $\overline{A0}$ output of the upper 74HC148 is asserted OR the $\overline{A0}$ output of the lower 74HC148 is asserted. The same logic is used to produce the B and C digits of the BCD code. The most significant bit of the desired BCD code is simply equal to the EO signal from the upper 74HC148 because this signal will be high for the 8 key and the 9 key and low for all others.

Finally, let's look at how the $\overline{GS}$ signals from the 74HC148s are used here. If a key connected to the upper device is pressed, the $\overline{GS}$ of this device will be asserted. Likewise, if the lower 74HC148 is enabled and a key connected to that device is pressed, the $\overline{GS}$ of that device will be asserted low. If no keys are pressed, the output of the gate connected to the two $\overline{GS}$ signals will be low. When any key is pressed this output will go high. A signal which indicates that a key has been pressed is often called a *key-pressed strobe*.

If no keys are pressed, the low on the key-pressed strobe line will assert the $\overline{BI}$ input of the 74LS47 and blank the LED display. We did this so that the display is different for the 0 key pressed and for no keys pressed. If we had used a single 74HC147 encoder for this system instead of two 74HC148s, the codes for 0 and for no keys pressed would be the same and the key-pressed signals would not be directly available. The point is that you have to think carefully about the requirements of a system rather than just grab what seems a simple solution at first glance.

Before we leave the 74HC148, let's take a closer look at the logic symbol for it in Figure 7-30b. If the ENα input is asserted, all the outputs identified with α will be enabled. The letter Z is used to show an internal logic connection. The Z10 on the 0 input of the symbol indicates that if the 0 input is asserted, the input of the internal OR block labeled 10 will be asserted through some internal logic connection. The internal OR block, identified with the characters ≤ 1, tells you that the $\overline{GS}$ will be asserted if any of the data inputs are asserted. The EO is a little messier to describe. The letter V, remember, is used to represent OR dependency. The V18 next to the EI input and the 18 next to the EO output tell you that the EO output will be asserted (high) if the $\overline{EI}$ input is not asserted OR if one of the data inputs is asserted (low).

In the preceding sections we have discussed some MSI devices which can be used for simple code conversions. For converting more complex codes such as ASCII to EBCDIC and for implementing logic functions which have more than four or five variables, these devices are not suitable. For these applications, we turn to logic devices which you can program to implement any desired logic function. In the next section we discuss some of these devices.

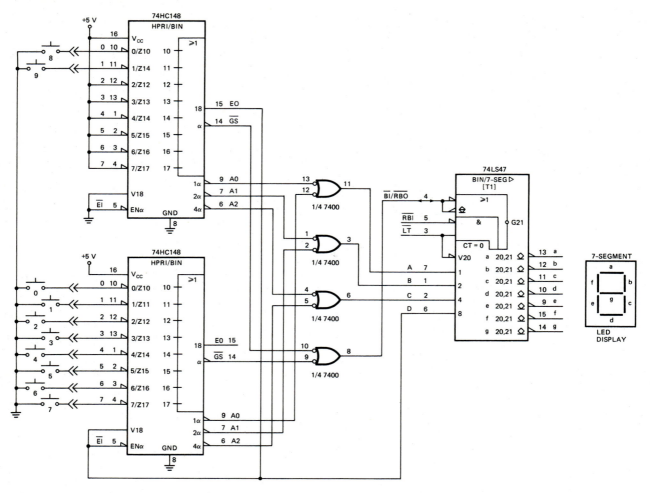

FIGURE 7-31 A partially completed schematic for a keyboard encoder and display circuit.

PLAS, PROMS, AND PALS

Introduction

There are several reasons for reducing the number of ICs required to implement a digital system. Usually fewer ICs mean less PC board space, faster and cheaper assembly, less power dissipation, smaller cabinets, less devices to keep in inventory, less chance of an IC failure, and easier troubleshooting in case of a malfunction. In this section we introduce you to some of the devices that allow you to implement several complex logic functions in a single IC package. A major purpose of the remainder of this chapter is to plant some seeds that we will fertilize and harvest in later chapters.

AND-OR-INVERT GATES

As we have shown you before, any combinational logic function can be expressed in a sum-of-products form. This sum-of-products expression can be implemented directly with AND gates, an OR gate, and perhaps some inverters. Based on these observations, a logical step would be to create an IC device which could implement any AND-OR function. Early attempts at this were devices such as the 74LS54, which is described in the TTL

data book as a 4-WIDE 2,3,3,2 INPUT, AND-OR-INVERT gate. The logic diagram for the device in Figure 7-32a should help you understand what this name means. The 4-wide means that the output OR gate has four inputs, so it can sum four product terms. The numbers 2,3,3,2 tell you the number of inputs on each of the four AND gates. INVERT in the name refers to the fact that the output is active low. As another example of this type of device, Figure 7-32b shows the logic diagram for a 74LS55, which is a DUAL, 2-WIDE, 4-INPUT, AND-OR-INVERT gate.

The difficulty with devices such as the 74LS54 and 74LS55 is that they are only useful if an expression happens to fit in the format of one of the available devices. None of the available devices, for example, will directly accommodate an expression which requires two 4-input AND gates and a 3-input AND gate.

MULTIPLEXERS

Another simple device which can implement any single AND-OR expression is a multiplexer. If you look at the logic diagram for the 74ALS151 multiplexer in Figure 7-5, you will see that it has a basic AND-OR structure. Earlier in the chapter we showed you how an 8-input

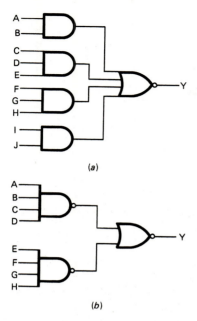

(a)

(b)

FIGURE 7-32   AND-OR-INVERT gates. *(a)* 74LS54.
(b) 74LS55.

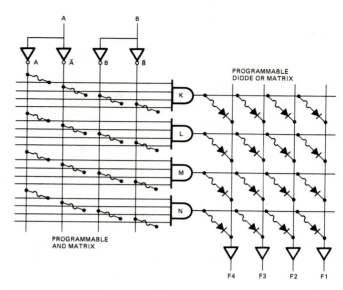

FIGURE 7-33   Programmable AND-OR circuit.

multiplexer such as the 74ALS151 can be used to implement any logic function of up to four input variables. The multiplexer approach is more flexible than the AND-OR-INVERT gate approach, but it still has one problem. The problem is that since a multiplexer has essentially only one output, you need one multiplexer for each output function. As the example in the preceding section showed, you would need seven multiplexers to implement a BCD-to-seven-segment decoder. Advances in semiconductor technology, however, have produced devices which contain many programmable AND-OR structures and can therefore implement almost any combinational logic functions such as this with a single IC. In the following sections we introduce you to some of these devices.

PLAs and FPLAs

Figure 7-33 shows a circuit for a simple AND-OR array which can be programmed to implement any four logic functions of the two input variables, A and B. Here's how it works.

Each of the input variables is applied to a buffer so that both the true and the complemented signal for that variable are available. The true and the complemented form of each variable are connected to inputs of an AND gate through fuses as shown. If all four fuses are intact on, for example, the K AND gate, then the output expression for that gate will be:

$$K = A \text{ AND } \overline{A} \text{ AND } B \text{ AND } \overline{B} = A\overline{A}B\overline{B}$$

Since the result of ANDing a variable with its complement is 0, the output of the K gate will be always low, regardless of the logic levels on A and B. This, of course, isn't very useful. Let's see what happens, however, if we

blast out some of the fuses on the inputs of the K AND gate. Suppose, for example, that we blow out the fuse on the $\overline{A}$ input and the fuse on the B input. These inputs of the AND gate will then be pulled high by internal pull-up resistors. The output expression for the gate will then be $K = A\overline{B}$.

If we had blown the $\overline{A}$ fuse and the $\overline{B}$ fuse, the resultant expression for the K AND gate would have been $K = AB$. The principle here is that the output expression for each AND gate depends on which fuses you blow and which fuses you leave. Since each AND gate has four inputs, you can program it by blowing fuses to produce any function of the two input variables. Now let's look at the OR array in the circuit in Figure 7-33.

For example, the input of the buffer on the F1 output is connected to each of the AND gate outputs with a diode and a fuse. If all the AND gate outputs are low, the output of the F1 buffer will be low. If any of the AND gate outputs is high, it will pull the F1 buffer input line high through a diode and a fuse. Since any one of the AND outputs can pull the line high, this connection acts like a 4-input OR gate with one input connected to each of the AND outputs, as shown in Figure 7-34. With all fuses on the F1 line intact, then, the expression for the F1 output is:

$$F1 = K + L + M + N$$

Since K, L, M, and N are product terms of A and B, F1 can be any sum-of-products relationship for A and B. If we blow a fuse on one of the AND outputs, that term will be removed from the expression. For example, if we blow the K fuse and the M fuse on the F1 line, the resultant expression for F1 will be $F1 = L + N$. By leaving intact or by blowing different fuses, the F1, F2, F3, and F4 outputs can each be programmed to produce a different logical function of the product terms from the K, L, M, and N AND gates. Incidentally, the diodes are necessary in this circuit to prevent a low on the output of one AND

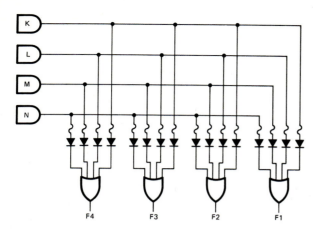

FIGURE 7-34   Gate equivalent of diode OR array.

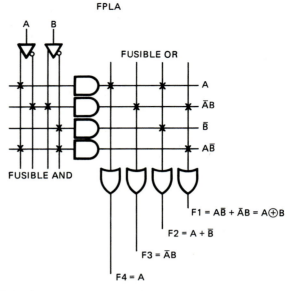

FIGURE 7-35   FPLA standard logic diagram.

gate from being connected directly to a high on the output of one of the other AND gates.

In summary, then, if a fuse is intact on the input of an AND gate, the variable on that input will be present in the product term for the AND gate. If a fuse on the output of an AND gate is intact, the product term produced by that AND gate will be present in the output OR expression.

The circuit in Figure 7-33 is a simple example of a device called a PLA or an FPLA, depending on how it is programmed. If it is programmed during manufacture, then programming is done by changing the "mask" used to produce diodes, transistors, etc., on the silicon die. The programming is done by simply making connections or not making connections on the AND inputs and by leaving out diodes or putting in diodes on the AND outputs. If the device is programmed in this way, it is called a *programmable logic array*, or *PLA*.

If the device is manufactured with fuses so that a user can custom program it, the device is called a *field programmable logic array*, or *FPLA*.

Even though it has only two input variables, the circuit diagram for the example FPLA in Figure 7-33 is somewhat messy. Manufacturers' data sheets commonly use a streamlined format to represent the internal structure of devices such as FPLAs. Figure 7-35 shows how our example FPLA can be represented in this streamlined form and how it can be programmed to implement four different logic functions.

First, the buffers on each input are condensed into a single buffer with both inverted and noninverted outputs. To simplify the AND gate connections, all the input lines are represented by a single line. Since you know from Figure 7-33 that the AND gates have one input for each input variable and for its complement, you can tell how many inputs each AND gate has by simply counting the number of lines which intersect the line on the input of the AND gate. Likewise, the inputs of the OR gates are sometimes represented by single lines to simplify the drawing. Again, you can determine the number of inputs on each OR gate in an FPLA by simply counting the number of AND gate outputs which intersect the line on the input of the OR gate.

An X at the intersection of two lines in Figure 7-35

means that the fuse there is intact. Therefore, those two lines are connected. If no X is present, it means that the fuse at that junction has been blown and there is no connection. In the top AND gate in Figure 7-35, for example, the A line has an X, so this signal is connected to one of the inputs of the AND gate. The $\overline{A}$ line, the B line, and the $\overline{B}$ line do not have an X, so they are not connected to the corresponding inputs of the AND gate. Internal resistors pull the disconnected AND inputs high so they do not interfere with the operation of the gate.

An X in the OR array also shows a fuse is present. The F4 OR gate, for example, has an X on the output line from the top AND gate but no Xs on the other lines. This means that the output of the top AND gate is connected to one of the inputs of the OR gate, but the other three are not. Unused OR gate inputs are internally pulled low, so that they do not interfere with the operation of the OR gate.

We have programmed the device in Figure 7-35 with some example logic functions. Here's how you determine the programmed functions.

The output expressions (product terms) generated by the AND gates are shown along the right-hand side of Figure 7-35. As we said before, only the A input line is connected to the top AND gate, so the expression for that gate is just A. Moving down from the top, one input of the second AND gate is connected to the $\overline{A}$ line and another is connected to the B line. The output expression for this AND gate then is $\overline{A}B$. In a similar manner, the expression for the third AND is $\overline{B}$, and the expression for the bottom AND gate is $A\overline{B}$.

Fuses in the OR array determine which product terms will be included in each of the final output expressions. The F4 OR gate is only connected to the A product term, so the expression for the F4 output is just F4 = A, as shown in Figure 7-35. Likewise, the F3 OR input is only connected to the $\overline{A}B$ product term, so the output expression for the F3 output is just F3 = $\overline{A}B$. F2 is a little more interesting. According to Figure 7-35, one input of the

F2 OR gate is connected to the A product term and another input of the OR gate is connected to the $\overline{B}$ product term. This produces $F2 = A + \overline{B}$. The F1 OR gate is also connected to two product terms. The output expression for F1 is $F1 = A\overline{B} + \overline{A}B$ which, of course, is equal to $A \oplus B$. These examples should convince you that an FPLA can implement any sum-of-products expression if its AND gates have enough inputs and its OR gates have enough inputs. Actual FPLAs are much larger than our simple example, so this is usually not a problem.

An example of an actual FPLA is the Texas Instruments TIFPLA839. This device is described in its data sheet as a $14 \times 32 \times 6$ programmable logic array. The number 14 in this name tells you that the device has 14 inputs to the AND array, so a product term can contain up to 14 variables. The number 32 in the name of the device shows that the AND array has 32 AND gates. This means that the device can be programmed to produce up to 32 different product terms. The number 6 in the name of the device tells you the number of output OR gates. This FPLA then can be used to implement up to six different logic functions, each containing up to 32 product terms of up to 14 variables each. Incidentally, each of the TIFPLA839 outputs can also be programmed to be active high or active low.

FPLAs can implement almost any sum-of-products expression, but the fact that they have two fuse arrays make them somewhat difficult to manufacture, program, and test. To program a TIFPLA839, for example, you have to program the output polarity, program and test the AND array, then program and test the OR array. Although these jobs are almost always done by a special instrument which does it automatically, FPLAs are not very convenient. In the next sections we discuss *ROMs* and *PALs*, which are much more commonly used than FPLAs. ROMs and PALs have the same basic AND-OR structure that FPLAs do, but since only one of the arrays is programmable in these devices, they are easier to manufacture, program, and test.

ROMs, PROMs, EPROMs, and EEPROMs

ROMs, PROMs, EPROMs, and EEPROMs are devices which can either be used to implement combinational logic functions in the same way as FPLAs or they can be used to store binary data words. The ROM in these names stands for *read-only memory*. The name comes from the fact that once these devices are programmed with a set of data words, the stored data words can be read out, but they cannot be easily changed or in some cases they cannot be changed at all. Once data words are programmed into a ROM, the data words remain stored there, even if the power is turned off.

The rest of the letters in these ROM names indicate how they are programmed and if they can be erased and reprogrammed. In a later chapter on memories we discuss these devices in detail, but as an introduction here's a brief explanation of the four types.

ROM—*Read-only memory.* Programmed during manufacture; ROMs cannot be reprogrammed.

PROM—*Programmable read-only memory.* Programmed by blowing fuses; PROMs cannot be reprogrammed.

EPROM—*Erasable programmable read-only memory.* Programmed by storing charges in insulated regions in device, can be erased by shining ultraviolet light through a quartz window in the top of the package. Once erased, they can be reprogrammed.

EEPROM—*Electrically erasable programmable read-only memory.* Programmed by storing charges in insulated regions in device; erased by a sequence of pulse signals. Once erased, they can be reprogrammed.

Figure 7-36 shows the structure of a PROM the same size as the FPLA we discussed in the preceding section. The dots on intersections of the AND array in Figure 7-36 show that these connections are *hardwired*, or fixed. Remember, in the FPLA these AND-gate connections are programmable. The OR array in the PROM is programmable by blowing or leaving fuses intact just as it is in an FPLA. An advantage of a PROM over an FPLA, then, is that the PROM has only one array of fuses to build, program, and test. To analyze this device, let's start by looking at the product terms produced by the hardwired AND array.

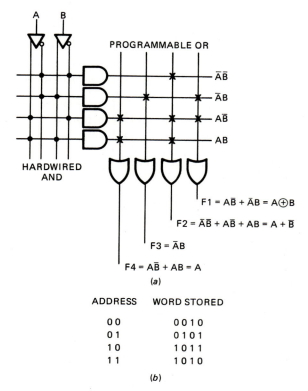

$$F1 = A\overline{B} + \overline{A}B = A\oplus B$$
$$F2 = \overline{A}\overline{B} + A\overline{B} + AB = A + \overline{B}$$
$$F3 = \overline{A}B$$
$$F4 = A\overline{B} + AB = A$$

(a)

ADDRESS	WORD STORED
0 0	0 0 1 0
0 1	0 1 0 1
1 0	1 0 1 1
1 1	1 0 1 0

(b)

FIGURE 7-36   PROM. *(a)* AND-OR representation of PROM structure. *(b)* Relationship between addresses and words stored at each address.

The top AND gate is connected to $\overline{A}$ and $\overline{B}$, so the product term for that gate is $\overline{A}\overline{B}$ as shown along the right-hand side of Figure 7-36. Moving down the array, the second AND gate is connected to $\overline{A}$ and B, so the product term for that gate is $\overline{A}B$. The two remaining product terms are $A\overline{B}$ and AB as shown in the figure. As you can see, these four product terms represent all the possible combinations of the input variables A and B. Now let's see how different combinations of these product terms can be ORed together to produce any desired sum-of-products function.

In the example PROM in Figure 7-36, the F4 OR gate inputs are connected to $A\overline{B}$ and to AB, so the output expression for F4 is $F4 = A\overline{B} + AB$. If A is high, F4 will be high, regardless of the state of B, so the expression reduces to $F4 = A$.

The F3 OR gate is connected only to the $\overline{A}B$ product term, so the expression for F3 is just $F3 = \overline{A}B$.

The expression for the F2 output is $F2 = \overline{A}\,\overline{B} + A\overline{B} + AB$. As a review of Boolean algebra, you can prove that this simplifies to $F2 = A + \overline{B}$. Finally, the expression for the F1 output is $F1 = \overline{A}B + A\overline{B}$, which is simply $F4 = A \oplus B$. As we mentioned before, a PROM such as the one in Figure 7-36 can also be used as a "memory" for a group of digital words. When used as a memory, the A and B inputs of the PROM in Figure 7-36 are called *address* inputs because the signal on these inputs addresses or selects the desired binary word stored in the device. Here's how this works.

Suppose for a start that the A and B inputs of the PROM in Figure 7-36 are both low. With the hardwired connections shown on the inputs of the AND gates, this will produce a high on the output of the top AND gate and lows on the outputs of the other three AND gates. The fuse on this AND output line will apply the high to one of the inputs of the F2 OR gate. This will then produce a high on the F2 output. The F1, F3, and F4 OR inputs have no fuses left in the top line, so the inputs of these three OR gates will be low. Therefore, the F1, F3, and F4 OR gate outputs will be low. Note that for input address 00, the outputs of the bottom three AND gates are low, so these outputs will have no effect on the F1, F3, or F4 outputs.

For input address 00, then, the PROM in Figure 7-36 produces and output word of 0010 on its F4 to F1 outputs. You can read this stored word directly from the top line in the PROM diagram, if you think of no fuse as a 0 and a fuse as a 1. As another example of this, let's see what happens if we apply an input address of 01 to the device. A = 0 and B = 1 will produce a high on the output of the second AND gate from the top in Figure 7-36. The output line from this AND gate has fuses on the F1 and F3 lines, so the F1 and F3 outputs will be high. Since the F2 and F4 OR gates have no fuses on this line, the F2 and F4 outputs will be low. For input address 01, then, the output word will be 0101. Another way of saying this is that the word stored in the device at address 01 is 0101.

Work out the words stored in the device at address 10 and address 11. You should find that 1011 is stored at address 10 and 1010 is stored at address 11 in this PROM. Figure 7-36b summarizes the relationship between the addresses and the words stored at each address in this device.

The simple example PROM in Figure 7-36a, then, can be used to implement any AND-OR function of two in-

put variables, or it can be used to store four binary words of 4 bits each. In either case the device functions the same, but we describe its operation in the way that best fits how it is being used in a specific application.

To introduce you to an actual PROM, Figure 7-37 shows a block diagram of National Semiconductor's DM74LS471. The device has eight data outputs labeled Q1 to Q8. These outputs have three-state buffers which will be enabled if the $\overline{E1}$ and $\overline{E2}$ enable inputs are asserted low. The eight pins labeled A0 to A7 on the device are the address inputs.

The 74LS471 is described in its data sheet as a 256×8 device. This means that it can store 256 binary words of 8 bits each. Since $256 = 2^8$, eight address inputs are required to select one of the 256 stored words. When the device is being used as a memory, then, the 8-bit binary address applied to the A0 to A7 inputs determines which one of the 256 stored words is present on the data outputs. For example, if 00000000 is applied to the A0 to A7 inputs, the byte stored at that address in the device will be present on the Q outputs. If 11111111 is applied to the A0 to A7 inputs, the byte stored at that address will be present on the Q outputs.

When a 74LS471 is being used to implement combinational logic functions, A0 to A7 are used as the variable inputs. The block diagram of the device in Figure 7-37 tells you that the device contains a 2048-bit (256×8) hardwired array which generates 256 different product terms from the eight input variables. Implied, but not stated, in the diagram is the fact that a programmable OR array on the outputs of the AND gates allows you to produce any eight OR functions of these 256 product terms. In other words, you can use this single device to implement any 8, AND-OR functions of up to 8 input variables. As an example application for this device, we will show you how it might be used as a simple code converter.

Suppose that you have a device which outputs ASCII codes and you need to convert these codes to EBCDIC so that they can be input to a particular computer system. If there is no readily available device which does this conversion, you can do it with a 74LS471 as follows.

First look at the ASCII codes and the corresponding EBCDIC codes in Table 7-2. For this design problem the ASCII codes represent the input variables and the EBCDIC codes represent the desired output codes. The two columns in Table 7-2 then represent the truth table you want to implement with the PROM. There are two different ways you can think about doing this.

One way is to write a sum-of-products expression for each bit of the output EBCDIC code in terms of the bits of the input ASCII code, simplify the expression, and implement the simplified expression with logic gates. This is the same approach you used previously to produce a BCD-to-seven-segment code converter. The difficulty with this approach here is that the ASCII codes contain 7 bits, so the product terms in the Boolean expressions contain up to 7 terms. This makes the expressions difficult to simplify with Boolean algebra or Karnaugh maps.

An easier way to solve this problem is to think of the device as a memory. If the ASCII data is applied to the address inputs of the PROM as shown in Figure 7-38,

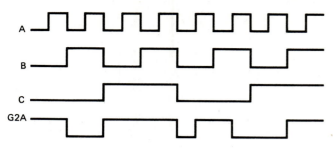

FIGURE 7-43 Input waveforms for Problem 10.

plexer if the waveforms shown in Figure 7-43 are applied to the indicated inputs.

11. Convert the following numbers to octal:
    a. 1010101_2 c. 111001011_2
    b. 10_{10} d. 24_{10}

12. Convert the following octal numbers to decimal:
    a. 314_8 c. 40_8
    b. 72_8 d. 24_8

13. Convert to hexadecimal:
    a. 42_{10} c. 01011101_2
    b. 798_{10} d. 11000011_2

14. Convert to decimal:
    a. A7H   c. 40H
    b. 3FFH   d. 10H

15. Convert to BCD:
    a. 72_{10} c. 37_{10}
    b. 59_{10} d. 15_{10}

16. a. Describe how an optical encoder produces a binary code which represents the position of a rotating shaft in a machine such as a lathe.
    b. What is the major advantage of gray code over standard binary for encoding the position of a shaft in a machine?

17. a. Draw a schematic for the internal circuitry of a common anode seven-segment LED display.
    b. Is a logic low or a logic high required to light the LEDs in a common-anode LED display?

c. What codes must be present on the a to g inputs of a common-anode seven-segment display to display a 4?
d. Why are resistors required in series with each segment of an LED display?

18. a. Define the term *parity*.
    b. Draw a logic circuit which will indicate whether the parity of a 4-bit binary word is odd or even.
    c. Describe how parity generator-checkers are used to detect an error in transmitted data.
    d. What is the major problem with the parity method of error detection?

19. The M key is depressed on an ASCII-encoded keyboard. What pattern of 1's and 0's would you expect to find on the seven parallel data lines coming from the keyboard? What pattern would a carriage return, CR, give?

20. Give the EBCDIC code for G.

21. Describe the difference between an encoder and a decoder.

22. Figure 7-26a shows a 74LS47 and a common-anode display.
    a. What effect will making the $\overline{LT}$ input low have on the display?
    b. What logic level will be on each of the segment outputs if the $\overline{BI}$ input becomes stuck low?
    c. List the logic levels that will be present on each of the segment outputs if 0101 is applied to the DCBA inputs.
    d. What character will be displayed if $\overline{RBI}$ is low and 0000 is applied to the DCBA inputs?

23. Write the sequence of displays that will be produced by the input waveforms to the left of the 74LS47 in Figure 7-44.

24. Figure 7-30b shows the logic symbol for a 74HC148 priority encoder.
    a. Is the enable input active low or active high?
    b. Determine the logic levels that will be present

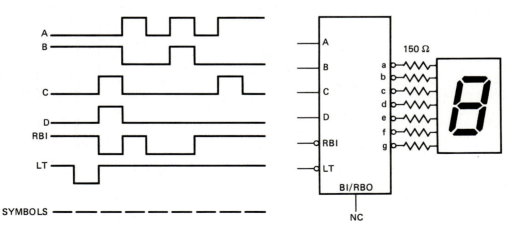

FIGURE 7-44 Circuit for Problem 23.

on the A2, A1, A0, EO, and GS outputs if the device is enabled and both the 2 data input and the 4 data input are low.

25. The 74S64 is described in the data book as a 4-wide 4,3,2,2 input AND-OR-INVERT gate. Draw a logic diagram for this gate.

26. Figure 7-45 shows the logic diagram for a simple FPLA.
    a. What is meant by the term *FPLA*?
    b. How is an FPLA different from a PLA?
    c. For the FPLA in Figure 7-45, how many inputs does each AND gate have?
    d. How many inputs does each OR gate have?
    e. What are the main disadvantages of FPLAs?

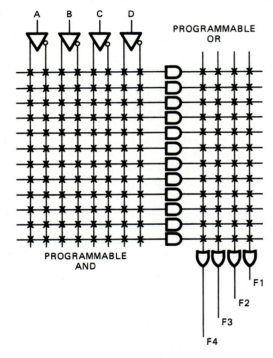

FIGURE 7-45   Logic diagram of FPLA for Problem 26.

27. Briefly define the terms *ROM*, *PROM*, *EPROM*, and *EEPROM*.

28. Figure 7-46 shows the logic diagram for a simple PROM.
    a. How many inputs does each AND gate have?
    b. Write the product term produced by each AND gate.
    c. Write the sum-of-products expression for each of the F outputs of the PROM.
    d. Show how this device could be programmed to convert a 4-bit standard binary input code to 4-bit gray code.

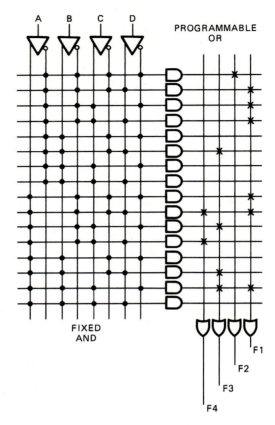

FIGURE 7-46   Logic diagram of PROM for Problem 28.

29. Figure 7-47 shows the logic diagram for a simple PAL.
    a. Define the term *PAL* and describe how a PAL is different from an FPLA.
    b. How many inputs does each AND gate have?
    c. How many inputs does each OR gate have?
    d. Write the product term produced by each AND gate.
    e. Write the sum-of-products expression for each F output.
    f. Based on the number of inputs, etc., how would this PAL be identified using the standard convention?

30. How are PROMs and PALs usually programmed?

31. Figure 7-48 shows a 6-bit binary counter that is started by pressing a switch. When the switch is released, the counters will run for a while and then stop. As the counter is clocked, it will step through the binary sequence of 000000 to 111111 over and over. Draw a circuit showing how some inverters and eight 74LS138s or four 74154s can be connected to the outputs of this counter to step a lit LED around a circle of 64 LEDs. If numbers are written next to each LED, the circuit will function as a "wheel of fortune" such as you might see at a carnival.

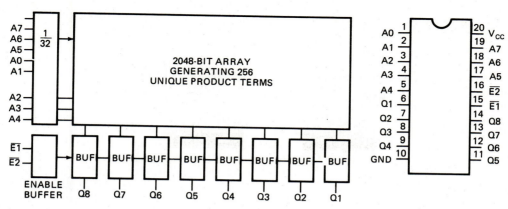

FIGURE 7-37 Block diagram of 256x8 DM74LS741 PROM.

each ASCII code will select a different address in the PROM. To produce the desired EBCDIC code on the outputs, then, all we have to do is program each address with the corresponding EBCDIC code. For an input ASCII code of 00110000, or 30H, the desired output EBCDIC code is 11110000, or F0H. To produce this, we program address 00110000 in the PROM with F0H, the desired EBCDIC code.

FIGURE 7-38 Using a DM74LS471 PROM as a decoder.

When the ASCII code 30H is present on the address inputs, the desired EBCDIC code, F0H, will be output on the Q lines of the PROM. As another example of this, we would program address 41H with C1H, because the EBCDIC code C1H corresponds to the ASCII code 41H. The principle here then is to simply treat each ASCII code as an address and program that address in the PROM with the corresponding EBCDIC code. Rather than write Boolean equations and attempt to simplify them, we just implement the entire truth table directly in the PROM! The advantages of this approach over the Karnaugh map and discrete gate approach should be clear.

To program a PROM we usually use a device called a *PROM programmer*. The desired sequence of data words is loaded into the PROM programmer and the programmer automatically blows fuses or stores charges as

needed to program all the addresses in the device. The whole operation only takes a few minutes.

In the next section we discuss another common AND-OR type device which, for some applications, has advantages over a PROM.

PALs and EPLDs

PAL is the acronym used to identify a *programmable array logic* device. Figure 7-39 shows the internal structure for a PAL which implements the same logic functions we used to demonstrate the operation of an FPLA and a PROM. Note that in a PAL the AND array is programmable by blowing fuses and the OR array is hardwired. This is the opposite of a PROM which has a fixed AND array and a programmable OR array.

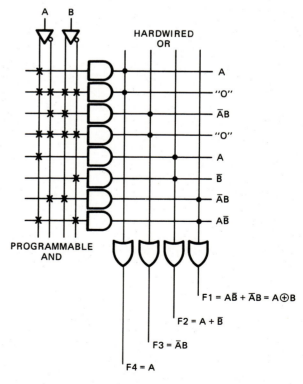

FIGURE 7-39 Internal structure of a simple PAL.

In a PAL the product terms are produced just as we described for an FPLA. If a fuse is left intact, the variable connected to that fuse will appear in the product term. If a fuse is blown, the variable associated with that input will not appear in the product term. If all the fuses on the inputs of an AND gate are left intact, the output of the AND gate will be a 0 as shown in Figure 7-39, regardless of the state of the input variables. The reason for this is that when a variable is ANDed with its complement, the result is always 0. To see how this product term of 0 is often used in PALs, let's take a look at the OR array of the PAL in Figure 7-39.

Remember from our discussion of a ROM that a dot at an intersection represents a hardwired connection. For our example PAL in Figure 7-39, then, the two inputs of each OR gate are hardwired to two AND gate outputs. For this device then the output expression for each OR gate will be just the sum of the two product terms connected to its inputs.

The F4 OR gate, for example, has the two product terms A and 0 connected to its inputs. Since a variable ORed with 0 reduces to just the variable, $F4 = A$. From this you can see that you remove a product term in a PAL by making that term a 0. The F3 OR gate shows another example of this. Here $F3 = \overline{A}B + 0 = \overline{A}B$. The F2 expression is just $F2 = A + \overline{B}$, and the F1 expression is $F1 = \overline{A}B + A\overline{B}$. Note that, since the OR input connections are fixed, you have to generate a particular product term, 0, for example, each time it is used in an output expression. Also note that the number of product terms that can be included in any output expression is limited by the number of inputs on each OR gate.

PALs come in a wide variety of formats, so it is usually not too difficult to find one that has enough product terms to solve a particular problem. To introduce you to a real PAL, Figure 7-40 shows the logic diagram for a Monolithic Memories PAL16P8A, which comes in a small, 20-pin package.

The first number in the name of a PAL tells you the number of data inputs on the device. The 16 in the name of our example device then tells you that it has 16 data inputs. This number does *not* include the complements of each input that are internally produced by the buffers. Look carefully at the 16P8A in Figure 7-40 to see if you can find the 16 data inputs. Note that six of the inputs are connected to outputs so that the output from one AND-OR section can be fed back into the AND array as an input signal to other AND-OR sections if needed to implement a complex logic function. The inputs are labeled 1 to 9, 11, and 13 to 18.

The number near the end of the PAL name tells you the number of outputs on the device. By now you have probably found the eight outputs on the 16P8A in Figure 7-40. The outputs are labeled 12 to 19.

The letter in the middle of a PAL name indicates the type of outputs it has. An H, for example, shows that the outputs are active high, an L shows the outputs are active low, and a P indicates that the outputs can be programmed to be active high or active low. In our example device in Figure 7-40, an output is active high if the fuse on the grounded input of its XOR gate is left intact. If this fuse is blown, the input of its XOR gate will inter-

nally be pulled high, and the logic signal passing through the gate will be inverted.

When it comes from the manufacturer, all the intersections in the AND array of the PAL in Figure 7-40 are connected with fuses. To make the logic diagram less cluttered, the fuses are not shown with Xs as we did in the preceding example. You can determine the number of inputs that each AND gate has by simply counting the number of lines which intersect the line on its input. For our example here, there are 32 intersecting lines, so each AND gate has 32 inputs. This number represents the 16 input variables and their complements. In other words, each product term can contain up to 16 variables in true or complemented form.

Since the OR array is hardwired in a PAL, the OR inputs are shown separately in Figure 7-40 so that you can directly see how many product terms are summed by each. Each of the OR gates in Figure 7-40 is connected to seven AND gate outputs, so the expression for each OR output can contain up to seven product terms.

The final parts of Figure 7-40 to look at are the three-state output buffers. The enable inputs for these buffers are connected to AND gates so that each buffer can be individually enabled when some desired logic conditions are present on the data input lines. An alternative might be to program the PAL so that when one input is asserted, it enables all the output buffers.

PALs such as the PAL16P8A are programmed by blowing fuses. This means that if you make a mistake while programming one, the device is only useful as a replacement for thumbtacks in sticking notes on the bulletin board. To eliminate this potential waste, several companies have developed PAL-type devices which can be erased and reprogrammed over and over. To distinguish these devices from the fuse-type PALs, they are called *erasable programmable logic devices*, or *EPLDs*. Some EPLDs use EPROM technology to make the erasable-programmable connections. These devices are erased with ultraviolet light as EPROMS are. Other EPLDs use EEPROM technology to make the erasable-programmable connections. These devices then are electrically erasable.

Having learned about ROM-type and PAL-type devices, the question that may occur to you is, "When do you use a PROM and when do you use a PAL?" The answer to this question depends on the application. A PROM produces all the possible product terms for the input variables, so for an application such as the ASCII-to-EBCDIC code converter where we want an output for each of the possible combinations of the input variables, we use a PROM. For applications where we have many input variables but we only need a limited number of product terms for these variables, a PAL is usually a better choice.

You will usually use a computer to generate the fuse map data for a PAL or the program data for a PROM and then transfer the data to a programmer which does the actual programming. In Chapter 12 we show you how to use PALs to implement a wide variety of combinational and clocked logic circuits. In Chapter 15 we discuss *gate arrays* and other VLSI devices which can be used to implement still larger amounts of logic circuitry with a single IC.

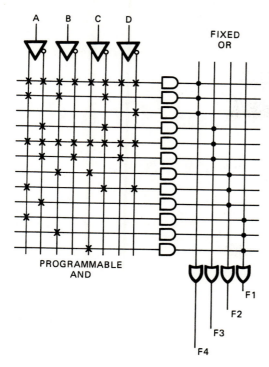

FIGURE 7-47 Logic diagram of a PAL for Problem 29.

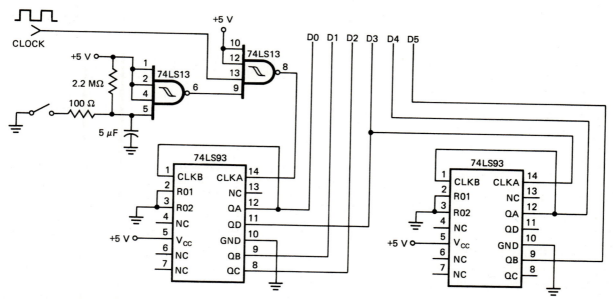

FIGURE 7-48 Binary counter for Problem 31.

CHAPTER 8

Latches, Flip-Flops, Counters, and Logic Analyzers

For the digital circuits we have discussed previously, a change on the inputs produces an almost immediate change on the outputs. The major topics of this chapter are devices and circuits whose outputs only change when a separate "clock" signal is applied to them. The 74LS93 binary counter, which we used to produce a binary count sequence in several previous circuits, is an example of one type of clocked device.

Many clocked circuits such as counters have several output signals. Unless you have a scope multiplexer such as the one in Figure 7-10, it is not very easy to observe these multiple waveforms with a standard dual-trace scope. Therefore, in this chapter we also describe the operation and use of a logic analyzer, which allows you to observe many signal waveforms on the same display.

OBJECTIVES

At the conclusion of this chapter you should be able to:

1. Describe how glitches or hazards are produced in combinational logic circuits.

2. Show how possible hazards can be predicted using Karnaugh maps and prevented by adding extra gates.

3. Write the truth tables for D latches, D flip-flops, JK flip-flops, and T flip-flops.

4. Draw the output waveforms produced when given input waveforms are applied to D latches, D flip-flops, and JK flip-flops.

5. Draw the timing diagram for a binary ripple counter of any modulo.

6. Draw a circuit diagram showing how an MSI counter such as a 7493 can be connected to divide an input frequency by any number.

7. Draw a circuit which will decode a specified state on the outputs of a counter.

8. Describe how ripple counters can be used for applications such as electronic organs and digital clocks.

9. Describe the operation of a logic analyzer.

10. Describe how a logic analyzer can be connected to a counter circuit to determine the sequence of states or measure propagation delays.

11. Describe how the controls of a logic analyzer should be set to produce a specified state or timing display.

12. Draw a block diagram showing how counters can be used to construct a digital clock.

13. Explain the major problem of ripple counters.

TIMING CONSIDERATIONS IN COMBINATIONAL LOGIC CIRCUITS

In the preceding chapters we have for the most part ignored the propagation delay of logic gates. In applications such as driving relays, LED displays, and other low-frequency devices, this creates few problems. However, in high-frequency applications, failure to take gate propagation delay into account can lead to the failure of an entire system. As a first example of the effect of propagation delay, we will use a simple OR gate.

Introduction to Hazards in Logic Circuits

To refresh your memory, Figure 8-1a shows the truth table for an OR gate. Figure 8-1b shows a circuit which ORs a variable, A, with its complement. This is not a very useful circuit because, according to the middle two lines of the truth table in Figure 8-1a, the output will always be high. However, when the circuit is built with real devices, the circuit does not perform as expected. The inverter in series with the top input of the OR gate delays the signal to that input by a few nanoseconds. This means that the A signal and the $\overline{A}$ signal do not get to the OR gate at the same time. Figure 8-1c shows the waveforms that will be produced on the output of the inverter and the output of the OR gate by a pulsed A signal. For these waveforms we have exaggerated the delay of the $\overline{A}$ signal somewhat so that it is more clearly visible. Also, to make it easier to see the effect of ORing

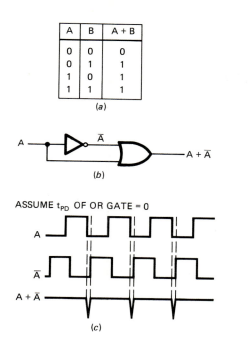

(a)

ASSUME t_{PD} OF OR GATE = 0

(c)

FIGURE 8-1   *(a)* OR gate truth table. *(b)* Circuit that ORs a variable with its complement. *(c)* Timing diagram for circuit.

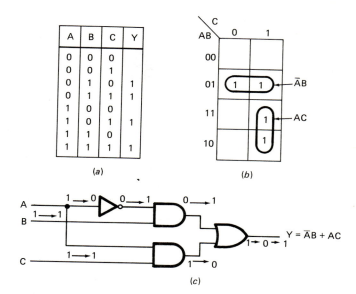

(a)   *(b)*

$Y = \overline{A}B + AC$

(c)

FIGURE 8-2   Simple combinational logic circuit. *(a)* Truth table. *(b)* Karnaugh map. *(c)* Circuit diagram for minimal expression.

the A and delayed $\overline{A}$ signals together, we have assumed in these waveforms that the propagation delay of the OR gate is zero. This is justified because the propagation delay of the two input signals through the OR gate is the same.

According to the truth table for an OR gate, if any input is high, the output will be high. When the A signal is high, the output of the OR gate will therefore be high. When the A signal goes low, the $\overline{A}$ signal will go high after a few nanoseconds. During the time it takes the $\overline{A}$ signal to go high, both inputs of the OR gate will be low. As shown in the waveforms in Figure 8-1c, this will cause the output of the OR gate to be low. The output of the OR gate will be low for the few nanoseconds it takes the change in the A signal to propagate through the inverter. A short, unwanted state such as the low on the output of the OR gate here is called a *hazard, glitch,* or *spike.* As you can see, hazards are caused by signal paths which have unequal propagation delay times through a logic circuit. As a more realistic example of how hazards can be produced, let's look again at how we typically implement a simple truth table with a logic circuit.

Predicting and Preventing Hazards in Combinational Circuits

AN EXAMPLE OF A LOGIC CIRCUIT HAZARD

Figure 8-2a shows a truth table for a simple 3-variable logic circuit, and Figure 8-2b shows a Karnaugh map for the truth table. The four 1's in the Karnaugh map can be circled with just two groups. The expressions for these two groups are $\overline{A}B$ and AC, so the simplified expression for the truth table is simply $Y = \overline{A}B + AC$. This expression can be easily implemented with an inverter,

two AND gates, and one OR gate, as shown in Figure 8-2c. To test a circuit such as this for hazards you theoretically have to predict the response of the circuit for each of the possible input signal transitions. For example, you have to check if a hazard is produced when the ABC inputs transition from 000 to 001, to 010, to 011, to 100, to 101, to 110, or to 111. Since the inputs can be changed from any one of eight states to any one of seven other states, there are 56 possible transitions to check! Checking all these transitions is a tedious and error-prone process, so after we show you how this circuit produces a hazard, we will show you a trick for identifying those transitions that may cause a hazard.

The transition we want to demonstrate here is the one between ABC = 111 and ABC = 011. We will first look at it intuitively, and then show an actual timing diagram.

According to the truth table for the circuit, the final output should be high for both of the input combinations, 111 and 011. Figure 8-2c shows a simple way to analyze the effect of a transition intuitively on the logic diagram of the circuit. The state of each input or output before a transition is shown to the left of an arrow, and the state of the input or output after the transition is shown to the right of the arrow. With input combination 111 the output of the inverter is low, as shown by the 0 on the left side of the arrow next to its output. With this low and the high on the B signal, the output of the top AND gate will be low. This is shown by the 0 on the left side of the arrow next to the upper AND output. For the input combination 111 the output of the lower AND gate will be high, as shown by the 1 on the left side of the arrow next to its output. This 1 will produce a 1 on the output of the OR gate.

If the inputs are changed to 011, as shown by numbers to the right of the arrows on the inputs, the output of the AC AND gate will go to a 0 after just the propagation delay of the NAND gate. The output of the $\overline{A}B$ AND gate will stay low until the change on the A input propagates through the inverter and through the AND gate. The propagation delay difference between the two paths is the delay of the inverter. In other words, it takes one propagation delay for the lower AND output to change to a 0, but it takes two propagation delay times for the upper AND output to change to a 1. For one propagation delay time, then, both inputs of the OR gate will be 0, and the output of the OR gate will be a 0. Once the output of the upper AND gate goes high, this high will cause the output of the OR gate to go high after one propagation delay time. The short duration 0 on the output is of course the hazard we were looking for. As you can see, the hazard is caused by a difference in propagation delay times for the two signal paths. Figure 8-3 shows, in a timing diagram, the effect of a transition from 011 to 111 and 111 to 011. Let's see how this corresponds to the intuitive approach.

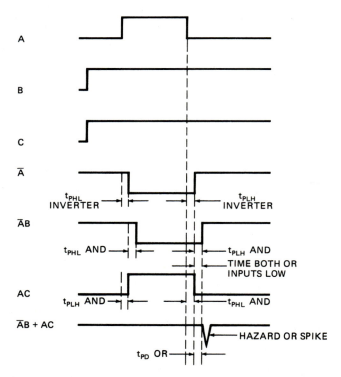

FIGURE 8-3   Timing diagram analysis of circuit in Figure 8-2c.

To simplify the timing diagram we have assumed that the propagations delay times (t_{PD}) of the inverter, the AND gates, and the OR gate are approximately equal. As you can see in the waveforms, a change in the A signal will be seen on the output of the inverter after one t_{PD}. A change in the $\overline{A}$ signal will produce a change in the $\overline{A}B$ signal after another t_{PD}. A change on the A signal then produces a change on the upper input of the OR gate after two t_{PD}s. Since the A signal is connected directly to one input of the lower AND gate, a change in the A signal will produce a change on its output after only one t_{PD}.

For the 011-to-111 transition, the AC signal from the lower AND gate goes high one t_{PD} before the $\overline{A}B$ signal from the upper AND gate goes low. In this case one of

the AND gates is always high, so the output is always high. For the 111-to-011 transition, the AC output goes low one t_{PD} before the $\overline{A}B$ output goes high. During this t_{PD}, both inputs of the OR gate are low. These lows will cause the output of the OR gate to go low after its t_{PD}. The width of the glitch is approximately equal to the difference in propagation delays for the two signal paths. In the example we are discussing here, the width of the glitch will be equal to the propagation delay of the inverter, or approximately 10 ns.

Now that we have demonstrated how a particular signal transition can cause a glitch or hazard, we will keep our promise to show you how to predict which transitions may cause this type of hazard.

PREDICTING POTENTIAL HAZARDS

The easiest way to predict a possible $1 \rightarrow 0 \rightarrow 1$ hazard in a logic circuit is to use the Karnaugh map for the circuit. Assuming only one input variable at a time is changed, the rule for predicting a potential $1 \rightarrow 0 \rightarrow 1$ hazard is:

> In-line adjacent 1's which are not circled in the same group may cause a hazard.

Let's try this rule on the preceding example.

Figure 8-4 again shows the Karnaugh map we used to implement the logic circuit in Figure 8-2c. The four 1's are circled in two groups. One group is represented by the expression $\overline{A}B$, and the other group is represented by the expression AC. The 1 representing the input combination 011 is adjacent to the 1 representing the input combination 111, but we did not group these two 1's together when we wrote the simplified expression for the map. The hazard rule then predicts that these adjacent 1's may cause a hazard on the output.

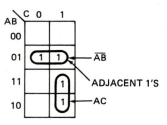

FIGURE 8-4   Karnaugh map for circuit in Figure 8-2c showing adjacent 1's not included in the same group.

According to the timing diagram in Figure 8-3, the transition from input state 011 to input state 111 does not cause a hazard, but the transition from 111 to 011 does produce a hazard, as predicted by the rule.

The hazard rule then quickly tells you which of the many possible signal transitions need to be checked to see if they produce a $1 \rightarrow 0 \rightarrow 1$ hazard. In the next section we show you one method that can be used to prevent this type of hazard.

PREVENTING $1 \rightarrow 0 \rightarrow 1$ HAZARDS

Hazards or glitches are caused by signal paths which have unequal propagation delays. One way to prevent hazards in combinational logic circuits then might be to equalize propagation delay times by adding inverters or noninverting buffers to the shorter signal paths. The difficulties with this approach are that it is hard to equalize the delays through all the different signal paths and that the resulting circuit is likely to be messy. A more efficient approach for some circuits is to use a Karnaugh map to help you redesign the circuit so it doesn't produce hazards. Here's how you eliminate the hazard we showed you in the preceding section.

In the preceding section we showed you that you can predict possible hazards by looking for adjacent 1's that are not circled in the same group on a Karnaugh map. To design a circuit which does not contain hazards, then, the first step is to draw some additional circles on the Karnaugh map so that all adjacent 1's are grouped together. You then add the terms for any additional groups to the sum-of-products expression for the map.

The circuit in Figure 8-2c, for example, produces a hazard because the 1 for input combination 111 is adjacent to the 1 for input combination 011, but it is not included in the same group. These two 1's can be circled as shown in Figure 8-5a and represented by the term BC. Sum-

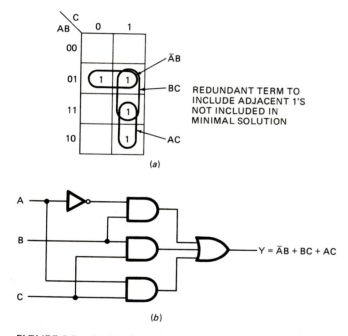

(a)

(b)

FIGURE 8-5 (a) Circling adjacent 1's in Karnaugh map to prevent hazard. (b) Resulting hazard-free circuit.

ming the three circled groups gives the expression Y = $\overline{A}B$ + BC + AC. As you know, this is not the minimal expression for the desired logic function, but when this expression is implemented with real devices, it will not produce a glitch during a 111-to-011 transition, as the minimum expression does.

The final step in the design process is to include in the circuit any gates needed to implement the additional terms. As a first step, Figure 8-5b shows how we have added another AND gate to implement the BC term of the new expression. Let's analyze the circuit to deter-

mine how it prevents the glitch produced by the 111-to-011 transition.

Intuitively you can see that if B = 1 and C = 1, the output of this added AND gate will be high, regardless of any change on the A input. The high on the BC output will cause the output of the OR gate to remain high as desired during a change on the A signal.

As an introduction to hazards we have only shown you how to predict and prevent 1→0→1 hazards produced by a change in one signal at a time on the inputs of a circuit. If more than one input signal changes at a time, additional 1→0→1 hazards, not predicted by the initial 101 hazard rule, may be produced.

In the circuit in Figure 8-5, for example, an input transition from 111 to 010 will produce a hazard on the output. This hazard, incidentally, can be predicted by observing that the diagonally adjacent 1's in boxes 010 and 111 of the Karnaugh map in Figure 8-5b are not included in the same circled group. The hazard can be eliminated by adding the term B(A XNOR C), which is the expression for a loop containing these two 1's.

Further analysis would show that additional hazards might be produced by more than two input signals changing at the same time. Also, combinational circuits with more than just an AND layer and an OR layer may have 0→1→0 hazards.

The point of all this is that it is very difficult to eliminate all the possible hazards from even a simple combinational circuit. For many digital systems, glitches which are a few nanoseconds wide are not a problem, so you can use the minimum solution from a Karnaugh map. For high-frequency or other digital circuits, we use several methods to overcome the hazard or glitch problem.

In some applications, such as the state machines we discuss in Chapter 12, we design the rest of the circuit so that only one input to the combinational circuit changes at a time. We then use the techniques we showed you in the preceding section to eliminate any glitches.

In many other applications we use devices called *latches* or devices called *flip-flops* to eliminate glitches in digital signals. In the next section we discuss these devices in detail and show you how they are used to do this.

LATCHES AND FLIP-FLOPS

Cross-Coupled NAND Gate Latch

The diagram in Figure 8-6a shows how the simplest type of latch circuit can be constructed with NAND gates. The circuit is shown with both types of NAND symbols, so you can use whichever one you are most comfortable with to analyze the circuit. Before we analyze the circuit, however, we need to introduce you to some labeling conventions commonly used with latches and flip-flops.

Latches and flip-flops usually have an active high output and an active low output. The active high output is identified with a Q, and the active low output is identified with a $\overline{Q}$ as shown in Figure 8-6a.

When the Q output of a latch or a flip-flop is high, the device is said to be *in the set state*, or just *set*. If the Q output is high, the $\overline{Q}$ output will, by definition, be low.

When the Q output of a latch or flip-flop is low, the de-

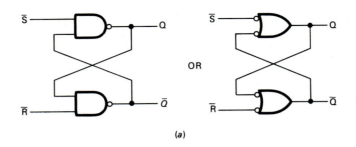

OR

(a)

STATE NAME	$\overline{S}$	$\overline{R}$	Q	$\overline{Q}$
INDETERMINATE	0	0	*	*
SET	0	1	1	0
RESET	1	0	0	1
LATCHED	1	1	Q	$\overline{Q}$

(b)

FIGURE 8-6   Cross-coupled NAND gate latch. (a) Circuits. (b) Truth table.

vice is said to be *in the reset state,* or just *reset.* If the Q output is low, then the $\overline{Q}$ output will, by definition, be high.

An input which directly causes a latch or flip-flop to go to the set state is labeled with an S or an $\overline{S}$, depending on whether the input is active low or active high. Other labels commonly used to identify the set input of a device are PRESET, or just PRE.

An input which directly causes a latch or flip-flop to go to the reset state is labeled with an R or $\overline{R}$, depending on whether the input is active low or active high. Other labels commonly used to identify the reset input on a device are CLEAR, or just CLR. Now, let's see how all this applies to the circuit in Figure 8-6a.

To start, let's predict the effect of putting a 0 on the $\overline{S}$ input and a 1 on the $\overline{R}$ input. A low on the $\overline{S}$ input causes the Q output to be high. The high from the Q output and the high on the $\overline{R}$ input will produce a low on the $\overline{Q}$ output of the lower NAND gate. The low on $\overline{Q}$ is fed back into one input of the upper NAND gate. If the Q output were not already high, this low would cause it to go high. With the $\overline{S}$ input low and the $\overline{R}$ input high, the circuit is in the set state. Now, let's see what happens if $\overline{S}$ is made high and $\overline{R}$ is made low.

A low on the $\overline{R}$ input causes the $\overline{Q}$ output to go high. This high, combined with the high on the $\overline{S}$ input, causes the Q output to go low. As shown by the truth table in Figure 8-6b, a low on $\overline{R}$ and a high on $\overline{S}$ causes the circuit to go to the reset condition.

The next case to consider is that where both $\overline{S}$ and $\overline{R}$ are made low. Based on the truth table for a NAND gate, $\overline{S}$ being low will cause Q to be high and $\overline{R}$ being low will cause $\overline{Q}$ to be high. This does not harm the devices, but it violates the definition of Q and $\overline{Q}$ being the complement of each other. Therefore, this input combination is referred to as *indeterminate, prohibited,* or *invalid.* This state is shown by using * in the truth table. In applications where we use this circuit, we try to make sure the input signals can never both go low at the same time.

Finally, let's predict the effect of making both $\overline{S}$ and $\overline{R}$ high at the same time. This case is a little more difficult,

because the output state depends on the logic levels that were present on $\overline{S}$ and $\overline{R}$ before they both went high. As a first example, suppose that initially the $\overline{S}$ input is low and the $\overline{R}$ input is high. These input conditions will cause the Q output to be high and the $\overline{Q}$ output to be low. The low from the $\overline{Q}$ output is fed back to one input of the upper NAND gate. This low from $\overline{Q}$ will cause the Q output to be high, regardless of the logic level on the $\overline{S}$ input. If the $\overline{R}$ input is kept high and the $\overline{S}$ input is made high, then the outputs will stay the same as they were before $\overline{S}$ was made high. A similar result occurs if the circuit is reset before $\overline{S}$ and $\overline{R}$ are both made high.

If $\overline{S}$ is high and $\overline{R}$ is low, then the Q output will be low and the $\overline{Q}$ output will be high. The low fed from the Q output to one input of the lower NAND gate will hold the output of this NAND gate high, regardless of the level on $\overline{R}$ input. Therefore, when $\overline{R}$ is made high, the circuit just remains in the reset state.

The condition produced by both inputs being high is called the *latched state.* As shown in the truth table in Figure 8-6b, a Q and a $\overline{Q}$ are used to indicate that the output levels during the latched state are the same as they were just before the latched state.

To further describe the operation of this simple latch circuit, Figure 8-7 shows how it will respond to some input waveforms. Initially the Q output is low which rep-

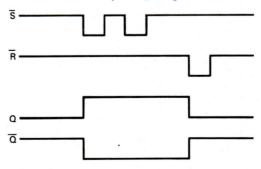

FIGURE 8-7   Waveforms showing effects of $\overline{S}$ and $\overline{R}$ pulses on Q and $\overline{Q}$.

resents the reset state. When the $\overline{S}$ input is pulsed low, the circuit goes to the set state. Beyond a certain minimum, the width of the $\overline{S}$ pulse doesn't matter because once the outputs change, feedback from the lower NAND gate holds the circuit in the set state. Likewise, the second $\overline{S}$ pulse has no effect because the circuit is already in the set state. When the $\overline{R}$ input is made low, the circuit will go to the reset state. Again, beyond a certain minimum, the width of the $\overline{R}$ pulse doesn't matter because, once the circuit is in the reset state, feedback from the upper NAND gate holds it in that state. A second $\overline{R}$ pulse would have no effect because the circuit is already reset. The only way to change the circuit to the set state is to pulse $\overline{S}$ low. Note in these waveforms that when $\overline{S}$ and $\overline{R}$ are both high, the outputs remain the same as they were at the time the inputs became high.

Now that you know how this simple latch circuit works, we will show you a common application for it.

A SIMPLE LATCH APPLICATION

When you flip them from one position to the other, most mechanical switches bounce for a few milliseconds before they make a solid contact. In many applications, such as light switches, this bouncing causes no problems. However, when we are using a mechanical switch to produce a logic signal, the bouncing can cause severe problems. Figure 8-8a shows how a mechanical switch might be connected to an inverter, and Figure 8-8b shows the waveform that would be produced on the output of the inverter when the switch was closed. As you can see, the output of the inverter shows a series of pulses, rather than the single low-to-high transition that we want.

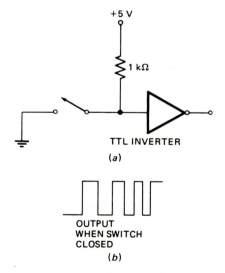

FIGURE 8-8   Output of TTL inverter showing effect of switch bounce. *(a)* Circuit. *(b)* Output waveforms.

Figure 8-9 shows how a double-throw switch can be connected to the simple latch circuit to produce a *debounced* output signal. To analyze the operation of this circuit, assume that the switch is initially in the down position, as shown in the figure. With the switch

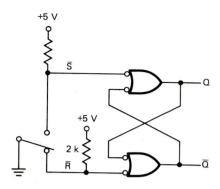

FIGURE 8-9   Cross-coupled NAND latch switch debouncer circuit.

in the down position, the Q output will be low. When the switch is thrown to the up position, the $\overline{R}$ input will be pulled high and the $\overline{S}$ input will be made low. This will cause the Q output to go high. The high on Q and the high on $\overline{R}$ will, within a few nanoseconds, produce a low on the $\overline{Q}$ output and on the other input of the upper NAND gate. As we described before, this low will hold the Q output high, regardless of any changes on the $\overline{S}$ input. If the switch bounces, the $\overline{S}$ input will be pulled high between bounces, but, because of the feedback from the lower NAND gate, this bouncing will have no effect on the Q output. The waveforms in Figure 8-7 demonstrate that once the Q output goes high, the only way to get it back to low again is to throw the switch to the $\overline{R}$ position. Likewise, once the Q output is made low, the only way to get it back to high again is to throw the switch to the $\overline{S}$ position. The circuit then produces a debounced output signal for either transition of the switch. If you need to debounce several switches, a 74LS279 contains four of these cross-coupled NAND gate, $\overline{S}\overline{R}$ latches.

You can also use cross-coupled NOR gates to build a latch circuit which functions similarly to the cross-coupled NAND circuit. A problem at the end of the chapter gives you a chance to analyze the operation of a NOR gate latch. In the next section here we discuss an improved version of the basic latch.

RS Latch with an Enable Input

The cross-coupled NAND latch we discussed in the preceding sections is useful for some applications, but it has two major limitations. The first of these is the indeterminate state caused by both inputs being made low at the same time. The second problem is that the output of the simple NAND circuit changes as soon as $\overline{S}$ or $\overline{R}$ is pulsed. In other words, the simple NAND circuit has no input which allows you to control when the outputs change to reflect new signals on $\overline{S}$ and $\overline{R}$.

Figure 8-10a shows a circuit which solves this second problem by providing an *enable* or *strobe* input. If the enable input of the circuit in Figure 8-10a is low, the outputs of the leftmost NAND gates are both high, regardless of the signals on the S and R inputs. The highs

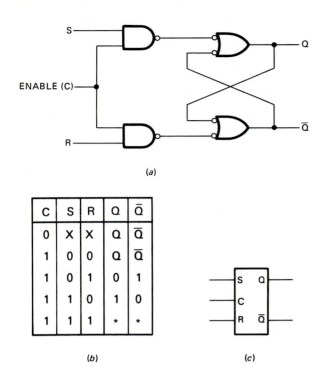

FIGURE 8-10 (a) RS latch circuit with enable input. (b) Truth table. (c) Traditional logic symbol.

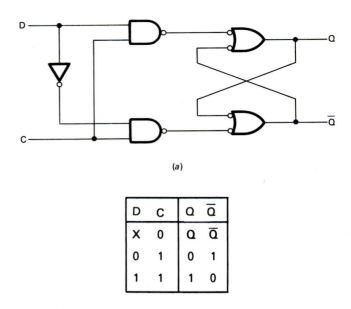

FIGURE 8-11 D latch. (a) Circuit. (b) Truth table.

on the outputs of these NANDs cause the rightmost NANDs to be in the latched state.

When the enable input is made high, the leftmost NAND gates will be enabled. The logic levels on the S and R inputs will then affect the output NANDs as shown in the truth table in Figure 8-10b. When the enable input is high, a high on S and a low on R cause the Q output to go to the set state. A low on S and a high on R cause the Q output to go to the reset state. Because of the signal inversion in the input NANDs, the S and R inputs are active high for this circuit.

If S and R are both low on the circuit in Figure 8-10a, the outputs will be in the latched state. Remember, this means the outputs will be held in the states they were in just before the inputs both became low. In this circuit, the output states will be indeterminate if S and R are both high at the same time.

The enable input on the circuit in Figure 8-10a allows you to control when the outputs change, but this circuit still has the problem that one set of inputs produces an indeterminate output state. The circuit we discuss next solves both of these problems.

D Latches

BASIC CIRCUIT ANALYSIS

Figure 8-11 shows one way to implement a circuit called a *D latch*. The circuit is essentially the same as the RS circuit in Figure 8-10a, except for the inverter connected between the input NAND gates. This inverter eliminates the indeterminate output state by making sure that the input signals to the two input NAND gates are always the complement of each other. If the D, or data, input is

high, for example, the corresponding input of the upper NAND gate will be high and the corresponding input of the lower NAND gate will be low. As shown by the truth table in Figure 8-11b, the operation of this circuit is quite simple.

If the enable input, C, is low, the signal on D will have no effect on the output state. The output will be held in whatever state it was in just before the C input was made low. If the C input is high, a high on D will produce a high on Q. A low on D will cause a low on Q. Another way of saying this is that if the C input is high, Q "follows" D. In data sheets, the term *transparent* is often used to indicate that when the C input is high, the Q output can "see" what is on the D input. When the C input is made low, the level present on D at that time will be latched on the Q output. When the enable input, C, is made high again, the Q output will again "follow" the D input.

To help fix in your mind the operation of this important circuit, Figure 8-12 shows some idealized (no propagation delay) waveforms for it. Work your way across

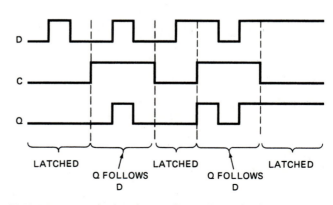

FIGURE 8-12 Idealized waveforms for D latch.

these waveforms from left to right and see if you agree with the Q output waveform shown.

Initially the C input is low, so the first pulse on the D input waveform has no effect on the Q output. For the section of the waveforms between the first set of vertical lines, C is high so the Q output waveform is the same as (follows) the waveform on the D input. As you can see in the waveforms, the second pulse on the D input waveform is transmitted to the Q output.

At the point in the waveforms where C first goes low, D is low. Therefore, a low will be latched on the Q output until C goes high again. When C goes high again, the D input is high, so after a short propagation delay, this high will be transmitted to the Q output. As shown by the waveforms between the second set of vertical lines, Q again follows D as long as the enable input is high. When C goes low the second time, D is high, so a high will be latched on the Q output.

A question that may occur to you at this point is, "What happens if the signal on D and the signal on C both change at the same time?" To answer this question, we will use the data sheet for a commonly available D latch device.

AN EXAMPLE D LATCH: THE 74LS75

The 74LS75 contains four D latches of the type we discussed in the preceding section. Figure 8-13*a* shows the traditional logic diagram for the device and Figure 8-13*b* shows the IEEE/IEC logic symbol or dependency notation symbol for the device. On older schematics, the enable inputs may be labeled with the letter G, but now enable inputs on latches are almost universally labeled with the letter C. As you can see in Figure 8-13*b*, a common enable input serves latch 1 and latch 2 in the device. Another common enable input serves both latch 3 and latch 4 in the device. Each latch has both a Q and a $\overline{Q}$ output. The truth table for each latch is the same as that in Figure 8-11*b*.

Now look at the timing waveforms for the 74LS75 in Figure 8-14*a*. After you use up your 5-min "freak-out" period, see if you can find any labeled timing parameters that you have met before. From discussions in earlier chapters, you should recognize t_{PHL} and t_{PLH}. The time t_{PLH} is the time between a change on an input and a low-to-high change on the corresponding output. t_{PHL} is the time between a change on an input and a high-to-low change on the corresponding output. In Figure 8-14*a*, these labels are shown several times because there are several different possible input and output transitions. We will use the two t_{PLH} labels just above the start of the Q waveform to show you how to interpret these.

The upper t_{PLH} in Figure 8-14*a* represents the time it takes for a high on the D input to cause Q to go high after C goes high. This parameter assumes that D has been high for a long enough time before the C transition that it has no effect. In the switching characteristics section of the data sheet in Figure 8-14*b*, this time is identified as t_{PLH} FROM C TO Q.

The lower t_{PLH} in Figure 8-14*a* represents the time between a change on the D input and the corresponding change on the Q output. Although the waveforms don't look that way, this parameter assumes that the C input

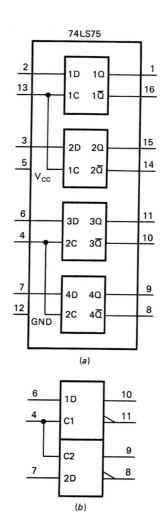

FIGURE 8-13 74LS75 quad D latch. *(a)* Traditional logic symbol. *(b)* IEEE/IEC logic symbol for one latch (one-half of 74LS75).

is already high. In Figure 8-14*b*, this time is referred to as t_{PLH} FROM D to Q. To help stick these data sheet conventions in your mind, find the t_{PHL} labels on the timing diagram and look up the corresponding parameters in Figure 8-14*b*.

The next timing parameter to look at for the 74LS75 is the minimum pulse width, $t_{w,MIN}$. The value for this parameter is listed in the Recommended Operating Characteristics section of the data sheet in Figure 8-14*b*. The value of 20 ns for this parameter tells you that C must be high at least 20 ns, or the logic level on the D input may not be transferred to the Q output.

Next, take a look at the setup time parameter labeled t_{su} in Figure 8-14*a*. This parameter tells you that the signal on the D input must be held stable for some amount of time before C is made low. If the signal on the D input is changed during this setup time, the Q output may not reflect the change in the D signal. According to the data sheet in Figure 8-14*b*, the output of a 74LS75 will be unpredictable if the D input is changed less than 20 ns before the 50 percent point of the C signal going low.

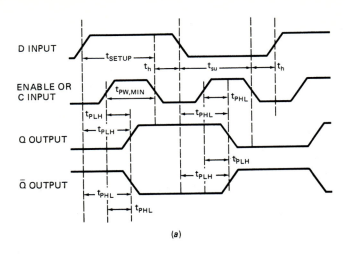

| D INPUT |
| ENABLE OR C INPUT |
| Q OUTPUT |
| Q̄ OUTPUT |

(a)

RECOMMENDED OPERATING CONDITIONS

	SN54LS75 SN54LS77			SN74LS75			UNIT
	MIN	NOM	MAX	MIN	NOM	MAX	
SUPPLY VOLTAGE, V_{CC}	4.5	5	5.5	4.75	5	5.25	V
HIGH-LEVEL OUTPUT CURRENT, I_{OH}			−400			−400	µA
LOW-LEVEL OUTPUT CURRENT, I_{OL}			4			8	mA
WIDTH OF ENABLING PULSE, t_W	20			20			ns
SETUP TIME, t_{su}	20			20			ns
HOLD TIME, t_h	5			5			ns
OPERATING FREE-AIR TEMPERATURE, T_A	−55		125	0		70	°C

SWITCHING CHARACTERISTICS, V_{CC} = 5 V, T_A = 25°C

PARAMETER	FROM (INPUT)	TO (OUTPUT)	'LS75			'LS77			UNIT
			MIN	TYP	MAX	MIN	TYP	MAX	
t_{PLH}	D	Q		15	27		11	19	ns
t_{PHL}				9	17		9	17	
t_{PLH}	D	Q̄		12	20				ns
t_{PHL}				7	15				
t_{PLH}	C	Q		15	27		10	18	ns
t_{PHL}				14	25		10	18	
t_{PLH}	C	Q̄		16	30				ns
t_{PHL}				7	15				

(b)

FIGURE 8-14   74LS75 D latch. (a) Timing waveforms. (b) Recommended operating conditions and timing parameters.

Finally, take a look at the hold time parameter labeled t_h in Figure 8-14a. This parameter represents the time that the signal on D must be held stable after the 50 percent point of C going low. Again, if the signal on D is changed during this time, the change may not be transferred to Q. As shown in the data sheet for the 74LS75 in Figure 8-14b, the minimum hold time for the device is 5 ns.

The setup time and hold time parameters answer the question we asked earlier about what happens if D and C both change at the same time. The setup time parameter of 20 ns tells you that if the D signal is changed during the 20 ns before C is made low, the output will be unpredictable. The hold time of 5 ns tells you that if D is changed sooner than 5 ns after C is made low, the

Q output will be unpredictable. When designing a system using these D latches, it is important to make sure the D signal remains stable during the time from 20 ns before C goes low until 5 ns after C goes low.

In Chapter 15 we will show you how to use a computer program called a *simulator*. A simulator can be used to check if the setup and hold times are met for all the devices in a circuit.

A SIMPLE D-LATCH APPLICATION

Mechanical switches and many other signal sources produce digital signals which have unwanted transitions. Figure 8-15a shows how a D latch, such as one of those in a 74LS75, can be used to remove unwanted transitions from a digital signal.

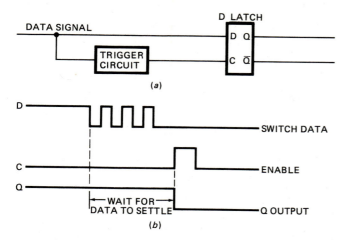

FIGURE 8-15 *(a)* Circuit to remove unwanted transitions from a digital signal. *(b)* Example timing waveform for the circuit.

The original digital signal is applied to the D input of the latch. A "trigger" circuit detects the first transition in the data signal, and after a delay it generates an enable pulse for the latch. In a later chapter we describe the operation of a "trigger" circuit such as this one, but for now all you need to know is that it generates an enable pulse after the data signal has stabilized. When the enable input goes high, Q will go to the level present on D. When the enable input goes low, the logic level present on D will be latched on Q. The point here is that by waiting until the D signal has stabilized before enabling the latch, you will eliminate the unwanted sections of the D signal.

Here's another application of a D latch which takes advantage of the fact that data on the output of a latch are held constant until the enable input is made high again.

USING A D-LATCH DEVICE TO SOLVE A DEMULTIPLEXER PROBLEM

In Figure 7-8 we showed you how a multiplexer such as a 74LS151 can be used to multiplex several digital signals on a single wire, and how a demultiplexer such as a 74LS138 can be used to separate the digital signals at the receiving end of the wire. As we explained in Chapter 7, the difficulty with using a 74LS138 as the demultiplexer is that when an output is not selected, it always goes high. In other words, as shown by the waveforms in Figure 7-17, the 74LS138 outputs only show the correct data for a channel while that channel is selected. They do not hold that data value stable until the next update. Now that you know how a D latch works, we can show you how this problem is easily solved.

Figure 8-16 shows the logic symbol and truth tables for a 74LS259 8-bit addressable latch. This device allows you to route data from a single input to the data input of any one of eight latches and latch the data on the Q output of that latch.

According to the latch selection table in Figure 8-16*b*, the 3-bit binary code on the S2, S1, and S0 select inputs determines which of the eight latches will be selected.

logic symbol

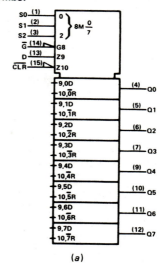

(a)

LATCH SELECTION TABLE

SELECT INPUTS			LATCH
S2	S1	S0	ADDRESSED
L	L	L	0
L	L	H	1
L	H	L	2
L	H	H	3
H	L	L	4
H	L	H	5
H	H	L	6
H	H	H	7

(b)

FUNCTION TABLE

INPUTS		OUTPUT OF ADDRESSED LATCH	EACH OTHER OUTPUT	FUNCTION
$\overline{CLR}$	$\overline{G}$			
H	L	D	Q_{i0}	Addressable Latch
H	H	Q_{i0}	Q_{i0}	Memory
L	L	D	L	8-Line Demultiplexer
L	H	L	L	Clear

H ≡ high level, L ≡ low level
D ≡ the level at the data input
Q_{i0} ≡ the level of Q_i (i = 0, 1 . . . 7, as appropriate) before the indicated steady-state input conditions were established.

(c)

FIGURE 8-16 74LS259 8-bit addressable latch. *(a)* Symbol. *(b)* Selection table. *(c)* Function table. *(Courtesy of Texas Instruments Inc.)*

The first line in the function table in Figure 8-16*c* tells you that if the enable input, $\overline{G}$, is low, the Q output of the selected latch will follow the data on the D input. The other Q outputs will stay the same. When the $\overline{G}$ input is made high, the level on the Q output of the selected latch will be held, rather than always going high as an output of a 74LS138 does when it is unselected.

Figure 8-17*a* shows how a 74LS259 can be connected to demultiplex data from a 74LS151 multiplexer, and Figure 8-17*b* shows the timing waveforms for this system. First note that the ABC select inputs for the two devices are connected in parallel so that the same channel is selected in each device by a 3-bit binary code from the 74LS193. As shown in the timing waveforms in Fig-

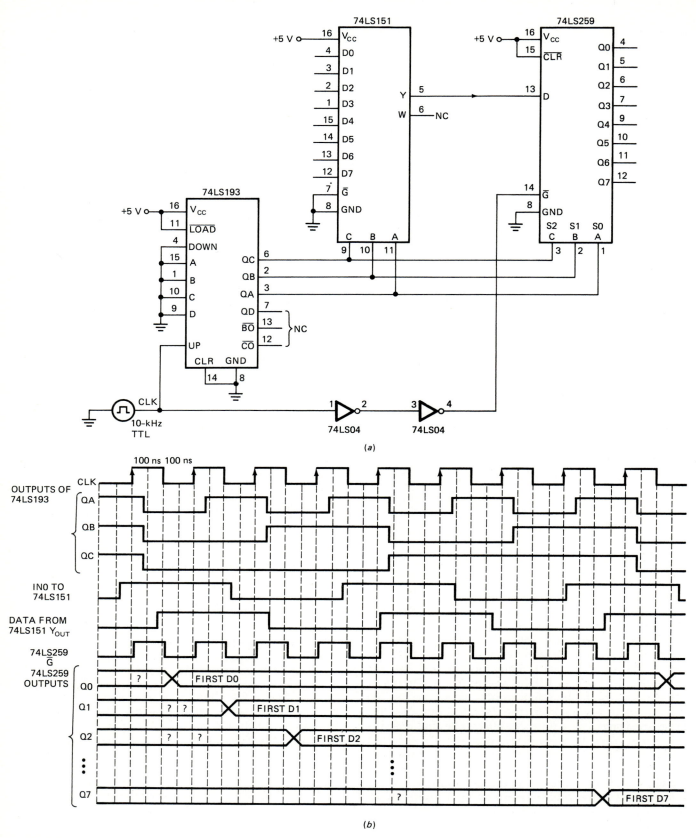

FIGURE 8-17 (a) Schematic showing how to connect a 74LS259 to demultiplex data from a 74LS151 multiplexer. (b) Timing waveforms for the system.

ure 8-17b, the signal used to clock the 74LS193 binary counter is delayed by two inverters and applied to the enable input, $\overline{G}$, of the 74LS259. This delayed clock signal will enable the 74LS259 each time a new address is applied to the select lines of the 74LS259 by the 74LS193. When the enable pulse goes low, data on the D input of the 74LS259 will be transferred to the addressed output. When the enable pulse goes high, the data will be latched on the selected output. As shown by the Q waveforms in Figure 8-17b, the data will be held constant on an output until that individual latch is addressed and enabled again. If the frequency of the signal applied to the CLK input of the 74LS193 is sufficiently higher than the frequency of the signals applied to the data inputs of the 74LS151, then the output signals from the 74LS259 will accurately represent the input data signals. The point here is that the latches in the 74LS259 hold the data values between updates rather than always returning the outputs to a high as the 74LS138 does.

As we will show you in a later chapter, D-latch devices such as the 74ALS373 are often used in microcomputer systems to demultiplex data from signal lines. In the next section we discuss D flip-flops, devices which are often needlessly confused with D latches.

D Flip-Flops

Figure 8-18 shows the traditional logic diagram for a 74LS175 which contains four D flip-flops. As you can see in the diagram, the symbol for a D flip-flop is similar to that for a D latch. However, the two devices differ as to when data on their D inputs is transferred to their Q outputs.

Remember from the previous section that the Q output of a D latch follows the D input during the time when the enable input, C, is high. When the enable input is made low, the data on the D input at that time is latched on the Q output.

For D flip-flops, a low-to-high transition of the signal on the CLOCK input causes data on the D input to be transferred to the Q output. Because an "edge" of the clock signal causes the transfer, a D flip-flop is called an *edge-clocked*, or *edge-triggered*, device. After data is transferred to the Q output of a D flip-flop, it will be held there until another rising edge occurs on the clock signal. Each time the CLOCK input goes high, the Q output will be updated to reflect the data present on the D input at that time. You might think of a D flip-flop as an "instant camera" which takes a "picture" of the data on the D input each time the CLOCK input goes high and displays that picture on the Q output until it takes the next picture.

In logic diagrams a small triangle is used to indicate that an input is activated by the transition or edge of an input signal. See if you can find the triangles next to the CLK labels in Figure 8-18. A bubble next to the triangle as in Figure 8-18 indicates that the negative, or falling, edge of a signal activates that input. A triangle with no bubble indicates that the input is activated by the positive, or rising, edge of an input signal. According to the logic diagram in Figure 8-18, the flip-flops in a 74LS175 are internally negative-edge-clocked. However, since the

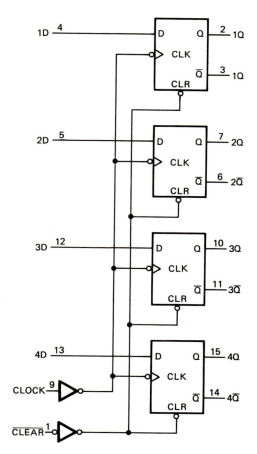

FIGURE 8-18   Traditional logic diagram for 74LS175, quad D flip-flop.

CLK inputs of the four flip-flops are driven by an inverting buffer, the four Q outputs will be updated when a rising edge is applied to the external CLOCK pin.

Besides CLOCK and D inputs, the 74LS175 also has an input labeled $\overline{\text{CLEAR}}$. If this input is asserted, all the Q outputs will go low, to the reset or clear state.

To further describe the operation of a typical D flip-flop, Figure 8-19 shows a truth table for one of the flip-

INPUTS			OUTPUTS	
RESET/ CLEAR	CLK	D	Q	$\overline{Q}$
H	↑	H	H	L
H	↑	L	L	H
H	H	X	Q	$\overline{Q}$
H	L	X	Q	$\overline{Q}$
L	X	X	L	H

FIGURE 8-19   Truth table for 74LS175 D flip-flop.

flops in a 74LS175. An upward arrow in a truth table such as this shows that the low-to-high edge of that signal activates the device. The top two lines in the truth table then tell you that, if the $\overline{\text{CLEAR}}$ input is not asserted, the signal present on the D input will be transferred to the Q output when the clock signal, CLK, makes a low-to-high transition. The third line in the truth table tells you that if the CLK input is at a steady high level, the signal level on D will have no effect on the Q and $\overline{Q}$ outputs. Likewise, the fourth line of the truth table tells you that if the CLK input is at a steady low level, the D input will have no effect on the Q and $\overline{Q}$ outputs. Finally, the last line in the truth table tells you that if the $\overline{\text{CLEAR}}$ is made low, the outputs will go to the reset state, regardless of the signals present on the D and CLK inputs. The $\overline{\text{CLEAR}}$ on the 74LS175 is an example of an *asynchronous*, or *direct*, input. The term asynchronous here means that this input causes the outputs to change immediately without waiting for a transition on the clock input. Incidentally, D flip-flops in some other common devices such as the 74LS74 have both a direct $\overline{\text{CLEAR}}$ input and a direct $\overline{\text{PRESET}}$ input. If the $\overline{\text{PRESET}}$ input is asserted on one of these flip-flops, the Q output will immediately go high, which of course is the set state. To see further how a D flip-flop operates, let's look now at some waveforms for one.

D FLIP-FLOP WAVEFORMS

Figure 8-20 shows some idealized (no propagation delay) waveforms for a positive-edge-triggered D flip-flop. The easiest way to predict the output waveform for a D flip-flop is to first mark the positive transitions of the CLOCK signal with arrows as shown in Figure 8-20. From the discussion above, you know that these are the points where a D flip-flop transfers the level on the D input to the Q output. Therefore, all you have to do is make a dot which represents the level on Q at each of these points and connect the dots together in the appropriate way. Here's how you do it for the example waveforms in Figure 8-20.

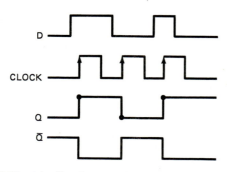

FIGURE 8-20 Idealized example waveforms for positive-edge-triggered D flip-flop.

The D input is initially low, so you draw a low across to the first arrow (positive clock edge). Since the D input is high at this point, you draw the Q output line high from this point to the next arrow. The D input is low at the second rising edge of the CLOCK signal, so draw a low on the Q waveform between this second rising edge, and the third rising edge. At the time of the third rising edge, the D input is high, so from that point on draw a high for the Q waveform. To summarize this process, then, all you have to do is update the Q waveform to agree with the level on D when a positive edge occurs on

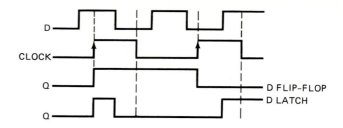

FIGURE 8-21 Timing diagram comparing D latch and D flip-flop output waveforms for same clock and D input signals.

the CLOCK signal. Once you have the Q waveform, you simply draw the complement of it to get the $\overline{Q}$ waveform.

To help fix in your mind the difference between the operation of a D latch and a D flip-flop, Figure 8-21 shows the idealized output waveforms that will be produced when the same D and clock signals are applied to each.

At the first rising edge of the clock signal, the D input is high, so the Q output of the D flip-flop will go high. The Q output will then stay high until the next rising edge of the clock signal. At this second rising edge, the D input is low, so Q will go low and stay low until the next positive edge of the clock signal. At that point, Q will be updated to agree with D.

For a D latch the Q output waveform will follow the D input waveforms during the times when the clock signal is high. In Figure 8-21 these times are represented by two sets of dashed vertical lines. If you check the D and Q waveforms during these times, you should find that they are the same.

The memory tricks for these two devices are:

D LATCH—Q follows D during entire time when clock input is high.

D FLIP-FLOP—Snapshot of D input every rising edge of clock.

D FLIP-FLOP TIMING PARAMETERS

Figure 8-22a shows the actual timing waveforms and Figure 8-22b shows the switching characteristics and recommended operating conditions for a 74LS175. The main points to observe in these waveforms are:

1. The time for a clear to occur is measured from the 50 percent point of CLEAR going low to the 50 percent point of a Q output going low. Find this parameter in the switching characteristics table to get an idea of its value, but don't memorize values such as this. This maximum value for the t_{PHL} output from clear parameter is 30 ns.

2. The time required for a signal on D to be transferred to the output is measured from the 50 percent point of the rising edge of the clock signal to the 50 percent point of the corresponding change on the Q or $\overline{Q}$ output. This is logical, because it is the rising edge of the clock signal that causes the transfer. Again for practice, find the Q and $\overline{Q}$ output from clock times

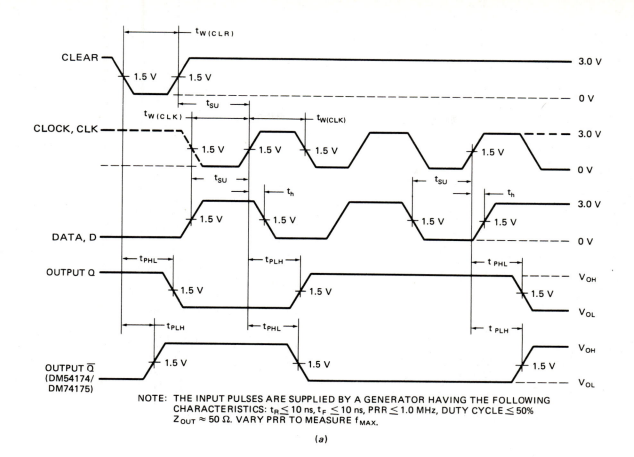

NOTE: THE INPUT PULSES ARE SUPPLIED BY A GENERATOR HAVING THE FOLLOWING CHARACTERISTICS: $t_R \leq 10$ ns, $t_F \leq 10$ ns, PRR ≤ 1.0 MHz, DUTY CYCLE $\leq 50\%$ $Z_{OUT} \approx 50\,\Omega$. VARY PRR TO MEASURE f_{MAX}.

(a)

SWITCHING CHARACTERISTICS, $V_{CC} = 5$ V, $T_A = 25°C$

PARAMETER	'LS174			'LS175			UNIT
	MIN	TYP	MAX	MIN	TYP	MAX	
f_{MAX} MAXIMUM CLOCK FREQUENCY	30	40		30	40		MHz
t_{PLH} PROPAGATION DELAY TIME, LOW-TO-HIGH-LEVEL OUTPUT FROM CLEAR					20	30	ns
t_{PHL} PROPAGATION DELAY TIME, HIGH-TO-LOW-LEVEL OUTPUT FROM CLEAR		23	35		20	30	ns
t_{PLH} PROPAGATION DELAY TIME, LOW-TO-HIGH-LEVEL OUTPUT FROM CLOCK		20	30		13	25	ns
t_{PHL} PROPAGATION DELAY TIME, HIGH-TO-LOW-LEVEL OUTPUT FROM CLOCK		21	30		16	25	ns

RECOMMENDED OPERATING CONDITIONS

		SN74LS174 SN74LS175			UNIT
		MIN	NOM	MAX	
SUPPLY VOLTAGE, V_{CC}		4.75	5	5.25	V
HIGH-LEVEL OUTPUT CURRENT, I_{OH}				−400	μA
LOW-LEVEL OUTPUT CURRENT, I_{OL}				8	mA
CLOCK FREQUENCY, f_{CLOCK}		0		30	MHz
WIDTH OF CLOCK OR CLEAR PULSE, t_W		20			ns
SETUP TIME, t_{su}	DATA INPUT	20			ns
	CLEAR INACTIVE-STATE	25			ns
DATA HOLD TIME, t_h		5			ns
OPERATING FREE-AIR TEMPERATURE, T_A		0		70	°C

(b)

FIGURE 8-22   74LS175 D flip-flop. (a) Timing waveforms. (b) Switching characteristics and recommended operating conditions.

in Figure 8-22b. You should find that the maximum value for the t_{PHL} is 25 ns.

3. The setup time, t_{su}, and the hold time, t_h, are also measured from the 50 percent point on the rising edge of the clock signal. According to the data sheet, the minimum t_{su} for a 74LS175 is 20 ns. This means that the signal on the D input must be stable at least 20 ns before the 50 percent point of the rising edge of the clock signal. If the signal on D is allowed to change within the 20 ns before the rising edge of the clock signal, the level on Q after the clock signal edge will be unpredictable. Failure to obey the setup time requirement is a common cause of problems in D flip-flop circuits.

4. The minimum value for t_h, according to the data sheet, is 5 ns. This means that the signal on the D input must be held stable at least 5 ns after the clock signal goes high to make sure that the data is properly transferred to the output.

COMMONLY USED D FLIP-FLOP DEVICES

For reference, here are some other commonly used D flip-flops that you might want to look up in a data book for further information:

74LS74, 74ALS74, 74HC74—Dual D with set and reset.

74LS174, 74ALS174, 74HC174—Hex D with reset.

74LS374, 74ALS374, 74HC374—Octal D with 3-state outputs.

74LS29821—10-bit D with 3-state outputs.

Throughout this book and its sequel, we show you many applications of D flip-flops. As a last point about D flip-flops here, let's take a quick look at the IEEE/IEC or dependency notation symbol for one.

D FLIP-FLOP IEEE/IEC SYMBOL

The indented box at the top of the IEEE/IEC symbol for a D flip-flop in Figure 8-23 represents a common control

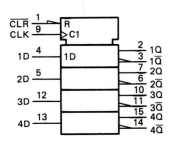

FIGURE 8-23   IEEE/IEC symbol for a D flip-flop.

block. Control signals that enter this box, unless otherwise indicated, affect all the elements below the control block. For example, the R next to the $\overline{CLR}$ input tells you that all the Q outputs will be reset when the $\overline{CLR}$ input is asserted. The letter C next to the CLK input represents control dependency. The 1 after the C tells you that

all the inputs labeled with a 1 will be activated when this signal is asserted. In this device, the CLK input affects all four flip-flops, but by convention only one of the flip-flops is labeled with 1D to show the dependency. As in the traditional symbol, the triangle next to the CLK input indicates that the four flip-flops are activated by the positive edge of the clock signal.

JK Flip-Flops

SYMBOLS AND TRUTH TABLE

A basic $\overline{SR}$ latch such as that in Figure 8-6 has the problem that one set of input conditions produces an indeterminate output state. In D latches and D flip-flops, this problem is solved by connecting an inverter between the two inputs so that the inputs to the SR part of the device are always the complement of each other. The fact that D latches and D flip-flops have only one data input, however, limits their usefulness for some applications. In a device called a *JK flip-flop*, the indeterminate output state problem is solved by another method which does not eliminate one data input. Here's how these devices work.

Figure 8-24 shows the logic symbols for a common JK flip-flop, the 74LS76A. For practice, let's see what we can determine about the operation of the device just from these symbols.

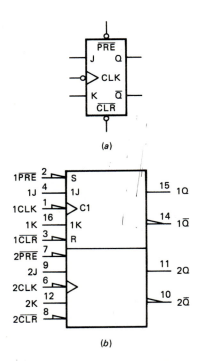

FIGURE 8-24   Logic symbols for 74LS76A dual JK flip-flop. (a) Traditional symbol. (b) IEEE/IEC symbol.

The device has two data inputs, labeled J and K, and a clock input labeled CLK. From the triangle and the wedge (or bubble) next to the CLK input, you can tell that the device is activated by the negative edge of the signal applied to the CLK input. Note that there are some positive-edge JK devices, such as the 74HC109, but most JK flip-flops are negative-edge-triggered. This JK flip-flop also has a set input labeled $\overline{PRE}$ and a reset input labeled $\overline{CLR}$. From the wedges or bubbles on these inputs, you can tell that they are active low. The complementary outputs are labeled Q and $\overline{Q}$, the same as they are on D latches and D flip-flops. To determine more details about the operation of a JK device, take a look at the truth table for it in Figure 8-25.

INPUTS					OUTPUTS	
PR	CLR	CLK	J	K	Q	$\overline{Q}$
L	H	X	X	X	H	L
H	L	X	X	X	L	H
L	L	X	X	X	H*	H*
H	H	↓	L	L	Q0	$\overline{Q}$0
H	H	↓	H	L	H	L
H	H	↓	L	H	L	H
H	H	↓	H	H	TOGGLE	
H	H	H	X	X	Q0	$\overline{Q}$0

FIGURE 8-25 Truth table for 74LS76A negative-edge-triggered JK flip-flop.

The $\overline{PRE}$ and $\overline{CLR}$ inputs of the 74LS76A are called *asynchronous*, or *direct*. As shown by the first line in the truth table, this means that, if $\overline{PRE}$ is asserted (low), the Q output will immediately go to a high, the preset state, regardless of the signals on the CLK, J, and K inputs. Likewise, as shown by the second line in the truth table, the Q output will immediately go low if the $\overline{CLR}$ input is asserted, regardless of the signals on the CLK, J, and K inputs. Asserting both $\overline{PRE}$ and $\overline{CLR}$ at the same time produces an indeterminate output state, just as it does on a basic SR device.

The downpointing arrows in the CLK column of the truth table tell you that this device is negative-edge-triggered. When the clock signal goes low, data on the J and K inputs determines the level on the Q output. The bottom line of the truth table in Figure 8-25 shows that the signals on J and K have no effect on the output when the clock signal is at a steady high level. This JK device then updates the Q output according to the rules for the JK inputs, each time the CLK input makes a high-to-low transition. Let's take a closer look at how the J and K inputs affect the output.

If J and K are both low when CLK goes low, the outputs stay the same as they were before the transition. This represents the latched, or hold, state for a JK device. If the J input is high and the K input is low when CLK goes low, the Q output will be high and the $\overline{Q}$ output will be low after the clock edge. In a similar manner, if the J input is low and the K input is high when CLK goes low, then the Q output will be low and the $\overline{Q}$ output will be high after the clock edge. An easy way to remem-

ber these two cases is to notice that if the states on J and K are different, the level on the Q output will be the same as the level on J after the falling edge of the clock.

The final input state to check out here is the case where J and K are both high when CLK goes low. For these input conditions, the Q output *toggles*. The term toggle means to change to the opposite state from the state present before CLK went low. For example, if J and K are both high and Q is low, the Q output will toggle to a high when CLK goes low. Likewise, if J and K are both high and Q is high when CLK goes low, Q will toggle to a low and $\overline{Q}$ will toggle to a high.

To summarize, then, the JK inputs of this device produce the following output responses:

J	K	Output
0	0	Q and $\overline{Q}$ remain the same
1	0	Q follows J
0	1	Q follows J
1	1	Q and $\overline{Q}$ each toggle to their opposite state

JK FLIP-FLOP TIMING WAVEFORMS

The next step is to use the JK flip-flop truth table to predict the output waveforms that will be produced by given input waveforms. We will use the waveforms in Figure 8-26 to show you how to do this.

The main points to keep in mind when predicting the output waveform for a JK flip-flop are:

1. If the device has direct $\overline{PRE}$ and $\overline{CLR}$ inputs, these signals override the effects of J, K, and CLK.

2. The effect of the J and K inputs will be transferred to the outputs on the negative edge of the clock signal.

The first step in predicting the output waveform for any edge active device is to mark the active edge of the CLK signal with arrows as shown in Figure 8-26. This allows you to easily identify the points where the Q output will be updated as predicted by the truth table if $\overline{PRE}$ or $\overline{CLR}$ is not asserted.

In the waveforms in Figure 8-26, Q is initially low. When the $\overline{PRE}$ input is asserted, the Q output immediately goes high. A low on $\overline{PRE}$ overrides the effect of J, K, and CLK, so when CLK goes low at point 1, the signals on J and K have no effect on Q.

The next event in our example waveforms in Figure 8-26 is $\overline{CLR}$ being asserted low. This will cause the Q output to immediately go low, regardless of the signal levels on J, K, and CLK. $\overline{CLR}$ is still low during the falling edge of CLK pulse 2, so J and K will have no effect on the Q output at this point. After $\overline{CLR}$ goes high, Q will remain low until $\overline{PRE}$ is asserted again or until J and K cause a change at a later falling edge of the clock signal. For the rest of the waveforms in Figure 8-26, $\overline{PRE}$ and $\overline{CLR}$ remain unasserted, so all you have to consider are the levels present on J and K at the times when CLK goes low.

At CLK edge 3 in the waveforms, J is high and K is low. According to the truth table in Figure 8-25, this will cause Q to go high. At CLK edge 4, J is low and K is high, so Q

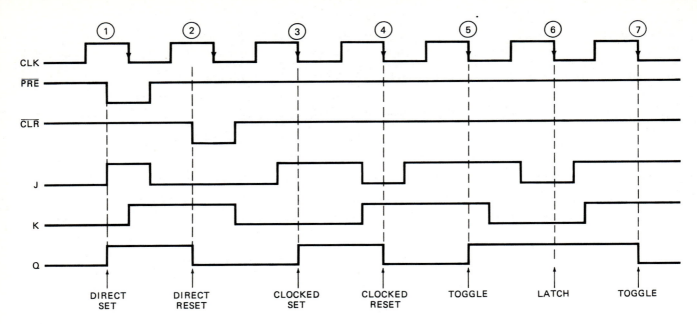

FIGURE 8-26 Example signal waveforms for 74LS76A negative-edge-triggered JK flip-flop with direct preset and clear inputs.

will go low. Since J and K are both high at CLK edge 5, Q will toggle from a low to a high. J and K are both low at CLK edge 6, so Q will stay the same. Finally, at CLK edge 7 both J and K are high, so Q will toggle from a high to a low. All you have to do then is simply use the rules for J and K to update the Q output at each falling edge of CLK. If you are predicting the output waveform for a positive-edge-triggered JK flip-flop such as a 74HC109, you use the rules for J and K to update the Q waveform at each rising edge of the clock signal.

Incidentally, in older equipment you may encounter a now obsolete type of JK flip-flop called *pulse-triggered*, or *level-triggered*. Examples of this type are the standard TTL 7473, 7476, and 74107. The flip-flops in these devices are made in two sections called a *master* and a *slave*. During the rising edge of a clock pulse, the data on J and K is used to update the output of the first section or master. When the clock signal goes low, data on the output of the master is transferred to the output of the second section or slave. The Q and $\overline{Q}$ outputs then change just after the falling edge of the clock. These level-triggered JK flip-flops function similarly to a negative-edge-triggered JK, except that the J and K inputs must be held stable from before the CLK signal goes high until after the CLK signal goes low. The major problem with these older devices is that levels on Q and $\overline{Q}$ are unpredictable if J or K changes while the CLK signal is high. Newer devices, such as the 74LS76 or the 74HC76, do not have this problem because only the levels present on J and K just before the falling clock edge affect the output level.

COMMONLY USED EDGE-TRIGGERED JK FLIP-FLOPS

For reference, here are a few other commonly used, edge-triggered JK flip-flops that you might want to look up in a data book for further information:

74LS107A, 74HC107—Dual JK with reset.

74LS109A, 74ALS109A, 74HC109—Dual JK with set and reset.

74LS112A, 74ALS112A, 74HC112—Dual JK with set and reset.

74LS113A, 74ALS113A—Dual JK with set.

74LS114A, 74ALS114A—Dual JK with set, common reset, and CLK.

JK FLIP-FLOP TIMING PARAMETERS

The timing parameters for a JK flip-flop are similar to those we described for a D flip-flop in the preceding section. As an example, Figure 8-27a shows the recommended operating characteristics and switching characteristics section of the data sheet for a 74LS76A. Since the 74LS76A is a negative-edge-triggered device, all these times, except pulse widths, are measured from the negative edge of the clock signal.

The parameters listed under t_w in Figure 8-27a tell you the minimum times that CLK must be high or $\overline{PRE}$ and $\overline{CLR}$ must be low to be recognized by the device. The first of the three setup times listed for the device tells you that data on the J and K inputs must be stable for at least 20 ns before CLK goes low. The second setup time tells you that $\overline{CLR}$ must return high at least 20 ns before CLK goes low, or the data on J and K will not dependably determine the level on Q output when CLK goes low. Likewise, the third setup time tells you that $\overline{PRE}$ must be returned high at least 25 ns before CLK goes low in order for J and K to dependably control the level on Q when CLK goes low. As with most newer flip-flops, the 74LS76A has a hold time, t_h, of 0 ns, so J and K do not have to be held stable after CLK goes low.

The f_{max} parameter in the switching characteristics

		SN54LS76A			SN74LS76A			UNIT
		MIN	NOM	MAX	MIN	NOM	MAX	
V_{CC}	Supply voltage	4.5	5	5.5	4.75	5	5.75	V
V_{IH}	High-level input voltage	2			2			V
V_{IL}	Low-level input voltage			0.7			0.8	V
I_{OH}	High-level output current			− 0.4			− 0.4	mA
I_{OL}	Low-level output current			4			8	mA
f_{clock}	Clock frequency	0		30	0		30	MHz
t_w	Pulse duration	CLK high 20			20			ns
		PRE or CLR low 25			25			
t_{su}	Setup time before CLK↓	data high or low 20			20			ns
		CLR inactive 20			20			
		PRE inactive 25			25			
t_h	Hold time-data after CLK↓	0			0			ns
T_A	Operating free-air temperature	− 55		125	0		70	°C

(a)

PARAMETER	FROM (INPUT)	TO (OUTPUT)	TEST CONDITIONS		MIN	TYP	MAX	UNIT
f_{max}			$R_L = 2\,k\Omega$,	$C_L = 15\,pF$	30	45		MHz
t_{PLH}	PRE, CLR or CLK	Q or Q̄				15	20	ns
t_{PHL}						15	20	ns

(b)

FIGURE 8-27 74LS76A dual JK flip-flop with preset and clear. (a) Recommended operating conditions. (b) Switching characteristics. (Courtesy of Texas Instruments Inc.)

section of Figure 8-27b refers to the maximum frequency CLK signal that can be applied to the device and still have it produce the correct output as predicted by its truth table. The manufacturer gives a typical value for this parameter, but as we have told you before, typical values are meaningless. You should always use the minimum or maximum value for a parameter, whichever represents the "worst-case" situation. For example, the worst-case values for t_{PHL} and t_{PLH} are usually the maximums, so the maximum values of these parameters are the ones you will look for on a data sheet.

At this point, you may wonder why we have described the timing parameters of D and JK flip-flops in such detail. The reason is that these parameters must be taken into account if you are designing a circuit using flip-flops, and failure to obey these times is a common reason for a circuit with flip-flops to malfunction.

Troubleshooting Flip-Flop Circuits

Malfunctioning flip-flop and latch circuits usually have one of three types of problems:

1. The Q output does not change to the state predicted by the D or by the J and K inputs when the device is enabled or clocked. This symptom is usually caused by a defective IC or by the Q output being shorted to V_{CC}, to ground, or to another signal line. If replacing the IC does not cure the problem, check if the output is shorted to something as we described in Chapter 3.

2. The Q output sets or resets at random times. This symptom is often caused by the PRE or CLR inputs of the device being left open. When left open, these inputs may pick up pulses from other signal lines. Unused PRE or CLR inputs should be tied to ground

or to V_{CC}, whichever level holds them in their unasserted state. If this is not the problem, use a scope to see if the PRE or CLR inputs are receiving random pulses from some other part of the circuit.

3. The Q output is sometimes correct and sometimes not correct. This symptom may be caused by a defective IC, or it may be caused by a violation of the minimum setup and/or hold time requirements for the device. Remember that if the input data changes too close to the active edge of the clock signal, the output may not be updated to reflect that data. Since these timing parameters vary with temperature and from IC to IC, marginal timing problems can be hard to find.

As an example of troubleshooting a malfunctioning flip-flop circuit, suppose that an instrument works fine with its cover off, but 5 min after the cover is put on, it shows incorrect data on its display. What you do here is connect one input of a scope to the data input of the suspect flip-flop, and another input of a scope to the CLK input. From the scope display, you can determine if, due to marginal circuit design, the data is changing too close to the CLK edge. If the circuit design obeys the minimum setup time for the device, next bring a heat source near the suspect IC to see whether this causes a malfunction. If it does, you can tell from the display whether the failure was caused by a change in timing or by a defective IC.

The reason you should not just replace an IC in a case like this is that you may by chance happen to find an IC that makes the instrument work fine until it gets to Phoenix.

T Flip-Flop Circuits

D or JK flip-flops can be connected to form circuits which are sometimes called T, or toggle, flip-flops. Figure 8-28a shows how a negative-edge-triggered JK flip-flop

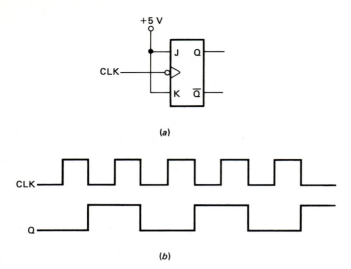

FIGURE 8-28   T flip-flop made from a negative-edge-triggered JK flip-flop. *(a)* Circuit. *(b)* Idealized signal waveforms.

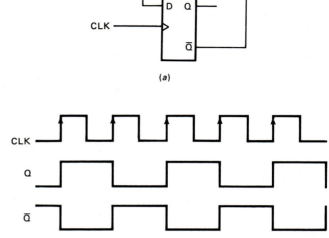

FIGURE 8-29   T flip-flop made from a positive-edge-triggered D flip-flop. *(a)* Circuit. *(b)* Idealized signal waveforms.

can be connected to form a T flip-flop. Let's look at the waveforms in Figure 8-28b to see how the circuit works.

The J and K inputs are both tied high, so according to the truth table for a JK, the Q output will change state, or toggle, each time CLK goes low. Q is initially low, so at the first negative edge of CLK, Q toggles high. At the second negative edge of CLK, Q toggles low, and at the third negative edge of CLK, Q toggles high again.

A quick look at the CLK and Q waveforms in Figure 8-28b should show you that the Q output signal is a square wave with a frequency that is one-half the frequency of the input clock signal. This circuit then functions as a frequency divider. In the next major section of the chapter, we show you how several T flip-flop circuits can be connected in series, or *cascaded*, to divide an input frequency by other numbers. To finish this section, however, we will show you how a D flip-flop can be connected so that it functions as a T flip-flop.

A positive-edge-triggered D flip-flop can be converted to a T flip-flop by connecting its $\overline{Q}$ output to its D input as shown in Figure 8-29a. Figure 8-29b shows the timing waveforms for this circuit. At the first rising edge of the CLK signal, the high on $\overline{Q}$ is transferred through the flip-flop to the Q output. When Q goes high, $\overline{Q}$, of course, goes low. On the next rising edge of the CLK signal, the low now on $\overline{Q}$ is transferred through the flip-flop to the Q output. When Q goes low, $\overline{Q}$ goes high. The Q and $\overline{Q}$ outputs then toggle at each positive edge of the CLK signal.

If you compare the frequency of the CLK signal and the frequency of the Q signal in Figure 8-29b, you should see that this circuit also produces a square-wave signal which is one-half the frequency of the CLK input signal. As we show you in the next section, either of the toggle flip-flop circuits in Figure 8-29 can be cascaded to form a circuit which divides an input frequency by powers of 2 such as 4, 8, 16, 32, etc.

ASYNCHRONOUS BINARY COUNTERS AND FREQUENCY DIVIDERS

Introduction

In the previous section of this chapter we showed you how D flip-flops and JK flip-flops can be connected to function as toggle flip-flops. As shown by the waveforms in Figure 8-28, the frequency on the Q output of a toggle flip-flop is one-half the frequency applied to the CLK input. Another way of saying this is that a T flip-flop divides an input frequency by 2. If the Q output from one T flip-flop is connected to the CLK input of another T flip-flop, the first flip-flop will divide the input frequency by 2 and the second flip-flop will divide that frequency by 2. The two flip-flops in series then divide an input frequency by 4. Connecting devices in series in this way is often referred to as *cascading* them. Three T flip-flops can be cascaded to divide an input frequency by $2 \times 2 \times 2$, or 8, and four T flip-flops can be cascaded to divide an input frequency by $2 \times 2 \times 2 \times 2$, or 16. For any number of T flip-flops cascaded in this way, the input frequency will be divided by 2^N, where N is the number of T flip-flops cascaded. The number that a circuit divides an input frequency by is often called the *modulo* of the circuit. A circuit which divides an input frequency by 8 is called a modulo-8 divider. After we introduce you to the 74LS93 4-bit binary counter, frequency divider IC, we will show you how to design circuits with any desired modulo.

74LS93 Operation and Waveforms

Figure 8-30a shows a simple circuit using a 74LS93. As you can see from the internal diagram for the 74LS93, the device contains four JK flip-flops. Although it is not shown in the diagram, all the J and K inputs are tied high, so the flip-flops all function as T-type circuits. The

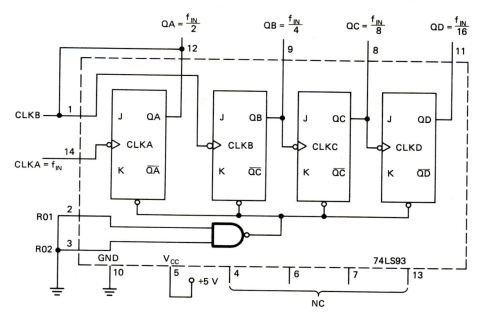

NOTE: ALL J AND K INPUTS TIED HIGH INTERNALLY, R01 AND R02
TIED LOW TO COUNT. A HIGH ON BOTH RESETS ALL Q OUTPUTS.

(a)

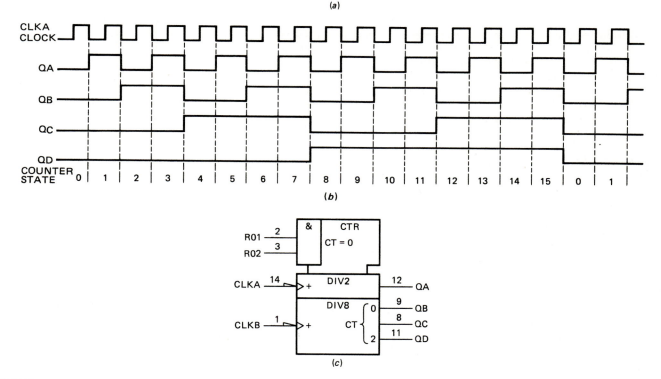

(b)

(c)

FIGURE 8-30   74LS93 (a) Circuit showing connections for a four-stage binary divider or counter. (b) Clock and output waveforms for the circuit. (c) IEEE/IEC symbol.

R01 and R02 inputs of the 74LS93 allow you to reset the flip-flips. If the R01 and R02 inputs of the device are both asserted high, the four Q outputs will all be forced to the low state.

If a clock signal is connected to the CLKA input of the A flip-flop in the 74LS93, the frequency of the signal on its Q output (QA) will be one-half the frequency of the signal on its CLKA input. The B, C, and D flip-flops in the 79LS93 are internally cascaded. If a clock signal is

applied to the CLKB input, the B flip-flop will divide this input frequency by 2, the C flip-flop will divide the output frequency from B by 2, and the D flip-flop will divide the output frequency from C by 2. The result of all this is that the frequency on the QD output is one-eighth the frequency applied to the CLKB input of the B flip-flop. If the QA output is connected to the CLK input of the B flip-flop as shown in Figure 8-30a, then the frequencies on the four Q outputs will be as follows:

$$QA = \frac{f_{IN}}{2}$$

$$QB = \frac{f_{IN}}{4}$$

$$QC = \frac{f_{IN}}{8}$$

$$QD = \frac{f_{IN}}{16}$$

The circuit in Figure 8-30a functions as a modulo-2, modulo-4, modulo-8, or modulo-16 divider, depending on which output you use. To get a better idea of how the circuit functions, let's take a look at the idealized waveforms for it in Figure 8-30b.

As shown by the bubble next to the CLK input of each flip-flop in Figure 8-30a, these flip-flops toggle on the negative edge of the signal applied to the CLK input. The QA output then will toggle on each negative edge of the CLKA input signal. As you can see in Figure 8-30b, the frequency of the QA output is one-half the frequency of the CLKA input signal.

The QA output of the A flip-flop is connected to the CLKB input of the B flip-flop, so each negative edge of the QA signal will cause the B flip-flop to toggle. In the same way, each negative edge of the QB output signal will cause the C flip-flop to toggle and each negative edge of the QC output signal will cause the D flip-flop to toggle. As you can see in Figure 8-30b, the result is that the frequency of the QD output is one-sixteenth the frequency of the signal applied to the CLKA input.

Besides its use for dividing an input frequency by a power of 2, the circuit in Figure 8-30a functions as a 4-bit binary counter. We introduced you to this use of a 74LS93 in the "automatic" logic tester circuit in Figure 5-17, the scope multiplexer circuit in Figure 7-10, and the multiplexer data link in Figure 8-17a. If you rotate the waveforms in Figure 8-30b 90° clockwise, you should see that these waveforms step through a binary count sequence from 0000 to 1111 over and over. The QA output represents the LSB of the count and the QD output represents the MSB of the count. To help you see this, the number of each counter state is written below that state in the figure.

When used as a counter, the circuit in Figure 8-30a is called a *ripple*, or an *asynchronous*, counter. The term ripple comes from the fact that the input CLKA signal propagates sequentially or ripples down through the chain of flip-flops. For the circuit in Figure 8-30a, then, the QD output will not change until four propagation delays after a negative edge of the CLKA signal. The term asynchronous here is just another way of showing that the four Q outputs do not all change in "sync" with the negative edge of the signal applied to the CLKA input.

Now that you have some understanding of the operation of a 74LS93, let's see how the IEEE/IEC symbol for the device represents its operation. The IEEE/IEC symbol in Figure 8-30c shows that the device is made up of two counting sections, a divide by 2 and a divide by 8. The wedge, triangle, and + symbols next to the CLK

inputs tell you the count on the Q output(s) increments by 1 on each negative edge of the CLK signal. The common control block at the top of the symbol tells you that the count on the Q outputs will be reset to 0000 if the R01 and R02 inputs are both asserted (high).

Designing Modulo-n Dividers with 74LS93s

In the preceding section we showed you how a 74LS93 can be connected to divide an input frequency by 2, 4, 8, or 16. If the QD output from one 74LS93 circuit is connected to the CLKA input of another 74LS93 connected in the same way, the outputs of the second device will produce signals which are equal to the initial CLKA frequency divided by 32, 64, 128, and 256. More 74LS93 circuits can be added to divide the input frequency by still higher powers of 2.

To calculate the resultant modulo of cascaded counters, you simply multiply the modulos of the cascaded counters. The modulo of a counter produced by cascading a modulo-8 and a modulo-4 counter is 4 × 8 or 32. For ripple counters produced by simply cascading T flip-flops, the modulo will always be some integral power of 2 such as 2, 4, 8, 16, 32, etc. Some counter applications, however, require frequency dividers with modulos such as 6, 10, and 12, which are not integral powers of 2. A digital clock, for example, requires counters with modulos of 6, 10, and 12.

To shorten the count of a ripple counter such as a 74LS93, you use digital feedback which resets the Q outputs to 0000 when the counter outputs reach the desired count. As a first example of how you do this, Figure 8-31a shows a 74LS93 connected to function as a modulo-10 counter/divider.

As pulses are applied to the CLKA input, the counter outputs step through a binary count sequence as shown in Figure 8-30b. When the counter outputs reach state 1010, the high on the QD output will assert the R01 input and the high on the QB output will assert the R02 input of the 74LS93. Asserting the two R0 inputs will cause all four counter outputs to go to 0 after a maximum of 40 ns. State 1010 then will only be present on the outputs for the time it takes all the flip-flops to reset to zero.

Figure 8-31b shows the waveforms that will be produced by the circuit in Figure 8-31a. Note that the waveforms step through 10 full states, 0 through 9. State 10 or 1010 is also present on the outputs, but only long enough to cause the outputs to be reset to state 0. For many applications such as driving LED displays, the unwanted extra state or "glitch" is present for such a short time that it causes no problems.

After the outputs are reset to 0000, the falling edge of the next pulse on the CLKA input causes the outputs to increment to state 0001. The count sequence then repeats as pulses are applied to the CLKA input. Now let's check out the frequency relationships between the input signal and the output signals.

As you can see from the waveforms in Figure 8-31b, it takes 10 pulses on the CLKA input to produce one cycle on the QD output. The frequency of the QD output then is one-tenth the input frequency as desired. The QD out-

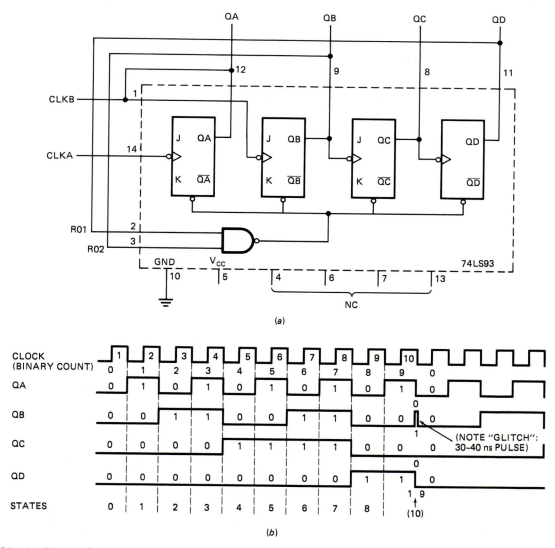

FIGURE 8-31 (a) Circuit showing 74LS93 converted to a modulo-10 counter. (b) Clock and output waveforms.

put waveform is high for two input CLKA cycles and low for eight CLKA cycles, therefore:

$$\text{Duty cycle of QD} = \frac{2}{2 + 8} \times 100 \text{ percent} = 20 \text{ percent}$$

The QC output of the circuit in Figure 8-31a also produces one output cycle for each 10 CLKA input pulses, so its frequency is also one-tenth the CLKA input frequency. However, the QC signal is high for four CLKA cycles and low for six CLKA cycles. Its duty cycle then can be calculated as follows:

$$\text{Duty cycle of QC} = \frac{4}{4 + 6} \times 100 \text{ percent} = 40 \text{ percent}$$

The QB output has a nonsymmetrical waveform with two cycles for each 10 CLKA input cycles. Therefore, its frequency is equal to the CLKA input frequency divided by 5. The frequency of the QA waveform is simply equal to one-half the frequency of the CLKA input frequency.

A scheme such as that in Figure 8-31a can be used

with a 74LS93 to divide an input frequency by any number between 2 and 16. The trick is to decode the state that is equal to the desired modulo and use the output of the decoder to reset the counter outputs to 0000. For a modulo-10 counter, the internal R01 and R02 NAND gate is all that is needed to decode state 1010. The reason for this is because state 1010 only has two 1's and 1010 is the first state which has 1's in the QB and QD bit positions. For some other modulo circuits, an external gate is needed as part of the decoding. As an example of this, Figure 8-32 shows how a 74LS93 can be connected as a modulo-11 counter.

To produce a modulo-11 counter, we need to decode state 11 or 1011 and use the decoder output to reset the counter outputs to zero with the R01 and R02 inputs. In the circuit in Figure 8-32a, we use an AND gate to detect the 1's on QA and QB. When QA and QB are both high, the R01 input of the 74LS93 will then be asserted. The R02 input will be asserted when QD is high. It's a minor point, but we connected the QA and QB outputs to the AND gate because these outputs are

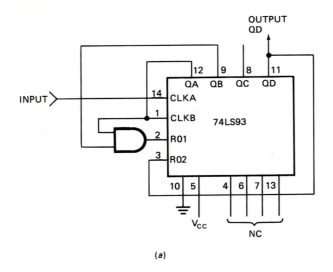

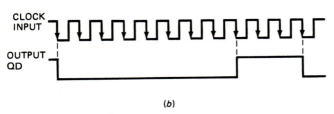

(b)

FIGURE 8-32 *(a)* Circuit for 74LS93 as modulo-11 counter. *(b)* Clock and QD output.

produced before the QD output signal. This connection allows the effects of the QA and QB signals to propagate through the AND gate while the CLK signals are rippling through flip-flop C and flip-flop D in the counter.

An easy way to predict the waveforms for any modulo counter made with a 74LS93 is to just modify the full-count waveforms in Figure 8-30b. Here's how you predict the output waveforms for the modulo-11 counter in Figure 8-32a.

Start by drawing a dotted vertical line through the waveforms in Figure 8-30b at the point where the counter outputs are reset to 0000. For the modulo-11 counter, this point is just after the start of state 1011. If it is not already 0, return each Q waveform to a low along the dotted vertical line. The outputs are now all back to 0. Figure 8-32b shows the clock and resultant QD output waveforms for a divide-by-11 counter. If you want to draw more of the waveforms, you simply repeat the count sequence from 0000 to 1011 again.

For dividing input frequencies by modulos greater than 16, you can cascade two or more 74LS93s. To do this you connect the QD output from the least significant counter to the CLKA input of the next most significant counter, etc. With external gates you decode the desired state and use the output signal from the decoder to reset all the Q outputs with the R01 and R02 inputs.

Now that you are familiar with the operation of a 74LS93, we will briefly introduce you to some other commonly available ripple counters.

OTHER COMMONLY AVAILABLE RIPPLE COUNTERS

The need for some modulo counters is great enough that manufacturers have implemented them directly in commonly available devices. As a first example, Figure 8-33a shows the logic diagram for a 74LS90 decade counter. The A flip-flop divides a signal applied to its CLK input by 2. The B, C, and D flip-flops are internally connected so that the signal on the QD output is one-fifth the frequency applied to the CLKB input. If the QA output is connected to the CLKB input, the outputs will show a binary count sequence from 0000 to 1001. When connected in this way, the device is called a *decade*, or *BCD*, counter. The frequency of the signal on the QD output will be one-tenth the frequency of the signal applied to the CLKA input. Because of the way the divide by 5 is internally produced, the output waveforms do not have the output glitch that you get when you make a divide-by-10 circuit with a 74LS93.

If an external signal is connected to the CLKB input and the QD output is connected to the CLKA input, the signal on the QA output will be a square wave one-tenth the frequency of the CLKB input signal.

Incidentally, the 74LS90 has R0 inputs which will cause the outputs to go to 0000 if they are both asserted, and it has R9 inputs which will cause the outputs to go to 1001 if they are both asserted.

As another example of a ready-made modulo counter, Figure 8-33b shows the internal logic diagram for a 74LS92. This device consists of a divide-by-2 flip-flop and three flip-flops internally connected to divide by 6. If the QA output is connected to the CLKB input, the Q outputs will cycle through the binary sequence 0000 to 1011 as CLKA is pulsed. The signal on the QD output will be one-twelfth the frequency of the CLKA input signal. As an alternative connection, the QD output can be connected to the CLKA input. With this connection the signal on the QA output will be a square wave with a frequency one-twelfth the frequency of the signal applied to the CLKB input.

Finally, Figure 8-33c shows the logic diagram for a 74HC4060 14-stage binary ripple counter. Note that in this device, T flip-flops are produced from D flip-flops by connecting the Q' outputs around to the D inputs as we showed in Figure 8-29. As we will show you in a later chapter, a resistor and capacitor can be connected on the R_{TC} and C_{TC} inputs so that the internal inverter functions as an *oscillator*. An oscillator is a device or circuit which produces a pulsed signal without the need for an external signal generator. If the T_{RC} and C_{RC} inputs of 74HC4060 are left open, a signal applied to the RS input will be divided by successive powers of 2 as it ripples through the series of flip-flops. If the master reset input, MR, is simply tied low, the frequency of the signal on the final Q output will be equal to the input frequency divided by 2^{14}. To produce modulos which are not powers of 2, you can decode a desired state and use the decoder output to assert the MR input. This will reset all the flip-flop outputs to 0, and the count will start over. The technique is the same one we described for a 74LS93.

In the preceding sections we have spent quite a bit of

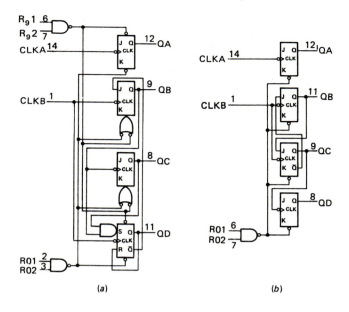

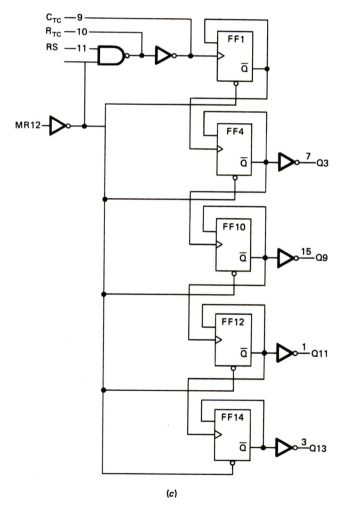

FIGURE 8-33 Internal logic diagrams of other common ripple counters. *(a)* 74LS90. *(b)* 74LS92. *(c)* 74HC4060. *(Courtesy of Texas Instruments Inc.)*

time showing you how ripple counters work. In the next section we show you some uses for them.

Some Applications of Ripple Counters

INTRODUCTION

Ripple counters are relatively inexpensive to produce, and they are suitable for many low-frequency applications. In the following sections we first show you some applications of ripple counters as frequency dividers and then some applications where they are used as counters.

AN ELECTRONIC ORGAN

Western music is based on an equally tempered scale of 12 notes per octave. The letters A through G are used to identify the notes in an octave. Notes within an octave have precise mathematical relationships to each other, and the frequency of a particular note in one octave is exactly half the frequency of the same note in the next higher octave. For example, an A note in one octave has a frequency of 440 Hz. The A in the next lower octave has a frequency of 220 Hz, the A in the next lower octave has a frequency of 110 Hz, and the A in the next lower octave has a frequency of 55 Hz.

Figure 8-34a shows the frequencies for the 12 notes in the highest commonly used octave. As also shown in Figure 8-34a, close approximations to these 12 frequencies can be produced by dividing a 2119.040-kHz signal by the appropriate modulos. The term *CENT ERROR* in the rightmost column of the table is a musical term used to express the accuracy of a note's frequency.

A single Sanyo LM8071 Equal Tempered Scale Generator will generate all the top octave frequencies shown in Figure 8-34b when driven by a two-phase, 2119.040-kHz clock signal. The frequencies needed for notes in lower octaves are equal to these frequencies divided by powers of 2. Therefore, all the frequencies needed for a simple electronic organ can be produced by connecting a string of binary dividers on each of the outputs of the LM8071 as shown in Figure 8-35a. In this circuit we used Sanyo LM3216 six-stage binary dividers to do the job because these devices use the same PMOS logic levels as the LM8071 device.

The circuit in Figure 8-35a generates square-wave signals with all the frequencies required for a simple electronic organ. The block diagram in Figure 8-35b shows the additional parts needed for the rest of the organ.

First, keyboard switches select the basic frequencies for the desired notes. The outputs from the tone generator part of the circuit are all square waves which, if amplified and played through a speaker, sound somewhat like a clarinet. To produce other sounds, the basic square-wave signal for a note is mixed with the same notes from higher octaves (harmonics) and the result passed through some filters. The output from the filter is then amplified by an audio amplifier and used to drive a speaker.

Research has shown that a simple divider system such as that in Figure 8-35a works, but it sounds somewhat "dull" to the ear because all the output frequencies are precisely derived from the one basic input clock fre-

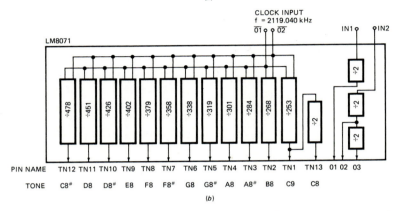

DIVISION	RATIO	(Hz) OUTPUT FREQUENCY	(Hz) TRUE FREQUENCY	CENT ERROR
C9	253	8375.652	8372.018	+0.751
B8	268	7906.866	7902.133	+1.037
A8#	284	7461.408	7458.620	+0.647
A8	301	7040.000	7040.000	0.000
G8#	319	6642.759	6644.875	−0.551
G8	338	6269.349	6271.927	−0.712
F8#	358	5919.106	5919.911	−0.235
F8	379	5591.135	5587.652	+1.079
E8	402	5271.244	5274.041	−0.918
D8#	426	4974.272	4978.032	−1.308
D8	451	4698.537	4698.636	−0.036
C8#	478	4433.138	4434.922	−0.697

C SCALE OUTPUT FREQUENCY (CLOCK FREQUENCY = 2119.040 kHz)

(a)

FIGURE 8-34 *(a)* Frequencies of top octave of equally tempered scale. *(b)* Sanyo LM8071 scale generator block diagram.

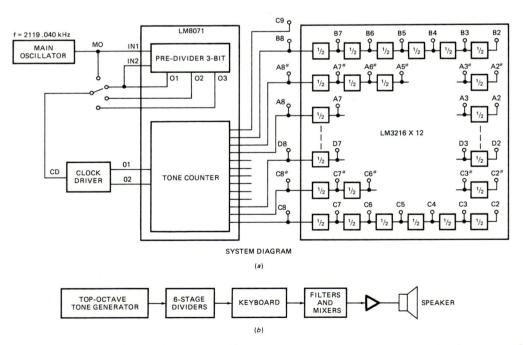

SYSTEM DIAGRAM

(a)

FIGURE 8-35 *(a)* Circuit using Sanyo LM8071 scale generator and LM3216 six-stage ripple dividers to produce all the frequencies needed for a simple electronic organ. *(b)* Block diagram showing additional parts needed for simple electronic organ.

quency. One solution to this problem is to apply a slight vibrato, or frequency modulation, to the 2119.040-kHz oscillator. Another solution is to produce two identical, parallel circuits with slightly different frequencies and mix the signals from the two together just before the audio amplifier. This latter approach is similar to the way the strings of a piano or guitar are slightly detuned to give the sound more body.

RHYTHM GENERATORS

Another interesting application of the frequency divider function of ripple counters is in the rhythm generators and drum machines commonly found in electronic music synthesizers. As an example of this, Figure 8-36 shows an internal block diagram for a Sanyo LM8471 Rhythm Generator IC connected to produce a complete rhythm generator system.

The LM8471 divides down the frequency of an input clock signal to produce signals at the right frequency for a desired rhythm. The outputs from the basic frequency dividers are decoded and used to select the actual pattern of pulses required on the outputs. The rhythm pattern generator in the LM8471 is a ROM which outputs a programmed sequence of signals as it is addressed by the decoder outputs. Let's take a look at the system block diagram in Figure 8-36 to get a better idea of how the device operates.

The LM8471 is a PMOS device, so if V_{SS} is ground, V_{DD} will be a negative voltage of perhaps -10 V. A clock signal is generated by the RC circuit on the left edge of the figure. This circuit is adjustable so that the basic rate or tempo can be changed as desired. The desired rhythm pattern is selected by connecting one of the RS inputs to V_{SS} as shown in the figure. If RS1 is connected to V_{SS}, for example, the circuit will generate a waltz rhythm, and if RS4 is connected to V_{SS}, the circuit will generate a rock-and-roll rhythm. The actual outputs for each rhythm are pulses on the RM1 to RM8 lines and the AC1 to AC3 lines. These pulses are used to turn on the corresponding tone generator in the tone generator section of the circuit. A pulse on RM1, for example, will turn on the cymbal generator for a time equal to the width of the pulse. A pulse on the AC2 or the AC3 output will turn on the bass drum tone generator for a time.

The outputs from the tone generators are combined (mixed), amplified, and used to drive a speaker. An additional feature is a tempo indicator LED which is turned on at each beat. This rhythm generator device uses ripple counters both as frequency dividers and as counters. In the next section we discuss applications where the use of ripple counters as counters is more obvious.

DECIMAL COUNTING UNITS

Figure 8-37 shows how 74LS90 decade counters, 74LS47 BCD-to-seven-segment decoders, and LED displays can be connected to make a decimal counting unit. The fall-

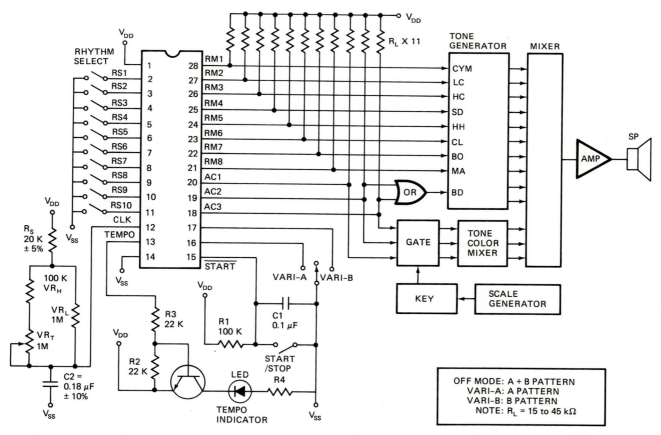

FIGURE 8-36 Block diagram of complete electronic rhythm generator system using an LM8471 rhythm generator IC.

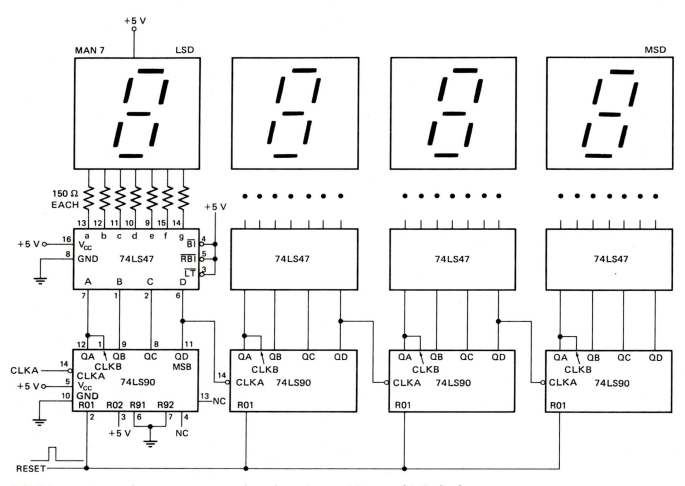

FIGURE 8-37 Decimal counting units made with 74LS90s, 74LS47s, and LED displays.

ing edge of each pulse applied to the CLKA input of the leftmost counter will cause the BCD count on its outputs to increment by 1. The 74LS47 will convert this BCD count to the corresponding seven-segment code so that the count will be displayed on the LED display. When the count on the leftmost counter reaches 1001 or 9, the next CLKA pulse will cause the count on that counter to roll back to 0000. As the QD output of the first counter falls back to 0, it clocks the second counter. This increments the count on that counter by 1. When the second counter rolls over from 1001 to 0000, the falling edge of its QD output clocks the third counter and increments its count by 1.

The circuit in Figure 8-37 then functions similarly to the odometer in a car. When each digit reaches its maximum count of 9, the next pulse rolls that digit back to 0 and increments the count on the next most significant digit. When all the digits reach a count of 9, the next pulse rolls them all back to 0 and the count starts over. More digits can be added to the basic circuit in Figure 8-37 to extend the maximum count as far as is needed for a given application.

In the circuit in Figure 8-37, the LSD is on the left and the MSD is on the right. This is reversed from the way you usually write numbers, but the circuit is drawn this way to follow the convention that signals enter a schematic on the left and proceed from left to right through the circuit.

You might use the circuit in Figure 8-37 to count the number of people who go into a stadium, the number of cans of applesauce that pass by on a conveyer belt, or the number of turns that a shaft rotates. To produce the pulses which correspond to people passing through a gate, you might use an LED and a phototransistor such as we showed for the optical encoder in Figure 7-18*b*. In a later chapter on interfacing, we show you other methods to produce pulses from physical events.

DIGITAL CLOCKS

If you connect a clock signal with a frequency of one pulse per second to the input of the counter circuit in Figure 8-37, the display will be incremented each second and the number displayed will represent the number of seconds that have passed. The circuit then functions as a digital readout clock, but the display is not in the familiar format of hours, minutes, and seconds. Figure 8-38 shows a block diagram for a digital clock circuit which will produce a traditional display of hours, minutes, and seconds. For the divide-by-10 sections of this clock, you could use a 74LS90 decade counter or a 74LS93 modified to divide by 10. For the divide-by-6 sections, you

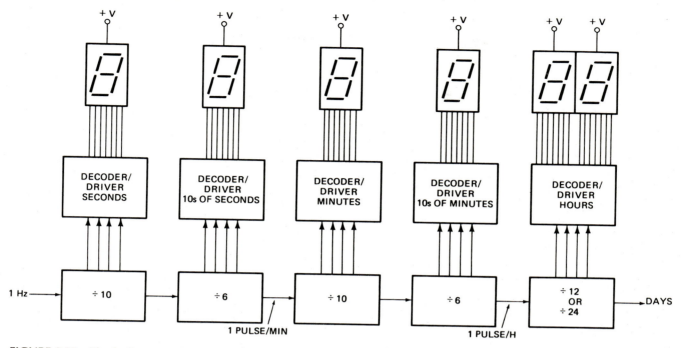

FIGURE 8-38 Block diagram of a digital clock with seven-segment readout.

could use the divide-by-6 section of a 74LS92 or a 74LS93 modified to divide by 6. Finally, for the divide-by-12 section, you could use a 74LS92 or a 74LS93 modified to divide by 12. An accurate 1-Hz input signal for the clock circuit can be produced from the 60-Hz ac power line. To do this, the power line signal is first converted to a TTL level signal by a circuit we discuss in a section of Chapter 10 on logic gate input interfacing. This 60-Hz signal is divided by 10 and then by 6 to give the desired 1-Hz signal.

If a digital clock circuit such as that represented by the block diagram in Figure 8-38 is built with discrete TTL or CMOS devices, it requires a relatively large number of IC packages to implement. LSI technology has made it possible to put all the circuitry for a complete digital clock or watch in a single MOS devices such as a Sanyo LM8460 Alarm Clock IC.

Figure 8-39a shows the block diagram of an LM8460 clock IC, and Figure 8-39b shows how the device can be connected with a power supply, a few switches, and a fluorescent display module to produce a complete digital alarm clock such as you may have next to your bed.

As shown in Figure 8-39b, the LM8460 uses the 60-Hz ac power line signal as a time base. This frequency is internally divided by 60 to produce a 1-Hz signal for the seconds counter. The output from the seconds counter is a one pulse per minute signal which is used to drive the minutes and hours counters. A digital comparator compares the time on the outputs of the minutes and hours counters with the values you load into the alarm counter. When the two counts match, the internal alarm signal will be activated. Depending on the signal level on the Alarm Select input, this will cause the Radio Control output to be turned on, the buzzer output signal to be turned on, both of these signals to be turned on, or neither of these to be turned on. Additional counters in the

LM8460 allow you to select a 10-min snooze time or up to a 59-min sleep time.

The segment decoder-driver in the LM8460 produces independent segment drive signals for a four-digit seven-segment vacuum fluorescent display. This type of display, commonly used in digital alarm clocks and microwave ovens, is typically blue in color.

We showed the circuit for a digital alarm clock in Figure 8-39 to help you get a better feeling for how a common digital device around your house operates. However, you would probably not want to build one of these because mass production has reduced the price of digital clocks and watches to the point where they are given away as prizes in cereal boxes and for renewing magazine subscriptions. At your local discount store, you can buy a complete digital alarm clock with a built-in AM/FM radio for about $10. Therefore, in accord with our policy of giving you as many real circuits that you can build and experiment with, we will show you a large display stopwatch circuit that uses a commonly available LSI device. You might use this stopwatch circuit to display the times at the finish line of a 10-km run or to time the events in a swim meet.

Figure 8-40 shows the complete circuit for our stopwatch. As you can see in the figure, the heart of the circuit is the Intersil ICM7215 four-function stopwatch device. Additional parts consist of the display LEDs, three D cell 1.5-V batteries, some switches, and a crystal.

The 3.2768-MHz crystal connected between pins 1 and 24 is used by an internal oscillator circuit to produce a 3.2768-MHz signal. Binary counters inside the ICM7215 divide this frequency by 32,768 to produce a precise 100-Hz signal. The 100-Hz signal drives the hundredths of seconds counter. The output of the hundredths counter drives the tenths of seconds counter. Additional counters in the counter chain count seconds, tens of seconds, min-

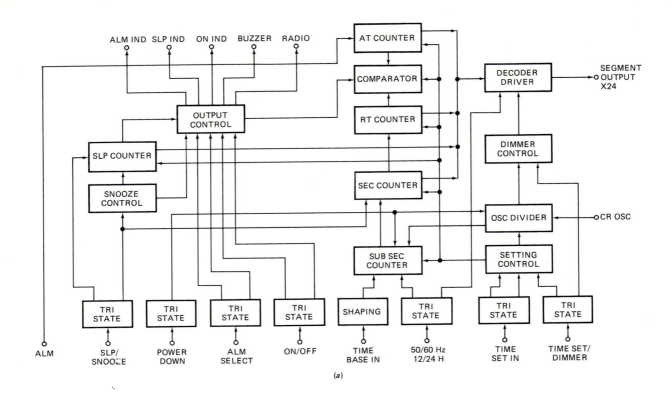

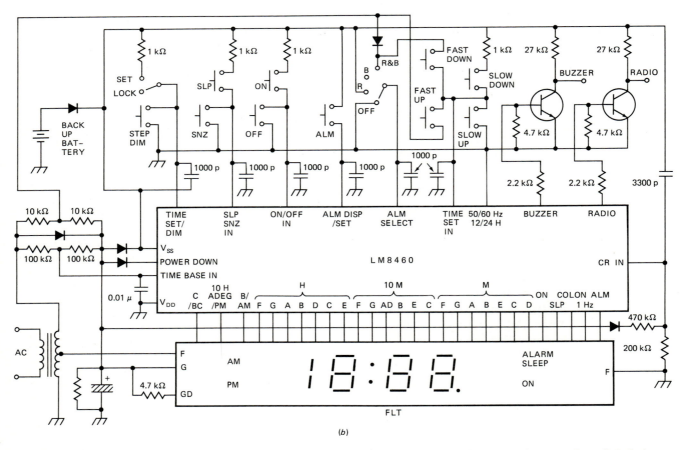

FIGURE 8-39 (a) Internal block diagram of Sanyo LM8460 digital alarm clock IC. (b) Circuit for complete digital alarm clock using Sanyo LM8460.

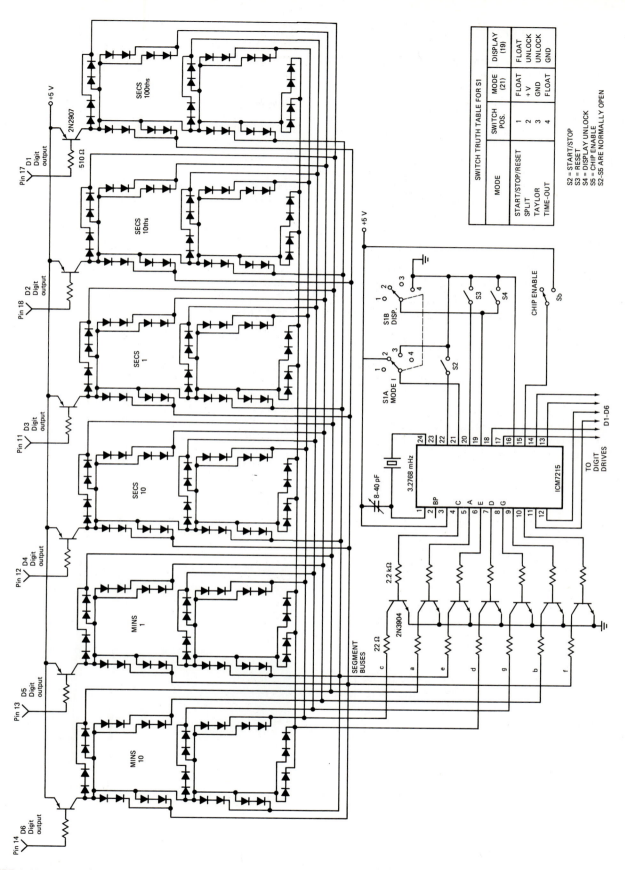

FIGURE 8-40 Large display digital timer using Intersil ICM7215 stopwatch IC.

utes, and tens of minutes. The circuit then can count up to 59 min and 59.99 s.

Instead of using standard common cathode seven-segment displays which the ICM7215 was designed to drive, we used discrete, high-intensity LEDs for the display. These LEDs can be mounted by drilling holes in paneling or Masonite to form a seven-segment display which can be seen from a considerable distance. Each segment consists of four LEDs connected in a series-parallel arrangement as shown.

To save power, these displays use a scheme called *display multiplexing*. In the next chapter we show you in detail how a multiplexed display works, but here's an introduction to the basic principle.

When they are lit, the current through each pair of LEDs in the circuit in Figure 8-40 is about 25 mA. Each "segment" then requires about 50 mA. Multiplying the 50 mA per segment times seven segments per display gives a current of 350 mA per digit. If all the digits were lit at the same time, then the total current for the displays would be 350 mA per digit times 6 digits, which is 2100 mA, or 2.1 A.

To drastically reduce the required amount of power, we use the same trick we used to produce multiple scope traces with a single beam. Basically we send out the seven-segment code for one digit on the segment bus and turn on that digit for a few milliseconds. Then we send out the seven-segment code for the next digit on the segment bus and turn that digit on for a few milliseconds. We continue the process until all six digits have had a turn and then start over with the first digit. At any given time, then, only one digit is lit, so the maximum current is 350 mA. If we cycle through the digits fast enough, the digits will all appear lit with the proper numbers because the human eye responds too slowly to see the multiplexing.

In the circuit in Figure 8-40, the "segment" LEDs for all six digits are connected in parallel on common lines. When the 7215 sends out a seven-segment code, it then goes out in parallel to all the digits. However, as we said before, only one digit will ever be turned on at any given time. Current for each digit is supplied by a digit driver transistor connected between +5 V and the anodes of the LEDs for that digit. The digit driver transistors are turned on one at a time by signals from the digit driver outputs of the ICM7215.

To complete our discussion of this circuit, we need to tell you a little bit about the different modes the circuit can operate in. The table in the bottom of Figure 8-40 shows the four stopwatch functions that are selected by the different positions of switch 1 in the circuit. If the switch is in position 1, the circuit operates in start/stop/reset mode. In this mode, the count is reset to 0 by pressing the reset switch once. Pressing the start/stop switch once starts the counters counting. As the counters are counting, the changing count will be displayed on the displays. Pressing the start/stop switch once again stops the count. The accumulated time is held on the displays until the reset switch is pressed again.

If the Taylor, or sequential, mode is selected, pressing the start/stop button will start the counters. When the start/stop switch is pressed again, the accumulated time will be displayed, the counters reset, and the count started from zero again. When the start/stop button is pushed again, the count since the last start is displayed, the counters are reset, and the count is started again. This mode might be useful for timing successive laps around a track.

The *split* mode of operation is similar to the Taylor mode except that the display shows the total time accumulated since pushing the reset button and then the start button. If the display is unlocked, the display will follow the counters. When the start/stop switch is pressed, the current count will be held.

In the *time out* mode of operation, pressing the start/stop switch alternately starts and stops the counters. The count displayed is the accumulated count since the last reset.

LOGIC ANALYZERS

Introduction

For the most part a simple dual-trace scope is sufficient to observe the waveforms of the circuits and devices we have discussed so far in this book. To observe the waveforms or sequence of states produced by many of the circuits and devices throughout the rest of the book, however, we need an instrument which can simultaneously display the data present on several signal lines. One instrument which fills this need is a *logic analyzer*.

Figure 8-41 shows a common logic analyzer, the Tektronix 1230. The least expensive version of the Tektronix 1230 can sample and display the data on up to 16 signal lines. Expansion boards can be added to increase the number of input signal lines to 32, 48, or 64. Hewlett Packard, Gould, and several other companies make comparable logic analyzers.

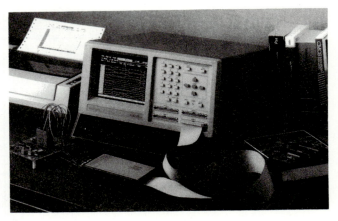

FIGURE 8-41 Tektronix 1230 logic analyzer. *(Courtesy of Tektronix Inc.)*

Personal computers can be adapted to function as logic analyzers by adding plug-in units such as the MicroCase Inc. (formerly Northwest Instruments) μAnalyst 2000 or Bitwise Designs Inc. Logic-20.

One method of connecting an analyzer to a circuit is

with a *pod* and test leads such as those shown in front of the analyzer in Figure 8-41. Here the test leads are attached to a "glomper clip," which is clipped onto one of the major ICs in the test circuit. Another method of connecting the pod to a circuit is with a special plug which is inserted in a socket on the board being tested. An important point to stick in your mind from the start is:

Turn off the power to the analyzer *before* you plug pods into the analyzer or unplug pods from the analyzer!

Failure to observe this simple precaution may damage the circuitry in the pod and/or the analyzer.

Most modern logic analyzers are essentially a specialized microcomputer or a microcomputer adapted for use as a logic analyzer. You program the analyzer to take a desired measurement by selecting choices from an on-screen menu with front panel keys. After we discuss the general operation of a logic analyzer, we will show you how to set up a logic analyzer to make some specific types of measurements.

Logic analyzers allow you to display data samples from the input channels in one of two basic formats. One format is a *timing diagram* format such as that shown in Figure 8-42a. This format is useful when you are using an analyzer to determine, for example, the time between transitions on one group of signal lines and transitions on another group of signal lines. As part of the setup parameters, most analyzers allow you to choose any one of a large number of times per division, similar to the way the TIME/DIVISION control on a scope operates.

The second major logic analyzer display format is a *state table* format such as that shown in Figure 8-42b. This format is most useful when you want to see clearly the sequence of states that a clocked circuit steps through. As part of the logic analyzer setup procedure, you can usually specify whether you want the sequence of states displayed in binary, octal, decimal, hexadecimal, ASCII equivalents, or some other code. Before we show you how to set up a logic analyzer for various measurements, we will describe how an analyzer works. Hopefully this will help make the analyzer setup parameters more understandable to you.

Logic Analyzer Operation

INTRODUCTION

Figure 8-43 shows a block diagram for the major sections of a simple logic analyzer. The overall principle of a logic analyzer is that it takes a sequence of samples of the signals on its data inputs and stores these samples in an internal memory. When the desired number of samples have been taken, the analyzer displays the stored sequence of samples on the CRT screen in the specified format. In the following sections we discuss how each part of the analyzer functions.

INPUT COMPARATORS

The first parts of the block diagram in Figure 8-43 to look at more closely are the adjustable threshold com-

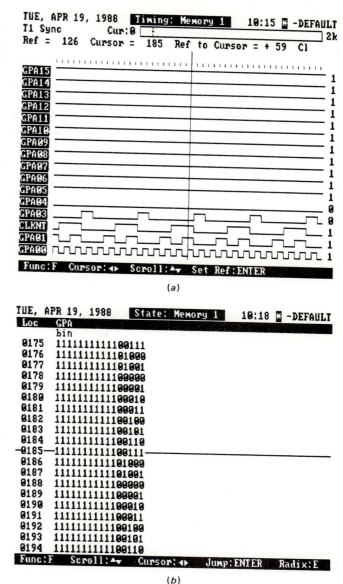

FIGURE 8-42   Common logic analyzer display formats. *(a)* Timing diagram format. *(b)* State table format.

parators on the left side of the figure. A *comparator* is a device which outputs one logic level if an input voltage is less than some reference voltage, and outputs the other logic level if an input voltage is greater than the reference voltage. If, for example, the reference voltage is set for 1.4 V and the voltage applied to the signal input of this circuit is a small fraction of a volt more positive than 1.4 V, the output of the comparator will be high. If the voltage on the signal input is a small fraction of a volt less than 1.4 V, the output of the comparator will be low.

In a logic analyzer you set the reference voltage on all the input comparators to the threshold voltage of the logic family you are working with. When looking at 74LS signals, for example, you will set the comparator thresholds for 1.0 to 1.4 V. Incidentally, some analyzers group all TTL-compatible logic families under one setting of TTL

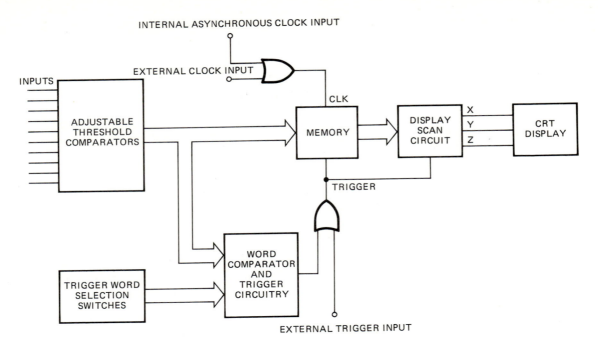

INTERNAL ASYNCHRONOUS CLOCK INPUT

EXTERNAL CLOCK INPUT

INPUTS

ADJUSTABLE
THRESHOLD
COMPARATORS

CLK

MEMORY

DISPLAY
SCAN
CIRCUIT

X
Y
Z

CRT
DISPLAY

TRIGGER

WORD
COMPARATOR
AND
TRIGGER
CIRCUITRY

TRIGGER WORD
SELECTION
SWITCHES

EXTERNAL TRIGGER INPUT

FIGURE 8-43   Block diagram of simple logic analyzer.

with a threshold of 1.4 V. For ECL signals you will set the comparator thresholds for about −1.3 V. The point here is that if you set the thresholds appropriately, the comparators will convert the input signals from any logic family to the logic levels required by the internal circuitry of the analyzer.

MEMORY

The next part of the analyzer block diagram to look at is the section labeled *memory*. Remember from a previous discussion that when a D flip-flop is clocked, the logic level on its D input is transferred to its Q output and held there until the flip-flop is clocked again. In other words, a D flip-flop functions as a 1-bit memory, because it "remembers" the value stored on its output until a new value is written to it. You can think of the memory in a logic analyzer as a very large array of D flip-flops which are used to store a series of samples taken from the input signals. Depending on the particular analyzer, each input channel has perhaps 2000 memory locations associated with it. An analyzer configured in this way can take and store up to 2000 sequential samples of the data on each input channel.

Each time the memory control circuitry receives a clock signal, it takes a set of samples of the logic levels on the inputs and stores these logic levels in a set of memory locations. The memory is configured so that the first set of data samples taken from the inputs is written to the first set of locations in the memory, the second set of data samples is stored in the second set of memory locations, the third set of data samples is stored in the third set of memory locations, etc. When the memory is full or the specified number of samples has been taken, the display circuitry will use the sets of data stored in memory to produce a display on the CRT.

CLOCK SELECTION

A key point in the preceding paragraph is that a clock signal determines the times when the analyzer takes a set of samples of the signal levels present on the data inputs and stores those samples in memory. Each pulse of the clock signal causes the analyzer to take a new set of samples and store the samples in memory. As shown in Figure 8-43, the clock signal can come from some *external* source or from a precise *internal* oscillator which is part of the analyzer circuitry.

To produce a display of the sequence of states that a clocked circuit such as a counter steps through, you usually set the analyzer for *external clock* and connect the counter's clock signal to the analyzer's *external clock* input. The analyzer will then take a new set of samples each time the counter is clocked to a new state. The sequence of samples stored in memory and displayed will then represent the sequence of states that the counter steps through. In a later section we will describe how you decide which edge of the external clock signal to use to take a set of samples.

To produce a timing diagram display for a circuit so that you can measure times between points on the waveforms, you usually use the analyzer's *internal clock*. The internal clock consists of a high-frequency oscillator and a chain of frequency dividers. As part of the setup you specify the frequency from the divider chain that you want the analyzer to use as a clock signal. The internal clock frequency you select then determines how often the analyzer takes samples of the data inputs. For example, if you select an internal clock frequency of 1 MHz, the analyzer will take a sample every $\frac{1}{1,000,000}$s, or every 1 μs. Since you know precisely the rate at which samples are taken, all you have to do to determine the time between two events on displayed waveforms is to count the number of sample points between the two events. Most newer logic analyzers, incidentally, will give you a direct readout of the time between two cursors you position on the display.

In summary, then, you usually use the *external* clock to produce a trace of the sequence of states that a circuit steps through, and you usually use the *internal* clock to make precise measurements of the time between points on waveforms.

TRIGGER CIRCUITRY

If a logic analyzer is receiving a clock signal, it will be continuously taking samples of the signals on the data inputs and storing the samples in its memory. Once the memory is full, the oldest set of samples in the memory will be written over when a new set of samples is taken. For an analyzer which stores 2048 samples per channel, the memory will at any point in time contain the latest 2048 samples taken. Since the analyzer is continuously taking new samples, the contents of the memory are constantly changing. If the display scan circuitry used this changing data to produce a display on the CRT, the display would, in most cases, be changing too fast to see anything useful in it. What is needed to solve this problem is a signal which tells the analyzer to stop taking samples and display the samples stored in memory over and over on the CRT. Remember from the discussion of scope operation in Chapter 4 that on most CRTs, the display must be traced out over and over or "refreshed" in order for it to remain visible.

The signal which tells the analyzer when to stop taking samples and display the contents of memory is the *trigger* signal. As shown in the block diagram in Figure 8-43, the trigger signal can come from some external source or from a word recognizer circuit inside the analyzer.

If the circuit you are looking at with the analyzer produces some appropriate signal at the end of a count sequence, for example, you can connect that signal to the *external trigger* input of the analyzer and set up the analyzer to trigger on this signal. In this case, when the analyzer receives the trigger signal, it will stop taking samples of the input data and display the samples taken before the trigger occurred.

If the circuit you are looking at does not have any signal which is appropriate to use as a trigger signal, you can set the analyzer for *internal trigger*. In this mode the analyzer uses an internal *word recognizer* circuit to generate a trigger signal. The word recognizer circuit compares the binary word present on the data inputs of the analyzer with a binary word specified as part of the analyzer setup. When the two binary words are equal, the word recognizer circuit generates a signal which is used to trigger the analyzer.

If you need to refresh your memory of the operation of a word recognizer circuit, take a quick look at Figures 5-12 and 5-19. The output of the XNOR gate circuit in Figure 5-12 will go high when the word on the A inputs is equal to the word on the B inputs. For use in an analyzer, the data input signals might be connected to the A inputs of the circuit and a word specified during the analyzer setup procedure might be sent to the B inputs of the circuit. When the two words match, the output of the circuit will

go high and trigger the analyzer. Figure 5-19 shows a circuit which will produce a low trigger signal when a word selected by switches is present on its inputs.

To summarize, then, if a logic analyzer is set for *internal trigger*, the word recognizer compares each set of samples with a word specified during setup. When the words match, the word recognizer produces a trigger signal. Most analyzers allow you to tell the analyzer to produce a trigger signal when any one of several different words is present on the data inputs or when a specified word is *not* present on the data lines. Later we will show you how to use some of these modes.

During setup you can tell an analyzer to respond in any one of several ways when it receives a trigger signal. One option is to tell the analyzer to *stop* taking samples when it receives a trigger signal. Here the display will show the 2000 or so samples that it took before the trigger occurred. This mode is sometimes called *pretrigger recording*. Another option is to tell the analyzer to *start* taking a new set of samples when it receives a trigger signal, continue taking samples until the memory is full, and then display the contents of memory. In this case the display will show the 2000 or so samples taken after a trigger occurs. This mode is sometimes called *posttrigger recording*. A third option is to tell the analyzer to take some specified number of samples after the trigger occurs and then display the contents of memory on the CRT. In this case the display will show some samples that occurred before the trigger and some samples that occurred after the trigger. If, for example, you tell the analyzer to continue taking samples for 1000 clocks after the trigger, the display will show about 1000 samples that were taken before the trigger and 1000 samples that were taken after the trigger occurred. The reason that you can display events that occurred before the trigger pulse is because the analyzer is continuously taking samples of the data inputs and storing them in memory as long as it is receiving a clock signal.

SUMMARY OF LOGIC ANALYZER GENERAL OPERATION

Before we show you how to set up and make measurements with a specific logic analyzer, here's a summary of the functions of the major blocks of a simple logic analyzer and the basic parameters you set for each:

1. Comparators on the data inputs convert logic signals from any logic family to the levels required by the analyzer internal circuitry. You must set the threshold parameter for the logic family being measured.

2. A clock signal tells the analyzer when to take samples of the signals on the data inputs and store the samples in the internal memory:

   a. EXTERNAL CLOCK is usually used to see sequence of states stepped through by a clocked circuit. You set the analyzer to take a sample on each positive edge of the input clock signal or a sample on each negative edge of the input clock signal, whichever edge occurs when valid data is present on the data inputs.

    b. INTERNAL CLOCK is usually used to make timing measurements on waveforms. You select the internal clock frequency so that samples are taken at appropriate intervals to make the desired measurement, just as you set the horizontal time base control on a scope.

3. A trigger signal tells the analyzer when to display the samples stored in memory:

    a. EXTERNAL TRIGGER uses a signal connected from some external source such as the system being observed. You must specify the edge of the external signal you want the analyzer to trigger on.

    b. INTERNAL TRIGGER uses a signal produced by a word recognizer as a trigger. The word recognizer compares the word on the data inputs with a word specified as part of the setup. You must specify the word you want the word recognizer to look for.

    c. TRIGGER DELAY or TRIGGER POSITION tells the analyzer to *start* taking a new memory full of samples when it receives a trigger, *stop* taking samples when it receives a trigger, or *delay* some specified number of samples after it receives a trigger signal. How you specify these depends very much on the particular analyzer you are using.

4. Display format controls determine whether the stored samples will be displayed in binary state table mode, in hexadecimal state table mode, in timing diagram mode, etc. Other parameters you set in this category include the number of samples to be displayed on the screen at a time, the number of channels to be displayed on the screen, and whether you want glitches displayed on waveforms or not.

Setting Up and Making Measurements With a Logic Analyzer

INTRODUCTION TO THE TEKTRONIX 1230 LOGIC ANALYZER

We chose the Tektronix 1230 logic analyzer as an example for this section because it is low in cost, can be expanded to as many signal lines as needed, and has personality modules which adapt it to work with a variety of microprocessors. Also, the Tektronix 1230 uses menus for its setup, so it is quite easy to use. If you are using some other analyzer, you should, with the help of the discussion above, be able to translate the following section as needed to apply to the controls on your analyzer.

As we mentioned previously, the Tektronix 1230 logic analyzer uses a series of on-screen menus to display the available setup parameters. To set up the analyzer for a measurement you select the desired parameters from each menu with the front panel keypad shown in Figure 8-41.

When you first turn on the power to a Tektronix 1230, it does an internal self-test and displays an introductory message which tells you about the NOTES key on the front panel. If you need an explanation of any of the parameters in a menu as you are working your way through it, you press the NOTES key. The analyzer will then display on the screen a brief description of the function of the parameters in that menu. Once you learn the basic operation of the analyzer, this "smart" feature will in most cases save you from having to dig around in the manual. When you have read the desired notes, you just press the MENU key to get back to the menu you are working with.

After you have read the introductory message, you press the MENU key to continue. The analyzer will then display the MAIN menu shown in Figure 8-44. Listed in

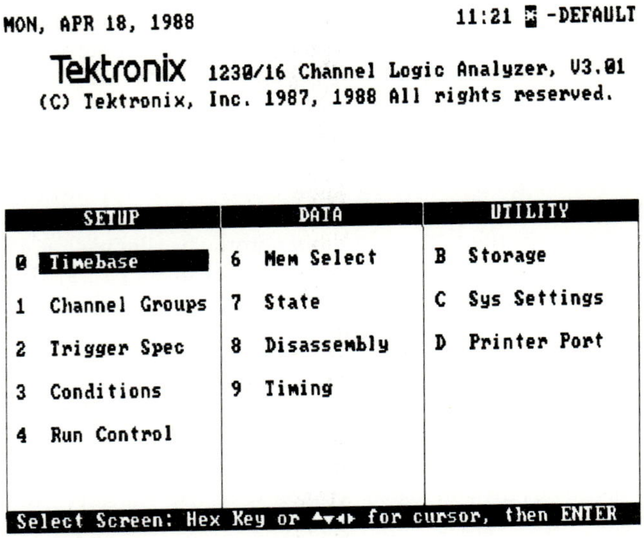

MON, APR 18, 1988   11:21 -DEFAULT

Tektronix 1230/16 Channel Logic Analyzer, V3.01
(C) Tektronix, Inc. 1987, 1988 All rights reserved.

SETUP		DATA		UTILITY	
0	Timebase	6	Mem Select	B	Storage
1	Channel Groups	7	State	C	Sys Settings
2	Trigger Spec	8	Disassembly	D	Printer Port
3	Conditions	9	Timing		
4	Run Control				

Select Screen: Hex Key or ▲▼◄► for cursor, then ENTER

FIGURE 8-44   Tektronix 1230 logic analyzer MAIN menu.

this MAIN menu are 12 submenus. Note that the submenus are arranged in three groups, SETUP, DATA, and UTILITY. Setting up the analyzer for a particular measurement mostly involves working your way through the SETUP submenus. The DATA menus are used to select the desired display format, and the UTILITY menus are used to specify memory usage, screen brightness, printer configuration, etc.

To get to a desired submenu you press the MENU key and then the number key associated with the desired menu. Another way to get to a desired submenu is to use the arrow keys to move the highlighted box to the desired menu in the list and press the ENTER key. Note that one of the submenus is a printer option. This option allows you to copy the display on the screen at any time to a simple graphics-type printer. With this option, you can use an inexpensive printer to produce printouts such as those shown in the following sections.

In each submenu you select the desired parameters by pressing the indicated keys on the front panel. This menu/keyboard approach greatly reduces the number of knobs required on the front panel. Also, with this approach you can display the help notes if you get confused about the meaning of a particular parameter. In the following sections we show you how to work through the menus for several types of measurements.

The most important step in setting up a logic analyzer is to first think carefully about what type of measurement you want to make and how you want to make it. You need to ask yourself questions such as:

1. How many signals do I want to look at?

2. Do I want to do a state or a timing measurement?

3. Do I want to use an internal or an external clock?

4. If I am using an external clock, do I want to take samples when a rising edge occurs on the external clock signal or when a falling edge occurs on the external clock signal?

5. Should I use an external or internal trigger source?

6. If I use the internal trigger source, what word should I set the word recognizer to trigger on?

7. Do I want to display all the samples before or after the trigger, or some before and some after?

8. If I am doing a state listing, do I want the display in decimal, binary, or hexadecimal format?

9. Do I just want to display a single sequence of states, or compare this sequence of states with another sequence stored in the reference memory of the analyzer?

Once you answer these questions it is usually quite easy to work your way through the setup menus.

SETTING UP A 1230 FOR TRACING A SEQUENCE OF STATES

Whenever you are first learning how to use a new instrument, it is wise to start by working with signals that you are already familiar with. Therefore, for our first example we will show you how to set up the analyzer to display the sequence of states that a 79LS93 4-bit binary counter, connected to divide by 10, steps through as it is clocked. The circuit for this is shown in Figure 8-31a and the output waveforms for it are shown in Figure 8-31b. For most of the examples in this section, we applied a 12-kHz square-wave signal to the CLKA input of the 74LS93.

In this first logic analyzer example, we will start by telling you the answers to the questions we told you to think about when you use a logic analyzer. Then we will step you through setting up the analyzer to make some common types of measurements. These examples are intended to give you an overview of the steps involved in setting up a logic analyzer, not to make you an expert user. Some exercises in the accompanying laboratory manual give you more detailed instructions on logic analyzer setup and use:

1. The 74LS93 has four outputs, so we only need four data inputs on the logic analyzer for now. The basic Tektronix 1230 has one 16-bit pod called *pod A*. For this measurement we simply connect the four counter outputs in order to the four LSB inputs of pod A. In other words, connect QA to A0, QB to A1, QC to A2, and QD to A3. Connect the unused inputs of pod A to ground so they do not pick up electrical noise signals and cause confusing results on the display. Connect the ground input on the pod to a ground point in the circuit.

2. We want to do a display of a sequence of states, so we will select the state mode.

3. For doing a trace of a sequence of states, we usually use an external clock signal. For this example we will use the signal which clocks the 74LS93 to also clock the analyzer. For the P6444 pod on a Tektronix 1230, there are two external clock inputs, CLK1 and CLK2. If you want the analyzer to take samples on falling edges of the clock signal, you connect the external clock signal to one of the CLK inputs and set the appropriate CLK polarity switch on the probe for $-$. If you want the analyzer to take samples on the rising edges of the external clock signal, then you connect the external clock signal to one of the CLK inputs and set the appropriate CLK polarity switch on the probe for $+$. Connect the GND input of the pod next to the CLK input you choose to a ground point in your circuit.

4. If the circuit you are looking at changes states on the positive edge of the applied clock, then you usually tell the analyzer to take samples or "clock" on the negative edge of the external clock signal. Likewise, if the circuit you are looking at changes states on the negative edge of the clock signal, then you usually tell the analyzer to "clock" on the positive edge of the applied signal. A quick look at the 74LS93 counter waveforms in Figure 8-31b should help you see the reason for this.

As shown by the waveforms in Figure 8-31b, the flip-flops in the 74LS93 change states on the negative edge of the applied clock signal. If you clock the analyzer on the same negative edges of this clock signal, the counter outputs may not have settled to their new state when the analyzer takes a sample. This depends on the propagation delays in the counter and in the logic analyzer. As you can see in the waveforms in Figure 8-31b, the new states are clearly stable on all the outputs at the times when the clock signal goes positive. For this counter, then, clocking the analyzer on the positive edges of the counter clock signal makes sure the data is stable when a set of samples is taken.

5. For this trace we will use the word recognizer to generate an internal trigger signal. This is the usual choice when we want to see the sequence of states that a circuit steps through.

6. As you know, the counter circuit in Figure 8-31a steps through the states 0000 to 1001 over and over when clocked. To see the sequence from the start, then, 0000 is a good word to trigger on.

7. Since the count sequence repeats over and over for this circuit, the position of the trigger in the stored

samples doesn't make much difference. Because we have to make a decision for the setup, we will choose to display mostly the samples that are taken after the trigger word occurs.

8. Since the outputs of this counter cycle through the decimal states 0 to 9, it is convenient to display the output states in decimal or hexadecimal format.

9. For the first trace we just want to look at the sequence of states the counter steps through, so we don't have to worry about comparing our trace with a trace stored in a reference memory.

Now let's work our way through setting up the 1230 menus to produce a state table trace for this counter.

The first menu you want to set up in the 1230 is the TIMEBASE menu. To get to this menu you first press the MENU key and then the 0 key. Figure 8-45 shows

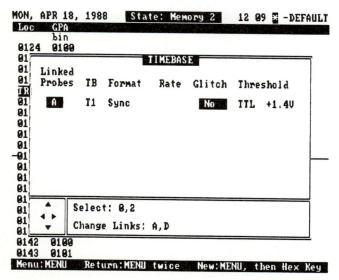

FIGURE 8-45   Tek 1230 TIMEBASE menu.

the TIMEBASE menu. As shown by the arrows in a box at the bottom of the menu, you use the arrow keys on the front panel to move a highlighted box to a parameter you want to change in the menu. Once you have the highlighted box (cursor) positioned on a menu item that you want to change to a new value, you use the 0 key or the 2 key to select the desired value for the highlighted parameter. As you press the 0 key or the 2 key, the available values for that parameter will be displayed in the highlighted box. When the desired value is displayed you stop pressing the key and move the cursor to the next parameter to set it. This is equivalent to turning a knob on older instruments to select a desired range.

As a first example of this, putting the cursor on the Format parameter in the TIMEBASE menu and pressing the 0 key several times would show you that there are two options for this parameter, Sync and Async. Selecting the Sync option in this menu tells the analyzer that you are using an external clock signal. The frequency of the external clock signal must be ≥25 MHz. Selecting Async for this option tells the analyzer that you

want to use the internal clock to take samples. If you use the Async option, you can specify an internal clock frequency as high as 100 MHz. For this first measurement we want to see a sequence of states so we use the Sync mode. We leave the Glitch option as No and leave the Threshold set for TTL. In a later example we show how to use the Async mode and Glitch display feature.

The next menu you want to look at for the analyzer setup is the CHANNEL GROUPING menu shown in Figure 8-46. To get to this menu you press the MENU key

FIGURE 8-46   Tek 1230 CHANNEL GROUPING menu.

and then the 1 key. The main parameters to note in this menu are Radix and Pol. You use the Radix parameter to tell the analyzer what number system you want a state listing in. The choices are octal, hex, and binary. We want our state listing in hex, the default, so we leave this parameter as it is. You use the Pol parameter to tell the analyzer whether you want the display in positive logic format or negative logic format. We want a standard positive logic display, which is the default, so we leave this parameter at + and go on to the next menu.

Figure 8-47 shows the TRIGGER SPEC menu which you get to by pressing the MENU key and the 2 key. The Condition box in this menu is used to tell the analyzer what internal or external condition you want it to trigger on. With this box you can, for example, tell it to trigger on a signal applied to a trigger input on the pod, an external trigger signal applied to the TRIGIN connector on the rear of the analyzer, or when the signal inputs match a specified word. The default here is the letter A. The A here does not represent pod A, it represents a trigger word you specify in the CONDITIONS menu. The box at the bottom of the TRIGGER SPEC menu gives you a look ahead at this CONDITIONS menu. For our first example here, we want to trigger when a specified word is on the data inputs, so we will leave the default A in place.

The Count box in the TRIGGER SPEC menu is used to tell the analyzer how many times you want the specified trigger condition to occur before it triggers the analyzer. This feature is useful when tracing execution

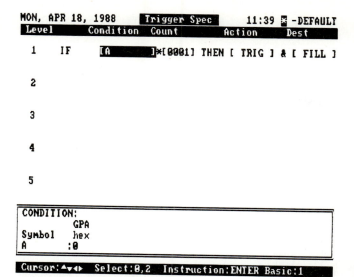

FIGURE 8-47   Tek 1230 TRIGGER SPEC menu.

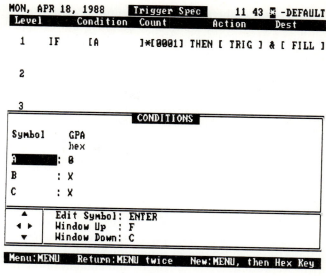

FIGURE 8-48   Tek 1230 CONDITIONS menu.

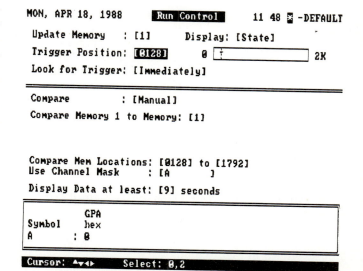

FIGURE 8-49   Tek 1230 RUN CONTROL menu.

through program loops in a microcomputer. In our example we want to trigger on the first occurrence of the trigger condition, so we leave the default 0001 in place.

The final two boxes in the first line of the TRIGGER SPEC menu tell the analyzer what action to take in response to the trigger. The most commonly used options are TRIG and FILL memory as shown. For complex trigger situations you can use an IF-THEN-ELSE structure in this menu to set up a list of alternative trigger conditions and the desired action for each. For example:

IF [A]*[0012] THEN [TRIG] & [FILL]
    ELSE [B]*[0001] THEN [STROFF]

tells the analyzer to trigger and fill the memory if condition A occurs 12 times and to turn off the storage if condition B occurs once. For basic digital observations you don't usually need this feature, but it is handy to know it is available.

The next menu to look at is the CONDITIONS menu shown in Figure 8-48. You get to this menu by pressing the MENU key and then the 3 key. This is the menu you use to specify the condition represented by the letter A in the TRIGGER CONDITION menu. In other words, you use this menu to tell the word recognizer the word you want it to look for. For this measurement we are only using the lower 4 bits of the pod, so we don't care what logic levels are present on the upper 12 lines. Therefore we leave X's in these positions of the A word to represent don't cares. The count on the lower four lines will cycle from 0000 to 1001 over and over, so any value in this range will work. We put a hex 0 in to represent these 4 bits so that the analyzer triggers at the start of the count sequence.

The final menu you have to set to get a basic display is the RUN CONTROL menu, which you get to by pressing the MENU key and then pressing the 4 key. Figure 8-49 shows the format for this menu.

The Update Memory line in the RUN CONTROL menu allows you to select one of four internal memories to store

the sequence of samples you take. This feature allows you to store a set of samples from a working system, store a set of samples from a nonworking system, and compare the two to help find the source of the problem in the nonworking system. For our first trace we will use Update Memory : [1].

The Display box in the RUN CONTROL menu is used to indicate whether you want the display in timing-diagram format or state-table format. We will use the state-table format for this first measurement.

The Tektronix 1230 stores about 2000 samples of each input channel. You use the Trigger Position line in the RUN CONTROL menu to tell the analyzer where you want the 2000 samples taken with respect to the trigger event. If you put the cursor on this box and press the 2 key several times, you will see the values in the cursor box step through values of 128, 256, 384, 512, 640, 768, 896, 1024, 1152, 1280, 1408, 1536, 1664, 1792, and

1920. These numbers represent the number of pretrigger samples you want the analyzer to keep. For our example we will select 128 for this parameter. The total number of samples that can be stored for each input is around 2000, so choosing 128 tells the analyzer to take 2000 − 128, or about 1872, samples after the trigger occurs and then stop. Most of the stored samples then will represent events that occurred after the trigger, so this is essentially posttrigger recording. Increasing the number selected here tells the analyzer to keep more samples that occurred before the trigger.

The Look For Trigger line in this menu is used to tell the analyzer whether you want it to trigger [Immediately] when it finds the specified trigger condition and then stop when it has completed the desired action such as FILL, or whether you want it to [Continuously] look for the trigger conditions and update the memory each time it finds the specified trigger condition. For this measurement we want analyzer to trigger Immediately and stop when the memory is full. We don't need the Compare section of this menu for our first example, so we will leave a discussion of it for a later section.

You now have all the parameters set that you need to take a set of samples. To tell the analyzer to take a set of samples, you simple press the START key. The analyzer should take the specified samples, give you an acquisition complete message, and then show a state listing of the signals on the A pod. You can use the cursor move keys to scroll the display forward or backward to look at the rest of the samples in the sample memory.

If you want to quickly find the trigger word in the display, press the F key until you see the message:

Search For: 0,2 [   ] Do Search:1

at the bottom of the screen. Press the 0 key until Trigger appears in the box after Search For: and then press the 1 key. The display will show the trigger word at sample number 128 in the middle of the screen with a cursor line as shown in Figure 8-50.

If you want to see the sequence of states displayed in binary instead of the default hexadecimal mode, press the E key until the line at the bottom of the screen says:

Select Group:0[GPA]   Select Radix:2[hex]   Exit:Enter

Press the 2 key until bin appears in the Select Radix box and then press the ENTER key. The list on the screen should now be in binary.

The point of the preceding discussion is not for you to memorize the controls on a Tektronix 1230 analyzer, but to give you an introduction to setting up an analyzer and to give you an idea of how easy it is to manipulate the display to a desired format. In the next section we introduce another very useful logic analyzer feature.

COMPARING ANALYZER TRACES

When troubleshooting a digital instrument, it is often helpful to compare a trace taken on a working system with one taken on the malfunctioning system. You can do this on a 1230 as follows.

In a 1230, Memory 1 is used as the reference memory

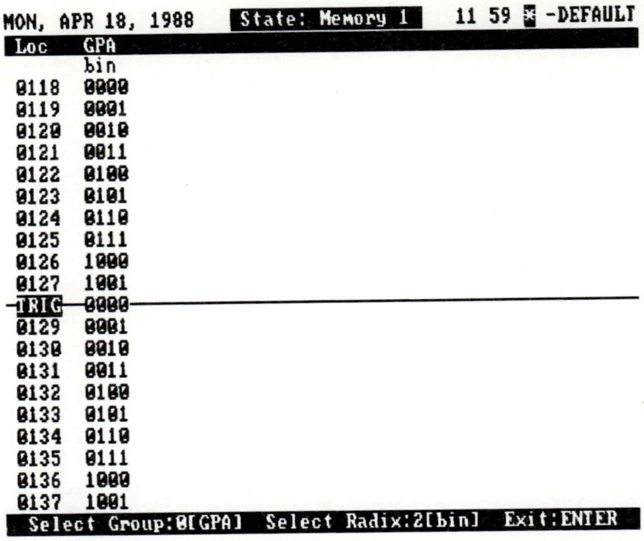

FIGURE 8-50   State listing of circuit in Figure 8-31a on A pod of 1230 analyzer.

for a compare, so as described you first store a trace from the "good" system in this memory.

For our example malfunctioning system, we removed input lead A1 from the QB output of the counter and connected the lead to ground to simulate a shorted line.

To do a trace for this malfunctioning system, the settings in the first four menus can be left as they were for the trace of the good system. The only changes we have to make are in the RUN CONTROL menu.

The first change in the RUN CONTROL menu is to change the Update Memory box from 1 to one of the other three available memories. For this first example we will use memory 2 to store the samples from the malfunctioning system. The other changes in the RUN CONTROL menu are as follows:

1. Set the Compare box in the menu to [Auto] so the analyzer will automatically compare the samples from the malfunctioning system with those from the working system.

2. Set the Compare Memory box for Memory 2 to Memory [1].

3. Set the When Memories Equal box to [Display and Stop].

4. Set the Compare Mem Locations box to [0128] to [1872].

5. Set the Display Data at least to [9] seconds.

6. When the RUN CONTROL menu is all set up, you press the START button to do a trace. The analyzer will automatically take a set of samples to fill memory 2, compare these samples with the set from the good system in memory 1, and display the samples from memory 2. Any differences between the reference samples in memory 1 and the new samples in memory 2 will be highlighted in the display.

The differences are more obvious if the display is in binary form. To get the display in binary, press the F key until the line containing Radix appears at the bottom of the screen, press the E key to get to the Radix box, and then press the 2 key until the Radix box shows bin. Press ENTER to carry out the command and get back to the display. The display should now be in binary as shown in Figure 8-51. In this form you should easily be able to see that bit A1 is always a 0. This leads you to the problem intentionally introduced on the A1 line.

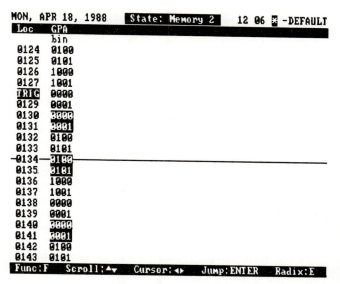

FIGURE 8-51 Listing showing samples that differ from good system samples in reference memory.

FINDING GLITCHES WITH A LOGIC ANALYZER

Remember from our previous discussions that a *glitch* is a short-duration, unwanted pulse on a signal line. One of the major uses of a logic analyzer is to find glitches on signal lines. If you are using normal internal or external clocking to take samples with a logic analyzer, a glitch that occurs between clock pulses will not be sampled by the analyzer. Therefore, it will not be displayed. One solution to this problem is to put a latch on each line to "catch" glitches. The glitch sets or resets the latch. At the next clock pulse, the glitch is stored in memory so that it can be displayed along with the normal samples.

As we discussed earlier in the chapter, our 74LS93 divide-by-10 counter circuit shown in Figure 8-31a produces a glitch on its QB output during the transition from state 9 to state 0. The reason for this is that state 1010 is decoded and used to reset the counter outputs to 0000 to start the count over. State 1010 is only present on the outputs for 20 to 40 ns, the time required to reset the counters to 0000. Since this 1010 state is present for such a short time, it does not show up in a normal state listing. The Tektronix 1230, however, can easily be set up to show glitches such as this. Only three minor changes from the previous setup are required:

1. Connect pod input A1 back to the QB output of the counter.

2. Press the MENU key and then the 0 key to get to the TIMEBASE menu. Move the cursor to the Glitch column and use the 0 key to turn the glitch option on.

3. Press the MENU key and then the 4 key to get to the RUN CONTROL menu. In this menu change the Update Memory to [1], the Display to [Timing], and the Compare to [Manual].

Now press the START button to make a new trace. You should see a display similar to the one in Figure 8-52. Notice that we have changed the labels on the traces so that they are the same as the counter output names. We have also moved the traces so that they appear in the same order as shown in data books.

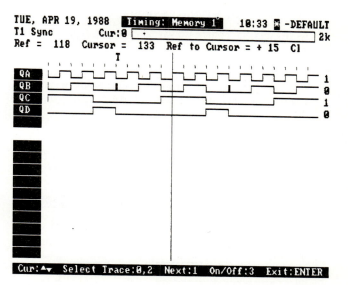

FIGURE 8-52 Display of state listing in timing diagram format to show glitches.

If the display is more compressed or more expanded than this, you can press the F key until you see the Resolution option in the line at the bottom of the screen. Then press the 4 or 5 key to change the number of samples displayed on the screen until the display does look like this one. Note that the glitch between state 1001 and state 0000 is clearly displayed on the QB line.

To see the sequence of states in this display, you can use the arrow keys to move the vertical cursor line across the display. As you move the cursor line across the display, the logic levels at the point where the cursor is located are displayed on the right side of the screen.

MAKING TIMING MEASUREMENTS WITH A LOGIC ANALYZER

The display in Figure 8-52 is in timing diagram format, but it is just a listing of the sequence of states that the counter steps through as it is clocked, not a precise timing diagram. In other words, the display cannot easily be used to accurately determine the time between two events on the display.

To make timing measurements with a logic analyzer, you use the precise internal clock so that the interval of time between samples is accurately known. The time between two events on the display can then be determined by simply counting the number of sample times between the two events and multiplying the number of samples by the time per sample.

The 1230 can be set up for making timing measurements on the 74LS93 counter outputs as follows:

1. In the TIMEBASE menu, set the Format for Async, the Rate for 10 μs between samples, and the Glitch for Yes.

2. In the CHANNEL GROUPING menu, set for the defaults of GPA HEX + T1 AAAA.

3. In the TRIGGER SPEC menu, set for the default IF [A]*[0001] THEN [TRIG] & [FILL].

4. In the CONDITIONS menu, set the A trigger word for X0.

5. In the RUN CONTROL menu, set for Update Memory [1], Display [Timing], Trigger Position [0128], Look For Trigger [Immediately], and Compare [Manual].

When you finish the setup, press the START button to do a trace. You should then see a display such as the one in Figure 8-53. Each dot in the line of dots above

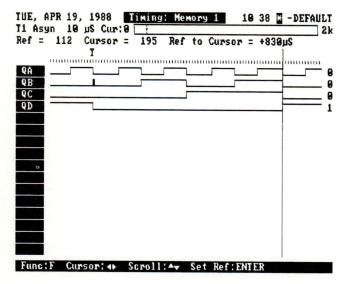

FIGURE 8-53   Trace of timing waveform using internal 100-kHz clock.

the waveforms in Figure 8-53 represents a sample. Since you know the time between samples is 10 μs, you can determine the time between two events on the display by simply counting the number of dots between them and multiplying this number by the sample interval of 10 μs. An easier way to measure times on these waveforms is to use the two vertical lines or cursors. You use the cursor move keys to position the movable cursor to one of the edges or events that you want to measure the time from. Press the ENTER key. This will set a fixed cursor on that point as a reference. Then use the cursor

move keys to move the movable cursor to the other edge or event that you want to measure to. The time between the two cursors will be displayed at the top of the screen. In Figure 8-53 we set the cursors to measure the time for one cycle of the clock signal connected to the A3 data input of the logic analyzer. Since we set the frequency of the clock signal applied to the counter at 12 kHz, and since the A3 signal is equal to this clock frequency divided by 10, the displayed value of 830 μs is correct. Other times can easily be measured by simply moving the two cursors around.

The question that may have occurred to you at this point is, "How did we know to set the internal clock frequency at 100 kHz for this measurement?" The answer to the question is that we tried several values until we found a sampling rate which produced a display of about one count sequence on the screen. A rule of thumb is to use a sampling clock frequency at least 10 times the highest frequency signal you want to look at, if possible. For our measurement, the frequency of the clock signal connected on the A3 input is 10 kHz, so a sampling clock rate of 100 kHz is an appropriate first choice.

Another important point to consider when choosing the frequency of the internal sampling clock is *resolution*. With a sampling clock frequency of 100 kHz, the analyzer takes a sample every 10 μs. If a signal changes between samples, the display won't reflect the change until the next sample time. With a 100-kHz sampling frequency, then, you can only measure or resolve any time interval to the nearest 10 μs. Another way of saying this is that the resolution of the measurement is 10 μs. Figure 8-53 shows one effect of this.

If you look very carefully at the QA waveform in Figure 8-53, you can see that the pulses vary in width by one sample interval. The reason for this is that if a signal changes between samples, the change will not show in the display until the next sample is taken.

Increasing the frequency of the sampling clock will decrease the time between samples and therefore increase the resolution of measurements. Usually you have to compromise with this because the number of samples that can be stored is limited. If you choose too high a sampling frequency, the memory will fill up before the desired section of the signal is sampled. With a sample interval of 10 μs and a memory which can hold 2000 samples, for example, the analyzer can only sample 2000 × 10 μs or 2 ms before the memory is full.

A CLOSE LOOK AT A GLITCH

Throughout the rest of this book and its companion laboratory manual we discuss many applications for a logic analyzer. As a final example of logic analyzer measurements here, we will show you how to use pretrigger recording and the 100-MHz internal clock to take a close look at the glitch produced on the outputs of our 74LS93 counter circuit as it is reset back to 0000 to start the count over again. To do this, all you have to do is make a few minor changes from the basic timing measurement setup in the previous section:

1. Increase the frequency of the signal applied to the CLKA input of the 74LS93 to 10 MHz, so the appropriate output waveforms can easily be displayed on the screen.

2. In the TIMEBASE menu, set the Format for Async, the Rate for 10 ns, and the Glitch for No.

3. In the RUN CONTROL menu, set the Trigger Position to its maximum value of 1920 so that the memory will be filled mostly with samples taken before the trigger. This is necessary in order to see the signal levels which occur just before the counter outputs return to 0000.

When you complete the setup, press the START button to do a trace. You should then see on the screen a display such as that in Figure 8-54. Each dot along the

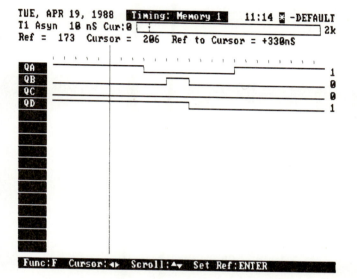

FIGURE 8-54 Glitch measurement using internal clock.

top of the screen represents one sample interval of 10 ns. If you look carefully at the waveforms, you should see that starting from state 9, the outputs go to state 8

for 20 ns, to state 10 for 20 ns, and then to state 0 when the flip-flops are reset. From our previous discussion of ripple counters, you know that the A flip-flop toggles to a 0 first. This produces state 8 on the outputs. The change on the QA output causes the QB output to change from a 0 to a 1, producing state 10 on the outputs. With the circuit connected as shown, state 10, or 1010, causes the counter outputs to reset to 0000 after a few nanoseconds.

The same logic analyzer setup can be used to see the rest of the glitches in the 74LS93 count sequence. To do this we first increase the frequency of the CLKA signal to about 10 MHz so that a complete count cycle will take place during the time the analyzer is taking its 2000 samples with a 100-MHz internal clock. We then press the START button to do a trace. Figure 8-55 shows the resultant display. To make the waveforms easier to interpret we have written the decimal equivalent of the count under each section of the waveform.

As you can see in the figure, many states besides the desired ones are present on the counter outputs. These extra states are present when the counter is operated at any frequency, but for many low-frequency applications such as clocks, they are not a problem. For many high-frequency applications, however, these extra states make ripple counters unusable. If, for example, you try to decode state 0000 on the 74LS93 we have used here, the waveforms in Figure 8-55 tell you that the decoder output will be asserted four times during a count sequence, rather than once as desired. At low-frequency operation the width of the glitches is very short compared to the actual state times, so the glitches often don't matter. At higher operating frequencies the glitches and the state times are more nearly the same, so the output count sequence is not meaningful. In the next chapter we discuss synchronous devices and circuits which for the most part do not have this problem. We suggest that if you have a logic analyzer available, you use it to observe the outputs of the circuits we discuss there. Besides giving you a more thorough understanding of the operation of the circuits, this will give you more proficiency in using a logic analyzer.

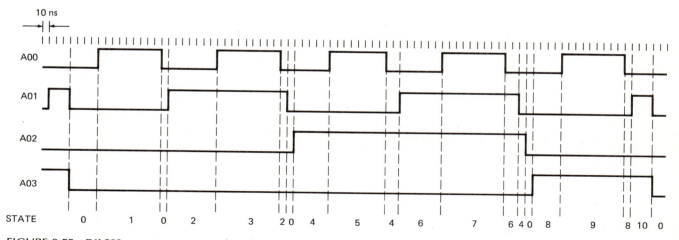

FIGURE 8-55 74LS93 count sequence showing unwanted states caused by propagation delay through flip-flops.

CHECKLIST OF IMPORTANT TERMS AND CONCEPTS IN THIS CHAPTER

Hazard, glitch, spike
Cross-coupled NAND gate latch
Set and reset state
Indeterminate, prohibited, invalid state
Latched state
Debounced signal
Latch with enable input
Enable or strobe input
D latch, D flip-flop
Edge-clocked, edge-triggered devices
Asynchronous or direct input
JK flip-flop
Toggle
Pulse- or level-triggered devices

T flip-flop
Cascading counters
Counter modulo
Ripple or asynchronous counter
Decade or BCD counter
Oscillator
Display multiplexing
Logic analyzer
Logic analyzer pod
Timing diagram format
State table format
Threshold or reference voltage
Internal and external clock signals
Trigger signal
Word recognizer circuit
Pretrigger recording
Posttrigger recording

REVIEW QUESTIONS AND PROBLEMS

1. Draw a timing diagram for the circuit in Figure 8-56. Assume the inverters and the AND gate each have a maximum propagation delay of 10 ns.

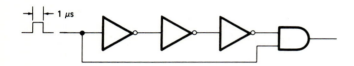

FIGURE 8-56   Circuit for Problem 1.

2. *a.* The circuit in Figure 8-57 is often called an *edge detector.* Draw a timing diagram for the circuit to see if you can determine why. Assume the propagation delay time of each of the inverters is 10 ns, and the propagation delay of the XOR gate is 15 ns.
   b. Using the timing diagram from part *a*, compare the output frequency of the circuit with the input frequency.

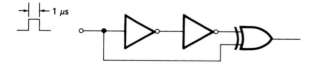

FIGURE 8-57   Circuit for Problem 2.

3. The circuit in Figure 8-58 contains a potential hazard.
   a. Use a Karnaugh map to determine the signal transitions which may cause the hazard.
   b. Try each of these signal transitions to determine which transition produces the hazard.
   c. Use the Karnaugh map to help you redesign the circuit so that the hazard is eliminated.

4. *a.* What logic level is present on the Q output of a latch or flip-flop when it is in the "set" condition?

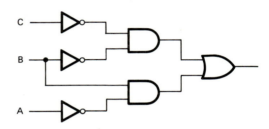

FIGURE 8-58   Circuit for Problem 3.

   b. What level is present on the Q output when a latch or flip-flop is "reset"?

5. *a.* Write the truth table for the simple latch circuit in Figure 8-59*a*.
   b. Draw the waveform that will be produced on the Q output of this latch circuit by the input waveforms shown in Figure 8-59*b*.
   c. Once this latch is in the set condition, what is the only way it can be changed to the reset state?

6. *a.* Write a truth table for the simple latch circuit in Figure 8-60.
   b. Use the truth table you wrote in part *a* to determine which input of the circuit is the set input and which is the reset input.

7. You have been assigned the task of designing a simple alarm system for a store which has six windows and two doors. Each window has a metal foil strip glued on it such that if the window is broken, the foil will tear and break the connection through the foil. Each door has a magnetic sensor, which is a short circuit if the door is closed and becomes an open circuit if the door is opened. Design a simple circuit which will assert a logic high on the input of the alarm controller if any of the foil strips are broken or a door is opened. Once a foil strip is broken or a door is opened, the alarm output should stay on until a

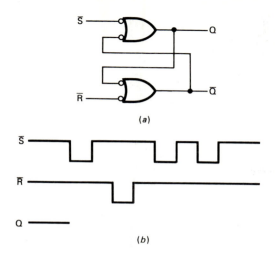

FIGURE 8-59　Problem 5. *(a)* Circuit. *(b)* Input waveforms.

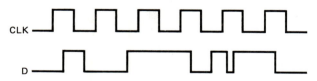

FIGURE 8-60　Circuit for Problem 6.

separate reset button is pressed. *Hint:* To simplify the logic, connect the sensors in series.

8.　*a.*　Draw the schematic symbol used to represent a D latch.
　　b.　Write the truth table for a D latch, assuming the enable input is active high.
　　c.　Show the waveform that will be produced on the Q output of a D latch by the input waveforms shown in Figure 8-61.
　　d.　The data sheet for a 74HC75 indicates that the D latches in this device have a minimum setup time of 20 ns. Describe where this parameter is measured from and what it tells you.
　　e.　The data sheet for a 74HC75 shows that the D latches in that device have a maximum hold time of 0 ns. Describe where this parameter is measured from and what it tells you.

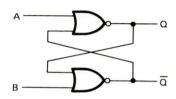

FIGURE 8-61　Waveforms for Problem 8.

9.　Figure 8-62*a* shows the IEEE symbol for a D latch and the IEEE symbol for a D flip-flop.
　　a.　Use the symbols to help you describe the difference between the two devices.

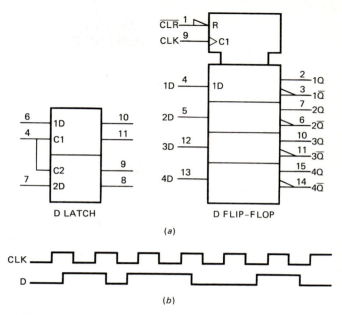

FIGURE 8-62　*(a)* IEEE symbols for a D latch and a D flip-flop. *(b)* D and CLK waveforms for Problem 9.

　　b.　Write a truth table for the D flip-flop. In the truth table, make sure you indicate which edge of the clock signal updates the Q output.
　　c.　Given the D and CLK waveforms shown in Figure 8-62*b*, draw a timing diagram which shows the waveform these inputs will produce on the Q output of a D latch. On the same diagram, show the waveform these inputs will produce on the Q output of a positive-edge-triggered D flip-flop. Compare the waveforms and note the vast difference between them.
　　d.　For a D flip-flop, where are the setup and hold times measured from?
　　e.　Why is it important to make sure that the input D signal doesn't change during the setup and hold time of a flip-flop?

10.　Figure 8-63*a* shows the traditional and the IEEE symbols for a JK flip-flop with direct preset and clear inputs.
　　a.　Write a truth table for the device represented by these symbols. In the truth table make sure to indicate which edge of the CLK signal activates the device.
　　b.　Draw a timing diagram showing the waveform that will be produced on the Q output of this flip-flop by the J, K, and CLK waveforms shown in Figure 8-63*b*. Assume Q is initially low.
　　c.　Draw a timing diagram showing the waveform that will be produced on the Q output of a positive-edge-triggered JK flip-flop by the J, K, and CLK waveforms in Figure 8-63*b*.

11.　*a.*　Draw a circuit diagram showing how a D flip-flop can be converted to a T flip-flop.
　　b.　Draw a timing diagram showing the output waveform the circuit will produce when clocked. Assume a positive-edge-triggered D flip-flop.

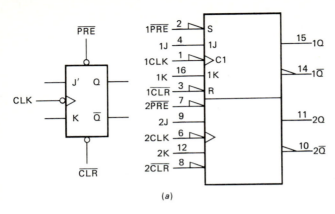

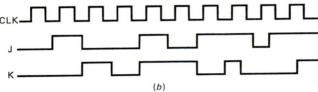

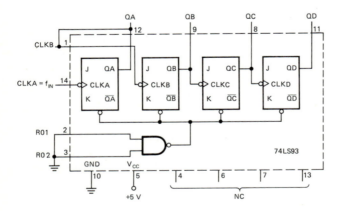

FIGURE 8-63 *(a)* IEEE/IEC symbol for JK flip-flop with preset and clear inputs. *(b)* Waveforms for Problem 10.

12. *a.* Draw a circuit showing how a JK flip-flop can be connected so that it acts as a T flip-flop.
    b. Draw a timing diagram for a JK flip-flop connected as a T flip-flop. Assume the flip-flop is negative-edge-triggered.

13. Figure 8-64 shows the circuit for a 4-bit binary counter made with a 74LS93.
    a. Draw a timing diagram for the circuit showing the input clock signal and the four Q output signals. Assume the flip-flops in the 74LS93 change on the negative edge of the CLK signal. Under each state, write its decimal value.
    b. For an input clock frequency of 480 kHz, calculate the frequency of the signal on each of the Q outputs.

FIGURE 8-64 Four-bit binary counter circuit for Problem 13.

14. What is the final output frequency of an eight-stage binary divider which has an input frequency of 10,240 Hz?

15. What is the maximum count for an 8-bit binary counter, for a 12-bit binary counter, and for a 16-bit binary counter?

16. *a.* Draw a circuit showing how a 74LS93 can be connected to make a modulo-12 counter.
    b. Draw a timing diagram showing the CLK and QD waveform for this counter.
    c. Calculate the duty cycle of the QD output waveform.

17. Draw a circuit showing how a 74LS93 can be connected to make a modulo-14 divider.

18. *a.* Draw a circuit showing how a 74LS93 can be connected to function as a modulo-6 divider.
    b. Draw a timing diagram showing the CLK and Q waveforms for the circuit.
    c. Which Q output has a signal which is one-sixth the frequency of the input CLK signal?

19. Show the logic gate circuits you would use to decode the following states from a 4-bit binary counter. Assume an active high decoder output is desired.
    a. State 4.
    b. State 9.
    c. State 13.

20. *a.* Show how two 74LS93 counters can be connected to divide an input clock signal by 24.
    b. Assuming the clock signal applied to the circuit in part *a* has a frequency of one pulse per hour, design a decoder which will output a high for 1 h starting at 12 o'clock.

21. Describe the symptoms you would observe if the QC output of the counter circuit in Figure 8-64 became shorted to ground.

22. Describe how a crystal oscillator and frequency dividers can be used to produce all the different frequencies needed for a simple electronic organ.

23. *a.* Draw a block diagram for a simple digital clock made with counters, decoders and seven-segment displays.
    b. Draw a circuit that you could add to the basic digital clock circuit in part *a* to toggle an a.m./p.m. LED on and off every 12 h.
    c. Show circuitry you could add to the basic clock in part *a* to produce an active high alarm signal when a preset count is present on the outputs of the counters.

24. Standard 7447s could be used for the decoder/drivers for the seconds and minutes displays in the digital clock circuit in Figure 8-38. A 7447 by itself, however, is not enough to provide the decoding for the hours display. Figure 8-65 shows a "black box" which represents the decoder needed to interface a divide-by-12 counter to a 7447 and a transistor to drive the hours display. As a review of Karnaugh maps, design the logic circuits that must be in the black box so that as the counter counts from 0000 to 1011, the two seven-segment displays show 1, 2, 3, 4, 5, 6, 7, 8, 9, 10, 11, 12.

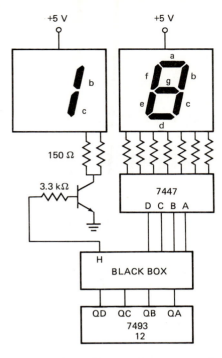

FIGURE 8-65 An hours-display driver circuit for a digital clock for Problem 24.

25. *a.* Describe in general terms how seven-segment LED displays are multiplexed.
    b. What is the major advantage of multiplexing LED displays?

26. *a.* What is the purpose of the comparators on the data inputs of a logic analyzer?
    b. What is the purpose of the memory in a logic analyzer?
    c. What is the purpose of the clock signal in a logic analyzer?
    d. For what type of measurements do you use an external clock signal to tell a logic analyzer to take samples of the input data signals?
    e. For what type of measurements do you use an internal clock signal to tell a logic analyzer to take samples of the input data signals?

27. *a.* What is the purpose of the trigger signal in a logic analyzer?
    b. Describe the operation of a word recognizer in a logic analyzer.
    c. How is it possible for a logic analyzer to display events which occurred before a trigger occurred?
    d. Describe what is meant by the terms *pretrigger*, *posttrigger*, and *center-trigger* recording.

28. *a.* Describe how you would set the clock, trigger, and display to see the sequence of states that an 8-bit counter steps through as it is clocked. Assume that the counter increments on the positive edge of the clock.
    b. Describe what you would add to the basic setup to see a glitch that may be present on the outputs of the counter.

29. Describe how you would set the clock, trigger, and display of a logic analyzer to measure the CLK to Q propagation delay times of the latches in a 74LS373 octal latch device.

30. Describe how the compare feature of a logic analyzer can be used to help troubleshoot a malfunctioning digital system.

31. Describe how you would set up a logic analyzer to see the details of a glitch on the outputs of a counter circuit.

32. Why are asynchronous counters not suitable for many high-frequency applications?

33. *a.* Given the timing parameters shown below, draw a detailed timing diagram for the data multiplexer-demultiplexer circuit in Figure 8-17*a*.
    b. Use the detailed timing diagram you drew in part *a* to determine if the timing of the signal applied to the $\overline{G}$ input of the 74LS259 obeys the setup and hold time requirements for the device.

Timing parameters:

74LS193	From UP or DOWN input to Q output:	
	$t_{PLH} = 38$ ns	$t_{PHL} = 47$ ns

74LS151	From DATA input to Y output:	
	$t_{PLH} = 32$ ns	$t_{PHL} = 26$ ns

74LS151	From ADDRESS input to Y output:	
	$t_{PLH} = 43$ ns	$t_{PHL} = 30$ ns

74LS04	$t_{PLH} = 22$ ns	$t_{PHL} = 15$ ns

74LS259B

Pulse duration	$\overline{G}$ low	$t_w = 17$ ns
Pulse duration	$\overline{CLR}$ low	$t_w = 10$ ns
Setup time	Data before	$t_{su} = 20$ ns $\overline{G} \uparrow$
Setup time	Address before	$t_{su} = 17$ ns $\overline{G} \uparrow$
Setup time	Address before	$t_{su} = 0$ ns $\overline{G} \downarrow$
Hold time	Data after	$t_h = 0$ ns $\overline{G} \uparrow$
Hold time	Address after	$t_h = 0$ ns $\overline{G} \uparrow$

CHAPTER

9 Synchronous Devices, Circuits, and Systems

Figure 8-30 in the last chapter shows the clock and output waveforms for a 74LS93 asynchronous counter. As shown in the figure, the output waveforms have out-of-sequence states. For example, state 0000 is present at several points in the waveforms rather than just at the start of the count sequence. These brief, out-of-sequence states are caused by the fact that the flip-flop outputs do not all change at the same time. Remember from the discussions in the last chapter that the effect of the input clock signal "ripples down through" the flip-flops in an asynchronous counter such as the 74LS93.

For some applications, such as those we described in the last chapter, the brief, out-of-sequence states do not cause any problems. For many applications, however, these extra states would cause severe problems. In those applications we use *synchronous* devices and circuits.

A synchronous circuit is designed so that all the devices in the circuit are clocked at the same time. Therefore, all the outputs change at very nearly the same time. In this chapter we show you how to analyze and work with a wide variety of synchronous circuits.

OBJECTIVES

At the conclusion of this chapter you should be able to:

1. Predict the sequence of states that a synchronous circuit made from D, T, or JK flip-flops will step through as it is clocked.

2. Describe the operation of a synchronous counter device and draw the input/output waveforms for it.

3. Draw a schematic for a circuit which uses MSI synchronous counter devices to divide an input clock signal by any specified number and draw the input/output waveforms for the circuit.

4. Describe the function and operation of each part of a frequency counter made with MSI devices.

5. Describe the operation of a multiplexed display circuit.

6. List a series of steps you would take to troubleshoot a digital instrument such as a frequency counter.

7. Describe how control signals can be multiplexed into a digital instrument such as an LSI frequency counter.

8. Describe the operation of SISO, SIPO, PISO, and PIPO shift registers.

9. Use the data sheet for an MSI shift register to determine the connections for a desired mode of operation.

10. Calculate the delay produced by a SISO shift register with a given number of stages and a given clock frequency.

11. Predict the output waveforms that will be produced by a variety of shift register counter circuits such as a ring counter and a Johnson counter.

12. Describe how a ring counter can be used as part of a keyboard encoder circuit.

13. Draw a diagram showing the format in which asynchronous serial data is often sent.

14. Describe how shift registers are used in UARTs to send and receive asynchronous serial data.

ANALYZING SYNCHRONOUS COUNTER CIRCUITS AND DEVICES

Synchronous Counter Circuits Made with D or JK Flip-Flops

D FLIP-FLOP EXAMPLE

Figure 9-1 shows a circuit for a 2-bit binary synchronous counter made with D flip-flops. In this circuit the clock signal is connected in parallel to the CLK inputs of the two flip-flops. This means that when a rising edge occurs on the CLK signal, the outputs of the two flip-flops will be updated at the same time, or *synchronously.*

As you know from the discussion in the previous chapter, the level that will be present on the Q output of a D flip-flop after a clock pulse is the level that was present on the D input just before the rising edge of the clock pulse. In the circuit in Figure 9-1, the signals for the D

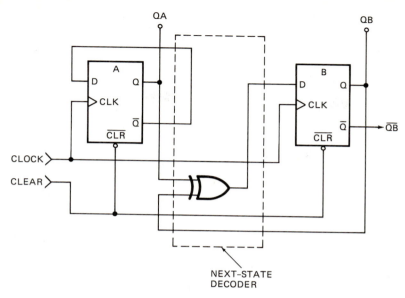

NEXT-STATE
DECODER

FIGURE 9-1   Circuit for 2-bit synchronous counter made with D flip-flops.

inputs are produced by decoding the flip-flop output signals. The circuitry used to do this decoding is often called the *next-state decoder*. The name comes from the fact that the circuit decodes the present-state output signals to produce the D input signals that will cause a transition to the desired next state when a CLK pulse occurs.

The first step in determining the count sequence for a circuit such as this is to write the Boolean expression for each flip-flop input. For the circuit in Figure 9-1, the expression for the A flip-flop is DA = $\overline{QA}$ and the expression for the B flip-flop is DB = QA $\oplus$ QB.

The next step in determining the count sequence is to draw two tables such as those shown in Figure 9-2. In

STATE	QB	QA	CLOCK	DB	DA
0					
1			1		
2			2		
3			3		
0			4		

FIGURE 9-2   Template for Q and D tables used to analyze the count sequence of a synchronous circuit made with D flip-flops.

the Q table you write the QB and QA values for each state as you determine them. In the D table you write the logic levels that each set of Q levels will produce on the DB and DA inputs. The Boolean expressions you wrote for the flip-flop inputs help you to determine the D levels that will be produced by the present-state levels on the Q outputs. The principle here is that if you know the level on D at a particular time, you know that, on the next rising edge of the clock, the level on D will be transferred to the Q output. Determining the D values produced by one output state then allows you to easily predict the state that the outputs will go to when the circuit is next clocked.

To fill in the charts such as those in Figure 9-2, start with a state that you know is part of the count sequence. The circuit in Figure 9-1 has a direct clear input, so a logical state to start the table with is the QB = 0 and QA = 0 state produced by asserting the $\overline{CLR}$ input.

If QA = 0, $\overline{QA}$ will be a 1. According to the expression for DA, DA will then be a 1, so write a 1 under DA on the top line of the table. With QA = 0 and QB = 0, the XOR gate will produce a 0 on the DB input, so write a 0 under the DB column in your table.

Now, when the positive edge of a clock pulse occurs, the 1 on the D input of the A flip-flop will be transferred to the QA output. At the same time, the 0 on the D input of the B flip-flop will be transferred to the QB output. After a clock pulse, then, the new output state will be QB = 0 and QA = 1. You write these values in the second line of the Q columns in the table.

To predict the state that the circuit outputs will go to when the circuit is clocked again, you again use the Boolean expressions to predict the D levels that will be produced by the next-state decoder. Since QA = 1, $\overline{QA}$ and DA are both 0's. The 1 on the QA output and the 0 on the QB output are decoded by the XOR gate to produce a 1 on the D input of the B flip-flop. DB = 1 and DA = 0, so on the positive edge of the next clock pulse, the QB output will go to 1 and the QA output will go to 0.

An output state of 10 produces a 1 on the DA input and a 1 on the DB input. Therefore, on the rising edge of the next clock pulse, the outputs will go to 11 or 3. QB = 1 and QA = 1 produces a 0 on the D input of the A flip-flop and a 0 on the D input of the B flip-flop, so on the next clock pulse, the outputs transition back to QB = 0 and QA = 0. The circuit then steps through the count sequence shown in Figure 9-3. If additional clock pulses are applied to the circuit, the outputs will step through the count sequence again.

To summarize, then, the principle of a synchronous circuit is that the output signals during the present state are decoded to produce the input signals re-

STATE	QB	QA	CLOCK	DB	DA
0	0	0	1	0	1
1	0	1	2	1	0
2	1	0	3	1	1
3	1	1	4	0	0
0	0	0			

FIGURE 9-3 Completed Q and D tables for 2-bit synchronous counter in Figure 9-1.

quired to cause a transition to the desired next state. To predict the next state of a circuit such as that in Figure 9-1, then, you use the next-state decoder expressions to determine the input signals that will be produced by the present-state Q levels. If you know the levels that are present on the inputs of the flip-flops, you can easily predict the levels that will be present on the outputs after a clock pulse. As another example of this process let's look at a synchronous counter circuit built with JK flip-flops.

JK FLIP-FLOP SYNCHRONOUS COUNTER CIRCUIT

Figure 9-4 shows a circuit for a 4-bit binary synchronous counter made with JK flip-flops. Note that in this circuit the input clock signal is also connected in parallel to all the flip-flop CLK inputs. As indicated by the bubble and the wedge on each CLK input, these flip-flops are negative-edge-triggered. On the falling edge of the input clock signal, then, the outputs of all the flip-flops will be updated, based on the levels on their J and K inputs. As in the previous circuit example, the present-state Q output signals are decoded to produce the flip-flop input signals required to cause the flip-flops to transition to the desired next states when a clock pulse occurs. The next-state decoder in the circuit in Figure 9-4 consists of the wire connecting the JA and KA inputs to +5 V, the wire connecting the QA output to the JB and KB inputs, and the two AND gates.

Again, the first step in determining the count sequence for the circuit is to write the Boolean expression for each flip-flop input. The J and K inputs of each flip-flop are tied together, so you only have to write one expression for each flip-flop. The four expressions are as follows:

$$JKA = 1$$

$$JKB = QA$$

$$JKC = QA \text{ AND } QB$$

$$JKD = QA \text{ AND } QB \text{ AND } QC$$

Since the J and K inputs of each flip-flop are tied together, J and K must both be high or J and K must both be low. Remember from the truth table for a JK flip-flop in Chapter 8 that, if J and K are both high, the Q output will toggle when a clock edge occurs. If J and K are both low, the Q output will remain the same when a clock edge occurs. For each of the flip-flops in Figure 9-4, then, there are only two possibilities:

Toggle if its JK inputs are high.

Keep the same states on Q and $\overline{Q}$ if its JK inputs are low.

These are the rules you will use to predict the level on each flip-flop after a clock pulse.

The J and K inputs of the A flip-flop in Figure 9-4 are permanently tied high so that the output of the A flip-flop will toggle on the negative edge of each clock pulse. As you will see when we work through the count sequence, this flip-flop then produces the LSB of the count.

The J and K inputs of the B flip-flop are tied to the QA output of the A flip-flop, so the QB output will toggle if QA is high when a negative clock edge occurs. If QA is low when a negative clock edge occurs, the QB output will stay in the same state.

According to the equations above, the J and K inputs of the C flip-flop will be high if the QA output AND the QB output are both high. The QC output then will only toggle if both QA and QB are high. Otherwise, QC will stay the same.

In the same manner, the equation for JKD above tells you that the D flip-flop will only toggle if QA is high AND QB is high AND QC is high. For all other combinations of QA, QB, and QC, the QD output will remain the same.

The next step in determining the count sequence of the circuit is to set up a table where you will write the sequence of Q output states and a table for the J and K

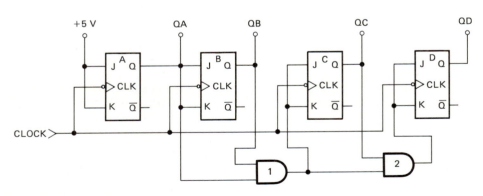

FIGURE 9-4 Circuit for 4-bit binary synchronous counter.

STATE	QD	QC	QB	QA	CLOCK	JKD	JKC	JKB	JKA
0	0	0	0	0	1				
1					2				
2					3				
3					4				
4					5				
5					6				
6					7				
7					8				
8					9				
9					10				
10					11				
11					12				
12					13				
13					14				
14					15				
15					16				
0									

FIGURE 9-5   Q and JK template for analyzing synchronous circuits made with JK flip-flops connected as T flip-flops.

values that will be produced by each output state. Figure 9-5 shows how to make a table for this example.

As in the previous example, let's assume the Q outputs are initially all 0's as shown by the top line of the table in Figure 9-5. Then you use the equations above to determine the JK levels that will be produced by these Q levels. For outputs of 0000 JKA will be high, JKB will be low, JKC will be low, and JKD will be low, as shown in the top line of the JK columns of Figure 9-6.

Once you know the levels that are present on the JK inputs of each flip-flop, you can use the rules for a JK flip-flop to predict the Q output level that each flip-flop will have after it is clocked. Remember, the rules for this circuit are simplified to toggle or no toggle because J and K are tied together.

The 1 on JKA will cause the QA output to toggle from a 0 to a 1 when clocked. Since the JKB, JKC, and JKD inputs are all 0's, the outputs of these flip-flops will all remain the same when the flip-flops are clocked. After a clock pulse, then, Q outputs will be 0001 as shown in the line labeled state 1 in Figure 9-6.

STATE	QD	QC	QB	QA	CLOCK	JKD	JKC	JKB	JKA
0	0	0	0	0	1	0	0	0	1
1	0	0	0	1	2	0	0	1	1
2	0	0	1	0	3	0	0	0	1
3	0	0	1	1	4	0	1	1	1
4	0	1	0	0	5	0	0	0	1
5	0	1	0	1	6	0	0	1	1
6	0	1	1	0	7	0	0	0	1
7	0	1	1	1	8	1	1	1	1
8	1	0	0	0	9	0	0	0	1
9	1	0	0	1	10	0	0	1	1
10	1	0	1	0	11	0	0	0	1
11	1	0	1	1	12	0	1	1	1
12	1	1	0	0	13	0	0	0	1
13	1	1	0	1	14	0	0	1	1
14	1	1	1	0	15	0	0	0	1
15	1	1	1	1	16	1	1	1	1
0	0	0	0	0					

FIGURE 9-6   Completed Q and JK table for circuit in Figure 9-4.

To predict the levels that will be produced on the Q outputs by the next clock pulse, you determine the levels that the new Q levels of 0001 produce on the JK inputs. As you determine the level present on each JK input, it is important that you write them in a table such as the one on the right side of Figure 9-6. If you try to remember these JK values in your head, it is very easy to make a mistake and get incorrect results from there on.

Again, once you determine the levels present on each JK input, you can simply use the JK rules to predict the Q levels that will be produced when the next clock pulse occurs.

With 0001 on the Q outputs, JKA as always is high. The QA output is connected to the JKB input, so the high on QA will also be present on the JKB input. Since QB is low, the outputs of both AND gates will be low. Therefore, both JKC and JKD will be low. The second line on the right-hand side of the table in Figure 9-6 summarizes these JK conditions.

When the next clock pulse occurs, JKA is high, so QA will toggle from a 1 to a 0. JKB is high, so QB will toggle from a 0 to a 1. JKC and JKD are low, so QC and QD will both remain 0's. After a clock pulse, then, the Q outputs will have 0010 on them as shown by the line labeled state 2 in Figure 9-6.

To predict the next output state, you again determine the levels that will be produced on the JK inputs by Q outputs of 0010. Then, as before, you use the JK rules to predict the effect that these JK levels will have on the outputs when a clock pulse occurs. JKA is, of course, high, so QA will toggle from 0 to a 1 on the next clock pulse. With Q outputs of 0010, QA is low, so JKB, JKC, and JKD will all be low. Therefore, the QB, QC, and QD outputs will remain the same when the next clock pulse occurs. After the next clock pulse, then, the outputs will contain 0011, which is shown as state 3 in Figure 9-6.

The determination of the JK inputs for the next transition shows the effect of one of the AND gates. In state 3 (0011), the high on QA and the high on QB assert the inputs of AND gate 1 to produce a high on JKC. As shown on the state 3 line in Figure 9-6, then, JKA, JKB, and JKC are all high, so QA, QB, and QC will all toggle when the next clock pulse occurs. However, with 0011 on the Q outputs, AND gate 2 will assert a low on JKD. Therefore, QD will not toggle when a clock pulse occurs. As shown in Figure 9-6, the outputs will go from 0011 to 0100 when the next clock pulse occurs.

For practice, work your way through the transitions from states 4 to 5, 5 to 6, and 6 to 7 in Figure 9-6. The next interesting transition we want to talk about here is the one from state 7 to state 8.

In state 7 (0111), the QA, QB, and QC outputs are all high. These three high signals will cause the outputs of the two AND gates to be high. The result of this is that the JKC and JKD inputs will be asserted high. Since JKA and JKB are also high, all the flip-flops will toggle on the next clock pulse. The resultant Q outputs will be 1000.

Another interesting transition to look at in Figure 9-6 is the one from state 15 (1111) to state 0. In state 15 all the Q outputs are 1's. According to the Boolean expressions above, then, all the JK inputs will be high. Since the JK inputs all have 1's on them, the Q outputs will all toggle on the next clock pulse. In this case, the 1111 on the outputs will toggle to 0000, which brings the outputs back to the start of the count sequence again.

As you can see from the completed count sequence in Figure 9-6, the circuit in Figure 9-4 is a 4-bit binary upcounter. The difference between this counter and a ripple counter such as the 74LS93 is that the outputs are all updated at the same time, or synchronously, rather than one after the other, as they are in an asynchronous counter such as the 74LS93.

Again, the main principle of a synchronous counter is that the levels on the Q outputs during one state are used to produce signals for the D or J and K inputs such that when the circuit is clocked, it transitions to the desired next state. As we will show you in detail in a later chapter, you can easily design a synchronous counter circuit using essentially the reverse of the technique you use to analyze one. In general, the design process for a circuit with JK flip-flops involves the following steps:

1. Making a table which lists the desired sequence of output states.

2. Making another table which shows the levels required on each J input and each K input to produce the desired transitions.

3. Designing the next-state decoder by developing a simplified Boolean expression for each J input and each K input based on the present-state Q output levels.

4. Implementing the simplified Boolean expressions with gates.

The AND gates in Figure 9-4 are an example of this. To help you with the design discussion in Chapter 12, keep a little bit of this design process in the back of your mind as you analyze the final example in this section.

JK FLIP-FLOP CIRCUIT WITH INDEPENDENT J AND K INPUTS

Figure 9-7 shows an interesting synchronous circuit made with JK flip-flops. As you can see, the next-state decoder in this circuit consists simply of some wires connecting outputs to inputs. Since the J and K inputs are

independent in this circuit, you have to write a Boolean expression for each. The four expressions are:

$$JA = \overline{QB}$$

$$KA = QB$$

$$JB = QA$$

$$KB = \overline{QA}$$

Figure 9-8 shows you how to set up the tables for this circuit. Note that because the J and K inputs are not tied together, you have to make separate columns for each J input and each K input as shown in Figure 9-8a.

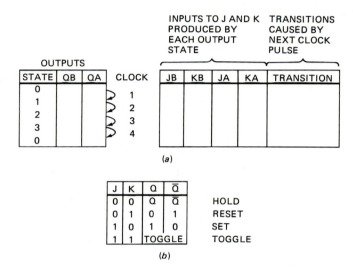

OUTPUTS				INPUTS TO J AND K PRODUCED BY EACH OUTPUT STATE				TRANSITIONS CAUSED BY NEXT CLOCK PULSE
STATE	QB	QA	CLOCK	JB	KB	JA	KA	TRANSITION
0			1					
1			2					
2			3					
3			4					
0								

(a)

J	K	Q	$\overline{Q}$	
0	0	Q	$\overline{Q}$	HOLD
0	1	0	1	RESET
1	0	1	0	SET
1	1	TOGGLE		TOGGLE

(b)

FIGURE 9-8 (a) Template for Q and JK tables used to analyze synchronous circuit with independent J and K inputs. (b) JK flip-flop truth table refresher.

To fill in the tables in Figure 9-8a, let's again assume the count sequence starts with QB = 0 and QA = 0. Since this circuit has no external decoding gates, predicting the J and K levels produced by these Q levels is quite easy. According to the expressions above, Q levels of 00 will give JB = 0, KB = 1, JA = 1, and KA = 0, as shown in the top line of the JK table in Figure 9-9.

Once you have determined the levels present on each J and each K input, the next step is to use the JK truth table to predict the effect these inputs will have when a clock pulse occurs. To refresh your memory, Figure 9-8b shows the truth table for a JK flip-flop. According to the truth table, the high on KB will cause the QB output to stay a 0 and the high on the JA input will cause the QA output to go to a 1. The Q outputs for state 1 then are QB = 0 and QA = 1.

If you work your way through the circuit or use the expressions above with Q levels of 01, you should determine that the levels they produce on the J and K inputs are JB = 1, KB = 0, JA = 1, and KA = 0, as shown in the second line of the JK table in Figure 9-9. The truth table for a JK flip-flop tells you that with these levels on

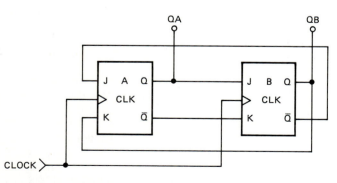

FIGURE 9-7 JK flip-flop synchronous circuit with independent J and K inputs.

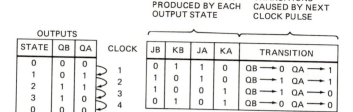

OUTPUTS				INPUTS TO J AND K PRODUCED BY EACH OUTPUT STATE				TRANSITIONS CAUSED BY NEXT CLOCK PULSE
STATE	QB	QA	CLOCK	JB	KB	JA	KA	TRANSITION
0	0	0	1	0	1	1	0	QB → 0 QA → 1
1	0	1	2	1	0	1	0	QB → 1 QA → 1
2	1	1	3	1	0	0	1	QB → 1 QA → 0
3	1	0	4	0	1	0	1	QB → 0 QA → 0
0	0	0						

FIGURE 9-9   Completed Q and JK tables for circuit in Figure 9-7.

their inputs, the QB output will go to a 1 and the QA output will stay a 1 when a clock pulse occurs. This result is shown as state 2 in the Q table in Figure 9-9. Note that the number of a state in the sequence does not always correspond to the binary value of the state.

For practice, work your way through the rest of the JK table and the Q table for this circuit to see if you agree with the values in Figure 9-9. If you look carefully at the output sequence, you may recognize it as a 2-bit gray code sequence such as you use along the edge of a Karnaugh map.

The circuit in Figure 9-7 did not require any external gates to decode the Q output signals for the J and K inputs, as most JK synchronous circuits do. To analyze a circuit which does have external decoding gates, you use the same technique we have just described for the circuit in Figure 9-7. Some exercises at the end of the chapter give you more practice predicting the count sequence for a variety of synchronous circuits made with D or JK flip-flops.

MSI SYNCHRONOUS COUNTERS AND COUNTER APPLICATIONS

Introduction

Figure 9-4 shows the circuit for a 4-bit binary synchronous upcounter. A major advantage of a synchronous counter over a ripple counter is that the outputs of a synchronous counter all change at very nearly the same time. This means that an output state can be decoded with few if any glitches. Also, the fact that the clock signal does not have to ripple down through a chain of flip-flops in a synchronous counter means that synchronous counters will count correctly with much higher input clock frequencies. If you try to clock a ripple counter at too high a frequency, the least significant flip-flop may toggle twice before the most significant flip-flop gets the message to toggle once.

Fortunately, you do not have to build a synchronous counter with flip-flops and gates every time you need one. Figure 9-10 lists a few of the commonly available synchronous counters. We have shown the parts with ALS family numbers, but most of them are available in other logic families such as 74LS and 74HC. Counters identified in Figure 9-10 as decade upcounters cycle through a BCD count of 0000 to 1001. Decade up- and

SYNCHRONOUS COUNTER TYPE	PARALLEL LOAD	PART NUMBER
DECADE UP	SYNCHRONOUS	74ALS160
	SYNCHRONOUS	74ALS162
DECADE UP/DOWN	SYNCHRONOUS	74ALS168
	ASYNCHRONOUS	74ALS190
	ASYNCHRONOUS	74ALS192
	SYNCHRONOUS	74ALS568
4-BIT BINARY	SYNCHRONOUS	74ALS161
	SYNCHRONOUS	74ALS163
	SYNCHRONOUS	74ALS561
4-BIT BINARY UP/DOWN	ASYNCHRONOUS	74ALS191
	ASYNCHRONOUS	74ALS193
	ASYNCHRONOUS	74ALS569

FIGURE 9-10   Commonly available synchronous counter types.

downcounters can be connected to count from 0000 to 1001 or from 1001 to 0000, depending on the need of a specific application. Counters identified in Figure 9-10 as 4-bit binary, count through the sequence 0000 to 1111, or, if operating as a downcounter they count through the sequence 1111 to 0000.

All the counters listed in Figure 9-10 have parallel load capability. This means that a binary number applied to the data inputs of the counter will be transferred to the Q outputs when a special *load* input on the counter is asserted. The parallel load column in Figure 9-10 indicates whether the load operation is synchronous or asynchronous. If the load operation is asynchronous, the binary number applied to the data inputs will be transferred to the Q outputs as soon as the load input is asserted. For a counter with a synchronous load, the data will not be transferred to the Q outputs until the next clock pulse occurs. Either type of counter will continue counting from the loaded count when it receives additional clock pulses. To give you a better understanding of how one of these synchronous counter devices works, the next section takes a close look at the 74ALS193.

The 74ALS193 4-Bit Binary Counter

SYMBOLS AND WAVEFORMS

Figure 9-11a shows a traditional logic symbol and Figure 9-11b shows the IEEE/IEC symbol for a 74ALS193 4-bit binary synchronous counter. As you can see in the diagrams, the counter has four data inputs and four Q outputs. When the $\overline{LOAD}$ input is asserted low, the logic levels on the four data inputs will be transferred to the corresponding Q outputs. When the clear input, CLR, is asserted (high), all four Q outputs will be reset to zero regardless of the signal levels present on the data, load, and clock inputs.

The 74ALS193 has two clock inputs labeled UP and DOWN. If the DOWN input is tied high and a clock signal is connected to the UP input as shown in Figure 9-12a, the counter will count up from 0000 to 1111 and back to 0000 as it receives clock pulses. Alternatively, if the UP input is made high and a clock signal is con-

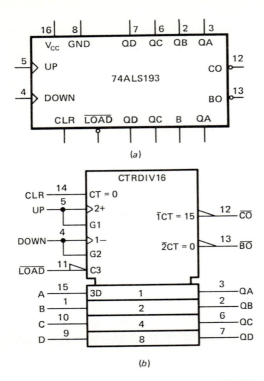

(a)

(b)

FIGURE 9-11 Logic symbols for 74ALS193 4-bit binary up-and-down counter. (a) Traditional symbol. (b) IEEE/IEC symbol.

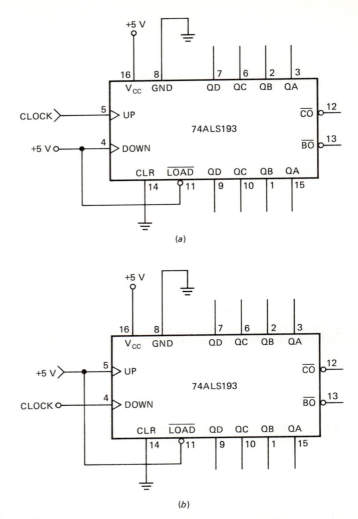

(a)

(b)

FIGURE 9-12 74ALS193 counter. (a) Connected to count up. (b) Connected to count down.

nected to the DOWN input, as shown in Figure 9-12*b*, the counter will count down from 1111 to 0000 as it receives clock pulses.

If a 74ALS193 is connected for counting up, the carry output, $\overline{CO}$, will go low when the count reaches 1111 or 15. If a 74ALS193 is connected for counting down, the borrow out, $\overline{BO}$, will go low when the output count reaches 0000. As we will show you a little later, these $\overline{CO}$ and $\overline{BO}$ signals can be used to clock the next counter in a cascaded chain of counters.

To help you understand the basic operation of the 74ALS193, Figure 9-13 shows a typical clear, load, and count sequence from a data sheet for the device. Let's work our way across these waveforms.

The 74ALS193 CLR input is asynchronous, so as soon as it is made high, the Q outputs all go low without waiting for a clock pulse. The next operation shown on the waveforms is LOAD or PRESET. Some data books refer to counters such as the 74ALS193 as *presettable*, because the Q outputs will be preset to the values present on the DATA inputs when the LOAD input is asserted. The LOAD input on the 74ALS193 is asynchronous, so as shown in Figure 9-13, the levels on the DATA inputs are transferred to the Q outputs as soon as LOAD is asserted (low). In this case the loaded count is 1101 or 13.

The next section of the waveforms in Figure 9-13 shows how the counter counts up from 13 to 15 and how it starts its count sequence over at 0. Remember that for counting up, the DOWN input is held high and the clock signal is applied to the UP input, as shown in the waveforms. The rising edge of the input clock signal then

causes the counter outputs to change states. For the example in Figure 9-13, the first rising edge causes the output count to increment to 14 and the next rising edge causes the count to increment to 15. When the clock signal returns low during state 15, the 74ALS193 asserts its carry output, $\overline{CO}$. On the next rising edge of the clock signal, the output count rolls over to 0000 and the $\overline{CO}$ output returns high. As you will soon see, the rising edge of this $\overline{CO}$ signal can be used to clock the UP input of another counter when cascading counters.

The final section of the waveforms in Figure 9-13 shows a count-down sequence. For counting down, the UP input is held high and the clock signal is applied to the DOWN input. Again, the rising edge of the applied clock signal causes the Q outputs to change to the next state. In the example waveforms, the first rising edge of the clock signal causes the output count to decrement from 2 to 1, and the second rising edge causes the output count to decrement from 1 to 0. When the clock signal returns low during state 0, the 74ALS193 asserts its borrow output, $\overline{BO}$. The next rising edge of the clock signal causes the count to roll over to 1111 or 15 and the $\overline{BO}$ output to return high. One of the uses of this $\overline{BO}$ signal

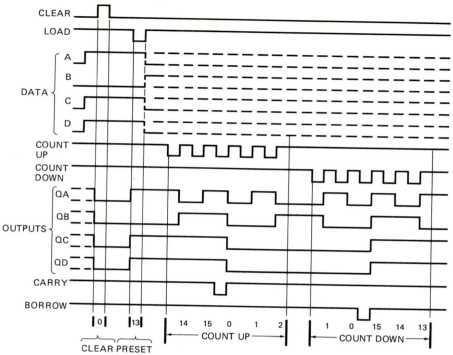

CLEAR

LOAD

DATA
- A
- B
- C
- D

COUNT UP

COUNT DOWN

OUTPUTS
- QA
- QB
- QC
- QD

CARRY

BORROW

|0| |13| 14 15 0 1 2 1 0 15 14 13

CLEAR PRESET ⟶COUNT UP⟶ ⟵COUNT DOWN⟶

SEQUENCE:

(1) CLEAR OUTPUTS TO ZERO.

(2) LOAD (PRESET) TO BINARY THIRTEEN.

(3) COUNT UP TO FOURTEEN, FIFTEEN, CARRY, ZERO, ONE, AND TWO.

(4) COUNT DOWN TO ONE, ZERO, BORROW, FIFTEEN, FOURTEEN, AND THIRTEEN.

NOTES:

(A) CLEAR OVERRIDES LOAD, DATA, AND COUNT INPUTS.

(B) WHEN COUNTING UP, COUNTDOWN INPUT MUST BE HIGH; WHEN COUNTING DOWN, COUNT–UP INPUT MUST BE HIGH.

FIGURE 9-13   Clear, load, and count sequences for 74ALS193 counter.

is to clock the DOWN input of another counter when cascading counters for longer counts.

As with most logic devices, there are some restrictions on when the input signals to a 74ALS193 can be changed with respect to each other. To acquaint you with these parameters, Figure 9-14 shows the recommended pulse widths, setup times, and hold times from a 74ALS193 data sheet. The main thing for you to remember from these parameters is that there must be 20 ns or so between operations. For example, according to these parameters, data must be stable on the data inputs for at least 20 ns before $\overline{LOAD}$ is asserted. As another example, CLR must be returned low at least 20 ns before a rising edge on the UP input or the DOWN input. Note that the device is guaranteed to count correctly with an input clock frequency as high as 30 MHz.

USING A 74ALS193 TO MAKE MODULO-n DIVIDERS

In Chapter 8 we showed you how a 74LS93 can be connected to divide an input frequency by any desired modulo. A synchronous counter such as the 74ALS193 can also be connected to divide an input clock signal by any desired modulo. As a first example, Figure 9-15a shows

a 74ALS193 circuit which divides an input clock signal by decimal 11.

The CLR input is not used in this circuit, so it is just tied low. The UP input is tied high, and the clock signal is connected to the DOWN input so that the counter will count down. The real key to the operation of the circuit is the connection from the borrow output, $\overline{BO}$, to the $\overline{LOAD}$ input. Remember from the waveforms in Figure 9-13, that when the counter counts down to 0000, the $\overline{BO}$ output goes low. With the connection in Figure 9-15a, then, $\overline{LOAD}$ will be asserted low when $\overline{BO}$ goes low. This will cause the signal levels on the data inputs (DCBA) to be transferred to the Q outputs. For the circuit in Figure 9-15a we have wired 1011 or decimal 11 on the data inputs, so these are the values that will be loaded on the Q outputs. The rising edge of the next pulse on the DOWN input will decrement this count to 1010, the next rising edge will decrement it to 1001, etc. Every time the count gets down to 0, $\overline{BO}$ will go low and load 1011 on the outputs again.

To see how this circuit functions as a modulo-11 frequency divider, you need to work your way through the waveforms in Figure 9-15b. There are three important points to observe in these waveforms.

RECOMMENDED OPERATING CONDITIONS FOR 74ALS193		MIN MAX	UNIT
f_{clock} CLOCK FREQUENCY		0 30	MHz
t_W PULSE DURATION	CLR HIGH LOAD LOW UP OR DOWN HIGH OR LOW	10 20 16.5	ns
t_{su} SETUP TIME	DATA BEFORE LOAD↑ CLR INACTIVE BEFORE UP↑OR DOWN↑ LOAD INACTIVE BEFORE UP↑OR DOWN↑	20 20 20	ns
t_h HOLD TIME	DATA AFTER LOAD↑ UP HIGH AFTER DOWN↑ DOWN HIGH AFTER UP↑	5 0 0	ns

FIGURE 9-14   Recommended operating conditions for 74ALS193 counter.

First, the $\overline{BO}$ is connected directly to $\overline{LOAD}$, so as soon as $\overline{BO}$ goes low, $\overline{LOAD}$ will be made low. After a few nanoseconds propagation delay, this will load 1011 on the Q outputs and make $\overline{BO}$ go high again. The result of this is that $\overline{BO}$ will only be low for 30 or 40 ns.

Second, if you count the number of different states

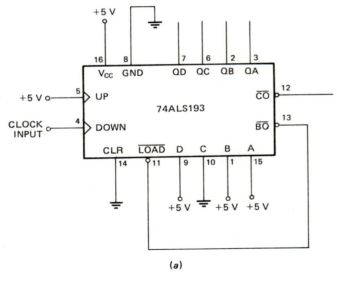

(a)

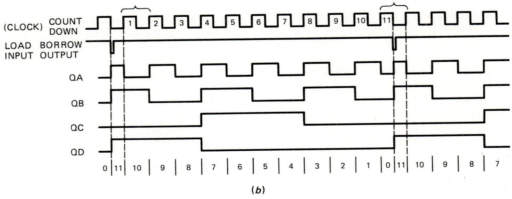

(b)

FIGURE 9-15   74ALS193 modulo-11 divider. (a) Circuit. (b) Waveforms.

that are present on the outputs during one count cycle, you will find that there are 12 states, 0 through 11. However, since state 0 and state 11 are each only present for one half of an input clock cycle, the total number of input clock cycles per count cycle is only 11. As you can see in the waveforms, the $\overline{BO}$ output goes low once for each 11 input clock pulses.

Third, the QC and QD outputs both produce signals with frequencies equal to the frequency of the input clock divided by 11. For some applications, these signals may be more useful than the 30- to 40-ns-wide $\overline{BO}$ pulses. Note that this circuit functions very well as a frequency divider, but it cannot ordinarily be used as a counter because 0000 and 1011 are both present on the outputs during one clock cycle.

The circuit in Figure 9-15a can be programmed to divide an input clock signal by any modulo between 2 and 15 by simply hardwiring the binary equivalent of the desired modulo on the data inputs. If you have a logic analyzer, you can connect it to the clock and Q signals to observe the count sequences and timing waveforms for different modulos

You can cascade two or more 74ALS193s for counting up or down by simple connecting the $\overline{BO}$ from one counter to the DOWN input of the next, and connecting $\overline{CO}$ from one counter to the UP input of the next, as shown in Figure 9-16. The $\overline{BO}$ and $\overline{CO}$ signals act as clock signals for the next counter. If the $\overline{LOAD}$ inputs of all the cascaded counters are simply tied high, the counters will count through a normal binary sequence as they are clocked.

To make a modulo-n divider circuit, you connect the UP input of the first counter high, apply the clock signal to the DOWN input of the first counter, and tie the $\overline{BO}$ output of the most significant counter to the $\overline{LOAD}$ inputs of all the counters, as shown in Figure 9-16. When the output count reaches all zeros, the $\overline{BO}$ will go low and load the count on the data inputs into the counters. The loaded number will be counted down by succeeding clock pulses. To make a divider of any modulo, then, all you have to do is apply the binary equivalent of the desired modulo to the data inputs. You can do this by hardwiring the data inputs to V_{CC} and ground, as shown in Figure 9-15, or you can connect some switches to the data inputs so that you can easily change the modulo of the circuit.

Now that you have a feeling for how synchronous counters are used, the next step is to show you some applications for them. As a first example, we will show you how synchronous counters, latches, decoders, and a few other common devices can be used to build a frequency counter.

SYNCHRONOUS COUNTER APPLICATIONS

An MSI Frequency Counter

INTRODUCTION

In Chapter 4 we described how you make measurements with a frequency counter. Most frequency counters allow you to make at least three basic types of measure-

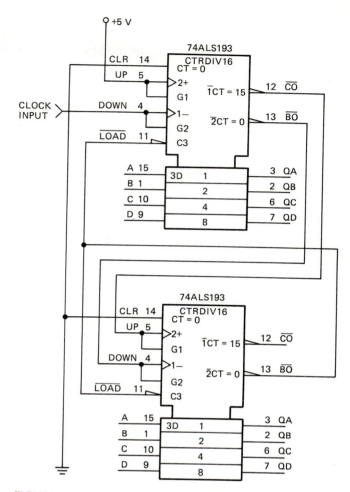

FIGURE 9-16 Cascading 74ALS193 counters for modulos greater than 16.

ments: total counts, frequency, and period. Here's an overview of how the circuitry in a frequency counter makes each of these three types of measurements.

To determine a total number of counts, the simple counter-decoder-display circuit we showed you in Figure 8-37 is mostly all that is needed. The only additional part needed is a push-button switch which you can use to reset the outputs of the counters to zero after each measurement.

To measure frequency, you have to count the number of pulses that occur within some time. A frequency counter does this by gating the unknown frequency signal into a chain of counters for a precise time such as 1 s. The time interval during which the counters are counting is often called the *count window*. The count accumulated during the count window is proportional to the frequency of the input signal and can simply be displayed on some seven-segment LEDs.

Period is simply the time for one cycle of a waveform. To measure period you need some way to measure the time between, for example, one rising edge on a signal and the next rising edge. To mark off intervals of time, you might use a precise 1-MHz oscillator. Each pulse of the 1-MHz oscillator represents a time of 1 μs. All you have to do to measure period, then, is use a chain of counters to count the number of 1-MHz pulses that occur between one rising edge of the unknown signal and the next rising edge of the unknown signal. The count accumulated on the counters will be equal to the period of the input signal in microseconds.

In the following sections, we show you some circuitry which can be used to make these measurements.

A FIRST LOOK AT THE CIRCUIT DIAGRAM

Figure 9-17 shows the complete schematic for a basic frequency counter made with readily available TTL and CMOS MSI parts. Since this schematic is more complex than the ones we have shown you before, you may at first glance feel overwhelmed. As we have told you before, the first step in dealing with this feeling is to use the 5-min rule. After you have used up your 5-min "freakout" period, you look for something you recognize in the schematic as a starting point.

In the schematic in Figure 9-17, probably the first part you will recognize is the seven-segment displays across the top of the page. These of course are used to display the measured frequency. Below each seven-segment display is a 74LS47 BCD-to-seven-segment decoder which produces the seven-segment codes for that display. Next, under each 74LS47 is a 74HC175 quad D flip-flop, the operation of which we described in Chapter 8. As we will discuss in greater detail later, the purpose of the 74HC175s in this circuit is to hold the results of one count stable on the displays while a new count is being taken.

Finally, under each 74HC175 is a 74HC160. According to the list of counters in Figure 9-10, a 74160 is a synchronous decade counter. The six 74HC160s are cascaded so that they form a six-decade counter. The top two-thirds of the circuit in Figure 9-17, then, functions as a six-digit counter module which can count, latch, decode, and display a count up to 999,999 pulses. Note that the $\overline{CLR}$ inputs of all the 74HC160 counters are connected to a common line so that the counters can all be reset to 0 by a control signal before each new count is taken.

In the introduction above, we told you that we measure frequency by *gating* the unknown signal into a counter chain for a precisely known time called a *count window*. The single NAND gate (U25c) on the left-hand side of the circuit in Figure 9-17 is used to gate the unknown signal into the counter-display block. If the enable input of the NAND gate is high, the unknown signal will pass through the gate to the CLK inputs of the 74HC160 counters. If the enable input (pin 10) of the NAND gate is low, the NAND gate will be disabled and no pulses will reach the counter block. To produce an output count proportional to the input frequency, we need to make the enable input of the NAND gate high for some precise time such as 0.1 s, 1 s, or 10 s.

To produce the desired count window pulses for the enable input of the NAND gate, we start with a precise 1-MHz crystal oscillator. The 1-MHz output signal from the oscillator is divided down by another chain of 74HC160 decade counters to produce signals of 10 Hz, 1 Hz, and 0.1 Hz. Switch SW1b selects one of these signals and routes it through a 74HC76 JK flip-flop to the

enable input of the NAND gate. An oscillator and divider chain such as this is sometimes called a *timebase generator*.

DETAILED OPERATION AND TIMING FOR FREQUENCY MEASUREMENTS

Once you have figured out the function of the major blocks of circuitry in a schematic, you can then look at the "glue" pieces that tie these large blocks together and work out the detailed operation of the circuit. As a start, let's take a closer look at the part that the 74HC76 flip-flop plays in producing the count window signal for the NAND gate.

Suppose, for example, that we want a count window of 1 s. A convenient signal to do this is a square wave which is high for 1 s and low for 1 s. The period of this signal is 2 s, so it has a frequency of 0.5 or ½ Hz. The J and K inputs of the 74HC76 in Figure 9-17 are tied high, so it produces a square-wave output signal which has half the frequency of its input CLK signal. If the 1-Hz signal from the 1-MHz divider chain is connected to the CLK input of the 74HC76, the signal on the Q output of the device will be the desired ½-Hz square wave.

The only part left to analyze in the frequency counter circuit is the 74LS121. You probably are not familiar with this device because we have not discussed it in any of the previous chapters. What you do when you find a part that you are not familiar with is to look up the data sheet for the device in a data book and/or figure out the operation of the device from how it is connected in the circuit.

You don't need to look up the data sheet for a 74LS121 right now because in the next chapter we discuss in detail the operation of such devices. All you need to know about a 74LS121 at this point is that when the device is triggered, it produces a single high pulse on its Q output and a single low pulse on its $\overline{Q}$ output. The output pulse width is determined by the value of the capacitor connected between pins 10 and 11, and the value of the resistor connected from pin 11 to +5 V. The values shown in Figure 9-17 give an output pulse width of about 1 μs. The 74LS121 can be triggered by applying a rising edge to its B input or by applying a falling edge to either its A1 input or its A2 input. Now let's see how this device fits into the operation of the frequency counter.

Figure 9-18 shows a timing diagram for the count window signal and the pulses from the 74LS121. During the 1 s that the count window signal is high, the NAND gate is enabled and the unknown signal pulses are counted by the main counter chain. When the count window signal goes low, the NAND gate is disabled, so the accumulated count stays constant on the outputs of the counters. As the count window signal on the Q output of the 74HC76 goes low, the $\overline{Q}$ output of the 74HC76 goes high. This low-to-high transition triggers the 74LS121 to output a positive-going pulse on its Q output and a negative pulse on its $\overline{Q}$ output.

The Q output of the 74LS121 is connected through the RUN/HOLD NAND gates to the CLK inputs of the 74HC175 D flip-flops. Since these D flip-flops are positive-edge-triggered, the rising edge of the Q pulse from the

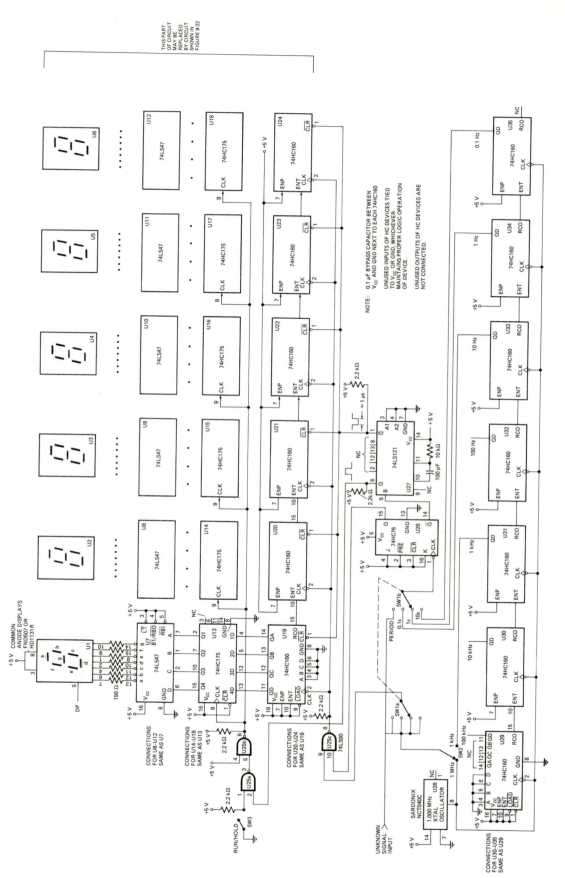

FIGURE 9-17   Complete schematic for an MSI frequency counter.

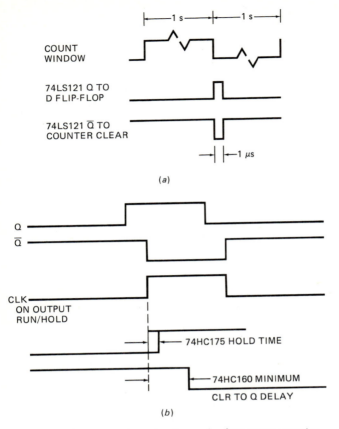

COUNT
WINDOW

74LS121 Q TO
D FLIP-FLOP

74LS121 Q̄ TO
COUNTER CLEAR

(a)

Q

Q̄

CLK
ON OUTPUT
RUN/HOLD

74HC175 HOLD TIME

74HC160 MINIMUM
CLR TO Q DELAY

(b)

FIGURE 9-18 *(a)* Timing waveforms for frequency counter control circuitry of Figure 9-17. *(b)* Expanded timing diagram.

74LS121 clocks them. When clocked, these flip-flops transfer the accumulated count from the outputs of the 74HC160 counters to the inputs of the 74LS47 decoders. The flip-flops then hold the count stable on the inputs of the 74LS47 decoders until the flip-flops are updated again by the next Q pulse from the 74LS121. Without these flip-flops in the circuit, the display would alternate between 1 s of garbage as the counters made a new count and 1 s of stable reading during the low time of the counter window signal. This would make the display very hard to read. The 74HC175 flip-flops hold the display data constant except for a few nanoseconds every 2 s when the flip-flops are clocked to transfer a new count to the decoders. If the RUN/HOLD switch is in the ground position, NAND gate U25a will be disabled, so the update pulse cannot get to the CLK inputs of the flip-flops. As long as this switch is in the ground position, the most recent count will be held on the display. When the switch is moved to the RUN position, the display will again be updated at the end of each count window.

The Q̄ signal from the 74LS121 is connected to the CLR inputs of the counters in the main divider chain. This signal will reset all the counter outputs to zero, so the counters are ready to do the next count of the unknown signal.

If you think a little about the timing diagram in Figure 9-18a, you may wonder how we can clock the flip-

flops to transfer the count from the counter outputs at the same time as we send a signal to reset the outputs of the counter. We will use the expanded timing diagram in Figure 9-18b to help explain how this works.

The Q pulse appears on the output of the 74LS121 about 10 ns before the Q̄ pulse does. The Q pulse, however, is delayed by the two 74LS00 NAND gates in the RUN/HOLD control for about 10 ns, so the Q pulse arrives at the 74HC175s at nearly the same time that the Q̄ pulse appears on the CLR inputs of the 74HC160s. The 74HC160 has a CLR-to-output delay time of about 20 ns, and the 74HC175 has a maximum hold time of only 5 ns. This means that by the time the CLR signal propagates through the 74HC160 counters, the counter output data will have been safely clocked into the flip-flops. This simple example again shows the importance of data sheet timing parameters.

The final device in the circuit in Figure 9-17 that we want to dig a little deeper into is the 74HC160 counter. The easiest way to show you what we need to know about this device is with the waveforms for the 74LS160A and 74LS162 decade counters in Figure 9-19.

The main difference between these two devices is that the 74LS160A has an asynchronous CLR input and the 74LS162A has a synchronous CLR input. Look at the waveforms to see the difference this makes in the time when the outputs go to zeros. For the frequency counter in Figure 9-17, we used the 74160 because we wanted the outputs to reset as soon as CLR was asserted, rather than waiting for a clock pulse.

Both of these counters load synchronously. This means, as shown in the waveforms, that if the LOAD input is made low, data on the data inputs will not be transferred to the Q outputs until the next rising edge of the CLK signal.

A 74LS160A has an Enable Parallel input, ENP, and an Enable Trickle input, ENT. As shown in the waveforms, both of these inputs must be high in order for the device to count when clocked. A 74LS160A also has a Ripple Carry output, RCO. The RCO on a counter will go high if its ENT input is high and if the count on its outputs is 9. To cascade 74LS60As, the RCO signal from one device is connected to the ENT input of the next most significant counter as shown in Figure 9-17. When the count on the less significant counter reaches 9, then the more significant counter will be enabled to increment on the next clock pulse. As an example of this, suppose that the leftmost (least significant) counter in Figure 9-17 has a 9 on its outputs and the next counter to the right has a 7 on its outputs. Since the leftmost counter has a 9, and its ENT input is high, its RCO output will be high. The high on RCO will enable the second counter. On the next clock pulse the leftmost counter will roll back to 0, and the next counter will increment to 8. These actions represent the count changing from 79 to 80. When all the counters reach a count of 9, all the RCO signals will be high. This will enable all the counters so that on the next clock pulse they will all roll back to 0.

Now that you have an understanding of the basic operation of a frequency counter, let's see why we need a count window other than 1 s for some frequency measurements.

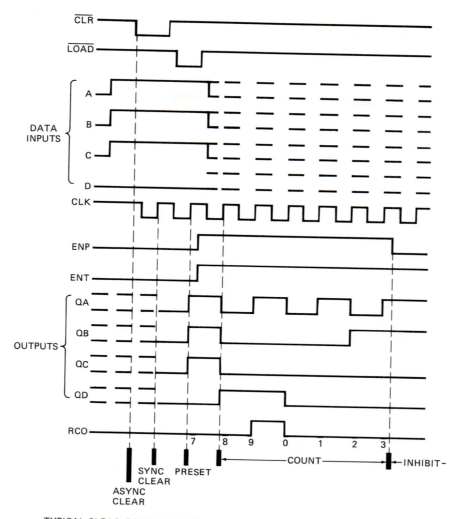

FIGURE 9-19 Typical clear, preset, count, and inhibit sequence for the 74LS160A.

A counter such as the one in Figure 9-17 has six digits, so with a 1-s window it can only measure a frequency up to 999,999 Hz before the counters overflow. On most frequency counters, a decimal point is inserted in the display so that this maximum value will be displayed as 999.999 kHz.

If the count window is reduced to 0.1 s, then a frequency of 999.999 kHz will only produce a count of 099999 on the displays. With a count window of 0.1 s, then, we can measure a frequency up to 9999.99 kHz or 9.99999 MHz before the counters overflow. Besides changing the count window, all you have to do is move the decimal point one digit position to the right in the display. A section of the same switch that changes the

count window can be used to turn off one decimal point and turn on the other.

For measuring low frequencies, a 1-s window presents another type of problem. With a 1-s window, the resolution of a measurement is 1 Hz. This means that if, for example, the display shows a reading of 20 Hz, the actual frequency might be anywhere between 20 and 21 Hz. This uncertainty represents a possible error of 1 part in 20, or 5 percent. The accuracy of low-frequency readings can be improved by increasing the count window to 10 s and moving the decimal point accordingly. In 10 s the counters will count 200 cycles of a 20-Hz signal and display it as 20.0 Hz. Because the counters cannot count fractions of counts, the actual count could be anywhere between 200 and 201 counts. This uncertainty of one count in 200 is a possible error of only 0.5 percent.

The only difficulty with using a relatively long count window such as 10 s is that you have to wait what seems like a long time for a new reading. With a count window signal that is high for 10 s and low for 10 s, you have to wait 20 s for a new reading. Therefore, for low-frequency signals it is usually more efficient to measure the period and use it to calculate the frequency. In the next section we discuss how the circuit in Figure 9-17 can be used to measure the period of an input signal.

DETAILED OPERATION AND TIMING FOR PERIOD MEASUREMENTS

At the start of this section we told you that we measure the period of a signal by counting how many pulses of a known frequency occur during one cycle of the unknown frequency signal. If, for example, we count how many pulses of a precise 1-MHz signal occur during one cycle of the unknown signal, then the count on the display will represent the period of the unknown signal in microseconds. If we count the number of cycles of a precise 1-kHz signal that occur during one cycle of the unknown input signal, then the display will represent the period of the unknown signal in milliseconds.

With the switches in Figure 9-17 in the positions shown, the circuit is set up to measure frequency with a 1-s window. If switch SW1 is thrown to its PERIOD position, then the circuit will be set up to measure the period of an input signal. Since SW2 is connected to a 1-MHz signal, the display produced will represent the period of the signal in microseconds.

The key to the operation of the circuit in this mode is that the unknown signal is used to produce the count window, and the counters actually count cycles of the 1-MHz timebase signal. A look at the example waveforms for the circuit in Figure 9-20 should help you see how this works.

Each negative edge of the unknown frequency signal toggles the Q output of the 75HC76 flip-flop. The Q output of the JK flip-flop, then, is high for one cycle of the unknown frequency signal and low for the next cycle, as shown in Figure 9-20. The high time of this Q signal functions as a count window, just as it did for the operation of the circuit as a frequency counter. During the

high time of the count window signal the 74HC160 counters are counting cycles of the 1-MHz signal now connected to their inputs through SW2, SW1, and NAND gate U25c. At the end of the count window time, the 74LS121 produces a CLK pulse to the 74HC75 D flip-flops and a reset pulse to the $\overline{\text{CLR}}$ inputs of the 74HC160 counters.

With a reference clock of 1 MHz, the system in Figure 9-17 can measure a signal with a period as long as 999,999 μs, or about 1 s. If SW2 is moved to the 1-kHz position, the system can measure a signal with a period of up to 999,999 ms, or about 1000 s.

IMPROVEMENTS ON THE BASIC FREQUENCY COUNTER

The basic frequency counter in Figure 9-17 can be improved in several ways. The first way we want to talk about is increasing the maximum frequency signal the system can count.

In a previous discussion we showed that the maximum input frequency for the system could be increased from about 1 MHz to about 10 MHz by using a 0.1-s count window instead of a 1-s count window. From this you might think that we could increase the maximum input frequency still further by decreasing the count window to 0.01 s or 0.001 s. In theory this works, but in reality it doesn't, because the maximum count frequency for the 74HC160 counter chain in Figure 9-17 is about 25 MHz.

The most effective way to increase the maximum input frequency of the system is to use a *prescaler*. A prescaler is a high-frequency divider device which is used to divide an input signal down to a frequency that the counter chain can count. Since a prescaler is simply used to divide down the frequency of the input signal, it does not have a latch, a decoder, and a display, as the other counters do. The switch which connects a prescaler in series with the unknown frequency input signal also switches the position of the decimal point in the main display so that it gives the correct reading.

As one example of a prescaler, a 74S196 is a decade counter/divider which can be used with an input clock signal as high as 100 MHz. A gallium arsenide device such as the 10G070 from Gigabit Logic can be used to

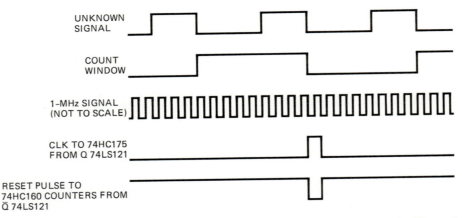

UNKNOWN SIGNAL

COUNT WINDOW

1-MHz SIGNAL (NOT TO SCALE)

CLK TO 74HC175 FROM Q 74LS121

RESET PULSE TO 74HC160 COUNTERS FROM $\overline{\text{Q}}$ 74LS121

FIGURE 9-20   Timing waveforms during measurement of input signal period with system in Figure 9-17.

divide down an input signal with a frequency of up to 3.0 GHz. (Remember 1 GHz = 1000 MHz.)

The second improvement we need to make on the basic frequency counter in Figure 9-17 is input signal conditioning. For the circuit as shown in Figure 9-17, the unknown frequency input signal must have TTL signal levels to properly assert the input of the NAND gate or the input of the JK flip-flop. Many signals, such as those in a radio receiver or transmitter, do not have TTL-compatible levels, so we could not measure them with the basic counter circuit.

A device called a *comparator* can be used to convert small sine-wave and other non-TTL signals to TTL levels. Figure 9-21 shows how a high-speed comparator,

9-17 extends its capability to measuring sine-wave signals of up to 100 MHz.

In the next section we discuss an improvement which reduces the amount of power that a frequency counter such as the one in Figure 9-17 consumes.

MULTIPLEXING THE FREQUENCY COUNTER DISPLAY

One problem with the frequency counter circuit in Figure 9-17 is that the displays dissipate a relatively large amount of power. If you assume a current of 20 mA for each segment and seven segments per display, each digit uses 7 × 20 mA or 140 mA. Multiplying this by six dig-

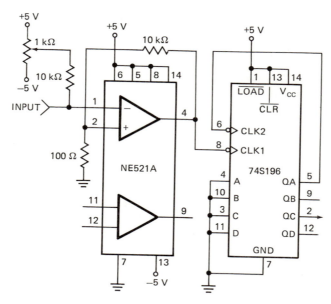

FIGURE 9-21   Circuitry to add input signal conditioning and divide-by-10 prescaling to the input of the basic frequency counter in Figure 9-17.

the Signetics NE521A, and a 74S196 high-speed decade counter can be connected on the input of a frequency counter to provide signal conditioning and prescaling. In the circuit shown, the trigger level of the comparator can be adjusted to a voltage anywhere between −5 V and +5 V. The comparator output levels are TTL-compatible.

Since the input signal is applied to the inverting input of the comparator, the output of the comparator will be inverted from that of the input signal. If the input voltage is a few millivolts above the applied threshold voltage, the output of the comparator will be low. If the input signal is a few millivolts below the applied threshold voltage, the output of the comparator will be high. A small sine-wave signal applied to the input of the comparator then will produce a TTL-compatible square-wave signal on the output of the comparator. The NE521A and the 74S196 counter can both operate correctly with an input frequency as high as 100 MHz. Connecting this circuit on the input of the frequency counter in Figure

its gives a maximum current of 6 × 140 mA or 840 mA, just for the displays. At 35 mA each, the six 74LS47s add another 210 mA to the total current. The decoders and displays of the circuit in Figure 9-17 then use a total maximum current of 1050 mA or about 1.1 A!

Both the number of ICs in the circuit and the amount of power dissipated can be reduced by multiplexing the display. The 8-channel scope multiplexer circuit in Figure 7-10 is one example of a multiplexed display. The trick in that circuit is that if you chop or alternate fast enough between the different traces, your eyes see the results as eight continuous displays.

Another example of a multiplexed display is that in the LSI stopwatch circuit in Figure 8-40. As we explained in Chapter 8, the principle of this display is that only one display digit is lit at a time. The segment code for one digit is sent out on a segment bus to all the display digits, but only one display digit is turned on, so only that one display will be lit. After a few milliseconds, that digit is turned off and the segment code for the next digit is sent out on the segment bus. The next display is then turned on to display that segment data. After a few milliseconds, this second display is turned off and the segment code for the third display is sent out on the segment bus. The third digit is then turned on to display that segment data. The process is repeated until all digits have had a turn, then the process cycles around to refresh the first display again. The response of your eyes is slow enough that, if the displays are refreshed at a rate of 100 to 1000 times per second, they all appear to be lit with the desired digit.

Figure 9-22 shows a counter, flip-flop, and multiplexed display circuit which can be substituted for the nonmultiplexed display in Figure 9-17. Note that we added a seventh counter to this circuit so that the basic counter can count to 9999999. The overall operating principle of this circuit is that the BCD code from one 74HC175 is sent to the inputs of the 74LS47 by the four 74HC151 multiplexers. The 74LS47 outputs seven-segment code for that count to the segment inputs of all the seven-segment displays. The 74LS138 turns on the transistor which supplies current to the common anode of the display we want to show that count. The 74LS93 counter cycles the multiplexers and the 74LS138 through all the displays over and over so that each display is lit for a few milliseconds in turn. Let's look a little more closely at how this circuitry works.

First note in Figure 9-22 that, for example, the a segments of all the displays are connected to a common bus line. Likewise, the other segments are all connected on common lines. Each display will then receive the same segment code from the 74LS47. Next note that the anode of each display is connected to V_{CC} with a transistor. To light a digit, this transistor is turned on by applying a logic low signal to its base with a 74LS138.

As you may remember from Chapter 7, one name for a 74LS138 is a one-of-eight low decoder. This means that if the device is enabled, only one of its eight Y outputs will ever be low at a time. The 3-bit binary code applied to the C, B, and A select inputs determines which of the eight outputs is low at a given time. If the code on the CBA inputs is 000, for example, the Y0 output will be low. If the code on the CBA inputs is 001, the Y1 output will be low, etc. In the circuit in Figure 9-22, the signals for the CBA inputs come from a 74LS93 binary counter. Since the 74LS93 counter is clocked by a 1-kHz signal from the timebase divider chain, the 74LS93 will output a binary count sequence over and over. This repeating binary count on the CBA inputs of the 74LS138 will cause it to step a low through its Y outputs over and over. The stepping low on the outputs of the 74LS138 will turn on one transistor at a time and thus light one display at a time.

The only part left to explain here is how the data from a counter gets to the segments of a display at the proper time. This is the job of the four 74HC151 multiplexers.

Remember from Chapter 7 that if a 74HC151 is enabled, it sends the signal from one of its eight inputs to its Y output. The input signal sent to the Y output is determined by the 3-bit code applied to the C, B, and A select inputs. If the CBA inputs have 000 on them, for example, the D0 input will be effectively connected to

the Y output. The C, B, and A select inputs of the four multiplexers are all connected in parallel to the outputs of the 74LS93 counter, so that at any particular time the four multiplexers will all be selecting the same input channel. The channel selected will depend on the 3-bit binary count present on the outputs of the 74LS93.

The data inputs of the multiplexers are connected to the 74HC160 counter outputs in parallel. The D0 inputs, for example, are all connected to the least significant counter; the D1 inputs are all connected to the second counter; the D2 inputs are all connected to the third counter; etc. If the outputs of the 74LS93 counter have 000 on them, then the D0 inputs of the multiplexers will be selected. This will route the data from the Q outputs of the least significant counter through the multiplexers to the inputs of the 74LS47. The 74LS47 will output the seven-segment code for this data on the segment bus to all the displays. The 000 on the outputs of the 74LS93 will also cause the 74LS138 to output a low on its Y0 output. This low on Y0 will turn on transistor Q7 and thereby light the rightmost digit of the display. The rightmost digit will then display the data from the least significant 74HC160 counter as desired.

When the 74LS93 outputs increment to 001, the multiplexers will route data from the second counter to the 74LS47 and the 74LS138 will output a low on its Y1 output to light the second display from the right as desired. As the 74LS93 increments, it causes the multiplexers to step across from counter to counter and the 74LS138 to light the corresponding display. The display refresh process starts over each time the 74LS93 output count rolls back to 000.

For the multiplexing system in Figure 9-22, we connected a 1-kHz signal to the CLK input of the 74LS93 counter. The rate at which the QC, QB, and QA outputs of the counter will cycle through a count of 000 to 111, then, is 1 kHz/8, or 125 Hz. This means that each display is getting refreshed 125 times a second. Another way of saying this is that the display refresh rate or multiplex rate is 125 Hz. Each display is only lit for one-eighth of the time. A common way of indicating this is to say that the displays have a 12.5 percent duty cycle.

The multiplexed display circuit in Figure 9-22 uses several hundred milliamps less than the nonmultiplexed display circuit in Figure 9-17. The current saving would be greater, except that when displays are multiplexed, the segment currents must be increased so that the displays appear as bright as nonmultiplexed displays. You might think that since each display is only on for one-eighth of the time, we would have to increase the current by 8 to get equivalent brilliance. Due to the fact the human eye "remembers" the average brilliance between refreshes, doubling or tripling the segment currents is usually sufficient.

One problem with a multiplexed display is that if the display scanning stops for some reason, the large segment currents will flow through one display continuously, rather than one-eighth of the time. Since the multiplexed segment current is usually larger than the display can stand on a continuous basis, this may cause the display to burn out. The 150-Ω current-limiting resistors in the circuit in Figure 9-22 limit the segment

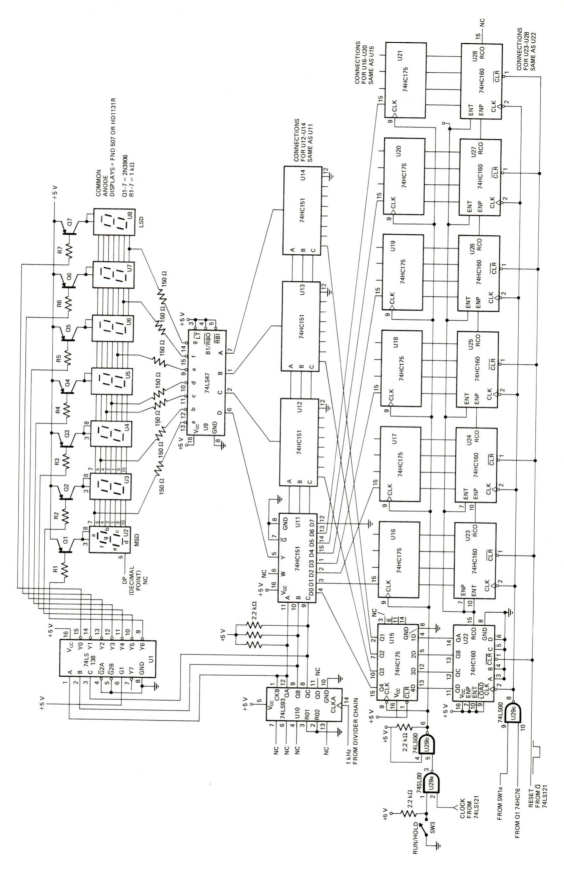

FIGURE 9-22 Schematic of MSI frequency counter with multiplexed displays.

currents to about 20 mA. This current value produces a somewhat dim display, but if you build the circuit to experiment with, it assures that the displays will not burn out if the scanning part of the circuit doesn't work.

BUILDING AND TESTING AN MSI FREQUENCY COUNTER

The timebase and control circuitry from Figure 9-17 can be combined with the counters and multiplexed display circuitry in Figure 9-22 to make a functioning MSI frequency counter. We will use this frequency counter to refresh and expand your knowledge of how you build, test, and troubleshoot a digital system.

The details of every system are different, but the basic approach to building, testing, and debugging is usually about the same. The key point is to break the system into functional modules which can be individually built, tested, and debugged, instead of building the whole system at once. It is usually much easier to find a problem in a small section of circuitry than in a large circuit. As you get each part working, you can connect it to previously tested sections and test the combined circuit.

The first step in building any circuit is to draw a schematic for the whole system, complete with all pin numbers and connections. Do not try to build a circuit from a collection of data books scattered around your bench.

Once you have a complete schematic, plan the layout of the circuit on protoboard, vector board, or whatever you are building the circuit on. Then analyze the schematic to see which blocks of circuitry can be built and tested separately. Here is one possible sequence of steps you might use to build our frequency counter with a multiplexed display:

1. Wire up the 1-MHz crystal oscillator and verify that it works correctly.

2. Wire up the first of the 74HC160 counters in the timebase divider chain and verify that it works correctly.

3. Add the rest of the counters in the divider chain, quickly testing the operation of each counter as you add it.

4. Build the 74HC76 count window circuit and test it.

5. Build and test the 74LS121 CLK and $\overline{CLR}$ pulse generator.

   NOTE: If you trigger the 74LS121 with a 10-kHz TTL signal from a signal generator instead of the $\overline{Q}$ signal from the 74HC76, you will be able to easily see the 74LS121 output pulses with a standard, dual-trace scope.

6. Next build the section of the circuit containing the seven-segment displays, the digit driver transistors, and the 74LS47 decoder. For initial testing, temporarily connect the $\overline{LT}$ input of the 74LS47 to ground. This will assert all the segment outputs of the 74LS47.

7. Connect the open end of the base resistor on one digit driver transistor at a time to ground. Since all

the segment outputs of the 74LS47 are asserted, this should cause an 8 to be displayed on that display. Connecting the open end of R7 to ground, for example, should cause an 8 to be displayed on the rightmost display. Test all the displays in this way.

8. Next add the 74LS138 and the 74LS93 counter parts of the circuit. With a 1-kHz clock signal connected to the CLK input of the 74LS93, the displays should all show 8's.

9. To determine whether the segments are wired in the correct order, connect the $\overline{LT}$ input of the 74LS47 high and hardwire a BCD number such as 0010 or 0101 on the DCBA inputs of the 74LS47. The displays should then all show the decimal equivalent of the number. Try a few different numbers to make sure there are no errors.

10. Build the main counter chain of 74HC160 counters, one at a time. Test each one after you wire it.

11. Add the 74HC175 D flip-flops to the circuit. To see if the flip-flops are being updated, check if the values on the outputs of each flip-flop agree with the values on the outputs of the corresponding counter.

12. Add the 74HC151 multiplexers to the circuit. You can now test if the entire circuit is working correctly. Apply a known frequency signal source to the input of the counter and set the counter switches for measuring frequency. If you have another frequency counter, you can connect it in parallel with the one you just built so that you can compare readings. If the readings agree, congratulate yourself. Then try making other frequency and period measurements.

13. If your frequency counter doesn't give a correct frequency reading, you can troubleshoot this final section in one of two ways.

One way is to connect four data inputs from a logic analyzer to each of the 74HC175 flip-flop outputs and four analyzer data inputs to the outputs of the multiplexers. Connect the analyzer external clock input to the 1-kHz clock signal going to the 74LS93. Set the analyzer to clock on the rising edge of the external clock signal. When you do a trace in state mode, the data on the outputs of the multiplexers should sequentially agree with the data on each of the flip-flops. In other words, for one analyzer sample, the multiplexer output data will agree with the data on the first flip-flop; on the next sample, the multiplexer output data will agree with the data on the second flip-flop; etc. By examining the state listing carefully, you should be able to determine the section of the multiplexer circuitry causing the problem.

If you don't have a logic analyzer handy, you can troubleshoot the multiplexer section on a static basis as follows.

Lift out the 74LS93 and jumper the C, B, and A inputs of the 74LS138 and the multiplexers to ground. With these connections, the rightmost display should be lit and the multiplexers should all be selecting the leftmost 74HC175.

To get a count other than 0 on the outputs of the 74HC160 counters, disconnect the $\overline{CLR}$ signal line from the $\overline{Q}$ output of the 74LS121 and connect it to +5 V. This will prevent the counter from being reset at the end of a count window. Connect a 1-MHz signal to the input of the frequency counter and let it run for a few seconds, then disconnect the 1-MHz signal. This should leave a count on the outputs of the counters.

The logic level now present on the output of each multiplexer should be the same as the logic level on the corresponding output of the 74HC175. These logic levels are static, so you can check them with a scope, voltmeter, or logic probe.

If the levels on the multiplexer outputs are not the same as the corresponding 74HC175 Q outputs, look for a wiring error. If the levels are the same, then jumper 001 on the C, B, and A select inputs. The outputs of the multiplexers should now be the same as the corresponding Q outputs of the second 74HC175, and the second display from the right should be displaying that number. Quickly work your way across the 74HC175s until you find the source of the problem.

The major point here is that if you understand the operation of a system, you can usually devise simple tests to help you narrow the source of a problem down to a small section of circuitry.

TROUBLESHOOTING AN MSI FREQUENCY COUNTER THAT PREVIOUSLY WORKED

As we show you in the next section of the chapter, most new frequency counters implement the majority of their circuitry in a single chip. However, since by now you are quite familiar with the operation of our MSI frequency counter, we will use it to show you more about how you troubleshoot a digital system which previously worked.

Start by analyzing the schematic and/or manual to determine the function of the major blocks of circuitry. The point here is to understand the operation of the system well enough to associate the symptom with a particular section of circuitry so that you know where to start looking for the problem. It is not necessary to initially determine the function of every IC. For most systems this would take far too much time. Once you isolate a problem to a particular section, you can then focus on the detailed operation of that section and devise tests to locate the problem. Here are a few examples.

If the counter counts but shows erratic counts, a first thing to check is that the power-supply voltage is within specifications. TTL devices function erratically with low supply voltages.

If one digit of the display is brightly lit but the others are dark, the 74LS138 is most likely not scanning the display digits. Knowing how the scanning circuitry works should help you see that there are several possible causes of this symptom:

1. The 74LS138 may be defective.

2. The 74LS138 may not be receiving a count sequence from the 74LS93 counter.

3. The 74LS93 counter may not be outputting a count sequence because it is defective or because it is not

receiving a 1-kHz signal from the timebase counter chain.

4. The 1-kHz signal may not be produced because the 1-MHz crystal oscillator has stopped working.

With a scope you can quickly determine the source of the problem in this signal chain.

If all the digits are lit equally, but the displayed numbers do not change when the input frequency is changed, the problem may be that the 74HC175 flip-flops are not receiving a CLK pulse to update them with a new count. If the display digits only show 0's, the count window may not be getting produced to let the input signal into the main counter chain. If the count on the displays constantly changes to random numbers, the 74HC160 counters may not be getting a $\overline{CLR}$ pulse to reset them after each measurement. These problems all point to the 74HC76 and the 74LS121. You can quickly check with a scope to see if the count window, CLK, and $\overline{CLR}$ signals are being produced.

If the lower digits of the display change as the input frequency is changed, but the more significant digits don't, this points to a problem in the main divider chain. Remember that the RCO from one 74HC160 enables the ENT of the next counter, so if one counter is bad, the counters after that one may also not work.

We could go on with more symptoms and their likely causes, but by now you should get the major point. The major point of systematic troubleshooting is to analyze the symptoms and the schematic, and then to think of signals you can check and tests you can perform to localize the cause of a problem to a small enough area that it is easily found.

An LSI Frequency Counter

The circuits shown in Figures 9-17 and 9-22 work well, but they are really a 1970's-style frequency counter. Since most parts of the circuit operate at relatively low frequencies, CMOS technology has made it possible to put most of the circuitry needed for a frequency counter such as this in a single IC. Figure 9-23 shows the schematic for a 100-MHz multifunction counter using the Intersil ICM7226A. This unit can measure total counts, frequency, period, time intervals, and the ratio of two input frequencies.

The ICM7226A contains the timebase oscillator, timebase divider chain, two main counter chains, latches, decoders, segment drivers, digit driver transistors, control circuitry, and multiplex circuitry for an eight-digit common-anode counter system. The A input of the ICM7226A can only accept an input frequency up to 10 MHz, and the B input of the ICM7226A can only accept an input frequency up to 2 MHz. Therefore, as you can see in Figure 9-23, we connected 74S196 high-speed decade counters in series with each of these inputs. The 74S196s function as prescalers and extend the maximum input frequency to 100 MHz for the A input and 20 MHz for the B input. The NE521A comparators on the inputs allow the unit to measure sine-wave and other non-TTL-compatible signals. Now that you have

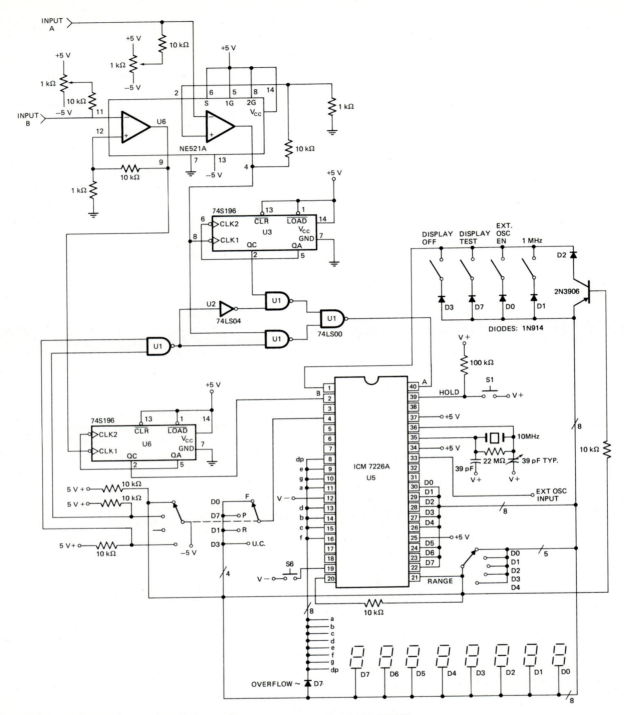

FIGURE 9-23 A 100-MHz multifunction counter using the Intersil ICM7226A.

an overview of the operation of the system in Figure 9-23, let's take a closer look at the operation of the ICM7226A.

Figure 9-24 shows the pin descriptions for the ICM7226A because they wouldn't easily fit on the symbol in Figure 9-23. You will need to refer to these periodically as we describe the operation of the circuit. To start, let's look at how the timebase oscillator is implemented.

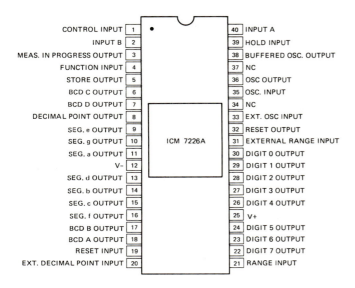

FIGURE 9-24 Pinouts for the ICM7226A.

The ICM7226A has internal circuitry for an oscillator, but it requires an external 10-MHz quartz crystal, two capacitors, and a resistor to operate, as shown on pins 35 and 36 in Figure 9-23. Many LSI devices require these external oscillator parts because they are difficult, if not impossible, to integrate directly in the IC. In the next chapter we will explain in detail how an oscillator such as this works. The next part of the circuit in Figure 9-23 to discuss here is the multiplexed display.

The ICM7226A has seven-segment outputs, a to g, and a decimal point output, dp. Note in Figure 9-23 that as soon as these eight lines leave the IC, they are drawn as a single line. This does not mean that the eight lines are all connected! Buses and other parallel signal lines are often drawn this way to make the schematics less cluttered. A / followed by a number is used to indicate how many lines are represented by a bus line such as this. If you look around the schematic in Figure 9-23, you should see several examples of this.

When a bus reaches a point where the signals are used individually, they are separated into individual lines again. An example of this is where the segment bus is separated into individual lines again to the left of the displays in Figure 9-23. Each segment driver in the ICM7226A can sink up to 25 mA. Note that the ICM7226A also has multiplexed BCD outputs so that you can use an external BCD-to-seven-segment decoder such as a 74LS47 if you need larger segment currents for particular displays.

The ICM7226A has eight digit driver lines labeled D0

to D7. One of these lines is connected to the common anode of each of the seven-segment displays in Figure 9-23. These lines correspond to the collectors of the digit driver transistors in the MSI display circuit in Figure 9-22. Each digit driver of the ICM7226A can source up to 170 mA from V_{CC}.

The operation of the display multiplex circuitry of the ICM7226A is essentially the same as we described for the operation of the circuitry in Figure 9-22.

Another important feature of the ICM7226A that we need to discuss is the way that function, range, and control signals are multiplexed into the device. Here's how this scheme works.

During the display refresh operation, the D0 to D7 lines are pulsed high one at a time in sequence. When the rightmost digit is being refreshed, D0 will be high; when the second digit is being refreshed, the D1 line will be high; etc. Look at the RANGE switch in the lower right corner of Figure 9-23, and note that the inputs of the switch are connected to some of the digit strobe lines, D0 to D4. The output of the switch is connected to the range input of the ICM7226A. Circuitry inside the ICM7226A constantly monitors the range input. Based on when it finds a high on that input, the internal circuitry sets the count window accordingly. The range input section of Figure 9-25 shows that if the internal circuitry finds a high during the D2 digit strobe time, it will set the count window to 1 s for frequency measurements.

In the same manner, the desired type of measurement is multiplexed from the FUNCTION switch into the FUNCTION pin (pin 4) of the ICM7226A. The top section of Figure 9-25 shows the digit strobe which corresponds to each function.

As shown by the circuitry in the upper right corner of Figure 9-23 and the bottom section of Figure 9-25, the same technique is used to multiplex some control signals into the control input of the ICM7226A.

This scheme used to multiplex RANGE, FUNCTION, and CONTROL signals into the ICM7226A greatly reduces the number of pins required on the device. Furthermore, if the mechanical switches are replaced with multiplexers such as 74HC151s, the function of the instrument can be programmed by the binary codes applied to the select inputs of the multiplexers. In other words, the measurement functions are programmable with binary words output from, for example, a microcomputer. Programmable instruments such as this are an important part of automatic test equipment (ATE) used to test systems during production.

Incidentally, if you want to build a counter such as the one shown in Figure 9-23, all the components for the basic counter are available in kit form (Intersil ICM 7226A EV/KIT) from your local electronics distributor.

REGISTERS, SHIFT REGISTERS, AND SHIFT REGISTER APPLICATIONS

Introduction

A register is a parallel group of flip-flops, usually D type, used for the temporary storage of binary data words. As

INPUT	FUNCTION	DIGIT
FUNCTION INPUT	FREQUENCY	D0
PIN 4	PERIOD	D7
	FREQUENCY RATIO	D1
	TIME INTERVAL	D4
	UNIT COUNTER	D3
	OSCILLATOR FREQUENCY	D2
RANGE INPUT	0.01 s/1 CYCLE	D0
PIN 21	0.1 s/10 CYCLES	D1
	1 s/100 CYCLES	D2
	10 s/1K CYCLES	D3
EXTERNAL RANGE INPUT PIN 31	ENABLED	D4
CONTROL INPUT	BLANK DISPLAY	D3 & Hold
PIN 1	DISPLAY TEST	D7
	1 MHz SELECT	D1
	EXTERNAL OSCILLATOR ENABLE	D0
	EXTERNAL DECIMAL POINT ENABLE	D2
	TEST	D4
EXTERNAL DECIMAL POINT INPUT, PIN 20	DECIMAL POINT IS OUTPUT FOR SAME DIGIT THAT IS CONNECTED TO THIS INPUT.	

FIGURE 9-25   Digit strobe connections for range, function, and control inputs of the ICM7226A.

we discuss in a later chapter, most microprocessors contain several registers which are used for this temporary data storage. Registers may be any number of bits wide, but common register sizes are 4, 8, 16, and 32 bits.

Figure 9-26 shows four D flip-flops connected as a simple data storage register. When the circuit is clocked, data on the D inputs is transferred to the Q outputs and is held or "stored" there until another clock pulse occurs.

If the output of each flip-flop in a register is connected to the input of the adjacent flip-flop, then the circuit is called a *shift register*. The name comes from the fact that each successive clock pulse moves or "shifts" the data bits one flip-flop to the left or to the right, depending on how the flip-flops are connected.

Figure 9-27a shows four D flip-flops connected as a *serial-in, serial-out shift register*. As you can see in the figure, the Q output of one flip-flop is simply connected to the D input of the next. The CLK, $\overline{PRE}$, and $\overline{CLR}$ signals are connected in parallel to all four flip-flops, so that they are all clocked, all set, or all reset at the same time. The shift register in Figure 9-27a is further referred to as a shift-right register because data bits are shifted to the right through the circuit as it is clocked. The example waveforms for the circuit in Figure 9-27b should help you see how this works.

Assume that the $\overline{CLR}$ input has been asserted at the left edge of the waveforms so that the Q outputs are all 0's. Then assume that a 1 data bit is applied to the D input of the first flip-flop as shown on the DATA waveform. When the next positive edge of the clock pulse occurs, the 1 on the D input of the A flip-flop will be

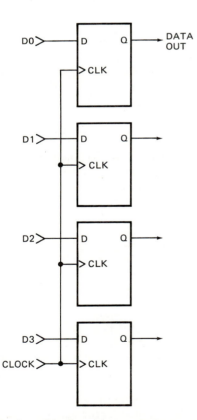

FIGURE 9-26   Four-bit data storage register made with D flip-flops.

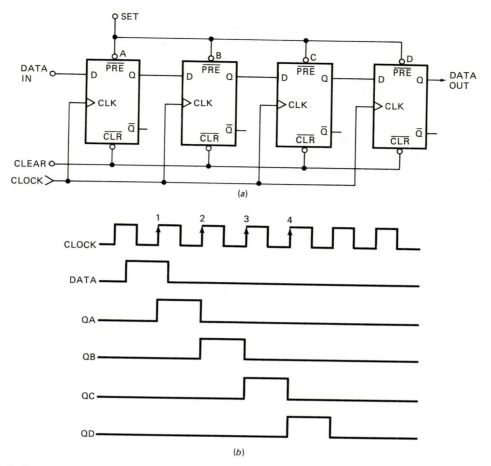

FIGURE 9-27 *(a)* D flip-flops connected as a shift register. *(b)* Example waveforms for shift register in Figure 9-27a.

transferred to the QA output. On the next rising edge of the clock signal, the 1 now present on the QA output will be transferred to the QB output. The D input of the A flip-flop is low at the time of the second rising edge of the clock, so this low will be transferred to the QA output. On the third rising edge of the clock, the 1 on QB will be transferred to QC, the 0 on QA will be transferred to QB, and the 0 on the DA input will be transferred to QA. Successive clock pulses will shift the 1 data bit along through the rest of the flip-flops.

As you can see in the waveforms, the 1 data bit will appear on the QD output after four clock pulses. When the fifth clock pulse occurs, the 1 data bit is shifted off the end of the register and is lost.

There are four basic types of shift register. The names for these four types tell you how data is loaded into the register and how data is read out from the register. We discuss the operation and uses of each type in the following sections, but here is an overview of the four types:

SISO—*Serial-in, serial-out.* Input data is applied one bit at a time to the first flip-flop in a chain and read out from the last flip-flop in a chain one bit at a time.

SIPO—*Serial-in, parallel-out.* Input data is applied one bit at a time to the D input of the first flip-flop in a chain and read out from the Q outputs in parallel after a data word is all shifted in.

PISO—*Parallel-in, serial-out.* A data word is loaded into all the flip-flops in parallel then shifted out the left end or the right end of the register one bit at a time.

PIPO—*Parallel-in, parallel-out.* A data word is loaded into all the flip-flops in parallel. The data word is read out from all the Q outputs in parallel after the desired number of shifts.

An Example MSI Shift Register

Figure 9-28a shows the internal logic diagram for a 74AS194 universal 4-bit shift register. We chose this device as an example because it can be connected to operate as any of the four shift register types described above. It will shift left, shift right, parallel load, or do nothing when clocked, depending on the logic levels present on its S1 and S0 inputs. For a refresher of tracing gate circuits, work your way through the circuit in Figure 9-28a to determine which levels on S1 and S0:

1. Route the signal from QA to the inputs of the B flip-flop for shift-right operation.

2. Route the signal from QD to the input of the C flip-flop for shift-left operation.

3. Route the data from the A, B, C, and D parallel inputs to the inputs of the flip-flops.

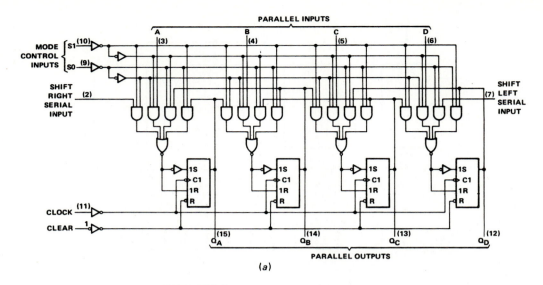

FIGURE 9-28 74AS194 4-bit-wide shift register. (a) Internal logic diagram. (b) Function table. (c) IEEE/IEC symbol. (Courtesy Texas Instruments Inc.)

When you finish working out the effects of S1 and S0, compare your results with those in the function table in Figure 9-28b.

The fourth and fifth lines in the function table tell you that if S1 is low and S0 is high, the data on the SHIFT RIGHT SERIAL INPUT will be shifted to the Q_A output on the rising edge of the CLOCK signal. The previous Q_A, Q_B, and Q_C output levels will be shifted one bit position to the right.

The sixth and seventh lines in the function table tell you that if S1 is high and S0 is low, the data on the SHIFT LEFT SERIAL INPUT will be shifted to the Q_D output on the rising edge of the CLOCK signal. The previ-

ous Q_D, Q_C, and Q_B output levels will all be shifted one bit position to the left through the register.

The third line in the function table tells you that if the S1 and S0 inputs are both high, the data on the A, B, C, and D parallel data inputs will be transferred to the corresponding Q outputs on the rising edge of the next CLOCK pulse. This is an example of a synchronous load operation.

The remaining lines in the function table tell you that the $\overline{CLEAR}$ input is direct or asynchronous, that CLOCK is inactive in the low state, and that the register outputs will not change if S1 and S0 are both low when the CLOCK goes high. An important footnote to the operation of the 74AS194 is that S1 and S0 must only be changed while the CLOCK input signal is high. As we will show you a little later, this is usually done by synchronizing the S0 and S1 signals with the CLOCK signal.

While the function table for a 74AS194 is fresh in your mind, take a look at the IEEE/IEC symbol for it in Figure 9-28c to see how this symbol represents the operation of the device.

The SRG4 at the top of the control block in the symbol identifies the device as a 4-bit shift register. The characters $\}M_3^0$ next to the S1 and S0 inputs indicates that the mode of operation of the device is determined by the binary number, 0 to 3, on these inputs. If binary 3 (11) is on these S1 and S0 inputs, for example, the data on the inputs labeled with 3's will be transferred to the corresponding Q outputs on the next CLK pulse. In mode 1 (S1 = 0, S0 = 1), the SR SER input will be activated. To indicate this, a 1 is put next to that input. Likewise, in mode 2, the SL SER input will be activated, so a 2 is put next to that input. None of the inputs is identified with a 0 because, in mode 0, none of the inputs has any effect on the outputs.

The wedge and the C4 on the CLK input tells you that the inputs labeled with a 4 affect the outputs on the rising edge of the CLK signal. Note that the A, B, C, and D inputs are labeled with 4's to tell you that the load operation occurs synchronously on the rising edge of the CLK signal. The characters $1\rightarrow/2\leftarrow$ shown on the lower CLK line tell you that if the device is in mode 1, the data on the registers will be shifted right on the rising edge of the CLK signal. If the 74AS194 is in mode 2, the data on the registers will be shifted left on the rising edge of the CLK signal. Finally, the R next to the $\overline{CLR}$ input tells you that the Q outputs will all be reset to 0 if that input is asserted.

In a later section we show you some applications for a 74AS194. We are using the 74AS194 device as an example because we want to remind you that AS devices can be used for high-speed applications. The connections and logic diagrams are the same for the LS and HC devices.

Using a Shift Register for Digital Delay

One common application for SISO shift registers is to delay digital signals by some desired amount of time. A signal applied to the input of the shift register in Figure 9-27a, for example, will appear on the QD output after four clock pulses. In other words, passing through the shift register delays a signal by four clock pulses.

The amount of delay produced by a shift register depends on the rate at which the register is clocked and the number of flip-flops in the register. The delay time is simply equal to the number of flip-flops times the period of the clock signal. As an example, a 256-bit shift register clocked by a signal with a period of 4 ms will delay a signal by 256 × 4 ms = 1024 ms, or about a second.

As it is clocked, a SISO shift register essentially takes samples of the signal on its input and passes the samples along to its output. The clock frequency, of course, determines the rate at which the input signal is sampled. In order to accurately reproduce the input signal

at the output, the clock frequency must be considerably higher than the frequency of the input signal. The trade-off here is that increasing the clock frequency increases the number of shift register stages required to produce a given delay. Fortunately, any number of SISO shift registers can be cascaded to produce a needed delay. Examples of commonly available longer SISO shift registers are the GE/RCA CD4031BE, which has 64 CMOS stages, and the Micrel MIC2833C, which has 1024 NMOS stages and can be clocked at up to 5 MHz. Here's an example of how you might use these devices.

Figure 9-29 shows in block diagram form how an analog-to-digital (A/D) converter, some long SISO shift registers, and a digital-to-analog (D/A) converter can be connected to replace the mechanical spring reverb system found in many guitar amplifiers and organs. The system works as follows.

The A/D converter takes samples of the audio signal on its inputs and converts each sample to an 8-bit binary value which is proportional to the amplitude of the signal. Each data bit from the A/D converter is connected to the input of a long SISO shift register. As the shift registers are clocked, the binary words representing the audio samples are shifted down through the registers. A D/A converter connected to the outputs of the shift registers converts the binary values back to a proportional voltage. If the clock frequency is enough higher than the frequency of the input audio signal, the signal on the output of the D/A will be just a delayed version of the audio signal applied to the input of the A/D converter. The amount of delay again depends on the clock frequency and the number of stages in each shift register.

An echo or reverb effect is created by tapping off a fraction of the delayed output signal with a voltage divider and mixing this fraction with the input signal. The mixed signal is then sampled by the A/D converter, and the binary result is shifted through the registers. Each time a fraction passes around through the A/D, shift register, D/A, and voltage divider loop, it gets smaller, just as a sound bouncing back and forth between the walls of a cathedral gets smaller on each echo. By adjusting the voltage divider, you can control the amount of the signal that is fed back to the input. The more you feed back, the stronger the echo will be and the longer it will last. Incidentally, Chapter 11 contains a detailed discussion of the operation of A/D and D/A converters.

Using Shift Registers for Waveform and State Sequence Generators

BASIC RING COUNTERS

Figure 9-30a shows a circuit commonly called a *ring counter* or a *recirculating shift register*. This second name comes from the fact that the QD output is connected to the D input of the A flip-flop so that data on the flip-flops circles around and around as the register is clocked.

Pressing the START switch loads the desired data word on the flip-flop outputs. For the circuit in Figure 9-30a, pressing the START switch will set the A flip-flop and

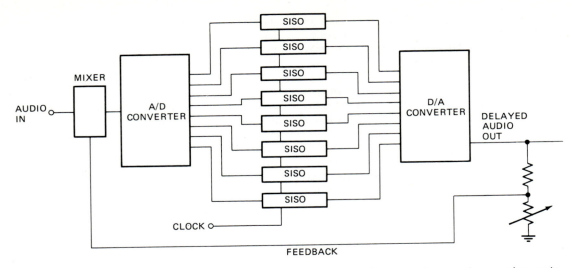

FIGURE 9-29 SISO shift register used with A/D and D/A converters to make an audio reverb, or echo, unit.

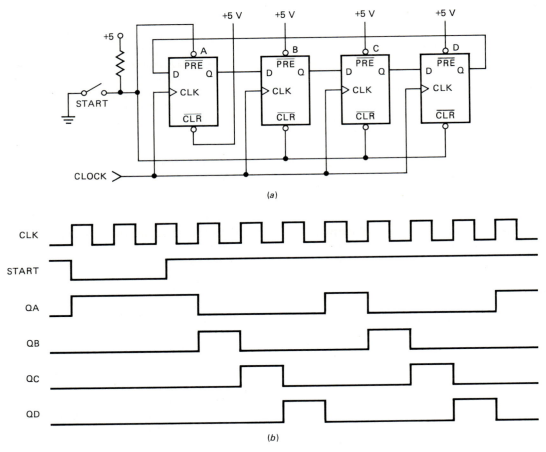

(a)

(b)

FIGURE 9-30 (a) Shift register connected as a ring counter. (b) Waveforms produced by ring counter loaded with QA = 1, QB = 0, QC = 0, and QD = 0.

reset the B, C, and D flip-flops. When the START switch is released, the loaded data word will be shifted around and around the register as it is clocked.

Figure 9-30b shows the waveforms that will be produced by the circuit in Figure 9-30a. The 1 initially loaded on QA is shifted down the chain until it reaches QD. On the next clock pulse, the 1 is transferred from QD to QA to begin its journey around the ring again.

If the circuit in Figure 9-30a is modified so that QA is reset and QB, QC, and QD are set when the START switch is pressed, a low will be circulated around the ring as the flip-flops are clocked. The waveforms produced by this circling low are the complements of those in Figure 9-30b. Other sets of useful waveforms can be produced by connecting the $\overline{PRE}$ and $\overline{CLR}$ inputs so that 1100 or 1010 is loaded into the flip-flops when the START switch is pressed.

You can use a shift register and the same principle to generate almost any digital waveform. As an example of this, suppose that you want to produce the waveform shown in Figure 9-31. The first step is to mark off a section of the waveform that repeats. For the waveform in

output to SL SER (shift-left serial input) so that a loaded pattern can be circulated to the left or to the right. To load the binary word you want to rotate, you simply hardwire the word on the A, B, C, and D inputs and set both switches in the up position. The next clock pulse will transfer the levels on the data inputs to the Q outputs. After the data word is loaded, throw SW2 down to rotate the word to the right or throw SW1 down to rotate the word left. To hold the word in place at any time, put both switches in the down position.

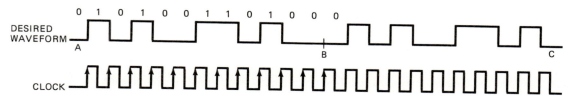

FIGURE 9-31 Generating a digital waveform by loading a bit pattern in ring counter made with PISO shift register.

Figure 9-31, the section between A and B is an example. The next step is to divide this section into equally spaced intervals as shown. From this you can see that the waveform consists of a 12-bit pattern, 010100110100, repeated over and over. You can create this waveform directly then with 12 flip-flops. The 12-bit pattern is loaded into the 12 flip-flops and shifted out serially with the appropriate clock signal. The clock signal must be such that it causes a shift at each desired transition of the output waveform. For this example, the clock waveform shown under the output waveform in Figure 9-31 will produce the required shifts, assuming the shift register consists of positive-edge-triggered D flip-flops.

Instead of using individual flip-flops to make a ring counter as we did in the circuit in Figure 9-30a, you can use any shift register which has parallel inputs and parallel outputs. Figure 9-32, for example, shows how a 74AS194 can be connected to operate as a 4-bit universal ring counter. Here's how it works.

Remember from the function table for the 74AS194 in Figure 9-28b that this device will shift left, shift right, load, or hold, depending on the levels on the S0 and S1 inputs. In the circuit shown in Figure 9-32a, the position of the debounced switches, SW1 and SW2, determine the operation for which the device is programmed. However, note that the outputs of the debounce circuits are not connected directly to the S inputs of the 74AS194. The reason for this is that according to the data sheet for the 74AS194, the S inputs can only be changed while the CLK input is high! To make sure the S inputs can only change when the clock signal is high, we connected a couple of flip-flops from a 74HC175 between the outputs of the debounce circuits and the S inputs of the 74AS194. The signal used to clock the 74AS194 is also used to clock the flip-flops. When the clock signal goes high, the switch signals will be transferred to the S inputs of the 74AS194. On the next clock pulse, 74AS194 will start performing the programmed action.

In the circuit in Figure 9-32a we connected the QD output to SR SER (shift-right serial input) and the QA

In the next chapter we show you how this circuit can be used to generate the waveforms needed to rotate a special type of motor called a *stepper motor*. As the name implies, a stepper motor rotates in small increments when pulsed rather than spinning around as a regular motor does. Stepper motors are often used in printers to advance the paper or position the print head.

You can cascade any number of 74AS194s to produce a longer shift register by:

1. Connecting the S0, S1, and CLK inputs of the 74AS194s in parallel.

2. Connecting the QD output of the rightmost device to the SR SER input of the leftmost device.

3. Connecting the QA output of the leftmost device to the SL SER input of the rightmost device.

WALKING RING COUNTERS OR JOHNSON COUNTERS

Figure 9-33a shows a shift-register circuit commonly called a *walking ring* or *Johnson* counter. Note in this circuit that the $\overline{Q}$ output of the rightmost stage of the shift register is connected to the D input of the leftmost stage. Assuming the flip-flops are all reset to start the sequence, the QC output will initially be a low and the $\overline{QC}$ output will be a high. On the next clock pulse, the high from $\overline{QC}$ will be transferred to QA. Succeeding clock pulses will shift highs down through the flip-flops until QC goes high and $\overline{QC}$ goes low. On the next clock pulse the low on $\overline{QC}$ will be transferred to QA. Succeeding clock pulses will shift lows down through the flip-flops until QC goes low and $\overline{QC}$ goes high again. The cycle then repeats as the counter is clocked further.

Figure 9-33b shows the waveforms produced on the Q and the $\overline{Q}$ outputs of this circuit as it is clocked. Each output has a square-wave signal with a frequency equal to one-sixth the frequency of the input clock. The circuit then functions as a modulo-6 divider. More flip-flops can be added to increase the modulo of the circuit. Four

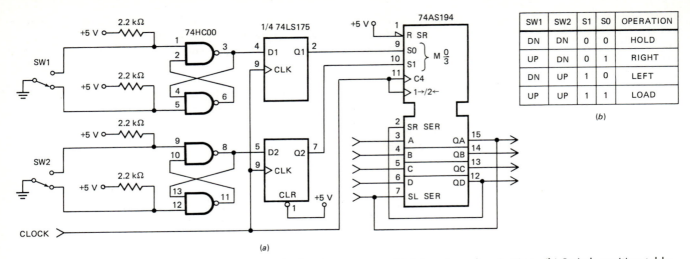

SW1	SW2	S1	S0	OPERATION
DN	DN	0	0	HOLD
UP	DN	0	1	RIGHT
DN	UP	1	0	LEFT
UP	UP	1	1	LOAD

(b)

FIGURE 9-32 (a) 74AS194 connected as universal 4-bit ring counter with 5-input synchronization. (b) Switch position table.

flip-flops will produce a modulo-8 circuit, and five flip-flops will produce a modulo-10 circuit.

Since only one output of a walking ring counter changes at a time, output states can be decoded with no glitches. A two-input AND gate is all that is required to produce a unique high pulse for each state. For the circuit in Figure 9-33, the AND gate connection to produce six decoded outputs are as follows:

State	AND input connections
0	$\overline{QA}$, $\overline{QC}$
1	QA, $\overline{QB}$
2	QB, $\overline{QC}$
3	QA, QC
4	$\overline{QA}$, QB
5	$\overline{QB}$, QC

The waveforms produced by the outputs of the six AND gates are stepped high pulses similar to those for the simple ring counter in Figure 9-30. A decoded walking ring counter circuit, however, requires only half as many shift register stages as a standard recirculating shift register does to produce the same number of stepped pulses. The 74HC4017 is a Johnson decade counter with ten active high decoded outputs. As described in most 4017 data sheets, these devices can be cascaded to produce a pulse which steps along as many outputs as needed.

SHIFT REGISTER CIRCUITS WITH FEEDBACK FROM INTERMEDIATE STAGES

A very wide variety of count sequences and/or output waveforms can be produced by decoding the outputs of a shift register and feeding the results back to the input. Figure 9-34a shows an example of this type of circuit. Assuming this circuit is reset to start the count sequence, it will step through the sequence of states shown in Figure 9-34b over and over as it is clocked. The sequence of output states from this circuit might be used to control some machine. Also, if you rotate Fig-

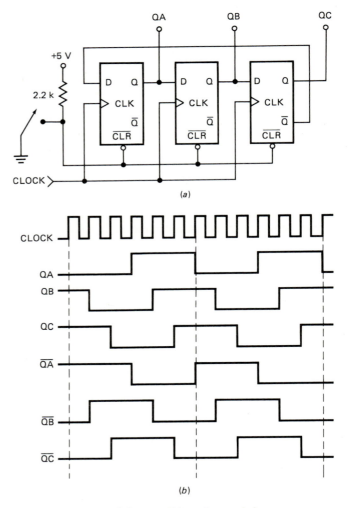

FIGURE 9-33 Modulo-6 walking ring or Johnson counter. (a) Circuit. (b) Waveforms.

ure 9-34b counterclockwise 90°, you can see that the parallel outputs produce stepped waveforms which are useful in some applications. By designing the feedback decoder differently, you can produce a circuit which steps

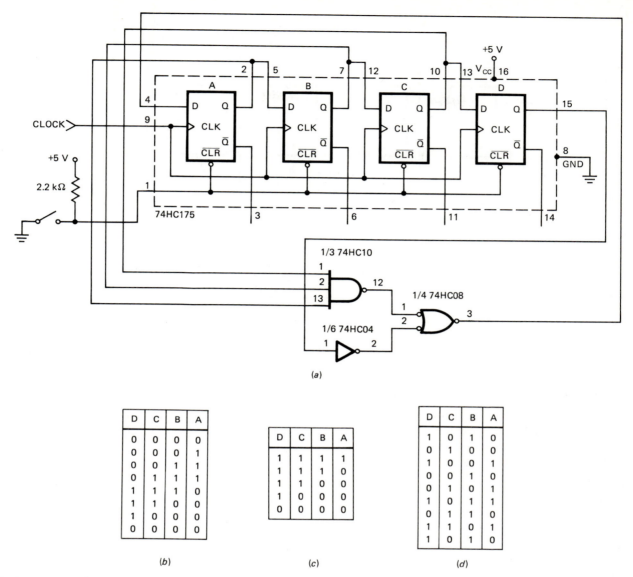

FIGURE 9-34 Shift register sequence generator with feedback from intermediate stages. *(a)* Circuit. *(b)* Sequence with 0000 starting state. *(c)* Sequence with 1111 starting state. *(d)* Sequence with 1010 starting state.

through any desired sequence and/or generates any desired waveform. The advantage of this feedback approach over the direct load approach described earlier is that the feedback approach requires fewer shift register stages to produce a given waveform. However, shift register circuits with feedback may have a problem unless they are carefully designed.

To illustrate this problem, let's first suppose that someone forgets to clear the circuit and it starts in state 1111 instead of in state 0000. In this case the circuit will step through the sequence shown in Figure 9-34c as it is clocked. Once the circuit reaches state 0000, it will continue through the main sequence shown in Figure 9-34b as it is clocked. This case usually doesn't cause a problem because the circuit gets into the main sequence after a few clock cycles. Now, let's see what happens if the circuit accidentally starts in state DCBA = 1010.

Figure 9-34d shows the sequence of states that the circuit will step through if it starts in state 1010. As you

can see, the circuit cycles through a sequence of eight states and returns to state 1010. Once it gets to state 1010 again, it will repeat this unwanted sequence over and over as it is clocked. In other words the circuit will never get to the desired main sequence states! These unwanted states or sequences of states are sometimes called *hang states*. For now we simply want you to be able to determine the sequence of states that a given shift register counter such as this will step through when it is clocked. In Chapter 12 we will show you how to design clocked circuits which avoid the hang state problem.

PSEUDO-RANDOM SEQUENCE GENERATORS

Another interesting application of shift register circuits with decoded feedback is a "pseudo"-random sequence generator. The term *random* here means that the outputs do not cycle through a normal binary count sequence. The term *pseudo* here refers to the fact that the sequence is not truly random because it does cycle

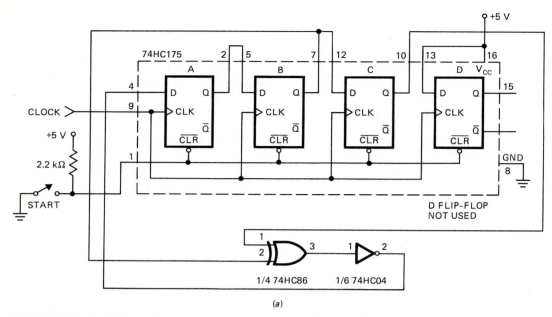

QC	QB	QA
0	0	0
0	0	1
0	1	1
1	1	0
1	0	1
0	1	0
1	0	0
0	0	0

(b)

FIGURE 9-35   A 3-bit random sequence generator. (a) Circuit. (b) Output sequence.

through all the possible combinations once every 2^n-1 clock cycles. In this case, n represents the number of shift register stages. The random sequence generator in Figure 9-35a, for example, has three stages, so it generates a sequence of 2^3-1 or 7 values before repeating. Figure 9-35b shows the sequence of values it will produce on its outputs starting from a reset.

Figure 9-36 shows a circuit for a random sequence generator which uses an 8-bit shift register, three XOR gates, and an inverter to produce a random sequence of 255 values. A 16-bit random sequence generator can be produced by XORing the outputs from the 4th, 13th, 15th, and 16th flip-flops and feeding the result back through an inverter to the D input of the first flip-flop. The 16-bit circuit will cycle through 65,535 values.

Pseudo-random number generators are used for testing digital circuits. They are used in encrypting digital data so that unauthorized persons cannot easily read it, and they are also connected to D/A converters to produce random noise used to test audio systems.

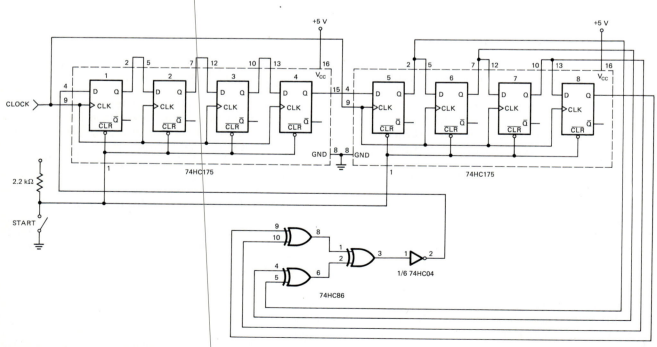

FIGURE 9-36   Circuit for 8-bit random number sequence generator.

A Keyboard Encoder using a Shift Register Ring Counter

The repetitive stepped waveforms produced by a ring counter make them useful in many applications. As an example, we will show you how a ring counter can be used as part of the circuitry for encoding a computer keyboard. This example also reviews some devices you have met before and gives you more practice working your way through a digital system.

Figure 9-37 shows a diagram for a standard computer-style keyboard and circuitry which will output a unique 8-bit code for each key when it is pressed. Getting useful information from a keyboard involves three major tasks:

1. Detecting that a key has been pressed.

2. Debouncing the keypressed signal.

3. Encoding the keypress signal, or in other words producing a compact code such as ASCII which uniquely represents the pressed key.

Let's see how the circuit in Figure 9-37 accomplishes these three tasks.

To start we will discuss how the 74199 part of the circuit operates as a ring counter. A 74199 is an 8-bit shift register which can be loaded in serial or in parallel. To help you understand the operation of the device, Figure 9-38 shows the function table for it.

The 74199 has a $\overline{CLR}$ input which can be used to reset all eight outputs. We don't need to reset in this circuit, so we just tied the $\overline{CLR}$ input high. The 74199 has a clock inhibit input, CLK INH, which, as shown by the last line in the function table, will prevent the device from shifting when it is clocked. We didn't need this function, so we tied CLK INH low, its inactive state.

The third line of the function table in Figure 9-38 shows that if the load input (SH/$\overline{LD}$) of the 74199 is asserted, the data present on the A to H inputs will be transferred to the corresponding Q outputs on the next rising edge of the clock input signal. The fourth line in the function table shows that when SH/$\overline{LD}$ is made high again, succeeding clock pulses will shift the loaded data to the right through the register. The fifth and sixth lines of the function table tell you that if the QH output is connected around to the J and $\overline{K}$ inputs, as it is in Figure 9-37, the level on QH will be transferred to QA when the register is clocked. This connection then makes the circuit a recirculating shift register or ring counter which will shift a loaded data word around and around as it is clocked.

To initially load the desired word into the shift register, we used a special circuit which is sometimes called a *power-on reset* or *load circuit*. Here's how it works. When the power is first turned on, the 33-μF capacitor has no charge, so the input of the 74HC14 inverter is low. The effects of the two inverters cancel, so this produces a low on the SH/$\overline{LD}$ input of the 74199. The next rising edge of the clock signal then loads the A to H data on the QA to QH outputs. When the capacitor charges to a voltage higher than the threshold of the 74HC14, the SH/$\overline{LD}$ input will be made high. Succeeding clock pulses will then shift the loaded data word around the register.

In the circuit in Figure 9-37, we connected the A to G data inputs high and the H input low, so that 11111110 will be loaded on the outputs. As the register is clocked, it will step a low around its outputs over and over. We use this stepping low to detect and encode a pressed key. Let's look at the rest of the circuit in Figure 9-37 to see how this is done.

The key switches in Figure 9-37 are wired as a matrix of rows and columns. If no keys are pressed, there are no connections between the rows and the columns, so all the column lines are held high by the 2.2-kΩ pull-up resistors on them. Pressing a key connects the row line of the key to the column line of the key. When the low circling the 74199 outputs reaches the row containing the pressed key, it will pull the column line containing the key low also. This low on the column line will cause several things to happen.

With no keys pressed, all the inputs of the 74HC30 (U7) will be high, so its output will be low. A low on any column will cause the output of the 74HC30 to go high. Because of the bouncing of the mechanical switch, however, the output of the 74HC30 may go alternately high and low for a few milliseconds as shown next to the 74HC30 in Figure 9-37. We debounce the pressed key by simply waiting until we are sure the bouncing has stopped before we encode the pressed key.

The first rising edge of the signal from the 74HC30 triggers the 74121 (U4). Remember from the discussion of the frequency counter in Figure 9-17, that a 74121 produces a single pulse on its Q output when triggered. The $\overline{Q}$ output produces the complement of the Q pulse. As we discuss in detail in the next chapter, the width of the output pulses is determined by the values of the resistor and capacitor connected to pins 10 and 11. For the values in Figure 9-37, the output pulse widths are about 10 ms.

As shown in the figure, the $\overline{Q}$ output of the U4 74121 is connected to one input of the U9 74HC10 which acts as a gate for the 5-kHz clock signal to the shift register. When the $\overline{Q}$ signal from the 74121 goes low, this gate is disabled so clock signal pulses cannot reach the shift register. The shift register outputs then will remain constant. This holds a stable low on the row containing the pressed key during the 10-ms debounce time.

Note in Figure 9-37 that we also connected the U7 74HC30 output signal through an inverter to one input of the U9 74HC10. We did this so that if a key is held down for more than 10 ms, the clock is prevented from starting the ring counter again until the key is released. Without this addition, the system would, after 10 ms, circle the low around the ring counter again and generate another keypressed signal from the 74HC30.

In summary of the operation so far, then, the 74HC30 detects a pressed key and the 74121 debounces the pressed key by stopping the clock signal for about 10 ms. Now let's see how the circuit produces the ASCII code for the pressed key.

Remember from a discussion in Chapter 7 that a 74HC148 is an eight-input priority encoder. If the device is enabled by making its enable input ($\overline{EI}$) low, it will output a 3-bit binary code which represents the high-

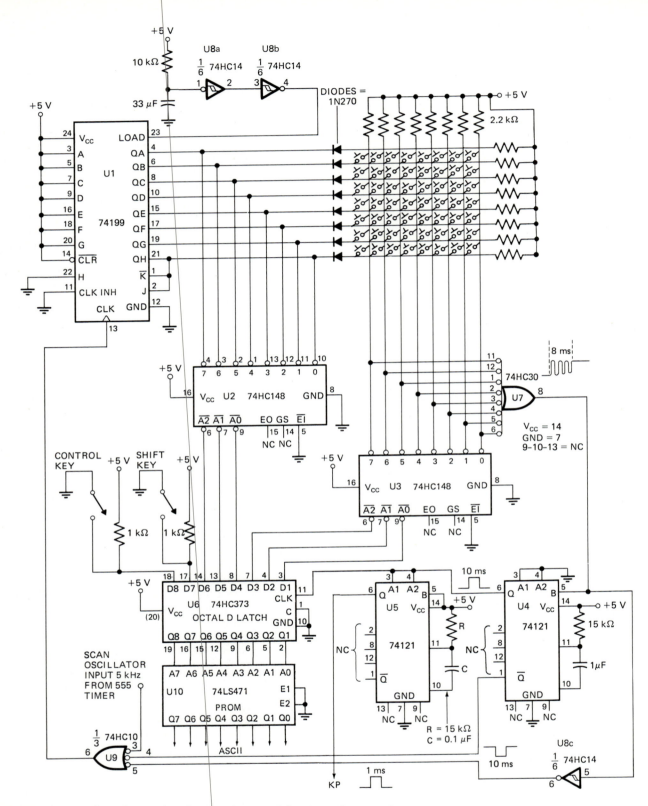

FIGURE 9-37 Keyboard encoder circuit using an eight-stage ring counter.

est numbered data input which has a low on it. The $\overline{A2}$, $\overline{A1}$, and $\overline{A0}$ are active low, so if the 3 input has a low on it, for example, the $\overline{A2}$, $\overline{A1}$, and $\overline{A0}$ outputs will have 100 on them. In the circuit in Figure 9-37, the U2 74HC148

produces a 3-bit code which represents the row which has a low on it. The U3 74HC148 produces a 3-bit code that represents the column which has a low on it when a key is pressed. The two 3-bit codes from the 74HC148s

INPUTS						OUTPUTS			
CLEAR	SHIFT/LOAD	CLOCK INHIBIT	CLOCK	SERIAL J K̄	PARALLEL A...H	QA	QB	QC	... QH
L	X	X	X	X X	X	L	L	L	L
H	X	L	L	X X	X	QA0	QB0	QC0	QH0
H	L	L	t	X X	a...h	a	b	c	h
H	H	L	t	L H	X	QA0	QA0	QBn	QGn
H	H	L	t	L L	X	L	QAn	QBn	QGn
H	H	L	t	H H	X	H	QAn	QBn	QGn
H	H	L	t	H L	X	Q̄An	QAn	QBn	QGn
H	X	H	t	X X	X	QA0	QB0	QB0	QH0

FIGURE 9-38 Function table for the 74199.

are connected to the inputs of a 74HC373 octal D latch along with the SHIFT key switch and the CONTROL key switch. Each key on the keyboard will then produce a different 8-bit code on the inputs of the 74HC373. The code produced, however, will not be the desired ASCII, so we have to go one step more. The outputs of the 74HC373 are connected to the address inputs of a programmable read-only memory (PROM) which converts the 8-bit code to ASCII. As you may remember, a section near the end of Chapter 7 showed you how a PROM can be used to do code conversions such as this.

The only functions of the circuit in Figure 9-37 left to discuss are those of the 74HC373 octal latch and the U5 74121. Remember that when the output of the 74HC30 goes high, it triggers the U4 74121 to produce a 10-ms-wide positive-going pulse on its Q output. This Q output is connected to the CLK input of the 74HC373 latch so that when Q goes high, the latches are enabled. This allows the codes from the 74HC148 encoders to pass through to the PROM. When the Q signal from the 74121 falls low again at the end of the 10-ms debounce time, the code present on the outputs of the 74HC373 will be latched there until another key is pressed. This provides stable input data for the PROM and thereby assures a stable ASCII code on the outputs of the PROM.

The falling edge of the Q signal from the U4 74121 is also used to trigger the U5 74121. When triggered, this 74121 produces a 1-ms-wide pulse which functions as a *keypressed strobe*. The purpose of this strobe is to tell an external system that a valid ASCII code is present on the outputs of the ROM. A microcomputer or some other external system can monitor this strobe signal so that it knows when to read a new ASCII code from the keyboard.

The circuit in Figure 9-37 is a good example of a digital system and shows you how a keyboard encoder works. However, if you need to encode a keyboard, all this circuitry and more is available in single LSI devices, such as the General Instruments AY5-2376 or KB3600. Besides detecting, debouncing, and encoding a keypress to an ASCII code, these devices provide a feature called *two-key rollover*. This means that if two keys are pressed at very nearly the same time, both keypresses will be detected and debounced. The code for the first key pressed will be output with a keypressed strobe, and then the code for the second pressed key will be output with a second keypressed strobe.

Using Shift Registers for Serial Data Transmitting and Receiving

INTRODUCTION TO SERIAL DATA TRANSMISSION

Most digital systems such as microcomputers work with digital data in parallel form. To reduce the number of wires needed to send data from one system to another, however, data words are often transmitted in serial form. In other words, data bits are sent one after another on a single wire or optical fiber.

Figure 9-39 shows one format that is commonly used to send data such as ASCII characters from one system to another in serial form.

If no data characters are being sent, the transmitter holds the signal line in a continuous high or *marking* state. To begin transmitting a character, the transmitter first makes the signal line low for one bit time. The receiver will use this low to identify the start of a character, so this always low bit is appropriately called a *start* bit. After sending the start bit, the transmitter will send the LSB of the data word. After the LSB is sent, each of the remaining bits of the data word are sent in order as shown. The shading in Figure 9-39 indicates that these bits may be high or low, depending on the specific ASCII character or other data word being sent. A parity bit may

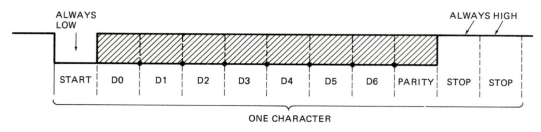

FIGURE 9-39 Commonly used asynchronous serial data transmission format.

or may not be sent after the data bits. After the data and parity bits are all sent, the transmitter makes the signal line high for one or two bit times. These bit(s) are called *stop* bits because they indicate the end of the data word. As you can see in Figure 9-39, it takes 10 or 11 bit times, depending on the number of stop bits, to send one 7-bit ASCII character plus a parity bit.

The rate at which data is being sent in this serial format is usually expressed in units called *baud* (Bd). The baud rate for a serial data transmission is simply 1 divided by the time per bit. For example, if a system is transmitting such that the time per bit is 3.333 ms, the baud rate is 1/3.333 ms = 300 Bd. Commonly used baud rates are 300, 600, 1200, 2400, 4800, 9600, and 19,200.

In systems using the format shown in Figure 9-39, data is transmitted asynchronously. This means that the time per bit within each transmitted character is fixed, but the time between characters is variable. The receiver detects the beginning of each character individually by looking at the line over and over until it finds a start bit. When the receiver finds a start bit, it then reads in the data bits which follow. A person pressing keys on a keyboard at random times is a good example of asynchronous data transmission. Incidentally, the alternative to asynchronous data transmission is *synchronous* transmission. In synchronous systems data words are sent immediately after each other, so start and stop bits are not needed for each data word. The receiver in a synchronous system locks onto the start of a block of transmitted data words and simply counts off bits from that point to identify individual words in the block. In the sequel to this book we discuss synchronous serial data transmission in detail.

To transmit data in the asynchronous serial form shown in Figure 9-39, the data must be converted from parallel to serial at the sending end and then converted from serial back to parallel at the receiving end. Shift registers can be used to perform each of these basic operations.

To perform the parallel-to-serial conversion, the data word, a start bit, a parity bit, and one or two stop bits are first parallel-loaded into an 11-bit parallel-in, serial-out (PISO) shift register. Then the loaded bits are simply shifted out the end of the register onto the transmission line by 11 pulses with a frequency equal to the desired baud rate.

The serial-to-parallel conversion can be done at the receiving end of the line by a serial-in, parallel-out (SIPO) shift register. After some control circuitry detects a start bit, the data and parity bits are clocked serially into a shift register. When all the bits have been shifted in, the data word can be read in parallel from the Q outputs of the shift register. The D0 bit came down the line first, so it will be shifted to the rightmost end of the register.

You seldom have to build serial data transmission circuitry with discrete shift registers because many commonly available LSI devices contain all the required transmit, receive, and control circuitry. In the next section we take a look at one of these LSI devices.

AN EXAMPLE LSI SERIAL TRANSMITTER/RECEIVER

Figure 9-40 shows a block diagram of the transmit section of the General Instruments AY-3-1015D UART. The name UART stands for *universal asynchronous receiver-transmitter*. As the name implies, UARTs contain all the circuitry required for both transmitting and receiving serial data, so putting one of these devices in each of two systems allows bidirectional data transfer between them.

As expected, the heart of the transmit section is a PISO shift register. The data word to be sent is applied to the DB1 to DB8 data inputs. When the $\overline{\text{DATA STROBE}}$ input is asserted, the data word is loaded into the PISO shift register.

Control input pins shown in the upper left corner of Figure 9-40 allow you to program the AY-3-1015D for 5 to 8 data bits, parity or no parity, odd or even parity, and one or two stop bits. Incidentally, a parity bit does not count as a data bit when setting the number of data bits. A parity generator circuit such as those we showed you in Chapter 7 is used to calculate the desired parity bit and insert it with the serial data bits being transmitted. The control bits programmed for transmit also affect the receive operation. For example, if you set the control inputs for even parity, the receiver part of the AY-3-1015D will generate an error signal if the parity of a received data word is not even.

The AY-3-1015D will send and receive data at any rate up to 30,000 Bd. The baud rates are determined by the frequency of the clock signal applied to the 16 × T CLOCK input and the 16 × R CLOCK input shown in Figure 9-41. The 16× in these pin names tells you that the frequency of the applied clock signal must be 16 times the desired baud rate. For example, if you want to transmit data at 4800 Bd, you must apply a signal of 16 × 4800 Hz or 76.800 kHz to the 16 × T CLOCK input. A counter in the AY-3-1015D divides down this 16× frequency to produce the actual clock signal for the shift register. The 16× input signal is required in the transmit section so that other internal operations can be clocked during bit times.

When the transmitter section has completed shifting out the data, parity, and stop bit(s) for a character, it sets the TRANSMITTER BUFFER EMPTY flip-flop. This is a signal to the rest of the world that a new character can be loaded on the DB1 to DB8 inputs to be sent. Pulsing the $\overline{\text{DATA STROBE}}$ input to load the new character resets the TRANSMITTER BUFFER EMPTY flip-flop.

Figure 9-41 shows a block diagram for the receive section of an AY-3-1015D UART. As expected, the heart of this section is the receive shift register, which is a SIPO device. The serial data line is connected to the SERIAL INPUT shown on the left side of the diagram. After the control circuitry detects a valid start bit, the data bits which follow are clocked into the RECEIVER SHIFT REGISTER. When all the bits are clocked in, the data word is checked for errors, the data available flip-flop (DAV) is set, and the data word is transferred to the RECEIVER HOLDING REGISTER BUFFER.

If the $\overline{\text{RECEIVE DATA ENABLE}}$ input is asserted, the RD1 to RD8 three-state outputs will be enabled, and the data word will be present on the RD1 to RD8 pins. If the $\overline{\text{STATUS WORD ENABLE}}$ input is asserted, the three-state status outputs will be enabled. The data available

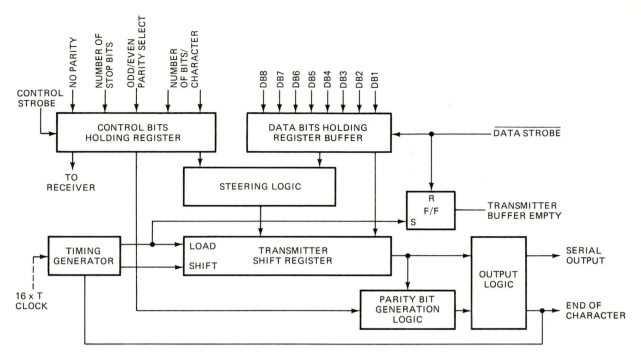

FIGURE 9-40　Internal block diagram for the transmit section of General Instruments AY-3-1015D UART.

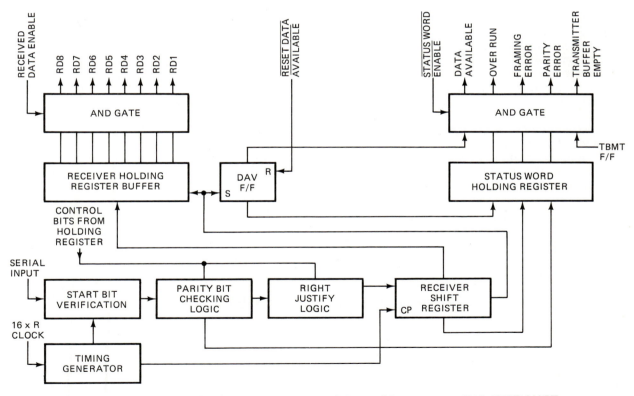

FIGURE 9-41　Internal block diagram for the receive section of General Instruments AY-3-1015D UART.

signal and the error signals will then be asserted on the indicated output pins.

The purpose of the data available signal is to let the receiving system know that a data word is ready to be read on the RD1 to RD8 outputs. When the receiving system reads the data word from RD1 to RD8, the receiving sys-

tem will assert the $\overline{\text{RESET DATA AVAILABLE}}$ input to reset the DAV flip-flop for the next operation.

Signals output on pins 13 to 15 of the device show the incoming data had a parity error, framing error, or an overrun. A *parity error* indicates that the word read in has different parity than the UART is programmed for. A *framing error* shows that the word read in has a different number of bits than the UART is programmed for. An *overrun error* indicates that a second data word was shifted into the UART before a previous word was read out from the UART by the receiving system.

The receive section of the AY-3-1015D also requires an input clock signal to its $16 \times R$ input which is 16 times the desired baud rate. The UART counts off eight pulses of this higher-frequency clock pulse to determine the center of each bit time. For example, when the UART detects a start bit, it counts off eight clock pulses to determine the center of the start bit time. At this time it checks to see if the input line is still low, indicating a valid start bit. If the line is not still low, the UART assumes that the previously detected low was just a noise pulse and it goes back to looking for a start bit again. If the UART finds the line still low at the center of the start bit time, it counts clock pulses until the center of the first data bit time. At this time the data bit is clocked into the RECEIVER SHIFT REGISTER. Reading the data bit at the center of the bit time assures that the data bit has had time to settle to a stable value. To read in the remaining data bits, the UART clocks the receive shift register at the center of each of the remaining bit times.

The $16\times$ clock signals for a UART must be stable in frequency, so they are usually produced by dividing down the signal from a higher-frequency crystal oscillator. Baud-rate generator ICs such as the GE/Intersil IM4702, for example, require only an external crystal to produce the clock frequencies needed for eight common baud rates up to 19,200 Bd. Some UARTs, such as the General Instruments AY2661A, now have a built-in baud rate generator.

A device similar to a UART is a *USART*, or *universal synchronous-asynchronous receiver-transmitter*. A USART can be programmed to send and receive data in asynchronous format or send and receive data in synchronous format. Since UARTs and USARTs are usually used with microcomputers, we wait until the sequel to this book to show the system connections and programming required to work with these devices.

Using Shift Registers in Mathematical Operations

As a final example in this chapter, a parallel-in, parallel-out shift register can be used to shift a loaded number left or right. For now, experiment with some example numbers to prove to yourself that shifting a binary number right one bit position divides it by 2 and shifting a binary number left one bit position multiplies the number by 2. As a start, shifting the binary number 111 or 7 left one bit position gives 1110 or 14.

In a later chapter we show you how two binary numbers can be multiplied by a series of add and shift-right operations. We also show you how division of a binary number by another binary number can be performed with a series of subtract and shift-left operations.

CHECKLIST OF IMPORTANT TERMS AND CONCEPTS IN THIS CHAPTER

Synchronous device

Presettable counter

Count window

Timebase generator

Prescaler

Multiplexed display

Multiplex rate

Comparator

Shift register

SISO, SIPO, PISO, PIPO

Ring counter, recirculating shift register

Stepper motor

Walking ring, Johnson counter

Power-on reset or load circuit

Keypressed strobe

Two-key rollover

Asynchronous serial data transmission format

Marking state

Start and stop bits

Parity bit

Baud, Bd

Baud rate

Synchronous transmission

UART, USART

Framing error

Overrun error

REVIEW QUESTIONS AND PROBLEMS

1. For the circuit in Figure 9-42:
   a. Set up D tables and determine the output sequence starting with state 00. Note that the A flip-flop produces the LSB of the count.
   b. Draw the clock and output waveforms for one count sequence.

2. For the circuit in Figure 9-43:
   a. Set up D tables and determine the output sequence starting with state 00. Note that the A flip-flop produces the LSB of the count.
   b. Draw the clock and output waveforms for one count sequence.

3. For the circuit in Figure 9-44:
   a. Set up JK tables and determine the output sequence starting from state 0000. Note that the A flip-flop produces the LSB of the count.
   b. Draw the clock and output waveforms for one count sequence.

4. For the circuit in Figure 9-45, set up the JK tables and predict the output count sequence starting from state 000.

5. For the circuit in Figure 9-46:
   a. Set up the JK tables and predict the output count sequence starting from state 0000.

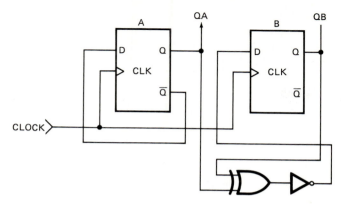

FIGURE 9-42   Circuit for Problem 1.

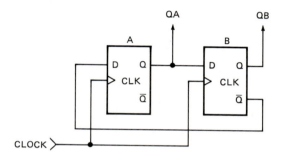

FIGURE 9-43   Circuit for Problem 2.

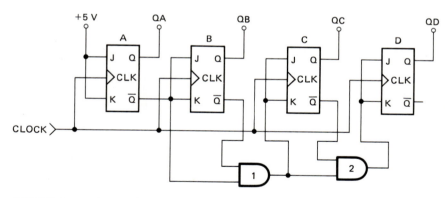

FIGURE 9-44   Circuit for Problem 3.

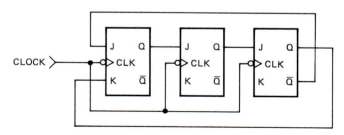

FIGURE 9-45   Circuit for Problem 4.

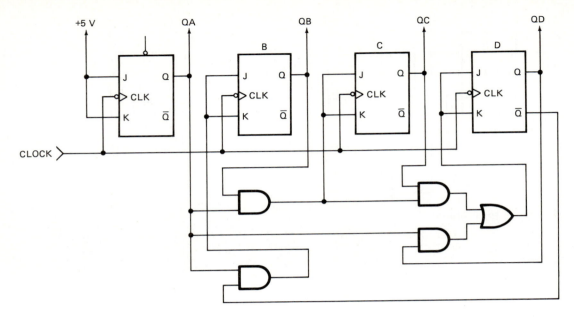

FIGURE 9-46   Circuit for Problem 5.

b.   Give the common name of this circuit.

6.   For the circuit in Figure 9-47, set up the JK tables and predict the output count sequence starting from state 000.

7.   For the circuit in Figure 9-48:
   a.   Set up the JK tables and predict the output count sequence starting from state 000.
   b.   Identify the code that this circuit produces as it is clocked.

8.   What are the major advantages of synchronous counters over ripple counters?

9.   Use the waveforms for a 74ALS193 in Figure 9-13 to help you answer the following questions.
   a.   The CLR input is active _____ and the LOAD input is active _____ .
   b.   On which edge of the input clock signal do the Q outputs of a 74ALS193 change?
   c.   Is the CLR operation synchronous or asynchronous for this counter?

d.   Is the LOAD operation synchronous or asynchronous for this counter?
e.   Draw a circuit diagram showing how a 74ALS193 is connected so that it counts down when clocked.
f.   At what point in a count sequence does the $\overline{BO}$ signal go low? What causes it to go back high again?

10.   *a.*   Draw a circuit diagram showing how two 74ALS193s can be connected to make an 8-bit upcounter.
   b.   Show how you would connect a logic analyzer to the circuit in part *a* to see the sequence of states produced when the circuit is clocked.
   c.   Describe how you would set the clock, trigger, and display controls on the logic analyzer to see the sequence of binary states the 8-bit counter steps through as it is clocked.
   d.   Describe how you connect and set up a logic analyzer to measure the delay time between a

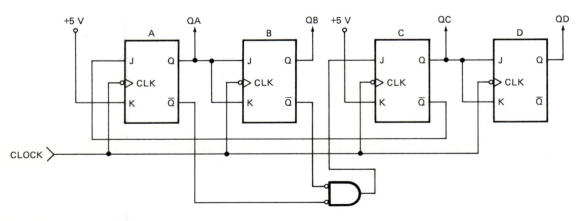

FIGURE 9-47   Circuit for Problem 6.

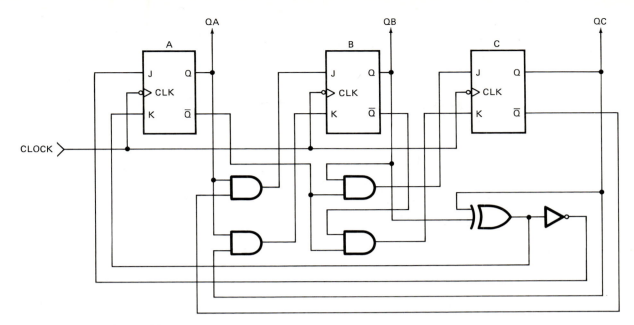

FIGURE 9-48 Circuit for Problem 7.

clock signal on the UP input and a transition on the $\overline{CO}$ on the most significant counter.

11. *a.* Draw a circuit showing how a 74ALS193 can be connected to make a modulo-14 counter.
    b. Use Figure 9-13 to help you draw the clock and Q waveforms for the circuit in part *a.*

12. *a.* Design a simple NAND gate circuit which will route the input clock signal to the *up* input or the *down* input of a 74ALS193, depending on the signal level on a control input. (Refer to the switch section of Chapter 7 if you need a review.)
    b. Figure 7-29*b* shows a circuit which uses a 4-bit binary counter and two 74LS138s to cycle a lit LED across 16 LEDs over and over. With the circuit in Figure 7-29, the lit LED cycles across to the highest LED, then starts back at the first as the 7493 counts up over and over. A 74ALS193 can count up or down, so if a 74ALS193 is connected to the inputs of the two 74LS138s in the circuit in Figure 7-29*b*, the lit LED can easily be made to sweep back and forth.

      Make the required additions to the circuit from part *a* so that it will count up from 0000 to 1111, down to 0000, and up to 1111 over and over. You can use the completed circuit to represent a scanning eye on the front of your robot.

13. *a.* Show the waveforms that will be produced on the outputs of a 74ALS193 if the signals shown in Figure 9-49 are applied to its inputs.
    b. According to the timing parameters for a 74ALS193, what is the minimum time that data must be present on the data inputs before $\overline{LOAD}$ is asserted?

14. Use the frequency counter schematic in Figure 9-17 to help you answer the following questions.

    a. Describe the purposes of the crystal oscillator and chain of decade dividers connected to it.
    b. What is the purpose of the JK flip-flop connected between an output of the divider chain and the enable input of the NAND gate?
    c. What improvement does adding D flip-flops on the outputs of the counters in the main counter chain make in the operation of the circuit?
    d. With a 1-s count window the maximum frequency that this frequency counter can measure is _____ Hz.
    e. With a 0.1-s count window, the maximum frequency that this counter can measure is _____ Hz.
    f. Why is a 10-s count window needed to make more accurate measurements of low frequencies such as 50 Hz?

15. Describe how the frequency counter in Figure 9-17 measures the period of an input signal.

16. *a.* What is the purpose of a prescaler in a frequency counter?
    b. Why is a comparator often used as part of the input circuitry of a frequency counter?

17. *a.* What is the major advantage of a multiplexed display such as that in Figure 9-22?
    b. Describe how the data from the least significant counter is sent to the least significant LED in the multiplexed display in Figure 9-22.
    c. Draw the waveforms that will be present on the A1, A2, and A3 inputs and the waveforms that will be present on the Y0 to Y7 outputs of the 74LS138 in Figure 9-22 as the displays are scanned.

18. Three students combined their protoboards to build the frequency counter from Figure 9-17 with the multiplexed display shown in Figure 9-22. Describe

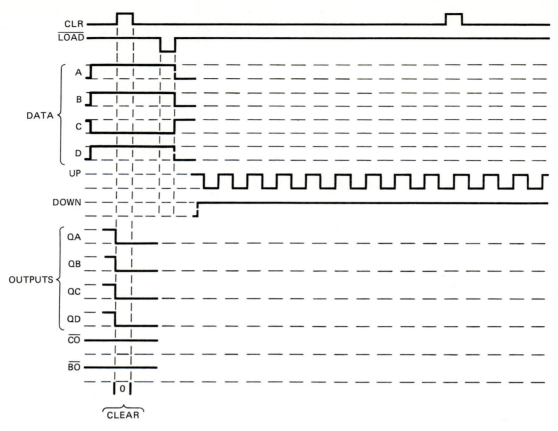

FIGURE 9-49 Circuit for Problem 13.

a series of steps they should take to systematically build and test the circuit.

19. Often when prototyping a digital system such as the frequency counter described in the preceding problem, it does not work correctly the first time. For each of the following symptoms, describe the possible cause(s) and describe how you would go about finding the actual cause of the symptom.
    a. The displayed count never changes when a new frequency is input.
    b. One digit of the display is brightly lit, but the others are all off.
    c. The displayed digits are out of order.
    d. The least significant digits are correct for an input frequency, but the three most significant digits are incorrect.
    e. One digit of the display is brightly lit and always shows a number 8. The other digits are dimly lit but show the correct numbers.
    f. The displayed frequency is off by a factor of 10.

20. Describe how you could use a logic analyzer to determine if the multiplexer section of the frequency counter in Figure 9-22 is working correctly.

21. a. Draw a circuit for a 4-bit SISO shift register made with D flip-flops.
    b. Explain why D latches cannot be used to make a shift register.

22. Use the 74AS194 data sheet portions in Figure 9-28 to help you answer the following questions.
    a. Is the $\overline{\text{CLEAR}}$ on the 74AS194 synchronous or asynchronous?
    b. What connection must be made so that data shifted to the right through the register is recirculated back to the leftmost flip-flop?
    c. What control input conditions are required to transfer the data word on the parallel data inputs to the Q outputs?
    d. Draw a circuit diagram showing how a 74AS194 can be connected as a shift-right ring counter which shifts the word 0011 around and around after loading.
    e. Draw the output waveforms that will be produced as the pattern 1000 is shifted around through the circuit in part d.

23. A 200-state SISO shift register is clocked at 10 kHz. How long is digital data delayed when it passes through this register?

24. a. Why are the 74HC75 D flip-flops required on the S inputs of the 74AS194 in Figure 9-32?
    b. What operation will the circuit in Figure 9-32 perform if SW1 is UP and SW2 is UP?
    c. Draw the output waveforms that will be produced by the circuit in Figure 9-32 if SW1 is UP and SW2 is DN after 1011 is loaded on the Q outputs.

25. The circuit shown in Figure 9-50 is used to control the S1 input of a 74AS194 shift register.
    a. Given the clock waveform shown, draw the waveform that will be produced on the S1 input when SW1 is thrown to the UP position.
    b. Describe the effect that this S1 waveform will have on the operation of the 74AS194.
    c. What advantage does this synchronization circuit have over the synchronization circuit shown in Figure 9-32? *Hint:* How soon does this circuit start shifting after SW1 is thrown?

26. Figure 9-51 shows a circuit for a 4-bit Johnson counter made with JK flip-flops.
    a. Draw the Q and $\overline{Q}$ output waveforms that will be produced by this circuit, starting with 0000 on the Q outputs.
    b. Eight unique states can be decoded from the Q and $\overline{Q}$ outputs of this circuit with a single two-input AND gate as described in the text. Determine the required connections for the eight AND gates. *Hint:* Each state is decoded with a signal which goes high at the start of that state

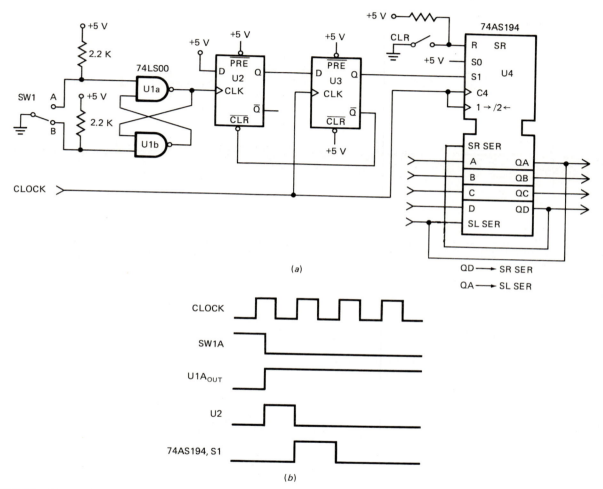

(a)

(b)

FIGURE 9-50   Circuit for Problem 25.

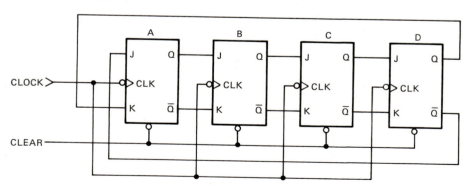

FIGURE 9-51   Circuit for Problem 26.

and a signal which goes low at the end of that state. State 0, for example, is decoded with $\overline{QA}$ and $\overline{QD}$; state 1 is decoded with QA and $\overline{QB}$; etc.

27. a. Determine the count sequence produced by the Johnson counter shown in Figure 9-51, starting with a clear of all the flip-flops.
   b. Suppose that the operator forgets to clear the registers and the circuit starts with 1001 on the Q outputs. Will the circuit ever get to the desired sequence you worked out in part *a*?

28. Figure 9-52 shows the circuit for a 6-bit pseudo-random sequence generator.

c. How does the circuit debounce the keypressed operation?
d. What will be the effect of holding a key down longer than 10 ms?
e. Show the logic levels that will be on the outputs of the two 74HC148 encoders if the third key from the left in the second row from the top is pressed.
f. What is the purpose of the 74HC373 octal latch in this circuit?
g. What is the purpose of the 1-ms-wide pulse generated by the U5 74121 in the circuit?
h. What symptom would you observe in this circuit if a key became stuck closed?

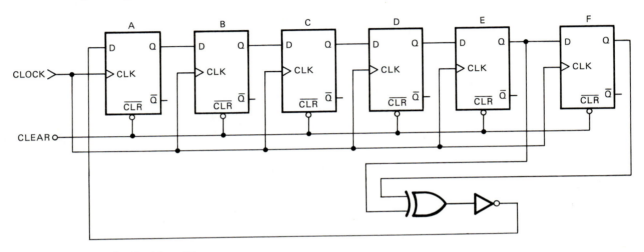

FIGURE 9-52   Circuit for Problem 28.

a. How many unique states does the circuit have?
b. Show how this circuit could be used with the switch and decoder-driver circuit from Figure 7-29b to make a "wheel of fortune" circuit which functions somewhat like a roulette wheel.

29. List the three tasks required to get useful information from a matrix keyboard.

30. Use the keyboard encoder circuit in Figure 9-37 to help you answer the following questions.
   a. What binary word is shifted around the 74199 shift register?
   b. Describe how the keyboard encoder circuit detects when a key is pressed.

31. a. Draw a waveform which shows how the data word 1101110 would be sent in asynchronous serial format with odd parity and one stop bit.
   b. Data is being sent serially at 2400 Bd. What is the time for each bit?
   c. Describe the conversions required to send and receive data in serial form.
   d. What tasks does a UART perform in addition to parallel-to-serial and serial-to-parallel conversions?
   e. Why does a UART require an input clock signal whichhas a frequency 16 times the desired baud rate?

10 Digital Signal Generation, Transmission, and Interfacing

This is the first of two chapters on digital signal generation and interfacing.

As you learned in the preceding chapters, most digital systems require one or more clock signals to sequence their operation. One of the goals of this chapter is to show you how we produce clock signals with the specific characteristics required for a given system. The second major section of this chapter shows you how digital signals are transmitted from one part of a system to another part, and from one system to another system. In the final section of this chapter, we show you how to condition signals so that they have the necessary characteristics to interface with digital inputs, and we show you how to interface digital device outputs to a variety of devices.

In the next chapter we show you how an analog-to-digital (A/D) converter can be used to produce a binary value proportional to the amplitude of an analog signal, and how a digital-to-analog (D/A) converter can be used to convert a binary value to a proportional analog voltage or current.

OBJECTIVES

At the conclusion of this chapter you should be able to:

1. Describe the operation of an astable multivibrator such as a 555.

2. Use a design chart and/or formula to determine the R and C values needed to produce a specified output frequency from a 555 astable multivibrator.

3. Describe the operation of a voltage-controlled oscillator circuit such as that in an MC14046 or a 74LS624.

4. Draw a circuit for a basic crystal oscillator and describe the basic operation of a crystal oscillator.

5. Draw a block diagram for a phase-locked loop circuit and describe its operation.

6. Draw a block diagram and explain how a phase-locked loop circuit can be used to produce a signal

with a frequency which is a multiple of the input frequency.

7. Describe the operation of an astable multivibrator and use a design chart and/or formula to calculate R and C values for a specified output pulse width.

8. Explain why transmission lines must be terminated with a resistor and draw waveforms which show the results of improper termination.

9. Describe how signals are transmitted on RS-232C, RS-422A, and RS-423 transmission lines.

10. Describe the three major methods of modulation used by modems to send digital signals over standard telephone lines.

11. Describe how fiber-optic lines are used to transmit digital signals.

12. Describe the operation of a Schmitt trigger device and explain why it is often used on a standard logic device input.

13. Draw a circuit diagram showing how a comparator can be used to convert a variety of input signals to acceptable logic level signals.

14. Draw circuits showing how logic gate outputs can be interfaced to LEDs, relays, and other power devices.

15. Describe the operation and the advantages of a solid-state relay.

DIGITAL SIGNAL GENERATION

An Astable Multivibrator Signal Generator—the 555

For many applications such as the scope multiplexer in Figure 7-10, the frequency of the required clock signal is relatively low and the exact frequency isn't critical. One possible choice for applications such as this is a 555 timer IC. This device is available from several manufacturers and is relatively inexpensive.

Figure 10-1 shows the internal block diagram and the

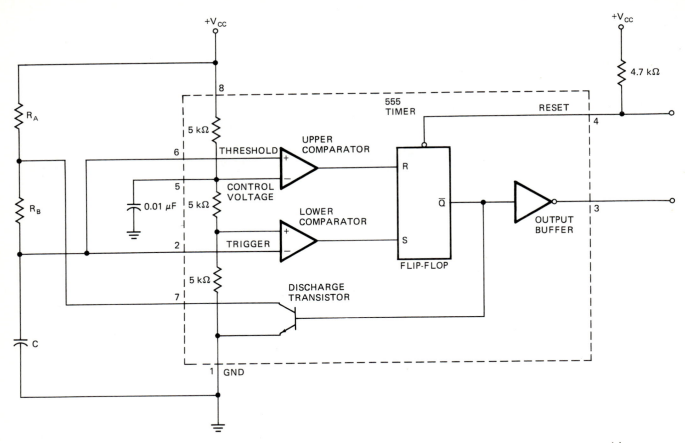

FIGURE 10-1   555 timer internal block diagram and external connections required for operation as an astable multivibrator.   *(Courtesy of National Semiconductor.)*

external components required to cause a 555 to operate as an astable multivibrator. The term *astable multivibrator* simply means that the output will continuously alternate or "vibrate" back and forth between a low and a high. For simplicity, we will refer to the circuit in Figure 10-1 with just the term *astable*. The frequency and duty cycle of the signal output by this astable circuit is determined by the values of resistors R_A and R_B and by the value of the capacitor labeled C. Let's look first at how the internal parts of the device work and then at how the external parts control the operation of the circuit.

The internal voltage divider of three 5-kΩ resistors sets a threshold voltage of $\frac{1}{3}V_{CC}$ on the noninverting (+) input of the lower comparator and a threshold voltage of $\frac{2}{3}V_{CC}$ on the inverting input (−) of the upper comparator. If the voltage applied to the signal input (pin 2) of the lower comparator is less than $\frac{1}{3}V_{CC}$, then the output of the lower comparator will be high. This will set the flip-flop. With the flip-flop in the set condition, the $\overline{Q}$ output will be low and the discharge transistor will be off. The inverting buffer connected to $\overline{Q}$ will produce a high on the final output.

If the voltage on the signal input of the upper comparator is greater than $\frac{2}{3}V_{CC}$, the output of the upper comparator will be high and the flip-flop will be reset. This will cause the discharge transistor to be turned on and the final output on pin 3 to be low.

In the circuit in Figure 10-1, the comparator inputs are both connected to the timing capacitor, C. When the power is first turned on, the capacitor has no charge, so the voltage on it is 0 V. Since this is less than $\frac{1}{3}V_{CC}$, the flip-flop will be set, the discharge transistor will be off, and the final output will be high. As time passes, the timing capacitor is charged through R_A and R_B. When the voltage on the capacitor passes $\frac{2}{3}V_{CC}$, the output of the upper comparator will go high and reset the flip-flop. This will turn on the discharge transistor and make the final output go low. When the discharge transistor is on it acts like a very low resistance, so the charge on the timing capacitor is drained off to ground through R_B. When the voltage on the capacitor drops below $\frac{1}{3}V_{CC}$ again, the output of the lower comparator will go high and set the flip-flop. This will cause the discharge transistor to turn off and the final output to go high again. The cycle repeats over and over as the capacitor is alternately charged to $\frac{2}{3}V_{CC}$ and discharged to $\frac{1}{3}V_{CC}$.

To determine the appropriate resistor and capacitor values for a desired output frequency, f, you can use the chart in Figure 10-2 or the relationship:

$$f = \frac{1.49}{(R_A + 2R_B)C}$$

The duty cycle, DC, of the output waveform is calculated using the formula:

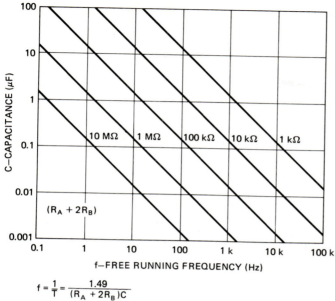

$$f = \frac{1}{T} = \frac{1.49}{(R_A + 2R_B)C}$$

DUImE CYCLE = $\dfrac{\text{TIME OUTPUT HIGH}}{T}$

$= \dfrac{R_A + R_B}{R_A + 2R_B}$

R_A OR R_B MINIMUM = 1 kΩ

$R_A + R_B$ MAXIMUM = 6.6 MΩ

FIGURE 10-2 Chart for determining R and C values for 555 timer operating as an astable. *(Courtesy of National Semiconductor.)*

$$DC = \frac{R_A + R_B}{R_A + 2R_B} \times 100\%$$

It usually takes a couple of trials to determine commonly available values which give the desired frequency and duty cycle. Note that for operation with a V_{CC} of 5 V the minimum values for R_A and R_B are 1 kΩ and the maximum value of $R_A + R_B$ is 6.6 MΩ.

As an example of how you determine the required values, suppose that you want to produce a 1-kHz signal with about a 50 percent duty cycle. To use the chart in Figure 10-2, move along the bottom of the chart until you find the vertical line which represents 1 kHz. Go up this vertical line until you reach the horizontal line which represents a convenient capacitor value and make a dot. For this example, suppose that we use a 0.01-μF capacitor. The diagonal line nearest this dot now gives you an approximate value for $(R_A + 2R_B)$. The dot for this example is between the 100-kΩ line and the 1-MΩ line, but it is closer to the 100-kΩ line, so a first guess at the value for $(R_A + 2R_B)$ might be 150 kΩ.

Since we want an output signal with approximately a 50 percent duty cycle, we want R_A to be as small as possible compared with R_B. For 5-V operation, the minimum recommended value for R_A is 1 kΩ, so we will use this value. Subtracting an R_A value of 1 kΩ from the $(R_A + 2R_B)$ value of 150 kΩ leaves a value of 149 kΩ for $2R_B$. This gives an approximate value of 74 kΩ for R_B. Once you get an approximate value from the chart, you can use the formula $f = 1.49/(R_A + 2R_B)C$ to get a more

precise value. With a frequency of 1 kHz, capacitor of 0.01 μF, and an R_A of 1 kΩ the formula gives a value of 74 kΩ for R_B. Since we made a lucky guess at reading the chart in Figure 10-2, this is the same value we determined from the chart. As shown by the chart in Table 1-3, the standard 5% resistor value nearest to this is 75 kΩ, which differs from the desired value by less than the tolerance of the resistor, so this is an appropriate value to use.

As we said above, the duty cycle of the output waveform is DC = [($R_A + R_B$)/($R_A + 2R_B$)] × 100 percent. Substituting the R_A value of 1 kΩ and the R_B value of 75 kΩ in this relationship gives a duty cycle of 50.3 percent, which is very close to the desired value of 50 percent.

In most applications where you decide to use a 555 as a signal generator, you probably don't care about the exact output frequency. However, if you need to improve the accuracy of the circuit, there are several things you can do. The first is to use a stable capacitor such as a Mylar or polycarbonate type rather than a ceramic disk type such as you use for V_{CC} bypassing. Second, use 1 percent metal film resistors, which change value less with changes in temperature than common carbon film resistors. Third, reduce the value of R_B and add a variable resistor in series with it so that you can adjust the output frequency. For example, instead of using a 75-kΩ fixed resistor for R_B, you might use a 62-kΩ fixed resistor with a 20-kΩ variable resistor in series with it.

NOTE: Don't use just a variable resistor for R_B because, if the variable resistor accidentally gets set at 0 Ω, the 555 may be destroyed!

Before we go to the next type of signal generator, here are a few additional notes about using a 555 as an astable.

The external RESET input (pin 4) can be used to gate the circuit on and off. If the external RESET input of a 555 is made low, the output immediately goes low and stays low. When RESET is made high, the circuit starts pulsing again.

The CONTROL VOLTAGE input (pin 5) can be used to vary the output frequency of the circuit. A voltage applied to this input will change the comparator threshold voltages and thereby change the frequency of oscillation. When operating with a V_{CC} of + 5 V, the CONTROL INPUT voltage can be varied from 1.7 to 5 V. Varying the CONTROL INPUT voltage over this range varies the output frequency about 25 percent around the value predicted by the chart in Figure 10-2. For applications where you need to vary the output frequency, however, the voltage-controlled oscillators we discuss in the next section are a better choice. If you are not using the CONTROL INPUT pin on a 555 to vary the output frequency, then you should connect a 0.01-μF capacitor between pin 5 and ground as shown in Figure 10-1 to prevent random noise from modulating the output frequency.

Voltage-Controlled Oscillators

In many applications we need an oscillator whose output frequency can be easily varied by changing an externally applied voltage. A device designed to do this is called a *voltage-controlled oscillator*, or *VCO*. As a first example of a VCO we will use the one contained in the CMOS 4046 device. This device is available in standard CMOS as the CD4046 or MC14046, and in high-speed CMOS as the 74HC4046. The devices are pin for pin compatible. The standard CMOS versions will operate with a V_{CC} of 3 to 18 V, but for the HC device, the V_{CC} must be between 3 and 6 V. With a V_{CC} voltage of 5 V, the standard CMOS devices have a maximum frequency of about 300 kHz and the HC devices have a maximum frequency of about 15 MHz.

Figure 10-3a shows a block diagram of a 74HC4046. For now we will just discuss the operation of the VCO in the device. In a later section of the chapter on phase-locked loops we will discuss the use of the phase detectors in the device.

Figure 10-3b shows the connections for a simple 4046 VCO circuit. For this discussion we will assume a V_{CC} of 5 V, so that the output signal level is compatible with common CMOS and TTL inputs.

A 74HC4046 VCO oscillates by charging and discharg-ing the capacitor connected between the C1A and C1B inputs. The output frequency, f, is determined by the value of capacitor C1, the values of resistors R1 and R2, and by the control voltage applied to the VCO input. When the VCO control input voltage is at ground, the circuit will oscillate at its minimum frequency. The approximate minimum frequency of oscillation is determined by the relationship:

$$f_{MIN} = \frac{1}{R2(C1 + 32 \text{ pF})}$$

When the VCO control input voltage is at V_{CC}, the circuit will oscillate at its maximum frequency. The approximate maximum output frequency is expressed by the relationship:

$$f_{MAX} = \frac{1}{R1(C1 + 32 \text{ pF})} + f_{MIN}$$

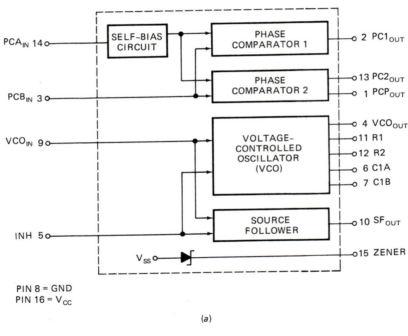

PIN 8 = GND
PIN 16 = V_{CC}

(a)

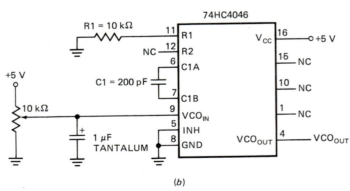

(b)

FIGURE 10-3   (a) Block diagram of CMOS 4046 device with VCO. (b) Circuit connections for 74HC4046 VCO to produce a 0-Hz to 1-MHz sweep range.

NOTE: These formulas are for estimation only, they may be off by as much as a factor of 4. Circuits should be built and values adjusted as needed to achieve desired frequencies.

For the 74HC4046, the value of capacitor C1 can be any value greater than 40 pF. Resistors R1 and R2 must each be between 3 and 300 kΩ. R2 should always be larger than R1. With a V_{CC} of 5 V, the maximum operating frequency of the 74HC4046 is about 15 MHz.

For a standard CMOS 4046, the value of capacitor C1 must be in the range 100 pF to 0.01 µF. Resistors R1 and R2 must each be in the range of 10 kΩ to 1 MΩ. R2 should always be larger than R1. With a V_{CC} of 5 V, the maximum operating frequency of the CD4046 is about 300 kHz.

If resistor R2 is omitted from the circuit, the minimum frequency of oscillation will be near 0 Hz. The output frequency will then vary from 0 to f_{MAX} as the VCO input voltage is varied from ground to V_{CC}. If resistor R2 is included in the circuit, the output frequency will vary from some f_{MIN} or *offset* frequency to a frequency of $f_{MAX} + f_{MIN}$ as the VCO input voltage is varied from ground to V_{CC}. With the values shown in Figure 10-3b, for example, the circuit can be adjusted from 0 to 1 MHz if R2 is left out. If R2 is included in the circuit, the frequency can be adjusted from approximately 100 kHz to 1.1 MHz.

When operating with a V_{CC} of 5 V, a 74HC4046 has a maximum output frequency of about 15 MHz. If you need a VCO which is capable of operating at much higher frequencies, you might use a 74S124, which can operate at frequencies as high as 85 MHz.

A fun application of a 4046 is to produce the sound of a siren. Figure 10-4 shows a circuit to do this. The transistor buffer on the output supplies the current necessary to drive a small speaker. When the switch is in the down position, the voltage on the 2.2-µF capacitor (C2) and the VCO input (pin 9) is held at ground by the 2.2-MΩ resistor (R4). The VCO output frequency then is near 0 Hz. When the switch is moved to the up position, the large capacitor charges slowly through the 2.2-MΩ resistor (R3). As the capacitor charges, the voltage on the VCO input increases until it reaches V_{CC}. This increasing voltage causes the output frequency to sweep from 0 Hz to the maximum frequency determined by C1 and R1. When the switch is moved to the down position again, the capacitor discharges slowly through the 2.2-MΩ resistor, causing the output frequency to drop down to 0 again.

The output frequency of a VCO such as a 74HC4046 or a 74S124 is not exactly predictable, it varies from device to device, and it changes appreciably with changes in temperature. In a later section of the chapter on phase-locked loops, we show you how these problems are overcome. In the next section here we discuss circuits which produce very stable output frequencies.

Crystal Oscillators

BASIC THEORY OF OPERATION

When we need a signal source with a very accurate and stable frequency, we usually use a *crystal-controlled oscillator*. Complete crystal oscillators are available in a variety of packages. Some units such as the Motorola K1149AM, shown in Figure 10-5a, plug into a standard IC socket. Other packages are designed for direct surface mounting, such as the STATEK CX-IV-SM shown in Figure 10-5b.

The crystal in most crystal oscillators consists of a thin slice of quartz mounted between two metal plates with connecting leads. Figure 10-5c shows a picture of a common crystal package. If a voltage is suddenly applied to a crystal, the crystal will vibrate at a specific frequency,

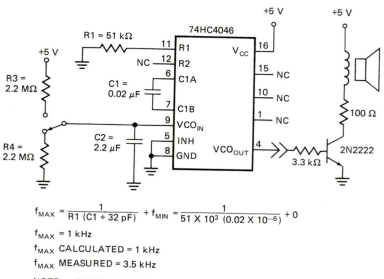

$$f_{MAX} = \frac{1}{R1\,(C1 + 32\,pF)} + f_{MIN} = \frac{1}{51 \times 10^3 \,(0.02 \times 10^{-6})} + 0$$

$f_{MAX} = 1\ kHz$

f_{MAX} CALCULATED = 1 kHz

f_{MAX} MEASURED = 3.5 kHz

NOTE: THESE FORMULAS ARE FOR ESTIMATION ONLY, THEY MAY BE OFF BY AS MUCH AS A FACTOR OF 4. CIRCUITS SHOULD BE BUILT AND VALUES ADJUSTED AS NEEDED TO ACHIEVE DESIRED RESULTS.

FIGURE 10-4 74HC4046-based siren circuit.

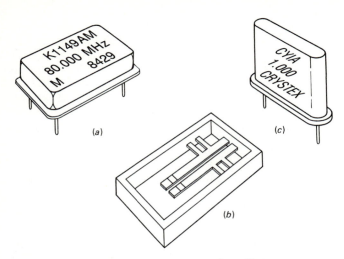

FIGURE 10-5 *(a)* DIP package crystal oscillator. *(b)* Surface-mounted crystal oscillator. *(c)* Common package used for crystal only.

just as a guitar string will vibrate at a specific frequency when it is strummed. The specific frequency that the crystal vibrates at is called its *resonant frequency.* The resonant frequency of the crystal is determined mostly by the size and thickness of the crystal slice. Available crystal frequencies range from a few kilohertz to above 100 MHz.

To explain how a crystal oscillator works, we will use the simple gate circuits shown in Figures 10-6 and 10-7.

As a crystal vibrates, it produces an ac voltage with a frequency equal to its resonant frequency. If the generated voltage is amplified and fed back to the crystal with the right phase, the crystal will continue to vibrate at this resonant frequency. The process is about the same as pushing a playground swing at just the right point in each cycle so that the swing keeps going at its resonant frequency.

Depending on how the crystal is connected in a cir-

cuit, it may vibrate in its *series resonant mode* or in its *parallel resonant mode.* If the crystal is vibrating in its series resonant mode, the generated voltage will be in phase with the externally applied voltage. What this means is that at a frequency equal to its series resonant frequency, the crystal will act simply as a small-value resistor. Here's how we use this to make an oscillator circuit.

The voltage dividers on the inputs of the 74LS04 inverters in Figure 10-6 adjust the voltages on the inputs to about 1.0 V, which is equal to their threshold voltages at 25°C. When *biased* or voltaged in this way, the inverters act as amplifiers. If you look at the transfer curve for an inverter in Figure 3-18, you can see that at this threshold voltage a small change on the input voltage causes a large change on the output voltage of the inverter. Now, suppose that we apply a small ac signal to the input of the first inverter. This signal will be amplified and shifted in phase by 180°. The second inverter will amplify the signal further and shift its phase by another 180°. The sum of the two phase shifts is equal to 360°, which is the same as 0°, so the final output signal is in phase with the original input signal. This is just another way of saying the effects of two inverters cancel. If the frequency of the signal we apply to the input is equal to the series resonant frequency of the crystal, then most of the output signal from the second inverter will be fed back through the low resistance of the crystal to the input of the first inverter. Since the signal fed back is in phase with the input signal, the two signals will add together and the sum will pass through the inverters again. Each time the signal passes around the loop it will get larger, until the outputs of the inverters are going from a logic low to a logic high on each cycle. The process is then self-sustaining, and the circuit will output a pulsed logic level signal with a frequency equal to the series resonant frequency of the crystal.

In actuality we don't have to apply a signal to get the process started because small noise signals which are

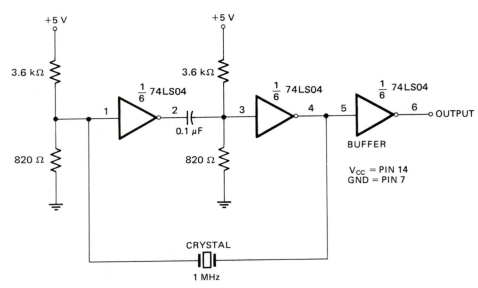

FIGURE 10-6 TTL circuit for crystal oscillator using series resonance mode of the crystal. *(Courtesy of Motorola.)*

always present will do the job for us. You may have seen an example of this feedback oscillator principle if you have heard a public address sound system produce a loud "howl" when sound from the speakers bounces off a wall and feeds back into the amplifier through a microphone.

The third inverter in the circuit in Figure 10-6 simply buffers the oscillator part of the circuit so that the capacitance of any gate inputs connected to the oscillator will not affect the output frequency. Note that if you build a circuit such as the one in Figure 10-6, you may have to adjust the resistor values somewhat so that the circuit will start oscillating.

If a crystal is operating in its parallel resonant mode, the ac voltage generated by the vibrating crystal is 180° out of phase with the applied voltage. Figure 10-7 shows

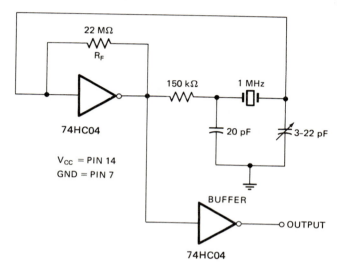

FIGURE 10-7 Circuit for CMOS oscillator using parallel resonance mode of the crystal. *(Courtesy of Motorola.)*

a CMOS oscillator circuit which uses the parallel resonant mode of a crystal. The 22-MΩ resistor from the output of the CMOS inverter to its input biases the input near its threshold voltage. A small ac signal applied to the input of the inverter then will be amplified and its phase shifted by 180° as it passes through the inverter. The crystal will shift the phase of the signal by another 180° and feed the result back to the input of the inverter. Because of the two 180° phase shifts, the signal fed back to the input of the inverter will be in phase with the input signal. The two signals will add and the result will be passed through the inverter again. Each time the signal loops around through the inverter it is amplified until it is swinging from a logic low to a logic high. The circuit is then self-sustaining and produces a stable output signal with a frequency equal to the parallel resonant frequency of the crystal.

As with the previous circuit, we don't have to insert a signal to get the process started because random noise signals will do it. The two capacitors form a voltage divider which determines the fraction of the output signal which is fed back to the input. The capacitor on the right is adjusted so that the oscillator starts quickly and holds a stable frequency. As with the previous crystal oscilla-

tor circuit, an output buffer is used to prevent the input capacitance of following devices from affecting the oscillating frequency.

For a given crystal, the parallel resonant frequency is slightly higher than the serial resonant frequency. Manufacturers usually specify whether a crystal is intended for serial operation or for parallel operation, so if you are building a crystal oscillator such as the ones in Figure 10-6 or 10-7, it is best to choose a crystal which is designed to operate in the proper mode.

CRYSTAL OSCILLATOR APPLICATIONS

As we said before, we use a crystal oscillator in any application where we want a very precise and stable signal frequency. One common application is in a digital watch or clock. Digital watches commonly use a 32,768-Hz crystal oscillator. This frequency is divided by 2^{15} or 32,768 to give a precise 1-Hz signal which can be used to count off seconds.

Crystal oscillators are used to produce the timebase for digital instruments such as frequency counters. The crystal oscillator and the divider chain in Figure 9-17 are examples of this.

Another use of crystal oscillators is to generate the baud-rate clocks needed for UARTs. Remember from Chapter 9 that UARTs need a transmit clock and a receive clock, each with a frequency 16 times that of the desired baud rate. An IC called a *baud-rate generator* divides the signal from a crystal oscillator by several modulos to produce the clock signals needed for several common baud rates. The Intersil IM4702 shown in Figure 10-8, for example, requires only an external 2.4576-MHz crystal, two capacitors, and a resistor to produce any of the commonly needed baud-rate clock frequencies. The binary word applied to the S0 to S3 inputs of the IM4702 determines which output from the internal dividers is routed to the Z output. As shown in Figure 10-8, a simple switch can be used to program the device to output the clock frequency needed for several of the most common baud rates.

Still another application for crystal oscillators is to produce stable frequency clock signals for microprocessors.

Monostable Multivibrators

The oscillators we described in the preceding sections generally produce square-wave signals. For many applications, we need to modify the duty cycle of the pulses to some value other than 50 percent. For applications where the exact pulse width is not critical, a monostable multivibrator is an inexpensive way to do this. Basically a monostable circuit functions as follows.

The output is normally in a low state. When the circuit receives a trigger pulse, the Q output goes high for a time and then returns to a low state. The time the Q output remains in the high state is determined by the value of a resistor and the value of a capacitor. The prefix "mono" in the name of the circuit tells you that the circuit has only one stable state to which it returns after the RC-determined time. Here are a few examples of devices which can be used to produce a variety of monostable circuits.

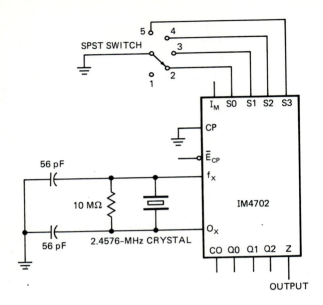

SWITCH POSITION	Z OUTPUT = 16 X BAUD RATE SHOWN
1	110 Bd
2	150 Bd
3	300 Bd
4	1200 Bd
5	2400 Bd

FIGURE 10-8   Circuit for crystal-controlled baud rate generator using an Intersil IM4702.

THE 555 TIMER USED AS A MONOSTABLE MULTIVIBRATOR

Figure 10-9a shows a monostable multivibrator circuit made with a 555 timer device. The 555 in this circuit is triggered by applying a low-going pulse to its TRIGGER input (pin 2). When triggered, the 555 output (pin 3) will go high for a time equal to $1.1R_A \times C$. With a V_{CC} of 5 V, R_A can have a value from 1 kΩ to about 6.6 MΩ. The value of the capacitor can be anywhere between 500 pF and several microfarads. For best results, use low-leakage types of capacitors and avoid extreme values of resistors and capacitors. Figure 10-9b shows a table you can use to determine reasonable values of R and C for a desired output pulse width. To use this chart, first find the desired output pulse width on the horizontal axis. Go vertically up from this value until you intersect with a horizontal line which represents a standard capacitor value. Use the slanted lines nearest to the intersection of these two lines to estimate the required value of resistor. Since common capacitors have a tolerance of 10 or 20 percent and common resistors have a tolerance of at best 5 percent, it is not usually worth agonizing too much about exact values when working with monostables.

The trigger pulse width for a 555 must be shorter than the desired time high for the output pulse. If the available trigger pulse is too wide, it can be coupled to the

TRIGGER input with a capacitor as shown by the ALTERNATE TRIGGER INPUT in Figure 10-9a.

A 555 is inexpensive and works well for frequencies up to 100 kHz or so and pulse widths down to about 10 µs. For higher-frequency and shorter-output pulse width applications, we turn to TTL family devices such as the 74121 and the 74LS122 described in the following sections.

THE 74121 MONOSTABLE MULTIVIBRATOR

A 74121 is the device we used to produce the clock pulses for the 74HC175 flip-flops and the reset pulses for the

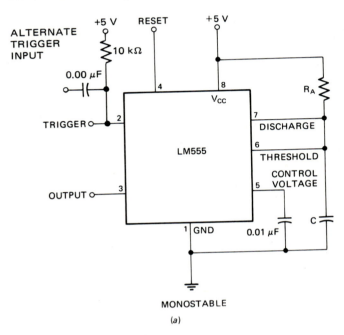

MONOSTABLE

(a)

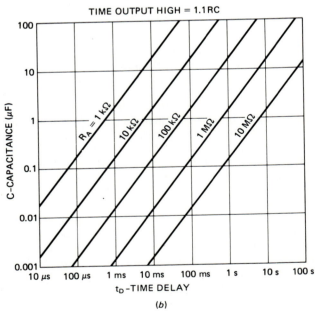

FIGURE 10-9   555 monostable multivibrator. (a) Circuit. (b) Chart used for determining R and C values needed to produce a desired output pulse width. (Courtesy of National Semiconductor.)

74HC160 counters in the frequency counter circuit in Figure 9-17. Here's a little more detailed look at how the device works.

Figure 10-10 shows a logic diagram, a function table, and an example circuit for a 74121. As you can see in the logic diagram in Figure 10-10a, the 74121 has Q and $\overline{Q}$ outputs. When the device is triggered, the Q output will go high and the $\overline{Q}$ output will go low. As you can also see in the logic diagram, a 74121 has three trigger inputs, A1, A2, and B. The A1 and A2 inputs are active low and the B input is active high. The arrows in the truth table in Figure 10-10b tell you that these inputs are edge-active. If, for example, the A1 input is low and the A2 input is low, a low-to-high transition on the B input will trigger the device. As another example, if the B input is high and the A1 input is high, a high-to-low transition on the A2 input will trigger the device.

Since the 74121 is an edge-triggered device, the width of the triggering pulse doesn't matter. Once the device is triggered, Q will go back low after a time determined by the R and C values, even if the trigger input is still active. Also, if a second trigger pulse occurs while Q is high, it will be ignored. Remember this response so that you can compare it with the response of the 74LS122 we discuss in the next section.

The output pulse widths for a 74121 are determined by the relationship:

$$t_{PW} = 0.7RC$$

As shown in the logic diagram in Figure 10-10a, the 74121 contains an internal resistor of about 2 kΩ that you can use for the timing resistor. If this internal resistor is not accurate enough or not a reasonable value for a desired pulse width, you can use an external resistor between 2 and 40 kΩ. The timing capacitor value can be in the range of 10 pF to 10 μF. Figure 10-11 shows a chart which you can use to help determine reasonable R and C values for a desired output pulse width.

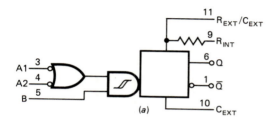

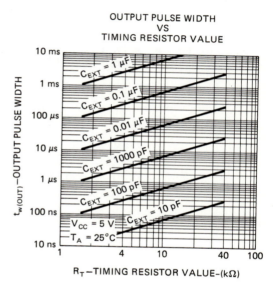

FIGURE 10-11 Design chart for 74121 monostable multivibrator. (Courtesy of Texas Instruments Inc.)

THE 74LS122 RETRIGGERABLE MONOSTABLE

Figure 10-12 shows the logic diagram, function table, and an example circuit for a 74LS122 monostable multivibrator. The logic diagram for this device is almost the same as that for the 74121, except that the 74LS122 has an additional active high trigger input and it has an asynchronous $\overline{CLR}$ input. As shown in the function table in Figure 10-12b, a 74LS122 can be triggered in any one of several different ways. For example, if the A1 or the A2 input is low and the B1 input is high, a rising edge on the B2 input will trigger the device. As with the 74121, the Q output will go high and the $\overline{Q}$ output will go low when the device is triggered. If the $\overline{CLR}$ input is made low, the Q output will immediately go low and stay low, regardless what point it was at in an output pulse. Also, as shown in the logic diagram, a low on $\overline{CLR}$ disables the trigger input gating, so the device will not respond to any new trigger pulses as long as $\overline{CLR}$ is low.

FUNCTION TABLE

INPUTS			OUTPUTS	
A1	A2	B	Q	$\overline{Q}$
L	X	H	L	H
X	L	H	L	H
X	X	L	L	H
H	H	X	L	H
H	↓	H	⊓	⊔
↓	H	H	⊓	⊔
↓	↓	H	⊓	⊔
L	X	↑	⊓	⊔
X	L	↑	⊓	⊔

(b)

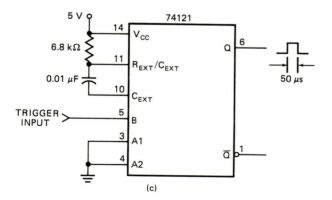

FIGURE 10-10 74121 monostable multivibrator. (a) Logic diagram. (b) Function table. (c) Example circuit. (Courtesy of Texas Instruments Inc.)

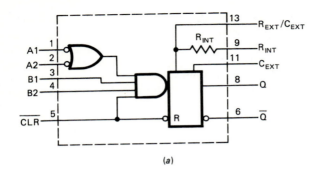

(a)

INPUTS					OUTPUTS	
CLEAR	A1	A2	B1	B2	Q	$\overline{Q}$
L	X	X	X	X	L	H
X	H	H	X	X	L	H
X	X	X	L	X	L	H
X	X	X	X	L	L	H
H	L	X	↑	H	⊓	⊔
H	L	X	H	↑	⊓	⊔
H	X	L	↑	H	⊓	⊔
H	X	L	H	↑	⊓	⊔
H	H	↓	H	H	⊓	⊔
H	↓	↓	H	H	⊓	⊔
H	↓	H	H	H	⊓	⊔
↑	L	X	H	H	⊓	⊔
↑	X	L	H	H	⊓	⊔

(b)

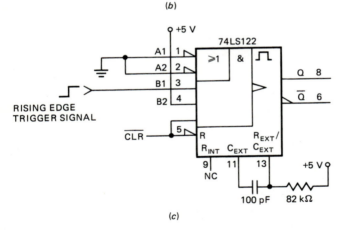

(c)

FIGURE 10-12   74LS122 monostable multivibrator. (a) Logic diagram. (b) Function table. (c) Circuit connections for positive edge trigger. (Courtesy of Texas Instruments Inc.)

The output pulse widths for a 74LS122 are again determined by the value of a timing resistor and the value of a timing capacitor. You can use the internal 10-kΩ resistor or an external resistor between 5 and 260 kΩ as shown in Figure 10-12c. The timing capacitor can be any value from a few picofarads to several microfarads. For capacitor values greater than 1000 pF, the width of the output pulse is approximately equal to $0.35 \times R_T \times C_{EXT}$. For capacitor values less than 1000 pF, Figure 10-13 shows a design chart to help you determine reasonable values of R and C for a desired pulse

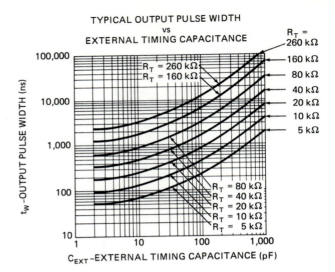

FIGURE 10-13   Design chart for 74LS122 monostable multivibrator. (Courtesy of Texas Instruments Inc.)

width. The minimum pulse width for a 74LS122 is 40 ns. For versions of this device in other logic families, such as a 74HC122, consult the data sheet for that specific device because the parameters vary somewhat from family to family.

A major feature of a 74LS122 is that it is *retriggerable*. This means that if a new trigger pulse arrives while the output is high, the 74LS122 will start the output pulse time over again at that point. The waveforms in Figure 10-14 show how this works. Figure 10-14 compares the responses of a 74LS122 and a 74121 to the same trigger signals. To make the comparison easier, the circuits which produced these waveforms were designed to have the same output pulse widths.

Both circuits were designed to be positive-edge-triggered, so when the first positive edges of the trigger signal occur, both Q outputs go high. When the 74121 receives a second trigger pulse while the Q is still high, it ignores this second trigger pulse. Its Q output simply goes low again after the time determined by the timing R and C values. When the 74LS122 receives a second trigger pulse while its Q output is still high, it starts the output high pulse time over again. When this time runs out, the Q output of the 74LS122 goes low again because it did not receive another trigger pulse

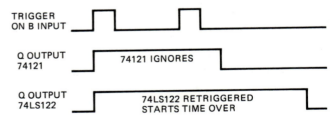

FIGURE 10-14   Comparison of output waveforms for 74121 nonretriggerable monostable and 74LS122 retriggerable monostable.

during the time it was high. If the 74LS122 had received another trigger pulse before the Q output went low, the Q output would have remained high for another time period. If a series of pulses is applied to the TRIGGER input so that it is continuously retriggered before Q goes low, the output will remain continuously high. This feature can be used to detect if a pulse is "missing" in a train of pulses. The output pulse width is calculated so that if a pulse is missing on the train of pulses on the trigger input, the Q output will go low to indicate this. In a problem at the end of the chapter we show you an application of this feature. Next here we discuss a type of circuit which can "track" or produce a multiple of the frequency of an external signal.

Phase-Locked Loops

BASIC OPERATION

In applications such as digital television tuning systems, digital compact audio disk players, and magnetic disk data storage systems, we need a signal generator whose frequency will follow or "lock-on" the frequency of some external signal. For these applications we use a circuit called a *phase-locked loop* or *PLL*. We will first give an overview of the operation of a PLL, and then we will use a specific device to discuss the detailed operation of a PLL circuit.

Figure 10-15 shows a block diagram for a basic PLL.

frequency to increase. The VCO frequency will keep increasing until it is equal to the frequency of the external signal. At this equilibrium point, the filtered output voltage from the phase comparator is just right to hold the VCO oscillating at a frequency equal to the frequency of the external signal. When the circuit reaches this equilibrium condition, it is said to be *locked*.

If the VCO frequency is initially greater than the external frequency, the filtered output from the phase comparator will be such that it causes the VCO frequency to go down until the two frequencies are equal. To summarize, then, the output of the phase comparator will always drive the loop toward the locked condition. In order for the circuit to lock on an external signal, the frequency of the external signal must be within a certain range called the *capture range*.

Once the loop is locked on the frequency of the external signal, it will follow or *track* the external signal. In other words, if the frequency of the signal applied to the external input is increased, the VCO output frequency will increase to match it. If the frequency of the external signal is decreased, the VCO output frequency will decrease to match it. The range of frequencies over which a particular loop can remain locked on the external signal is called its *lock range*.

Now, let's see how the loop functions if we throw switch SW1 in Figure 10-15 to the down position so that a divide-by-N counter is inserted in the feedback line from the VCO output. For a first example, suppose that the

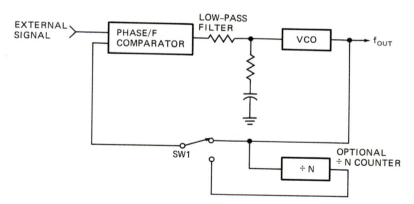

FIGURE 10-15 Block diagram of phase-locked loop circuit with switch to select ×1 or ×N mode.

The actual signal generator in the circuit is a VCO, such as the ones we discussed earlier in the chapter. With switch SW1 in the up position as shown, the phase comparator compares the phase and frequency of a signal from some external source with the phase and frequency of the signal output by the VCO. The phase comparator outputs a series of pulses with a duty cycle proportional to the difference in phase and frequency between the two signals. The pulses from the phase comparator output are filtered to near a dc voltage by a low-pass filter and applied to the control input of the VCO. The phase comparator is designed so that if the VCO output frequency is less than the external frequency, the voltage on the output of the filter will cause the VCO oscillating

divider is programmed to divide the frequency of the signal from the VCO by 10. The phase comparator then will compare the external signal with the VCO signal divided by 10. The pulses produced on the output of the phase comparator will drive the frequency of the VCO up until the frequencies on the two inputs of the phase comparators are the same. Since the signal from the VCO is divided by 10, the frequency on the output of the VCO must be 10 times the frequency of the external signal in order for the two inputs of the phase comparator to be equal. This circuit then effectively multiplies the frequency of the externally applied signal by 10. The circuit can be used to produce a signal whose frequency is any integral multiple of the external signal, as long as the desired output frequency is within the capture and lock ranges of the loop. This circuit configuration is often called a *digital frequency synthesizer*. As we show you later, this method of frequency multiplication with a PLL has many applications. For now, let's look at a specific PLL device so that we can discuss the operation of a PLL circuit in greater detail.

THE CD4046/74HC4046 PLL DEVICES

There are several types of PLL devices commonly available. Devices in the Signetics NE560 family are examples of analog PLLs. The Motorola MC4024 VCO and MC4044 phase detector can be used to make a digital PLL circuit. The device we have chosen to use as an example here is the 4046 CMOS PLL which contains a VCO and two different types of phase detectors. The device is available in standard CMOS as the RCA CD4046 or the Motorola MC14046. It is also available in high-speed CMOS as the 74HC4046. For the examples here we will use the 74HC4046 because it has a higher maximum operating frequency. For applications where the maximum VCO output frequency is below 100 kHz, you can substitute a standard CMOS 4046 device.

Figure 10-16 shows how a 74HC4046 is connected for

$1/[R2(C1 + 32 \text{ pF})]$. If R2 is omitted, the minimum VCO output frequency will be about 0 Hz. The maximum VCO frequency, or the frequency the VCO will oscillate at with a VCO input voltage of approximately 5 V, is expressed by the relationship $f_{MAX} = 1/[R1(C1 + 32 \text{ pF})]$. The circuit in Figure 10-16 has a maximum operating frequency of about 1.5 MHz. With the optional resistor R2 omitted, the minimum frequency is about 0 Hz. If the control input (VCO_{IN}) is left unconnected, the VCO will oscillate at half the maximum frequency, or 0.75 MHz. This frequency is often called the *center frequency*.

A 74HC4046 contains two different types of phase comparators. The comparator used depends on the particular application. We will explain a little bit about how each works and then show you a comparison of the advantages and disadvantages of each.

As shown in the Phase Comparator section in Figure 10-16, phase comparator 1 (PC1) is simply an XOR gate. The input signals to PC1 must both be symmetrical square waves for proper operation. As a start in explaining how this comparator works, suppose that we apply an external square-wave signal with a frequency equal to half the maximum frequency of the VCO to the reference input of PC1. Figure 10-17*a* shows the waveforms

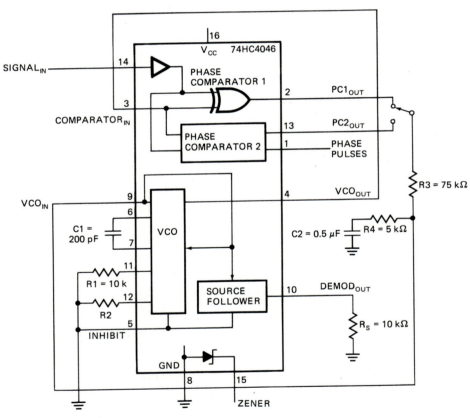

FIGURE 10-16   74HC4046 nonmultiplying phase-locked loop circuit. *(Courtesy of National Semiconductor.)*

operation as a simple, nonmultiplying PLL. As we described in a previous section on VCOs, the frequency the VCO will oscillate at, with the VCO input voltage at 0 V, is determined by the relationship $f_{MIN} =$

that will be present on the output of the VCO (VCO_{OUT}), the output of the phase comparator ($PC1_{OUT}$), and the output of the low-pass filter (VCO_{IN}) when the loop locks on this frequency. Note that VCO_{OUT} has the same fre-

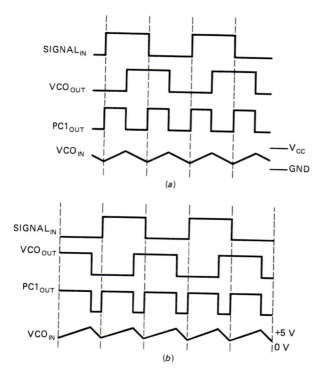

FIGURE 10-17   74HC4046 phase comparator 1
waveforms. *(a)* In lock at center frequency. *(b)* In lock
above center frequency.

quency as the external signal, but it is out of phase by
90°. As shown in the waveforms, this 90° phase difference between the inputs produces a square wave with a
frequency twice that of the input signals on the output
of the phase comparator. When this square wave is filtered, it produces a voltage which averages out to half
V_{CC}. As we said before, a voltage of half V_{CC} applied to
the control input of the VCO will cause it to oscillate at
its center frequency.

Now, suppose that we increase the frequency of the
external signal above the center frequency of the VCO.
When the loop acquires lock, the waveforms present on
the VCO, phase comparator, and low-pass filter will be
something like those in Figure 10-17b. The VCO frequency and the external frequency are the same, but the
two signals differ in phase now by more than 90°. The
waveform produced on the output of the phase comparator is now high for a longer time than it is low. When
this waveform is filtered by the low-pass filter, it averages out to a voltage greater than half V_{CC}. This voltage,
applied to VCO_{IN}, will hold the VCO oscillating at a frequency equal to the higher external frequency.

If a frequency lower than the center frequency is applied to the reference input of the phase comparator, the
phase difference between the external signal and VCO_{OUT}
will be less than 90° when the loop acquires lock. The
average voltage on the output of the low-pass filter will
then be less than half V_{CC}. This voltage, applied to VCO_{IN},
will hold the VCO oscillating at a frequency equal to the
new lower external frequency.

To summarize, then, in a PLL circuit using phase comparator 1, the input signal and the VCO output signal

will have the same frequency, but the two signals will be
out of phase by between 0 and 180°, depending on
whether the input frequency is below or above the VCO
center frequency.

Phase comparator 2 (PC2) in the 74HC4046 only looks
at the rising edges of the input signals, so the duty cycle
of the input signals doesn't matter. When a PLL circuit
using PC2 is in lock, the frequency of the VCO output
signal will be the same as the input frequency, and the
phase of the two signals will be the same. If the VCO
frequency is lower than the input signal, the output of
the phase comparator will be continuously high. This
will supply V_{CC} to the control input of the VCO through
the low-pass filter and drive the VCO frequency up. When
the frequency of the VCO becomes equal to the input
frequency, but the phase of the signals is still different,
the PC2 will output pulses which have a duty cycle proportional to the difference in phase between the two signals. After passing through the low-pass filter, these
pulses will cause the VCO output signal to shift in phase
until it is in phase with the input signal. The loop is
then locked in phase and frequency. Figure 10-18 shows

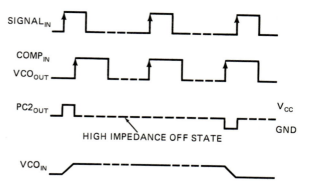

FIGURE 10-18   74HC4046 phase comparator 2 waveforms
with loop locked at center frequency.

some typical waveforms for a locked loop using PC2. Note
that since the phase detector uses the rising edges of
the signals to lock the phase, these edges are lined up in
the two waveforms. Also note that when the loop is
locked, the PC2 output only turns on for a very short
time every once in a while. These narrow pulses on the
output of PC2 are just enough to keep the low-pass filter
capacitor charged to a voltage which will keep the VCO
locked on the external signal. The dashed line on the
PC2 output waveform represents a condition where the
phase comparator output is turned off. During this time
the control voltage for the VCO is just held on the low-
pass filter capacitor. You should never connect a scope
or voltmeter to the output of phase comparator 2 because
even a 10-MΩ load at that point will drain some of the
charge off the capacitor and affect the operation of the
circuit. If you need to monitor the approximate
waveforms present on the VCO input, connect your scope
probe to the source follower buffer output on pin 10 of
the 74HC4046.

To summarize, then, when a PLL circuit using phase
comparator 2 is in lock, the VCO output will have the
same frequency and phase as the external signal. Here,

as promised, is a comparison of the characteristics of PLL circuits using each type of phase detector:

Phase Comparator 1	*Phase Comparator 2*
Maximum tracking range is ±30% of center frequency	Tracks input frequency change over 1000:1 range
Excellent noise immunity	Poor noise immunity
Input signal must be a square wave	Duty cycle of input signal does not matter
With no VCO input signal, output frequency goes to center frequency	With no VCO input, output frequency goes to f_{MIN}
May lock on harmonics of input signal	Not affected by harmonics of input signal frequency
Output is 90° out of phase with input when locked at center frequency	Output is in phase with input for all input frequencies

The design details of the low-pass filter between the output of the phase comparator and VCO_{IN} are beyond what we have space to do in this text. Basically, the values of R3 and C2 determine how quickly the loop can acquire lock after the input frequency is changed. If the product of R3 and C2 is too small, the output frequency will tend to vary because there is not enough filtering of the phase comparator output signal. If the product of R3 and C2 is too large, the loop will take a long time to acquire lock because it takes a long time for C2 to charge or discharge to the voltage needed to produce the new VCO frequency.

The purpose of R4 in the low-pass filter circuit is damping. This resistor serves the same purpose here as a shock absorber does in a car. It controls how smoothly the loop acquires lock when the frequency of the input signal is changed. If R4 is too small or left out, the VCO output frequency may drastically overshoot the new frequency and continue to change above and below the new frequency, just as a car with no shock absorbers will continue to bounce after it hits a bump in the road. If the value of R4 is too large, the loop will take a long time to acquire lock. If you want to build a circuit such as this, start with the values given in Figure 10-16. Apply a signal to the input and use a scope to see if the loop acquires lock properly. Then change the frequency of the input signal and watch how quickly and smoothly the loop acquires lock on this new frequency. Based on your observations, you can alter the values of R3 and R4 until the loop has the desired response to signal changes. A properly adjusted loop will usually overshoot the new frequency at first and then settle down after a few excursions above and below the new frequency.

Phase-Locked Loop Applications

One application for a PLL is for a digital frequency synthesizer such as the 74HC4046 circuit shown in Figure 10-19. The VCO in this circuit has a maximum frequency of 150 kHz and a minimum frequency of 0 Hz. The reference input of the phase comparator is fed with a stable 1-kHz signal from the crystal oscillator and a divider chain. The output of the VCO is fed back to the other phase comparator input through a divider that can be programmed to divide by any number from 1 to 999.

The loop will lock when the frequency fed back through the divider is equal to 1 kHz. If the divider is programmed to divide by 1, the loop will acquire lock when the VCO output frequency is equal to 1 kHz. If the divider is programmed to divide by 4, the phase comparator output will drive the VCO output frequency up until the signal fed back to the phase comparator is 1 kHz. At this point the VCO output frequency will be 4 kHz. The output frequency of the VCO then will always be N×1 kHz, where N is the number the divider is programmed for. This circuit then will produce a stable signal with a frequency of 1 to 100 kHz, depending on the BCD number applied to the P inputs of the MC14522 dividers.

A circuit such as this might be used as part of an automatic test system. The ATE computer might be programmed to output a sequence of numbers to the divider inputs. The PLL then would output a stepped sequence of frequencies to the device or system being tested.

Another very important use of PLLs is in digitally tuned radio and TV receivers. As an example of this application, we will show you how a digital FM receiver works.

The standard broadcast FM band in the United States uses the frequency band from 88 to 108 MHz. Each station is assigned a 200-kHz-wide band of frequencies within this range. A 20-MHz band divided by 200 kHz per station means that there are 100 possible stations in the standard FM band. The center, or *carrier*, frequency for the first station in the band is 88.1 MHz, the carrier frequency for the second station is 88.3 MHz, the carrier frequency for the third is 88.5 MHz, etc. Audio signals are transmitted by changing, or *modulating*, the frequency of the carrier signal at a rate equal to the audio signal.

Figure 10-20 shows a block diagram for an FM receiver such as you might have in your home. Here's how it works.

The signal coming in from the antenna is first amplified by a radiofrequency or *RF amplifier*. The amplified RF signal is then combined or *mixed* with the signal from a local oscillator (LO). When you tune the radio, you adjust the frequency of this local oscillator so that it is always 10.7 MHz lower than the center frequency of the desired station. Due to the way a mixer operates, the output from a mixer will contain four different frequency signals. The frequencies of the four signals are equal to:

1. The local oscillator input signal

2. The radiofrequency input signal

3. The sum of the frequencies of the two input signals

4. The difference between the frequencies of the two input signals

It is the difference signal that we are interested in. Since tuning in a station involves adjusting the local oscillator so that its frequency is 10.7 MHz below the carrier frequency of the station, the difference output from

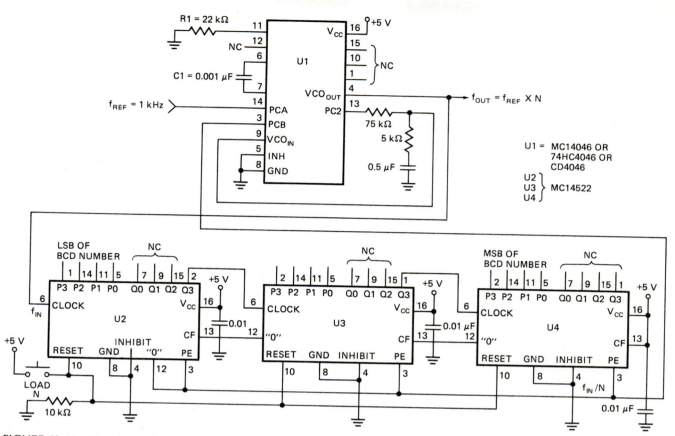

FIGURE 10-19   Circuit for frequency synthesizer using a PLL.

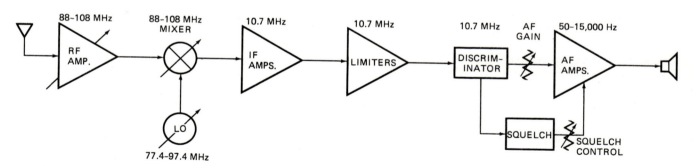

FIGURE 10-20   Block diagram of standard broadcast band FM radio receiver.

the mixer will always have a frequency of 10.7 MHz, regardless of the station you are tuned to. This difference frequency signal is filtered out from the others and passed through some intermediate frequency or IF amplifiers. The advantage of this scheme is that these IF amplifiers only have to be designed to amplify a narrow band of frequencies around 10.7 MHz, rather than amplify frequencies over the whole FM band.

The 10.7-MHz signal passed through the IF amplifiers contains the desired audio modulation, so after amplification this signal is passed through a detector or discriminator circuit to extract the audio signal. The audio signal is then amplified by an audiofrequency (AF) amplifier and used to drive a speaker.

One use of a PLL in a receiver such as this is to digi-

tally control the frequency of the local oscillator as you tune in a station. Figure 10-21 shows a block diagram of how this can be done. It is basically an extension of the frequency synthesizer application we showed you in Figure 10-19.

The difference in frequency between stations is 200 kHz, so we apply a stable 200-kHz signal to the external input of the phase comparator of the PLL. The output of the phase detector is filtered and applied to the control input of a VCO designed to oscillate over the range of 77.4 to 97.4 MHz. The 77.4-MHz frequency is 10.7 MHz below the 88.1-MHz carrier frequency of the lowest station of the band, and the 97.4-MHz frequency is 10.7 MHz lower than the 107.9-MHz carrier frequency of the highest station in the band.

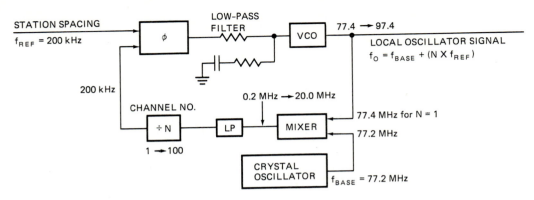

FIGURE 10-21 Block diagram of digitally programmed phase-locked loop circuit used as local oscillator in FM radio receiver.

The output of the VCO is mixed with a 77.2-MHz signal from a stable crystal oscillator. As we described before, a mixer output will contain the two original signals, a signal with a frequency equal to the sum of the two frequencies, and a signal with a frequency equal to the difference in frequency between the two input signals. For this application, we use a low-pass filter to filter out all but the difference frequency. The difference frequency signal is passed back to the other input of the phase comparator through a programmable divider.

To tune in the lowest station in the FM band, the LO signal has to have a frequency of (88.1 − 10.7) MHz or 77.4 MHz. This frequency is 200 kHz or 0.2 MHz higher than the signal from the crystal oscillator, so the difference output signal from the mixer will have a frequency of 200 kHz. If the divider is programmed to divide by 1, this 200-kHz signal will be applied to the other input of the phase comparator. With these connections the phase comparator output will drive the VCO until the loop is locked and its output frequency is 77.4 MHz.

Now, suppose you change the divider programming to a 3. The phase comparator will drive the VCO output frequency up until the difference signal from the mixer is equal to 3×200 kHz. The VCO frequency required to do this is 77.2 MHz + 3×200 kHz or 77.8 MHz. This frequency is 10.7 MHz below 88.5 MHz, the frequency of the third station in the FM band. In general terms the VCO output frequency of this circuit will always be:

$$f_O = f_{BASE} + (N \times f_{REF})$$

This means that you tune from one station to another by simply changing the number applied to the program inputs on the divider. To help you tune in a desired station, the number applied to the divider is usually converted to codes which will display the frequency of the selected station on some seven-segment displays. This conversion is easily done with a small ROM.

Much of the circuitry needed to implement digital-based PLL tuning systems is now available in single ICs. The Motorola MC6196, for example, contains much of the circuitry required to build a digitally tuned standard broadcast TV receiver. The MC6195 contains much of the circuitry needed to implement a digitally tuned cable TV converter. For further information on this application of PLLs, consult the data sheets for these devices.

DIGITAL SIGNAL TRANSMISSION

Introduction

If the distance between the output of one gate and the input of another gate is short, a simple jumper wire or printed-circuit board trace will transmit a signal from one to the other with no problems. However, if the distance between the gates increases beyond a certain length, which is different for each logic family, there are several potential problems we have to be concerned about.

One possible problem is the dc resistance of the line. This value depends on the size of the wire and its length. A given size of wire has a certain number of ohms of resistance per foot of length. As an example, 24-gage copper wire has a resistance of 0.02617 Ω per foot of length. To find the total resistance of a line, you simply multiply the resistance per foot times the length of the line in feet. For dc signals, the main consideration is that the voltage drop in the line, or *line drop*, is not so large that the signal levels at the receiving end are illegal. To calculate the voltage drop for a given connection distance, simply multiply the total resistance of the signal wire and the return wire by the current flowing through the wire. For commonly used current levels and voltage levels, the dc resistance of the line is not usually a problem until the line length exceeds 1000 ft.

Another potential problem is *crosstalk* between adjacent signal lines. An ac or pulsed signal on one conductor will induce a signal on nearby signal lines. The induced crosstalk may produce unwanted signal transitions on these adjacent lines. The amount of crosstalk between two lines is determined by how close together the lines are and how fast the signals on the lines are changing. As you might expect, the closer together the signal lines are to each other, the more chance there is for crosstalk. Faster rise and fall time signals produce more crosstalk because they allow a greater amount of a signal to capacitively couple from one line to others.

Crosstalk can be reduced by making the rise and fall times of the transmitted signals longer. As we show you in detail later, however, increasing the rise and fall times reduces the maximum frequency that can be transmitted. Another way to reduce crosstalk is to use twisted-pair wire or coaxial cable to transmit the signal. With twisted-pair or coaxial lines, the electric field of the sig-

nal is mostly between the two conductors, so it does not couple as much to adjacent lines.

Still another potential problem with longer signal lines is *electromagnetic interference,* or EMI. Transmitting alternating or pulsed signals on lines causes them to emit radiofrequency electromagnetic waves. Since these emitted waves may interfere with the operation of radios, TVs, and other communication equipment, the Federal Communications Commission (FCC) sets standards for the maximum amount of radiation permitted from any signal line or system. The longer a signal line is, the better antenna it makes for this emitted radiation. Also, the higher the frequency and the faster the rise and fall times of the transmitted signals, the more EMI the lines radiate.

EMI can be reduced by using shielded, twisted-pair wire or coaxial cable to transmit signals. The grounded shield or the outer conductor of the coaxial cable helps contain the emitted radiation. Increasing the rise and fall times of the transmitted signals also reduces the amount of EMI produced, but again, increasing the rise and fall times reduces the maximum frequency signal that can be transmitted.

Still another potential problem with long signal lines is reflections of the transmitted signal back and forth on the line. In the next sections we show you the problems that signal reflections can cause and how reflections are prevented.

Basic Transmission Line Characteristics

The term *transmission line* is a name given to the wire, PC board trace, or coaxial cable used to transmit a signal from one point to another. For ac or pulsed signals, a transmission line will act like a "load" equal to its *characteristic impedance,* Z_o, when the line is connected to the output of a gate. Numerically:

$$Z_o = \sqrt{\frac{L_o}{C_o}}$$

where L_o is the inductance per foot for the line and C_o is the capacitance per foot between the wire and ground or between the signal line and the return line. Since L_o and C_o both contain "per foot" in their units, this unit cancels out in the expression. This means that Z_o has the same value whether the transmission wire is 5 or 500 ft long! The value of Z_o is expressed in ohms.

For open connecting wires such as you might use when making protoboard connections, Z_o will have a value of 30 to 600 Ω, depending on how far the signal wire is separated from the ground conductor. Round coaxial cables such as you may have used to get a signal from a function generator to a circuit are manufactured with standard fixed impedance values such as 50, 62, 75, and 93 Ω. The Z_o of a common coaxial cable, RG-58/AU, for example, is 50 Ω. As another example, RG-59/U, the coaxial cable type often used for cable TV installations, has an impedance of 75 Ω.

Uniform impedance transmission lines are produced on double-sided PC boards by leaving a copper layer on

one side of the board where possible as a *ground plane.* In multilayer PC boards, one or more ground planes are sandwiched in the board so that signal traces can be on both outer sides of the board, and in some cases, on inner layers of the board. The impedance of a signal line on a PC board is determined by the width of the trace and the thickness of the dielectric between the trace and the ground plane layer inside or on the other side of the board. As an example, a 0.105-in- (0.27-cm-) wide trace on a 0.062-in-thick fiberglass-epoxy board with a ground plane on the other side will act as a transmission line with a Z_o of 50 Ω.

The speed at which electrical signals travel along a transmission line is determined by the dielectric material between the signal line and the return line or ground plane. Typical values are 6 to 8 in/ns. The speed of light in a vacuum is 11.8 in/ns, so these values represent about half to two-thirds the speed of light. Now, let's see why you need to know all this about transmission line impedance and signal propagation times.

If you apply a pulse to one end of an infinitely long transmission line, as shown in Figure 10-22a, the pulse will just propagate along the line forever. At any point in the line the pulse will appear undistorted when viewed with a scope.

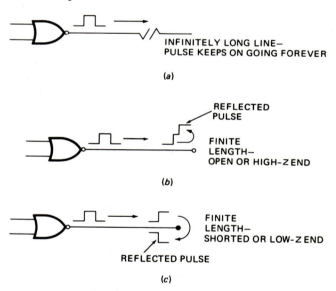

FIGURE 10-22 *(a)* Pulse traveling on an infinitely long transmission line. *(b)* In-phase reflection of pulse from open end of finite-length transmission line. *(c)* Out-of-phase reflection of a pulse at the end of finite-length transmission line with a short or low impedance connected across it.

If you apply a pulse to an open-ended transmission line which is finite in length, the pulse will reflect from the open end as shown in Figure 10-22b. One way to describe what happens at the open end is to think of charge piling up at the open end and creating a pulse which travels back toward the source. It is as if the reflected pulse climbs over the top of the incoming pulse, rolls down the back side of the incoming pulse, and travels back to the source. Since the reflected pulse adds to the incoming pulse during the duration of the pulse, the signal seen at the open end of the line will have twice the actual amplitude during this time. Since the incoming pulse and the reflected pulse add, we say the reflected pulse is *in phase* with the incoming pulse. To describe what will usually happen to the pulse when it gets back to the source again, we need to discuss what happens if you send a pulse out on a transmission line of finite length which has a short or a low impedance connected across its end.

When a pulse is sent out on a transmission line which has a short or an impedance less than the characteristic impedance of the line connected across its end, the pulse will be inverted and reflected back toward the source as shown in Figure 10-22c. The reason for this is that, at the shorted end of the line, the sum of the incoming and outgoing voltages must be zero at any instant in time. In order for this to be true, the reflected pulse must be inverted or 180° out of phase with the incoming pulse. If the line has an impedance lower than Z_o, but not a short connected across its end, the reflected pulse will be smaller in amplitude, but will still be 180° out of phase. Now let's see what happens when a pulse is reflected back to the output of a gate.

Figure 10-23a shows a circuit which sends a pulse with a 1-ns risetime and a 1-ns falltime over an 8-in open wire from one gate to another. The input of the receiving gate has a relatively high input impedance, so the receiving end of the line is essentially open. The upper half of Figure 10-23b shows the waveform that you will see if you look at the input of the receiving gate with a scope. Here's how this waveform is produced.

When the leading edge of the pulse reaches the open end of the transmission line, it is reflected back toward the source in phase with the incoming pulse. The signal present on the input of the receiving gate is then the sum of the incoming pulse and the reflected pulse, or about twice the amplitude of the transmitted pulse. The point labeled 1 in Figure 10-23b shows the overshoot produced by addition of the incoming and outgoing signals. As you can see, the waveform drastically overshoots the desired logic high input level for the receiving gate. An initial overshoot such as this may be enough to cause a CMOS input to go into a latchup condition! (See Chapter 3 for discussion of CMOS latchup.)

The output impedance of the driving gate is less than 50 Ω, so when the reflected pulse reaches this output, it will be inverted and reflected back toward the receiving gate again. When this inverted reflection reaches the receiving gate, it will subtract from the initial pulse level, as shown at the point labeled 2 in the top of Figure 10-23b. This will most likely cause the signal level present at the input of the receiving gate to drop below the level required to produce a logic high on the output. This will cause the output of the receiving gate to temporarily go low, as shown in the output waveform in the bottom half of Figure 10-23b.

The signal will bounce back and forth on the line until the reflections damp out, just as a sound wave will bounce back and forth between the walls of a canyon until it dies out. A lattice diagram such as that in Figure 10-23c is often used to keep track of the reflections on the signal line. As you can see in this diagram, the third

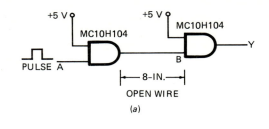

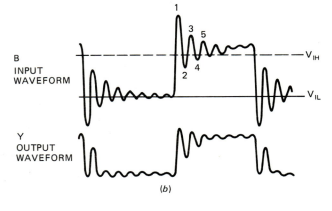

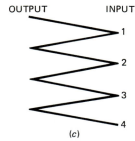

FIGURE 10-23 *(a)* Pulse sent out on a 50-Ω transmission line to MC10H104 input. *(b)* Signal observed at input and output of MC10H104. *(c)* Transmission line reflection diagram.

time the signal arrives at the gate input it causes the overshoot labeled 3 in Figure 10-23b. The fourth time the signal arrives at the gate input it causes the undershoot labeled 4 in Figure 10-23b.

The point of the preceding discussion is to show you that an open-ended transmission line can cause a pulse to bounce back and forth several times on the line. The reflections can cause the signal on the input of the receiving gate to overshoot and "ring" across the logic threshold before finally settling down to the desired logic level. As shown in Figure 10-23b, this ringing on the input can cause the output of the receiving gate to undergo unwanted transitions.

Figure 10-24 shows the effect that will be produced by the same pulse if the receiving end of the transmission line has a resistor less than the Z_o of the line connected across it. In this case, the initial reflection from the receiving end of the line subtracts from the incoming pulse. This causes the incoming signal to undershoot the desired logic level. After several reflections, the signal on the input of the receiving gate finally reaches the desired logic high level. Here again, the impedance mismatch at

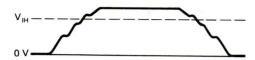

FIGURE 10-24 Waveform seen at the input of MC10H104 if a resistance less then Z_o is connected between the transmission line and ground.

the receiving end of the line causes reflections which distort the incoming signal.

A simple way to prevent these reflections on a transmission line is to trick the pulse going out on the line into thinking that it sees an infinite length line. You do this by connecting a resistor with a value equal to Z_o between the transmission line and ground at the receiving end, as shown in Figure 10-25. This resistor dissipates the energy of the pulse and prevents reflections. Connecting a matched-value resistor across the end of a transmission line in this way is called *parallel termination*. This trick is analogous to putting sound-absorbing material on the walls of a theater to prevent echoes.

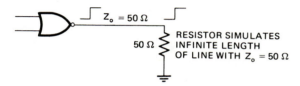

FIGURE 10-25 Proper method of terminating a 50-Ω transmission line to prevent reflections.

Determining the Maximum Length for an Unterminated Line

The question that may occur to you at this point is, "How long does a circuit connection have to be before I have to use a terminated transmission line or some other technique?" The answer to this question depends on the risetime and falltime of the signal being transmitted.

If the major reflections on the signal line take place during the risetime or the falltime of an output pulse, they will not usually cause any problems. The definition of what represents major reflections depends on the particular logic family, but as you can see in Figure 10-23b, the signal on the input of the MC10H104 remains clearly high after one overshoot and one undershoot. Likewise on the falling edge of the pulse, the signal remains clearly low after one overshoot and one undershoot. Some references indicate that the maximum line length should be such that the signal travels twice the length of the line during the risetime of the signal. Our experience has shown that it is safer to limit the maximum unterminated line length so that the signal has time to travel the length of the line four times during the risetime or falltime of the signal being transmitted. The reason for this is that, as shown in Figure 10-23c, the third trip of the signal on the line produces the undershoot labeled 2 in Figure 10-23b. By allowing four line-propagation delay times, we make sure that both the initial overshoot and

this potential undershoot are within the risetime of the transmitted signal. Here's how you determine the maximum length of unterminated connecting wire you can use for a particular device:

1. From its data sheet determine the output risetime and falltime for the device that is sending the pulse.

2. Divide the lesser of these two times by 4 to determine the maximum allowable time for signal propagation between the output of the one gate and the input of the next.

3. Multiply this time by a worst-case signal velocity of 6 in/ns to convert it to an equivalent distance.

As a first example of this determination, the data sheet for a 74LS00 gate shows a risetime of about 13 ns and a falltime of about 6 ns. The falltime of 6 ns represents the "worst-case" situation, so we use that value. The result for the maximum length, L, of unterminated connecting wire, then, is:

$$L = \frac{6}{4} \text{ ns} \times 6 \text{ in/ns} = 9 \text{ in}$$

Dividing the falltime of 6 ns by 4 gives a value of 1.5 ns, which is the maximum signal propagation time for an unterminated line with this device. In 1.5 ns the signal can travel a distance of 1.5 ns $\times$ 6 in/ns = 9 in. This tells you that when you are building a circuit with 74LS family devices, you can use connecting wires of up to about 9 in long without having to worry about signal reflections causing problems. Since we used "worst-case" values in each part of the calculations for this example, you can mostly use longer unterminated connecting wires before encountering problems. However, to use the old cliché, "Better safe than sorry."

As another example of these calculations, devices in the Fairchild F100K ECL family have risetimes of about 0.7 ns when operated with a supply voltage of -4.5 V. The unterminated line length, L, for these devices then is calculated as follows:

$$L = \frac{0.7}{4} \text{ ns} \times 6 \text{ in/ns} = 1.05 \text{ in}$$

From this calculation, you can see that when working with this family of ECL devices, a terminated transmission line must be used to make any connection longer than 1 in! For reference, here are the maximum unterminated line lengths you can use for common logic family devices:

Device Family	Risetime, ns	Falltime, ns	Maximum Unterminated Line Length, in
7400	12	5	7.5
74AS	5	3	4.5

(continued)

(continued)

Device Family	Risetime, ns	Falltime, ns	Maximum Unterminated Line Length, in
74S	6	3	4.5
74LS	13	3	4.5
74ALS	10	6	9.0
74HC	6	7	9.0
74HCT	6	7	9.0
ECL10KH	0.8	0.8	1.2

As you can see from the preceding examples, the slower the risetime of a signal, the longer the line you can use without worrying about signal reflections. The question that may occur to you at this point is, "Why don't we limit the rise- and falltimes of signals so that we can use longer unterminated connecting wires?" The answer is that in some systems we do just that, but in most systems increasing the rise- and falltimes would cause a multitude of timing problems. One obvious problem is that increasing the rise- and falltimes of a gate output reduces the maximum frequency that the gate can output. Here's how this works.

Figure 10-26a shows the output signal produced when a 20-MHz square-wave signal is applied to the input of a logic gate with 10-ns rise- and falltimes. As you can see, the signal reaches legal low and legal high logic levels. Figure 10-26b shows what happens if you apply the same 20-MHz signal to the input of a gate with output rise- and falltimes greater than 50 ns. As you can see in this second case, the signal does not have time to reach a legal high or a legal low level before it is switched in the other direction. The point here, then, is that if you want to transmit high-frequency signals, you must use devices which have short propagation delay times, risetimes, and falltimes. A general rule is that to transmit a square-wave signal of a given frequency, the output risetime of the transmitting device should be no greater than 0.1 to 0.3 times the time the signal is high each cycle. For a 20-MHz square wave which is high for 25 ns, then, the transmitting device should have an output risetime of 2.5 to 7.5 ns.

Now that you know about the risetime-frequency trade-off and how to determine the maximum length unterminated line you can use for a given signal risetime, the next step is to show you some of the techniques that are used to transmit digital signals for distances greater than those allowed for unterminated lines.

FIGURE 10-26 Output signal produced when a 20-MHz square-wave signal is applied to a gate with output rise- and falltimes of (a) 10 ns (b) greater than 50 ns.

ECL Transmission Lines

As we showed above, the risetimes and falltimes of ECL devices are so short that any connecting wire over an inch or so must be a terminated transmission line to avoid signal reflection problems. The transmission line can be a trace going to another IC on the PC board, or a coaxial cable going to another board or system. For an ECL signal we usually use a transmission line with a low impedance such as 50 Ω. The reason for this is that a low-impedance line can charge or discharge the input capacitance of the receiving gate more quickly than a higher impedance line would. The output of an ECL device can easily supply enough current to directly drive a 50-Ω terminated transmission line.

As you may remember from Chapter 3, ECL devices are usually operated with a negative supply voltage. In order to maintain the correct dc voltage levels on device outputs and also provide proper termination, one of two schemes is commonly used to terminate ECL signal lines.

The first scheme in Figure 10-27a uses a 50-Ω resistor connected to a voltage of −2 V. This connection supplies the required current path from the output to a negative voltage. Also, since the −2 V is fixed, it acts like a ground for logic signals transmitted on the line. Therefore, the 50-Ω resistor to −2 V acts as a proper termination for pulses on the transmission line. The disadvantage of the scheme shown in Figure 10-27a is that it requires a separate, −2-V supply.

The second ECL termination scheme shown in Figure 10-27b has the advantage that the only voltage it requires is −5.2 V, the same voltage used to power the gate. This scheme has the disadvantage, however, that it requires two resistors. Using Thevenin's theorem, which we don't have time to explain here, it can be shown that the voltage divider formed by the 82-Ω resistor and the 130-Ω resistor functions identically to a 50-Ω resistor connected to −2 V. From a dc standpoint, the voltage divider produces a dc voltage of 2.0 V at its center. For pulse signals on the signal line, the −5.2 V is fixed and ground is fixed; to a signal on the line, the two resistors act as if they are in parallel between the line and ground. An 82-Ω resistor in parallel with a 130-Ω resistor is equivalent to about 50 Ω.

TTL and CMOS Level Transmission Lines

Ohm's law tells you that a current of 40 mA is required to produce a 74XX or 74HCTXX minimum input high voltage of 2.0 V across a 50-Ω termination resistor. As shown in Table 3-2, standard TTL and CMOS device outputs will not output enough current to directly drive a low-impedance terminated transmission line such as this. Therefore, when you need to transmit a TTL or CMOS level signal for a distance greater than those we showed you above, you need to use a special device called a *buffer* or *line driver*. A buffer or line driver can output enough current to produce legal TTL logic levels across the terminating resistor at the end of the line. One example of a line driver device is a 74128. Each of the four NOR gates in a 74128 is capable of driving a 50-Ω terminated line. Another common line driver, the 74S140,

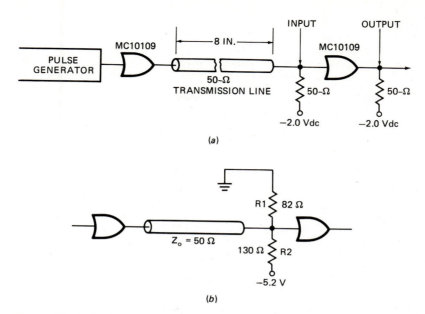

(a)

(b)

FIGURE 10-27 Termination methods for ECL transmission lines. (a) Single resistor to −2 V. (b) Equivalent two-resistor method which does not require separate −2-V supply.

contains two 4-input NAND gates which are each capable of driving a 50-Ω terminated line.

For applications where you don't need the speed of a 50-Ω transmission line, you can use a buffer such as those in a 74XX240, 74XX241, or 74XX244. These devices each contain eight buffers which are capable of driving a terminated transmission line with an impedance of 130-Ω or more. The buffers in the 74XX240 devices are inverting and the buffers in the 74XX241 and 74XX244 devices are noninverting. All these devices have three-state outputs so that several device outputs can be connected on a common line. Figure 10-28 shows a

the amount of current that the buffer output has to source or sink. For a pulse traveling down the line, the fixed + 5 V acts like a ground, so the 330- and 220-Ω resistors act as if they are in parallel between the signal line and ground. This produces a terminating resistance of about 130 Ω, which is approximately the impedance of an open wire separated from ground by a half inch or so. Buffers such as the 74XX244 are often used to transmit signals on microcomputer signal buses. Incidentally, in a later section of this chapter on Schmitt triggers we will explain the meaning of the symbol shown inside the 74LS244 buffer in Figure 10-28.

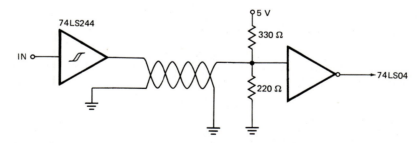

FIGURE 10-28 Logic diagram for 74LS244 octal buffer showing the two-resistor termination scheme commonly used on output signal lines.

logic diagram for one-half of a 74LS244, and the termination scheme often used for transmission lines driven by these buffers. The voltage divider formed with the 330-Ω resistor and the 220-Ω resistor produces a voltage of 2 V at the junction of the resistors. The buffer in the 74LS244 either sources some additional current to pull this junction to a logic high or sinks the additional current needed to pull the junction to a logic low level. The advantage of this two-resistor scheme is that it reduces

RS-232C Signals

The line drivers and buffers described in the preceding section all transmit signals with TTL/CMOS levels. Often we convert logic signals from these levels to some other levels in order to transmit them from one system to another. In this section and the next two sections we describe the RS-232C, RS-423, and RS-422A signal standards which use different voltage levels. Signals using one of these voltage standards are often used to transmit asynchronous serial data from remote CRT terminals to a central computer. Signals using one of these standards are also used to communicate between computers and devices called modems, which we discuss a little later in this chapter.

The most common of these three voltage standards is part of the RS-232C standard developed by the Electronic Industries Association (EIA). In this standard a logic high is defined as a voltage between −3 and −15 V. A logic low is defined as a voltage between +3 and +15 V. Obviously these signal voltages are not compatible with the levels of standard logic families, so devices are needed to convert standard logic signals to and from these levels. The devices most commonly used to do this are the MC1488 quad TTL-to-RS-232C line driver and the MC1489 quad RS-232C-to-TTL line receiver. Figure 10-29 shows how these devices are connected.

Note in Figure 10-29 that there are no termination resistors on the signal line. RS-232C uses a somewhat sneaky method of avoiding the need for a terminated transmission line. The capacitor connected between the output of the MC1488 driver and ground increases the

rise- and falltimes of the transmitted signal to the point where the signal can travel a considerable distance before reflections become a problem. The RS-232C standard specifies a *minimum* rate of change of voltage, dV/dT, of 30 V/μs! This rate of change of voltage gives rise- and falltimes of about 1 μs for a signal going between −15 and +15 V. These long rise- and falltimes are specified to help reduce crosstalk between adjacent signal lines and to allow a connecting wire to be many feet long before reflections become a problem. A major disadvantage of the long rise- and falltimes, however, is that they limit the rate at which data bits can be sent on the line.

As we explained in a discussion of UARTs in Chapter 9, the term baud rate is used to describe the rate at which asynchronous serial data is transmitted. Baud rate, remember, is defined as 1 divided by the time per bit. Commonly used baud rates are 300, 1200, 2400, 4800, 9600, and 19,200 Bd. Since each cycle of a square-wave signal can be thought of as two data bits, one high bit and one low bit, the equivalent frequency of a data signal transmitted in the simple asynchronous format shown in Figure 9-39 is half the baud rate. Transmitting a signal at 9600 Bd, for example, is equivalent to transmitting a 4800-Hz square wave.

For a signal line a few feet long, RS-232C specifies a maximum data rate of 20,000 Bd. At lower baud rates RS-232C specifies a maximum line length of 50 ft. The baud rate and distance specifications are both too limiting for many current systems. In the next two sections we discuss a couple of newer standards which permit data to be transmitted at much higher rates.

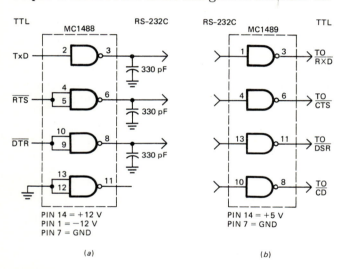

FIGURE 10-29 TTL-to-RS-232 and RS-232-to-TTL signal conversion. *(a)* MC1488 used to convert TTL to RS-232C. *(b)* MC1489 used to convert RS-232C to TTL.

RS-423 Signals

RS-423 is another EIA standard which is somewhat of an improvement over RS-232C. A logic high in this standard is represented by a voltage between −4 and −6 V. A logic low is represented by a voltage between +4 and +6 V. As shown in Figure 10-30, a line driver such as the MC3488A can be used to translate standard TTL/ CMOS signal levels to these levels, and an MC3486 line receiver can be used to translate RS-423 signals back to TTL/CMOS levels. Note that the signal ground is only

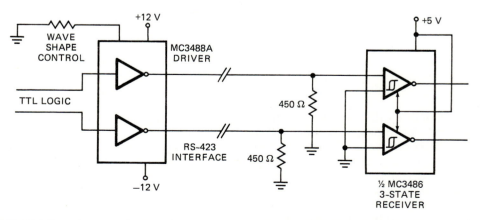

FIGURE 10-30 MC3488A line driver and MC3486 line receiver used for RS-423 signal transmission.

connected to the system ground at the driver end of the line. This prevents large currents from flowing through the signal ground wire and affecting the logic signal between it and the signal wire.

RS-423 uses two methods to help solve the signal reflection problem. First, the RS-423 driver can output enough current to drive a load resistor as small as 450 Ω at the end of the line. This load resistor serves as a partial termination for the line.

Second, as shown in Figure 10-30, the MC3488A line driver has a wave shape control input. The value of the resistor connected from this input to ground determines the risetime and the falltime of output signal pulses. A resistor value of 10 kΩ, for example, will produce output rise- and falltimes of 1.4 μs, and a resistor value of 100 kΩ will produce output rise- and falltimes of about 14 μs. As we discussed earlier, the longer the risetime and falltime of the pulses being transmitted, the greater the unterminated line length can be without reflection problems. Increasing the rise- and falltimes, however, decreases the maximum data rate or the maximum frequency signal you can transmit on the line. The point here is that with RS-423 line you adjust the risetime, falltime, and data rate for best performance in a given application.

The RS-423 standard allows a maximum data rate of 100,000 Bd over a 40-ft or shorter line. The maximum specified line length for RS-423 is 4000 ft. With a line this long, the maximum data rate is only 1000 Bd.

RS-422A Signals

Another improved EIA standard, RS-422A, specifies that signals will be sent differentially on two adjacent wires in a ribbon cable, or on a twisted pair of wires as shown in Figure 10-31a. To create the differential signals from TTL/CMOS logic signals we use devices such as the MC3487 quad TTL-to-RS-422A line driver. To translate differential signals back to standard TTL/CMOS signal levels we use devices such as the MC3486 quad RS-422A/423-to-TTL line receiver.

The term *differential* in this standard means that the signal voltage is developed between the two signal lines, rather than between a signal line and ground as in RS-232C and RS-423. In RS-422A a logic high is transmitted by making the "b" line more positive than the "a" line. A logic low is transmitted by making the "a" line more positive than the "b" line. The voltage difference between the two lines must be greater than 0.4 V but less than 12 V. Typical drivers produce a differential voltage of about 2 V. The center or common-mode voltage on the lines must be between −7 V and +7 V. RS-422A specifies signal rise- and falltimes of 20 ns or 0.1 × the time for 1 bit, whichever is greater.

Figure 10-31b shows the relationship between maximum cable length and baud rate for RS-422A line. As you should be able to see in this graph, the maximum data rate for RS-422A lines ranges from 10 million Bd on a line 40 ft long to 100,000 Bd on a 4000-foot line. The reason that the data rates are so much higher than for RS-423 lines is that the differential line functions as a fully terminated transmission line. Common 24-gage

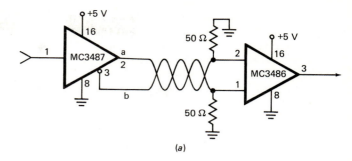

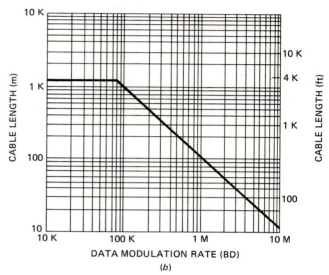

FIGURE 10-31 (a) MC3487 driver and MC3486 receiver used for RS-422A differential signals. (b) Specified maximum line length versus baud rate for RS-422A signal lines.

twisted-pair wire has a Z_o of about 100 Ω, so the line can be terminated with a matching 100-Ω resistor connected between the signal lines. A more common termination method, however, is to use a 50-Ω resistor from each signal line to ground as shown in Figure 10-31a. This method helps keep the two signal lines balanced.

A further advantage of differential signal transmission is that any electrical noise induced in one signal line will be induced equally in the other signal line. The operation of a differential line receiver is such that it only responds to the voltage difference between its two inputs. Any noise voltage that is induced equally on the two inputs will not have any effect on the output of the differential receiver.

RS-422A drivers and receivers allow you to transmit digital signals for distances up to 4000 ft, but often this is still not enough. In the next sections we show you some methods used to transmit digital signals for longer distances over standard phone lines.

Modems

INTRODUCTION

A common reason for transmitting digital data is to communicate between a terminal and a central computer or between two computers. Computers internally work with data in parallel form, but to reduce the number of wires needed to transmit data to another computer the data is converted to serial form. This is usually done by a UART such as we described in Chapter 9. To send the resultant serial data for long distances, such as from Portland, Oregon, to New York City, telephone lines are convenient because they are already in place.

To transmit signals over long distances, telephone companies usually convert signals to a digital form and transmit them on wide-bandwidth microwave links, fiberoptic cables, or satellite links. For a variety of reasons, however, most local phone lines are designed to transmit signals only in the frequency range of 300 to 3000 Hz. If you try to transmit digital signals directly on standard phone lines such as these, the transmitted signals will be severely distorted. Local phone lines may in the future all be converted to a wide-bandwidth digital system known as the *integrated services digital network* or *ISDN*. ISDN will allow direct, high-speed digital communication over phone lines. For now, however, we are stuck with two, less desirable ways of using phone lines to transmit digital data.

The first alternative is to lease a specially conditioned line which has enough bandwidth to allow high-speed digital signals to be transmitted directly. For digital communication between two fixed points, such as a branch office and the main office of a company, leased lines are often a good approach. The disadvantages of leased lines, however, are the cost and the fact that the connection is fixed between two points. You have to lease a separate line for each desired destination.

The second alternative is to convert the digital signals to a form which can be transmitted over standard phone lines. This alternative has the advantage that you can just dial up anyone you want to communicate with.

To transmit digital signals over standard phone lines we use a device called a *modulator-demodulator*, or *modem*. At the transmitting end of the line the modulator part of the modem converts the digital signals to sine waves which have frequencies that can be transmitted over phone lines without distortion. At the receiving end of the line the demodulator part of the modem converts the transmitted sine waves back to the equivalent digital signals. To represent digital ones and zeros the transmitted sine waves are modulated in one of three ways. The three methods are *amplitude modulation*, *frequency-shift keying*, and *phase-shift modulation*. Here's an overview of how these methods work.

MODEM MODULATION METHODS

An amplitude-modulated signal consists of a single-frequency tone of perhaps 387 Hz. The tone is turned on to represent a 1 and turned off to represent a 0, as shown in Figure 10-32. Amplitude modulation is only used for very low frequency communication because it is adversely affected by amplifier gains and line resistance. A temporary increase in the resistance of the line, for example, might drop the amplitude of a "one" signal below the threshold of the detector in the receiver. The receiver will then erroneously output the bit as a zero.

As shown in Figure 10-33, frequency-shift keying or

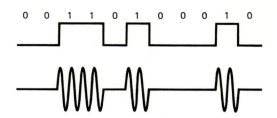

FIGURE 10-32   Representation of digital ones and zeros with amplitude-modulated sine-wave signal.

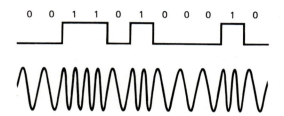

FIGURE 10-33   Representation of 1's and 0's with frequency-modulated sine-wave signal (FSK). Lower frequency represents 0, higher frequency represents 1.

FSK modulation uses a tone of one frequency to represent a digital 1 and a tone of another frequency to represent a digital 0. A common FSK modem type, the Bell 202, for example, uses a frequency of 1200 Hz to represent a 0 and a frequency of 1700 Hz to represent a 1. In order to allow communication in both directions at the same time on a line, four different frequencies are often used. An example of this type of FSK modem is the old standard 300-Bd Bell 103A type, which uses 2025 Hz for a 0 and 2225 Hz for a 1 in one direction, and 1070 Hz for a 0 and 1270 Hz for a 1 in the other direction.

FSK modulation is easy to implement and easy to detect, but for standard phone lines the maximum baud rate is limited to 1200 Bd. To communicate at higher baud rates such as 2400, 4800, and 9600 Bd on standard phone lines, we use phase-shift modulation.

In the simplest form of *phase-shift modulation* (PSK), the phase of a constant-frequency sine-wave carrier of perhaps 1700 Hz is shifted by 180° to represent a change in the data from a 1 to a 0 or from a 0 to a 1. Figure 10-34*a* shows an example of this. As the digital data changes from a 0 to a 1, near the left edge of the figure, the phase of the signal is shifted by 180°. When the data changes from a 1 to a 0, the phase of the carrier is again shifted by 180°. For the next section of the digital data where the data stays 0 for 3 bit times, the phase of the carrier is not changed. Likewise, in a later section of the waveform where the data remains at a one level for 2 bit times, the phase of the carrier is not changed. The phase of the carrier then is only shifted by 180° when the data line changes from a 1 to a 0 or from a 0 to a 1.

The simple phase-shift modulation shown in Figure 10-34*a* has no real advantage over FSK as far as maximum bit rate, etc. However, by using additional phase angles besides 0 and 180°, 2 or 3 bits can be sent in one baud time. Two bits sent in one baud time are called *dibits*, and three bits sent in one baud time are called *tribits*. Here's how dibits and tribits are encoded.

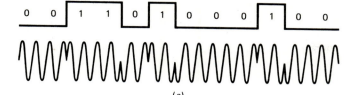

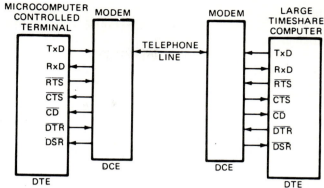

DTE – DATA TERMINAL EQUIPMENT
DCE – DATA COMMUNICATION EQUIPMENT

FIGURE 10-35 Serial data communication between remote terminal and a central computer using modems and standard phone lines.

GRAY CODE TRIBIT VALUE	DEGREES OF PHASE SHIFT
0 0	0
0 1	90
1 1	180
1 0	270

(b)

GRAY CODE DIBIT VALUE	DEGREES OF PHASE SHIFT
0 0 1	22.5
0 0 0	67.5
0 1 0	112.5
0 1 1	157.5
1 1 1	202.5
1 1 0	247.5
1 0 0	292.5
1 0 1	337.5

(c)

FIGURE 10-34 Phase-shift modulation (PSK).
(a) Waveforms for simple phase-shift modulation.
(b) One common set of phase shifts used to represent four possible dibit combinations. *(c)* One common set of phase shifts used to represent the eight possible tribit combinations.

For dibit encoding, each pair of bits in the data stream is treated together. The value of these two bits determines the amount that the phase of the carrier will be shifted. Figure 10-34b shows the angle that the phase of the carrier will be shifted for the four possible values of a dibit. If, for example, the value of 2 bits taken together is 01, the phase of the carrier will be shifted 90° to represent that dibit. The trick here is that the phase of the carrier only has to shift once for each two transmitted bits. Remember from a previous discussion that baud rate is the rate at which the carrier is changing. In this case it is not the same as the number of bits per second. Bell 212A–type modems use this scheme to transmit 1200 bits/s at a baud rate of only 600 Bd. Two carrier frequencies, 1200 and 2400 Hz, are used to permit full-duplex operation. For tribit encoding, the data stream is divided into groups of 3 bits each. Figure 10-34c shows one possible set of angles that the phase of the carrier might be shifted to represent the eight different values that the tribit can have. The Bell 208 modem uses this tribit scheme to transmit data at 4800 bits/s.

Another similar scheme, called *quaternary amplitude modulation* (QAM), encodes 4 bits in each baud time. This makes it possible to transmit data at 9600 bits/s over conditioned standard phone lines. The QAM scheme uses eight different angles to represent three of the bits in each group of 4, and two different amplitudes to represent the fourth bit in each quadbit. A baud rate of 2400 Bd and 4 bits/Bd produces a data rate of 9600 bits/s.

MODEM INTERFACE SIGNALS

Probably the most common use of modems is for communication between a distant terminal and a computer or between two computers. Figure 10-35, for example,

shows how two modems can be used to communicate between a terminal and a large time-share computer over a standard phone line. Modems and other equipment used to send serial data over long distances are referred to in the literature as *data communication equipment*, or DCE. The computers and the terminals that are sending or receiving the serial data are called *data terminal equipment*, or DTE.

The lines labeled T×D in Figure 10-35 are the lines used to transmit serial data bytes from a terminal or computer to a modem. The lines labeled R×D represent the lines used to receive serial data bytes from a modem. The arrowheads should help you see the direction of signal flow on these and the other lines.

Interfacing a terminal or computer with a modem requires some control signal lines as well as the T×D and R×D lines. The function of these control signals is to make sure that each part of the data link is ready before any attempt is made to send data. The data and control signal names shown in Figure 10-35 are another part of the RS-232C standard which we introduced you to in an earlier section. After we give you an overview of how these control signals function, we will take a closer look at this part of the RS-232C standard.

The term *handshake* is often used to describe the process that the system goes through to make sure that all parts are ready before data is sent. Here's an overview of how the handshake takes place, starting from the terminal end of the link.

When the power to the terminal is turned on, the terminal runs some internal tests and then asserts the *data-terminal-ready* ($\overline{DTR}$) signal to tell the modem it is ready. When the modem is powered up and ready to transmit or receive data, it will assert the *data-set-ready* ($\overline{DSR}$) signal to the terminal. Under manual or terminal control the modem then dials up the computer.

If the computer is available, it will send back a specified tone. Now, when the terminal has a character actually ready to send, it will assert a *request-to-send* ($\overline{RTS}$) signal to the modem. The modem will then assert its *carrier-detect* ($\overline{CD}$) signal to the terminal to indicate that it has established contact with the computer. When the modem is fully ready to transmit data, it asserts the *clear-to-send* ($\overline{CTS}$) signal back to the terminal. The terminal then sends serial data characters to the modem. When the terminal has sent all the characters it needs to, it makes its $\overline{RTS}$ signal high. This causes the modem to unassert its $\overline{CTS}$ signal and stop transmitting. A similar handshake occurs between the modem and the computer at the other end of the data link. The important point now is that a set of handshake signals is defined for transferring serial data to and from a modem. Now that you have an overview of RS-232C handshake signals, we will take a closer look at this part of the RS-232C standard, and show you how this standard is used to connect equipment other than modems.

A Second Look at RS-232C

SIGNAL AND PIN DEFINITIONS

As we told you in a previous section, RS-232C standard specifies a logic low signal voltage of +3 to +15 V and a logic high signal voltage of −3 to −15 V. The RS-232C standard also specifies 25 signal pins and it specifies that the DTE connector should be a male type and the DCE connector should be a female type. No specific connector is specified, but the most commonly used connectors are the DB-25P male shown in Figure 10-36a and the mating DB-25S female.

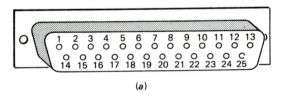

(a)

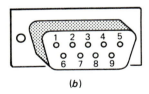

(b)

FIGURE 10-36   Connectors commonly used for RS-232C connections. *(a)* DB-25P 25-pin male connector. *(b)* DE-9P 9-pin DIN connector.

In systems where many of the 25 pins are not needed, a nine-pin DIN connector such as the DE-9P male shown in Figure 10-36b and the corresponding female connector are commonly used. When you are wiring up one of these connectors, it is important to note the unusual order in which the pins are numbered.

Figure 10-37 shows the signal names, signal direction, and a brief description for each of the 25 pins defined for RS-232C. For most applications only a few of these pins are used, so don't get overwhelmed. Here are a few additional notes about these signals.

First note that the signal direction is specified with respect to the DCE. This convention is part of the standard. We have found it very helpful to put arrow-

heads on all signal lines as shown in Figure 10-35 when we are drawing circuits for connecting RS-232C equipment.

Next observe that there is both a chassis ground (pin 1) and a signal ground (pin 7). To prevent large ac-induced ground currents in the signal ground, these two should be connected only at the power supply in the terminal or the computer, but not both places.

The T×D, R×D, and handshake signals, shown with common names in Figure 10-37, are the ones most often used for simple systems. We gave an overview of their use in the introduction to this section of the chapter. These signals control what is called the *primary* or *forward* communications channel of the modem. Some modems allow communication over a *secondary* or *backward* channel, which operates in the reverse direction from the forward channel and at a much lower baud rate. Pins 12, 13, 14, 16, and 19 are the data and handshake lines for this backward channel. Pins 15, 17, 21, and 24 are used for synchronous data communication. Next we want to show you some of the tricks in connecting RS-232C-"compatible" equipment.

CONNECTING RS-232C EQUIPMENT

A major point we need to make right now is that you can seldom just connect two pieces of equipment described by their manufacturers as "RS-232C compatible" and expect them to work the first time. There are several reasons for this. To give you an idea of one of the reasons, suppose that you want to connect the terminal in Figure 10-35 directly to the computer rather than through the modem-phoneline-modem link. The terminal and the computer probably both have DB-25-type connectors so that, other than a possible male-female mismatch, you might think you could just plug the terminal cable directly into the computer. To see why this doesn't work, hold your fingers over the modems in Figure 10-35 and refer to the pin numbers for the RS-232C signals in Figure 10-37. As you should see, both the terminal and the computer are trying to output data (T×D) from their number 2 pins to the same line. Likewise, they are both trying to input data (R×D) from the same line on their number 3 pins. The same problem exists with the handshake signals. RS-232C drivers are designed so that connecting the lines together in this way will not destroy anything, but connecting outputs together is not a productive relationship. A solution to this problem is to make an adapter with two connectors so that the signals cross over as shown in Figure 10-38. This crossover connection is often called a *null modem*. Again, we put arrowheads on the signals in Figure 10-38 to help you keep track of the direction for each. As you can see in the figure, the T×D from the terminal now sends data to the R×D input of the computer. Likewise, the T×D from the computer now sends data to the R×D input of the terminal as desired. The handshake signals are also crossed over so that each handshake output signal is connected to the corresponding input signal.

A second reason that you can't just plug RS-232C-compatible equipment together and expect it to

PIN NUMBERS FOR 9 PINS	PIN NUMBERS FOR 25 PINS	COMMON NAME	RS-232C NAME	DESCRIPTION	SIGNAL DIRECTION ON DCE
3	1		AA	PROTECTIVE GROUND	—
2	2	TXD	BA	TRANSMITTED DATA	IN
7	3	RXD	BB	RECEIVED DATA	OUT
8	4	$\overline{RTS}$	CA	REQUEST TO SEND	IN
	5	$\overline{CTS}$	CB	CLEAR TO SEND	OUT
6	6	$\overline{DSR}$	CC	DATA SET READY	OUT
5	7	GND	AB	SIGNAL GROUND (COMMON RETURN)	—
1	8	$\overline{CD}$	CF	RECEIVED LINE SIGNAL DETECTOR	OUT
	9		—	(RESERVED FOR DATA SET TESTING)	—
	10		—	(RESERVED FOR DATA SET TESTING)	—
	11			UNASSIGNED	—
	12		SCF	SECONDARY RECEIVED LINE SIGNAL DETECTOR	OUT
	13		SCB	SECONDARY CLEAR TO SEND	OUT
	14		SBA	SECONDARY TRANSMITTED DATA	IN
	15		DB	TRANSMISSION SIGNAL ELEMENT TIMING (DCE SOURCE)	OUT
	16		SBB	SECONDARY RECEIVED DATA	OUT
	17		DD	RECEIVER SIGNAL ELEMENT TIMING (DCE SOURCE)	OUT
	18			UNASSIGNED	—
	19		SCA	SECONDARY REQUEST TO SEND	IN
4	20	$\overline{DTR}$	CD	DATA TERMINAL READY	IN
	21		CG	SIGNAL QUALITY DETECTOR	OUT
9	22		CE	RING INDICATOR	OUT
	23		CH/CI	DATA SIGNAL RATE SELECTOR (DTE/DCE SOURCE)	IN/OUT
	24		DA	TRANSMIT SIGNAL ELEMENT TIMING (DTE SOURCE)	IN
	25			UNASSIGNED	—

FIGURE 10-37 RS-232C pin names and signal descriptions.

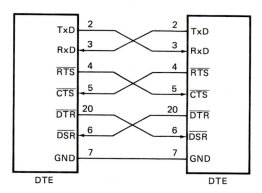

FIGURE 10-38 Null-modem method for connecting two RS-232C systems which are both set up as DTE.

work is that a partial implementation of RS-232C is often used to communicate with printers, plotters, and other computer peripherals besides modems. These other peripherals may be configured as DCE or as DTE. Also, they may use all, some, or none of the handshake signals.

The point here is that whenever you have to connect RS-232C-compatible devices such as terminals or serial printers, get the schematic for each and work your way through the connections one pin at a time. Make sure that an output on one device goes to the appropriate input on the other device. Sometimes you have to look at the actual drivers and receivers on the schematic to determine which pins on the connector are outputs and which are inputs. This is necessary because some manufacturers label an output pin connected to pin 2 as R×D, indicating that this signal goes to the R×D input of the receiving system.

Fiber-Optic Signal Lines

INTRODUCTION

All the data communication methods we have discussed so far use metallic conductors. The systems we describe here transfer data through very thin glass or plastic fibers with a beam of light and have no conducting electrical path. Therefore, they are called *fiber-optic* systems. Figure 10-39 shows the connections for a basic fiber-optic data link you can build and experiment with.

The light source here is a simple infrared LED. Higher-performance systems use an *infrared injection laser diode* (ILD) or some other laser driven by a high-speed, high-current driver.

NOTE: If you are working on a fiber-optic system, you should never look directly into the end of the fiber to see if the light source is working because the light beam from some laser diodes is powerful enough to cause permanent eye damage. Use a light meter, or point the cable at a nonreflecting surface to see if the light source is working.

Digital data is sent over the fiber by turning the light beam on for a 1 and off for a 0. Data rates for some currently available systems are as high as a gigabit per second. Current systems use one of three light wavelengths, 0.85, 1.3, or 1.500 μm. These wavelengths are used because the absorption of light by the optical fibers is minimum at these wavelengths.

The fibers used are made of special plastic or glass. Depending on the desired operating mode, bandwidth, and transmitting distance, different-diameter fibers are used. Fiber diameters used range from 2 to 1000 μm. As

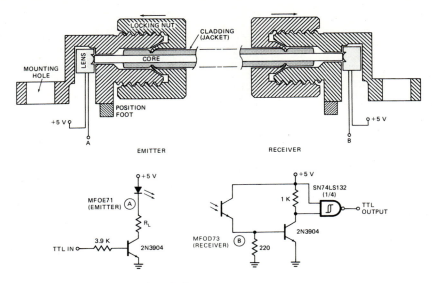

FIGURE 10-39   Diagram of simple fiber-optic data link.

THE OPTICS OF FIBERS

shown in Figure 10-39, the fiber-optic cable consists of three parts. The optical fiber core is surrounded by a *cladding* material which is also transparent to light. We will explain the function of this cladding later. The cladding layer is enclosed in a sheath which protects the cladding and does not allow external light to enter.

To convert the light signal back into an electrical signal at the receiving end, Darlington photodetectors such as the MFOD73 shown in Figure 10-39, p-i-n FET devices, or avalanche photodiodes (APDs) are used. APDs are more sensitive and operate at higher frequencies, but the circuitry for them is more complex. A special gate called a *Schmitt trigger* is usually used on the output of the detector to "square up" the output pulses. Now that you have an overview of an optical fiber link, we will briefly discuss some of the optics involved.

THE OPTICS OF FIBERS

The path of a beam of light going from a material with one optical density to a material of different optical density depends on the angle at which the beam hits the boundary between the two materials. Figure 10-40 shows the path that will be taken by beams of light at various angles going from an optically dense material such as glass to a less-dense material such as a vacuum or air. If the beam hits the boundary at a right angle, it will go straight through as shown in Figure 10-40a. When a beam hits the boundary at a small angle away from the perpendicular, or *normal*, it will be bent away from the normal when it goes from the more dense to the less-dense material, as shown in Figure 10-40b. A light beam going in the other direction would follow the same path. A quantity called the *index of refraction* is used to describe the amount that the light beam will be bent. Using the angle identifications shown in Figure 10-40b, the index of refraction, n, is defined as:

$$n = \frac{\text{sine of angle B}}{\text{sine of angle A}}$$

A typical value for the index of refraction of glass is 1.5. The larger the index of refraction, the more the beam will be bent when it goes from one material to another.

Figure 10-40c shows a unique situation that occurs when a beam going from a dense material to a less-dense material hits the boundary at a special angle called the *critical angle*. The beam will be bent so that it travels parallel to the boundary after it enters the less-dense material.

A still more interesting situation is shown in Figure 10-40d. If the beam hits the boundary at an angle greater than the critical angle, it will be totally reflected from the boundary at the same angle on the other side of the normal. This is somewhat like skipping stones across water. Here the light beam will not leave the more-dense material. The boundary essentially functions as a mirror.

To see how all this relates to optical fibers, take a look at the cross-sectional drawing of an optical fiber in Figure 10-40e. If a beam of light enters the fiber parallel to the axis of the fiber, it will simply travel through the fiber. If the beam enters the fiber such that it hits the fiber-cladding layer boundary at the critical angle (as shown for beam Y in Figure 10-40e), it will travel through the fiber-optic cable in the cladding layer. If the beam enters the cable such that it hits the fiber-cladding layer boundary at an angle greater than the critical angle, it will bounce back and forth between the walls of the fiber as shown for beam X in Figure 10-40e. The glass or plastic used for fiber-optic cables has very low absorption, so the beam can bounce back and forth along the fiber for several feet or miles without being reduced in intensity.

MULTIMODE AND SINGLE-MODE FIBERS

If an optical fiber has a diameter many times larger than the wavelength of the light being used, then beams which enter the fiber at different angles will arrive at the other end of the fiber at slightly different times. The different angles of entry for the beams are referred to as *modes*. A

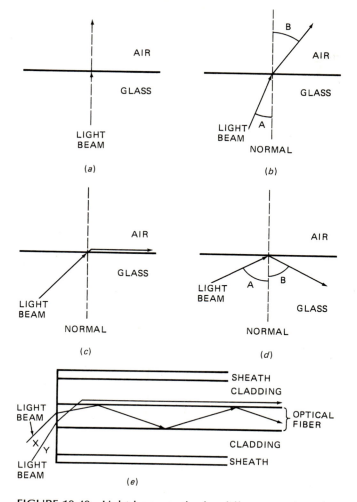

FIGURE 10-40 Light-beam paths for different angles of incidence with boundary of material having a lower optical density. *(a)* Right angle. *(b)* Angle less than critical angle. *(c)* At critical angle. *(d)* Angle greater than critical angle. *(e)* Angle greater than critical angle in optical fiber.

fiber with a diameter large enough to allow beams with several different entry angles to propagate through it is called a *multimode* fiber. Since multimode fibers are larger, they are easier to manufacture, are easier to work with manually, and can use inexpensive LED drivers. However, the phase difference between the output beams in multimode fibers causes problems at high data rates. One partial solution to this problem is to dope the glass of the fiber so that the index of refraction decreases toward the outside of the fiber. Light beams travel faster in the region where the index of refraction is lower, so beams which take a longer path back and forth through the faster outer regions tend to arrive at the end of the fiber at the same time as those that take a shorter path through the slower center region.

A better solution to the phase problems of the multimode fiber is to use a fiber that has a diameter only a few times the wavelength of the light being transmitted. With a small-diameter fiber, only light beams which are very nearly parallel to the axis of the fiber can be

transmitted. This is called *single-mode* operation. Single-mode systems currently available can transmit data a distance of over 60 km at a rate of nearly 1 Gbit/s. An experimental system developed by AT&T multiplexes 10 slightly different wavelength laser beams onto one single-mode fiber. The system can transmit data at an effective rate of 20 Gbits/s over a distance of 68 km without amplification.

One of the main problems with single-mode fibers is the difficulty in making low-loss connections with the tiny fibers. Another difficulty is that in order to get enough light energy into the tiny fiber, relatively expensive laser diodes or other lasers, rather than inexpensive LEDs, must be used.

FIBER-OPTIC CABLE APPLICATIONS

Fiber-optic transmission has the advantages that the signal lines are much smaller than the equivalent electrical signal lines, the signal lines are immune to electrically induced noise, and signals can be sent much longer distances without repeater amplifiers. Since a large number of fibers can be put in a single cable, one of the first major uses of fiber-optic transmission systems has been for carrying large numbers of phone conversations across oceans and between cities. The specifications of currently available fiber-optic systems are impressive, but relatively primitive as compared to the possibilities shown by laboratory research. In the future it is possible that the high data rate of fiber-optic transmission may make picture phones a household reality, replace TV cables, replace satellite communication for many applications, replace modems, and provide extensive computer networking.

LOGIC GATE INPUT AND OUTPUT INTERFACING

Introduction

So far in this chapter we have shown you how to generate logic level signals with oscillators and astables, how to modify the characteristics of logic signals with monostables and PLLs, and how to transmit logic level signals from one system to another with buffers and transmission lines. In the first part of this section we want to show you how we deal with signals which are too small or have too slow a risetime to be applied directly to the input of a logic gate. In the final part of this section we will show you how to interface high-power devices such as lights, relays, and motors to the outputs of logic gates.

Input Interfacing

SCHMITT TRIGGER INPUT DEVICES

If the risetime and/or falltime of a signal applied to the input of a standard logic gate is greater than about 500 ns, small noise voltages on the signal may cause the output of the device to produce unwanted transitions as shown in Figure 10-41a. The unwanted transitions are produced because the noise voltage on the input signal causes the input voltage to cross the threshold voltage several times during the slow risetime.

If a slow rise- and falltime signal is applied to the input of a *Schmitt trigger* inverter such as one of the six in a 74LS14 or 74HC14, the output is a clean signal as shown in Figure 10-41b. The hysteresis symbol inside the basic inverter shape in Figure 10-41b indicates that the device has a different threshold voltage for a positive-going signal than for a negative-going signal.

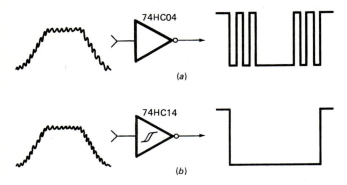

FIGURE 10-41 Output signal produced by applying noisy, slow-rise- and falltime signal to (a) input of standard inverter, (b) Schmitt trigger inverter such as that found in a 74LS14 or 74HC14.

Remember from Chapter 3 that the input threshold voltage for a standard 74LS04 inverter is about 1 V. As shown by the transfer curve in Figure 3-18, this means that when a rising input voltage reaches about 1 V, the output changes from a high to a low. Likewise, when a falling input signal voltage reaches about 1 V, the output goes from a low to a high.

Figure 10-42 shows the typical transfer curve for a 74LS14 Schmitt trigger inverter. The arrows on the center sections of this transfer curve tell you that the typical threshold voltage for a positive-going signal is about 1.6 V, and that the typical threshold voltage for a negative-going signal is about 0.8 V. The difference in voltage between the two thresholds is called *hysteresis*. The practical implication of this hysteresis is that a rising input signal has to reach about 1.6 V to cause the output to go low, and in order to cause the output to go back high, the input voltage has to drop past the low-going threshold voltage of about 0.8 V. This means that noise on the input signal would have to be greater than the hysteresis of 0.8 V to cause the output to make unwanted transitions. The minimum guaranteed hysteresis for 74LS14 and 74HC14 devices is actually only 0.4 V, but this still means that the noise on a signal has to be quite large to cause unwanted transitions on the output.

There are many applications for Schmitt trigger devices. Line receivers such as the MC3486s shown in Figures 10-30 and 10-31a have Schmitt trigger inputs to help overcome the effects of noise induced on signal transmission lines. Figure 10-43 shows how Schmitt trigger inverters and a zener diode can be used to produce a 60-Hz, TTL/CMOS level signal from the 120-V ac

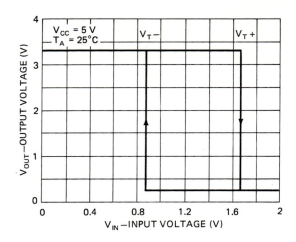

FIGURE 10-42 Output voltage versus input voltage transfer curve for 74LS14 Schmitt trigger inverter. *(Courtesy of Texas Instruments Inc.)*

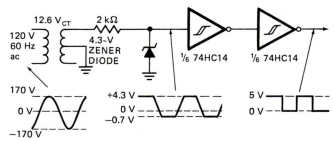

FIGURE 10-43 Circuit using Schmitt trigger devices to produce 60-Hz TTL/CMOS signal from 120-V ac line. Note waveforms produced at indicated points in the circuit.

power line. The zener diode limits the amplitude of the signal from the transformer to levels that are reasonably acceptable for a gate input, but the risetime and falltime of the signal produced by the zener are much too long for a standard gate input. The 74HC14 inverters "square up" the edges of the signal to the point where they are acceptable for standard gate inputs.

NOTE: Device diagrams in older books show the hysteresis symbol drawn in the same direction as the transfer curve in Figure 10-42. However, the ANSI/IEEE Std 91-1984 specifies the hysteresis symbol as shown in Figure 10-41b. Therefore, most newer data books will show it in this way.

COMPARATORS

Another problem commonly encountered when interfacing signals to gate inputs is that the available signal is at the wrong level or has too small an amplitude to be applied directly to a gate input. A device called a *comparator* can often be used to solve both of these problems.

Figure 10-44a shows how a common comparator, the LM319, can be used to convert a small sine-wave signal to TTL/CMOS input levels. When the voltage on the

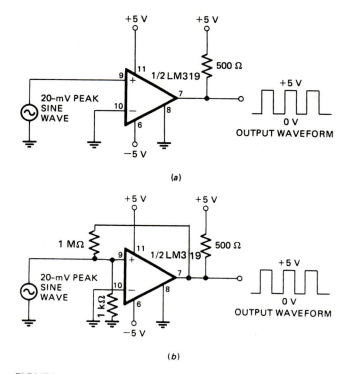

(a)

(b)

FIGURE 10-44 *(a)* Simple comparator circuit used to convert small-amplitude sine-wave signal to TTL/CMOS logic level signal. *(b)* Comparator circuit with hysteresis added to prevent oscillations on output.

noninverting (+) input is made a small fraction of a volt more positive than the inverting (−) input, the output of the comparator will go to a logic high level. When the noninverting (+) input is made a small fraction of a volt more negative than the voltage on the inverting input, the output of the comparator will go to a logic low state. The negative part of the 20-mV sine wave then causes the comparator output to go low and the positive part of the sine wave causes the comparator output to go high.

The comparator circuit in Figure 10-44a has two possible problems. One problem is that if a slow-risetime signal is applied to the input of the comparator, noise on the input of the comparator may cause the comparator output to oscillate high and low. This problem can be solved by adding feedback resistors as shown in Figure 10-44b. These resistors introduce some hysteresis in the comparator response so that it functions similarly to the Schmitt trigger devices we described in the preceding section.

The second potential problem with the circuit in Figure 10-44a is that its frequency response is limited to a few megahertz for small-amplitude signals. This problem is solved by using high-speed comparators such as the NE521A shown in Figure 9-21. This device can handle input signals with a frequency of up to 100 MHz. The circuit in Figure 9-21 also shows how the threshold or reference voltage for the comparator can be adjusted to a value other than 0 V so that a wide variety of signals can be converted to TTL/CMOS levels.

As a further example of this, Figure 10-45 shows how a comparator can be used to change the dc level of an

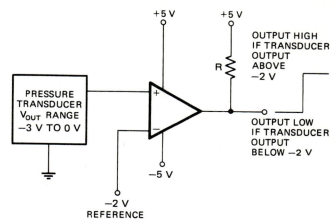

FIGURE 10-45 Comparator circuit used to determine if boiler pressure exceeds safety limit. The comparator shifts the negative voltage output signal from pressure transducer to TTL/CMOS logic levels.

input signal so that it is compatible with TTL/CMOS inputs. The voltage output by a pressure transducer ranges from −3 to 0 V, depending on the pressure applied to it. The reference input of the comparator is connected to −2 V, so if the output from the pressure transducer is below −2 V, the comparator output will be low. If the output from the pressure transducer is above −2 V, the output of the comparator will be high. This circuit then produces a logic signal which tells you if the pressure applied to the transducer is above or below some specified value.

Now that we have shown you some examples of logic gate input signal conditioning, we will show you how to interface the outputs of logic gates to a variety of devices.

Output Interfacing

The outputs of standard logic gates will not supply enough current to drive high-power devices such as lamps, heaters, solenoids, and motors. In the following short sections we show you some of the techniques you can use to interface to these high-power devices.

TRANSISTOR BUFFERS

In Chapter 3 we showed you how small NPN or PNP transistors can be used to drive simple LEDs from logic gate outputs. To refresh your memory, we will again show you some basic transistor circuits and how you quickly determine the parts values you need in these circuits for a particular application.

First determine how much current you need to flow through the LED, lamp, or other device. For our example here, suppose that we want 20 mA to flow through an LED. Next determine whether you want a logic high on the gate output to turn on the device or whether you want a logic low to turn on the device. If you want a logic high to turn on the LED, then use the NPN circuit shown in Figure 10-46a. If you want a logic low to turn on the LED, use the PNP circuit shown in Figure 10-46b. For a

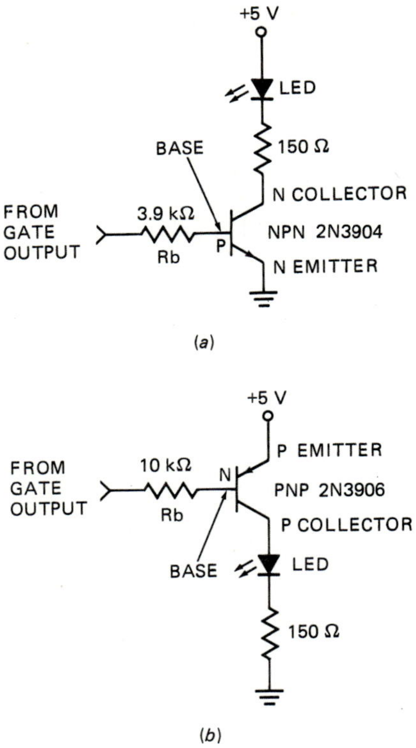

+5 V

LED

150 Ω

BASE

FROM
GATE
OUTPUT

3.9 kΩ

Rb

N COLLECTOR

NPN 2N3904

N EMITTER

(a)

+5 V

FROM
GATE
OUTPUT

10 kΩ

Rb

N

P EMITTER

PNP 2N3906

P COLLECTOR

BASE

LED

150 Ω

(b)

FIGURE 10-46   Circuit for driving an LED or small lamp.
(a) NPN buffer circuit. *(b)* PNP buffer circuit.

first example, suppose that you decide to use an NPN circuit.

Next look through your transistor collection to find an NPN transistor which can carry the required current, has a collector-to-emitter breakdown voltage (VBCEO) greater than the applied supply voltage, and can dissipate the power generated by the current flowing through it. We usually keep some inexpensive 2N3904 NPNs and some 2N3906 PNPs on hand for low-current switch applications such as this. Some alternatives are the 2N2222 NPN and the 2N2907 PNP.

When you decide what transistor you are going to use, look up its current gain, h_{FE}, on a data sheet. If you don't have a data sheet, assume a value of 50 for the current gain of a small signal transistor such as this. Remember, current gain, or β as it is commonly called, is the ratio of collector current to the base current needed to produce that current. To produce a collector current of 20 mA in a transistor with a $\beta = 50$ then requires a base current of 20 mA/50 or 0.4 mA. To drive this buffer transistor, then the gate output only has to supply the 0.4 mA. A look at Table 3-2 shows that most TTL and CMOS gate devices can supply this current in the high state with no difficulty.

The value of the base current-limiting resistor, Rb, is not very critical as long as it is small enough to let through enough base current to drive the transistor. To calculate a value for this resistor, let's assume the voltage on the output of the gate is 2.4 V. The base of the NPN transistor will be at about 0.7 V when the transistor is conducting, so this leaves a voltage of 1.7 V across Rb (2.4 V − 0.7 V). Dividing the 1.7 V across Rb by the

desired current of 0.4 mA gives an Rb value of 4.25 kΩ. A 3.9-kΩ or 4.3-kΩ resistor will work fine.

For the PNP circuit in Figure 10-46*b*, the gate end of Rb will be at a voltage near 0 V when the gate output is low. The other end of Rb will be at a voltage of about 4.3 V (5 V − 0.7 V). The voltage across Rb then is about 4.3 V. Assuming the PNP transistor requires a base current of 0.4 mA, the value of Rb for this circuit is then equal to 4.3 V/0.4 mA, or about 10 kΩ.

To calculate the value of the LED current-limiting resistor for either circuit you subtract the voltage across the lit LED from +5 V and divide the result by the current you want through the LED. The voltage across an LED is a function of the color of the LED and the amount of current through it. Red LEDs have a voltage drop of about 1.5 V, yellow LEDs a voltage drop of about 2.4 V, and green LEDs a voltage drop of up to 3 V. Subtracting an LED voltage drop of 2 V from the supply voltage of 5 V leaves 3 V across the LED current-limiting resistor. Dividing 3 V by a current of 20 mA gives a value of 150 Ω for the LED current-limiting resistor.

When you need to switch currents larger than about 50 mA on and off with the output of a gate, a single transistor does not have enough current gain to do this dependably. One solution to this problem is to connect two transistors in a Darlington configuration. Figure 10-47 shows how we might do this to drive a small solenoid-controlled valve or a small solenoid in the print head of a dot-matrix printer.

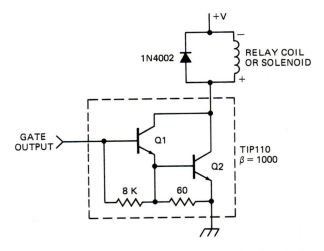

FIGURE 10-47   Darlington transistor used to drive relay coil or solenoid.

The dashed lines around the two transistors in Figure 10-47 indicate that both devices are contained in the same package. Here's how this configuration works. The gate output supplies base current to transistor Q1. This base current produces a collector-emitter current that is β times as large in Q1. The collector current of Q1 becomes the base current of Q2 and is amplified by the current gain of Q2. The result of all this is that the device acts like a single transistor with a current gain of (βQ1 × βQ2) and a base-emitter voltage of about 1.4 V. The internal resistors help turn off the transistors. The

TIP110 device we show here has a minimum β of 1000 at 1 A. If we assume that we need 400 mA to drive the solenoid, then the worst-case current that must be supplied by the gate output is about 400 mA/1000, or 0.4 mA, which it can easily do. If the drive current required for the Darlington base is higher than a gate output can supply, you can add a resistor from the transistor base to +5 V to supply the added current. When the gate output is in the low state, it can easily sink the added current supplied by the resistor. Note that since the V_{BE} of the Darlington is about 2 V, no Rb is needed here.

If you need several Darlington switches such as this, the ULN2068B manufactured by Sprague and Motorola contains four NPN Darlingtons which can each sink up to 1.5 A if the device is mounted on a heat sink. Each Darlington in the device has a built-in diode which can be used with inductive loads as shown in Figure 10-47.

When working with large currents and high-power devices, it is very important to make sure that the maximum power ratings of the devices are not exceeded. To give you an idea of how to do this, let's check out the power dissipation for the Darlington circuit in Figure 10-47.

According to the data sheet for the TIP110, it comes in a TO-220 package that can, with no heat sink, dissipate up to 2 W at an ambient (room) temperature of 25°C. With 400 mA flowing through the device, it will have a collector-to-emitter saturation voltage of about 2 V. Multiplying the current of 400 mA times the voltage drop of 2 V gives us a power dissipation of 0.8 W for the TIP110 in this circuit. This value is well within the limit for the device. A rule of thumb that we like to follow is: If the calculated power dissipation for a device such as this is more than half of its 25°C no-heat-sink rating, mount the device on a metal chassis or a heat sink to make sure it will work on a hot day. If mounted on an appropriate heat sink, a TIP110 will dissipate 50 W with an ambient temperature of 25°C.

One more important point to mention about the circuit in Figure 10-47 is the reverse-biased diode connected across the solenoid coil. You must remember to put in this diode whenever you drive an inductive load such as a solenoid, relay, or motor. Here's why.

As we explained in Chapter 4, the basic principle of an inductor is that it fights against a change in the current through it. When you apply a voltage to the coil by turning on the transistor, it takes a while for the current to start flowing. This does not cause any major problems. However, when you turn off the transistor, the collapsing magnetic field in the inductor keeps the current flowing for a while. This current cannot flow through the transistor because it is off. Instead, this current develops a voltage across the inductor with the polarity shown by the + and − signs on the coil in Figure 10-47. This induced voltage, sometimes called inductive "kick," will usually be large enough to break down the transistor if you forget to put in the diode. When the coil is conducting, the diode is reversed-biased, so it doesn't conduct. However, as soon as the induced voltage reaches 0.7 V, this diode turns on and supplies a return path for the induced current. The voltage across the inductor is clamped at 0.7 V, which saves the transistor.

Figure 10-48 shows how a power MOSFET transistor

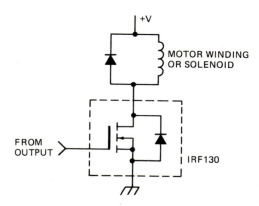

FIGURE 10-48 Power MOSFET circuit for driving solenoid or motor winding.

can be used to drive a solenoid, relay, or motor winding. Power MOSFETs are more expensive than bipolar Darlingtons, but they have the advantage that they only require a voltage to drive them. The Motorola IRF130 shown here, for example, only requires a maximum gate voltage of 4 V to turn on a drain current of 8 A. Note that a diode is placed across the coil in this circuit to protect the transistor from inductive kick.

INTERFACING TO A STEPPER MOTOR

During a discussion of shift registers in Chapter 9, we told you that the pattern of pulses output by a ring counter is useful for rotating a stepper motor. A stepper motor is a unique type of motor often used for moving things in small increments. If you have a dot-matrix printer, such as the Epson FX-80, look inside and you will probably see one small stepper motor which is used to advance the paper to the next line position, and another small stepper motor which is used to move the print head to the next character position. Stepper motors are also used to position the read/write head over the desired track of a hard or floppy disk, and to position the pen on X-Y plotters.

Instead of rotating smoothly around and around as most motors do, stepper motors rotate or "step" from one fixed position to the next. Available step sizes range from 0.9° to 30°. A stepper motor is stepped from one position to the next by changing the currents through the fields in the motor. The two common field connections are called two-phase and four-phase. We will discuss *four-phase steppers* here because their drive circuitry is much simpler.

Figure 10-49a shows a circuit you can use to interface a small four-phase stepper such as the Superior Electric MO61-FD302, IMC Magnetics Corp. Tormax 200, or a similar, nominal 5-V unit to the outputs of the ring counter in Figure 9-30. If you build up this circuit, bolt some small heat sinks on the MJE2955 transistors and mount the 10-W resistors where you aren't likely to touch them.

Since the 7406 buffers are inverting, a high on one of the inputs produces a low on the corresponding output. This low turns on the PNP power transistor and supplies current to one of the windings in the motor.

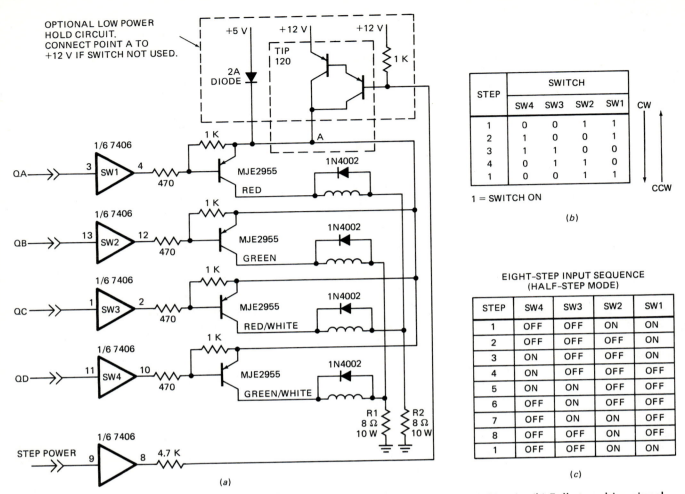

Optional low power hold circuit. Connect point A to +12 V if switch not used.

STEP	SWITCH				
	SW4	SW3	SW2	SW1	CW
1	0	0	1	1	
2	1	0	0	1	
3	1	1	0	0	
4	0	1	1	0	
1	0	0	1	1	CCW

1 = SWITCH ON

(b)

EIGHT-STEP INPUT SEQUENCE (HALF-STEP MODE)

STEP	SW4	SW3	SW2	SW1
1	OFF	OFF	ON	ON
2	OFF	OFF	OFF	ON
3	ON	OFF	OFF	ON
4	ON	OFF	OFF	OFF
5	ON	ON	OFF	OFF
6	OFF	ON	OFF	OFF
7	OFF	ON	ON	OFF
8	OFF	OFF	ON	OFF
1	OFF	OFF	ON	ON

(c)

FIGURE 10-49   Four-phase stepper motor interface circuit and switch sequence. *(a)* Circuit. *(b)* Full-step drive signal order. *(c)* Half-step drive signal order.

Figure 10-49*b* shows the switching sequence required to step a motor such as this clockwise, as you face the motor shaft, or counterclockwise. Here's how this works. Suppose that SW1 and SW2 are turned on. Turning off SW2 and turning on SW4 will cause the motor to rotate one step of 1.8° clockwise. Changing to SW4 and SW3 on will cause the motor to rotate another 1.8° clockwise. Changing to SW3 and SW2 on will cause another step. After that, changing to SW2 and SW1 on again will cause another step clockwise. You can repeat the sequence until the motor has rotated as many steps clockwise as you want. To step the motor counterclockwise, you simply work through the switch sequence in the reverse direction. In either case, the motor will be held in its last position by the current through the coils. Figure 10-49*c* shows the switch sequence that can be used to rotate the motor half steps of 0.9° clockwise or counterclockwise.

A close look at the switch sequence in Figure 10-49*b* shows an interesting pattern. To take the first step clockwise from SW2 and SW1 being on, the pattern of 1's and 0's is simply rotated one bit position around to the right. The 1 from SW1 is rotated around into bit 4. To take the next step, the switch pattern is rotated one more bit po-

sition. To step counterclockwise, the switch pattern is rotated left one bit position for each step desired. As we showed you in Chapter 9, this pattern can easily be produced by loading 0011 into a ring counter. To step the motor clockwise, you just rotate this pattern right one bit position. To step counterclockwise, you rotate the switch pattern left one bit position. After you output one step code you must wait a few milliseconds before you output another step command because the motor can only step so fast. Maximum stepping rates for different types of steppers vary from a few hundred steps per second to several thousand steps per second. To achieve high stepping rates, the stepping rate is slowly increased to the maximum, then it is decreased as the desired number of steps is approached. In the sequel to this book we show you how a stepper motor can be controlled by a microprocessor, as is done in printers and plotters.

Before we go on, here are a couple of additional points about the circuit in Figure 10-49*a*, if you want to add a stepper to your robot or some other project. First, don't forget the clamp diodes across each winding to save the transistors from inductive kick. Second, we need to explain the function of the current-limiting resistors, R1 and R2. The motor we used here has a nominal voltage rating of 5.5 V. This means that we could have designed the circuit to operate with a voltage of about 6.5 V on the emitters of the driver transistors (5.5 V for the motor plus 1 V for the drop across the transistor). For low stepping rates, this would work fine. However, for higher stepping rates and more torque while stepping, we use a higher supply voltage and current-limiting resistors as shown. The point of this is that by adding series resistance, we decrease the L/R time constant. This allows the current to change more rapidly in the windings. For the motor we used, the current per winding is 0.88 A. Since only one winding on each resistor is ever on at a time, 5.5 V/0.88 A gives a resistor value of 6.25 Ω. To be conservative, we used 8-Ω, 10-W resistors. The optional transistor switch and diode connection to the +5-V supply are used as follows. When not stepping, the switch to +12 V is off so the motor is held in position by the current from the +5-V supply. Before you send a step command, you turn on the transistor to +12 V to give the motor more current for stepping. When stepping is done, you turn off the switch to +12 V and drop back to the +5-V supply. This cuts the average power dissipation.

INTERFACING TO AC POWER DEVICES

To turn 110-, 220-, or 440-V ac devices on and off under logic gate control, we usually use *mechanical* or *solid-state relays* because the control circuitry for both of these devices is electrically isolated from the actual switch. This is very important because if the 110-V ac line gets shorted to the V_{CC} line of a digital system, it usually destroys most of the ICs in the system.

Figure 10-50a shows a picture of a mechanical relay. This relay has both normally open and normally closed contacts. When a current is passed through the coil of the relay, the switch arm is pulled down, opening the top contacts and closing the bottom set of contacts. The contacts are rated for a maximum current of 25 A, so this relay could be used to turn on a 1- or 2-hp motor or a large electric heater. When driven from a 12-V supply, the coil requires a current of about 170 mA. The Darlington driver circuit shown in Figure 10-47 could easily drive this relay coil from a logic gate output.

Mechanical relays, sometimes called contactors, are available to switch currents from milliamps up to several thousand amps. Mechanical relays, however, have several serious problems. When the contacts are opened and closed, arcing takes place between the contacts. This causes the contacts to oxidize and pit, just as the ignition points in some cars do with age. As the contacts become oxidized, they make a higher-resistance contact and may get hot enough to melt. Another disadvantage of mechanical relays is that they can switch on or off at any point in the ac cycle. Switching on or off at a high-voltage point in the ac cycle can cause a large amount of electrical noise called *electromagnetic interference*, or EMI. To a large extent, the solid-state relays discussed next avoid these problems.

Figure 10-50b shows a picture of a solid-state relay which is rated for 25 A at 25°C if it is mounted on a suitable heat sink, and Figure 10-50c shows a block diagram of the circuitry in the device. Figure 10-50c also

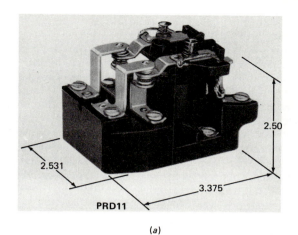

PRD11

(a)

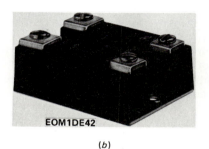

EOM1DE42

(b)

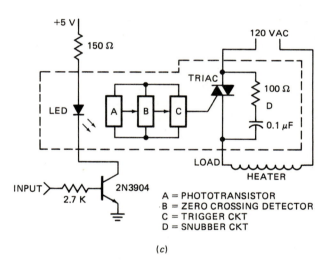

(c)

FIGURE 10-50   Relays for switching large currents. *(a)* Mechanical. *(b)* Solid-state. *(Courtesy of Potter and Brumfield.) (c)* Internal circuitry for solid-state relay.

shows how the relay is connected to a gate output and to a high-power ac load such as a heater.

The input circuit is essentially an LED and a current-limiting resistor. To turn the device on you simply apply a logic low to the negative end of the internal LED. This pulls current through the LED and turns it on. The light from the LED is focused on a phototransistor which turns on when the light falls on it. The phototransistor then activates the circuitry which turns on the ac power. Since the only connection between the input circuit and the output circuit is a beam of light, there are several thousand volts of electrical isolation between the input circuitry and the output circuitry.

The actual ac switch in a solid-state relay is a device called a *triac*. When triggered, this device conducts in either direction, so it works for both halves of the ac cycle. In order for the triac to be triggered, two conditions must be present. First, the LED and phototransistor must be on. Second, the ac line voltage must be at or near one of its zero crossing points. The zero-voltage detector in the device makes sure that the triac is only triggered when the ac line voltage is very close to one of its zero-voltage crossing points. This zero-point switching eliminates most of the EMI that would be caused by switching the triac on at high-voltage points in the ac cycle.

Triacs automatically turn off when the current through them drops below a small value called the *holding current*. If the input control signal is on, the trigger circuitry will automatically retrigger the triac for each half cycle. When the input control signal is turned off, the triac will turn off the next time the current drops to zero and will not be retriggered.

Solid-state relays then have the advantages that they produce less EMI, they have no mechanical contacts to arc, and they are easily driven from logic gate outputs. Their disadvantages are that they are more expensive than an equivalent mechanical relay and there is a voltage drop of a couple of volts across the triac when it is on. Another potential problem with solid-state relays occurs when driving a large inductive load such as a motor.

As you may remember from basic ac theory, the voltage waveform leads the current waveform in an ac circuit which has inductance. A triac turns off when the current through it drops to near zero. In an inductive circuit, the voltage waveform may still be at several tens of volts when the current is at zero. When the triac is conducting, it has perhaps 2 V across it. When the triac turns off, the voltage across the triac will quickly jump to the several tens of volts. This large dV/dT may turn on the triac again at a point you don't want it turned on. To keep the voltage across the triac from changing too rapidly, a *RC snubber* circuit is connected across the triac as shown in Figure 10-50c.

CHECKLIST OF IMPORTANT TERMS AND CONCEPTS

Astable multivibrator
Voltage-controlled oscillator, VCO
Offset frequency
Crystal-controlled oscillator
Resonant frequency

Series- and parallel-resonant mode
Baud rate generator
Monostable multivibrator
Retriggerable
Phase-locked loop
    Locked, lock range
    Capture range
    Phase-frequency comparator
Digital frequency synthesizer
Center frequency
Carrier frequency
Modulation
RF amplifier
Mixing
Line drop
Crosstalk
Electromagnetic interference, EMI
Characteristic impedance
Ground plane
In phase, out of phase
Parallel termination
Buffer, line driver
RS-232C, RS-423, RS-422A
Modem, modulator-demodulator
Integrated services digital network, ISDN
Amplitude modulation
Frequency-shift keying, FSK
Phase-shift modulation
Dibit, tribit, quadbit
Quaternary amplitude modulation, QAM
Data communication equipment, DCE
Data terminal equipment, DTE
Handshake signals,
    Data terminal ready, $\overline{DTR}$
    Data set ready, $\overline{DSR}$
    Request to send, $\overline{RTS}$
    Carrier detect, $\overline{CD}$
    Clear to send, $\overline{CTS}$
Null modem
Fiber-optic signal lines
Infrared injection laser diode, ILD
Index of refraction
Critical angle
Multimode and single-mode optical fibers
Schmitt trigger
Hysteresis
Comparator
Four-phase stepper
Relay, mechanical and solid-state
Triac
RC snubber circuit

REVIEW QUESTIONS AND PROBLEMS

1. *a.* Starting with the point where the timing capacitor has no charge, describe the operation of the 555 astable multivibrator circuit in Figure 10-1 for one charge-discharge cycle. Assume the final output is initially high.

   b. During the first cycle after power-up, the output time high for the circuit in Figure 10-1 is longer than the time the output is high during later cycles. Explain why this is the case.

2. Determine the approximate frequency and duty cycle of the 555 astable multivibrator circuit in Figure 10-51.

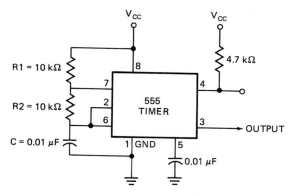

FIGURE 10-51   555 astable multivibrator circuit for Problems 2 and 3.

3. Given a 0.01-µF timing capacitor, use the table and formulas in Figure 10-2 to determine values for R_A and R_B that will cause a 555 circuit in Figure 10-51 to output a 100-Hz signal that is approximately a square wave. Choose resistor values from the standard 5 percent resistor values shown in Table 1-3.

4. *a.* Explain how the voltage applied to the control input of the 74HC4046 VCO circuit in Figure 10-3b affects the output frequency.

b. Explain how the circuit in Figure 10-4 produces an increasing frequency sound when the switch is flipped to the up position and why the circuit produces a decreasing frequency sound when the switch is flipped to the down position.

c. What symptom would you observe if the 2.2-µF capacitor in Figure 10-4 became shorted?

5. For the oscillator circuit in Figure 10-6
a. What is the purpose of the voltage dividers on the inputs of the inverters?
b. What starts this oscillator oscillating?
c. What determines the frequency of oscillation for the circuit?

6. Draw a circuit which uses a 2.45-MHz crystal oscillator and some other devices to produce the 16× baud rate clock for a UART which is transmitting serial data at 2400 Bd.

7. Is the 555 in Figure 10-52 operating as an astable multivibrator or as a monostable multivibrator? Explain how you arrived at your conclusion.

8. Use the table in Figure 10-9b to help you determine the output time high for the 555 circuit in Figure 10-52.

9. *a.* Use the chart in Figure 10-11 to help you determine the resistor value needed with a 100-pF capacitor to produce a 500-ns-wide pulse on the output of a 74121 when the device is triggered.

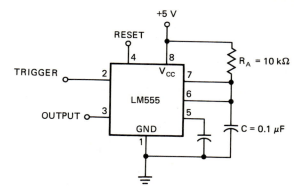

FIGURE 10-52   555 timer circuit for Problems 7 and 8.

b. Draw a 74121 circuit which includes the R and C values from part *a* and sets the 74121 up to trigger on the rising edge of the trigger signal.

c. Draw the waveform that will be produced on the output of the circuit in part *b* if the waveform shown in Figure 10-53 is applied to the trigger input of the 74121.

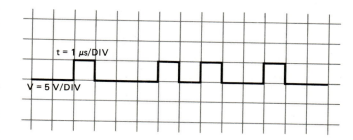

FIGURE 10-53   Input waveform for Problem 9.

10. *a.* Use the graph in Figure 10-13 to help you calculate the output pulse width for the 74LS122 circuit in Figure 10-54a
b. Is this circuit triggered by a rising edge on the trigger signal or by a falling edge on the trigger signal?
c. The data sheet for a 74LS122 says that the device is "retriggerable." What does this mean?
d. Show the waveform that will be produced on the output of the circuit in Figure 10-54a if the waveform shown in Figure 10-54b is applied to the trigger input.

11. One use of a monostable multivibrator is a "missing pulse detector." Design and draw the schematic for a 74HC122 circuit that will output a low pulse if any cycles are missing in a 1-kHz square-wave signal applied to its trigger input. Set up the 74HC122 to trigger on the rising edge of the 1-kHz input signal.

12. Another common use of monostable multivibrators is to change the width of a pulse. Suppose that a system outputs positive pulses which are 1 ms wide, and you need these pulses converted to positive pulses which are only 1 µs wide.

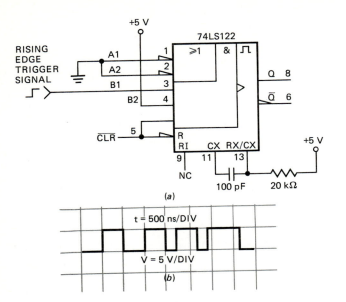

FIGURE 10-54 (a) 74LS122 circuit for Problem 10.
(b) Input waveform for Problem 10d.

    a. Draw a circuit which shows how a 74LS122 can be used to do this. (Assume positive edge triggering.)

    b. Calculate the required R and C values for the circuit.

13. Figure 10-55 shows a block diagram of a basic phase-locked loop.

VCO output and the input of the phase comparator, what frequency will the VCO output when a 100-kHz signal is applied to the external input of the phase comparator?

14. Four major problems encountered when transmitting digital signals over long wires are line drop, crosstalk, EMI, and signal reflections.

    a. Explain the meaning of each of these terms.

    b. Describe a method used to reduce each of these potential problems.

15. Figure 10-56 shows a positive-going pulse with fast rise- and falltimes being sent down a transmission line. Assume that the time required for the signal to travel the length of the line is several times longer than the risetime of the applied signal.

    a. Sketch the waveform that will be produced at the end of the line if the line is open.

    b. Sketch the waveform that will be seen at the end of the line if it has a resistance less than Z_o connected across it.

    c. Sketch the waveform that will be seen at the end of the line if the line is terminated with a resistance equal to the Z_o of the line.

16. a. How far does an electrical signal typically travel on a signal line in 1 ns?

    b. Assuming a logic family has output rise- and falltimes of 2 ns, use the 4× rule to determine the longest unterminated connecting wires that can safely be used with devices in that family.

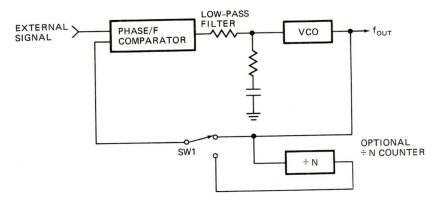

FIGURE 10-55 Block diagram of basic phase-locked loop for Problem 13.

    a. Describe the response that this system will make if the frequency of the external signal applied to the phase comparator is greater than the frequency of the VCO.

    b. Describe the response that the system will make if the frequency of the external signal is changed to a frequency below the VCO frequency.

    c. Describe the response that the system will make if the external frequency is increased to a frequency greater than the lock range of the circuit.

    d. If a divide-by-5 counter is inserted between the

    c. GaAs logic devices have output rise- and falltimes of about 150 ps. What is the maximum unterminated connecting line length for devices in this family?

    d. If a logic family has output rise- and falltimes of 3 ns, what is the maximum frequency square-wave signal that simple devices in this family can work with?

17. Signal reflections are often used to find shorts or opens in metallic and fiber-optic cables. The technique is called *time domain reflectometry* (TDR),

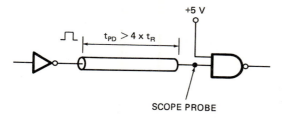

FIGURE 10-56   Fast-rise- and falltime pulse applied to transmission line with transmit time longer than 4 × risetime.

or "cable radar." Instruments such as the Tektronix OF235A, for example, are used to do TDR on fiber-optic cables, and instruments such as the Tektronix 1502B are used to do TDR on metallic cables.

To find a short or open on a line, a narrow pulse is sent out on the line as shown in Figure 10-57a.

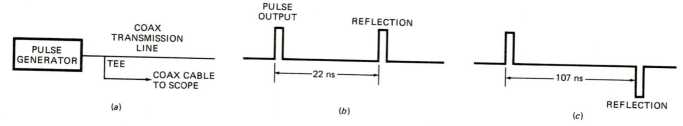

FIGURE 10-57   Time domain reflectometry waveforms for Problem 17.

The time between the transmitted pulse and the reflected pulse is proportional to the distance to the problem in the line. The shape of the reflected pulse indicates whether the cable has a short or an open.

For the TDR displays shown in Figure 10-57b and c, identify the type of problem and determine the distance from the source to the problem. (Assume the pulse travels 8 in/ns on the cable.)

18. *a.* Why are ECL transmission lines usually terminated with a resistor connected to a negative voltage or with an equivalent voltage divider, rather than with a resistor connected to ground?

b. Why will most standard logic gates not drive a 50-Ω terminated transmission line?

c. What method is often used to transmit TTL/CMOS level signals for distances longer than those permitted for open lines?

d. What is a typical impedance value for the twisted-pair wire commonly used for transmission lines?

19. *a.* On an RS-232C signal line a voltage of −12 V is a logic _____ and a voltage of +12 V is a logic _____ .

b. How are TTL signal levels commonly converted to and from RS-232C levels?

c. RS-232C signal lines are not terminated. How then is it possible to send signals for distances

of up to 50 ft on RS-232C lines without serious reflection problems?

20. Figure 10-58 shows a buffer and a receiver commonly used for RS-422 signal transmission.

a. How are a logic low and a logic high defined in this standard?

b. What is the purpose of the resistors between the signal lines and ground in this circuit?

c. What are the major advantages of RS-422 over RS-232C?

d. Use the graph in Figure 10-31b to determine the maximum recommended RS-422 line length for transmission at 4800 Bd.

21. *a.* Why are standard phone lines not suitable for transmitting digital signals directly?

b. What is meant by the term *modem*?

c. List the three types of modulation commonly used by modems.

d. Figure 10-59 shows a section of a waveform observed on the output of a modem. Identify the type of modulation used and the sequence of 1's and 0's represented.

e. Describe how phase-shift modulation allows two data bits to be encoded and transmitted during one bit time.

22. *a.* Why are "handshake" signals necessary between a terminal and a modem and between a computer and a modem?

b. Describe the functions of the RS-232C $\overline{\text{RTS}}$ and $\overline{\text{CTS}}$ signals.

c. The two RS-232C connectors in Figure 10-60 both come from equipment configured as DTE. Show how you would connect wires between the two connectors so that the two systems can communicate properly.

23. *a.* Draw a diagram that shows the construction of a fiber-optic cable, and label each part.

b. Identify two types of devices that are used to produce the light beam for a fiber-optic cable and two devices which are commonly used to detect the light at the receiving end of the fiber.

c. Why should you never look into the end of a fiber-optic cable to see if light is getting through?

d. Describe the difference between a multimode

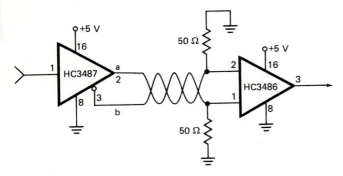

FIGURE 10-58   Line driver and line receiver connections for RS-422 signal transmission for Problem 20.

A ◁\/\/\/\/\/\/\/\/\/\/\/\/\/\/\▷ B
ASSUME 1200 BD

FIGURE 10-59   Section of modem output waveform for Problem 21d.

T x D	2	2	T x D
R x D	3	3	R x D
$\overline{RTS}$	4	4	$\overline{RTS}$
$\overline{CTS}$	5	5	$\overline{CTS}$
$\overline{DSR}$	6	6	$\overline{DSR}$
GND	7	7	GND
$\overline{DTR}$	20	20	$\overline{DTR}$

DTE   DTE

FIGURE 10-60   RS-232C connectors for Problem 22.

fiber and a single-mode fiber. Give a major advantage and a major disadvantage of each type.

e.   What are the major advantages of fiber-optic cables over metallic conductors?

24.   The circuit in Figure 10-61 is an inexpensive way to make an oscillator for applications where the exact frequency does not matter.

a.   What is the meaning of the boxlike symbol in the inverters in the circuit in Figure 10-61.

b.   Assuming the capacitor has 0 V on it when the power is first turned on, describe the operation of this oscillator for one cycle of operation. (Assume the + threshold is 2.5 V and the − threshold is 1.6 V.)

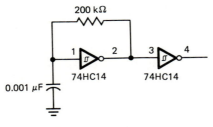

FIGURE 10-61   Inexpensive oscillator circuit using Schmitt trigger inverters for Problem 24.

c.   Why will a standard 74HC04 inverter not work in this circuit?

d.   Why is a Schmitt trigger device such as a 74HC14 used to interface slow-risetime signals to the inputs of standard logic gates?

25.   a.   Draw the waveform that will be produced on the output of the comparator circuit in Figure 10-62a if a 1-V peak ac sine-wave signal is applied to the + input.

b.   Predict the logic level that will be on the output of the comparator in Figure 10-62b with the input voltages shown.

c.   Why are resistors which provide hysteresis often added to a comparator circuit?

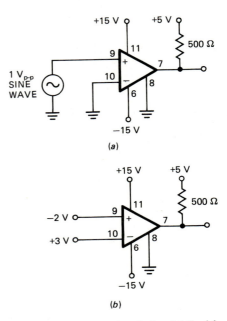

FIGURE 10-62   Comparator circuit for (a) Problem 25a, (b) Problem 25b.

26.   The transistor buffer circuit in Figure 10-63 is intended to interface a 74HC00 gate output to a relay coil.

a.   Is this transistor an NPN or a PNP?

b.   Will a logic high or a logic low turn on this transistor?

c.   What is the purpose of the diode connected across the relay coil?

d.   If the coil requires a current of 100 mA and the transistor has a $\beta = 60$, how much current must be supplied to the base of the transistor by the gate output.

e.   According to Table 3-2, can a 74HC00 output source enough current in the high state to supply the current you calculated in part d?

f.   What voltage will be present on the base of the transistor when it is conducting?

g.   Assuming a voltage of 2.5 V on the output of the 74HC00 gate, calculate the value of R_b which will allow the needed base current to flow.

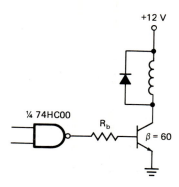

FIGURE 10-63   Transistor buffer circuit for Problem 26.

27. What is the major advantage of a power MOSFET device over a bipolar Darlington device when interfacing from a gate output to, for example, a large solenoid?

28. a. Why can't a standard bipolar or MOSFET transistor be used to turn ac power devices off and on?

    b. How is a mechanical relay activated?

29. a. Describe the operation of the optical coupling circuit on the input of the solid-state relay circuit in Figure 10-64.

    b. Describe the operation and purpose of the "zero crossing detector" circuit found in most better quality solid-state relays such as that in Figure 10-64.

    c. Give the major advantages that solid-state relays have over mechanical relays.

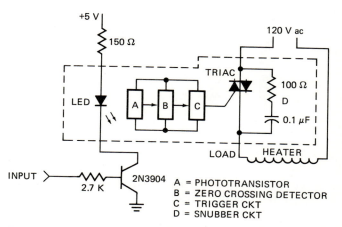

FIGURE 10-64   Solid-state relay block diagram for Problem 29.

A = PHOTOTRANSISTOR
B = ZERO CROSSING DETECTOR
C = TRIGGER CKT
D = SNUBBER CKT

11 A/D and D/A Converter Circuits and Applications

Most physical quantities such as pressure, temperature, and flow are analog in nature. There are usually several steps in producing electrical signals which represent the values of these variables and in converting the electrical signals to a digital form that can, for example, be used to drive an LED display or be stored in the memory of a microcomputer.

The first step involves a *sensor* which produces a current or voltage signal that is proportional to the amount of the physical pressure, temperature, or other variable. The signals from most sensors are quite small, so they must be amplified and perhaps filtered to remove unwanted noise. Amplification is usually done with some type of operational-amplifier (op-amp) circuit. The final step is to convert the signal to a proportional binary word with an analog-to-digital (A/D) converter.

In this chapter we introduce you to some basic op-amp circuits used in these steps, describe some common sensors, and then discuss the operation of several types of A/D and D/A converters. Finally in the chapter we discuss A/D and D/A applications such as digital multimeters, digital thermometers, digital storage oscilloscopes, and digital audio systems.

OBJECTIVES

At the conclusion of this chapter you should be able to:

1. Recognize several common op-amp circuits, describe their operation, and predict the voltages at key points in each.

2. Describe the operation and interfacing of several common sensors used to measure light intensity, temperature, and pressure.

3. Describe the operation of a weighted resistor D/A converter and predict its output voltage for a given input word.

4. Define A/D and D/A data-sheet parameters such as *resolution, settling time, accuracy, quantizing error,* and *linearity.*

5. Determine the number of bits required for a specified resolution with an A/D or D/A converter.

6. Given a block diagram of each, describe the operation of flash, single-ramp, dual-ramp, counter, and successive approximation type A/D converters.

7. Describe the operation of the circuits used to convert a basic A/D converter to a digital multimeter.

8. Given a block diagram for a digital storage oscilloscope, explain how this instrument produces a display of an input signal.

9. Describe how audio signals are digitally recorded and played back from optical disks or digital audio tapes.

INTRODUCTION TO OPERATIONAL-AMPLIFIER CHARACTERISTICS AND CIRCUITS

Basic Operational-Amplifier Characteristics

Figure 11-1 shows the schematic symbol for an operational amplifier commonly called an *op amp.* The basic points for you to remember about an op amp are:

1. The pins labeled +V and −V represent the power-supply connections. These pins are usually connected

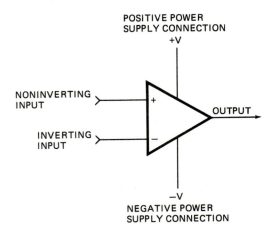

FIGURE 11-1 Schematic symbol for basic operational amplifier, or op amp.

to voltages such as $+15$ V and -15 V, or $+12$ V and -12 V.

2. The amplifier has two signal inputs labeled $-$ and $+$, and a single output. The input labeled with a $-$ sign is called the *inverting input* and the one labeled with the $+$ sign is called the *noninverting input*. The $+$ and $-$ on these inputs have nothing to do with the power-supply voltages. These signs show the phase relationship between a signal applied to that input and the result that signal produces on the output. If, for example, the noninverting input is made more positive than the inverting input, the output will move in a positive direction, which is in phase with the applied input signal.

3. The signal inputs of the amplifier have very high input impedances, so almost no current flows into or out of either one.

4. The amplifier amplifies the difference in voltage between the two signal inputs by 100,000 or more.

Now let's see how these rules enable you to predict how an op amp works in a variety of common circuits.

Op-Amp Circuits and Applications

Figure 11-3 shows an overview of some commonly used op-amp circuits. We put these all in one figure so that you can easily compare them. Refer to these circuits as we discuss them in the following sections.

Op Amps as Comparators

We said previously that the op amp amplifies the difference in voltage between its inputs by 100,000 or more (the number is variable with temperature and from device to device). Now, suppose that you power an op amp with $+15$ V and -15 V, tie the inverting input of the op amp to ground, and apply a signal of $+0.01$ V dc to the noninverting input as shown in Figure 11-2a. The op amp will attempt to amplify this signal by 100,000 and produce the result on its output. An input signal of 0.01 V times a gain of 100,000 predicts an output voltage of 1000 V. However, the op-amp output can only go positive to a voltage that is a volt or two less than the positive supply voltage, perhaps 13 V, so this is as far as it goes. Now suppose that you connect the inverting input to ground and apply a signal of -0.01 V to the noninverting input as shown in Figure 11-2b. The output will now try to go to -1000 V as fast as it can. With a negative supply voltage of -15 V, however, the output can only go to about -13 V, so this is where it stops.

In this circuit the op amp effectively compares the signal applied to the noninverting input with the voltage on the inverting input and gives a high or low output depending on the result of the comparison. If the input is more than a few microvolts above the reference voltage on the inverting input, the output will be high ($+13$ V). If the input voltage is a few microvolts more negative than the reference voltage, the output will be low (-13 V). An op amp used in this way is called a *comparator*. Figure

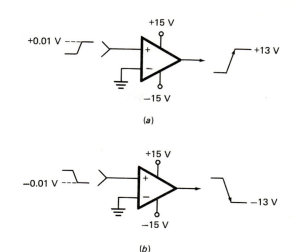

FIGURE 11-2 *(a)* Response of op amp with input signal voltage of $+0.010$ V. *(b)* Response of op amp with input signal voltage of -0.010 V.

11-3a shows how a comparator is usually labeled. The reference voltage applied to the inverting input does not have to be ground (0 V). As we showed you in a section on comparators in the last chapter, the threshold voltage can be set to any voltage within the input signal range specified for the particular op amp.

As you will see throughout this chapter, comparators are used in many circuits. The output levels of the op-amp comparator in Figure 11-3a are not TTL/CMOS-compatible; so in most circuits we use devices such as the LM319, which we showed you in Figure 10-44, because these devices have TTL/CMOS-compatible outputs. Figure 11-3b shows another commonly used comparator circuit. Note in this circuit that the reference signal is applied to the noninverting input, and the input voltage is applied to the inverting input. This connection simply inverts the output states from those in the previous circuit. Note also in Figure 11-3b the positive-feedback resistors from the output to the noninverting input. This feedback gives the comparator a characteristic called *hysteresis*, which we discussed in a section of the last chapter on Schmitt triggers. Remember, hysteresis is the difference between the threshold for a positive-going signal and the threshold voltage for a negative-going signal. Hysteresis prevents the noise on a slowly rising input signal from causing the comparator output to oscillate as the input signal passes the reference voltage.

To determine the amount of hysteresis in a circuit such as that in Figure 11-3b, assume $V_{REF} = 0$ V and $V_{OUT} = 13$ V. A simple voltage-divider calculation will tell you that the noninverting input is at about 13 mV. The voltage on the inverting input of the amplifier will have to go more positive than this before the comparator will change states. Likewise, if you assume $V_{OUT} = -13$ V, the noninverting input will be at about -13 mV, so the voltage on the inverting input of the amplifier will have to go below this to change the state of the output. The hysteresis of this comparator is then ±13 mV, or 26 mV.

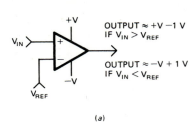
COMPARATOR

$$\text{OUTPUT} \approx +V - 1\,V$$
IF $V_{IN} > V_{REF}$

$$\text{OUTPUT} \approx -V + 1\,V$$
IF $V_{IN} < V_{REF}$

(a)

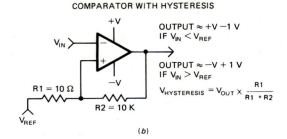
COMPARATOR WITH HYSTERESIS

$$\text{OUTPUT} \approx +V - 1\,V$$
IF $V_{IN} < V_{REF}$

$$\text{OUTPUT} \approx -V + 1\,V$$
IF $V_{IN} > V_{REF}$

$$V_{HYSTERESIS} = V_{OUT} \times \frac{R1}{R1 + R2}$$

(b)

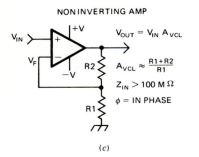
NONINVERTING AMP

$$V_{OUT} = V_{IN}\,A_{VCL}$$

$$A_{VCL} \approx \frac{R1 + R2}{R1}$$

$$Z_{IN} > 100\,M\Omega$$

$$\phi = \text{IN PHASE}$$

(c)

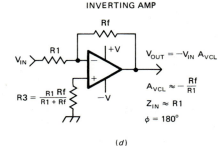
INVERTING AMP

$$V_{OUT} = -V_{IN}\,A_{VCL}$$

$$A_{VCL} \approx -\frac{Rf}{R1}$$

$$Z_{IN} \approx R1$$

$$\phi = 180°$$

(d)

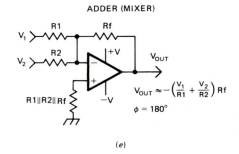
ADDER (MIXER)

$$V_{OUT} \approx -\left(\frac{V_1}{R1} + \frac{V_2}{R2}\right)Rf$$

$$\phi = 180°$$

(e)

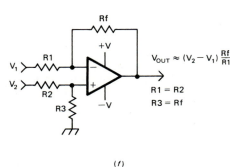
DIFFERENTIAL AMP

$$V_{OUT} \approx (V_2 - V_1)\frac{Rf}{R1}$$

$$R1 = R2$$
$$R3 = Rf$$

(f)

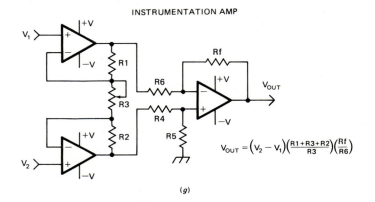
INSTRUMENTATION AMP

$$V_{OUT} = \left(V_2 - V_1\right)\left(\frac{R1 + R3 + R2}{R3}\right)\left(\frac{Rf}{R6}\right)$$

(g)

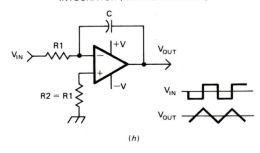
INTEGRATOR (RAMP GENERATOR)

$$R2 = R1$$

(h)

FIGURE 11-3 Commonly used op-amp circuits. *(a)* Op amp used as a comparator. *(b)* Alternative op amp comparator circuit. *(c)* Basic noninverting amplifier. *(d)* Basic inverting amplifier circuit. *(e)* Adder or mixer circuit. *(f)* Basic differential amplifier circuit. *(g)* Instrumentation amplifier. *(h)* Integrator or ramp generator circuit.

Noninverting Amplifier Op-Amp Circuit

When operating in open-loop mode (no feedback to the inverting input), an op amp has a very high, but unpredictable gain. This unpredictable gain is acceptable when you are using an op amp as a comparator, but not when you want to use it to amplify a signal by some precise amount.

To produce an op-amp circuit with stable and predictable gain, we use a technique called *negative feedback*. The principle of negative feedback is to send a fraction of the output signal back to the inverting input of the amplifier. The *noninverting amplifier* circuit in Figure 11-3c shows one way to do this. First, notice in this circuit that the input signal is applied to the noninverting input. The output signal then will be in phase with the input signal. Second, notice that a voltage divider from the output to ground feeds a fraction of the output signal back to the inverting input. Here's how this feedback controls the circuit gain.

To start, assume that $V_{IN} = 0$ V, $V_{OUT} = 0$ V, and the voltage on the inverting input is 0. Now, suppose that you apply a +0.01-V dc signal to the noninverting input. Since the 0.01-V difference between the two inputs will be amplified by 100,000, the output will head toward 1000 V as fast as it can. However, as the output goes positive, some of the output voltage will be fed back to the inverting input through the resistor divider. This feedback to the inverting input will decrease the difference in voltage between the two inputs. To make a long story short, the circuit quickly reaches a predictable balance point where the voltage on the inverting input (V_F) is very, very close to the voltage on the noninverting input (V_{IN}). For a 1.0-V dc output, this equilibrium voltage difference might be about 10 μV.

If you assume that the voltages on the two inputs are equal at this equilibrium point, then predicting the output voltage for a given input voltage is simply a voltage-divider problem:

$$V_{OUT} = V_{IN} \frac{R1 + R2}{R1}$$

If R2 = 99 kΩ and R1 = 1 kΩ, then $V_{OUT} = V_{IN} \times 100$. For a 0.01-V input signal, the output voltage will be 1.00 V. The gain of an amplifier circuit which uses negative feedback is called its *closed-loop gain* and is represented by the abbreviation A_{VCL}. The closed-loop gain for this noninverting op-amp circuit is equal to the simple resistor ratio:

$$A_{VCL} = \frac{V_{OUT}}{V_{IN}} = \frac{R1 + R2}{R1}$$

To see another advantage of feeding some of the output signal back to the inverting input, let's follow what happens when the load connected to the output of the op amp changes and draws more current from the output. If we increase the amount of current drawn from the output, the output voltage will drop because of the increased load. As the output voltage drops, the voltage fed back to the inverting input of the amplifier will also drop. This will increase the difference in voltage between the two inputs. The increased voltage difference between the two inputs will cause the op amp to output more current. The op amp will come to a new equilibrium point, which very nearly compensates for the increased load. Because of the negative feedback, then, the op amp will work day and night to keep its output stabilized and its two inputs at nearly the same voltage. This is probably the most important point you need to know to analyze or troubleshoot an op-amp circuit with negative feedback, so draw a box around this point in your mind so you don't forget it!

The noninverting circuit we have just discussed is used mostly as a *buffer*, because it has a very high *input impedance* (Z_{IN}). The term input impedance refers to the amount that the input of the amplifier circuit will "load"

whatever it is connected to. Think of the input impedance as a resistor connected from the output of the sensor or other device to ground. The higher the value of this resistor, the smaller the amount of current that will be drawn from the sensor or other device. This is important because the sensor output may not be accurate if it has too much current drain put on it. Op amps with bipolar transistor input stages will have a dc input impedance greater than 100 MΩ. FET input op amps such as the National LF356 have an input impedance of 10^{12} Ω.

Inverting Amplifier Op-Amp Circuit

Figure 11-3d shows another common amplifier circuit which uses negative feedback. Note that in this *inverting amplifier* circuit, the noninverting input of the op amp is tied to ground with a resistor and the signal you want to amplify is applied to the inverting input through a resistor. Since the input signal is applied to the inverting input, the output signal will be 180° out of phase with the input signal.

In this circuit the resistor Rf supplies negative feedback. As we explained for the previous circuit, an op amp with negative feedback will work day and night to keep the voltage on its two inputs very nearly the same. Since the noninverting input is tied to ground, the op amp will sink or source the current to hold the inverting input also at 0 V. In this circuit the junction of Rf and R1 at the inverting input is called *virtual ground* because the op amp holds this point in the circuit at 0 V, or ground.

The voltage gain of this inverting amplifier circuit is also determined by the ratio of two resistors. The A_{VCL} for this circuit at low frequencies is equal to −Rf/R1. You can derive this for yourself by just thinking of the two resistors as a voltage divider with V_{IN} at one end, 0 V in the middle, and V_{OUT} on the other end. The minus sign in the gain expression simply indicates that the output is inverted from the input. The input impedance (Z_{IN}) of this circuit is approximately equal to the value of R1. The op amp holds the amplifier end of R1 at 0 V, so an input signal "sees" an impedance of R1 to ground.

Now that you know how to determine the closed-loop gain of the two most common amplifier circuits, we can introduce you to the concept of *gain-bandwidth product*. As we indicated previously, an op amp may have an open-loop dc gain of 100,000 or more. As the frequency of an applied signal increases, the gain decreases until, at some frequency, the open-loop gain drops to 1. Figure 11-4 shows an open-loop voltage gain-versus-frequency graph for a common op amp such as a 741. The frequency at which the gain is 1 is referred to on data sheets as the *unity-gain bandwidth*, or the *gain-bandwidth product*. A common value for this is 1 MHz.

The bandwidth of an amplifier circuit with negative feedback times the low-frequency closed-loop gain will be equal to the unity-gain bandwidth value. For example, if an op amp with a gain-bandwidth product of 1 MHz is used to build an amplifier circuit with a closed-loop gain of 100, the bandwidth of the circuit (f_c) will be about 1 MHz/100, or 10 kHz, as shown in Figure 11-4b.

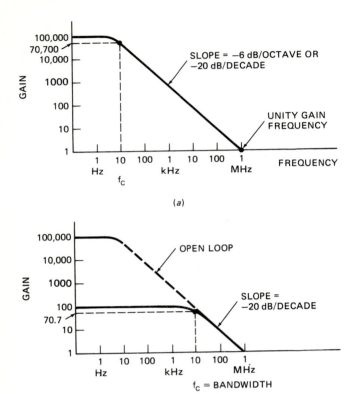

(a)

(b)

FIGURE 11-4   (a) Open-loop gain-versus-frequency response of 741 op amp. (b) Gain-versus-frequency response of 741 op-amp circuit with closed-loop gain of 100.

Op-Amp Adder Circuit

Figure 11-3*e* shows a commonly used variation of the inverting amplifier described in the previous section. This *op-amp adder* circuit adds together, or mixes, two or more input signals. Another way of saying this is that it produces an output voltage which is proportional to the sum of the input voltages. Here's how it works.

Remember from the previous discussion of an inverting amplifier circuit, that the op amp holds the inverting input at virtual ground. The input voltage, V_1, produces a current through R1 to this point. The input voltage, V_2, produces a second current through R2 to this point. The two currents add together at the virtual ground point. In order to hold the virtual ground at 0 V, the op amp pulls the sum of these two currents through Rf. The voltage on the output of the amplifier, then, is equal to the sum of the currents times the value of Rf, or:

$$V_{OUT} = (I_1 + I_2)Rf = \left(\frac{V_1}{R1} + \frac{V_2}{R2}\right)Rf$$

Since R1 and R2 are constants, the output voltage is proportional to the sum of the input voltages. The virtual ground point in an adder circuit is often called a *summing point*. An adder circuit such as this is used to "mix" audio signals in an amplifier and to sum binary-weighted currents in a D/A converter. For simplicity, we have shown an adder with two inputs, but an adder can have any number of inputs.

Simple Differential-Input Amplifier Circuit

As we will show later, many sensors have two output signal lines with a dc voltage of several volts on each signal line. The dc voltage present on both signal leads is referred to as a *common-mode voltage*. The actual signal you need to amplify from these sensors is a few millivolts difference in voltage between the two signal lines. If you connect one of these signal lines to the input of a standard inverting or noninverting amplifier circuit in an attempt to do this, the large dc voltage will be amplified along with the small difference voltage you need to amplify. Figure 11-3*f* shows a simple *differential amplifier* circuit which, for the most part, solves this problem without using coupling capacitors to block the dc. The analysis of this circuit is beyond the space we have here, but basically the resistors on the noninverting input hold this input at a voltage near the common-mode voltage. The feedback to the inverting input of the amplifier holds the inverting input at the same voltage. If the resistors are matched carefully, the result is that only the difference in voltage between V_2 and V_1 will be amplified. The voltage output then will represent only the amplified difference. One way to describe the operation of this circuit is to say that the differential signal is amplified and the common-mode voltage is rejected. The output voltage for the circuit is:

$$V_{OUT} = (V_2 - V_1) \times \frac{Rf}{R1}$$

An Instrumentation Amplifier Circuit

Figure 11-3*g* shows an op-amp circuit used in applications needing a greater rejection of the common-mode signal than is provided by the simple differential circuit in Figure 11-3*f*. The first two op amps in this circuit provide high input impedance buffers for the differential-input signal and provide some amplification for the differential signal. The output op amp provides additional amplification and converts the differential signal to a signal referenced to ground. *Instrumentation amplifier* circuits such as this are available in single packages.

An Op-Amp Integrator Circuit

Figure 11-3*h* shows an op-amp *integrator*, or *ramp generator*, circuit that can be used to produce a voltage signal which is a linear ramp. A dc voltage applied to the input of this circuit will cause a constant current of $V_{IN}/R1$ to flow into the virtual-ground point. Since it cannot flow into the input of the op amp, this current flows onto one plate of the capacitor. In order to hold the inverting input at ground, the op-amp output must pull an equal current from the other plate of the capacitor. The capacitor is then getting charged by the constant current $V_{IN}/R1$. Basic physics tells you that the voltage across a capacitor being charged by a constant current is a *linear ramp*. Note that because of the inverting amplifier connection, the output of the op amp will ramp negative for a positive input voltage. For a negative input voltage, the output of the op amp will ramp in a positive direction. In an integrator circuit such as this, some provision must be made to stop the ramp before the output of the amplifier reaches its output saturation voltage.

The circuit is called an *integrator* because it produces an output voltage proportional to the integral or "sum" of the current produced by an input voltage over a period of time. The waveforms in Figure 11-3h show the circuit response for a pulse-input signal.

Now that we have introduced you to some basic op-amp circuits, we will next discuss some of the different types of sensors you can use to determine the values of temperatures, pressures, position, etc.

SENSORS AND TRANSDUCERS

It would take a book many times the size of this one to describe the operation and applications of all the different types of available sensors and transducers. What we want to do here is introduce you to a few of these and show how they produce electrical signals which are proportional to the size of physical quantities such as light, temperature, and pressure. After we introduce you to some sensors, we show you in a later section of the chapter how the analog voltages from these sensors are converted to a proportional digital values with A/D converters.

Light Sensors

One of the simplest light sensors is a light-dependent resistor such as the Clairex CL905 shown in Figure 11-5a. A glass window allows light to fall on a zigzag pattern of cadmium sulfide or cadmium selenide whose resistance depends on the amount of light present. The

(a)

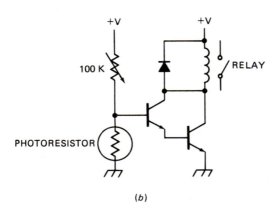

(b)

FIGURE 11-5 *(a)* Cadmium sulfide photocell. *(Clairex Electronics.)* *(b)* Light-controller relay circuit using a photocell.

resistance of the CL905 varies from about 15 MΩ when in the dark to about 15 kΩ when in a bright light. *Photoresistors* such as this do not have a very fast response time and are not stable with temperature, but they are inexpensive, durable, and sensitive. For these reasons they are usually used in applications where a precise measurement of the amount of light isn't needed. The devices mounted on top of street lights to turn them on when it gets dark, for example, contain a photoresistor, a transistor driver, and a mechanical relay as shown in Figure 11-5b. As it becomes dark, the resistance of the photoresistor goes up. This increases the voltage on the base of the transistor until, at some point, the transistor turns on. When the transistor turns on, it pulls current through the relay and turns on the street light.

Another device used to sense the amount of light present is a *photodiode.* If light is allowed to fall on a specially constructed silicon diode, the reverse-leakage current of the diode increases linearly as the amount of light falling on it increases. A circuit such as that shown in Figure 11-6a can be used to convert this small leakage

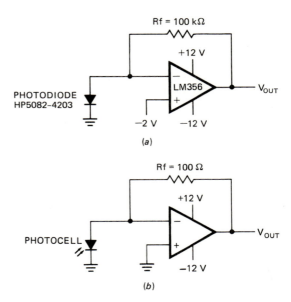

FIGURE 11-6 *(a)* Photodiode circuit to measure infrared light intensity. *(b)* Photocell circuit to measure light intensity.

current to a proportional voltage. Note that in this circuit a negative reference voltage is applied to the noninverting input of the amplifier. The op amp will then produce this same voltage of −2 V on its inverting input. Applying this −2 V to the positive end of the diode *reverse-biases* the photodiode. This reverse bias produces a leakage current through the diode to the inverting input of the amplifier. The op amp will pull the photodiode leakage current through Rf to produce a proportional voltage on the output of the amplifier. For a typical photodiode such as the HP 5082-4203 shown in Figure 11-6a, the reverse-leakage current varies from near 0 μA to about 100 μA, so with the 100-kΩ Rf, an output voltage of about 0 V to −10 V will be produced as the amount of light on the diode is varied from dark to light. A high-input impedance LM356 FET op amp is used in the circuit here to minimize the effect of the amplifier on the diode sensor.

A photodiode circuit such as this might be used to determine the amount of smoke being emitted from a smokestack. To do this, a gallium arsenide infrared LED is put on one side of the smokestack and the photodiode circuit put on the other. Since smoke absorbs light, the amount of light arriving at the photodiode is a measure of the amount of smoke present. An infrared LED is used here because the photodiode is most sensitive to light wavelengths in the infrared region.

Still another useful light-sensitive device is a *solar cell*. Common solar cells are simply large, very heavily doped, silicon PN junctions. Light shining on the solar cell causes a reverse current to flow, just as in the photodiode. Because of the large area and the heavy doping in the solar cell, however, the current produced is in milliamps rather than microamps. Thus, a solar cell functions as a light-powered battery. Solar cells can be connected in a series-parallel array to produce a solar power supply.

Light meters in cameras and photographic enlargers often use solar cells to determine light intensity, since the current from the solar cell is a linear function of the amount of light falling on the cell. A circuit such as that in Figure 11-6b can be used to convert the output current to a proportional voltage. Because of the larger output current, we decrease Rf to a much smaller value, depending on the output current of the cell. We also connect the noninverting input of the amplifier to ground because we don't use reverse biasing with solar cells.

The point here is that a photodiode or a solar cell and a simple op-amp circuit will give a signal voltage which is proportional to the amount of light falling on the sensor. Let's see how you can get a signal voltage which is proportional to the temperature of an object.

Temperature Sensors

Again, there are many types of temperature sensors. The two types we discuss here are semiconductor devices that are inexpensive and can be used to measure temperatures over the range of −55 to 100°C, and thermocouples that can be used to measure very low temperatures and very high temperatures.

SEMICONDUCTOR TEMPERATURE SENSORS

The two main types of semiconductor temperature sensors are *temperature-sensitive voltage sources* and *temperature-sensitive current sources.* An example of the first type is the National LM35, for which we show the circuit connections in Figure 11-7a. The voltage output from this circuit increases by 10 mV for each degree Celsius that its temperature is increased. By connecting the output to a negative reference voltage (V_S) as shown, the sensor will give a meaningful output for a temperature range of −55 to +150°C. You adjust the output to 0 V for 0°C. The output voltage can be amplified to give the voltage range you need for a particular application. In a later section of this chapter we show another circuit using the LM35 temperature sensor. The accuracy of this device is about ±1°C.

Another common semiconductor temperature sensor is a temperature-dependent current source such as the Analog Devices AD590J. The AD590J produces a current of 1 µA/°K. Figure 11-7b shows a circuit which converts this current to a proportional voltage. In this circuit the current from the sensor (I_T) is passed through a 1-kΩ resistor to ground. This produces a voltage which changes by 1 mV/°K. The AD580 is a precision voltage reference used to produce a reference voltage of 273.2 mV. With this voltage applied to the inverting input of the amplifier, the amplifier output will be at 0 V for 0°C. The advantage of a current-source sensor is that voltage drops in long connecting wires do not have any effect on

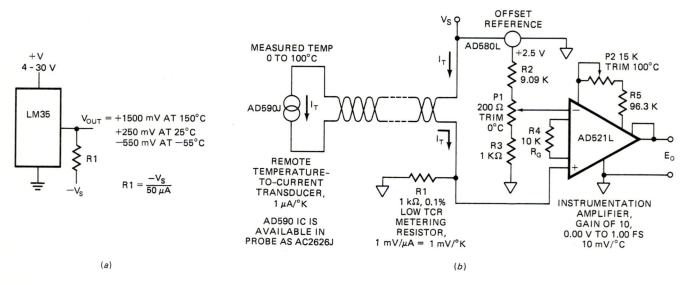

(a)   (b)

FIGURE 11-7 Semiconductor temperature-sensor circuits. *(a)* LM35 temperature-dependent voltage source. *(b)* AD590J temperature-dependent current source. *(Analog Devices Inc.)*

the output value. If the gain and offset are carefully adjusted, the accuracy of the circuit in Figure 11-7b is ±1°C using an AD590J part.

THERMOCOUPLES

Whenever two different metals are put in contact, a small voltage is produced between them. The voltage developed depends on the types of metals used and the temperature. Depending on the metals, the developed voltage increases between 7 and 75 μV for each degree Celsius increase in temperature. Different combinations of metals are useful for measuring different temperature ranges. A thermocouple junction made of iron and constantan, commonly called a *type-J thermocouple*, has a useful temperature range of about −184 to +760°C. A junction of platinum and an alloy of platinum and 13 percent rhodium has a useful range of 0 to about 1600°C. Figure 11-8a shows common thermocouple packages. Thermocouples can be made small, rugged, and stable; however, they have three major problems which must be overcome.

First, the output is very small and must be amplified a great deal to bring it up into the range where it can, for example, drive an A/D converter. Second, in order to make accurate measurements, a second junction made of the same metals must be included in the circuit as shown in Figure 11-8b. Adding this second junction is referred to as *cold-junction compensation*.

The first thing to notice in the circuit in Figure 11-8b is that the reference junction is connected in the reverse direction from the measuring junction. This is done so that the output connecting wires are both constantan. The thermocouples formed by connecting these wires to the copper wires going to the amplifier will then cancel out. The resultant output voltage will be the difference between the voltages across the two thermocouples. If we simply amplify the output of the two thermocouples, however, there is a problem when the temperature of both thermocouples is changing. The problem is that it is impossible to tell which thermocouple caused a change in output voltage. One cure for this is to put the reference junction in an ice bath or a small oven to hold it at a constant temperature. This solution is usually inconvenient, so instead a circuit such as that in Figure 11-8b is used to compensate electronically for changes in the temperature of the reference junction.

As we discussed in a previous section, the AD590 shown here produces a current proportional to its temperature. The AD590 is attached to the reference thermocouple so that they are both at the same temperature. The current from the AD590, when passed through the resistor network, produces a voltage which compensates for temperature changes in the reference thermocouple. The output amplifier for this circuit is a differential amplifier such as that shown in Figure 11-3f or the instrumentation amplifier shown in Figure 11-3g.

The third problem with thermocouples is that their output voltages do not change linearly with temperature. This can be corrected with analog circuitry which changes the gain of an amplifier according to the value of the signal. Another method of doing the correction is with a lookup table stored in a ROM. For this second

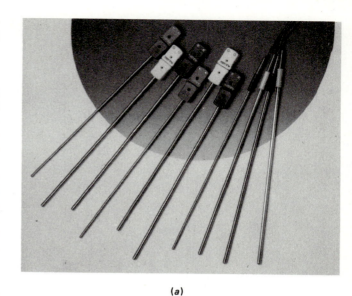

(a)

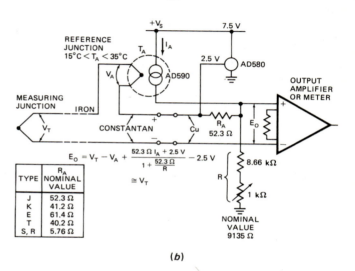

TYPE	R_A NOMINAL VALUE
J	52.3 Ω
K	41.2 Ω
E	61.4 Ω
T	40.2 Ω
S, R	5.76 Ω

$$E_O = V_T - V_A + \frac{52.3\,\Omega\,I_A + 2.5\,V}{1 + \frac{52.3\,\Omega}{R}} - 2.5\,V$$

$$\cong V_T$$

(b)

FIGURE 11-8 *(a)* Photograph of thermocouples. *(Courtesy of Omega Engineering.)* *(b)* Circuit showing amplification and cold-junction compensation for thermocouple. *(Analog Devices Inc.)*

method, an A/D is used to first convert the analog voltage to a digital value. The digital value is then used to address a ROM location which contains the correct temperature value for that reading.

Force and Pressure Transducers

To convert force or pressure (force/area) to a proportional electrical signal, the most common method involves some type of *strain gage*. Here's how strain gages work.

A strain gage is a small resistor whose value changes when its length is changed. It may be made of thin wire, thin foil, or semiconductor material. Figure 11-9a shows a simple setup for measuring force or weight with strain gages. One end of a piece of spring steel is attached to a fixed surface. A strain gage is glued on the top of the flexible bar. The force or weight to be measured is ap-

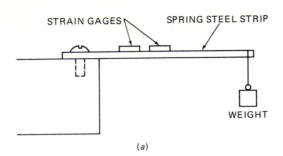

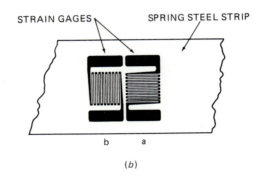

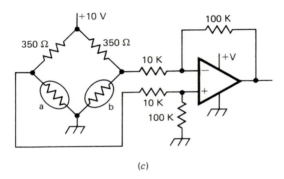

FIGURE 11-9 Strain gages used to measure force. (a) Side view. (b) Top view (expanded). (c) Circuit connections.

plied to the unattached end of the bar. As the applied force bends the bar, the strain gage is stretched, increasing its resistance. Since the amount that the bar is bent is directly proportional to the applied force, the change in resistance will be proportional to the applied force. If a constant current is passed through the strain gage, then the change in voltage across the strain gage will be proportional to the applied force.

Unfortunately, the resistance of strain-gage elements also changes with temperature. To compensate for this problem, two strain-gage elements mounted at right angles as shown in Figure 11-9b are often used. Both of the elements will change resistance with temperature, but only element "a" will change resistance appreciably with applied force. When these two elements are connected in a balanced-bridge configuration as shown in Figure 11-9c, any change in resistance of the elements due to temperature will have no effect on the differential output of the bridge. However, as force is applied, the resistance of the element under strain will change and produce a small differential-output voltage. The full-scale

differential-output voltage is typically 2 or 3 mV for each volt applied to the top of the bridge. For example, if 10 V is applied to the bridge as shown in Figure 11-9c, the full-load output differential voltage will only be 20 or 30 mV. This small signal can be amplified with a differential amplifier or an instrumentation amplifier. The strain-gage bridges are used in many different forms to measure many different types of force and pressure. If the strain-gage bridge is connected to a bendable beam structure as shown in Figure 11-9a, the result is called a *load cell* and is used to measure weight. Figure 11-10 shows a 10-lb load

FIGURE 11-10 Photograph of load-cell transducer used to measure weight. *(Transducers Inc.)*

cell that might be used in a microprocessor-controlled delicatessen scale or postal scale. Larger versions can be used to weigh barrels being filled, or even trucks.

If a strain-gage bridge is mounted on a movable diaphragm in a threaded housing, the output of the bridge will be proportional to the pressure applied to the diaphragm. If a vacuum is present on one side of the diaphragm, then the value read out will be a measure of the absolute pressure. If one side of the diaphragm is open, then the output will be a measure of the pressure relative to atmospheric pressure. If the two sides of the diaphragm are connected to two other pressure sources, then the output will be a measure of the differential pressure between the two sides. Figure 11-11 shows a Sensym pressure transducer which measures pressures between 0 and 15 lb/in^2.

Flow Sensors

One method of determining the flow of a liquid or gas through a pipe is to put a paddle wheel in the flow as shown in Figure 11-12a. The rate at which the paddle wheel turns is proportional to the rate of flow of a liquid or gas. An optical encoder consisting of an LED and a phototransistor can be attached to the shaft of the paddle wheel to produce one or more pulses each time the shaft rotates. The number of pulses per second then is proportional to the flow of liquid through the pipe. A

FIGURE 11-11 Strain-gage pressure transducer used to measure gas or liquid pressure. *(Sensym Inc.)*

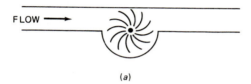

(a)

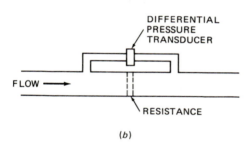

(b)

FIGURE 11-12 Flow sensors. *(a)* Paddle wheel. *(b)* Differential pressure.

simple frequency counter circuit can be used to count the number of pulses per second.

A second common method of measuring flow is with a *differential-pressure transducer*, as shown in Figure 11-12b. A wire mesh or screen is put in the pipe to create some resistance. Flow through this resistance produces a difference in pressure between the two sides of the screen. The pressure transducer gives an output proportional to the difference in pressure between the two sides of the resistance. In the same way that the voltage across an electric resistor is proportional to the flow of current through the resistor, the output of the pressure transducer is proportional to the flow of a liquid or gas through the pipe.

Other Sensors

As we mentioned previously, the number of different types of sensors is very large. Besides the types we have

discussed, there are sensors to measure pH, concentration of various gases, thickness of materials, and just about anything else you might want to measure. Often you can use commonly available transducers in creative ways to solve a particular application problem you have. Suppose, for example, that you need to accurately determine the level of a liquid in a large tank. To do this you could install a pressure transducer at the bottom of the tank. The pressure in a liquid is proportional to the height of the liquid in the tank, so you can easily convert a pressure reading to the desired liquid height.

DIGITAL-TO-ANALOG CONVERTERS

In the preceding sections of the chapter we showed you how to use sensors to get electrical signals which are proportional to the amount of light, temperature, and pressure. We also showed you how op-amp circuits can be used to amplify the small signals from these sensors. The next logical step would be to show you how we use an A/D converter to get these amplified signals to a form that can be stored in a memory or used to drive a display. However, since D/A converters are simpler in operation, and several types of A/D converters use D/A converters as part of their circuitry, we will first show you how D/A converters work.

D/A Converter Operation

A BINARY-WEIGHTED RESISTOR D/A CONVERTER

Figure 11-13 show a circuit which will produce an output voltage proportional to the binary word applied to its inputs by the four switches. Since this converter has four data inputs, it is called a 4-bit converter.

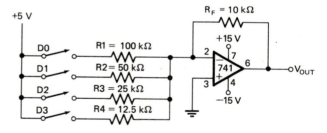

FIGURE 11-13 Binary-weighted resistor D/A converter.

The circuit in Figure 11-13 is just a 4-input op-amp adder circuit such as we discussed earlier in the chapter. To help you understand the operation of the circuit, let's review some basic op-amp characteristics.

The noninverting input of the op amp is tied to ground, so that input will be at 0 V. There is feedback from the output of the op amp to the inverting input, so the op amp will work continually to hold this input at 0 V. Remember that in this circuit configuration, the inverting input is referred to as a *virtual ground*, or *summing point*.

If none of the data switches are closed, there will be no current through any of the input resistors and no current through R_F. If there is no current through R_F, there will be no voltage across R_F. The output of the op amp, then, will be at the same voltage as the inverting input, 0 V.

Now, suppose the D0 data switch is closed. This will apply a voltage of +5 V to one end of R1. The other end of R1 will be held at 0 V by the op amp, so R1 has a voltage of 5 V across it. Ohm's law tells you that the positive current through R1 then is 0.05 mA (5 V/100 kΩ). In order to hold the inverting input at 0 V, the op amp pulls this current through R_F. The voltage across R_F produced by this current will be 0.050 mA × 10 kΩ = 0.5 V. Since one end of R_F is at 0 V, the output of the op amp must be at −0.5 V in order to keep the current flowing from +5 V through R1 to 0 V and on through R_F.

To summarize, then, closing the D0 data switch produces a voltage of −0.5 V on the output of the converter circuit. Now let's see what happens if you open switch D0 and close switch D1.

Since R2 is only half the value of R1, twice as much current, or 0.1 mA, will flow through R2 to the summing point and on through R_F into the output of the op amp. In order to pull a current of 0.1 mA though R_F, the op amp will assert a voltage of −1 V (−0.1 mA × 10 kΩ) on its output. The D1 data switch then produces an output voltage of −1 V, or twice as much as that produced by the D0 switch.

If switch D1 is opened and switch D2 is closed, a current of 5 V/25 kΩ, or 0.2 mA, will flow through R3 to the summing point and on through R_F into the output of the op amp. This current will produce a voltage of −2 V on the output of the op amp. Likewise, if switch D2 is opened and switch D3 is closed, a current of 5 V/12.5 kΩ, or 0.4 mA, will flow into the summing point and on through R_F into the output of the op amp. This current will produce a voltage of −4 V on the output of the op amp.

Now, suppose that switches D2 and D3 are both closed. The 0.2-mA current through R3 will combine with the 0.4-mA current through R4 at the summing point to produce a total current of 0.6 mA. In order to pull this current through R_F, the op amp will assert a voltage of −0.6 mA × 10 kΩ, or −6 V, on its output. The output voltage, then, is proportional to the sum of the currents produced by the closed switches. The values of the four currents are related to each other in the same way that the weights of binary digits are. Therefore, the output voltage will be proportional to the binary word applied to the data inputs. D0 represents the least significant bit (LSB) because it produces the smallest current, and D3 represents the most significant bit (MSB) because it produces the largest current.

Since there are four inputs, there are 16 possible input words, 0000 to 1111. These words produce output voltages ranging from 0 V for an input word of 0000 to −7.5 V for an input word of 1111. If you connect the outputs of a 4-bit binary counter to the inputs of the D/A converter as shown in Figure 11-14a, the inputs of the D/A converter will be stepped through all the possible input values as the counter is clocked. This will produce the interesting stair-step waveform shown in Figure 11-14b on the output of the op amp.

Technically the waveform on the output of the op amp

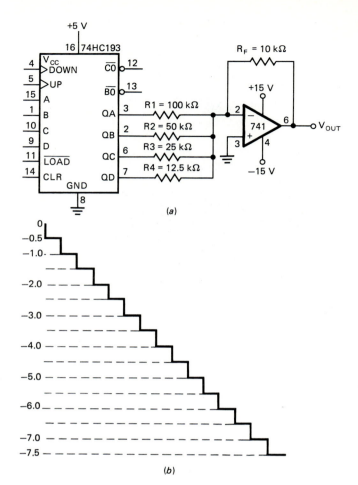

FIGURE 11-14 (a) 74HC193 binary counter connected to inputs of 4-bit binary-weighted resistor D/A converter. (b) Stepped output waveform.

is "digital" because it can only have certain fixed values, just like the value displayed on a digital voltmeter. However, since the output approximates a range of analog values, we refer to the output as analog. If you use a D/A converter with more input bits, then the output waveform will more closely resemble an analog waveform because the steps will be smaller. Later in this chapter we will show you how a D/A converter is used to produce high-quality audio from the digital words stored on a compact audio disc.

As you can see here, the heart of a D/A converter is a set of binary-weighted currents which are switched on or off based on the binary word applied to the inputs. There are several ways to produce the binary-weighted currents in a D/A converter, but since these weighted current sources are almost always buried inside an IC, we don't have to worry about the details of the different methods. Our main concern in the next section is the operation of D/A converters which have all their components in one package.

MONOLITHIC AND HYBRID D/A CONVERTER EXAMPLES

The term *monolithic* means "one stone." When the term is used to describe an electronic device such as a D/A converter, it means that all the parts for a circuit are contained on a single silicon chip. When it is not possible to put all the required parts on a single silicon chip, manufacturers often build a hybrid device such as the hybrid D/A converter shown in Figure 11-15. The term

FIGURE 11-15   Datel hybrid D/A converter.

hybrid means that the device contains one or more separate silicon chips and other microminiature components such as resistor networks and capacitors mounted on a ceramic substrate. The ceramic substrate is often put in an IC type package so that it can be plugged into an IC socket or soldered on a PC board. Often circuits start out being manufactured as hybrids, and then as technology permits, they are made into monolithic devices.

As an example of a commonly available monolithic device, Figure 11-16a shows a circuit using the MC1408 or DAC0808 D/A converter. The MC1408 contains eight binary-weighted current switches which are controlled by the TTL/CMOS logic signals applied to the data inputs, A1 to A8. Note for this converter A1 is the MSB and A8 is the LSB of the input data word. This is reversed from the way the data bits are labeled on counters and memories. Since some D/A converters label the input bits in this way and some label them in the same order as counters, you have to carefully check the data sheet when you work with a specific device.

The output signal on pin 4 of the MC1408 is a current which depends on the data word applied to the A8 to A1 inputs. Since there are eight data inputs, this current can have any one of 2^8, or 256, values. The op amp connected to pin 4 converts the output current to a proportional voltage. Here's how it works.

As shown by the arrow next to pin 4, the MC1408 causes a positive current, I_o, to flow into pin 4. The op-amp output will supply this current to the MC1408 output through R_O. In order to supply this positive current, the output of the op amp must be at some positive voltage. The op amp holds its inverting input at 0 V, so the output voltage, V_o, is simply equal to the current multiplied by R_o. Since the current depends on the data word

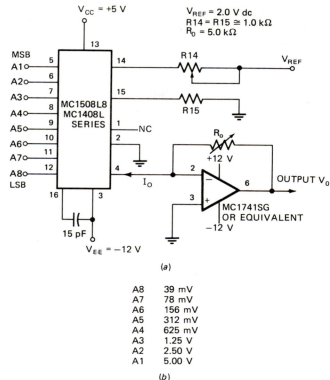

A8	39 mV
A7	78 mV
A6	156 mV
A5	312 mV
A4	625 mV
A3	1.25 V
A2	2.50 V
A1	5.00 V

(b)

FIGURE 11-16   *(a)* Circuit using MC1408 monolithic D/A converter. *(b)* Voltage equivalents for each bit of 0 to 10 V, 8-bit D/A converter.

on the inputs, we can write the complete expression for the output voltage as:

$$V_o = \frac{V_{REF}}{R14} (R_o) \left(\frac{A1}{2} + \frac{A2}{4} + \frac{A3}{8} + \frac{A4}{16} + \frac{A5}{32} + \frac{A6}{64} + \frac{A7}{128} + \frac{A8}{256} \right)$$

This expression looks messy, by it is quite simple. The A's in the fractions indicate the logic state on that data input. For example, if the A1 input is high, you put a 1 in the top of the A1 fraction. If the A2 input is low, you put a 0 in the top of the A2 fraction. A fraction with a 0 in its top is equal to 0, so it drops out of the expression. For a given data word, you just add all the fractions which have 1's as their top number. The sum of these fractions represents the fraction of the *reference current* that the MC1408 will output for that input data word. The reference current is determined by $V_{REF}/R14$. For the values in the circuit in Figure 11-16a, the reference current will be 2.0 V/1 kΩ = 2.0 mA. Let's see what output voltage will be produced by a few different input data words.

For an input data word of 00000000, all the fractions in the expression above are 0's, so the entire expression reduces to 0, and V_o = 0 V.

For an input data word of 1000000, the expression becomes:

$$V_o = \frac{V_{REF}}{R14} (R_o) \left(\frac{1}{2} + \frac{0}{4} + \frac{0}{8} + \frac{0}{16} + \frac{0}{32} + \frac{0}{64} + \frac{0}{128} + \frac{0}{256} \right)$$

which reduces to:

$$V_o = \frac{2\,V}{1\,k\Omega} \times 5\,k\Omega\left(\frac{1}{2}\right) = 5\,V$$

From this you can see that a 1 on the MSB input produces +5 V on the output of the converter.

For an input word of 00000001, the expression becomes:

$$V_o = \frac{V_{REF}}{R14}\,(R_o)\left(\frac{0}{2} + \frac{0}{4} + \frac{0}{8} + \frac{0}{16} + \frac{0}{32} + \frac{0}{64} + \frac{0}{128} + \frac{1}{256}\right)$$

which reduces to:

$$V_o = \frac{2\,V}{1\,k\Omega} \times 5\,k\Omega\left(\frac{1}{256}\right) = 0.039\,V$$

This tells you that a 1 on the LSB input produces an output voltage of about 39 mV. This also tells you that the difference between any output level and the next level is about 39 mV. For reference, Figure 11-16b shows you the voltage produced by each bit in the D/A converter circuit in Figure 11-16a.

Finally, for an input data word of 11111111, the expression becomes:

$$V_o = \frac{V_{REF}}{R14}\,R_o\left(\frac{1}{2} + \frac{1}{4} + \frac{1}{8} + \frac{1}{16} + \frac{1}{32} + \frac{1}{64} + \frac{1}{128} + \frac{1}{256}\right)$$

which reduces to:

$$V_o = 10\,V \times \left(\frac{255}{256}\right) = 9.961\,V$$

The result tells you that the full-scale output voltage for this converter is 9.961 V.

The D/A converter circuit in Figure 11-16a then has 256 output levels, ranging from 0.0 V for an input word of 00000000 to 9.961 V for an input word of 11111111. Although the circuit is described as a 10-V D/A converter, the full-scale output voltage is 9.961 V, rather than 10.0 V, as you might expect. The reason for this is that 0.0 V counts as one of the 256 levels. The full-scale output voltage for any D/A converter will always be 1 LSB less than the rated full-scale value when the circuit is properly adjusted. Now let's look at an example of a 12-bit D/A converter.

Figure 11-17 shows an internal block diagram for the Datel hybrid D/A converter we showed you a photograph of in Figure 11-15. This converter has 12 data inputs, so it is called a *12-bit converter*. One version of the device is set up to accept 12-bit binary data words, and another available version of the device is set up to convert 3-digit BCD numbers applied to its data inputs. Note the order of the data inputs.

Included in this hybrid package is a precision voltage source which can be used to control the full-scale current. To produce the precise currents needed for a 12-bit converter, the resistors in the thin-film resistor network are individually trimmed by a computer-controlled laser beam.

A current output signal is available directly on pin 20

of the device. To produce a voltage output signal, you connect either pin 18 or pin 19 to the output of the on-board op amp on pin 15. The op amp then functions as a current-to-voltage converter just as the op amp in the MC1408 circuit in Figure 11-16 does. With a 5-kΩ feedback resistor, the DAC HZ12BGC will have a 10-V output range, and if the two 5-kΩ resistors in series are used as the feedback resistor, the device will have a 20-V output range. Let's see what size the output voltage steps are for this converter.

The output voltage produced by the MSB of a D/A converter is always equal to half the rated voltage. For a 10-V converter, the MSB then will produce an output voltage of +5 V. The next most significant bit will produce an output voltage of half this, or 2.5 V. The voltage produced by each bit will be half the voltage produced by the preceding bit. For a 12-bit converter, the voltage produced by the LSB, V_{LSB}, is:

$$V_{LSB} = \frac{1}{2^{12}} \times \text{rated voltage}$$

$$= \frac{1}{4096} \times 10\,V = 2.44\,mV$$

The value of the LSB, or the voltage difference between output levels, is 2.44 mV. A 12-bit converter produces much closer levels than a converter with fewer bits. If a 12-bit binary counter is connected to the data inputs of this converter, the sawtooth waveform produced on the output will very closely approximate an analog sawtooth waveform.

From the preceding discussion you should see that an important characteristic of a D/A converter is the number of input data bits it has. The number of input data bits is one way of specifying the *resolution* of a D/A converter. In the next section we talk more about resolution and discuss some other D/A characteristics specified on their data sheets.

D/A Characteristics, Error Sources, and Specifications

RESOLUTION

Resolution is a specification of how many levels the output range is divided into. This is determined by the number of data inputs. A converter with eight data inputs, such as the one in Figure 11-16, has 2^8, or 256, output levels, so its resolution is 1 part in 256. A 10-bit converter has 2^{10}, or 1024, output levels, so its resolution is 1 part in 1024. The 12-bit converter in Figure 11-17 has 2^{12}, or 4096, output levels, so its resolution is 1 part in 4096.

Resolution is sometimes expressed as a percentage. The resolution of an 8-bit converter expressed in this way is $(\frac{1}{256}) \times 100$ percent, or 0.39 percent. In practice, most people express the resolution of a D/A converter simply as the number of input bits.

FULL-SCALE OUTPUT VOLTAGE

As we showed you in the preceding section, the full-scale output voltage of a D/A converter is determined by the reference current and the value of the feedback resistor on the current-to-voltage converter op amp. D/A convert-

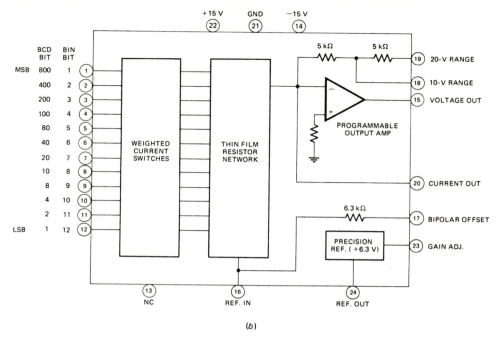

FIGURE 11-17   Internal block diagram for Datel DAC-HZ12BGC 12-bit hybrid D/A converter.

ers usually have full-scale output voltage ratings of +5 V, +10 V, or ±5 V, depending on the intended application. Remember from our previous discussion that the actual maximum output voltage for a D/A converter will always be 1 LSB less than the rated full-scale voltage of the converter. As another example of this, suppose that you have a 10-bit D/A converter with a rated full-scale output voltage of 5 V. The voltage for the smallest step or LSB is 5 V/1024, or 0.0049 V. The actual full-scale output voltage, then, will be 4.9951 V (5 V − 0.0049 V) when the D/A converter is properly adjusted.

ACCURACY

The accuracy specification for a D/A converter is a measure of how close the actual output voltage is to the predicted output voltage for a given input data word. Accuracy is specified as a percentage of the full-scale output voltage or as a fraction of the LSB. As an example of the first way, a data sheet may specify that an 8-bit D/A converter has a full-scale output voltage of 10 V and an accuracy of ±0.2 percent. The maximum error for any output level will be ±0.002 × 10.000 V = ±20 mV. For a predicted output voltage of 5.000 V, then, you know the actual output voltage will be between 4.980 and 5.020 V.

Ideally the maximum error for a D/A converter should be no greater than ±½LSB. If the error is greater than this, then an output may be closer to the next higher or lower value than to the desired value. Some data sheets indicate directly that the accuracy of the converter is ±½LSB or better.

To compare a percentage specification with a specification given as ±½LSB, determine the size of an LSB and compare this with the percentage converted to a voltage. As an example of this calculation, the size of the LSB for an 8-bit, 10-V D/A converter is 10 V/256, or 0.039 V. For this converter, then, an accuracy specification of

±½LSB or ±0.020 V is about the same as a specification of ±0.2 percent.

LINEARITY, GAIN, AND OFFSET SPECIFICATIONS

These three specifications are best described in terms of how a D/A converter output responds when its inputs are driven by a binary upcounter. Ideally the output should be a stair-step sequence which very nearly approximates a linear ramp that goes from 0 V to the maximum output voltage. However, there are several errors that may cause the actual output to deviate from the ideal.

The first of these is a *linearity error* such as that shown in Figure 11-18a. In this case the actual output is below the ideal at some points in the ramp and above the ideal at other points in the ramp. The linearity specification for a D/A is a measure of the maximum amount that the actual output differs from the predicted ramp. To be consistent with the accuracy, the maximum linearity error should be no greater than ±½LSB. However, many D/A converters which have greater linearity errors than this are commonly marketed. National Semiconductor, for example, markets the DAC1020, DAC1021, and DAC1022 series of 10-bit resolution converters. The linear specification for the DAC 1020 is 0.05 percent, which is appropriate for a 10-bit converter. The DAC1021 has a linearity specification of 0.10 percent, and the DAC1022 has a specification of 0.20 percent. The question that may occur to you at this point is, "What good is it to have a 10-bit converter if the linearity is only equivalent to that of an 8- or 9-bit converter?" The answer to this question is that for many applications the resolution given by a 10-bit converter is needed for small output signals, but it doesn't matter if the output value is somewhat nonlinear for large signals. Devices with poorer linearity are considerably cheaper, so it is advantageous to use them if they will do the job.

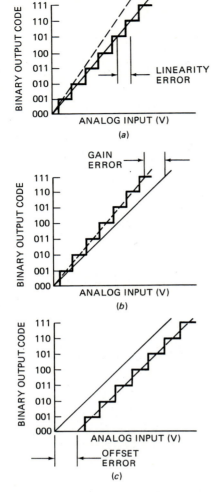

FIGURE 11-18   D/A error examples. *(a)* Linearity.
(b) Gain or scale factor. *(c)* Offset.

Besides giving a linearity specification, some data sheets may tell you that a device is "monotonic over its full temperature range." A D/A converter is said to be *monotonic* if each output step is higher than the last when the D/A is stepped through its entire range by a counter.

Another error that can be present in a D/A is a *gain* or *scale factor error* such as that shown in Figure 11-18*b*. Here the gain of the current-to-voltage converter amplifier is usually incorrect. A gain error can usually be corrected by adjusting the value of the feedback resistor on the current-to-voltage converter.

Still another error commonly found in D/A circuits is an *offset error* such as that shown in Figure 11-18*c*. The cause of this error is usually a leakage current which adds to the current from the switches such that the output is offset by a fixed amount for all input words. Correcting this error involves adding circuitry to cancel or compensate for the added current.

SETTLING TIME

Figure 11-19 shows the response that the output of a D/A converter will make if all its data bits are switched from 0's to 1's. As you can see, the output initially overshoots the desired level and rings back and forth across the desired level a few times. The amount of time required for the output to get to a point where the output level remains within $\pm\frac{1}{2}$LSB of the final value is called

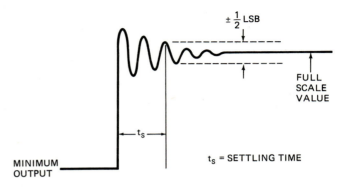

FIGURE 11-19   Illustration of D/A converter settling time.

the *output settling time*. As an example, the National DAC1020 10-bit converter has an output settling time of 500 ns for a full-scale change on the output. This specification is important because, if a D/A is operated at too high a frequency, it may not have time to settle on one value before it is switched to the next.

OUTPUT GLITCHES

When the binary word applied to the input of a D/A is changed, the current switches usually do not all switch at the same rate. This means that the output does not smoothly change from one value to the next. For example, if the word on the inputs of an 8-bit converter is changed from 00000111 to 00001000, the lower bits may change to 0's before bit 4 changes to a 1. This means that for a short time the value on the output of the D/A will represent the binary word 00000000. When viewed with a scope, the output may show a strong glitch such as that shown in Figure 11-20*a*. For some applications such as digital audio, these glitches are highly undesirable. The cure for these output glitches and a help for too long a settling time is the *sample-and-hold* circuit shown in Figure 11-20*b*. Here's how it works.

The normal position for the electronic switch is the open position, so there is no direct connection between the D/A and the final output. After a new word is applied to the inputs of the D/A, the output is allowed to settle down to a stable value. Then, the sample switch is closed for a short time to charge the hold capacitor to the new value on the output of the D/A. When the sample switch is opened, the D/A output voltage will he held on the capacitor, because the high-input impedance buffer amplifier does not drain off the stored voltage. Each time a new word is applied to the inputs of the D/A converter, the hold capacitor is updated to the new value. Since each new value is sampled after glitches and ringing have passed, these will not be passed on to the output of the buffer amp. We will discuss sample-and-hold circuits again in a later section of the chapter on A/D converters.

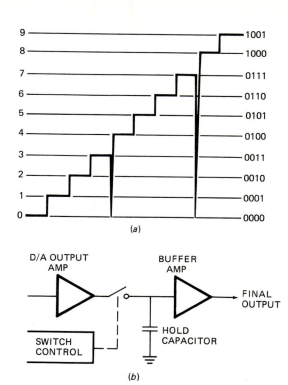

(a)

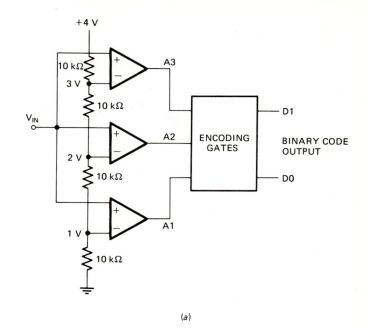

(a)

V_IN (VOLTS)	COMPARATOR OUTPUTS			BINARY OUTPUT	
	A3	A2	A1	D1	D0
0 to 1	0	0	0	0	0
1.001 to 2	0	0	1	0	1
2.001 to 3	0	1	1	1	0
3.001 to 4	1	1	1	1	1

(b)

FIGURE 11-20 (a) D/A converter waveform showing output glitches. (b) Sample-and-hold circuit used to eliminate output glitches.

FIGURE 11-21 (a) Circuit for parallel comparator or "flash" type A/D converter. (b) Comparator output codes and binary equivalents for various input signal ranges.

ANALOG-TO-DIGITAL CONVERTERS

The purpose of an A/D converter is to produce a digital word which best represents the value of an analog voltage or current. In other words, an A/D converter "digitizes" the signal applied to its input. Two major A/D characteristics that you need to know about before we start our discussion now are *resolution* and *conversion time*.

The resolution of an A/D converter is determined by the number of bits in the output word. An 8-bit A/D converter, for example, will represent the value of an input voltage with one of 256 possible words. Another way of putting this is to say that it will resolve an input voltage to the nearest one of 256 levels.

The conversion time for an A/D converter is simply the time it takes the converter to produce a valid output word after it is given a "start conversion" signal. When we refer to an A/D converter as "high speed" or "fast," we mean that it has a short conversion time.

There are many different ways to do an analog-to-digital conversion. The method chosen depends on the resolution, speed, and type of interfacing needed for a given application. In the following sections we describe some of the more common A/D methods and devices.

Parallel Comparator or Flash Converter

The simplest in concept and the fastest type of A/D converter is the *parallel converter,* or *"flash,"* type shown in Figure 11-21a. Here's how this converter works.

The resistor voltage divider sets a sequence of voltages on the reference inputs of the comparators as shown. Remember from our discussion of comparators in the last chapter, that if the voltage on the + input of a comparator is greater than the reference voltage on its − input, the output of the comparator will be high. In the circuit in Figure 11-21a the voltage to be converted is applied to the + inputs of all the comparators in parallel, so the number of comparators that have highs on their outputs indicates the amplitude of the input voltage. An input voltage of between 0 and 1 V, for example, is less than the reference voltage on any of the comparators, so none of the comparator outputs will be high. If the input voltage is between 1.001 and 2 V, only the A1 output will be high. For an input voltage between 2.001 and 3 V, both the A1 and A2 outputs will be high. Finally, for an input voltage greater than 3 V, all the comparator outputs will be high. Figure 11-21b summarizes the comparator output code that will be produced by input voltages between 0 and 4 V. The code produced on the outputs of the comparators is not binary, but with a simple gate circuit it can easily be converted to the binary equivalents shown in the rightmost column of Figure 11-21b.

Although the four binary codes represent values of 0, 1, 2, and 3 V, the full-scale input voltage for this converter is said to be 4 V. Just as with a D/A converter, the rated full-scale voltage for an A/D converter is the value of 1 LSB more than the highest actual level. Again, the reason for this is that 0 counts as one of the levels.

Doing a conversion with the A/D converter in Figure 11-21a is similar to making a measurement with a 4-in ruler that has only the inch marks on it. The converter will resolve an input voltage only to one of four levels. This is equivalent to just 2 bits. To increase the resolution, more comparators can be added. Seven comparators are required for 3-bit resolution and 15 comparators are required for 4-bit resolution. An N-bit converter requires $2^N - 1$ comparators, so an 8-bit converter requires 255 comparators, a 10-bit converter would require 1023 comparators, and a 12-bit converter would require 4095 comparators! As you can see from these numbers, the number of comparators needed increases rapidly as the desired number of bits increases.

The main advantage of a parallel comparator type A/D converter is its speed. The binary word is present on the outputs after just the propagation delay time of the comparators plus the delay time of the encoding gates. This is why a parallel comparator type A/D converter is often called a *flash* converter. Experimental GaAs units can produce a 2-bit output word in 1 ns. The AD9002, an 8-bit converter from Analog Devices, can do conversions at 150 MHz, or, in other words, a conversion every 6.7 ns. There are a wide variety of flash converters available ranging from 2 to 8 bits for high-speed applications. The number of comparators required for resolution greater than 8 bits is not currently practical, so other techniques are used.

Single-Slope A/D Converter

Figure 11-22a shows the circuit for an A/D converter which uses a ramp generator, a comparator, and a counter/display module to digitize an input voltage. Here's how this circuit works.

At the start of a conversion cycle, the ramp generator and the counters are each reset to zero. The signal to be digitized is applied to the + input of the comparator. Assuming the unknown input voltage is positive and the ramp voltage is at 0 V, the + input of the comparator will be more positive than the − input of the comparator. The output of the comparator will then be high. This will enable the AND gate and allow the 1-MHz clock signal to reach the counter chain.

As the ramp voltage goes positive, it will eventually exceed the unknown input voltage on the + input. This will cause the comparator output to go low. A low output on the comparator will disable the AND gate and shut off the 1-MHz clock signal to the counter chain. Since the ramp voltage increases at a constant rate, the count accumulated on the counters is proportional to the input voltage. Control circuitry such at that used in the frequency counter in Figure 9-22 can be used to latch the count, reset the counters, reset the ramp generator, and start the conversion cycle again.

A numerical example should help you further see how this circuit works. Assume the counter block contains four BCD counters and the unknown input voltage is 2.000 V. Further assume that the ramp has a slope of 1 V/ms as shown in Figure 11-22b and that the clock signal has a frequency of 1 MHz, or 1000 cycles/ms. Starting from 0 V, the ramp generator will take 2 ms to reach a voltage of 2 V and shut off the clock pulses to the counters. During this time the counters will have counted off 2000 clock pulses. The comparator output going low will strobe the latches and transfer the accu-

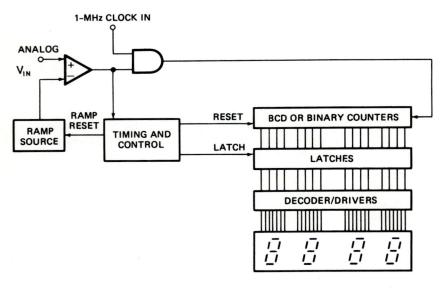

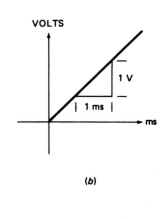

(b)

(a)

FIGURE 11-22   Single-slope A/D converter. (a) Block diagram. (b) Ramp slope.

mulated count to the displays. Inserting a decimal point at the proper place on the displays gives a reading of 2.000 V. With this circuit any input voltage up to 9.999 V can be converted to a BCD equivalent value and displayed on the LEDs. This circuit, then, is a simple digital voltmeter. To produce a binary output code instead of BCD, you just use binary counters instead of BCD counters in the circuit in Figure 11-22a.

For A/D applications such as a digital voltmeter which may require a resolution of 1 part in 20,000 or more, it is difficult to make a ramp generator which is stable enough. The dual-slope converter discussed in the next section overcomes the stability problems of the single-slope converter.

Dual-Slope A/D Converters

Figure 11-23a shows a block diagram for a dual-slope A/D converter. The first part of this circuit to examine is

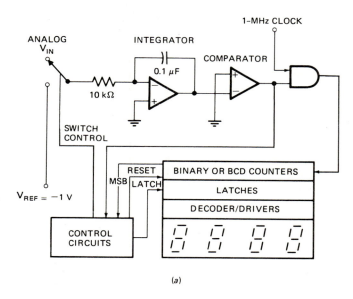

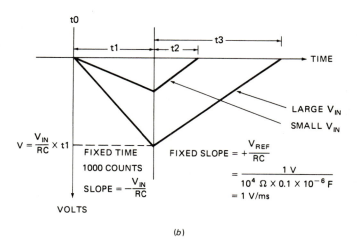

FIGURE 11-23 (a) Dual-slope A/D converter. (b) Integrator output waveform for two different input voltages.

the *ramp generator*, or *integrator*, which we introduced you to in an earlier section of the chapter on op amps. To refresh your memory, here's how it works.

Since there is feedback from the output of the op amp to the inverting input, the op amp will sink or source whatever current is required to keep the − input of the op amp at the same voltage as the + input of the op amp, 0 V. If a voltage of, for example, +2 V is applied to the input end of the 10-kΩ resistor, a constant positive current of 2 V/10 kΩ or 0.2 mA, will flow through the resistor to the summing point. Since this current cannot flow into the high-impedance op-amp input, it flows into one plate of the capacitor. In order to keep the − input of the op amp at 0 V, the op-amp output must pull an equal, constant current from the other plate of the capacitor. As the capacitor charges, the output of the op amp must go to a more negative voltage to keep this constant current flowing.

The voltage across a capacitor being charged by a constant current is a linear ramp. With a positive input voltage, the output of the integrator will ramp negative as shown by the waveform section labeled t1 in Figure 11-23b. If a negative voltage is applied to the input of the integrator, the output will ramp in a positive direction. Using some basic relationships, it can be shown that the slope of the integrator output ramp, $\Delta V/\Delta t$ is equal to $-V_{IN}/RC$. For the RC values shown and an input voltage of +2 V, the integrator output will ramp negative at 2 V/ms. With the reference voltage of −1 V connected to the input of the integrator, the output of the integrator will ramp positive at 1 V/ms. Now that you know how the integrator works, we can work our way through a conversion cycle.

A conversion cycle starts with the integrator output at 0 V, the counters reset to all zeros, and the input switch connected to the unknown input voltage. Assuming the input voltage is positive, the integrator output will start ramping in a negative direction. As soon as the integrator pushes the − input of the comparator a few microvolts negative, the output of the comparator will go high and enable the AND gate. This lets the clock signal into the counters.

The integrator output is allowed to ramp negative for a fixed number of counts. This is shown for two different input voltages as t1 in Figure 11-23b. After the fixed number of counts, the control circuitry resets the counters to zero and switches the integrator input to the negative reference voltage. A negative input voltage will cause the integrator output to ramp positive as shown by the sections labeled t2 and t3 in Figure 11-23b. When the integrator output goes a few microvolts above 0 V again, the comparator output will go low. This will disable the AND gate and shut off the clock signal to the counters. When the control circuitry detects this comparator transition, it latches the accumulated count, resets the counters to zero, and switches the integrator input back to the unknown input voltage to start another conversion cycle. The number of counts stored on the latches is proportional to the input voltage. Here's why.

In the fixed time, t1, the integrator output ramps down to a voltage equal to (V_{IN}/RC) × t1. In order for the output to return to 0 V, the integrator must ramp up the same amount of voltage during the reference integrate period, t2. The voltage the integrator output ramps during this time is $(V_{REF}/RC) \times t2$. Since the change in integrator output voltage is the same for both, the two expressions can be set equal to each other as follows:

$$\frac{V_{IN}}{RC} \times t1 = \frac{V_{REF}}{RC} \times t2$$

The RC term appears in each side of the equation, so it can be canceled out. The practical implication of this is that since the same R and C are used for the signal integrate ramp and the reference integrate ramp, variations in R or C with temperature will have no effect on the accuracy of the output reading. This is the major advantage of a dual-ramp converter over a single-ramp type.

The equation above can be simplified and shuffled around to give:

$$t2 = V_{IN} \times \frac{t1}{V_{REF}}$$

Since t1 and V_{REF} are constants, this equation tells you that t2 is directly proportional to V_{IN}. For the circuit shown in Figure 11-23a, t1 is 1000 counts of the 1-MHz clock, or 1 ms, and V_{REF} is −1 V. With a 2-V signal applied to the unknown input, t2 will be $(2 V/1 V) \times 1000$ counts, or 2000 counts. Putting a decimal point in the right place in the display will make this reading represent 2.000 V.

The inner ramps in Figure 11-23b represent the response of the circuit to an input voltage of +0.8 V. For this input voltage:

$$t2 = \frac{0.8 V}{1 V} \times 1000 \text{ counts} = 800 \text{ counts}$$

which will be displayed as 0.800 V on the LEDs. Incidentally, the values for R and C on the integrator are

selected so that the output of the integrator op amp does not go into saturation when the maximum specified input voltage is applied to the signal input. For the values shown in Figure 11-23a, the integrator will ramp down to −10 V with an input signal of +10 V. This is acceptable if the op amp is powered with ±15 V.

The advantages of a dual-slope converter are its resolution, low cost, and immunity to temperature-caused variations in R or C. The major disadvantage of a dual-slope converter is its slow speed. A converter which is able to resolve 1 part in 20,000 may take 100 ms or more to do a conversion. For use in a digital voltmeter or multimeter, however, this relatively long conversion time is not a problem. In a later section of the chapter, we show how this type of converter is used as the heart of a digital multimeter.

Charge Balance A/D Converters

The charge balance type of A/D converter uses essentially the same circuitry as the dual-slope converter in Figure 11-23a. However, instead of the input being switched back and forth between the unknown voltage and a reference voltage, the input is left connected to the unknown voltage. Pulses of a reference current are inserted directly into the summing junction of the integrator for a fixed time. The number of reference current pulses required to keep the summing junction at 0 V for this fixed time is proportional to the unknown input voltage. The advantage of this technique is that the voltage across the integrating capacitor is kept very near 0 V. Therefore, dielectric absorption by the capacitor is not a problem as it is in high-accuracy dual-slope converters.

Counter Type A/D Converters

One solution to the stability problems of a single-ramp converter is to use the dual-ramp approach we showed you previously. Another solution is to use a counter and a D/A converter to produce a stable ramp. Figure 11-24

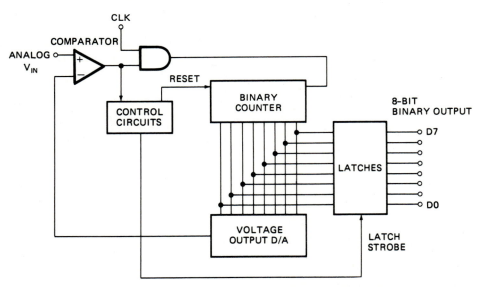

FIGURE 11-24 Binary upcounter and D/A converter used to make a single-ramp type of A/D converter.

shows a block diagram for a circuit which uses this method.

At the start of a conversion cycle, the counters are reset to 0, so the output of the D/A is at 0 V. A positive unknown voltage applied to the input of the converter will cause the output of the comparator to go high and enable the AND gate. This will let the clock pulses into the counter. Each clock pulse increments the counters by 1 and increases the voltage on the output of the D/A by one step. When the voltage on the output of the D/A converter passes the voltage on the unknown V input, the output of the comparator will go low and shut off the clock pulses to the counters. The count accumulated on the counters is proportional to the input voltage. The control circuitry then strobes the latches to transfer the count to the output and resets the counters to start the another conversion cycle.

This counter method is faster than the dual-ramp method, but it has the disadvantage that it requires a precision D/A converter. Another disadvantage of this type is that the counter has to start at 0 and count up until the D/A output passes the input voltage. For an 8-bit converter, then, a conversion may take as long as 255 clock cycles, and for a 10-bit converter, a conversion may take as many as 1024 clock pulses. The successive approximation type A/D converter we discuss in the next section uses a D/A converter in a much more efficient manner.

Successive Approximation A/D Converters

The successive approximation method of A/D conversion is similar to finding the height of a table with a set of eight binary-weighted blocks which are 128, 64, 32, 16, 8, 4, 2, and 1 cm in height. To start, the most significant block, 128 cm, is set next to the table. If the block is higher than the table, it is set aside and 0 is recorded for that bit position. If the block is not higher than the table, then it is kept and 1 is recorded for that bit position. The next most significant block, 64 cm, is then tried by itself or on top of the 128-cm block. If adding the 64-cm block makes the pile higher than the table, then it is set aside and 0 is recorded for that bit position. If adding the 64-cm block does not make the pile higher than the table, 1 is recorded in that bit position. The rest of the blocks are then tried in order. Any time an added block makes the pile higher than the table, the block is set aside and 0 is recorded in that bit position. If adding a particular block does not make the pile higher than the table, then the block is kept and 1 is recorded in that bit position. Each block requires only one trial or "clock" cycle, so eight trials produce an 8-bit word which represents the height of the table. The big advantage of a successive approximation A/D converter, then, is that it can do an N-bit conversion with only N clock pulses rather than the many clock pulses of a counter type.

Figure 11-25 shows a circuit for an 8-bit successive approximation A/D converter. As you will see later in the chapter, a complete successive approximation type A/D converter is available in a single IC. However, we chose a discrete component circuit for the example here so that you can build the circuit and explore its operation with a scope.

NOTE: The MC14549 is a CMOS device, so if you build the circuit, take appropriate precautions.

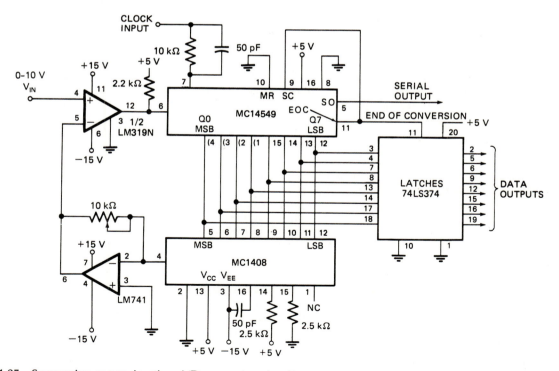

FIGURE 11-25 Successive approximation A/D converter circuit.

The heart of this type of A/D converter is a *successive approximation register* or *SAR* such as the MC14549. The purpose of the 10-kΩ resistor in series with the clock input of the SAR is to protect the device from damage if the power to the MC14549 is accidentally shut off before the clock signal.

The D/A converter used in this circuit is the MC1408 that we introduced you to in an earlier section of the chapter. As in the circuit in Figure 11-16, this circuit is set up for a 10-V full-scale output. Remember that the actual output voltage range of the D/A will be from 0.000 to 9.961 V when it is properly adjusted.

The LM319N comparator is used to determine whether the voltage output by the D/A converter is above or below the input voltage. If the D/A output is below the input voltage, the comparator output will be high. If the D/A output is above the unknown input voltage, the comparator output will be low.

The latches in the 74LS374 octal latch device are used to hold the data outputs constant while a new conversion is being done. The end-of-conversion, EOC, pulse from the SAR will update the latch outputs at the end of a conversion. Now that you have an overview of the parts in the circuit, we can explain how they all work together.

At the start of a conversion cycle the SAR outputs a 1 on its most significant data bit and 0's on the rest of its data bits. In response to this input word, the MC1408 D/A converter and the LM741 current-to-voltage converter will output a voltage equal to half the full-scale value for the converter. For the circuit in Figure 11-25, this is +5 V. If this voltage is higher than the input voltage, it will cause the comparator output to go low. A low on the comparator output tells the SAR to turn off that bit. A high on the output of the comparator would tell the SAR to keep that bit on in the data word output to the D/A.

In either case, the SAR will, on the next clock pulse, output a word with a high in the next most significant bit position. The SAR will keep this bit on or turn it off, depending on the output from the comparator. A high on the output of the comparator tells the SAR to keep the bit on, a low on the output of the comparator tells the SAR to turn the bit off before trying the next bit. In the same manner, the SAR proceeds down to the least significant bit, trying each bit in turn. If the output from the comparator is high when a bit is tried, the SAR keeps that bit on in the output word. If the output of the comparator is low when a bit is tried, the SAR turns that bit off in the output word. Since only one clock pulse is required to try each bit, the complete binary word will be present on the outputs of the SAR after just eight clock pulses. An EOC pulse produced by the SAR is used to transfer the binary word to the outputs of the 74LS374 latches. If the EOC signal is connected to the start conversion (SC) pin, as shown in Figure 11-25, the SAR will perform continuous conversion cycles.

To better understand the operation of this circuit, it is helpful to look at the waveforms produced on the output of the D/A as the circuit does a conversion on different input voltages. An exercise in the accompanying laboratory manual gives you some practice with this, but to give you a start, Figure 11-26 shows a few examples.

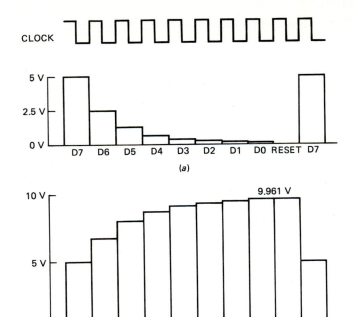

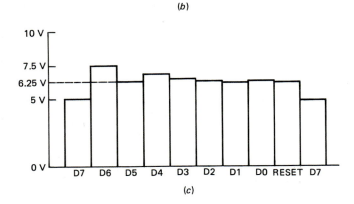

FIGURE 11-26 Waveforms present on output of D/A in a successive approximation A/D converter with different input voltages. *(a)* D/A output waveform with 0 V unknown voltage as input. *(b)* 10-V input. *(c)* 6.360-V input.

The first example in Figure 11-26*a* shows the D/A and comparator output waveforms that will be produced by an input voltage of 0.000 V. As you can see, the system tries and rejects all eight bits because all the D/A output voltage steps are greater than the V_{IN} of 0.000 V. Note that a complete conversion actually requires nine clock cycles instead of the theoretical eight. The additional clock cycle is required to reset the internal circuitry and start another conversion. Incidentally, the MC14549 outputs the data word 1 bit at a time as it is determined on the serial output, SO, pin. This SO format is useful in applications where you want to store the output data words on tape or disk.

The second example in Figure 11-26*b* shows the D/A output waveform that will be produced by an input voltage of 10.000 V. Since this input voltage is greater than the maximum output voltage of 9.961 V for the D/A converter, the SAR will keep all the output bits on for this conversion.

The final example in Figure 11-26c shows the D/A output waveform that will be produced by an input voltage of 6.360 V. To help you analyze this waveform, refer to Figure 11-16b, which shows the voltage steps that correspond to each bit of the D/A. As you can see in the waveform, the SAR keeps the MSB of 5.000 V. It rejects the 2.500-V bit, because the sum of 5.000 V and 2.500 V is 7.5 V which is greater than the input voltage. Proceeding on through the waveform it keeps the 1.250-V bit, and the 0.078-V bit, to give an output binary word of 10100010.

The successive approximation type A/D converter has the disadvantage that it requires a D/A converter, but it has the advantages of good resolution and relatively high speed. As an example, the GE/DATEL ADC-500 does a 12-bit conversion in 500 ns and the INTECH A8016-8 does a 16-bit conversion in 8 μs.

Voltage-to-Frequency A/D Converters

Another type of A/D converter commonly used in noisy industrial environments is the voltage-to-frequency (V-F) type. In this type of converter, the voltage to be digitized is applied to the control input of a VCO such as those we described in the last chapter. The frequency of the signal output by the VCO is proportional to the input voltage. A V-F converter can be used to convert the signal output by a remote sensor to digital pulses which are more resistant to noise than an analog signal. At the receiving system, the VCO frequency can be converted to an equivalent binary value by a simple frequency counter circuit. Good examples of this low-cost V-F type of converter are the National LM131/LM231/LM331 family of devices.

A/D Characteristics and Specifications

Before we show you some applications for A/D and D/A converters, we need to discuss the data sheet parameters of A/D converters and introduce you to some special considerations you have to take into account when you are using A/D converters.

RESOLUTION, ACCURACY, AND LINEARITY

Resolution, accuracy, and linearity specifications are the same for A/D converters as they are for D/A converters. To refresh your memory, resolution indicates the number of bits in the output word, or in other words how many decision levels the converter has. An 8-bit converter resolves an input voltage to 1 of 2^8 or 256 levels.

The accuracy of a converter is a measure of how closely the output represents the applied input voltage. As with a D/A converter, the accuracy of an A/D converter should be ±½LSB or better over the temperature range that you are going to use the converter.

The linearity of a converter indicates whether a stepped input signal produces an equally stepped value on the output. Again, the linearity specification for a converter should be no worse than ±½LSB for best performance. However, if you look at the data sheets for many A/D con-

verters, you will see that, as with D/A converters, devices are available which have worse linearity than this. These lower-specification devices are useful in applications where you need the resolution for small-amplitude signals but don't care if the output value is off somewhat for large input signals. The advantage of the lower-specification devices is that they are usually considerably less expensive.

QUANTIZING ERROR

As we said before, resolution of an A/D converter is a way of expressing how many different levels an A/D converter has to choose from in representing an input voltage. A 10-bit converter, for example, will represent an input voltage as one of 2^{10} or 1024 levels. *Quantizing error* is a way of expressing the maximum difference between the level used to represent an input voltage and the actual input voltage.

The 2-bit flash converter in Figure 11-21a, for example, has levels which are 1 V apart, so the value of an LSB is 1 V. If a voltage of 2.5 V is applied to this converter, the binary output code will be 10, or 2, which represents 2 V. This is an error of ½LSB. With the thresholds set as they are in the circuit in Figure 11-21a, an input voltage can be anywhere between 2.001 and 3.000 V and still give a binary output value of 2. Even if the converter has perfect accuracy, then, the output word can be in error by as much as about 1 V or one LSB!

In most applications the comparator thresholds in the circuit in Figure 11-21a would be shifted to 0.5, 1.5, 2.5, and 3.5 V, so that the maximum error or quantizing error would be ±0.5 V instead of +1 V. When set up in this way, the quantizing error for any converter is ±LSB. Quantization error, then, is a way of describing the best precision you can expect for a measurement taken by the A/D.

DYNAMIC RANGE

The dynamic range for an A/D converter is the ratio of the largest signal the converter can digitize to the smallest value it can resolve. Dynamic range, then, is a way of expressing the resolution of a converter in a form that is useful for some comparisons such as those in audio measurements. A unit called a *decibel*, or dB, is usually used to express dynamic range. If n is the number of bits for a converter, the dynamic range in decibels is:

$$DR = 20 \log 2^n = 20n \log 2$$

$$= 20n \times 0.301 = 6.02 \times n$$

An 8-bit A/D converter, then, can be said to have a dynamic range of 48.2 dB, a 12-bit converter a dynamic range of 72.2 dB, and a 16-bit converter a dynamic range of 96.3 dB. Some signals in industrial measurement applications require a converter with a dynamic range of 130 dB, or a resolution of about 22 bits.

CONVERSION TIME

As we told you at the start of this section on A/Ds, *conversion time* is simply the time required for a converter to produce a binary output word which represents an input voltage. The discussions in the preceding sections have shown you that conversion time ranges from a nanosecond to hundreds of milliseconds, depending on the method and the resolution. High-resolution, slow-speed applications usually use inexpensive, dual-slope types of converters. High-speed, high- and medium-resolution applications usually use a successive approximation type of converter. Very high speed applications are forced to use a parallel comparator type of converter.

Some A/D converter data sheets tell you the maximum clock frequency you can use with a converter rather than the conversion time. Since period = 1/frequency, you can easily convert between the two.

SAMPLE AND HOLDS

In applications such as digital storage scopes and digital audio recording where you need to digitize rapidly changing signals, a severe problem may occur. As shown in Figure 11-27, the problem is that by the time the conversion is complete, the input signal will have changed. Depending on the operation of the converter, the output word may represent the signal level present on the input at the end of the conversion cycle or at some intermediate point in the conversion cycle. Since the error voltage, ΔV, is a function of how fast the input signal is changing, the amount of error in the output digital words is variable. Therefore, it is difficult to correct for.

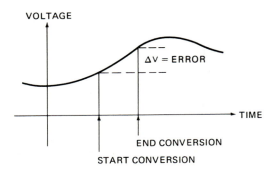

FIGURE 11-27   Error caused by input voltage changing during A/D conversion time.

This error can be eliminated or greatly reduced by connecting a sample-and-hold circuit such as that shown in Figure 11-28 on the input of the A/D. Before each conversion the electronic switch is very briefly closed to allow the capacitor to charge up to the voltage of the input signal and then quickly opened. After the signal is acquired on the capacitor, the A/D converter is signaled to perform a conversion on the stored sample. The principle here is that the switch is closed for such a relatively short time that the input signal doesn't change much during that time. Therefore, each sample accurately represents the voltage on the input when the sample was taken, regardless of the rate at which the signal is changing. Since the A/D digitizes these samples, the A/D output values will accurately represent sample points on the original waveform.

Two very important parameters for sample-and-hold circuits are *acquisition time* and *aperture time*. Acqui-

sition time is the time required for the switch to close and the capacitor to charge up to the input signal voltage after a sample signal is sent. Acquisition time depends on the output impedance of the input amplifier, the size of the hold capacitor, and the amount the capacitor has to charge or discharge from one sample to the next.

Aperture time is the time required for the sample switch to open or "disconnect" the input from the hold capacitor after a hold signal is sent. Aperture time, then, is a measure of how close a change on the input can be to the hold command without the input change affecting the output. Obviously you would like both acquisition time and aperture time to be as short as possible so that the input signal doesn't change during the time the sample is being taken and held.

For low-speed applications, the National LF198 has an acquisition time of about 6 μs to 0.01 percent and an aperture time of about 200 ns for an input change in voltage of 10 V. For high-speed applications, the Analog Devices HTS-0010 shown in Figure 11-28 has an acquisition time of 10 ns and an aperture time of 200 ps.

Sampling Rate and Aliasing

Another potential A/D problem becomes evident when you digitize, for example, a sine-wave signal with an A/D converter and then attempt to reconstruct the original sine-wave signal with a D/A converter as is done in digital storage scopes and digital audio recording/playback.

A theoretical analysis called the *sampling theorem* says that to accurately reconstruct a sine-wave signal, you need at least two nonzero samples from each cycle. In other words, the rate at which samples of the waveform are taken and digitized must be at least twice the frequency of the input signal. As you will see, a much higher sampling rate is necessary to get a good reproduction of the input signal. To see the effect of sampling rate on output fidelity, you might want to build and experiment with the circuit shown in Figure 11-29.

The circuit in Figure 11-29 is basically the successive approximation A/D converter from Figure 11-25 with a D/A converter connected to its outputs. The input range of the A/D converter is shifted from 0–10 V to ±5 V so that it can digitize signals such as sine waves which have both a negative part and a positive part. Likewise the output range of the D/A converter is set up for ±5 V. Before we show you the results produced by this circuit for different sampling rates, here's how the range is shifted for the D/A's in this circuit.

First note in the circuit in Figure 11-29 that the + input of each current-to-voltage op amp is connected to −5 V rather than to ground. Since the 10-kΩ resistor supplies negative feedback to its inverting input, the op amp will work day and night to hold its inverting input at −5 V also. If an input word of 00000000 is applied to the inputs of the D/A converter, the D/A output current will be zero and there will be no current through the feedback resistor. The output of the op amp, then, will be at the same voltage as the inverting input, −5 V. As the inputs of the D/A are incremented from 00000000 to 11111111, the voltage on the output of the

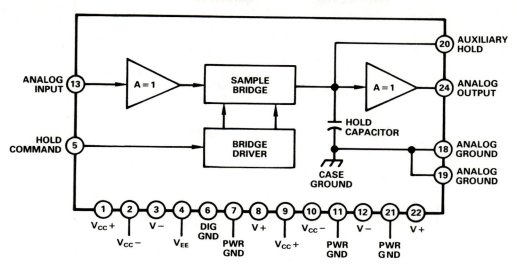

FIGURE 11-28 Block diagram of HTS-0010 sample-and-hold circuit used on the input of an A/D converter to provide a stable signal during the conversion cycle. *(Courtesy of Analog Devices.)*

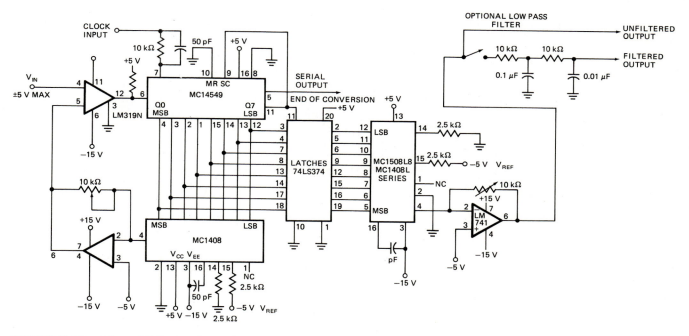

FIGURE 11-29 A/D and D/A circuit used to digitize and reconstruct analog signals such as sine waves.

op-amp will step up from −5 V to +4.961 V, instead of from 0 to +9.961 V as it did in the circuit in Figure 11-16a. Connecting the + input of the op amp to −5 V instead of to ground shifts the output voltage range to ±5 V instead of 0 to +10 V. This shifts the input range of the A/D converter to the same ±5-V range. Now let's look at some output waveforms.

Figure 11-30a shows the waveform that will be produced on the output of the D/A converter if a 100-Hz, 3-V peak sine-wave signal is applied to the input of the A/D converter and a 72-kHz clock signal is applied to the CLOCK input of the A/D converter. As you can see, the output waveform very closely resembles the input waveform. It takes nine clock cycles for the A/D converter

to do each conversion, so a 72-kHz clock produces 8000 (72,000/9) conversions or samples per second. Since one cycle of a 100-Hz signal takes 1/100 s, or 0.01 s, the converter then takes 80 samples during each cycle of the input waveform.

Figure 11-30b shows the input and output waveforms that will be produced if the frequency of the clock signal applied to the A/D converter is reduced to 7.2 kHz. This clock frequency causes the converter to take samples at a rate of 800 Hz (7.2 kHz/9). At this rate the converter takes eight samples for each cycle of the input waveform. As you can see, the output waveform is now a much more crude approximation to the input waveform. The approximation, however, is good enough so that if the signal is

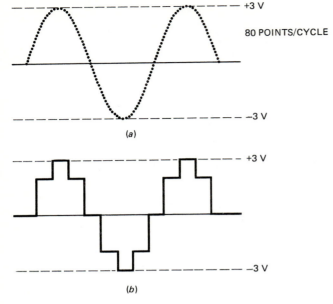

FIGURE 11-30   Output signal produced by A/D-D/A converter system with sine-wave input signal and different sampling rates. (a) Input signal sampled 80 times per cycle. (b) Input signal sampled eight times per cycle.

passed through a low-pass filter and an amplifier, the result will be very close to the input signal.

Finally, Figure 11-31 shows what happens if the frequency of the clock signal applied to the A/D converter is reduced to 720 Hz. This clock frequency produces a sampling rate of 80 Hz (720 Hz/9). Since this frequency is less than the frequency of the sine-wave signal applied to the A/D input, the converter will take only one sample

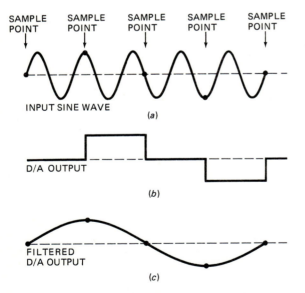

FIGURE 11-31   Effect of sampling rate less than frequency of input signal. (a) Input sine-wave signal showing sample points. (b) Unfiltered output from D/A converter. (c) Filtered output from D/A showing alias frequency produced.

during each input cycle. The signal these samples produce on the output of the D/A converter is shown in Figure 11-31b. As you can see, it doesn't look anything like the input signal. In fact, if this output from the D/A is amplified and filtered, the result will be the sine-wave signal shown under the D/A output waveform in Figure 11-31c. This output signal has a very different frequency from the original input frequency. An output frequency produced in this way is called an *alias frequency*, or *alias signal*. A simple way to see the alias frequency signal is to draw a line connecting the sampling points shown on the sine-wave input signal. The point for you to remember from all this is that aliasing is prevented by sampling at a frequency more than twice the frequency of the input signal! This principle has important implications in some of the A/D and D/A applications we discuss in the next section.

A/D AND D/A APPLICATIONS

Digital Voltmeters and Multimeters

One of the most common applications of A/D converters is in digital voltmeters (DVMs) and digital multimeters (DMMs). DVMs are commonly available as 3½-, 4½-, 5½-, 6½-, and 7½-digit units. The ½ digit here refers to the fact that the leading digit can only be a 0 or a 1. The full-scale reading then for a 3½-digit meter then is 1999, the full-scale reading for a 4½-digit meter is 19999, etc. Some available DVMs are described as 3¾ digit units. The ¾ digit in the name means that the leading digit can be a 0, 1, 2, or 3, so the maximum display is 3.999.

In this section of the chapter we first discuss an A/D circuit that can be used as the heart of a DVM, and then we show you circuits you can add to the basic A/D so that you can measure current, resistance, ac voltage, and temperature.

BASIC A/D CONVERTER SECTION

Figure 11-32 shows a circuit for a complete 3½-digit A/D converter with an input voltage range of ±1.999 V. We chose the Intersil ICL7137 for this example because it is inexpensive, digitizes both positive and negative voltages, and requires a minimum of external components.

As you can see in Figure 11-32, the ICL7137 is designed to directly drive common-anode LED displays that are cheap and readily available. For a battery-powered application, you might use an ICL7136, which functions the same as the ICL7137, except that it is designed to drive an LCD display. LCD displays have the advantage that they require almost no power to operate; however, they are considerably more expensive and more difficult to obtain than LED displays.

The ICL7137 uses a dual-slope conversion technique such as we described earlier in the chapter. The device is divided into two basic sections, an analog section and a digital section.

The digital section of the ICL7137 shown in Figure 11-33a consists mostly of a decade counter–latch–display module. The LED displays are driven directly rather than multiplexed. This eliminates any switching noise that would be produced on the V_{CC} line by digits

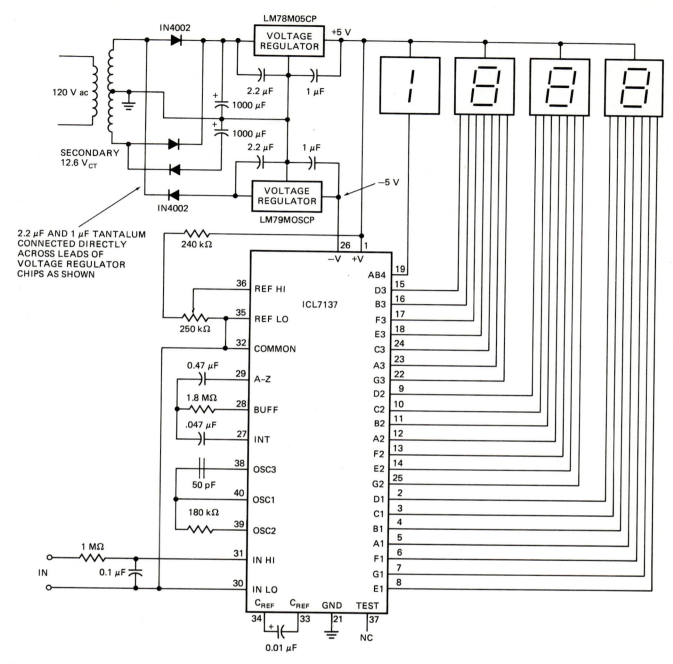

FIGURE 11-32 Schematic for complete 3-digit A/D converter using an Intersil ICL7137.

being turned on and off during multiplexing. The LED segment currents are internally limited to about 8 mA, so no external current-limiting resistors are needed. The digital section also includes circuitry that controls the integrator and an oscillator that produces the basic clock for the converter. The frequency of the clock is determined by the value of R and the C connected between pins 38 and 39. For the values shown in Figure 11-32, the clock frequency is 48 kHz. This produces three complete conversions per second.

The analog section of the converter shown in Figure 11-33b consists of an integrator, a comparator, and some MOS switches which are represented in the drawing by circled X's. These switches are used to determine which signal is applied to the input of the integrator and other parts of the circuit at various times in the conversion cycle. A conversion cycle for this converter is divided into four periods, as shown in Figure 11-33c, rather than just two, as we described for the basic dual-slope converter in Figure 11-23. The two additional periods or phases allow the converter to work with both positive and negative input voltages and cancel out the effects of offset voltages in the internal circuitry. Briefly, here's how the device does a conversion cycle.

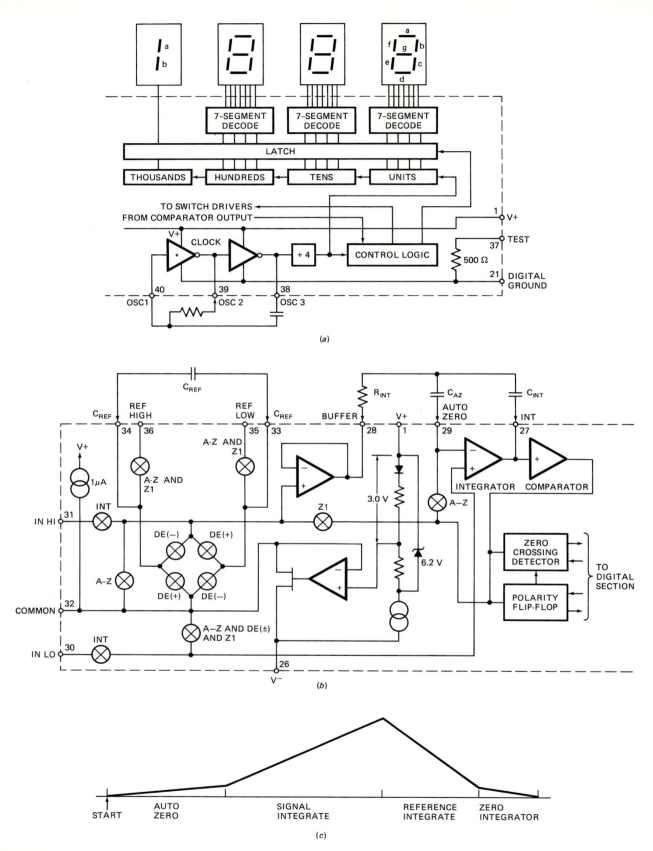

FIGURE 11-33   (a) Block diagram of digital section of ICL7137 A/D converter. (b) Block diagram of analog section of ICL7137 A/D converter. (c) Waveform showing four periods (phases) of ICL7137 conversion cycle for negative input voltage.

During the auto-zero phase, the reference capacitor, C_{REF}, is charged to a voltage equal to the reference voltage. Charging the reference capacitor to the reference voltage allows the converter to work with both positive and negative voltages. A second operation that is done during the auto-zero period is to charge the auto-zero capacitor, C_{AZ}, to a voltage which is equal to the sum of the offset voltages in the buffer amp, integrator, and comparator. This will cancel out the effects of any offsets in the circuit.

The second phase of a conversion is the signal integrate phase. During this phase the differential voltage between the IN HI and the IN LO pins is applied to the input of the integrator for 1000 clock cycles. For a positive input voltage, the integrator output will ramp down, and for a negative input voltage, the integrator will ramp up. The polarity flip-flop will be set or reset to show which way the integrator output ramped.

The third phase of the conversion cycle is the reference integrate phase. Depending on the setting of the polarity flip-flop, the reference capacitor is connected to the integrator such that the output of the integrator ramps back to the voltage stored on the auto-zero capacitor. As we explained in the initial discussion of a dual-ramp converter, the number of counts required for the reference integrate phase is proportional to the input voltage. For this circuit the reading displayed will be $1000 \times (V_{IN}/V_{REF})$.

The final phase of the conversion cycle is the zero integrate phase. The purpose of this phase is to return the integrator output back to zero so that everything is ready for the next conversion cycle. The labels next to the MOS switches in Figure 11-33b show which switches are on during each phase of a conversion, so if you want more detail of the operation of the circuit you can mentally close the appropriate switches and trace the current paths during each phase of a conversion.

As we told you before, the basic converter in Figure 11-32 will digitize voltages in the range of -1.999 to $+1.999$ V. A basic converter-display module such as this is available for use mostly in industrial applications as a *digital panel meter*. As we show you in the following section, interface circuits can be added to the basic converter to enable it to make a wide variety of measurements.

MEASURING RESISTANCE, CURRENT, AC VOLTAGE, AND TEMPERATURE

One useful addition to the basic A/D converter is a precision metal-film-resistor voltage divider such as that shown in Figure 11-34. A circuit such as this is often called an *attenuator* because it reduces the amplitude of an input signal to the point where it is in the range of the A/D. The attenuator in Figure 11-34 is designed so that the voltage on each output is one-tenth of the voltage on the next higher output. For example, If 500 V is applied to the input of the circuit, the top output will be at 500 V, the next output down will be at 50 V, the next output will be at 5 V, and the bottom output will be at 0.5 V. Since only the 0.5V output is in the range of the converter, the switch should be in this position to make a voltage measurement. Another section of the switch

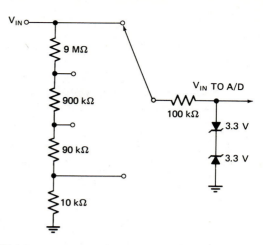

FIGURE 11-34 Attenuator circuit used to scale voltages for input to basic A/D converter.

that selects the input range can be used to light a decimal point at the proper position in the display.

Many digital voltmeters now have autoranging circuitry that automatically sets the meter to the correct range. These circuits use an electronic switch to connect the input of the A/D to one of the attenuator outputs. The control circuitry starts with the switch at the lowest voltage output and steps up the divider until it finds an output which has a voltage between 10 and 90 percent of the input range for the converter.

The 100-kΩ resistor and the back-to-back zener diodes in the circuit in Figure 11-34 protect the A/D converter input from overvoltage if the input switch is initially set for too high a range. When the voltage at the input of the converter reaches -4 V in the negative direction or reaches $+4$ V in the positive direction, the diodes will turn on. This prevents the input of the converter from being forced outside the safe range of ± 4 V. The 100-kΩ resistor limits the current through the protection diodes if they get turned on by overvoltage, but it does not attenuate the input signal. The input impedance of the converter is so high that normally almost no current flows through the 100-kΩ resistor, so there is no voltage drop across it.

Figure 11-35 shows a basic circuit which will convert the A/D converter in Figure 11-32 to an ohmmeter. When an unknown resistor is connected between the output

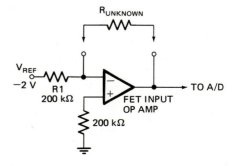

FIGURE 11-35 Circuit to convert basic A/D converter to an ohmmeter.

of the op amp and its inverting input, the op amp will hold the inverting input at 0 V. With the inverting input at 0 V, R1 will have a fixed voltage of −2 V across it, regardless of the value of $R_{UNKNOWN}$. The current through R1 then will always be −0.01 mA (−2 V/200 kΩ). The op-amp output will supply this current through the unknown resistor. The reference voltage is negative, so the output of the op amp will be positive. Since the current passed through the unknown resistor is constant, V = IR tells you that the output voltage is directly proportional to the value of the unknown resistor. For the values shown in Figure 11-35, the largest resistor that can be measured without going over the ±1.999-V input range of the A/D is 199.9 kΩ. Other values of R1 can be switched in to change the range of the circuit.

Figure 11-36 shows a circuit which converts the basic A/D converter in Figure 11-32 to an ammeter. The current to be measured is passed through a small *sense* resistor to produce a proportional voltage. The op-amp circuit amplifies the voltage drop across the sense resistor by 100. The amplified voltage is then digitized by the

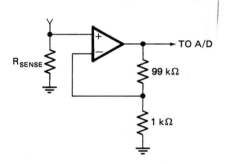

FIGURE 11-36   Circuit to convert current to a proportional voltage for input to A/D converter in Figure 11-32.

A/D converter. Amplifying the voltage across the sense resistor in this way allows a much smaller sense resistor to be used. In the circuit in Figure 11-36, a sense resistor of only 10 Ω will produce an output voltage of 1.999 V with a current of 1.999 mA. Different value sense resistors can be used to give different current ranges. One difficulty with the simple current adapter circuit in Figure 11-36 is that one end of the sense resistor must be tied to ground. A more versatile current adapter circuit can be built using a differential-input amplifier such as the instrumentation amplifier in Figure 11-3g.

Still another type of adapter you can add to the input of an A/D converter is one to convert an ac input signal to a dc voltage which can be digitized. Figure 11-37 shows an op-amp circuit which converts an input ac signal to a dc voltage which is equal to the RMS value of the ac input signal. We don't have space here to go into all the details of how the circuit works, but here's an overview. The first op amp is an active half-wave rectifier which produces the waveform shown above point A in the circuit. The second op amp is an adder which combines this waveform with the input ac waveform to produce a full-wave rectified dc waveform. Without C_F, the 10-μF capacitor, the output of the second op amp would look like the typical full-wave rectified waveform shown

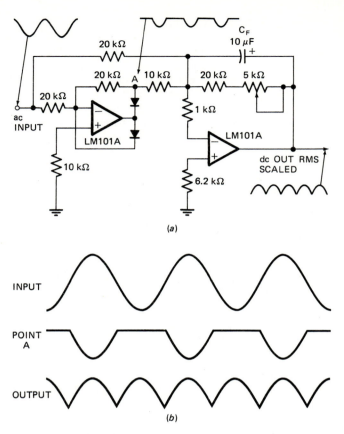

FIGURE 11-37   Circuit to convert ac input voltage to a dc voltage which is equal to the RMS value of the ac input voltage.

next to its output. With C_F in place, the signal on the output of the op amp is filtered to a pure dc waveform which can be digitized by the A/D. The 5-kΩ potentiometer is adjusted to scale the dc output voltage so that it is equal to the RMS value of sine-wave ac input signals. The A/D converter then converts this dc voltage to an equivalent binary word.

Besides measuring electrical quantities such as voltage, current, and resistance, the basic A/D converter in Figure 11-32 can be used to make a digital thermometer, scale, or pressure gage. For example, to make a digital thermometer you can connect the outputs of the LM35 temperature sensor circuit shown in Figure 11-7a to the HI input and the LO input of the ICL7137 A/D. The voltage output by this sensor circuit increases at 10 mV/°C from −0.550 V at −55°C to +1.500 V at 150°C. The output voltage range of the sensor, then, is within the conversion range of the A/D, so no additional circuitry is needed.

In a similar manner the differential output from a load cell or pressure transducer can be amplified and applied to the inputs of the ICL7137 to produce a digital readout scale or pressure gage. The possibilities are almost endless.

Digital Storage Oscilloscopes

BASIC OPERATION

In Chapter 4 we explained how a standard oscilloscope produces a trace by sweeping a beam of electrons across a phosphor coating on the front of a CRT. The phosphor used in standard scopes fades rapidly, so the beam must sweep out a trace over and over in order for the trace to remain visible. In cases where you are looking at repetitive signals of medium or high frequency, this presents few problems. Each time the beam comes back to the left side of the CRT, the scope will be retriggered and the beam will sweep out a trace of a later, identical section of the waveform. For very low frequency signals or signals that do not repeat, a standard scope doesn't work very well. You may have noticed that with a standard scope set at a very slow sweep speed, all you see is a bright dot moving across the screen. This moving dot display is difficult to analyze unless you have a photographic memory.

Joking aside, one way to observe low-frequency signals or nonrepetitive signals is to take a picture of the trace with a camera mounted on the front of the scope. If the camera shutter is opened for just the time the beam is sweeping across the screen, the picture will show the desired trace. This method, however, can be very time consuming and expensive.

A second way to observe low-frequency or nonrepetitive signals is with a special type of scope called a *storage scope*. As the name implies, a storage scope does a trace and in some way retains it for viewing. Storage scopes come in two basic types, analog and digital.

Analog storage scopes mostly use one of two methods to store a trace. One method is to use a phosphor which emits light for a relatively long time after the trace is swept out. The phosphor retains a visible trail where the electron beam swept across. This type of storage scope is good for some applications, but the storage time is very limited. A second analog storage method is to store the trace as charges on wire meshes between the electron gun and the phosphor. This mesh storage approach works at higher sweep speeds and permits much longer storage times than the phosphor storage type, but it is still very limited compared with digital storage scopes.

A digital storage oscilloscope, or DSO, allows you to store and display analog waveforms in a manner very similar to the way a logic analyzer allows you to store and display purely digital signals. Samples of the input signal are digitized and stored in memory. The stored values can then be read out and sent to a D/A converter over and over to produce a stable display on the CRT. The block diagram in Figure 11-38 shows the major parts of a digital storage scope. Let's take a closer look at how it works. Under the direction of the control section, the sample-and-hold (S/H) section takes a sample of the input signal and passes it along to the A/D converter. One A/D conversion technique used in DSOs is the "flash" technique we described earlier in the chapter. Another commonly used technique is to shift a sequence of perhaps 1000 samples into an analog MOS shift register with a high-frequency clock signal. Then, when the shift register is full, the stored analog samples are shifted out to a successive approximation converter at a lower clock rate to be digitized. The advantage of this two-step approach is that the A/D converter doesn't have to work at as high a frequency. The disadvantage of this approach is that the analog shift register introduces considerable noise into the signal.

The binary word produced by the A/D converter is stored in a memory. After one sample is taken, digitized, and stored in memory, the memory address counter is incremented by 1 to point to the next storage location in the memory. In *free-run mode*, the system will repeat the process. When the memory is full, the oldest sample word will be written over by the next word from the A/D converter. Depending on the particular instrument, memory may hold from 1024 words per input channel to 64K words per input channel. When the control circuitry receives a specified trigger signal from the input waveform or from an external source, it stops the sampling process and displays a selected group of sample words on the CRT.

The voltage ramp which sweeps the electron beam horizontally across the screen is produced by a counter and a D/A converter. A second D/A converter changes each word read out from memory to an equivalent analog voltage. The output from this D/A is used to deflect the electron beam vertically as it is swept across the screen by the horizontal amplifier.

The word read out from memory is determined by the memory address counter. The memory address counter and the horizontal timebase counter are clocked by the

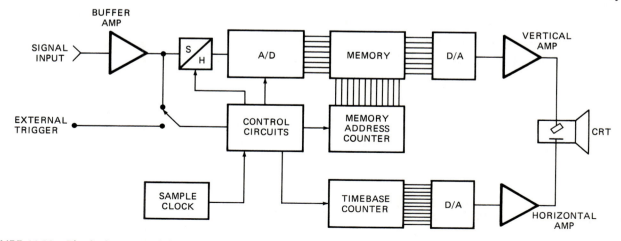

FIGURE 11-38 Block diagram of digital storage oscilloscope, or DSO.

same clock, so, as the timebase counter steps the beam horizontally to the next step, the memory address counter causes the memory to output the next stored word to the vertical D/A converter. The counters cycle around so that the stored samples are displayed on the screen over and over. The resultant display, then, is a dot-by-dot picture of the waveform applied to the scope input. There are many advantages of this system.

First, once the samples are digitized and stored in memory, they can be read out and displayed over and over as long as desired. This is just what is needed for a signal that occurs only once.

Second, the frequency of the clock used to take samples can be very different from the frequency of the clock used to produce a trace on the screen. To produce a trace of a breath waveform, for example, you might use a slow clock to take samples of the input waveform because one breath requires an interval of several seconds. Using this same clock to drive the display part of the circuitry would produce a slowly moving bright dot, such as the display that would be produced on a standard scope. Using a higher-frequency clock to drive the display circuitry refreshes the display more often and makes it clearly visible. From the horizontal time/division switch setting of the scope you know the "real" time represented by each horizontal division on the screen, so you can correctly determine the time between points on the signal waveform.

As another example of this second advantage, suppose you want to make a trace of a signal which only lasts a few hundred nanoseconds. You can use a sample frequency clock of 400 MHz to take a sample, digitize it, and store the result in memory. If you tried to use this frequency clock to drive the display circuitry, the beam would sweep across the screen so fast that it would be hardly visible. Using a lower-frequency clock for the display circuitry makes the display more easily visible.

A third advantage of the digital storage approach is that data which occurred before the trigger can be held in memory and displayed. Just as with logic analyzers, most sampling scopes allow you to display mostly the samples that occurred before the trigger, some samples before and some samples after the trigger, or mostly a group of samples taken after the trigger occurred.

Still another advantage of a DSO is that it can be operated in a variety of recording modes. In the next section we discuss some of these.

MAKING MEASUREMENTS WITH A DSO

There are a wide variety of DSOs available, ranging from relatively inexpensive general-purpose units such as the Tektronix 2230 that has a maximum sampling rate of 20 megasamples per second to the Hewlett Packard HP54111D with a maximum sampling rate of 1 gigasample per second.

Most of the controls on a general-purpose DSO such as the Tektronix 2230 are the same as those on a standard scope such as the Tektronix 2236 shown in Figure 4-3. DSOs commonly allow you to select either a nonstore mode where the unit operates as a standard scope, or a store mode where the unit operates in one of several recording modes.

The simplest DSO store/display mode is probably the *one-shot mode*. In this mode the DSO is allowed to take sam-ples continuously until a specified trigger slope and trigger level occur on the input signal. When the trigger event occurs, the DSO stops taking samples or continues to take a specified number of samples. It then displays the stored samples and stops until given further instructions. This mode is sometimes called "babysitting" mode because the DSO just sits and waits until some event such as an earth tremor or a laser pulse triggers it.

Another useful recording mode is the *scan mode*. In this mode the DSO takes enough samples for one display (typically 1000), and displays these samples on the CRT. It continues to take samples and store them in another section of memory until another 1000 samples are collected. Then the display is updated to show the new set of samples. The process continues over and over until you press the SAVE button on the front panel. The DSO then stops taking samples and holds the display of the latest set of samples.

Still another useful recording mode is *roll mode*. In this mode the DSO functions as a strip chart recorder. The DSO is continuously taking samples, and as each new sample is taken, the display is updated to show that sample on the left side of the screen. When you see the display you want, you press the SAVE button to hold that display.

DSOs such as the Tektronix 2230 allow you to store one set of samples in a reference memory and another set of samples in the current display memory. Both sets of samples can then be displayed on the screen at the same time and compared.

Most DSOs have intensified cursors which you can position on the displayed waveform to get a direct digital readout of the voltage between the two points or the time between the two points. If you think about the way the samples are taken and stored in a DSO, perhaps you can see how this time and voltage information is easily obtained. Often the voltage and time measurements determined from the cursor settings are more accurate than measurements you might make by interpreting the display on the screen. Most DSOs use 8-bit converters, so the values stored in memory should be accurate to at least 1 percent, whereas the accuracy of a CRT display is typically 3 to 5 percent.

Another powerful feature which is available in the higher-performance DSOs is the *sampling scope mode*, which allows the DSO to sample and display very high frequency repetitive signals. This capability is becoming more important as GaAs logic devices with clock frequencies in the gigahertz range become more commonplace. Figure 11-39 shows how this technique works.

The major principle of the sampling mode is that the display is made up from samples taken from widely separated sections of the signal waveform rather than from just one section, as is done in normal DSO operation. A typical sampling rate in this mode is about 50 kHz. This corresponds to one sample every 20,000 cycles of a 1-GHz input signal. Now, if the sample were taken at the same time in each cycle of the input waveform, the value stored for each sample would be the same, and the display would simply be a straight horizontal line. If, however, sample 1 is taken at time t, sample t1 is taken 20,000 cycles plus a small time (Δt) later, and sample 3

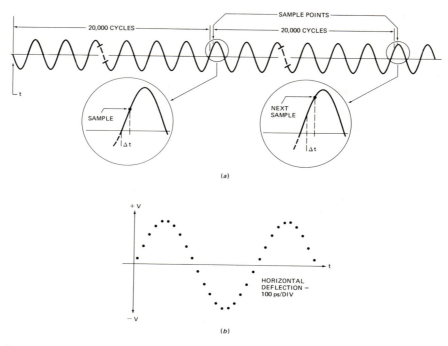

(a)

(b)

FIGURE 11-39 Sampling scope for observing very high frequency repetitive signals. *(a)* 1-GHz sine wave showing sample points for 50-kHz sampling rate. *(b)* Display produced on scope by collected samples.

is taken 20,000 cycles plus Δt after t2, then the sample points will step along the input waveform in increments of Δt as shown in Figure 11-39. The data values stored in memory, then, represent samples from many different cycles, but since the cycles are all the same, the display will accurately represent the desired section of the input waveform. Figure 11-39*b* shows the display that will be produced on the CRT by the stored samples. The horizontal timebase setting or an on-screen display will tell you the "real" time in nanoseconds or picoseconds represented by each division on the screen. Most DSOs with this sampling mode capability have on-screen cursors which you can position to determine the voltage or time between two points on an input waveform.

In the next section we describe how a personal computer can be set up to operate as a low-frequency DSO or data analysis system.

Data Acquisition Systems

HARDWARE

The term *data acquisition system* has at least two meanings. One use of the term is to describe a multiplexed input A/D converter such as the National ADC0808 shown in the circuit in Figure 11-40*a*. As you can see in the block diagram of the ADC0808, this device consists of an 8-bit successive approximation A/D converter and an 8-input analog multiplexer. The principal feature of this circuit is that it can digitize the signal on any one of the eight inputs of the multiplexer. The input selected to be digitized is determined by the 3-bit address, 000 to 111, applied to the address inputs of the multiplexer.

The op-amp circuit in the lower left corner of Figure 11-40*a* generates a precise reference voltage of 5.120 V

for the converter. This sets up the A/D converter for an input range of 0 V to 5.120 V, or 256 steps of 20 mV per step. Since the ADC0808 only requires 1 mA of supply current, the 5.12-V reference is also used to supply a "clean" V_{CC} to the converter. This keeps out any noise that may be present on the main +5-V supply of the converter. The inexpensive Schmitt trigger oscillator in the circuit supplies a 300-kHz clock to the converter. To help you see how you tell this circuit to do a conversion, Figure 11-40*b* shows the timing waveforms for it.

You start a conversion cycle by sending the 3-bit address for the input channel you want to digitize to the multiplexer address inputs and sending a pulse to the address latch enable (ALE) input. This selects the desired input channel and connects it to the output of the analog multiplexer. The next step is to send a start pulse to the successive approximation A/D converter. Pulsing high initiates a conversion cycle and causes the START EOC output signal to be reset. When the conversion is complete, the ADC0808 will assert the EOC signal high again. If the three-state outputs are enabled by a high on the OE input, then when EOC goes high the 8-bit binary word that represents the analog input value will be present on the outputs.

The ADC0808 data acquisition converter is designed with three-state outputs so that it can be interfaced to a microprocessor bus along with other devices. In the sequel to this book, we show you how to interface a converter such as this to a microprocessor and how it is used as part of a system which controls a small factory.

A second use of the term *data acquisition system* is to describe a system which includes both a microprocessor or microcomputer and a multiplexed input A/D converter such as an ADC0808. A common way to pro-

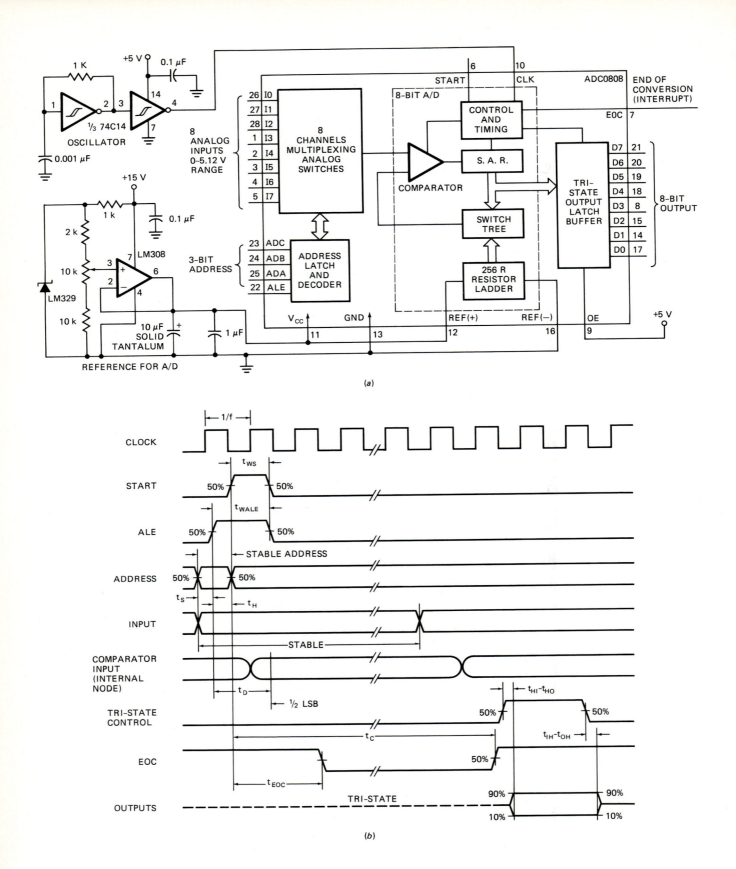

FIGURE 11-40 ADC0808 8-input, 8-bit data acquisition system. (a) Circuit for input range of 0 to 5.120 V (256 steps of 20 mV per step). (b) Timing waveforms for a conversion.

duce this type of system is with a PC board which plugs into an expansion slot in a microcomputer. Shown in the lower right corner of Figure 11-41 is the DAS-8 board from Metrabyte. This small board converts an IBM PC- or PC/AT-compatible microcomputer to an 8-input, 12-bit data acquisition system. The ribbon cable and connector box shown to the left of the PC board allow signals from sensors to be easily connected to the inputs of the A/D converter. Metrabyte and other companies make a wide variety of A/D, D/A, and digital interface boards for personal computers.

FIGURE 11-41   Metrabyte PC board, cable, and connector box used to convert an IBM PC-, XT-, or AT-compatible microcomputer to an 8-input data acquisition system. Binder contains Labtech software used to control the system.

SOFTWARE

Also shown in Figure 11-41 is the Labtech ACQUIRE software package. The programs in this package allow you to control the operation of the DAS-8 board from the keyboard of the microcomputer using simple commands. With these programs you can, for example, digitize as many as four channels simultaneously at sampling rates from a few hundred samples per second to one sample every 11 days. The data samples can be displayed on the screen of the computer and sent to files on a disk for later analysis and/or printing. The data files are stored on disk in a format that is compatible with popular spreadsheet programs such as LOTUS 123 and other analysis programs. These additional programs allow you to determine rate of change, total weight, minimum and maximum values, etc., for the input signals.

A data acquisition system like this might be used to record weather information such as barometric pressure, relative humidity, wind direction, wind velocity, and rainfall accumulation. There are many applications for a system such as this in an industrial environment where it can be used for measuring, analyzing, and controlling temperatures, pressures, flow rates, etc.

Another commonly available program called Unkelscope allows an IBM PC- or AT-compatible computer with a Metrabyte DAS-8 board to function as a low-speed DSO.

Digital Audio

Throughout this book we have tried to show you not just the theoretical side of digital electronics, but also practical applications such as digital clocks, frequency counters, and digital tuners. An area of electronics where A/D and D/A converters have had a tremendous impact is in audio recording. In this section we give you an overview of how A/D and D/A converters are used in recording and playing back digitally recorded audio from *compact audio disks* (CDs) and *digital audio tapes* (DATs).

Figure 11-42 shows a basic block diagram of the system used to record stereo digital audio on tape or CDs. First, in Figure 11-42 note the 16-bit successive approximation A/D converter for each channel. The system uses 16-bit converters so that it can digitize the loudest roar and still not miss the details of the softest whisper.

For CD recording, the A/D converters are operated at a 44.1-kHz sampling rate and for DAT recording, the converters use a 48-kHz sampling rate. Remember from our previous discussion of the sampling theorem that in order to reconstruct a sine-wave signal, you need to take samples at a frequency equal to or greater than twice the frequency of the sine wave. Sampling rates of 44.1 or 48 kHz, then, are sufficient to reconstruct signals in the audio range of 20 Hz to 20 kHz. A sample-and-hold circuit is used on the input of each A/D converter to make sure the signal level on the input of the converter doesn't change during a conversion.

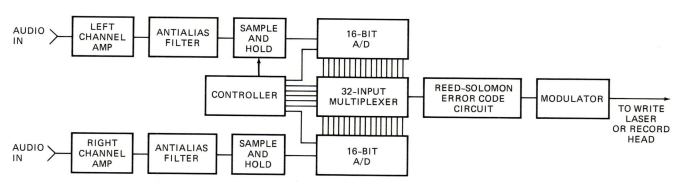

FIGURE 11-42   Block diagram of circuitry used to convert stereo audio signals to a form which can be recorded on a CD or DAT.

Remember also from our discussion of the sampling theorem that if you sample a signal at a rate less than twice its frequency, the reconstruction from a D/A converter will be a lower-frequency signal called an "alias" signal. If frequencies above about 20 kHz were allowed to get to the A/D converter, the output from the playback D/A converter would contain alias frequencies which were not part of the original signal. To prevent this aliasing, the audio signal to be digitized is passed through a low-pass filter which removes or sharply attenuates frequencies above 20 kHz.

The parallel binary output codes from the A/D converters are fed into the inputs of a multiplexer, which accomplishes two tasks. First, it converts the parallel data from the A/D converter to the serial form it must be in for writing to a track on a disk or tape. Second, it interleaves data bits from the two channels so that the data for both audio channels can be recorded on a single track on the disk or tape.

After converting the data to a serial bit stream, error detection/correction bits are added to the basic data words. In a previous section we explained how a parity bit is added to ASCII codes to detect errors introduced during transmission. Digital audio systems use a complex method called *cross-interleave Reed-Solomon Code* (CIRC) to add a group of calculated bits to the data bits. These added bits enable the playback system to detect and, in most cases, correct bit errors in the data words read out from disk or tape. After the error detection/correction bits are inserted in the data words, the data words are assembled into blocks and written to a master disk or tape.

On DAT the bits are recorded as magnetic flux changes on a Mylar tape coated with a magnetic material such as iron oxide. The same "head" used to record the data bits is also used to read the stored data bits for playback.

On a CD, a laser beam is used to write the binary data blocks along a track as tiny pits and lands. *Lands* are reflective sections between *pits*. The data track on the disk is a spiral which starts at the center of the disk and winds outward. The track is only about 20 µin wide, so it can wind around on the disk many times. Somewhere we read that the track would be about 3 miles long if it were stretched out in a straight line! It is also interesting to note that about a third of the bits stored on the disk are used for directory, formatting, synchronizing, and error detection/correction.

Data words are read out from a CD by focusing the beam

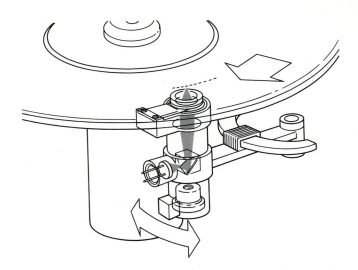

FIGURE 11-43 Laser mechanism to read data from compact audio discs.

from a laser diode on the data track from below, as shown in Figure 11-43. A phototransistor is used to detect the amount of reflection from the track. The lands between pits reflect better than the pits, so it is easy to tell the difference between them. The speed of the motor which spins the disk is varied from about 480 rpm when reading the track near the center of the disk to about 210 rpm when reading the track near the outer edge of the disk. This change in disk rotational speed is necessary so that the data track passes over the laser diode and phototransistor at a constant rate. If you look a little closer at the actual circuitry in a CD player, you will find that a PLL circuit is used to keep the drive speed synchronized with the data bits being read off a track. Data bits, clock bits, parity bits, and so forth are read out from the track at a rate of 4.3218 Mbits/s.

Once the stored bits are read out from a CD or DAT, the original data words must be reassembled, corrected if necessary, separated into left and right channel signals, and converted to the equivalent analog voltages. Figure 11-44 shows a block diagram of the circuitry used to do this. As you can see in the figure, a demultiplexer is used to separate the corrected stream of data bits into the left and right channel data words for the D/A converters. Sample-and-hold circuits are used on the outputs of each D/A converter to prevent glitches on the output waveforms. A low-pass filter on each output

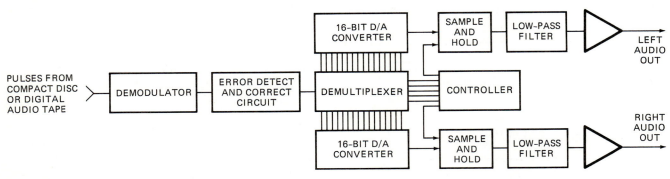

FIGURE 11-44 Block diagram of circuit to convert digital data from CD or DAT to stereo audio signals.

smoothes the steps in the output waveforms so that they very closely approximate the original audio waveforms. Tone control circuits then allow you to adjust the frequency response of each channel to match your tastes. Finally, amplifiers on each channel boost the power level of the output signals to the point where they can drive speakers and shake the walls.

If you want to learn about some of the specific devices that can be used to build a CD playback system, consult the data sheets for the Signetics TDA photodiode signal processor, the SAA7210 decoder, the TDA1541 dual 16-bit D/A converter, and the SAA7220 digital filter.

All circuitry and techniques described in the preceding section may seem to be a lot of trouble to get an hour or so of music recorded on and played back from a disk. However, the advantages of CD and DAT recording over standard analog records and tapes makes them well worth the trouble to most listeners. First, digital recordings have a frequency response of 20 Hz to 20 kHz ± 0.3 dB, which is much better than that of LP records.

Second, CDs and DATs have a much wider dynamic range than standard LP records. Dynamic range is a way of expressing the ratio between the loudest sound the system can produce without distortion and the softest sound the system can produce distinguishably. This ratio is usually expressed in decibels. As explained in an earlier section of the chapter, 16-bit A/D converters such as those used for digital audio give a theoretical dynamic range of $20 \log 2^{16} = 96.3$ dB. In practice the dynamic range of CDs and DATs is about 90 dB, which is at least 20 dB greater than that for traditional analog methods. On a system with poor dynamic range, a single violin playing will have nearly the same sound level as a cannon firing.

A whole group of other advantages of CDs involve the physical disk itself. The discs can hold more music than a standard LP record. CDs do not wear out because the read mechanism does not physically touch the recorded data and because the disk is enclosed in a plastic coating which can be easily washed if something is spilled on it. Unlike LP records, the error-correcting mechanism used with CDs prevents small scratches on the surface of the disk from affecting the output sound quality.

Still another advantage of CDs is that the audio information does not use all the available space on a disk. The data required for several hundred video images can be stored on the disc along with the audio information. On the appropriately designed player, this feature allows you to see a few pictures of your favorite band while you listen to their music.

Digital Voice Communications and CODECS

Analog signals such as video or voice communications get noisy when transmitted long distances. Since the induced noise is random, it is difficult, if not impossible, to filter out. Digital signals, however, are very resistant to noise because you only have to detect whether the signal is low or high. Therefore, if you are trying to get an intelligible signal back from the moon or from Mars, it is well worth the trouble to digitize the signal before sending it, and reconstruct the analog signal at the receiving end with a D/A converter.

Another application where it is well worth the trouble to convert signals to digital form is in telephone signal transmission systems. During a discussion of modems in Chapter 10, we told you that telephone signals are transmitted to and from your house as analog signals in a frequency range of about 300 Hz to 3 kHz. When the analog signal from your phone reaches the branch office, a *subscriber line interface circuit* (SLIC) converts the analog signals to digital form and performs a variety of control functions. The functions performed by the SLIC are summarized by the acronym *BORSCHT*, which stands for battery feed, overvoltage protection, ringing, signaling, coding, hybrid, and testing. Figure 11-45 shows a block diagram of an SLIC in which various BORSCHT functions are highlighted in this type of application.

The connection of the lines from your phone to the SLIC is called a *local subscriber* loop. The local loop signals to and from your phone connect to the lines labeled TIP and RING on the left side of the diagram. The terms TIP and RING are a holdover from the days when phone connections were made with phone plugs such as those used on electric guitar cords.

When it reaches the SLIC, the analog speech signal from your phone is passed through an antialiasing filter and then converted to a stream of 8-bit binary words by a section of the circuit called a *coder-decoder*, or *CODEC*. The stream of binary words leaves the SLIC in serial form on the line labeled TRANSMIT PCM. The term PCM stands for *pulse code modulation*, which is just another way of saying that pulses on the line represent the binary value of the sample taken by the A/D. After the PCM signal leaves the SLIC, it is multiplexed with control signals and the outputs from other SLICs onto a single line. A common system called the T1 transmission system uses a 1.544-MHz clock to multiplex 24 PCM lines on a single line to be transmitted to a central office. At the central office many of these 24 channel lines are multiplexed onto a single channel to be sent long distances over microwave links, satellite links, or fiber-optic cables.

Digital voice signals come back into the SLIC on the line labeled RECEIVE PCM. The CODEC in the SLIC converts the digital signal to an analog signal which is filtered and sent out on the RING and TIP lines to your phone.

From the preceding discussion, you can see that the major function of the CODEC is to convert the analog signal to a sequence of PCM codes to go out on the transmission line and convert the incoming PCM codes to equivalent analog values. To get a better understanding of how the CODEC works, let's take a look at the block diagram of the Texas Instruments TCM29C13 CODEC and filter device in Figure 11-46.

The successive approximation register, comparator, and D/A converter (DAC) in the upper half of Figure 11-46 should tell you that the converter uses a successive approximation technique to produce an 8-bit serial data word for each sample taken by the sample-and-hold circuit. The usual sampling rate used by this device is 8 kHz, which is more than twice the highest frequency in the 300-Hz to 3-kHz range allowed for analog phone signals. To make sure that no alias frequencies are produced, an antialiasing filter is put in front of the A/D converter's sample and hold. The decoder section of the CODEC in Figure 11-46 uses a

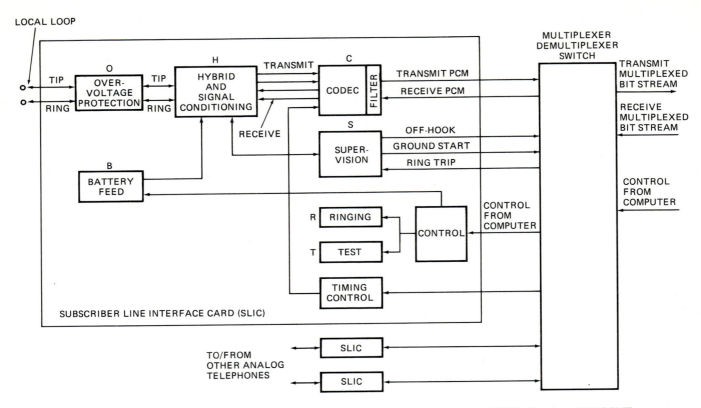

FIGURE 11-45 Block diagram for telephone system subscriber line interface circuit (SLIC) showing BORSCHT functions. *(Permission Howard W. Sams & Co., Indianapolis. Understanding Telephone Electronics, by Fike & Friend,* © *1983, 1984.)*

D/A converter, output sample and hold, and low-pass filter to regenerate the analog output signal.

The question that may have occurred to you is, "Why don't we just refer to the CODEC as an A/D-D/A, instead of giving it a special name?" To answer this question, we have to discuss the code produced by the A/D converter for each sample of the input waveform. For most applications, you want an A/D to be linear in its response. The A/D and D/A in a CODEC, however, are deliberately designed with nonlinear responses so that 8 bits can represent signals over a wide dynamic range. Remember that the A/D converter in a digital audio system requires 16 bits to accurately reproduce the wide dynamic range of signals encountered in music. Here's how a CODEC produces its unique output codes.

The A/D in the CODEC is set up to digitize both negative and positive signals. Figure 11-47a shows the format for the 8-bit code produced by the A/D. The MSB of this code is used to show whether the sample is positive or negative. A 1 in this bit indicates the sample is positive, and a 0 in this bit indicates that the sample is negative. The other 7 bits in the code are used to express the size or magnitude of the signal in a unique way. Figure 11-47b shows the output codes that will be produced by positive signals applied to the input of a 29C13 CODEC operating in what is called μlaw *mode*.

To determine the magnitude of an input signal the A/D converter divides the input range up into 16 segments, or *chords*, 8 positive and 8 negative. As we said before, bit 7 of the data word indicates whether the chord is negative or positive. Bits 6, 5, and 4 in the data word are used to represent the number of the chord. Each chord is further divided into 16 *intervals*, as shown in Figure 11-47b. Bits 3, 2, 1, and 0 of the output data word are used to represent the number of the interval within a chord. If you look closely at Figure 11-47b, you should see that only 16 codes are used to represent positive input voltages between full scale and ½ full scale, 16 codes are used to represent input voltages between ½ and ¼, and 16 codes are used to represent input voltages between ¼ and ⅛ of full scale, etc. The result of this nonlinear weighting scheme is that the resolution of the converter is finer for small signals than it is for large signals. This means that it can digitize both small signals and large signals with resolution sufficient for phone signals. This scheme wouldn't work for digital audio because the fine detail in large-amplitude waveforms would be lost. Incidentally, most systems in the United States and Japan use μlaw CODECs, but European systems more commonly use an A-law CODEC, which has a slightly different transfer curve. Some devices such as the 29C13 can be used for either mode, depending on the logic level applied to a pin on the device.

The D/A converter in the CODEC is designed with a complementary nonlinear response so that it accurately reproduces the original signal from the received pulse codes. The name CODEC then tells you that the A/D and D/A are working with the unique codes of PCM, rather than standard binary or BCD.

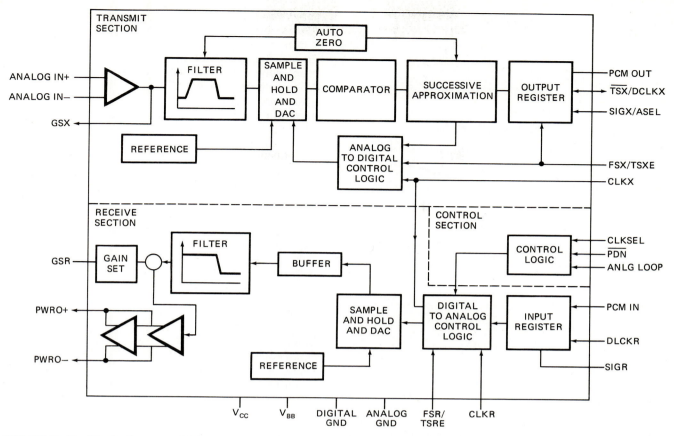

FIGURE 11-46 Block diagram of TCM29C13 synchronous CODEC and filter device for transmitting and receiving PCM signals. *(Courtesy of Texas Instruments Inc.)*

As the discussion of modems in Chapter 10 and the discussion of SLICs here should show you, the major roadblock to high-speed digital communication over telephone networks is the outmoded analog phone connections between your home and the local branch office. Perhaps, the integrated services digital network (ISDN) and/or fiber-optic technology will eventually remove this roadblock.

CHECKLIST OF IMPORTANT POINTS AND CONCEPTS IN THIS CHAPTER

Op amp
Comparator
Hysteresis
Noninverting amplifier
Negative feedback
Closed-loop gain
Buffer
Input impedance
Inverting amplifier
Virtual ground
Gain-bandwidth product, unity-gain bandwidth
Op-amp adder circuit
Summing point
Differential-input amplifier
Common-mode voltage
Instrumentation amplifier

Integrator, ramp generator, linear ramp
Light sensor, photoresistor
Reverse bias
Solar cell
Temperature-sensitive voltage source
Temperature-sensitive current source
Thermocouple
Cold-junction compensation
Strain gage
Load cell
Flow sensor
Differential-pressure transducer
Monolithic, hybrid
D/A converter
   Resolution
   Linearity error
   Monotonic
   Gain, scale factor error
   Offset error
   Output settling time
Sample-and-hold circuit
   Acquisition and aperture time
A/D converter
   Resolution, conversion time
   Flash A/D
   Single-slope A/D

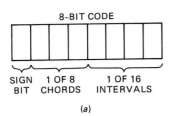

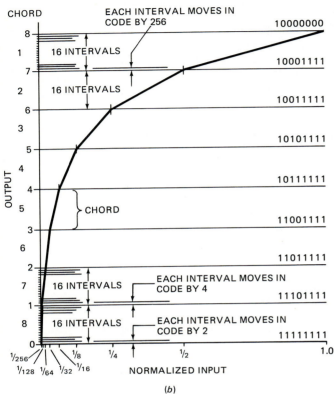

FIGURE 11-47 (a) Data word bit format for 29C13 CODEC. (b) PCM output words produced by 29C13 CODEC for positive input signals. (Permission Howard W. Sams & Co., Indianapolis. Understanding Telephone Electronics by Fike & Friend © 1983, 1984.)

Dual-slope A/D
Charge balance A/D
Counter type A/D
Successive approximation A/D
Successive approximation register, SAR
Voltage-to-frequency A/D
Quantizing error
Dynamic range
Sampling rate and aliasing
Alias frequency
Digital voltmeter, DVM
Digital multimeter, DMM
Digital panel meter
Attenuator
Digital storage scope, DSO
   Free-run mode
   One-shot mode
   Scan mode
   Roll mode
   Sampling scope mode
Data acquisition system
Compact audio disk, CD
Digital audio tape, DAT
Dynamic range
Digital voice communications
CODEC, coder-decoder
Subscriber line interface circuit, SLIC
Pulse code modulation (PCM)

REVIEW QUESTIONS AND PROBLEMS

1.   a. About what voltage will be present on the output of the op-amp circuit in Figure 11-48a?
    b. About what voltage will be present on the output of the comparator circuit in Figure 11-48b?

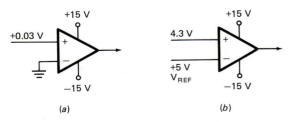

FIGURE 11-48 (a) Circuit for Problem 1a. (b) Circuit for Problem 1b.

2.   a. For the circuit in Figure 11-49a, calculate the closed-loop voltage gain and the voltage that will be produced on the output by the input voltage shown. What voltage would you measure on the inverting input of the amplifier with the input signal shown? What voltage gain would this circuit have if R2 is shorted out?
    b. For the circuit in Figure 11-49b, calculate the closed-loop voltage gain and the output voltage that will be produced by an input voltage of +0.56 V. What voltage should you always measure on the inverting input of this circuit?

3.   a. Calculate the voltage produced on the output of the differential amplifier in Figure 11-50 by the input voltages shown.

b. If the amplifier in the circuit in Figure 11-50 has a gain-bandwidth product of 1 MHz, what will be the closed-loop bandwidth of the circuit?

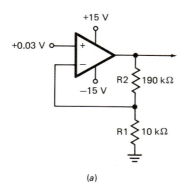

(a)

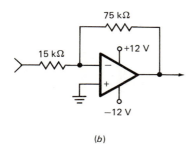

(b)

FIGURE 11-49   (a) Circuit for Problem 2a. (b) Circuit for Problem 2b.

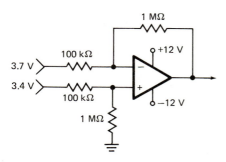

FIGURE 11-50   Circuit for Problem 3.

4. Draw a circuit showing how a light-dependent resistor can be connected to a comparator so that the output of the comparator changes state when the resistance of the LDR is 10 kΩ.

5. For the photodiode amplifier circuit in Figure 11-6a, what voltage will you measure on the inverting input of the amplifier? Why is it important to use an FET input amplifier for this circuit? Which direction are electrons flowing through the photodiode?

6. In what application might you use a temperature-dependent current device such as the AD590 rather than a temperature-dependent voltage device such as the LM35?

7. Why must thermocouples be cold-junction-compensated in order to make accurate measurements? How can the nonlinearity of a thermocouple be compensated for?

8. *a.* Why are strain gages usually connected in a bridge configuration?
   b. Why do you use a differential amplifier to amplify the signal from a strain-gage bridge?

9. *a.* Figure 11-51 shows a circuit for a 6-bit weighted resistor D/A converter. Assuming the resistor for the MSB has a value of 50 kΩ, calculate the values for the other five input resistors.
   b. How many discrete output steps does the 6-bit converter in part *a* have?
   c. Assuming an input is connected to +5 V when on, what output voltage will be produced by just the LSB?
   d. What will be the full-scale output voltage for this D/A converter?
   e. Sketch the output waveform that will be produced by this converter if a 6-bit binary CMOS counter is connected to the inputs. (You do not need to show all the steps.)

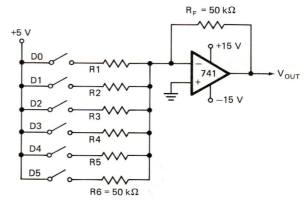

FIGURE 11-51   Circuit for 6-bit weighted resistor D/A converter in Problem 9.

10. *a.* What is the purpose of the op-amp circuit connected to the output of the MC1408 D/A converter in Figure 11-16a.
    b. Calculate the voltage that will be produced on the output of the 8-bit D/A converter circuit in Figure 11-16a if the binary word 10110101 is applied to the inputs.
    c. What voltage will be present on the inverting input of the op amp with this binary word on the inputs of the D/A?
    d. If the inputs of the D/A converter are being driven with an 8-bit binary counter and the feedback resistor on the op amp accidentally becomes adjusted to a value of 10 kΩ, what symptom(s) will you observe in the operation of the circuit?

11. a. What is the resolution of a 13-bit D/A converter?
    b. If the converter has a full-scale output of 10.000 V, how many volts is each step?
    c. What voltage should the output be adjusted to when the input data bits are all 1's?
    d. What accuracy should this converter have to be consistent with its resolution?

12. Assuming the inputs of a D/A converter are connected to a binary counter, draw a sketch which shows how each of the following sources of error will affect the output waveform.
    a. Linearity
    b. Gain error
    c. Offset error

13. a. What accuracy is required from a 10-V, 12-bit D/A converter to be consistent with its resolution?
    b. Why do manufacturers commonly make available D/A converters with linearity specifications less than that required to be consistent with their resolution?

14. a. Draw a waveform which shows what is meant by the term "settling time" for a D/A converter.
    b. What symptom will you see in a system with a D/A converter which has too long a settling time for the rate at which it is being operated?

15. Figure 11-20b shows a sample-and-hold circuit. Describe the operation of this circuit and explain why a circuit such as this is often connected on the output of a D/A converter.

16. Figure 11-52 shows a circuit for a 3-bit parallel comparator A/D converter.
    a. Describe the operation of this converter and explain why it is commonly called a "flash" converter.
    b. Determine the voltage that will be present on each of the comparator thresholds.
    c. What is the full-scale voltage for this converter?
    d. Show the logic levels that will be present on the outputs of the comparators if a 3.2-V signal is applied to the inputs of the circuit.
    e. What is the resolution of this converter?
    f. Search the earlier chapters of this book or a TTL data book to see if you can find a single commonly available device which will encode the signals on the outputs of the comparators to an equivalent 3-bit binary code.
    g. How many comparators are required to make a 6-bit flash converter?

17. Draw a block diagram of a single-ramp A/D converter and briefly describe its operation. What is a disadvantage of single-ramp types?

18. a. Draw a block diagram of a dual-slope A/D converter and briefly describe its operation.
    b. Explain the terms *reference integrate* and *signal integrate*.
    c. How does the dual-slope A/D converter overcome the main problem of the single-slope type?

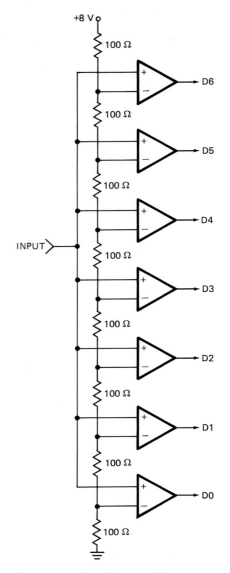

FIGURE 11-52 Circuit for 3-bit parallel comparator A/D converter for Problem 16.

19. a. For the integrator in Figure 11-23a, what slope will the output have for an input voltage of 1.5 V?
    b. What negative voltage will the output reach after 1000 counts of the 1-MHz clock?
    c. How long will it take to digitize the 1.5-V input with the circuit in Figure 11-23a?

20. Using a 100-kHz input clock, how long will it take the 8-bit counter type A/D converter in Figure 11-24 to digitize a full-scale signal?

21. a. Describe the operation of an 8-bit successive approximation type A/D converter such as the one in Figure 11-25.
    b. How many clock pulses are required for each conversion cycle an 8-bit successive-approximation type of A/D converter?
    c. With a 100-kHz clock, how long will a conversion take?

d. Compare this time with that for the 8-bit counter type.

22. For the successive approximation A/D converter in Figure 11-25:
    a. What is the purpose of the latches on the outputs of the SAR?
    b. What is the purpose of the 10-kΩ resistor on the clock input of the SAR?
    c. Draw the waveform that you would see on the output of the D/A op amp if 3.240 V is applied to the input of the converter.
    d. What symptom would you observe if, due to a static discharge, the D6 bit of the SAR was stuck high?

23. a. Define the term *quantizing error* that is often used in A/D data sheets and give an example of how quantizing error is expressed.
    b. What is the quantizing error for the 3-bit flash A/D converter in Figure 11-52?
    c. How could the circuit be modified so that the quantizing error is equally distributed around the comparator thresholds?

24. a. Why is it necessary to use a sample-and-hold circuit when you are digitizing rapidly changing signals?
    b. Define acquisition time and aperture time for a sample-and-hold circuit.

25. a. According to the sampling theorem, what is the minimum rate that a sine-wave signal must be sampled with an A/D if it is to be later reconstructed with a D/A?
    b. What effect will be produced by sampling an input waveform at a rate less than the minimum specified by the sampling theorem?

26. What is meant by the $\frac{1}{2}$ in the term $4\frac{1}{2}$-digit DVM?

27. Show the schematic for an attenuator network that might be used on the input of an A/D converter to extend its voltage range.

28. a. Figure 11-35 shows a circuit which enables an A/D converter to function as an ohmmeter. Describe how this circuit produces a voltage which is proportional to resistance connected between its terminals.
    b. Figure 11-36 shows a circuit which enables an A/D converter to function as an ammeter. Describe the operation of this circuit. As part of your discussion, explain the advantage of having the op amp in the circuit.

29. a. Describe how samples of an input waveform are taken, stored, and displayed in a DSO.
    b. How does a DSO make it possible to observe very slowly changing signals instead of trying to follow a bright dot moving slowly across the screen?
    c. How is it possible for a DSO to display events which occurred before a specified trigger.
    d. What DSO recording mode would you use to display a trace of a laser pulse which occurs only once?
    e. What DSO recording mode would you use to see a continuous display of a 2-Hz electrocardiograph (ECG) waveform?

30. a. Describe how a DSO makes it possible to see a 2.2-GHz pulse train when operated in "sampling scope mode."
    b. What is the major requirement of a signal so that you can see a trace of it in the sampling scope mode?

31. a. Draw a block diagram showing the major parts of an IC data acquisition system.
    b. What are the advantages of putting a data acquisition device on a PC board so that it interfaces with a microcomputer?

32. a. Why do digital audio systems use 16-bit A/D and D/A converters instead of less expensive converters with fewer bits of resolution?
    b. At about what sampling rates are the A/D converters in CD and DAT recorders operated? Why can't the converters be operated at lower frequencies?
    c. How do digital recorders prevent the introduction of alias frequencies into the output signal?
    d. Why are the digital words for the left channel signal multiplexed on a common line with the digital words for the right channel signal?
    e. What is the purpose of the CIRC bits which are added in with the data bits before they are recorded on a CD or DAT?

33. a. Describe how data bits are read off a CD and converted to the left and right channel audio signals.
    b. Describe the major advantages of digital audio recording over traditional analog methods such as LP records and magnetic tapes.

34. a. Describe the purpose of an SLIC in a telephone system.
    b. Define the term *CODEC*.
    c. Define the term *pulse code modulation*, or *PCM*.
    d. Why do the A/D and D/A converters in a CODEC use a nonlinear conversion scheme?
    e. What is the major advantage of sending telephone signals in digital form?

12 Programmable Logic Devices and State Machines

As we told you in Chapter 7, programmable logic devices such as FPLAs, PROMs, and PALs are now used to implement most of the "glue" logic between the large LSI devices in a system. The purpose of this chapter is to teach you the methods and tools used to program and test these devices.

In the first section of this chapter we show you how to use a microcomputer to program PALs, PROMs, and FPLAs for implementing combinational logic functions.

In the second section of this chapter we show you how to design synchronous circuits, or *state machines*, such as those we taught you to analyze in Chapter 9.

Finally, in the third section of this chapter we show you how to use a computer program to implement state machines with "register output" PALs.

OBJECTIVES

At the conclusion of this chapter you should be able to:

1. Describe the differences between a PAL, a PROM, and an FPLA.

2. Construct the fuse map required to implement a simple combinational logic function with a PAL, PROM, or FPLA.

3. Describe how a computer-based tool such as ABEL is used to automatically develop the fusemap for a programmable device.

4. Write test vectors for a PAL which implements a combinational logic function.

5. Describe how a programmable device is programmed and tested.

6. Draw a state diagram or an ASM chart for a simple state machine.

7. Draw block diagrams for a Mealy type state machine and for a Moore type state machine and explain the difference between the two types.

8. Design the next-state decoder for a simple state machine using D or JK flip-flops.

9. Describe how a computer-based tool such as ABEL can be used to develop the fusemap which implements a state machine in a register output PAL.

10. Write test vectors to determine if a PAL-based state machine follows the desired sequence of steps.

PROGRAMMABLE LOGIC DEVICES FOR COMBINATIONAL LOGIC FUNCTIONS

Review of Programmable Logic Devices

To refresh your memory, Figure 12-1 shows the basic structures of PLA (programmable logic array), PROM (programmable read-only memory), and PAL (programmable array logic) devices. As we discussed in Chapter 7, all these devices function as an AND array followed by an OR array. Therefore, each type of device can easily implement sum-of-product logic expressions such as you might derive from a Karnaugh map. If you remember that X's represent programmable connections and dots represent fixed connections in logic diagrams such as those in Figure 12-1, you can quickly see the difference between the structures of these three devices. In a PLA or an FPLA, both the AND array and the OR array are programmable. In a PROM the AND array has fixed connections, but the OR array is programmable. In a PAL the AND array is programmable, but the OR array has fixed connections. Now let's review some of the other characteristics of these devices.

A *PLA* is programmed during manufacture by putting in or leaving out connections in the masks. An *FPLA* is manufactured with fuses on the inputs of the AND gates and fuses on the inputs of the OR gates. You custom program the FPLA for a desired sum-of-products expression by blowing fuses on the lines that represent terms you don't want to be present in the output expression. Figure 12-2a shows a very small FPLA device programmed to implement some two-variable logic expressions. The X's in Figure 12-2 represent fuses left intact so that a connection remains between the intersecting wires. If no X is present, there is no connection between the intersecting wires.

FPLAs are very flexible, because both the AND matrix

360

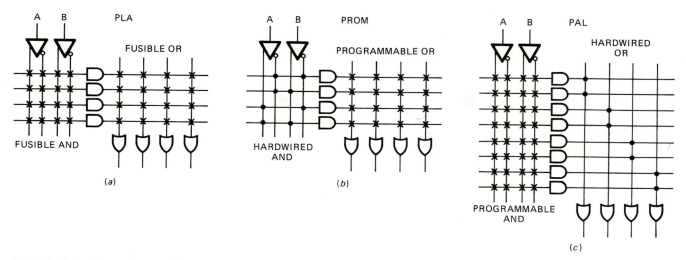

FIGURE 12-1   Comparison of basic structures of FPLA, PROM, and PAL. *(a)* FPLA. *(b)* PROM. *(c)* PAL.

and the OR matrix are programmable. The price of this flexibility is greater difficulty in programming. The AND matrix and the OR matrix must be separately programmed and tested. We use FPLAs mostly in applications where we have many input variables, but relatively few combinations of these variables.

As we described in Chapter 7, ROMs, PROMs, EPROMs, and EEPROMs can be used to implement combinational logic functions or to store a sequence of binary data words. *ROMs* are programmed during manufacture and cannot be reprogrammed. *PROMs* are programmed by blowing fuses and cannot be reprogrammed. *EPROMs* (erasable PROM) are programmed by storing charges in insulated regions in the device and can be erased by shining ultraviolet light through a quartz window in the top of the package. Once erased, they can be reprogrammed. *EEPROMs* (electrically erasable PROM) are programmed by storing charges in insulated regions in the device and can be erased by a sequence of pulse signals. Once erased, they can be reprogrammed.

The fixed connections on the AND matrix in ROMs and PROMs produce product terms for each combination of the input variables. Figure 12-2*b* shows how some simple two-variable logic functions can be implemented with a PROM. ROMs and PROMs are useful for applications such as ASCII-to-EBCDIC code converters that require an output for each of the possible combinations of the input variables. For applications where all the possible product terms are not needed, ROMs are not very efficient, so for these applications we usually use a PAL.

As shown in Figure 12-1*c*, the AND array in a PAL is programmable, but the OR array is hardwired. To refresh your memory, Figure 12-2*c* shows how you can implement some two-variable sum-of-product logic expressions with a PAL. Note that in a PAL the number of product terms ORed together to produce an output expression is limited by the number of inputs on each OR gate. This just means that when choosing a PAL to implement a specified logic expression, you have to choose one that has enough OR inputs to sum all the product terms in the expression.

PAL devices such as the PAL16P8A shown in Figure 7-40 are programmed by blowing fuses. The problem with this type device is that it cannot be reprogrammed if you make a mistake, and obviously the manufacturer cannot test the device to see if the fuses blow out properly. To solve this problem, several companies now produce PAL-type devices which can be erased and reprogrammed. To distinguish these type devices from the fuse-type PALs they are called *erasable programmable logic devices,* or *EPLDs.* Some EPLDs, such as the Altera and Intel EP1800 devices, use a technology similar to that in EPROMs to make them reprogrammable. As with EPROMs, these CMOS devices are erased by shining an ultraviolet light through a quartz window in the top of the package. The erasable version of the device can be used for prototyping, and then a cheaper, nonerasable version of the device can be used for production. Other available EPLDs such as the Lattice Semiconductor, National Semiconductor, and VLSI Technology GAL16V8 devices use a technology similar to that in EEPROMs to make them reprogrammable. These CMOS devices are erased electrically during the reprogramming process, so they are very easy to use for experimentation and prototyping.

For a variety of reasons PALs are currently the most commonly used devices for implementing logic functions which require the equivalent of up to a few hundred gates. Therefore, we will use a large part of this chapter to show you the programming and applications of the various types of PALs.

PAL Output Types and Numbering Conventions

PAL and PLD devices are available with several different types of output structures. The output may, for example, be active high, active low, or programmable as either active high or active low.

Unlike the numbering system used for standard logic devices, the numbering system used for most PALs and EPLDs tells you a considerable amount about the characteristics of the device. Figure 12-3 shows the naming

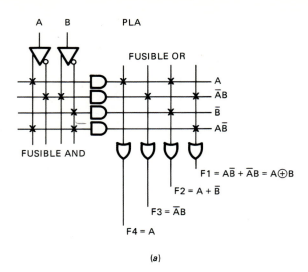

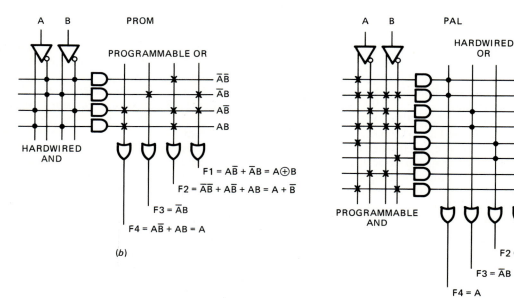

FIGURE 12-2   FPLA, PROM, and PAL programmed to implement some simple logic functions. (a) FPLA. (b) PROM. (c) PAL.

conventions which are generally followed throughout the industry to identify PAL-type devices.

The first number in the name indicates the maximum number of inputs to the device. The 16 in the name PAL16P8A, for example, indicates that the device has 16 data inputs. As you can see in Figure 7-40, the 16 data inputs produce 32 possible inputs to the AND array, since both the inverted and the true form of each input signal are generated by the input buffers. The next number in a PAL device name shows the maximum number of outputs from the device. The 8 in PAL16P8A, for example, shows that the device has a maximum of eight outputs. As we pointed out in Chapter 7, the maximum number of inputs plus the maximum number of outputs may be greater than the total number of pins on the device. This is possible because some pins such as

pins 13 through 19 on the PAL16P8A in Figure 7-40, for example, can be used both as inputs and as outputs.

The letter in the center of a PAL name indicates the type of output structure it has. Figure 12-3 lists the common types and the letter used to represent each. For reference and to help you understand the meaning of these letters, Figure 12-4 shows logic diagrams for some of the major output types.

Figure 12-4a shows a simple active high output. Note that the output buffer is enabled by a product term from the AND matrix. This means that you can enable the output buffer with a single input signal or with a function of several input variables.

The active low output of the buffer in the output of the PAL in Figure 12-4b will be asserted when the sum-of-products logic expression is high. Note that this out-

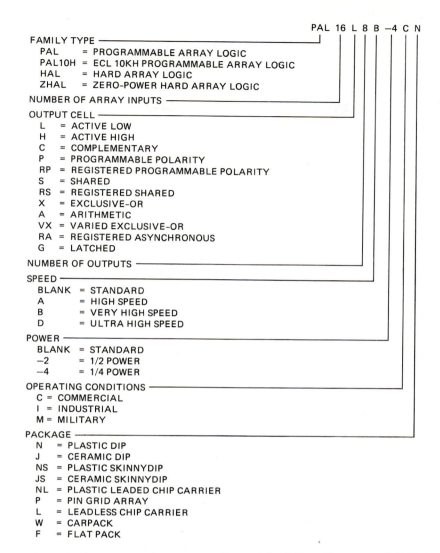

FAMILY TYPE
  PAL   = PROGRAMMABLE ARRAY LOGIC
  PAL10H = ECL 10KH PROGRAMMABLE ARRAY LOGIC
  HAL   = HARD ARRAY LOGIC
  ZHAL   = ZERO-POWER HARD ARRAY LOGIC
NUMBER OF ARRAY INPUTS
OUTPUT CELL
  L   = ACTIVE LOW
  H   = ACTIVE HIGH
  C   = COMPLEMENTARY
  P   = PROGRAMMABLE POLARITY
  RP = REGISTERED PROGRAMMABLE POLARITY
  S   = SHARED
  RS = REGISTERED SHARED
  X   = EXCLUSIVE-OR
  A   = ARITHMETIC
  VX = VARIED EXCLUSIVE-OR
  RA = REGISTERED ASYNCHRONOUS
  G   = LATCHED
NUMBER OF OUTPUTS
SPEED
  BLANK = STANDARD
  A   = HIGH SPEED
  B   = VERY HIGH SPEED
  D   = ULTRA HIGH SPEED
POWER
  BLANK = STANDARD
  −2   = 1/2 POWER
  −4   = 1/4 POWER
OPERATING CONDITIONS
  C = COMMERCIAL
  I = INDUSTRIAL
  M = MILITARY
PACKAGE
  N = PLASTIC DIP
  J = CERAMIC DIP
  NS = PLASTIC SKINNYDIP
  JS = CERAMIC SKINNYDIP
  NL = PLASTIC LEADED CHIP CARRIER
  P = PIN GRID ARRAY
  L = LEADLESS CHIP CARRIER
  W = CARPACK
  F = FLAT PACK

FIGURE 12-3   Naming and numbering convention commonly used for PALs. *(Courtesy of AMD/Monolithic Memories.)*

put buffer is likewise enabled by a product term from the AND matrix. Also note the active high output shown in Figure 12-4*a* and the active low output shown in Figure 12-4*b* are both fed back into the AND matrix. This is done so that one sum-of-products output expression can be used as one of the input terms for a more complex expression.

The output in Figure 12-4*c* can be programmed to be active high or active low, so this type output is represented with a P. The PAL16P8A in Figure 7-40 is an example of this type device.

The Exclusive OR output shown in Figure 12-4*d* produces an output which is the XOR of two sum-of-product expressions. In this particular example, the XOR output is also passed though a D flip-flop, so the output would be called a *Registered XOR output.* This output structure is useful for implementing some counters and digital arithmetic functions.

The simple Registered output shown in Figure 12-4*e* has a D flip-flop and a buffer between the OR output and the pin of the device. For this type output, the buffer is usually enabled by a single line connected to a pin on the device. A CLOCK pulse is required to transfer data from the OR output to the output of the D flip-flop. Later in the chapter, we show you how to implement synchronous circuits such as counters and shift registers with this type of PAL.

The Versatile, or V, type PAL output uses an output macrocell to make the output type programmable in several different ways. As an example of this type, Figure 12-5 shows one of the output macrocells from a Lattice Semiconductor Corporation GAL16V8.

The XOR gate in the center of the macrocell allows the output to be programmed as active high or active low. The OMUX in the macrocell allows the output to be programmed as purely combinational, or registered. The FMUX allows either the output from the register, the output from the buffer, or the output from an adjacent pin to be fed back into the AND matrix. The TSMUX at the at the top of Figure 12-5 allows the output buffer to be enabled either by a product term from the AND matrix or by a separate OE input pin. If this product term is not being used to enable the output, the PTMUX can be programmed to direct it into the output OR gate. This

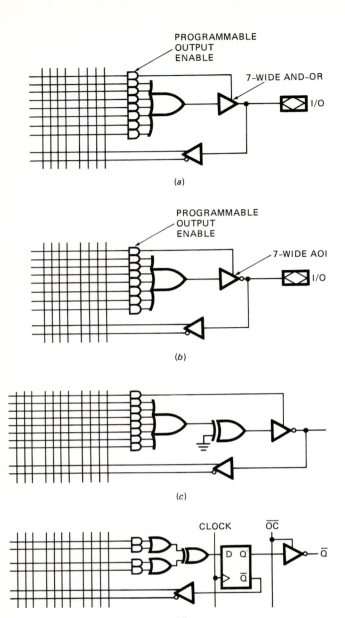

FIGURE 12-4 Logic diagrams for common types of PAL outputs. *(a)* Active high. *(b)* Active low. *(c)* Programmable as active low or active high. *(d)* Exclusive OR. *(e)* Register output.

gives the output OR an additional product term if needed. Because it is erasable and because its outputs can be programmed in all these different ways, we use the GAL16V8 for the application examples in this chapter.

The final part of the PAL numbering convention we want to comment on is the speed/power designation. For a given PAL device, manufacturers often offer versions with different speed/power tradeoffs. The letters and numbers in the indicated part of the device number tells the tradeoff for that particular device. For comparison, a PAL16L8A has a t_{PD} of 25 ns and a maximum current of 180 mA, a 16L8A-2 has a t_{PD} of 35 ns and a maximum current of 90 mA, and a 16L8A-4 has a t_{PD} of 55 ns and a maximum current of 50 mA. Incidentally, the fastest currently available PALs have a t_{PD} of about 7.5 ns.

Now that we have told you a little more about the basics of PALs, the next step is to show you how you program them to implement combinational logic functions.

Overview of the PLD Programming Process

In the early days of PLDs, we used to develop the fuse map for an FPLA or PROM by hand on a large sheet of paper and send this paper off to the manufacturer to be programmed into devices. Now we generate the fusemap or other programming data for a PLD with a microcomputer and download the data through a cable to a programmer such as the Data I/O model 60A shown in Figure 12-6. The programmer uses the downloaded data to automatically program the device and test to see that it is programmed correctly. Universal programmers such as the model 60A can program most common FPLAs, PROMs, or PALs, once you develop and download the appropriate programming information. In this chapter we show you how to develop the programming and test data for PALs, because they are best suited to the types of problems we want to show you how to solve. To develop the programming and test data for other devices such as EPROMs and FPLAs, you simply use a different development program.

There are several commonly available programs you can use to develop the fusemap and test information for PALs. Examples are PALASM from AMD/Monolithic Memories, CUPL from Logical Devices Incorporated, A+PLUS from Altera Corporation, and ABEL from Data I/O Corporation. All these programs have versions which will run on IBM PC type microcomputers. For the examples in this chapter, we have chosen to use ABEL version 3.00A because it is easy to use and has some very helpful features. If you are going to be working with one of the other programs, you should have little difficulty adapting most of our discussion to it because the programs are similar. Before we work through some specific examples of using ABEL, here's an overview of the development process.

As always, the first step in the design process for a combinational logic circuit is to carefully define the problem you want to solve and write a truth table which expresses the desired relationships between input signals and output signals.

Based on the number of input variables and the number of output functions, you then choose a PAL device which has the appropriate internal architecture to implement the desired functions.

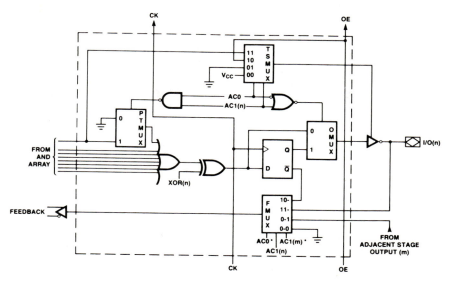

FIGURE 12-5 Output macrocell from Generic Array Logic device GAL16V8. The macrocell allows the device to be programmed for almost any type output. *(Courtesy of Lattice Semiconductor Corp.)*

FIGURE 12-6 Data I/O programmer for FPLAs, PROMs, and PALs.

The next step in the development process is to use a text editor program to write a *source file*, which describes the logic functions you want the programmed PAL to perform. This step is analogous to developing the source file for a Pascal or C program. You can use Wordstar, PC write, or some other editor in the "nondocument" mode to generate this source file. To distinguish it from other files, the extension ".ABL" is always given to an ABEL source file. As an example of this, we used BASICGTS.ABL as the name of the ABEL source file for the example in the following section. Along with the logic functions you want the PAL to perform, you also include *test vectors* in the source file. These test vectors are used to see if the PAL fusemap generated by ABEL correctly implements the desired logic functions. They are also used by the programmer to determine if the programmed PAL works correctly.

After you develop the source file, you process the file with the ABEL language processor program, ABEL.BAT. Figure 12-7 shows the sequence of six subprograms which ABEL.BAT runs to process the source file and also shows the four files generated by these programs.

The PARSE program reads the source file you wrote and checks for any illegal statements or other syntax errors. If the PARSE program finds syntax errors in your source file, it will display an error message on the screen, generate a list file which identifies the errors, and return control to DOS. You can print out the list file to determine the errors and then use your editor to correct the errors. After you correct the errors, you run ABEL again. When the PARSE program finds no syntax errors in the source file, ABEL generates the list file and automatically runs the TRANSFOR subprogram.

TRANSFOR translates Boolean expressions, truth tables, and other statements in the source file to forms that can be used by the REDUCE subprogram. When TRANSFOR is complete, ABEL automatically runs the REDUCE subprogram.

The REDUCE subprogram uses a Boolean algebra technique to reduce logic expressions in the same way that you have used Boolean algebra and Karnaugh maps to reduce logic expressions. When REDUCE finishes, ABEL automatically runs the FUSEMAP subprogram.

As the name implies, FUSEMAP generates the fusemap and test data that will be downloaded to the programmer to actually program and test the PAL. After ABEL runs FUSEMAP, it next runs the SIMULATE subprogram.

SIMULATE tests to see if the fusemap generated by the FUSEMAP program correctly implements the Boolean expression or truth table you specified in your source file. To do this test, SIMULATE uses test vectors which you included in your source file. If the fusemap fails simulation, SIMULATE displays an error message and generates a file which shows where the errors occurred during simulation. You can print out the simulation file to identify the errors, edit the source file to fix the errors, and then run ABEL on the source file again. After running SIMULATE or after finding an error when it ran one of the previous programs, ABEL runs the DOCUMENT subprogram.

With its default values, DOCUMENT produces a file

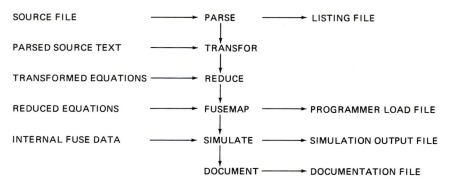

which contains a listing of the reduced equations and a pin diagram for the PAL showing the signal connections to each pin. If desired, the ABEL.BAT file can be modified so that DOCUMENT also prints out the transformed equations, the fusemap, and the test vectors. When DOCUMENT finishes, ABEL returns control to DOS.

After ABEL successfully runs through all its subprograms, you download the fusemap file to a programmer. You then connect an antistatic strap to your wrist, remove the PAL from its protective foam or tube, and insert the PAL in the programmer socket. Finally, you tell the programmer to program the device. When it is done programming and testing the programmed PAL, the programmer displays a PASS or FAIL message. If the PAL passes, you can be remove it from the programmer and see if it works in the prototype circuit.

NOTE: Most EPLDs are CMOS, so you should make sure you use an antistatic strap and turn off the power on the prototype before inserting or removing any of these devices.

Now that we have given you an overview of the PAL development process, we will show you a couple of specific examples. Keep in mind that the point of these examples is not to make you an expert with ABEL, but to show you how easy it is to implement almost any combinational logic circuit with a PAL and computer-based tools such as ABEL.

Implementing Basic Logic Gates in a PAL

DESIGN PROBLEM

Whenever you are first learning a new tool or test instrument, it is best to start with the simplest possible example so that you can concentrate on the basic operation of the tool rather than getting lost in the complexity of the example. Therefore, to introduce you to the format and syntax of ABEL source files, we will show you how to implement some simple 2- and 3-input gates in a PAL. The specific device we are using for this example is the Lattice GAL16V8 shown in Figure 12-8. This device is much more complex than needed for this simple example, but as we said before, it is erasable and can be used for the more complex examples we show you later.

The four logic functions we want to implement for this first example are:

$$R = AB$$

$$S = C + D$$

$$T = \overline{ABE}$$

$$U = \overline{GHL} + \overline{GH}\overline{L} + \overline{GH}\overline{L}$$

ABEL SOURCE FILE FOR BASIC GATES

Figure 12-9 shows an ABEL source file which will implement and test the four desired logic functions. Quickly read through this source file to see how much you can intuitively figure out before you read through the detailed explanation here. As you read through the file, you might highlight the key words MODULE, FLAG, TITLE, DEVICE, PIN, EQUATIONS, TEST VECTORS, and END.

The MODULE and END labels mark off a section of instructions and give a name to that section. This is similar to the use of BEGIN and END in a Pascal program or parentheses in an algebra expression.

The FLAG statement, which must come immediately after the MODULE statement, is used to tell ABEL any special processing instructions you want done. In this example we are using the FLAG statement to tell ABEL to use reduction level 1 when it reduces the Boolean expressions. A reduction level of 0 represents no reduction, and a reduction level of 3 represents the greatest amount reduction. As we explained in a discussion of hazards at the beginning of Chapter 8, reducing a combinational logic expression to its absolute minimum may give you a circuit which produces glitches for certain input signal transitions. Usually we try running ABEL on the module with a reduction level of 1, and if the resultant expressions are too complex to be implemented in the PAL, we change the reduction level to 2 or 3 and run ABEL on the module again.

The TITLE statement in Figure 12-9 is optional, but highly recommended. The text of the title can extend for more than one line. A single quotation mark is used to indicate the start of the title, and another single quotation mark is used to indicate the end of the title text.

The DEVICE statement gives a name to the PAL so you can identify it on a schematic and tells ABEL the device characteristics file to use for processing this module. When you purchase ABEL, you get a library of device files. Each file contains the characteristics of a specific programmable logic device. For the GAL16V8, ABEL has three device files which are identified as

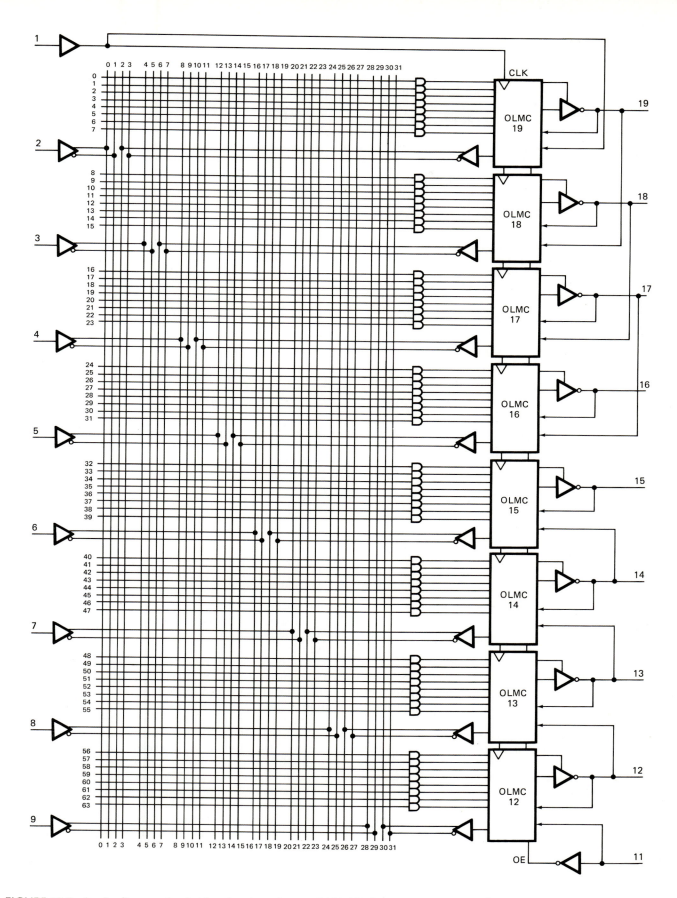

FIGURE 12-8 Logic diagram for Lattice Semiconductor GAL16V8 Generic Array Logic device.

```
MODULE BASICGTS
FLAG '-R1'
TITLE 'Implementing basic logic gates with ABEL and a GAL16V8 EPLD'
        U1   DEVICE  'P16V8S'; "S device file allows pin 12 as input
      A,B,C,D,E,F,G,H,L  PIN   2,3,4,5,6,7,8,9,12;
                R,S,T,U  PIN  19,18,17,16;
EQUATIONS
        R = A & B; " 2-input AND
        S = C # D; " 2-input OR
        !T = A & B & E; "3 input NAND - alternative is T = !(A&B&E)
        U = (!G & H & !L) # (G & !H & !L) # (G & H & !L); "sum-of-products

TEST_VECTORS  " For AND gate"
        ([A,B] -> R)
        [0,0] -> 0;
        [0,1] -> 0;
        [1,0] -> 0;
        [1,1] -> 1;
TEST_VECTORS " For OR gate
        ([C,D] -> S)
        [0,0] -> 0;
        [0,1] -> 1;
        [1,0] -> 1;
        [1,1] -> 1;
TEST_VECTORS " For 3-input NAND
        ([A,B,E] -> T)
        [0,0,0] -> 1;
        [0,0,1] -> 1;
        [0,1,0] -> 1;
        [0,1,1] -> 1;
        [1,0,0] -> 1;
        [1,0,1] -> 1;
        [1,1,0] -> 1;
        [1,1,1] -> 0;
TEST_VECTORS " For sum-of-products expression
        ([G,H,L] -> U)
        [0,0,0] -> 0;
        [0,0,1] -> 0;
        [0,1,0] -> 1;
        [0,1,1] -> 0;
        [1,0,0] -> 1;
        [1,0,1] -> 0;
        [1,1,0] -> 1;
        [1,1,1] -> 0;

END BASICGTS
```

FIGURE 12-9   ABEL source file to implement some basic logic gates in a GAL16V8.

P16V8C, P16V8S, and P16V8R. As shown in Figure 12-5, the GAL16V8 output macrocell can be programmed in several different ways. The "C" device file is used for combinational circuits where you want to use one product term to enable/disable the outputs. The "S" device file is used for applications where all eight product terms on each OR gate are to be used as part of the logic for the outputs. The "R" device file is used for applications that use the register in the output macrocells. For this example we will tell ABEL to use the P16V8S device file for processing our module. A semicolon (;) is used to terminate the DEVICE statement and other statements in ABEL programs.

Note that anything written on a line after the " character is treated as a comment by ABEL and does not have any effect on the programming of the PAL. A comment is terminated by another " character or by the carriage return at the end of the line.

The PIN statements are used to assign input and output signal names to the physical pins of the device. The first PIN statement, for example, tells ABEL that the A signal will be connected to pin 2, the B signal will be connected to pin 3, etc. To assign signals to pins, you use a logic diagram of the device such as the one for the GAL16V8 in Figure 12-8. Note in Figure 12-8 that pin 12 on the GAL16V8 can be programmed to function as either an input or an output, so we can assign the L input signal to it. The PIN statements and all the other statements in the file are terminated with a ; (semicolon).

The EQUATIONS directive tells ABEL that the statements which follow are the Boolean expressions you want to implement in the device. These are the same as the Boolean expressions you have written in the past, except that symbols more commonly found on a computer keyboard are used to represent logic functions. The symbols used to represent common logic operators are shown on the next page. Note our example expressions show that an input signal can be used in more than one expression. Also note that ABEL is "case-sensitive," so if you use lowercase letters for a variable name at one point in a file, you must use lowercase letters for that variable name throughout the file!

Operator	Function	Example
!	complement or NOT	!A
&	logical AND	A & B
#	logical OR	A # B
$	Exclusive OR	A $ B = (A&!B) # (!A&B)
!$	Exclusive NOR	A!$ B = (A&B) # (!A&!B)

The TEST_VECTORS sections of the example in Figure 12-9 are used to tell ABEL the desired output state for each combination of the input signals. A test vector consists of one set of input conditions and the desired output logic level(s) for that input combination. The SIMULATE subprogram uses these test vectors to determine whether the generated fusemap correctly implements the Boolean expressions for all the input combinations. Also, the programmer uses these test vectors to determine if the programmed PAL works correctly. The first line under each TEST_VECTOR heading is used to show the format for the test vectors in that section. The remaining lines show all the possible input combinations and the desired output for each. The square brackets are used to enclose a set of signals which are to be treated as a group.

RUNNING ABEL ON BASICGTS

When you run ABEL on the source file shown in Figure 12-9, the display shown in Figure 12-10 will appear on your computer screen, one section at a time. Note that it only takes a few seconds for ABEL to run each of the six subprograms.

If PARSE finds any syntax (grammatical) errors in your source file, it will create a list file which shows the errors it found and return control to you instead of proceeding on through the rest of the subprograms.

As we explained before, after PARSE runs successfully to completion, ABEL runs TRANSFOR to convert the equations to a form that REDUCE can work with, then runs REDUCE. REDUCE simplifies the Boolean equations using the rules for the R1 level of reduction specified in the FLAG 'R1' statement.

ABEL then runs FUSEMAP to produce the file which will be downloaded to a programmer to program and test the PAL. For this example, the file will have the name U1.JED. U1 is the name you used in the DEVICE statement to identify the device. JED is short for JEDEC (Joint Electron Device Engineering Council), an organization which has established a standard format for programmer load files. The programmer load file or JEDEC file contains the fusemap for the PAL, the test vectors to determine if the PAL is programmed correctly, and a check sum to determine if the file was downloaded to the programmer correctly.

The 7 of 64 terms line under FUSEMAP in Figure 12-10 tells you how many of the possible product terms were used to implement the specified logic functions. The GAL16V8 we used for this example has eight AND gates with eight inputs each, so the maximum number of prod-

```
A:\>echo off
+ abel BASICGTS

PARSE ABEL(tm) 3.00a Copyright 1983-1988 FutureNet Division, Data I/O Corp.
 module BASICGTS
PARSE complete.  Time: 4 seconds

TRANSFOR ABEL(tm) 3.00a Copyright 1983-1988 FutureNet Division, Data I/O Corp.
 module BASICGTS
TRANSFOR complete.  Time: 2 seconds

REDUCE ABEL(tm) 3.00a Copyright 1983-1988 FutureNet Division, Data I/O Corp.
 module BASICGTS
_device U1
Simple Reductions (Sum Of Products)
REDUCE complete.  Time: 3 seconds

FUSEMAP ABEL(tm) 3.00a Copyright 1983-1988 FutureNet Division, Data I/O Corp.
 module BASICGTS
_device U1 'P16V8S'
 7 of 64 terms used
FUSEMAP complete.  Time: 7 seconds

SIMULATE ABEL(tm) 3.00b Copyright 1983-1988 FutureNet Division, Data I/O Corp.
 module BASICGTS
_device U1 'P16V8S'
24 out of 24 vectors passed
SIMULATE complete.  Time: 7 seconds

DOCUMENT ABEL(tm) 3.00a Copyright 1983-1988 FutureNet Division, Data I/O Corp.
 module BASICGTS
_device U1
DOCUMENT complete.  Time: 4 seconds

A:\>
```

FIGURE 12-10 Computer screen display produced as ABEL processes a source file.

uct terms you can implement with this PAL is 8 × 8, or 64. If an application requires more product terms than a device can provide, you can use more reduction in ABEL, go to a larger device, or provide some reduction with external gates before supplying the signals to the PAL.

When FUSEMAP finishes, ABEL runs SIMULATE. SIMULATE uses the test vectors you provided in the source file to see if the fusemap implements the desired Boolean equations. SIMULATE generates a file called BASICGTS.SIM. If any vectors did not pass simulation, you can print out the .SIM file to find which ones didn't and correct the problem in the source file.

When everything else runs correctly, you can print out the Document file produced by ABEL. As you can see in Figure 12-11, the file produced by DOCUMENT shows you the reduced equations, and an actual pin diagram for the PAL which implements the specified logic functions. Since

we specified a reduction level of only R1 in the source file, the "reduced" equations are the same as those in the source file in Figure 12-9. For comparison, Figure 12-12 shows the reduced equations that are produced by using a reduction level of 3 instead of 1 in the source file in Figure 12-9. You might find it interesting to do a Karnaugh map for the U function in the source file to see the difference between these two results. Now let's look at a more complex combinational logic application for a PAL.

Implementing a Hexadecimal to Seven-Segment Decoder with a PAL

THE DESIGN PROBLEM

A common use of PALs in the "real" world is to implement various types of nonstandard encoders and decod-

```
ABEL(tm) 3.00a  -  Document Generator          22-Jul-88 01:07 PM
Implementing basic logic gates with ABEL and a GAL16V8 EPLD
Equations for Module BASICGTS


Device U1
- Reduced Equations:

    R = (A & B);
    S = (D # C);
    T = !(A & B & E);
    U = (G & H & !L # G & !H & !L # !G & H & !L);

                                                       Page 2
ABEL(tm) 3.00a  -  Document Generator          22-Jul-88 01:07 PM
Implementing basic logic gates with ABEL and a GAL16V8 EPLD
Chip diagram for Module BASICGTS


Device U1

                         P16V8S
              ------\  -----  /------
              |         -----         |
          A  |  1               20  | Vcc
          B  |  2               19  | R
          C  |  3               18  | S
          D  |  4               17  | T
          E  |  5               16  | U
          F  |  6               15  |
          G  |  7               14  |
          H  |  8               13  |
         GND |  9               12  | L
             |  10              11  |
              |                     |
              -----------------------


end of module BASICGTS
```

FIGURE 12-11 Document file produced by ABEL for source file in Figure 12-9.

```
ABEL(tm) 3.00a  -  Document Generator          23-Jun-88 12:12 PM
Implementing basic logic gates with ABEL and a GAL16V8 EPLD
Equations for Module BASICGTS


Device U1

- Reduced Equations:

    R = (A & B);
    S = (D # C);
    T = !(A & B & E);
    U = (G & !L # H & !L);
```

FIGURE 12-12 Reduced equations produced by using a reduction level of 3 in the source file in Figure 12-9.

ers. As a simple example of this, we will show you how to use a GAL16V8 to implement a decoder which converts a 4-bit binary code to a seven-segment code that will display the corresponding hexadecimal digit on a common-anode LED display. Figure 12-13a shows the circuit connections for this example. The PAL here functions similarly to the 74LS47 in Figure 7-26a, except that the symbols displayed on the LED are 0 to 9, A, b, C, d, E, and F, instead of just the 0 to 9 produced by a 74LS47. The GAL16V8 outputs can sink up to 25 mA in the output low state, so all that is needed to interface these outputs to the LED segment inputs are the 150-Ω current limiting resistors as shown.

Figure 12-13b shows the truth table needed to implement this binary-to-HEX decoder and the symbols that will be displayed for each binary input.

AN ABEL SOURCE FILE FOR THE DECODER

Logic design programs such as ABEL, A+PLUS, and PALASM allow you to enter the design information in several different formats. The most common forms for combinational logic designs are schematics, Boolean equations, and truth tables. ABEL allows you to enter the design information in any one of these three forms. The choice of entry form depends on whether you are implementing an existing circuit with a PAL or a new design with a PAL.

If you already have a logic design drawn with schematic capture programs such as OrCAD SDTIII and want to implement the circuit in a PAL, then schematic entry may be the easiest to use. If the PAL development program you are using does not have the schematic entry capability, you can just use the schematic diagram to write the Boolean equations for the circuit and enter the equations in the source file.

If you are doing a new design, you probably developed a truth table or Boolean expressions as one of the early steps in the design process. In that case, equations or truth tables are the easiest way to enter the design data.

In the previous example we showed you how to enter the design data in an ABEL source file in the EQUATIONS form. For this example we show you the format used to enter design data as a TRUTH_TABLE. This is much easier than writing a Boolean expression for each of the seven segment outputs and entering it in the equations form.

Figure 12-14 shows the ABEL source file for our binary-to-hexadecimal seven-segment decoder. Again, the MODULE and END statements bracket the statements of the file. When driving LEDs, we don't care about glitches, so in the FLAG statement we tell ABEL to use a reduction level of 3. The reduced equations will then use a minimum number of product terms. The leftover pins and product terms in the PAL might then be used to implement some other logic function needed in the system.

After the key word TRUTH_TABLE, the truth table for the PAL is entered in the same format as in Figure 12-13b. The top line of the truth table shows the meaning of each column in the truth table. Most text editors and word processors have a copy feature which makes it very easy to type in one line, and then just copy and modify it over and over to produce the rest of the lines in the table.

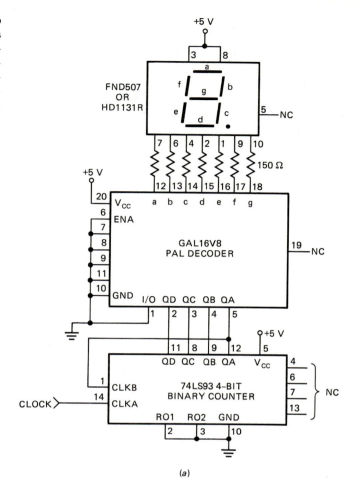

(a)

BINARY INPUT				PAL OUTPUT							SYMBOL DISPLAYED
QD	QC	QB	QA	a	b	c	d	e	f	g	
0	0	0	0	0	0	0	0	0	0	1	0
0	0	0	1	1	0	0	1	1	1	1	1
0	0	1	0	0	0	1	0	0	1	0	2
0	0	1	1	0	0	0	0	1	1	0	3
0	1	0	0	1	0	0	1	1	0	0	4
0	1	0	1	0	1	0	0	1	0	0	5
0	1	1	0	0	1	0	0	0	0	0	6
0	1	1	1	0	0	0	1	1	1	1	7
1	0	0	0	0	0	0	0	0	0	0	8
1	0	0	1	0	0	0	1	1	0	0	9
1	0	1	0	0	0	0	1	0	0	0	A
1	0	1	1	1	1	0	0	0	0	0	b
1	1	0	0	0	1	1	0	0	0	1	C
1	1	0	1	1	0	0	0	0	1	0	d
1	1	1	0	0	1	1	0	0	0	0	E
1	1	1	1	0	1	1	1	0	0	0	F

(b)

FIGURE 12-13 Binary-to-hexadecimal seven-segment display decoder. (a) Circuit diagram. (b) Truth table.

The TEST_VECTORS for this example are just a copy of the truth table, so the text editor block copy commands can be used to easily put these in the source file. Note again that each statement in the source file is terminated with a semicolon (;).

```
MODULE HEX7SEG
FLAG '-r3' "flag must be first after module"

TITLE 'HEXADECIMAL TO SEVEN SEGMENT DECODER - D.HALL, 1988'

            U2          DEVICE  'P16V8S'; "S gives 8 product terms/output
D,C,B,A         PIN     2,3,4,5;
a,b,c,d,e,f,g   PIN     12,13,14,15,16,17,18;

    TRUTH_TABLE     ([D,C,B,A] ->       [a,b,c,d,e,f,g])
                    [0,0,0,0] ->        [0,0,0,0,0,0,1]; "0
                    [0,0,0,1] ->        [1,0,0,1,1,1,1]; "1
                    [0,0,1,0] ->        [0,0,1,0,0,1,0]; "2
                    [0,0,1,1] ->        [0,0,0,0,1,1,0]; "3
                    [0,1,0,0] ->        [1,0,0,1,1,0,0]; "4
                    [0,1,0,1] ->        [0,1,0,0,1,0,0]; "5
                    [0,1,1,0] ->        [0,1,0,0,0,0,0]; "6
                    [0,1,1,1] ->        [0,0,0,1,1,1,1]; "7
                    [1,0,0,0] ->        [0,0,0,0,0,0,0]; "8
                    [1,0,0,1] ->        [0,0,0,1,1,0,0]; "9
                    [1,0,1,0] ->        [0,0,0,1,0,0,0]; "A
                    [1,0,1,1] ->        [1,1,0,0,0,0,0]; "b
                    [1,1,0,0] ->        [0,1,1,0,0,0,1]; "C
                    [1,1,0,1] ->        [1,0,0,0,0,1,0]; "d
                    [1,1,1,0] ->        [0,1,1,0,0,0,0]; "E
                    [1,1,1,1] ->        [0,1,1,1,0,0,0]; "F


    TEST_VECTORS    ([D,C,B,A] ->       [a,b,c,d,e,f,g])
                    [0,0,0,0] ->        [0,0,0,0,0,0,1];
                    [0,0,0,1] ->        [1,0,0,1,1,1,1];
                    [0,0,1,0] ->        [0,0,1,0,0,1,0];
                    [0,0,1,1] ->        [0,0,0,0,1,1,0];
                    [0,1,0,0] ->        [1,0,0,1,1,0,0];
                    [0,1,0,1] ->        [0,1,0,0,1,0,0];
                    [0,1,1,0] ->        [0,1,0,0,0,0,0];
                    [0,1,1,1] ->        [0,0,0,1,1,1,1];
                    [1,0,0,0] ->        [0,0,0,0,0,0,0];
                    [1,0,0,1] ->        [0,0,0,1,1,0,0];
                    [1,0,1,0] ->        [0,0,0,1,0,0,0];
                    [1,0,1,1] ->        [1,1,0,0,0,0,0];
                    [1,1,0,0] ->        [0,1,1,0,0,0,1];
                    [1,1,0,1] ->        [1,0,0,0,0,1,0];
                    [1,1,1,0] ->        [0,1,1,0,0,0,0];
                    [1,1,1,1] ->        [0,1,1,1,0,0,0];


    END HEX7SEG
```

FIGURE 12-14   ABEL source file for binary-to-hexadecimal display decoder.

RUNNING ABEL ON THE SOURCE FILE

As we described in the previous example, the next step after writing the source file is to process it with ABEL. ABEL will run each of its six subprograms in turn on the source file. If an error is found at any point, an error message will be displayed, and control returned to DOS, so you can correct the error and run ABEL again. When ABEL runs all the way through with no errors, a display such as that shown in Figure 12-10 will be written out on your computer screen. You can then download the JEDEC file to the programmer to program and test the PAL. To see the reduced equations and pinouts for the PAL, you can print out the document file, HEX7SEG.DOC, as shown in Figure 12-15.

One look at the reduced equations in Figure 12-15 should convince you that this PAL method is much easier than doing a Boolean map for each output expression and implementing the result with discrete logic gates. Incidentally, the simplified expressions generated by REDUCE may be somewhat different from those you get if you circle the 1's in a Karnaugh map because the default for the algorithm used by REDUCE is to write the expression you would get by circling adjacent 0's and writing the expression for an active low output. This is why all the right-hand sides of the expressions in Figure 12-15 start with an "!," which indicates an active low output expression. Remember from Chapter 7 that circling 0's to produce an active low output expression implements the same truth table as circling 1's to produce an active high expression. The resultant expressions are just the Boolean equivalent expressions for the same truth table. In the next example we show you how you can use an ISTYPE statement to generate the expression for an active high output if needed.

AN ALTERNATIVE SOURCE FILE FOR THE HEX DECODER

As with most high-level languages, ABEL allows you to represent constants in your source files with names so that the functions are more easily understood. As an example of this we will show you how you can write the HEX decoder source file in a more "Englishlike" form,

Page 1

ABEL(tm) 3.00a - Document Generator   22-Jul-88 08:30 AM
HEXADECIMAL TO SEVEN SEGMENT DECODER - D.HALL, 1988
Equations for Module HEX7SEG

Device U2

- Reduced Equations:

a = !(B & C # !A & D # !B & !C & D # A & C & !D # B & !D # !A & !C);

b = !(A & !B & D # !A & !C # A & B & !D # !A & !B & !D # !C & !D);

c = !(A & !B # !C & D # C & !D # A & !D # !B & !D);

d = !(!A & !B & D # !A & B & C # A & !B & C # A & B & !C # !A & !C & !D);

e = !(C & D # B & D # !A & B # !A & !C);

f = !(B & D # !C & D # !A & C # !B & C & !D # !A & !B);

g = !(A & D # !C & D # !A & B # !B & C & !D # B & !C);

Page 2

ABEL(tm) 3.00a - Document Generator   22-Jul-88 08:30 AM
HEXADECIMAL TO SEVEN SEGMENT DECODER - D.HALL, 1988
Chip diagram for Module HEX7SEG

Device U2

```
                        P16V8S
              ------\  /------
              |      -----      |
              |   1          20 |  Vcc
          D   |   2          19 |
          C   |   3          18 |  g
          B   |   4          17 |  f
          A   |   5          16 |  e
              |   6          15 |  d
              |   7          14 |  c
              |   8          13 |  b
              |   9          12 |  a
         GND  |  10          11 |
              |                 |
              -------------------
```

end of module HEX7SEG

FIGURE 12-15   Document file for binary-to-hexadecimal display decoder implemented with a GAL16V8.

and how you can add an enable input to the circuit so that the display can be blanked by applying a high to an input pin on the PAL. Figure 12-16a shows the source file for this second version.

The first item to notice in this version is that we tell ABEL to use the P16V8C device file so that one product term from each OR gate is available to control the output buffer. To help you see this, take another look at Figure 12-5. (The term from the AND array is routed through the TSMUX multiplexer to the enable input of the output buffer.) Also in this version we have added a third PIN statement to identify pin 6 as the enable input, ENA.

Now, look carefully at the next section of Figure 12-16 where we have given names to individual signals or sets of signals. The name BINARY, for example, is given to the set of input signals [D, C, B, A]. To see how we use this name, look at the format line next to TRUTH_TABLE. The name BINARY is just a replacement for the set of D, C, B, A input signals. In the actual truth table entries you can represent the value of this set with its decimal value as shown.

Another example of this symbolic representation is the statement which equates the word "on" with a logic 0 and the word "off" with a logic 1. When referring to the state of an LED segment, it is easier to think of segments as being on or off, rather than 0 or 1. The entries in the TRUTH_TABLE show how this is used to make the segment output part of the truth table more understandable.

Still another example of naming constants is the line in Figure 12-16 which equates the letter L with a logic 0 and the letter H with a logic 1. In some applications the terms low and high are more reasonable than 0 and 1. When describing the operation of an enable input, for example, it rolls more easily off your tongue to say the enable input must be asserted low to light the LEDs. The ENA columns of the truth table and the test vectors in Figure 12-16 show how we used these names.

Next in our naming, we set the letter X equal to .X. and the letter Z equal to .Z. In ABEL language, .X. is used to represent a "don't care" input signal and .Z. is used to represent an output signal that is in the high

```
MODULE SEVEN_SEG
FLAG '-R3'
TITLE 'Hexadecimal to 7-segment decoder with enable input'
            U3  DEVICE 'P16V8C'; "C = one product term for output enable
            D,C,B,A      PIN 2,3,4,5; "Binary input pins
            a,b,c,d,e,f,g PIN 12,13,14,15,16,17,18;
                    ENA PIN 6;

    BINARY =  [D,C,B,A];          "Equate name BINARY with set of inputs
    LED    =  [a,b,c,d,e,f,g];    "Equate name LED with segment outputs
    on,off =  0,1;                "Active levels for common anode segments
    L,H    =  0,1;                "Active levels for enable input
    X,Z    =  .X.,.Z.;            "Shorthand for don't care input and HiZ output

    a,b,c,d,e,f,g  ISTYPE 'POS'; "Generate active high expressions


TRUTH_TABLE        ([ENA, BINARY] -> LED)
"                          a     b     c     d     e     f     g
        [L,0]   -> [ on,   on,   on,   on,   on,   on,  off];   "0
        [L,1]   -> [off,   on,   on,  off,  off,  off,  off];   "1
        [L,2]   -> [ on,   on,  off,   on,   on,  off,   on];   "2
        [L,3]   -> [ on,   on,   on,   on,  off,  off,   on];   "3
        [L,4]   -> [off,   on,   on,  off,  off,   on,   on];   "4
        [L,5]   -> [ on,  off,   on,   on,  off,   on,   on];   "5
        [L,6]   -> [ on,  off,   on,   on,   on,   on,   on];   "6
        [L,7]   -> [ on,   on,   on,  off,  off,  off,  off];   "7
        [L,8]   -> [ on,   on,   on,   on,   on,   on,   on];   "8
        [L,9]   -> [ on,   on,   on,  off,  off,   on,   on];   "9
        [L,10]  -> [ on,   on,   on,  off,   on,   on,   on];   "A
        [L,11]  -> [off,  off,   on,   on,   on,   on,   on];   "b
        [L,12]  -> [ on,  off,  off,   on,   on,   on,  off];   "C
        [L,13]  -> [off,   on,   on,   on,   on,  off,   on];   "d
        [L,14]  -> [ on,  off,  off,   on,   on,   on,   on];   "E
        [L,15]  -> [ on,  off,  off,  off,   on,   on,   on];   "F

EQUATIONS
        a.OE = !ENA; b.OE = !ENA; c.OE = !ENA; d.OE = !ENA;
        e.OE = !ENA; f.OE = !ENA; g.OE = !ENA;


TEST_VECTORS       ([ENA,BINARY] -> LED)
"                          a     b     c     d     e     f     g
        [L,0]   -> [ on,   on,   on,   on,   on,   on,  off];   "0
        [L,1]   -> [off,   on,   on,  off,  off,  off,  off];   "1
        [L,2]   -> [ on,   on,  off,   on,   on,  off,   on];   "2
        [L,3]   -> [ on,   on,   on,   on,  off,  off,   on];   "3
        [L,4]   -> [off,   on,   on,  off,  off,   on,   on];   "4
        [L,5]   -> [ on,  off,   on,   on,  off,   on,   on];   "5
        [L,6]   -> [ on,  off,   on,   on,   on,   on,   on];   "6
        [L,7]   -> [ on,   on,   on,  off,  off,  off,  off];   "7
        [L,8]   -> [ on,   on,   on,   on,   on,   on,   on];   "8
        [L,9]   -> [ on,   on,   on,  off,  off,   on,   on];   "9
        [L,10]  -> [ on,   on,   on,  off,   on,   on,   on];   "A
        [L,11]  -> [off,  off,   on,   on,   on,   on,   on];   "b
        [L,12]  -> [ on,  off,  off,   on,   on,   on,  off];   "C
        [L,13]  -> [off,   on,   on,   on,   on,  off,   on];   "d
        [L,14]  -> [ on,  off,  off,   on,   on,   on,   on];   "E
        [L,15]  -> [ on,  off,  off,  off,   on,   on,   on];   "F
        [H,X]   -> [  Z,    Z,    Z,    Z,    Z,    Z,    Z];   "FLOAT
END SEVEN_SEG
```

(a)

FIGURE 12-16 Alternate ABEL source file for binary-to-hexadecimal display decoder showing how to implement an enable input and how to represent constants symbolically. *(Continued on next page.)*

impedance or floating state. We made this substitution simply because we don't like typing in all the periods when we use these terms.

The statement a,b,c,d,e,f,g ISTYPE 'POS'; in Figure 12-16 tells ABEL to generate active high expressions for the segment outputs instead of the default active low output expressions.

Finally, in Figure 12-16 note the EQUATIONS section. The a.OE = !ENA; statement tells ABEL to program the PAL so that the "a" signal output will be enabled when the ENA pin is asserted low. The "keyword" to notice in this statement is the .OE term. The rest of the equations tell ABEL to generate the fusemap so that the other segment outputs are also enabled when the ENA pin is

```
ABEL(tm) 3.00a  -  Document Generator                      Page 1
Hexadecimal to 7-segment decoder with enable input  22-Jul-88 08:25 AM
Equations for Module SEVEN_SEG

Device U3

- Reduced Equations:

  a = (A & !B & C & D & !ENA
        # A & B & !C & D & !ENA
        # !A & !B & C & !D & !ENA
        # A & !B & !C & !D & !ENA);

  b = (!A & C & D & !ENA
        # A & B & D & !ENA
        # !A & B & C & !ENA
        # A & !B & C & !D & !ENA);

  c = (B & C & D & !ENA # !A & C & D & !ENA # !A & B & !C & !D & !ENA);

  d = (!A & B & !C & D & !ENA
        # A & B & C & !ENA
        # !A & !B & C & !D & !ENA
        # A & !B & !C & !ENA);

  e = (A & !B & !C & !ENA # !B & C & !D & !ENA # A & !D & !ENA);

  f = (A & !B & C & D & !ENA
        # A & B & !D & !ENA
        # B & !C & !D & !ENA
        # A & !C & !D & !ENA);

  g = (!A & !B & C & D & !ENA
        # A & B & C & !D & !ENA
        # !B & !C & !D & !ENA);

  enable a = (!ENA);
  enable b = (!ENA);
  enable c = (!ENA);
  enable d = (!ENA);
  enable e = (!ENA);
  enable f = (!ENA);
  enable g = (!ENA);
```

(*b*)

FIGURE 12-16 *(continued)*

asserted low. By default, the segment outputs will float (high Z state) if the ENA input is high.

In this application we could have simply written an entry in the truth table that programmed the segment outputs to go high when the ENA input was high. However, many PAL applications require outputs which can be floated, so we used this example to show you how to do it. In a more complex application the !ENA on the right side of the ENABLE statement could be replaced with a sum-of-products expression.

The last entry in TEST_VECTORS shows you how to test the ENA input. The X in the statement indicates that if the ENA input is high, you don't care what logic level is on the BINARY inputs, because the outputs should be floating. The Z's in the segment columns indicate that you want SIMULATE to test if the segment outputs are programmed to float when ENA is high.

Figure 12-16*b* shows the reduced equations produced by the source file in Figure 12-16*a*. As you can see, the effect of the ENA input is included in the expressions and the ISTYPE statement has caused ABEL to generate active high expressions, similar to those you would get by circling 1's on a Karnaugh map.

Since the active low expressions produced by the ABEL default are just the Boolean equivalents of the active high expressions produced by including the ISTYPE 'POS' statement, you can usually use the default. However, if ABEL gives you an error message of "too many product terms" for a particular expression, you can try solving the problem by specifying that output as active high with an ISTYPE statement and running ABEL on the source file again. For the decoder in this example, either polarity fits easily in the PAL. The file in Figure 12-14 requires 35 of the 64 possible product terms to implement, and the file in Figure 12-16*a* happens to require only 32 of the 64 possible product terms to implement.

We don't expect you to remember all this, but this last example should give you some idea of the flexibility built into the development programs and the current programmable devices. PALs which have register outputs can also be used to easily implement clocked logic circuits. In the next section of the chapter we show you the general techniques used to design synchronous circuits, and then in a later section of the chapter we show you how to implement synchronous circuits with PALs.

DESIGNING SYNCHRONOUS CIRCUITS AND STATE MACHINES

Introduction

In Chapter 9 we showed you how to analyze the count sequence of synchronous circuits made with D, T, and JK flip-flops. The operating principle of these circuits is that the flip-flop outputs during one state are decoded to produce input signals which will cause the flip-flops to step to the desired next states when they are clocked. As we will start to show you here, this is a very general technique which can be used to design and implement many types of synchronous circuits.

It is not the major purpose of this book to teach you how to design digital circuits, but we do want to teach you enough about digital design so that you understand what is involved in the process. Also, we want to show you enough of the basics of digital design so that you can easily go further in that direction if you want or need to.

There are many books now available which discuss digital system design techniques. One difficulty with these books, however, is that each of them tends to use a somewhat different approach and different notation. In our discussions here we show you the techniques and notations used in most of the current manufacturers' literature. You should have little difficulty adapting to any minor differences you find in other books.

After introducing the basic techniques, we show you several complete design examples. Then we show you how synchronous circuits can be implemented with single programmable logic devices such as PALs.

Moore Machines and Mealy Machines

Digital designers often use the term *state machine* to refer to synchronous circuits such as the counters we discussed in Chapter 9. The term state machine comes from the fact that the outputs step from one binary state to another as the circuit is clocked. Traditionally digital designers have classified state machines into one of two major types: *Moore machines* and *Mealy machines*. It is important to differentiate between these two types of machines because they are designed differently, they function differently, and they are represented differently in the graphic drawings we use to help design state machines.

MOORE-TYPE STATE MACHINES

Figure 12-17a shows a block diagram of a Moore-type state machine, and Figure 12-17b shows a simple 2-bit binary counter circuit which is an example of this type machine. In a Moore-type machine the flip-flop output signals and any external control signals are decoded by the next state decoder to produce the flip-flop inputs required for the flip-flops to go to the next state. The out-

puts of the next-state decoder determine the response that the circuit will make when it is next clocked. In the example circuit in Figure 12-17b, the next-state decoder consists of the gates enclosed by the dashed box labeled NSD. A little later we will discuss the operation of this part of the circuit, but for now let's look at the output decoder section of the circuit in Figure 12-17b.

In a Moore-type state machine the output decoder uses only the flip-flop output signals to produce the system output signals. The level on the P output in Figure 12-17b, for example, is determined only by the levels on the Q outputs of the two flip-flops, not by any additional external signals. Specifically, the P signal will be asserted if the outputs of the two flip-flops are both low. An output signal such as the P signal is often called *unconditional* because its level depends only on the current output states of the flip-flops, not on the level of any external signal.

Since external signals are routed into the next-state decoder in a Moore machine circuit, any change of an external signal will not have any effect on the system output signals until after the next clock pulse occurs. For example, if the E input signal in Figure 12-17b changes, the P output signal will not be affected until the next rising edge on the clock signal updates the flip-flops.

The main point to stick in your mind about a Moore-type state machine then is:

> The signals present on the system outputs during a state are determined only by the decoding logic and the logic levels present on the outputs of the flip-flops, not by any external signals. Therefore, changes in external signals can only affect the system output signals after the next clock pulse occurs.

For contrast, let's look at the operation of a Mealy-type state machine.

MEALY-TYPE STATE MACHINES

Figure 12-18a shows a block diagram of a Mealy-type state machine and Figure 12-18b shows our 2-bit binary counter circuit with an addition to show you how a Mealy machine is different from a Moore machine.

As with a Moore machine, the output signals from the flip-flops and any appropriate external signals are decoded by the next-state decoder to produce the input signals required for the flip-flops to go to their next states. The difference between the two types of machines is in their output decoder sections.

In a Mealy-type state machine the system output signals are produced by decoding external signals *and* the outputs of the flip-flops. In the circuit in Figure 12-18b, for example, the level on the Z output is determined by the levels on the QA output, the QB output, *and* the level on the external M signal. The Z signal can be thought of as a *conditional* output signal because it depends not only on the output state of the flip-flops, but also on the "condition" or level of an external signal. If, for example, the flip-flop outputs are in state 11 and the M signal is low, the Z signal will be low. If the M signal goes high

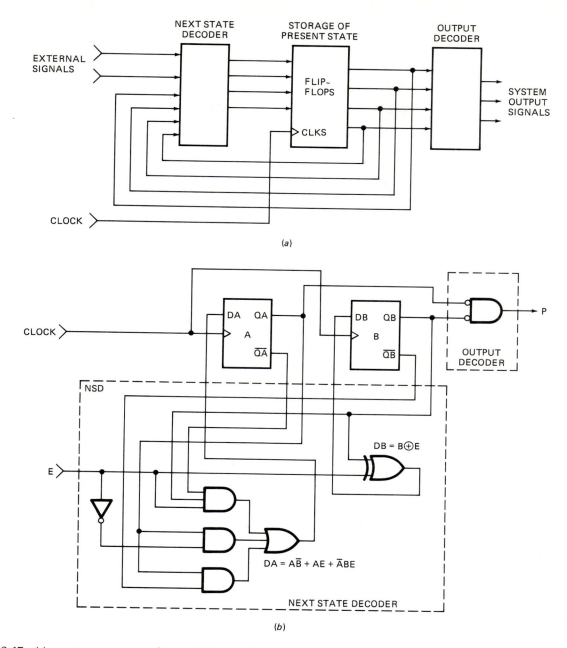

FIGURE 12-17 Moore-type state machine. *(a)* Block diagram. *(b)* Example circuit.

while the flip-flop outputs are still in state 11, the Z output will immediately go high.

The main point for you to stick in your mind about a Mealy machine then is:

> Signals on the system outputs are determined by the logic levels on the outputs of the flip-flops *and* by any external signals connected to the inputs of the output decoder. Therefore, these external signals can immediately affect the system output signals.

Many state machines have both conditional and unconditional system output signals, so in effect they function as both Moore machines and Mealy machines. In the circuit in Figure 12-18*b*, for example, the P output is unconditional because the P signal will be asserted whenever state

00 is on the Q outputs of the flip-flops. The P output, then, might be thought of as a "Moore output." The Z output of the circuit in Figure 12-18*b* is conditional because it will be asserted during flip-flop output state 11 if the external M signal is also asserted. The Z output, then, might be thought of as a "Mealy output." In the following sections we show you how these two types of signals are represented in the drawings we use to graphically represent the operation of state machines.

State Diagrams and ASM Charts

INTRODUCTION

Synchronous circuits or state machines should be designed in a *top-down* manner. This means that you

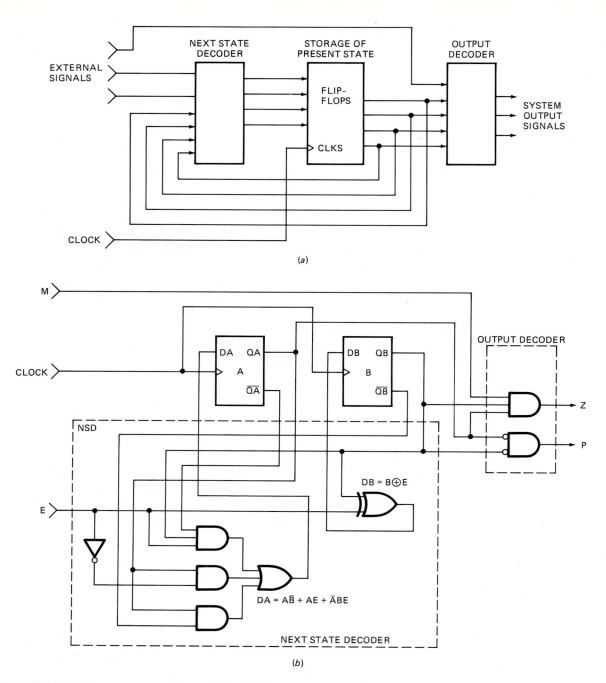

FIGURE 12-18   Mealy-type state machine. (a) Block diagram. (b) Example circuit.

should first think a great deal about the problem you want the circuit to solve and the best overall approach to solving the problem. Part of this thinking should be to make a list of the required input and output signals. The next big step is to express the desired relationships between the input and output signals in some convenient form.

In previous discussions you have used truth tables, Boolean expressions, and/or timing waveforms to represent the relationship between input and output signals of a circuit. To describe the operation of synchronous circuits other than simple counters and shift registers, you usually use *state diagrams* or *ASM charts*. These

two graphic techniques help you to more easily visualize the sequence of desired states, the input conditions required to go from one state to the next, and the output signals produced during each state. In the following sections we show you how to develop each of these for some existing circuits. Then we show you how to develop them for a circuit to be designed and how to use them to design the actual circuit. To start we will analyze the operation of the simple 2-bit synchronous binary counter in Figure 12-18b.

The basic flip-flop part of the circuit in Figure 12-18b is similar to the 2-bit binary upcounter circuit we showed you how to analyze in Figure 9-1. The count sequence for the flip-flops is 00, 01, 10, 11, 00, etc. The main addition to the circuit in Figure 12-18b is an enable input (E) which controls whether the counter steps to the next state or remains in the same state when clocked. If the E input is low when a clock pulse occurs, the flip-flop outputs will remain in the same state. As long as the E input is low, the flip-flop outputs will remain in the same state. If the E input is high when a clock pulse occurs, then the flip-flops will step to the next state when a clock pulse occurs.

As we explained previously, the NOR gate in the upper right corner of the circuit in Figure 12-18b decodes state 00, so when state 00 is present on the outputs of the flip-flops, the P output will be asserted. Also as we explained previously, the Z output of the 3-input AND gate in the upper right corner of Figure 12-18b will be asserted if state 11 is on the outputs of the flip-flops and the external M signal is asserted. Let's see how you can represent the operation of this circuit in a graphic way called a state diagram.

STATE DIAGRAMS

Figure 12-19 shows a state diagram for the example circuit in Figure 12-18b.

Each state is represented by a large circle or oval. These circles or ovals have led some engineers to refer to this type of state diagram as a "bubble diagram." Since we are used to using this term, we will probably slip and use it a few times as we work our way through the chapter. The name or number which identifies the state is written in the center or just above the center of the circle. In the example in Figure 12-19 we have used the decimal equivalent of each binary output state to iden-

tify that state. As you can see, the states are simply labeled as 0, 1, 2, and 3.

Arrows are used to indicate the response the machine will make when it is clocked. If the next-state decoder input signals are such that the machine transitions to another state when a clock pulse occurs, an arrow is drawn from the old state to the new state. An example of this is the arrow connecting state 0 with state 1 in Figure 12-19. Any external signal level that must be present for the transition to take place is written next to the tail of the arrow. The SYMBOL FORMAT statement at the top of the diagram identifies the signals represented in these labels. For our example here, the external E signal must be a 1 in order for the circuit to transition to the next state when it is clocked. To show this we write a 1 in the E position of the label next to the arrow connecting the two states. In the same manner we write 1's in the E positions of the labels next to the tails of the arrows which show the transitions from state 1 to 2, 2 to 3, and 3 to 0.

A short, looped arrow is used to indicate that the machine will remain in the current state as long as some condition is present on the inputs. The looped arrow to the left of the state 0 bubble in Figure 12-19 is an example. Symbol(s) next to the tail of this arrow show the input condition(s) which will cause the machine to remain in the same state. The machine shown in Figure 12-18b, for example, will stay in state 0 if E = 0. Again, the SYMBOL FORMAT statement at the top of the state diagram tells you the first value shown next to the tail of each arrow represents the E signal. Therefore, we write a 0 in the E position of the label next to the tail of the arrow which loops back to state 0. The looped arrows on the other three states indicate that if the machine is in any of these states, it will stay in that state as long as E = 0. It is assumed that a clock pulse is required to cause the machine to transition from one state to the next, so clock pulses are not shown on a bubble diagram.

The system output signals asserted during each state are also shown on a state diagram, but there is no standard for how to do this. We feel that the most reasonable convention is as follows.

Unconditional, or Moore-type, output signals that are asserted during a particular state are identified directly in the bubble representing that state. A horizontal line is drawn to separate the name of the state from the names of the signals that are asserted during that state. As an example, the P signal is unconditionally asserted during state 0, so we write a P in the bottom half of the state 0 bubble to show this.

Conditional, or Mealy-type, output signal levels are shown next to the tails of arrows along with the input signal levels. The SYMBOL FORMAT line at the top of Figure 12-19 shows the order in which the signals are represented. Input signal conditions are shown in front of a /, and Mealy (conditional) output signal levels produced during a state are shown after the /. Note that output conditions shown to the right of the / are for the state at the *tail* of the arrow. Let's look at a couple of examples of this representation.

The symbol format for the example in Figure 12-19 is E,M/Z. The 1X/0 next to the tail of the arrow from state

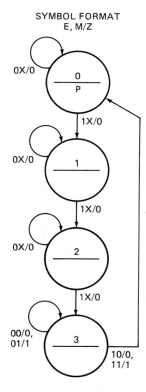

SYMBOL FORMAT
E, M/Z

FIGURE 12-19 State diagram for circuit in Figure 12-18b.

0 to state 1 then tells you that the E signal must be a 1 in order for this transition to take place when a clock pulse occurs. The X in the M position and the 0 in the Z position of the label tells you that in this state the Z signal output will be a 0, regardless of the signal level on the external M input. An alternative, less compact way of writing this label is 10/0, 11/0.

The 0X/0 label on the looped arrow next to state 0 tells you that the machine will stay in state 0 if E = 0. The X/0 in this label also tells you that if the machine is in state 0, the Z signal will be a 0, regardless of the level on the external M signal. The labels on the arrows next to state 3 tell a somewhat different story.

The 10/0, 11/1 label next to the transition arrow from state 3 to state 0 tells you that the E signal must be a 1 in order for the transition to take place. This label also tells you that if the machine is in state 3, the Z signal will be a 0 if M = 0 and Z will be a 1 if M = 1. The 00/0, 01/1 label next to the state 3 looped arrow tells you that the machine will stay in state 0 if E = 0. It also tells you that in state 3 the Z signal will be a 1 if the external signal M = 1. For practice work your way around the rest of the labels on the diagram in Figure 12-19. After we show you how to draw ASM charts in the next section, we will show you how to draw a state diagram for a desired machine and how to get from the state diagram to the circuit for it.

ASM CHARTS

Another common graphic way of representing the operation of a state machine is with an *algorithmic state machine*, or *ASM*, chart. An algorithm is the name given to a formula or sequence of steps used to solve a problem. An ASM chart represents the sequence of states that a state machine steps through when it is clocked.

ASM charts are similar to flow charts that you may have used when programming in BASIC or FORTRAN. A rectangular box is used to represent each state, a diamond-shaped box is used to represent a decision point, and a rounded-end box is used to represent conditional output signals.

As an example of how these boxes are used, Figure 12-20 shows an ASM chart for our example state machine circuit in Figure 12-18b. Four rectangular boxes are used to represent the four states. The identifying name or number for each state is written in a circle next to the upper left corner of the rectangle. For the example here we have again identified each state with the decimal equivalent of its binary output code. The binary code for each state is often written next to the upper right hand corner of each state rectangle as shown.

Transitions from one state to another are shown with arrows. If the transition from one state to another depends on the level of some external signal, then a diamond-shaped decision box is drawn between the state boxes. The name of the determining signal is written in the diamond-shaped box. For the circuit in Figure 12-18b, the E signal determines whether the machine will transition, for example, from state 0 to state 1, so we write an E in the decision box between state 0 and state 1. The arrows attached to the points of the decision box show how the machine will respond to each of

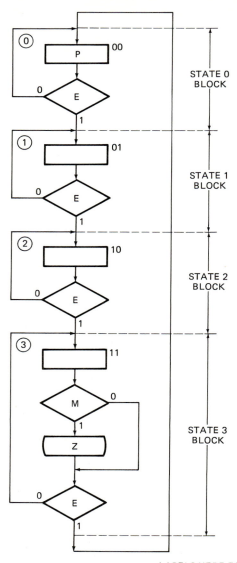

FIGURE 12-20 Algorithmic State Machine (ASM) chart for circuit in Figure 12-18b.

the possible values for the input signal. From the arrows on the top decision box in Figure 12-20, you can see that if E = 1, the machine will go from state 0 to state 1 when a clock pulse occurs. If E = 0, then the machine will stay in state 0 when a clock pulse occurs.

System output signals asserted during a state are shown in one of two ways depending on whether they are conditional or unconditional outputs. Signals that are unconditionally asserted during a state are identified by writing their names in the rectangular box which represents that state. The P signal in the circuit in Figure 12-18b is asserted during state 0, so a P is written in the state 0 box.

Conditional output signals that are asserted during a state are represented by rounded-end boxes in an ASM chart. A decision box is drawn in front of the rounded-end box to indicate the signal that must be present in order for that output signal to be asserted. The state 3 block in Figure 12-20 shows an example of this. The decision box just after the state 3 rectangle tells you that if M = 1 during state 3, the Z signal will be asserted. If M = 0 during state 3, the Z signal will not be asserted.

Work your way through the rest of the ASM chart in Figure 12-20 and compare it with the bubble diagram for the same circuit in Figure 12-19. Most books and manufacturer's application notes written during the last 10 years or so have used bubble diagrams, but now more designers seem to be using ASM charts. Therefore, you should become familiar with both methods. In the following sections we will show you more examples of each.

State Machine Design Examples

DESIGNING A 2-BIT BINARY SYNCHRONOUS UP-ONLY COUNTER

For our first design example we will develop a 2-bit binary counter such as the one in Figure 9-1 that we used to show you how to analyze synchronous circuits. We have chosen a very simple machine to start with so you can concentrate on learning the design steps rather than getting lost in the problem. The techniques we show you here can be extended to any type of state machine. In this section we show you how to design simple counter-type state machines; then, later in the chapter, we show you how to design a state machine which controls the operation of an elevator. Here is a list of the design steps and a description of how you carry out each step for our 2-bit counter example:

1. Think carefully about what you want the system to do.

   The circuit we want to design here is a 2-bit binary counter-type state machine which steps through the sequence 00, 01, 10, and 11 over and over as it is clocked. When the machine is in state 01, we want a signal named HOT to be unconditionally asserted, and when the machine is in state 10, we want a signal named COLD to be unconditionally asserted.

2. Draw a state diagram or an ASM chart for the machine.

   Figure 12-21a shows a bubble diagram for the machine, and Figure 12-21b shows an ASM chart for it. Note that since the machine simply steps from one state to the next when clocked, there are no looped arrows in the bubble diagram or decision boxes in the ASM chart. Also note that since the two desired output signals are unconditional, their names are written directly in the appropriate state bubbles or rectangles.

3. Make a *next-state*, or *transition-state*, *table* for the machine.

   Figure 12-22 shows the next-state table for the diagrams in Figure 12-21a and b. To make a table such as this, you first write all the possible flip-flop states in the Present State column. Then, for each present state, you write in the Next State column the state that the outputs will transition to when a

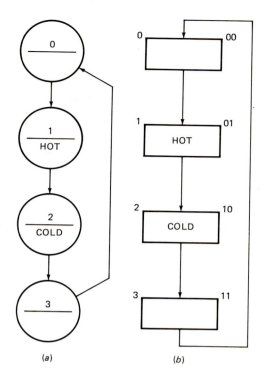

FIGURE 12-21 Algorithms for 2-bit binary synchronous upcounter with Hot and Cold outputs. *(a)* State diagram. *(b)* ASM chart.

PRESENT STATE		NEXT STATE		SYSTEM OUTPUTS	
QB	QA	QB	QA	HOT	COLD
0	0	0	1	0	0
0	1	1	0	1	0
1	0	1	1	0	1
1	1	0	0	0	0

FIGURE 12-22 Next-state table for 2-bit binary upcounter.

clock pulse occurs. For example, if the machine is in state 00, it will transition to state 01 when a clock pulse occurs. Likewise, if the machine is in state 01 when a clock pulse occurs, it will transition to state 10. For future reference also show the levels you want on the output signals such as Hot and Cold during each *present* state.

4. Decide how many flip-flops you need and whether you want to use D or JK flip-flops to implement your state machine.

   The required number of flip-flops, N, is related to the number of states that the machine has by the expression, States = 2^N. The machine here has four states, so we need two flip-flops. As a first example we will show you how to implement this machine with D flip-flops, and then we will show you how to do it with JK flip-flops.

5. For each flip-flop, determine the D or J and K input signals which are required for that flip-flop to tran-

sition from each present state to each next state in the table.

This sounds difficult, but all you really need is a truth table for the type of flip-flop you are using and a little patience. Here's how you do it for D flip-flops.

With a D flip-flop, the level present on D is transferred to the Q output when a clock pulse occurs. Therefore, if you want the next state of a D flip-flop to be a 0, you apply a 0 to the D input. If you want the next state of a D flip-flop to be a 1, you apply a 1 to the D input. To keep track of the inputs required on the two D flip-flops for each present-state to next-state transition, we add a DA input and a DB input column to the next-state table as shown in Figure 12-23a. Let's look at a few transitions to see how you fill in the D columns.

PRESENT STATE		NEXT STATE		SYSTEM OUTPUTS		FLIP-FLOP INPUTS	
QB	QA	QB	QA	HOT	COLD	DB	DA
0	0	0	1	0	0	0	1
0	1	1	0	1	0	1	0
1	0	1	1	0	1	1	1
1	1	0	0	0	0	0	0

(a)

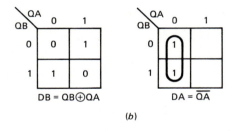

$DB = QB \oplus QA$ $DA = \overline{QA}$

(b)

FIGURE 12-23 (a) Next-state table with required D flip-flop input signals added. (b) Karnaugh maps and reduced equations for next-state decoder.

For the transition from state 00 to state 01, we want a 0 on the DB input and a 1 on the DA input. After a clock pulse, these levels on the D inputs will be transferred to the Q outputs to give the desired output state of 01. For the transition from state 01 to state 10, we want a 1 on DB and a 0 on DA. After a clock pulse, these levels will be transferred to the Q outputs to give the desired output state of 10. About this time you should notice that the DB and DA columns are identical to the next-state QB and QA columns! With D flip-flops this will always be the case, so writing the D columns is quite easy.

6. Design the next-state decoder.

Remember from our previous discussions that the next-state decoder uses the present-state Q levels to produce the D or JK signals required for the flip-flops to transition to the next state when clocked. The next step is to develop a simplified Boolean ex-

pression for each D signal, based on the *present state* QB and QA signal levels. Usually the easiest way to do this is to write a Karnaugh map for each flip-flop input and derive the simplified expression for that flip-flop from the map.

Figure 12-23b shows a Karnaugh map for the DB input and another Karnaugh map for DA input. From these maps you can see that the simplified expression for the DB signal is QB + QA, and the simplified expression for the DA signal is $\overline{QA}$.

7. Implement the simplified expressions with logic gates.

The simplified expressions for this example can be implemented with an XOR gate and a piece of wire as shown in Figure 12-24.

8. Test the response of the circuit with a few present states to see if it transitions to the desired next state when clocked.

9. Design the output decoder by developing the simplified Boolean expression for each system output signal in terms of the *Present-State* Q values. Again, Karnaugh maps are usually the best way to do this.

The Hot and Cold signals in our example here are each only present during one state, so you don't need a Karnaugh map to develop these simplified expressions. Directly from the next-state table, you can see that Hot = $\overline{QB}$ AND QA. Cold = QB AND $\overline{QA}$.

10. Implement the simplified output decoder expressions with gates.

Since the D flip-flops we chose have both Q and $\overline{Q}$ outputs, the simplified expressions can be implemented with two, 2-input AND gates as shown in Figure 12-24.

IMPLEMENTING A 2-BIT BINARY UPCOUNTER WITH JK FLIP-FLOPS

In the preceding section we showed you the steps involved in developing a state machine design using D flip-flops. The main difference in designing a state machine based on JK flip-flops is that the next-state decoder expressions are different. To help fix the 10 design steps in your mind, we will repeat them here as we show you how the state machine described by the bubble diagram and the ASM chart in Figure 12-21 can be implemented with JK flip-flops.

1. Think carefully about what you want the system to do.

This example is the same as the one in the previous section. To refresh your memory, you want a 2-bit binary counter with an unconditional output signal named Hot which is asserted during state 01 and an unconditional output signal called Cold which is asserted during state 10.

2. Draw a state diagram or an ASM chart which represents the desired operation of the entire machine.

The state diagram and the ASM chart for this example are the same as those in Figure 12-21.

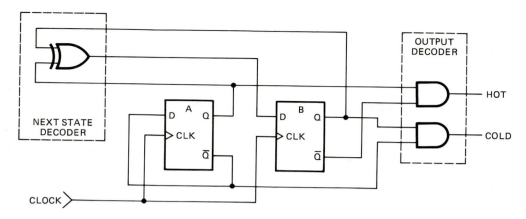

FIGURE 12-24   Circuit for a 2-bit binary upcounter with output decoder logic added.

3. Make a next-state table for the machine.

   Since this is the same basic design problem as the last example, Figure 12-22 shows the next-state table for it.

4. Decide how many flip-flops you need and what type you want to use.

   Here we have four states, so we need two flip-flops. The purpose of this example is to show you how to use JK flip-flops, so this is the obvious type to choose.

5. For each flip-flop, determine the input signals which are required for that flip-flop to transition from each present state to each next state when clocked.

   If you are using JK flip-flops, you need to determine the signal levels that are required on each J and K input for each present-state to next-state Transition and add columns for these to the next-state table. The process for this step is different from that for D flip-flops, so we will show you in detail how to do it.

   Figure 12-25 shows the *excitation table,* or what we like to call the *transition truth table,* for a JK flip-flop. This truth table shows the J and K signal levels that will cause a JK flip-flop to make each of the four possible transitions when clocked. For example, If Q = 0, this table shows that it will stay a 0 when the flip-flop is clocked if J = 0 and K = 0 or if J = 0 and K = 1. Since for this transition K can be either a 0 or a 1, it can be represented by an X as shown in the two rightmost columns of Figure 12-25.

   In a similar manner the Q output will transition from a 0 to a 1 if J = 1 and K = 0 or if J = 1 and K = 1. Since K can be either a 0 or a 1 for this transition, the two possibilities can be represented by J = 1 and K = X as shown in the two rightmost columns of Figure 12-25. As also shown in Figure 12-25, inputs of J = X and K = 0 will cause the Q output to stay a 1, and inputs of J = X and K = 1 will cause the Q output to transition from a 1 to a 0 when the flip-flop is clocked. You use the four resultant transition rules in Figure 12-25 to determine the J and K inputs required for each flip-flop transition of your state machine.

QN	→	QN+1	J	K	J	K
0	→	0	0	0		
0	→	0	0	1	0	X
0	→	1	1	1		
0	→	1	1	0	1	X
1	→	1	0	0		
1	→	1	1	0	X	0
1	→	0	1	1		
1	→	0	0	1	X	1

FIGURE 12-25   Transition truth table for a JK flip-flop.

Figure 12-26a shows the basic next-state table for this machine with columns added to show the J and K inputs required to cause each desired transition. Let's take a look at a few of the desired transitions to see how the values in the J and K columns were determined.

For the transition from state 00 to state 01, we want the QB output to stay a 0. According to the first rule in Figure 12-25, JB = 0 and KB = X will produce this result when the flip-flop is clocked. Therefore, we wrote a 0 in the JB column and an X in the KB column on the state 00 line of the table in Figure 12-26a.

For the 00 to 01 transition, we want the QA output to change from a 0 to a 1 when the flip-flop is clocked. According to the second rule in Figure 12-25, J = 1 and K = X will cause this transition. Therefore, we wrote a 1 in the JA column and an X in the KA column on the state 00 line of the table. In a similar manner we used the rules in Figure 12-25 to determine the J and K inputs required for each transition of each flip-flop and wrote the results in the table. For practice, work your way through the rest of the transitions.

6. Design the next-state decoder by developing a minimal Boolean expression for each flip-flop input based on the present-state QB and QA outputs.

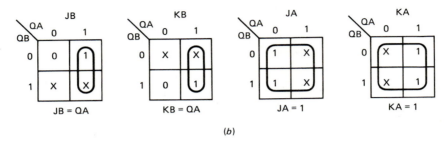

PRESENT STATE		NEXT STATE		SYSTEM OUTPUTS		FLIP-FLOP INPUTS			
QB	QA	QB	QA	HOT	COLD	JB	KB	JA	KA
0	0	0	1	0	0	0	X	1	X
0	1	1	0	1	0	1	X	X	1
1	0	1	1	0	1	X	0	1	X
1	1	0	0	0	0	X	1	X	1

(a)

(b)

FIGURE 12-26 *(a)* Next-state table with columns added to show required inputs for J and K flip-flops. *(b)* Karnaugh maps and reduced expressions for next-state decoder using JK flip-flops.

In most cases the easiest way to do this manually is with Karnaugh maps. Figure 12-26b shows the Karnaugh maps and the simplified expressions derived from each for the JB, KB, JA, and KA inputs. Because X's can be circled along with 1's in a Karnaugh maps, the resultant expressions are just JB = KB = QA and JA = KA = 1.

7. Implement the next-state decoder expressions with gates.

For the example here, the JA = KA = 1 means that JA and KA can simply be tied to +5 V. The JB = KB = QA means that JB and KB can simply be tied to the QA output. Figure 12-27 shows the completed circuit. This is the same circuit as the first two stages of the 4-bit binary synchronous counter we analyzed in Figure 9-4. As you can perhaps guess from this example, a state machine designed with JK flip-flops usually has a less complex next-state decoder than one designed with D flip-flops.

8. Test the response of the circuit with a few present states to see if the circuit will transition to the desired next state when it is clocked.

9. Design the output decoder by developing the simplified Boolean expression for each system output signal in terms of the present-state Q values.

The expressions for the output decoder for this example are the same as those in the previous section, Hot = $\overline{QB}QA$ and Cold = $QB\overline{QA}$.

10. Implement the simplified output decoder expressions with gates.

The output decoder circuitry for this example is the same as that for the circuit in Figure 12-24b, so we won't bother to repeat it here.

DESIGNING A COUNTER WITH AN UP/DOWN CONTROL INPUT

The previous design examples showed you how to develop a simple up-only counter with either D or JK flip-flops. In this section we show you how to design a 2-bit binary counter which will count up if a control input is high and will count down if the control input is low. This control signal is an example of an external signal that is fed into the next-state decoder of a Moore or Mealy machine along with the output signals from the flip-flops. To refresh your memory of this, review

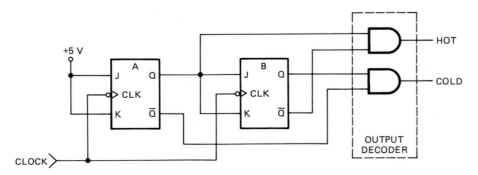

FIGURE 12-27 Circuit for 2-bit binary upcounter using JK flip-flops.

Figures 12-17a and 12-18a. Here's how you go about designing this circuit:

1. Think carefully about what you want the machine to do.

   For this example, we want a 2-bit binary counter which counts up when a control input signal, U, is high, and counts down when the U input is low. As in the previous examples, we want an unconditional output signal named Hot which is asserted during state 01 and an unconditional output signal named Cold which is asserted during state 10.

2. Draw a state diagram or an ASM chart for the machine.

   Figure 12-28a shows the bubble diagram for this machine, and Figure 12-28b shows an ASM chart for it.

3. Make a next-state table for the machine.

   Figure 12-29a shows a next-state table for our up/

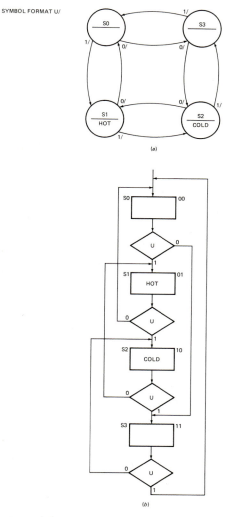

SYMBOL FORMAT U/

(a)

(b)

FIGURE 12-28 Algorithms for 2-bit binary up/down counter with Hot and Cold outputs. (a) Bubble diagram. (b) ASM chart.

down counter. Work your way through the entries in this table to see how they represent the state diagram and the ASM chart in Figure 12-28.

Since the U signal will be used as an input to the next-state decoder along with the present-state QB and QA signals, we could have made a column for U in the present-state section. However, the form shown in Figure 12-29a makes it easy to fill in Karnaugh maps and is consistent with that used by most authors.

4. Decide how many flip-flops you need and whether you want to use D, T, or JK flip-flops.

   This machine has four states, so we will use two D flip-flops to implement this design.

5. For each flip-flop, determine the input signals which are required for that flip-flop to transition from each present state to each next state in the table.

   Figure 12-29a shows the next-state table with columns added for the DB and DA inputs. Note that there are two sets of D columns because you have to determine the DB and DA inputs required for the case where U = 0 and for the case where U = 1.

6. Design the next-state decoder using Karnaugh maps.

   Since the next-state decoder for this example has three input variables, the Karnaugh maps each require eight squares. Figure 12-29b shows the Karnaugh maps for DB and DA. First note that, if you label the Karnaugh map as we have done here, the D columns in the next-state table correspond to the columns in the Karnaugh map. For example, the DB column under U = 0 in the next-state table contains the values for the U = 0 column of the Karnaugh map for DB. The D values are slightly out of order because of the way the variables are arranged in a Karnaugh map, but this is easily compensated for as you write values in the map.

   The simplified Boolean expression for DA taken from the map is $DA = \overline{QA}$. The map does not produce any simplification for the DB input, but from our discussion in Chapter 6 you may recognize the pattern in the DB Karnaugh map as that of a 3-input XNOR. The logic required for DB can then be represented as $DB = \overline{QA \oplus QB \oplus U}$.

7. Implement the simplified expression with logic gates.

   The expression for the DA input can be implemented by simply connecting a wire from the $\overline{QA}$ output to the DA input. The XNOR expression required for the DB input can be implemented with a single XOR gate device as shown in Figure 12-30. XOR gates 1 and 2 produce the XOR of QA, QB, and U. An XOR gate with one input tied high functions as an inverter, so XOR gate 3 inverts the XOR input to produce the desired XNOR function of QA, QB, and U.

8. Test the response of the circuit with a few present-state inputs to see if it transitions to the desired next state when clocked. We leave it to you to work a few values throughout the circuit in Figure 12-30.

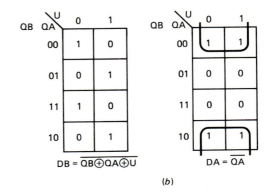

PRESENT STATE		NEXT STATE				SYSTEM OUTPUTS		FLIP–FLOP INPUTS			
		U = 0		U = 1				U = 0		U = 1	
QB	QA	QB	QA	QB	QA	HOT	COLD	DB	DA	DB	DA
0	0	1	1	0	1	0	0	1	1	0	1
0	1	0	0	1	0	1	0	0	0	1	0
1	0	0	1	1	1	0	1	0	1	1	1
1	1	1	0	0	0	0	0	1	0	0	0

(a)

$DB = \overline{QB \oplus QA \oplus U}$ $DA = \overline{QA}$

(b)

FIGURE 12-29 (a) Next-state table for 2-bit binary up/down counter with columns added for required D flip-flop inputs. (b) Karnaugh maps and reduced expressions for next-state decoder logic using D flip-flops.

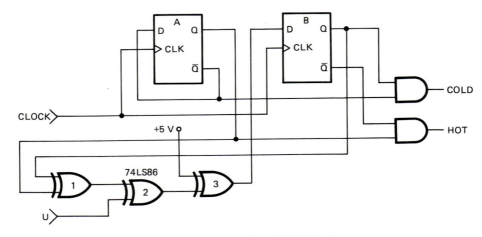

FIGURE 12-30 Circuit for 2-bit binary up/down counter made with D flip-flops.

9. Design the output decoder based on the present-state Q values.

The Hot and Cold output signals specified for this machine depend only on the output state. Therefore, the U signal does not enter into the output decoder expressions. The output decoder expressions for this machine, then, are the same as those for the machine in Figure 12-24b.

10. Implement the simplified output decoder expressions with gates.

The expressions for this circuit are the same as those for the previous example, so the output decoder circuit is the same as that in Figure 12-24b.

PREVENTING HANG STATES IN STATE MACHINES

The counter-type state machine examples we have used so far all cycle through the full binary count sequence. If a state machine does not use all the states possible with the number of flip-flops used, then you must take special precautions to make sure the machine does not get stuck in one or a series of the unused states.

As an example of this, suppose that we design a counter with four flip-flops that is intended to count from 0000 to 1001 and then return to 0000 and repeat the sequence when clocked. The unused states 1010, 1011, 1100, 1101, 1110, and 1111 represent possible problems. When the power is first turned on, for example, the circuit might initially come up in state 1101. Unless some provision is made to get the counter onto the main count sequence of 0000 to 1001, the counter may step through some unwanted sequence such as 1101, 1110, 1010, 1101, etc., over and over as it is clocked. As we explained in Chapter 9, these undesired extra sequence states are sometimes called *hang states*. Here are two common methods of preventing or overcoming hang states.

One method of attempting to solve the hang-state problem is to use a power-on set/reset circuit such as that in Figure 9-37 to force all the flip-flops to a state in the main sequence when the power is first applied. This approach gets the machine started properly, but if a noise pulse causes the machine to go to one of the hang states while it is running, the system will have to be reset to get it back to the main count sequence.

The second method of solving the hang-state problem is to design the machine such that it will step to one of the main sequence states on the next clock pulse if it accidentally gets into one of the unused states. Since it is usually not convenient to reset "real" circuits every time a power line spike occurs, the second method is the better one. The following example demonstrates the problem and shows you how to design a circuit so that it will always get back into the desired sequence of states.

Figure 12-31a shows you a simple synchronous circuit which we first introduced you to in Problem 4 in Chapter 9. As you perhaps determined then, if the outputs start at state 000, the circuit will step through the sequence of states shown in the top six lines of the next-state table in Figure 12-31b when it is clocked. The bottom two lines in Figure 12-31b show what will happen if the circuit happens to power up in state 101. As you can see, the outputs will transition to state 010 on the next clock pulse and then back to state 101 on the next clock

main sequence, so once the outputs get to this state, the circuit will stay in the main sequence from that point on. There is nothing particularly sacred about causing the circuit to transition from the hang states to state 000. We could have set up the J and K inputs to cause it to transition to any state in the main sequence. Sometimes choosing some other main sequence state to go to produces slightly simpler logic in the next-state decoder.

Figure 12-34a shows the Karnaugh maps we used to produce the simplified expressions for each of the J and K inputs, and Figure 12-34b shows one way to implement the simplified expressions. As you can see, this circuit

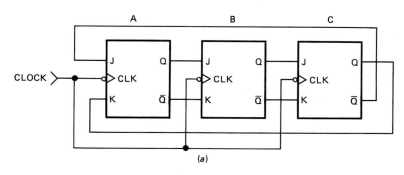

(a)

PRESENT STATE			NEXT STATE			FLIP-FLOP INPUTS					
QC	QB	QA	QC	QB	QA	JC	KC	JB	KB	JA	KA
0	0	0	0	0	1	0	1	0	1	1	0
0	0	1	0	1	1	0	1	1	0	1	0
0	1	1	1	1	1	1	0	1	0	1	0
1	1	1	1	1	0	1	0	1	0	0	1
1	1	0	1	0	0	1	0	0	1	0	1
1	0	0	0	0	0	0	1	0	1	0	1
0	1	0	1	0	1	1	0	0	1	1	0
1	0	1	0	1	0	0	1	1	0	0	1

(b)

FIGURE 12-31   (a) Simple synchronous circuit. (b) Sequence of states for circuit.

pulse. These two sequences are graphically represented in the bubble diagrams in Figure 12-32. From these diagrams you can see that if the outputs power up in any one of the main sequence states, the circuit will sequence through the main sequence states as desired. If the circuit powers up in one of the hang states, it will stay in the hang-state sequence and never get to the main sequence.

The cure for this problem is really very simple. You just design the circuit so that whatever state it powers up in, the next clock pulse will cause it to go to a state in the main sequence. Once the circuit is operating in the main sequence, it will remain there.

Figure 12-33 shows how you set up the next-state table for a redesign of the circuit in Figure 12-31a. Look carefully at the bottom two lines of the table because these are the only ones that are different from the original design. For each of these two states, we have set up the J and K inputs to produce a transition to state 000 when the next clock pulse occurs. State 000 is in the

is considerably more complex than the original, but the added gates are a small price to pay for the increased predictability of the circuit. As we show you in the next major section of the chapter, PLDs and PLD development tools such as ABEL make it possible to implement a hang-state free state machine almost as fast as you can write down the state diagram or ASM chart for it.

IMPLEMENTING STATE MACHINES WITH PROGRAMMABLE LOGIC DEVICES

In Figures 9-1 and 9-2 we showed you that a basic state machine consists of a next-state decoder, an output decoder, and flip-flops or registers which store the present-state levels on their outputs between clock pulses. The next-state decoder and the output decoder are simple combinational logic functions, so they can easily be implemented with the AND-OR part of a PAL as we showed you in an earlier section of the chapter. A register out-

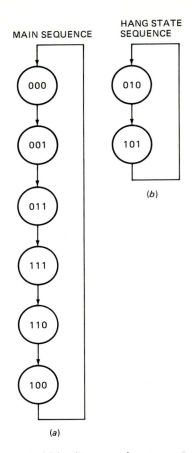

MAIN SEQUENCE

HANG STATE SEQUENCE

(a)

(b)

FIGURE 12-32   Bubble diagrams for state sequences in Figure 12-31b.

put PAL or a GAL such as the GAL16V8 then contains all the functions needed to implement a state machine. In this section of the chapter we teach you more about designing state machines and show you how to use a PLD programming language such as ABEL to easily implement state machines with register output PALS.

A 2-Bit Up/Down Counter with Synchronous Clear

DESIGN DESCRIPTION

As our first example of implementing a state machine with a PAL, we use a 2-bit up/down counter with a

CLEAR input and a single system output signal. We used a 2-bit circuit for this example to keep the size of the figures reasonable, but the techniques we show can be used for circuits with any number of bits.

The state machine we show you how to implement here is very similar to the one we described in Figure 12-28. The two major changes from that design are that the count sequence is different and we have added a synchronous clear capability. Figure 12-35a shows a block diagram of the circuit we want to implement for this example. Figure 12-35b shows the desired sequence of states and the truth table for the system output signal. Note that the sequence of binary states for this machine does not follow the normal binary count sequence.

If the UD input is high, we want the flip-flop outputs to step from state 0 to state 1, from state 1 to state 2, etc., as they are clocked. If the UD input is low, we want the flip-flop outputs to step through the listed states in the reverse order. If the CLR input is high, we want the QB and QA outputs to go to 0's synchronously, or, in other words, when the next clock pulse occurs. When the machine is in state 1 or state 3, we want the Y output to be unconditionally asserted.

Figure 12-36a shows a state diagram which summarizes how we want this machine to work. Note that if the UD input is high and CLR is low, the machine moves counterclockwise through the state sequence as it is clocked. If the UD input is low and the CLR input is low, the machine will step clockwise through the state sequence as it is clocked. The bubble labeled N in Figure 12-36a is a shorthand way to represent "any state." The arrow from this bubble and the X,1 next to the arrow tells you that if the CLR input is asserted, the machine will go to state 0 when the next clock pulse occurs, regardless of the level on the UD input. Using this N bubble simplifies the state diagram because you don't have to draw lines from each state bubble back to state 0 to represent the effect of the CLR input. The Y output signal is a Moore, or unconditional, output, so it is written inside the bubbles which represent the states where it is asserted.

Figure 12-36b shows an ASM chart for our state machine. Note in this diagram that we put the CLR decision box before the UD decision box to show that the CLR operation takes precedence over the count operation. If CLR = 1, the machine will transition to state S0, regardless of the level on the UD input. If CLR = 0, then UD = 1 will cause the machine to step through the state

PRESENT STATE			NEXT STATE			FLIP-FLOP INPUTS					
QC	QB	QA	QC	QB	QA	JC	KC	JB	KB	JA	KA
0	0	0	0	0	1	0	X	0	X	1	X
0	0	1	0	1	1	0	X	1	X	X	0
0	1	1	1	1	1	1	X	X	0	X	0
1	1	1	1	1	0	X	0	X	0	X	1
1	1	0	1	0	0	X	0	X	1	0	X
1	0	0	0	0	0	X	1	0	X	0	X
0	1	0	0	0	0	0	X	X	1	0	X
1	0	1	0	0	0	X	1	0	X	X	1

FIGURE 12-33   Next-state table for redesigned circuit in Figure 12-31.

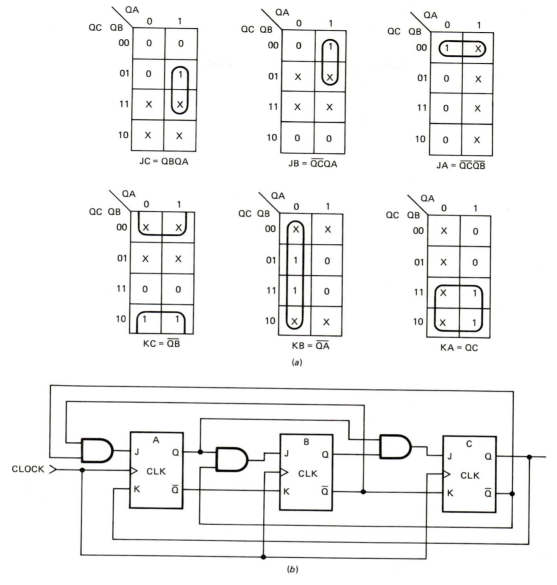

FIGURE 12-34 For next-state table in Figure 12-33: (a) Karnaugh maps. (b) Circuit.

sequence S0, S1, S2, S3, etc. If CLR = 0 and UD = 0, the machine will step through the state sequence S0, S3, S2, S1, etc. Again note that since the Y output signal is unconditional, we write it in the boxes which represent the states where it is asserted.

When you use a PAL development tool such as ABEL to implement a state machine, you can usually go directly from the state diagram to a source file for the PAL. However, for this first example, we will show you both the traditional method and the ABEL method so that you can compare them.

TRADITIONAL DESIGN METHOD

As we explained in the last major section of the chapter, the next step in the traditional method of designing a state machine is to develop a next-state table such as the one in Figure 12-37a for it. From this next-state ta-

ble you determine the logic level that must be present on each D input to cause the desired state transitions when a clock pulse occurs. Figure 12-37a also shows these required D input levels.

Now you develop a simplified logic expression for each D input in terms of the QA and QB output signals and the UD and CLR input signals. Figure 12-37b shows the Karnaugh maps and the reduced expressions for the two D inputs. Note that the Karnaugh maps must have 16 boxes to represent all the possible combinations of QB, QA, UD, and CLR, the four signals applied to the inputs of the next-state decoder. One of the reasons we did not show you a 4-bit up/down counter with clear here is that simplifying the next-state decoder equations would require six-variable Karnaugh maps. Working with maps this large is tedious and prone to errors.

Next, implement the simplified decoder expressions with gates. The reduced equations shown in Figure

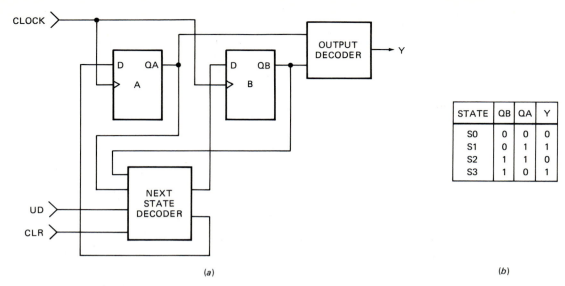

FIGURE 12-35   First ABEL state machine design example. *(a)* Block diagram. *(b)* State assignments and output decoder truth table.

STATE	QB	QA	Y
S0	0	0	0
S1	0	1	1
S2	1	1	0
S3	1	0	1

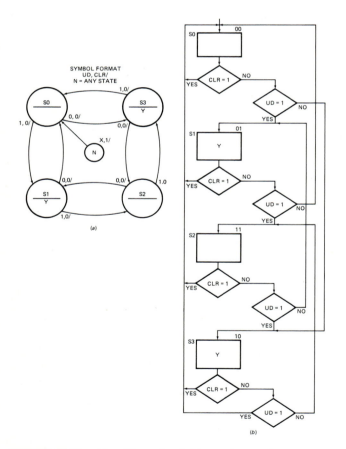

FIGURE 12-36   Algorithms for 2-bit up/down counter with synchronous clear. *(a)* State diagram. *(b)* ASM chart.

12-37*b* for the next-state decoder could easily be implemented with gates. As shown by the truth table in Figure 12-37*a*, the output decoder function is just QA $\oplus$ QB,

so it can also be implemented with simple gates. Now that we have reviewed the traditional design method, let's take a look at how you can do it with ABEL or a similar PAL development tool.

USING ABEL AND A PAL TO IMPLEMENT A STATE MACHINE

Figure 12-38 shows an ABEL source file which will implement our 2-bit state machine in a Lattice GAL16V8. When you run ABEL on this file, it will automatically produce the simplified expressions for both the next-state decoder and the output decoder, generate the fusemap to program the GAL16V8, and simulate the operation of the machine to make sure it steps through the desired sequence of states as it is clocked. Although this example uses only 2 bits, the basic format shown in Figure 12-38 can be expanded to implement larger state machines. An exercise in the accompanying laboratory manual gives you a chance to do this. For now let's work our way through the file to see how you tell ABEL what you want it to do.

Much of the format of the ABEL file in Figure 12-38 is the same as the format of the combinational examples we showed you earlier in the chapter. To help you pick them out, we have used uppercase letters for the key words, input signals, and output signals in the program.

The MODULE directive and the END directive give the module a name and "bracket" the rest of the statements in the file. The FLAG statement can be used to show several different processing options in a program. In this example, the FLAG '-R3' statement tells ABEL to use level 3 reduction, the maximum, on the equations for the next-state and output decoders.

The DEVICE statement tells ABEL the device file to use when processing the file. The R in P16V8R tells ABEL to use a device file which programs the output macrocells in the GAL16V8 as registered for those outputs which are clocked.

STATE NAMES	PRESENT STATE		NEXT STATE UD = 0 CLR = 0		NEXT STATE UD = 1 CLR = 0		NEXT STATE UD = X CLR = 1		SYSTEM OUTPUT
	QB	QA	QB	QA	QB	QA	QB	QA	Y
S0	0	0	1	0	0	1	0	0	0
S1	0	1	0	0	1	1	0	0	1
S2	1	1	0	1	1	0	0	0	0
S3	1	0	1	1	0	0	0	0	1

PRESENT		INPUTS FOR NEXT STATES					
QB	QA	DB	DA	DB	DA	DB	DA
0	0	1	0	0	1	0	0
0	1	0	0	1	1	0	0
1	1	0	1	1	0	0	0
1	0	1	1	0	0	0	0

(a)

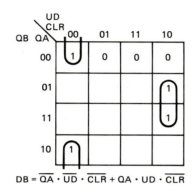 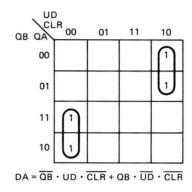

$$DB = \overline{QA} \cdot \overline{UD} \cdot \overline{CLR} + QA \cdot UD \cdot \overline{CLR}$$

$$DA = \overline{QB} \cdot UD \cdot \overline{CLR} + QB \cdot \overline{UD} \cdot \overline{CLR}$$

(b)

FIGURE 12-37 Traditional design method. (a) next-state table. (b) Karnaugh maps and simplified next-state decoder equations.

The PIN statements assign input and output signals to specific pins on the device. As before, you use the logic diagram for the GAL16V8 in Figure 12-8 to help decide these pin assignments. The main constraint here is that, for clocked designs, the clock signal must be applied to pin 1 of the GAL16V8. The GAL16V8 allows the output macrocells to be individually programmed, so that you can have both register outputs and combinational outputs from the same device. In this design, for example, we want QA and QB to be register outputs and Y to be a simple combinational output. The QB and QA outputs are internally fed back into the AND matrix and decoded to produce the Y output signal.

In ABEL state machine designs, the term .C. is used to represent a clock pulse in test vectors. To avoid typing the periods, we assigned the name ck to .C. and X to .X.

The next section of the program in Figure 12-38 assigns names to the four binary states of the machine. This process is often referred to as "making state assignments." Note that the state names identify the position of the state in the sequence of states, not the decimal equivalent of the Q outputs during that state. As you will see in a later example in this chapter, we usually assign a state name which is descriptive of the action which occurs during that state, rather than just a state number. The ^b in these state assignments shows

that the data is in binary. We like to put these assignments in binary so that we can clearly see the sequence of 1's and 0's. A fine point in state machine design is that if you assign the binary values so that just 1 bit at a time changes as you go from state to state, the next-state decoder will usually require less logic than other sequences.

The next section of the program in Figure 12-38 contains some substitutions which make the program less cluttered and more readable. The mode = [UD,CLR] statement tells ABEL the set of input signals that will be represented by the term mode in later sections of the program. The other three statements with this one give names to the different combinations of these input signals so that you can refer to each set of control signal values by name rather than having to type in the binary values. Note that in the clear = [X,1] statement, we use the X to indicate a "don't care" for the UD signal. In other words, if the CLR signal input is a 1, we want the outputs to go to zeros, regardless of the level on the UD input.

ABEL and other PAL development programs allow you to enter the design data for the next-state decoder in the form of Boolean expressions derived from the next-state table or in a state diagram form. We like the state diagram form because it can be derived directly from a bubble diagram or ASM chart such as those in

```
MODULE UP_DOWN   FLAG '-R3'
TITLE 'TWO BIT UP/DOWN COUNTER WITH SYNCHRONOUS CLEAR'

        U8        DEVICE      'P16V8R'; "note R for registered output
        CLK,UD,CLR PIN        1,2,3;
        QB,QA      PIN        19,18;
        Y          PIN        17;

        QB,QA     ISTYPE    'POS';

"Assign name to special constant
        ck = .C.;                    X = .X.;
"Assign state names to binary output states
        ST = [QB,QA];
        S0 = ^b00;                   S2 = ^b11;
        S1 = ^b01;                   S3 = ^b10;
"Assign names to modes produced by input control signals
        mode = [UD, CLR];       clear= [X,1];
        down = [0,0];           up =    [1,0];

STATE_DIAGRAM [QB,QA]
        "present state              mode            next state
        STATE S0:        CASE (mode==clear)     :S0;
                              (mode == down)    :S3;
                              (mode == up)      :S1;
                         ENDCASE;
        STATE S1:        CASE (mode==clear)     :S0;
                              (mode == down)    :S0;
                              (mode == up)      :S2;
                         ENDCASE;
        STATE S2:        CASE (mode==clear)     :S0;
                              (mode == down)    :S1;
                              (mode == up)      :S3;
                         ENDCASE;
        STATE S3:        CASE (mode==clear)     :S0;
                              (mode == down)    :S2;
                              (mode == up)      :S0;
                         ENDCASE;
EQUATIONS
        [QB,QA] := (mode == clear) & [0,0];      "clocked assignment
        Y  =   !QA&QB # QA&!QB;                   "combinational assignment

TEST_VECTORS ([CLK,mode] -> [QB,QA])
        [ck,clear] -> S0;       "Puts register outputs in known state
        [ck,up]    -> S1;       [ck,up]    -> S2;
        [ck,up]    -> S3;       [ck,up]    -> S0;
        [ck,down]  -> S3;       [ck,down]  -> S2;
        [ck,down]  -> S1;       [ck,down]  -> S0;
        [ck,down]  -> S3;

TEST_VECTORS ([CLK,ST] -> [ST, Y])        "ST = [QB,QA]
"      Load State on QB,QA              Test Y output with State
        [.P.,S0] -> [X,X];              [X,S0] -> [S0,0];
        [.P.,S1] -> [X,X];              [X,S1] -> [S1,1];
        [.P.,S2] -> [X,X];              [X,S2] -> [S2,0];
        [.P.,S3] -> [X,X];              [X,S3] -> [S3,1];

END UP_DOWN
```

FIGURE 12-38   ABEL source file to implement 2-bit state machine in GAL16V8 PAL.

Figure 12-36 without having to write equations. The STATE_DIAGRAM section of the ABEL program in Figure 12-38, for example, follows the bubble diagram in Figure 12-36*a* very closely, so you should be able to see how they correspond. Keep the assignments we explained before in the back of your mind as you read through these.

The CASE-ENDCASE structure is one way to enter a state diagram design. The == in these statements can be read as "is equal to." The first CASE-ENDCASE block in Figure 12-38 tells you that, for the case where the machine is in state S0 and the mode is equal to clear, the machine

will stay in state S0 when the next clock pulse occurs. It also tells you that for the case where the mode is equal to down, the machine will go to state S3, and for the case where the mode is equal to up, the machine will go to state S1 when the next clock pulse occurs. Remember, mode in these statements represents the levels present on the input control signals UD and CLR.

If you are not familiar with the CASE structure from other programming languages, then you can think of each of these CASE-ENDCASE sections in Figure 12-38 as an IF-THEN-ELSE statement. You might translate the response of the machine in a present state of S0 as follows:

IF mode is equal to clear THEN
    go to state S0 when a clock pulse occurs
ELSE IF mode is equal to down, THEN
        go to state S3 when a clock pulse occurs
    ELSE IF mode is equal to up THEN
        go to state S0

Incidentally, when you are designing a state machine with a tool such as ABEL, you can eliminate "hang states" in the design by just writing STATE__DIAGRAM entries which step the machine from unused states to main sequence states.

The CASE structure in the STATE__DIAGRAM doesn't indicate directly that the CLR input has priority as shown in Figure 12-36b, but defining clear = [X,1] will make sure that the next-state decoder equations implement clear in this way.

Note in Figure 12-38 that each CASE-ENDCASE block contains a statement which tells ABEL that if the UD and CLR inputs are in the clear mode, then we want the QB and QA outputs to go to state S0 when the next clock pulse occurs. An alternative is to use a single equation to tell ABEL to implement the desired clear function. The equation [QB,QA]:=(mode==clear) &[0,0] in the EQUATIONS section of Figure 12-38 does just that.

The syntax of this equation may seem a little strange, but it tells ABEL that if mode is equal to clear, assert the values shown to the right of the & on the QB and QA outputs when the next clock pulse occurs. The : = in the expression is what tells ABEL to assert to 0's on the outputs when the next clock pulse occurs. This is called a *clocked assignment*. In Figure 12-38 we show both the STATE__DIAGRAM method of implementing the clear function and the EQUATIONS method, but actually you should only use one or the other of these. If you use the equation method, you have to include the QB, QA ISTYPE 'POS'; statement shown in Figure 12-38 so that the actual output pins go to 00 when CLR is asserted. Otherwise, as shown in the GAL16V8 macrocell in Figure 12-5, the default for the programmable output buffers is inverting and the Q outputs will go to 11 when the register outputs are cleared to 00.

The second expression in the EQUATIONS section of the program is for the output decoder. This expression, Y = !QA & QB # QA & !QB, is an example of a purely combinational expression. The Y output will be asserted high as soon as the QA and QB have levels which make the right-hand side of the expression true. The single = in this expression tells ABEL to program the output macrocell associated with the Y output as a combinational output rather than as a register output.

Finally in Figure 12-38 we get to the TEST__VECTORS section. As we told you earlier, the ABEL Simulate subprogram uses the test vectors you supply to see if the fusemap implements the desired logic functions. The test vectors are also used by the PAL programmer to determine if the programmed PAL functions correctly.

To test the operation of a state machine, you have to specify the clock signal as one of the input signals along with the UD and CLR control signals. The ([CLK,mode] → [QB,QA]) line at the top of the TEST__VECTORS sec-

tion shows the format we used to test this machine. Remember that the name mode here represents the set of input signals [UD,CLR].

When a state machine is first powered up, it may start in any state. There are two ways of getting the machine into a known state so you can then step it from that state through a sequence of states with the test vectors. One way to get the machine into a known state is to write the first test vector so that it takes the machine to state S0 when the next clock pulse occurs. The first test vector, [ck,clear] → S0; (Figure 12-38) is an example of how to do this. Once the machine is in S0, the following test vectors in the first section step the machine up through the state sequence and then back down again.

Another method of getting a state machine to a known state is with the "register preload" option shown in the second set of test vectors in Figure 12-38. Most of the newer registered PALs allow you to preload a specified value on the outputs of the registers using .P. instead of .C. in the CLK position in a test vector. This preload option is useful for testing state machines which do not have an external pin that causes them to go to a known state when clocked. Another use of the preload feature is to put the Q outputs in known states for testing combinational outputs which are derived from the Q outputs as shown in Figure 12-38.

To test the Y output in our example state machine, for example, we preload the Q outputs with a known state and then write a test vector which determines if the Y output is correct for that state. Figure 12-38 shows how to do this.

The [.P.,S0] → [X,X] test vector of the second group in Figure 12-38 loads a 0 on QB and a 0 on QA (state S0). The [X,X] on the right side of the expression is required because the outputs are being changed to state S0 with a direct set/reset operation instead of the normal clocked operation. The [X,S0] → [S0,0] test vector in the lower group in Figure 12-38 is just a simple combinational one which tests whether the loaded state produces the correct result on the Y output. The rest of the vectors in Figure 12-38 load and test the effect of each of the remaining states.

The two sets of test vectors in Figure 12-38 can actually be combined into one set, but in more complex designs, such as the elevator controller at the end of the chapter, we often write them separately to make them easier to read. For cases where you want to combine them, Figure 12-39 shows how to do this for the test vectors in Figure 12-38.

Figure 12-40 shows the reduced equations produced by running ABEL on the source file in Figure 12-38 with a reduction level of -R3. Since we used the QB, QA ISTYPE 'POS'; statement, the equations are the same as those produced by circling 1's in the Karnaugh maps in Figure 12-37b.

We used the preceding example to show you how to implement a state machine using ABEL and a programmable logic device, but we didn't say anything about what you might use a state machine such as this for. In the next section we show you how to design a state machine which controls the operation of an elevator.

```
TEST_VECTORS  ([CLK,ST, mode ] -> [ST,Y])      "ST = [QB,QA]"
               [.P.,S0, X    ] -> [X, X];
               [ck, S0, up   ] -> [S1,1];
               [ck, S1, up   ] -> [S2,0];
               [ck, S2, up   ] -> [S3,1];
               [ck, S3, up   ] -> [S0,0];
               [ck, S0, down ] -> [S3,1];
               [ck, S3, down ] -> [S2,0];
               [ck, S2, down ] -> [S1,1];
               [ck, S1, down ] -> [S0,0];
               [ck, S0, down ] -> [S3,1];
               [ck, S3, clear] -> [S0,0]; "Test clear function
```

FIGURE 12-39   Format for combining state machine and combinational test vectors from Figure 12-38.

```
                                               Page 1
    ABEL(tm) 3.00a  -  Document Generator    22-Jul-88 09:32 AM
    TWO BIT UP/DOWN COUNTER WITH SYNCHRONOUS CLEAR
    Equations for Module UP_DOWN

    Device U8
    - Reduced Equations:

        QB := (!CLR & QA & UD # !CLR & !QA & !UD);
        QA := (!CLR & QB & !UD # !CLR & !QB & UD);
        Y = (QA & !QB # !QA & QB);
```

FIGURE 12-40   Reduced equations produced by running ABEL on the source file in Figure 12-38.

A State Machine Controller for a Two-Story Elevator

State machines are an important part of many digital systems. Microprocessors, which we introduce at the end of the next chapter, are essentially complex state machines. As a final state machine example in this chapter, we wanted to show you a complete state machine application example. However, most of the application examples we initially thought of were either boring or took too much space to explain. The basic idea for the two-story elevator controller example we have developed here came from an article by Dean Suhr in the *Lattice GAL Handbook*. We chose this example because it demonstrates most of the principles in the chapter and because we thought most people could relate to it from their personal experience. In the following sections we lead you through the actual process we used to design this controller. Because this is a relatively complete design example, you probably will need to read through this section a couple times to pick up the details. However, if you work your way through this section, you should have a good understanding of the design and use of a state machine.

DEVELOPING A STATE DIAGRAM

As we told you earlier in the chapter, the first step in designing anything is to do a great deal of thinking about what you want the end product to do. For this example we wanted to control a simple two-story elevator such as the one in Figure 12-41a. The controller needs to respond to input signals from users on either floor and in the elevator. It needs to produce output signals to control the motor that moves the elevator up or down, output signals that open or close the first floor door at the appropriate times, and output signals that open or close the second floor door at the appropriate times. Also the

controller needs output signals to drive some displays which show the position or motion of the elevator to users. To help you visualize this, Figure 12-41b shows a block diagram of the basic controller.

To make sure that the elevator proceeds from floor to floor in an orderly manner, and that the doors open and close at appropriate times, we will implement the controller with a state machine rather than with a combinational logic circuit. The state machine is clocked by the CLOCK input signal shown in Figure 12-41b.

The next step in designing the state machine controller is to think about the different states we want the controller to have and the action(s) we want performed in each state. Figure 12-42 shows our initial bubble diagram for the controller.

The definitions of the states are as follows:

FLOOR1 —Arrive at floor 1, open door, go to WAIT1 on next clock.

WAIT1 —Close door, wait on floor 1 for service request from floor 1, person in elevator, or floor2

GO__UP —Turn on elevator motor to go to FLOOR2

FLOOR2 —Arrive at floor 2, open door, go to WAIT2 on next clock.

WAIT2 —Close door, wait on floor 2 for service request from floor 2, person in elevator, or floor 1

GO__DOWN —Turn on elevator motor to go to FLOOR1

We decided that we would clock this controller every 10 seconds. This time was arrived at by riding an elevator and measuring the time it took to go from state to state. The state that the controller will go to when a clock

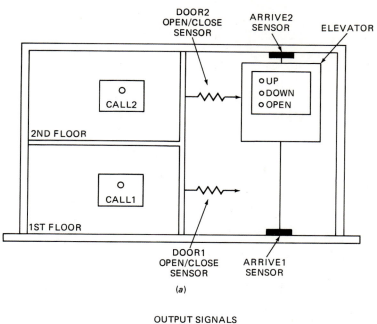

(a)

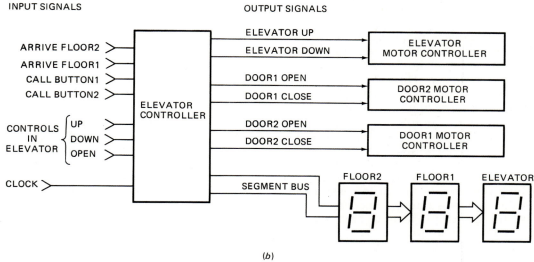

(b)

FIGURE 12-41   Two-story elevator. (a) Cutaway view showing control buttons and sensors. (b) Block diagram of controller.

pulse occurs is determined by the input control signals, so the next logical step was to determine the expected input signals. Once we had these signals, we could identify the signal(s) which produce each transition in the bubble diagram. To start, we made the following list of the needed user input signals:

CALL1   —Button on wall on FLOOR1
CALL2   —Button on wall on FLOOR2
UP   —Button inside elevator to send elevator up
DOWN   —Button inside elevator to send elevator down
OPEN   —Button inside elevator to open door

Next we made a list of the sensor signals we would need to determine whether the elevator was up or down and whether the door was closed. The initial list looked

something like this:

ARRIVE2   —Elevator on second floor
ARRIVE1   —Elevator on first floor
DOOROPEN1   —Door on floor 1 open
DOORCLOSED1 —Door on floor 1 closed
DOOROPEN2   —Door on floor 2 open
DOORCLOSED2 —Door on floor 2 closed

NOTE: The elevator has three doors, one on floor 1, one on floor 2, and one that travels up and down with the elevator. For this example we assume the inner door is mechanically opened and closed by the outer door on the floor where the elevator is located. Therefore we don't need a separate set of signals for it.

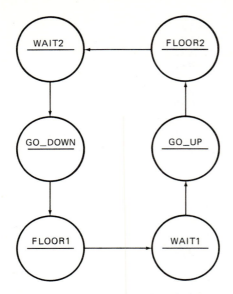

FIGURE 12-42 First state diagram for elevator controller.

The two lists of signals above seemed too long, so we thought about how we could combine signals and how we might get double usage from a single signal.

The first simplification came with the realization that the CALL2 button and the UP button both tell the controller to send the elevator up. These buttons then can be wired in parallel to produce a single signal which we called UPORCALL2. Likewise, the DOWN button and the CALL1 button both tell the controller to send the elevator down, so these buttons can be wired in parallel to produce a single signal which we called DNORCALL1.

The next simplification came to mind when we realized that the signals from the user buttons have to be latched. Latching is necessary so that the user doesn't have to stand there holding a button until the elevator comes, or until the elevator gets to the desired floor. Figure 12-43 shows the circuits we used to latch the signals from the UP, CALL2, OPEN, DOWN, and CALL1 buttons. As we will show you later, the latches are reset at the appropriate time by a signal from the controller. A fringe benefit of these latch circuits is that, as we explained in Chapter 8, they also debounce the switch signals.

The set and reset action of the latches in Figure 12-43 brought to mind a way to simplify the elevator position and door position signals. To determine if the elevator was up or down, we connected two spring-loaded switches to an SR latch shown in Figure 12-44a. If the elevator is in the down position, the bottom switch will be closed, the latch will be reset, and ARRIVE2 will be low. When the elevator reaches the second floor, the upper switch will be closed, the latch will be set, and the ARRIVE2 signal will be high. As shown in Figure 12-44b, we used the same scheme to produce a single signal which indicates whether each door is closed or open. If, for example, door 1 is open, the upper switch in the door 1 circuit will be closed, the latch will be set, and the DOOROPEN1 signal will be high. If door 1 is closed, the lower switch will be closed, the latch will be reset, and the DOOROPEN1 signal will be low. The six input con-

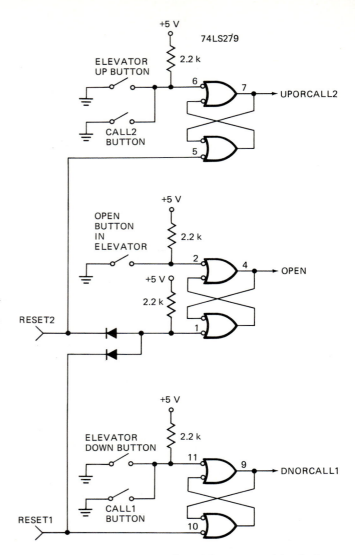

FIGURE 12-43 Circuits used to debounce and latch the UPORCALL2, DNORCALL1, and OPEN signals for the elevator controller.

trol signals we ended up with then were:

UPORCALL2	DNORCALL1
OPEN	ARRIVE2
DOOROPEN1	DOOROPEN2

After we worked out the details of the input control signals, we went back to the basic state diagram and labeled each transition with the expressions for the signal conditions which cause that transition. Figure 12-45 shows the result of this. We labeled the transitions with expressions rather than 1's and 0's because, when you have many input variables, expressions are more understandable and because the expressions can be directly used in an ABEL source file. Let's take a look at these transitions.

First note in Figure 12-45 that no exit conditions are shown for state FLOOR1. This means that when the next clock pulse occurs, the machine will unconditionally go from state FLOOR1 to state WAIT1. Likewise, a clock

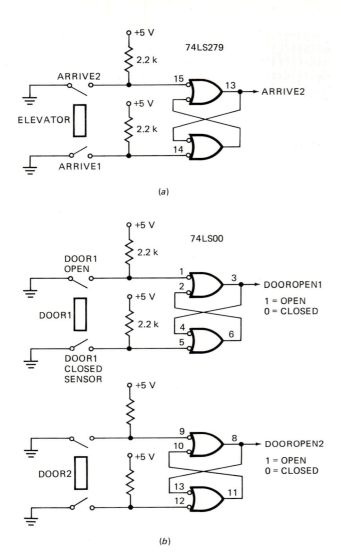

FIGURE 12-44 *(a)* Circuit for debouncing elevator position signal, ARRIVE2. *(b)* Circuits for producing door position signals DOOROPEN1 and DOOROPEN2.

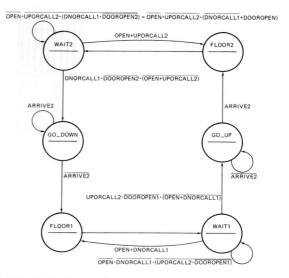

$$\overline{OPEN \cdot UPORCALL2 \cdot (DNORCALL1 \cdot \overline{DOOROPEN2})} = \overline{OPEN} \cdot \overline{UPORCALL2} \cdot (DNORCALL1 + DOOROPEN)$$

OPEN+UPORCALL2

DNORCALL1·DOOROPEN2·(OPEN+UPORCALL2)

UPORCALL2·DOOROPEN1·(OPEN+DNORCALL1)

OPEN+DNORCALL1

$$\overline{OPEN \cdot DNORCALL1 \cdot (UPORCALL2 \cdot \overline{DOOROPEN1})}$$

FIGURE 12-45 Second version of state diagram for elevator controller showing conditions which cause state transitions.

pulse is all that is needed to cause a transition from state FLOOR2 to state WAIT2.

Next in Figure 12-45 let's take a look at states GO__UP and GO__DOWN, which also have simple exit conditions. If, for example, the machine is in state GO__UP, it will transition to state FLOOR2 on the next clock pulse after the ARRIVE2 input signal goes high. The looped arrow on the GO__UP bubble shows that the machine will stay in state GO__UP as long as the ARRIVE2 signal is low. It may seem redundant to include this loop, but it is part of the state diagram convention and it reminds you to include the "stay in the same state" case in your ABEL source file.

Finally in Figure 12-45 let's look at state WAIT1, which has two exit routes and a stay in the same state loop. In state WAIT1, remember, the elevator simply stays on floor 1 with the door closed and waits for a user command. If the machine is in state WAIT1 and a user inside the elevator presses the OPEN button, or a user on floor 1 presses the CALL1 button, the machine will return to

state FLOOR1 and open the door so passengers can enter. The conditions for a transition from WAIT1 to FLOOR1 then are represented by the expression OPEN + DNORCALL1.

To prevent a possible conflict situation, the expression for a transition from WAIT1 to GO__UP is somewhat more complex. Basically the expression says that the controller will go to state GO__UP if a user inside the elevator presses the UP button or a user on floor 2 presses the CALL2 button, the door is closed, and no one has pressed the OPEN button or the CALL1 button.

The $\overline{DOOROPEN1}$ part of the expression is necessary to prevent the elevator from starting upward with the door still partially open. You can imagine the problems that might occur if the elevator started upward with someone only halfway in the door.

The $\overline{OPEN + DNORCALL1}$ part of the expression gives priority to a user request from floor 1. In other words, if the elevator is on floor 1 and receives a CALL1 signal and a CALL2 signal at the same time, it will respond to the CALL1 signal before it responds to the CALL2 signal. This is more efficient and gives you a little glimpse of the amount of thinking that must go into designing even the simplest real system.

The expression next to the looped arrow on the WAIT1 bubble in Figure 12-45 represents the conditions which will cause the controller to stay in state WAIT1. Using DeMorgan's theorem to split the long overbar in the expression gives the expression $\overline{OPEN} \cdot \overline{DNORCALL1} \cdot (\overline{UPORCALL2} + DOOROPEN1)$, which is perhaps more intuitive. It tells you that the controller will stay in state WAIT1 if no one has requested service or the door is open. If you look closely at the expressions around the WAIT2 bubble in Figure 12-45, you should see that these expressions have the same form as those for the WAIT1 transitions that we have just discussed.

Before we leave the discussion of the state diagram,

there is one more major point to consider. The state diagram in Figure 12-45 has six states, so it will require three flip-flops to implement. The maximum number of states for a machine with three flip-flops is eight, so if we are only using six, there are two undefined states. As we showed in an earlier discussion of hang states, you must include these undefined states in the design. In other words you must specify the response you want the machine to make if it powers up in one of these undefined states. If you don't do this in an instance where the machine is controlling a real system such as our elevator, a catastrophe may result.

Figure 12-46 shows our new, improved bubble diagram with these two undefined states added. Initially we were going to have each undefined state transition to FLOOR2 if ARRIVE2 was high and transition to FLOOR1 if ARRIVE2 was low. Then the thought occurred, what happens if the controller comes on in, for example, UNDEFND2, and the elevator is halfway between floors? Going directly to state FLOOR1, for example, would cause the door on floor 1 to open without the elevator in front of it. This would leave an opening which a child might fall through into the basement.

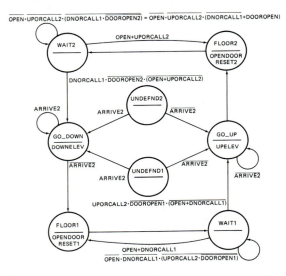

FIGURE 12-46   Final version of state diagram for elevator controller.

To solve this problem, we decided that if the controller powered up in one of the undefined states, it should close both doors and go to either state GO_DOWN or state GO_UP, depending on which floor it was last on. If the elevator was last on floor 1, ARRIVE2 will be low. We use this condition to cause a transition to state GO_UP. In state GO_UP the elevator motor will be activated to move the elevator to floor 2. Once the elevator is on floor 2, it is in the main sequence and will continue from there as specified by user commands.

If ARRIVE2 is high and the controller powers up in one of the undefined states, it will transition to state GO_DOWN. In state GO_DOWN the elevator motor will be activated to move the elevator to floor 1. Once the elevator reaches floor 1 and state FLOOR1, it is in the

main sequence and will remain there. You might think of what we have implemented here as a "mechanical reset" of the elevator to a known physical position. This must almost always be done in a controller which operates a mechanical system.

DESIGNING THE OUTPUT DECODER

As the next step in the design of our elevator controller, we started to think about the output decoder signals that would be required to control the motors and user request buttons in the system. The first signals that came to mind were the signals required to reset the UPORCALL2, DNORCALL1, and OPEN latches shown in Figure 12-43. We named these two signals RESET2 and RESET1 as shown.

The next output signals that came to mind were the signals required to control the motor which moves the elevator up and down. Figure 12-47a shows the actual circuit we used to do this on our model. If the UPELEV input is high and the $\overline{\text{DOWNELEV}}$ input is high, the upper output transistor will be on and the motor will rotate in a direction which takes the elevator up. If the UPELEV input is low and the $\overline{\text{DOWNELEV}}$ input is low, the lower output transistor will be on and the motor will rotate in a direction which takes the elevator down. A low on UPELEV and a high on $\overline{\text{DOWNELEV}}$ will turn off both output transistors and leave the motor unpowered. A high on UPELEV and a low on $\overline{\text{DOWNELEV}}$ will turn on both output transistors and short out the power supplies. Obviously this set of input conditions should be carefully avoided. Incidentally, this circuit is an example of how real systems often have both active high and active low signals. We could put an inverter on the $\overline{\text{DOWNELEV}}$ signal to make it active high, but there is no need to do this, because we can simply write the ABEL equations to produce the desired logic level on this input.

Finally, we thought about the signals to control the motors that open and close the doors on floor 1 and floor 2. As with the elevator motor, each of these motors must rotate in one direction to open the door, rotate in the other direction to close the door, or be off when the door is in the desired position. To control each door motor we used the same type of motor controller circuit that we used to control the elevator motor. As shown in Figure 12-47b, we named the four signals for the two door controllers CLOSEDOOR1, $\overline{\text{OPENDOOR1}}$, CLOSEDOOR2, and $\overline{\text{OPENDOOR2}}$.

Besides controlling the operation of the elevator and doors, we want the controller to display the status of the elevator on three, 7-segment LED displays. One display will be on floor 1, one on floor 2, and one in the elevator. Figure 12-48 shows the symbol we want to display for each state and the seven-segment code required to produce each symbol. We decided to use common anode displays, but since we have three displays to drive, we can't drive the segments directly off a PAL. The inverting buffers we used to drive the segments means that a high from the PAL lights a segment. Note in Figure 12-48 that the output levels for the d segment are the same as those for the e segment, so these two buffers can be driven by the same pin of the output decoder.

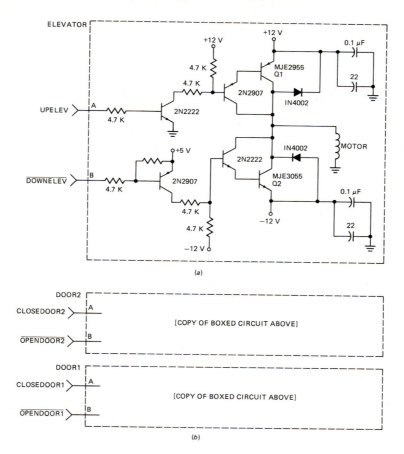

FIGURE 12-47   Interface circuits. *(a)* Elevator motor driver. *(b)* Door motor drivers.

STATE	a	b	c	d	e	f	g	SYMBOL	QC	QB	QA
FLOOR1	0	1	1	0	0	0	0	1	0	0	0
WAIT1	0	1	1	0	0	0	0	1	0	0	1
GO_UP	1	1	0	0	0	1	0	∩	0	1	1
FLOOR2	1	1	0	1	1	0	1	2	1	1	1
WAIT2	1	1	0	1	1	0	1	2	1	0	1
GO_DOWN	0	0	0	1	1	0	0	⊔	1	0	0
UNDEFND1	0	0	0	0	0	0	1	-	1	1	0
UNDEFND2	0	0	0	0	0	0	1	-	0	1	0

—— SAME SO ONLY ONE NEEDED

FIGURE 12-48   Symbols to be displayed on
seven-segment LED displays for each state of elevator
controller.

To help you visualize where we are in the design at
this point, Figure 12-49 shows a block diagram of the
major parts.

The next step in the design process was to determine
how we could most efficiently implement each of the signals from the output decoder. One approach might be to
write a large truth table which showed all the input signals and all the output signals. However, since there are
many input variables and since each of the output control signals is only asserted for very few combinations of
the input signals, it is much more efficient to intuitively

write the equations for each with the help of the state
diagram in Figure 12-46.

As an example of this, think about the conditions
when you would want the UPELEV signal to be asserted.
If you look at the bubble diagram in Figure 12-46, you
can see that you want UPELEV asserted only in state
GO__UP. Perhaps you can also see that you would want
the UPELEV signal to be turned off as soon as the
ARRIVE2 signal goes high and not wait until the next
clock pulse causes a transition to state FLOOR2. If you
don't turn off the motor immediately when ARRIVE2
goes high, the elevator could spend up to 10 s trying to
force its way through the roof! The expression for the
UPELEV signal then can be represented as:

$$\text{UPELEV} = (\text{state GO\_\_UP}) \cdot \overline{\text{ARRIVE2}}$$

In a similar manner the expression for the active low
DOWNELEV output can be written as:

$$\overline{\text{DOWNELEV}} = (\text{state GO\_\_DOWN}) \cdot \text{ARRIVE2}$$

Since these outputs depend on the level of an external
signal as well as the state of the machine, they are examples of Mealy-type outputs. This example should show
you why you often want Mealy-type outputs on a state
machine. To reduce the initial clutter in Figure 12-46,
we did not show these two Mealy output signals. You
can add them now if you want.

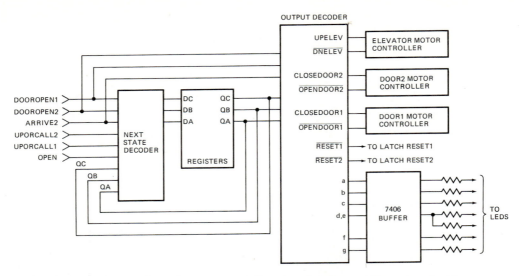

FIGURE 12-49 Block diagram of complete elevator controller showing Output Decoder signals.

The next output decoder signals to look at are the RESET1 and RESET2 signals. We want the RESET1 signal to be asserted low only during state FLOOR1 and the RESET2 signal to be asserted low only during state FLOOR2, so these two outputs can be represented with the expressions:

$$\overline{\text{RESET1}} = \text{state FLOOR1}$$
$$\overline{\text{RESET2}} = \text{state FLOOR2}$$

These signals are unconditional or Moore-type signals, so they are written directly in the appropriate bubbles in Figure 12-46, if you want.

Finally, let's look at the conditions we want to open and close each door. We want the CLOSEDOOR1 signal to be asserted in states WAIT1, UNDEFND1, and UNDEFND2. However, as soon as the DOORCLOSED1 signal goes high, we want to turn off the power to the motor that closes door 1. The expression for CLOSEDOOR1 then can be written as:

CLOSEDOOR1 = (state WAIT1 + state UNDEFND1
              + state UNDEFND2)·DOOROPEN1

Likewise, the expression for CLOSEDOOR2 can be written as:

CLOSEDOOR2 = (state WAIT2 + state UNDEFND1
              + state UNDEFND2)·DOOROPEN2

We only want to assert the OPENDOOR1 signal in state FLOOR1, and we want to turn the signal off as soon as the door is open. The expression for active low OPENDOOR1 signal then can be written as:

$$\overline{\text{OPENDOOR1}} = (\text{state FLOOR1}) \cdot \overline{\text{DOOROPEN1}}$$

and the expression for the active low OPENDOOR2 signal can be written as:

$$\overline{\text{OPENDOOR2}} = (\text{state FLOOR2}) \cdot \overline{\text{DOOROPEN2}}$$

A look at these last two signals shows that expression for $\overline{\text{OPENDOOR1}}$ is almost the same as the expression for $\overline{\text{RESET1}}$. Since the reset inputs of the latches only have to be pulsed low for a few nanoseconds, holding these inputs low for the time it takes the door to open is more than enough to reset the latches. This means that the $\overline{\text{OPENDOOR1}}$ signal can also be used for the $\overline{\text{RESET1}}$ signal. Likewise, the $\overline{\text{OPENDOOR2}}$ signal can also be used for the $\overline{\text{RESET2}}$ signal. This reduces the decoding logic and the number of PAL output pins required.

Once we had expressions or a truth table for all the output decoder signals, the next step was to assign binary values to each of the states in our controller. The question that may occur to you at this point is, "Why did we wait this late in the design process to make these state assignments?" The answer to this is that often you can simplify the next-state decoder by assigning the binary state codes such that one of the Q outputs corresponds directly to one of the required output decoder signals. Unfortunately, for this example we couldn't match up any of the Q outputs with any of the output decoder signals, but it is usually worth trying.

The rightmost columns in Figure 12-48 show the binary state assignments we choose for our controller. Note that we assigned the values so that only 1 bit at a time changes as you go from state to state in the main sequence. As we told you earlier, assigning states in this way usually simplifies the next-state decoder.

The GAL16V8 devices we wanted to use for this controller have only eight outputs, so the final step here was to determine which output signals we could implement along with the state machine in one PAL and which functions would have to be produced with a second PAL. The decisions we made were as follows:

PAL1	PAL2
CLOSEDOOR1	UPELEV
$\overline{\text{OPENDOOR1}}$	$\overline{\text{DOWNELEV}}$
CLOSEDOOR2	a
$\overline{\text{OPENDOOR1}}$	b
QC	c
QB	de
QA	f
	g

The QC, QB, and QA signals from PAL1 and the ARRIVE2 signal from the elevator position sensor will be connected to the inputs of PAL2. PAL2 will decode these signals to produce the elevator motor control signals and the segment driver output signals. As you can see from this list, all the basic logic for the controller and the output decoder is implemented with a couple of PALs.

THE ABEL SOURCE FILE FOR PAL1

Figure 12-50 shows the ABEL source file we wrote for PAL1 of our elevator controller. The basic format of this file is the same as the one we showed you in Figure 12-38 and discussed in an earlier section of the chapter. The main difference here is that we have used the GOTO and IF-THEN-ELSE structures instead of the CASE structure in the STATE__DIAGRAMS section of the program. With a little thought, you can write these statements directly from a state diagram such as the one in Figure 12-46. Note, for states where there are several possible exits it is important to include the final ELSE statement which tells the controller to stay in the same state if none of the exit conditions are met. If this is not done, the machine will go to the default value of all 0's or all 1's on the outputs of the registers, depending on the particular device.

ABEL uses the STATE__DIAGRAM section of the program in Figure 12-50 to develop the equations for the next-state decoder. The Equations section in Figure 12-50 tells ABEL the expressions for the four output decoder functions we are implementing in this PAL along with the basic state machine. Note that in these equations we put in the QC, QB, and QA values which represent the states where we want each signal to be asserted.

Both the state machine part of PAL1 and output decoder part of PAL1 must be tested. The first set of test vectors in Figure 12-50 shows how we test the operation of the basic state machine. The principle here is to load a known state on the outputs of the registers and then see if the controller transitions to the desired next state in response to the input signals. To be complete, each transition from each state must be tested. Here's how you develop the transition test vectors such as those we show in Figure 12-50.

Start with a known state such as WAIT1 and write a truth table which shows how you want the machine to transition for each combination of the input signals which affect transitions from that state. Figure 12-51a, for example, shows the desired response from state WAIT1. You don't have to write test vectors for all these combinations because, if you look closely at the truth table, some major simplifications should be obvious. If the DNORCALL1 signal is a 1, for example, the controller will go to state FLOOR1, regardless of the levels on the other inputs. For the eight input combinations where DNORCALL1 is a 1, then the other input signals are "don't cares" and can be represented by X's in the test vectors. Likewise, if the OPEN signal is a 1, the controller will go to state FLOOR1 regardless of the levels on the other inputs. The result of this is that we only have to test the effect of the five input combinations shown in Figure 12-51b.

To test the effect of these five combinations, you use a preload (.P.) test vector to put the controller in state WAIT1, then first try combinations which cause the controller to stay in state WAIT1. Then you try combinations which cause the controller to go to other states. Each time an input combination causes a transition to another state, you use a preload (.P.) vector to bring the controller back to WAIT1 for the next transition test. When you have tested all the transitions from WAIT1, allow the controller to transition to another state and test the transitions from that state. Note that there are no conditions for a transition from state FLOOR1 to WAIT1 or a transition from FLOOR2 to WAIT2, so only one vector is needed to test each of these transitions. Only ARRIVE2 determines the transitions from GO__UP and GO__DOWN, so only two vectors are needed to test each of these transitions.

In a similar manner the conditional outputs which control the door motors are tested by putting the machine in a known state and determining if the output signals are correct for each input combination.

As you can probably guess from looking at Figure 12-50, generating complete test vectors for even a relatively simple PAL such as our elevator controller is a tedious and error-prone task. To help with this problem, programs which automatically generate test vectors for a PAL design are now available. An example of this type program is PLDtest from Data I/O Corporation.

THE ABEL OUTPUT FILE

Figure 12-52 shows the equations produced by ABEL for the next-state decoder and the output decoder. Again notice the :=, which means that the specified value will occur after the next clock pulse, and the simple =, which means that the signal will be asserted as soon as the input conditions are met.

We were quite happy to let ABEL produce these simplified expressions, rather than spend a day drawing Karnaugh maps. Remember that ABEL also automatically generates a fusemap which implements all the reduced equations. All we had to do to produce the desired part was to download the JEDEC file to a programmer and program the PAL with it.

THE ABEL SOURCE FILE FOR PAL2

Figure 12-53a shows the ABEL source file for PAL2 in our design. The segment outputs are unconditional, so that all these functions need as inputs are the QC, QB, and QA signals from PAL1. The UPELEV and DOWNELEV signals depend on the level of the ARRIVE2 signal, so this signal must also be connected to one of the inputs of PAL2. As we showed you in Figure 12-14, the TRUTH__TABLE format is useful for implementing the segment outputs, and the EQUATIONS format is useful for implementing the elevator motor outputs. Figure 12-53b shows the reduced equations we got for PAL2.

THE COMPLETE SYSTEM

Once we had programmed the two PALs, it was a simple matter to connect the entire circuit as shown in Figure 12-54. As you can see, all that is needed besides the two

```
MODULE ELEV FLAG '-R3'

TITLE 'STATE MACHINE TO CONTROL TWO STORY ELEVATOR'

                U10    DEVICE  'P16V8R';
"INPUTS                                    OUTPUTS

CLK         PIN  1;                QA        PIN 19; "Register output
UPORCALL2  PIN  2;  "Buttons"      QB        PIN 18; "Register output
DNORCALL1  PIN  3;  "Buttons"      QC        PIN 17; "Register output
OPEN        PIN  4;  "Button"      CLOSEDOOR1 PIN 16; "Active high
ARRIVE2    PIN  5;  "Sensor"       OPENDOOR1  PIN 15; "Active low
DOOROPEN1  PIN  6;  "Sensor"       CLOSEDOOR2 PIN 14; "Active high
DOOROPEN2  PIN  7;  "Sensor"       OPENDOOR2  PIN 13; "Active low
JAM1        PIN  8;  "Dr1 jam'd"   OE         PIN 11; "Output enable"
JAM2        PIN  9;  "Dr2 jam'd"

"State assignments

FLOOR1      = ^B000; "Arrive floor 1, open door
WAIT1       = ^B001; "Close door, wait for CALL1, CALL2, UP or OPEN
GO_UP       = ^B011; "Elevator motor on until ARRIVE2 = 1
FLOOR2      = ^B111; "Arrive floor 2, open door
WAIT2       = ^B101; "Close door, wait for CALL1, CALL2, UP or OPEN
GO_DOWN     = ^B100; "Elevator motor on until ARRIVE2 = 0
UNDEFND1    = ^B110; "Undefined start up state #1
UNDEFND2    = ^B010; "Undefined startup state #2

CK, X, P = .C., .X., .P.;  ST = [QC, QB, QA];

STATE_DIAGRAM [QC,QB,QA]

"PRESENT STATE          CONDITION                       NEXT STATE

STATE FLOOR1:           GOTO                            WAIT1;

STATE WAIT1:            IF OPEN # DNORCALL1  THEN        FLOOR1
                        ELSE IF UPORCALL2 & !DOOROPEN1 &
                             !(OPEN # DNORCALL1)  THEN   GO_UP
                               ELSE                      WAIT1;

STATE UNDEFND1:         IF !ARRIVE2  THEN               GO_UP
                        ELSE                            GO_DOWN;

STATE GO_UP:            IF ARRIVE2 THEN                 FLOOR2
                        ELSE                            GO_UP;

STATE FLOOR2:           GOTO                            WAIT2;

STATE WAIT2:            IF OPEN # UPORCALL2   THEN       FLOOR2
                        ELSE IF DNORCALL1 & !DOOROPEN2 &
                             !(OPEN # UPORCALL2) THEN    GO_DOWN
                               ELSE                      WAIT2;

STATE UNDEFND2:         IF !ARRIVE2    THEN             GO_UP
                        ELSE                            GO_DOWN;

STATE GO_DOWN:          IF !ARRIVE2 THEN                FLOOR1
                        ELSE                            GO_DOWN;

"FLOOR1     = [!QC & !QB & !QA];  WAIT1     = [!QC & !QB & QA];
"GO_UP      = [!QC & QB & QA];    FLOOR2    = [QC & QB & QA];
"WAIT2      = [QC & !QB & QA];    GO_DOWN   = [QC & !QB & !QA];
"UNDEFND1  = [QC & QB & !QA];    UNDEFND2  = [!QC & QB & !QA];

EQUATIONS

CLOSEDOOR1 = (!QC & !QB & QA # QC & QB & !QA # !QC & QB & !QA) &
              DOOROPEN1; "State WAIT1 and Undefined states

!OPENDOOR1 =  !QC & !QB & !QA & !DOOROPEN1; "State FLOOR1

CLOSEDOOR2 = (QC & !QB & QA # QC & QB & !QA # !QC & QB & !QA) &
              DOOROPEN2; "state WAIT2 and Undefined states
```

FIGURE 12-50 ABEL source file for PAL which implements basic state machine and door motor signals for the elevator controller. *(Continued on next page.)*

```
!OPENDOOR2 = QC & QB & QA & !DOOROPEN2;  "State FLOOR2

"Test vectors for basic state machine

TEST_VECTORS
([CLK,ST,UPORCALL2,DNORCALL1,OPEN,ARRIVE2,DOOROPEN1,DOOROPEN2] -> ST)
"        STATE         DNORCALL1   ARRIVE2      DOOROPEN2
"CLK  |  UPORCALL2  |  OPEN   |  DOOROPEN1  |        STATE
[P,   FLOOR1,    X,    X,   X,   X,      X,      X ] -> X;
[CK,  FLOOR1,    X,    X,   X,   X,      X,      X ] -> WAIT1;
[CK,  WAIT1,     0,    0,   0,   X,      X,      X ] -> WAIT1;
[CK,  WAIT1,     1,    0,   0,   X,      1,      X ] -> WAIT1;
[CK,  WAIT1,     X,    1,   X,   X,      X,      X ] -> FLOOR1;
[CK,  FLOOR1,    X,    X,   X,   X,      X,      X ] -> WAIT1;  "BACK TO WAIT1
[CK,  WAIT1,     X,    X,   1,   X,      X,      X ] -> FLOOR1;
[CK,  FLOOR1,    X,    X,   X,   X,      X,      X ] -> WAIT1;  "BACK TO WAIT1
[CK,  WAIT1,     1,    0,   0,   X,      0,      X ] -> GO_UP;
[CK,  GO_UP,     X,    X,   X,   0,      X,      X ] -> GO_UP;
[CK,  GO_UP,     X,    X,   X,   1,      X,      X ] -> FLOOR2;
[CK,  FLOOR2,    X,    X,   X,   X,      X,      X ] -> WAIT2;
[CK,  WAIT2,     0,    0,   0,   X,      X,      X ] -> WAIT2;
[CK,  WAIT2,     0,    1,   0,   X,      X,      1 ] -> WAIT2;
[CK,  WAIT2,     1,    X,   X,   X,      X,      X ] -> FLOOR2;
[CK,  FLOOR2,    X,    X,   X,   X,      X,      X ] -> WAIT2;  "BACK TO WAIT2
[CK,  WAIT2,     X,    X,   1,   X,      X,      X ] -> FLOOR2;
[CK,  FLOOR2,    X,    X,   X,   X,      X,      X ] -> WAIT2;  "BACK TO WAIT2
[CK,  WAIT2,     0,    1,   0,   X,      X,      0 ] -> GO_DOWN;
[CK,  GO_DOWN,   X,    X,   X,   1,      X,      X ] -> GO_DOWN;
[CK,  GO_DOWN,   X,    X,   X,   0,      X,      X ] -> FLOOR1;
[P,   UNDEFND1,  X,    X,   X,   X,      X,      X ] -> X;
[CK,  UNDEFND1,  X,    X,   X,   1,      X,      X ] -> GO_DOWN;
[P,   UNDEFND1,  X,    X,   X,   X,      X,      X ] -> X;
[CK,  UNDEFND1,  X,    X,   X,   0,      X,      X ] -> GO_UP;
[P,   UNDEFND2,  X,    X,   X,   X,      X,      X ] -> X;
[CK,  UNDEFND2,  X,    X,   X,   1,      X,      X ] -> GO_DOWN;
[P,   UNDEFND2,  X,    X,   X,   X,      X,      X ] -> X;
[CK,  UNDEFND2,  X,    X,   X,   0,      X,      X ] -> GO_UP;

"Test vectors for output decoder

TEST_VECTORS
([CLK,ST,DOOROPEN1,DOOROPEN2] ->
          [ST,CLOSEDOOR1,OPENDOOR1,CLOSEDOOR2,OPENDOOR2])

"CLK  ST      DROPN1 DROPEN2       ST    CLSDR1 OPNDR1 CLSDR2 OPNDR2
[P,  FLOOR1,    X,    X]  -> [X,       X,    X,    X,    X ];
[X,  FLOOR1,    0,    X]  -> [FLOOR1,  0,    0,    0,    1 ];
[X,  FLOOR1,    1,    X]  -> [FLOOR1,  0,    1,    0,    1 ];
[P,  WAIT1,     X,    X]  -> [X,       X,    X,    X,    X ];
[X,  WAIT1,     0,    X]  -> [WAIT1,   0,    1,    0,    1 ];
[X,  WAIT1,     1,    X]  -> [WAIT1,   1,    1,    0,    1 ];
[P,  GO_UP,     X,    X]  -> [X,       X,    X,    X,    X ];
[X,  GO_UP,     X,    X]  -> [GO_UP,   0,    1,    0,    1 ];
[P,  FLOOR2,    X,    X]  -> [X,       X,    X,    X,    X ];
[X,  FLOOR2,    X,    0]  -> [FLOOR2,  0,    1,    0,    0 ];
[X,  FLOOR2,    X,    1]  -> [FLOOR2,  0,    1,    0,    1 ];
[P,  WAIT2,     X,    X]  -> [X,       X,    X,    X,    X ];
[X,  WAIT2,     X,    0]  -> [WAIT2,   0,    1,    0,    1 ];
[X,  WAIT2,     X,    1]  -> [WAIT2,   0,    1,    1,    1 ];
[P,  GO_DOWN,   X,    X]  -> [X,       X,    X,    X,    X ];
[X,  GO_DOWN,   X,    X]  -> [GO_DOWN, 0,    1,    0,    1 ];
[P,  UNDEFND1,  X,    X]  -> [X,       X,    X,    X,    X ];
[X,  UNDEFND1,  0,    0]  -> [UNDEFND1,0,    1,    0,    1 ];
[X,  UNDEFND1,  0,    1]  -> [UNDEFND1,0,    1,    1,    1 ];
[X,  UNDEFND1,  1,    0]  -> [UNDEFND1,1,    1,    0,    1 ];
[X,  UNDEFND1,  1,    1]  -> [UNDEFND1,1,    1,    1,    1 ];
[P,  UNDEFND2,  X,    X]  -> [X,       X,    X,    X,    X ];
[X,  UNDEFND2,  0,    0]  -> [UNDEFND2,0,    1,    0,    1 ];
[X,  UNDEFND2,  0,    1]  -> [UNDEFND2,0,    1,    1,    1 ];
[X,  UNDEFND2,  1,    0]  -> [UNDEFND2,1,    1,    0,    1 ];
[X,  UNDEFND2,  1,    1]  -> [UNDEFND2,1,    1,    1,    1 ];

END ELEV
```

FIGURE 12-50   *(continued)*

STATE	DNCALL1	OPEN	UPCALL2	DOOROPEN1	STATE
WAIT1	0	0	0	0	WAIT1
WAIT1	0	0	0	1	WAIT1
WAIT1	0	0	1	0	GO_UP
WAIT1	0	0	1	1	WAIT1
WAIT1	0	1	0	0	FLOOR1
WAIT1	0	1	0	1	FLOOR1
WAIT1	0	1	1	0	FLOOR1
WAIT1	0	1	1	1	FLOOR1
WAIT1	1	0	0	0	FLOOR1
WAIT1	1	0	0	1	FLOOR1
WAIT1	1	0	1	0	FLOOR1
WAIT1	1	0	1	1	FLOOR1
WAIT1	1	1	0	0	FLOOR1
WAIT1	1	1	0	1	FLOOR1
WAIT1	1	1	1	0	FLOOR1
WAIT1	1	1	1	1	FLOOR1

(a)

STATE	DNCALL1	OPEN	UPCALL2	DOOROPEN1	STATE
WAIT1	0	0	0	X	WAIT1
WAIT1	0	0	1	1	WAIT1
WAIT1	1	X	X	X	FLOOR1
WAIT1	X	1	X	X	FLOOR1
WAIT1	0	0	1	0	GO_UP

(b)

FIGURE 12-51   Truth tables used to develop test vectors for transitions from state WAIT1. (a) Full truth table. (b) Simplified truth table.

Page 1
ABEL(tm) 3.00a – Document Generator   23-Jun-88 12:07 PM
STATE MACHINE TO CONTROL TWO STORY ELEVATOR
Equations for Module ELEV

```
Device U10

- Reduced Equations:

    QC := !(!ARRIVE2 & !QC # !ARRIVE2 & !QA # !QB & !QC);

    QB := !(QA & QB & QC
          # ARRIVE2 & !QA
          # DOOROPEN1 & !QB & !QC
          # !OPEN & !QB & !UPORCALL2
          # DNORCALL1 & !QB & !QC
          # OPEN & !QB & !QC
          # !QA & !QB);

    QA := !(!QA & !QB & QC
          # DNORCALL1 & !DOOROPEN2 & !OPEN & !QB & QC & !UPORCALL2
          # ARRIVE2 & !QA & QB
          # DNORCALL1 & QA & !QB & !QC
          # OPEN & QA & !QB & !QC);

    CLOSEDOOR1 = (DOOROPEN1 & !QA & QB # DOOROPEN1 & QA & !QB & !QC);

    OPENDOOR1 = !(!DOOROPEN1 & !QA & !QB & !QC);

    CLOSEDOOR2 = (DOOROPEN2 & !QA & QB # DOOROPEN2 & QA & !QB & QC);

    OPENDOOR2 = !(!DOOROPEN2 & QA & QB & QC);
```

FIGURE 12-52   Reduced equations produced by ABEL for elevator controller state machine.

```
MODULE ELEVPAL2 FLAG '-r3'
TITLE 'DOOR MOTOR AND SEVEN SEGMENT DECODER FOR ELEVATOR - D.HALL, 1988'

            U2          DEVICE 'P16V8C';   "Combinational only
ARRIVE2,QC,QB,QA    PIN     2,3,4,5;
a,b,c,de,f,g        PIN     12,13,14,15,16,17;
UPELEV,DOWNELEV     PIN     18,19;

TRUTH_TABLE     ([QC,QB,QA] -> [a,b,c,de,f,g])
                [0,0,0] ->      [0,1,1,0,0,0]; "1
                [0,0,1] ->      [0,1,1,0,0,0]; "1
                [0,1,1] ->      [1,1,0,0,1,0]; "^
                [1,1,1] ->      [1,1,0,1,0,1]; "2
                [1,0,1] ->      [1,1,0,1,0,1]; "2
                [1,0,0] ->      [0,0,1,1,0,0]; "|_|
                [1,1,0] ->      [0,0,0,0,0,1]; "-
                [0,1,0] ->      [0,0,0,0,0,1]; "-
EQUATIONS
                UPELEV = !QC & QB & QA & !ARRIVE2;
                !DOWNELEV = QC & !QB & !QA & ARRIVE2;
TEST_VECTORS
        ([ARRIVE2,QC,QB,QA] ->      [a,b,c,de,f,g,UPELEV,DOWNELEV])
                [0,0,0,0] ->        [0,1,1,0,0,0,0,1]; "1
                [0,0,0,1] ->        [0,1,1,0,0,0,0,1]; "1
                [0,0,1,1] ->        [1,1,0,0,1,0,1,1]; "^
                [0,1,1,1] ->        [1,1,0,1,0,1,0,1]; "2
                [0,1,0,1] ->        [1,1,0,1,0,1,0,1]; "2
                [0,1,0,0] ->        [0,0,1,1,0,0,0,1]; "|_|
                [0,1,1,0] ->        [0,0,0,0,0,1,0,1]; "-
                [0,0,1,0] ->        [0,0,0,0,0,1,0,1]; "-
                [1,0,0,0] ->        [0,1,1,0,0,0,0,1]; "1
                [1,0,0,1] ->        [0,1,1,0,0,0,0,1]; "1
                [1,0,1,1] ->        [1,1,0,0,1,0,0,1]; "^
                [1,1,1,1] ->        [1,1,0,1,0,1,0,1]; "2
                [1,1,0,1] ->        [1,1,0,1,0,1,0,1]; "2
                [1,1,0,0] ->        [0,0,1,1,0,0,0,0]; "|_|
                [1,1,1,0] ->        [0,0,0,0,0,1,0,1]; "-
                [1,0,1,0] ->        [0,0,0,0,0,1,0,1]; "-
END ELEVPAL2
```

(a)

```
                                                        Page 1
ABEL(tm) 3.00a  -  Document Generator        21-Jun-88 05:00 PM
DOOR MOTOR AND SEVEN SEGMENT DECODER FOR ELEVATOR - D.HALL, 1988
Equations for Module ELEVPAL2

Device U2

- Reduced Equations:

    a = !(!QA # !QB & !QC);
    b = !(!QA & QB # !QA & QC);
    c = !(QA & QC # QB);
    de = !(!QA & QB # !QC);
    f = !(!QA # QC # !QB);
    g = !(!QA & !QB # QA & !QC);
    UPELEV = (!ARRIVE2 & QA & QB & !QC);
    DOWNELEV = !(ARRIVE2 & !QA & !QB & QC);
```

(b)

FIGURE 12-53   (a) ABEL source file for seven-segment and elevator motor decoder PAL. (b) Reduced equations for decoder.

PALs are a 74LS00 and a 74LS279 to condition the input signals, a 555 timer to produce the 10-s clock, some power transistors to interface to the motors, and a 7406 buffer to interface to the LED segments. The simplified equations for the PALs in Figures 12-52 and 12-53b should show you that if we had implemented this design with standard SSI parts, the circuit would have taken either one large PC board or a couple of small ones. Incidentally, many PALs have a *security fuse* which you can tell the programmer to "blow" so that no one can easily determine the functions implemented in your PAL. Also, some PALs now have "buried" state registers. The outputs of these registers feed back into the AND-OR matrix but do not connect to pins on the package. For applications where the register outputs are not needed, this architecture makes more package pins available for other inputs or outputs.

As we mentioned earlier in the chapter, a PAL or an

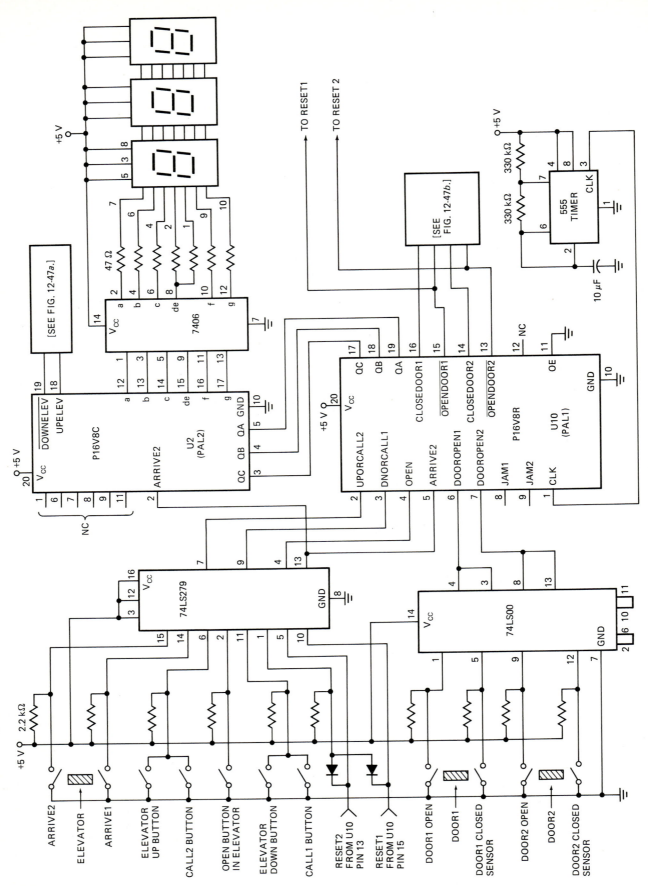

FIGURE 12-54 Complete schematic for elevator controller using PALs.

EPLD is useful for implementing circuits of up to a few hundred gates. In the next two chapters we show you some other types of digital systems, and then in Chapter 15 we show you how custom *Application Specific Integrated Circuits*, or *ASICs*, can be used to implement circuits equivalent to 100,000 or more basic gates in a single IC package.

CHECKLIST OF IMPORTANT TERMS AND CONCEPTS IN THIS CHAPTER

PLA, FPLA, ROM, PROM, EPROM, EEPROM, PAL, EPLD
Source file
Test vector
PAL outputs
  Active high
  Active low

Programmable
Exclusive OR
Register
Macrocell
State machine
  Moore machine
  Mealy machine
  Next-state decoder
  Output decoder
State diagrams
Algorithmic state machine (ASM) charts
Next-state or transition-state tables
JK excitation table or transition truth table
Hang state
Mechanical reset

REVIEW QUESTIONS AND PROBLEMS

1. *a.* Briefly describe the differences between an FPLA, a PROM, and a PAL.

   b. Define the term *EPLD* and explain the major advantage of this type device over a standard PAL.

2. For the simple PAL shown in Figure 12-55:

   a. Mark with X's the junctions where you would leave the fuses intact in the input AND matrix to implement the following functions:

$$F1 = A \oplus B$$

$$F2 = A{\cdot}B{\cdot}C + \overline{A}{\cdot}\overline{B}{\cdot}C$$

$$F3 = \overline{A}{\cdot}B{\cdot}\overline{C}{\cdot}D + A{\cdot}\overline{B}{\cdot}C{\cdot}\overline{D}$$

$$F4 = \overline{A}{\cdot}\overline{B}{\cdot}\overline{C}{\cdot}\overline{D}$$

   b. Explain why you cannot directly implement the function:

$$Y = A{\cdot}B{\cdot}C{\cdot}D + \overline{A}{\cdot}\overline{B}{\cdot}\overline{C}{\cdot}\overline{D} + \overline{A}{\cdot}B{\cdot}\overline{C}{\cdot}D + A{\cdot}\overline{B}{\cdot}C{\cdot}\overline{D}$$

     with this PAL.

3. *a.* Explain the meaning of the numbers and letters in each of the following PAL-type devices:
     i. PAL10L8AJ
     ii. PAL20R4A-2N
     iii. PAL16P6
     iv. GAL20V8A
     v. PAL64R32

   b. How can the specified number of inputs plus the specified number of outputs for a PAL be greater than the number of pins on the actual device?

4. Briefly describe the steps involved in developing the fusemap for a PAL and programming it.

5. *a.* Write the standard Boolean expression for the following ABEL equations:

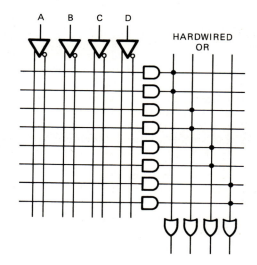

FIGURE 12-55   PAL for Problem 2.

     i. Z = !A & B & !C # !B & C
     ii. X = !(R & !S & P) # R & S & P
     iii. J = A $ B $ C

   b. What are the purposes of test vectors in an ABEL source program?

   c. Write the ABEL TEST__VECTORS for the equation J = ABC.

6. *a.* Use a Karnaugh map to develop the simplified Boolean expression for the d segment of the seven-segment to hexadecimal decoder in Figure 12-13.

   b. Show how the reduced equations can be implemented with simple gates.

   c. To appreciate how much you can implement with a single, small PAL, estimate how many gates would be required to implement the simplified expressions for the decoder as shown in Figure 12-16.

7. In Figure 7-18 we showed you how an optical encoder is used to produce a 4-bit gray code which represents the position of a rotating shaft. For many applications this gray code must be converted to a standard 4-bit binary code. For ABEL or whatever development program you have, write a source file which will produce the fusemap to implement this conversion in a GAL16V8. (Use U15 DEVICE 'P16V8S' statement to configure a macrocell for eight product terms per output.)

8. a. Draw a block diagram which shows the major parts of a Moore-type state machine.
   b. Draw a block diagram which shows the major parts of a Mealy-type state machine.
   c. Describe the major difference between a Moore-type state machine and a Mealy-type state machine.

9. For a 2-bit binary upcounter with synchronous CLEAR and an output decoder which asserts a signal called OPEN for count 10:
   a. Draw a state diagram.
   b. Draw an ASM chart.
   c. Make a next-state table.
   d. Add columns to the next-state table showing the inputs required to implement the basic state machine with D flip-flops.
   e. Use Karnaugh maps to produce the simplified expressions for the next-state decoder.
   f. Draw the circuit for the complete state machine, including the next-state decoder and the output decoder.

10. Design a 3-bit synchronous counter which counts up or down through the sequence 000, 001, 011, 010, 110, 111, 101, 100, depending on the logic level on an input labeled $U/\overline{D}$. To the basic counter add an output decoder which produces a high on a signal called PARITY each time a code with an even number of 1's appears on the outputs of the flip-flops. As part of the design:
   a. Draw a state diagram.
   b. Draw an ASM chart.
   c. Make a next-state table.
   d. Add columns to the next-state table showing the inputs required to implement the basic state machine with JK flip-flops.
   e. Use Karnaugh maps to produce the simplified expressions for the next-state decoder.
   f. Develop the equation for the output decoder.
   g. Draw the circuit for the complete state machine, including the next-state decoder and the output decoder.

11. Describe two methods that are used to solve the "hang-state" problem in synchronous circuits.

12. Design a 4-bit divide-by-13 binary synchronous counter which automatically goes to state 0000 if it powers up in any of the unused states. Design the circuit to be implemented with D flip-flops. Include in your design a bubble diagram or ASM chart, a next-state table with D input columns,

Karnaugh maps, and the completed circuit diagram.

13. Write an ABEL or other PAL development tool source file for the 2-bit counter with synchronous clear described in Problem 9. If you do not already have one, draw a state diagram or an ASM chart for the machine first.

14. Write an ABEL or other PAL development tool source file for the 3-bit synchronous up/down counter with PARITY output described in Problem 10. If you do not already have one, draw a state diagram or an ASM chart for the machine first.

15. Write an ABEL or other PAL development tool source file for the 4-bit, divide-by-13 counter described in Problem 12. If you do not already have one, draw a state diagram or an ASM chart for the machine first.

16. Each door in the elevator controller system we described in the last section of the chapter has a sensor which detects if the door jams as it attempts to close. To keep our example in the text as simple as possible and to leave some of the fun for you, we did not include these signals in the discussion there:
   a. Redraw the state diagram from Figure 12-46.
   b. Relabel the appropriate transition arrows to show the desired effect of JAM1 and JAM2 on these transitions.
   c. In the elevator controller state diagram in Figure 12-46, we showed you how to implement a "mechanical reset" from the two undefined states UNDEFND1 and UNDEFND2. Show how you could use a RESET input signal to cause a mechanical reset from any state in case the elevator powers up in some other state and it is physically between floors.
   d. Rewrite the STATE_DIAGRAM section of the ABEL source program in Figure 12-50 to include the improvements you made in the preceding steps.

17. a. Why was it necessary to latch the signals from the CALL and OPEN button in the elevator controller in Figure 12-50?
   b. Why do we usually assign binary values to states so that only 1 bit at a time changes as the machine goes from state to state?
   c. What is the purpose of the security fuse found in some PALs?

18. As we showed you in Chapter 9, the 74LS193 is a 4-bit binary counter which can count up or down, be preloaded with a count applied to its data inputs, or be cleared. One difficulty with the 74LS193 is that it has two CLOCK inputs, one for counting up and one for counting down. For this exercise, we want you to design a 4-bit binary counter which functions similarly to the 74LS193 except that it has only one CLOCK input and the counting direction is determined by the level on a pin labeled U/D. If the U/D input is high, we want the counter to

count up, and if the U/D input is low, we want the counter to count down.

a. Draw a state diagram for this counter. (You can use the shorthand of a circled N to represent "any state" for the clear and load operations.)

b. Write an ABEL or other PAL development tool source file for the basic counter transitions up and down.

c. To your source file, add equations which implement a synchronous clear function and a synchronous load function similar to the way we showed in Figure 12-38.

d. Write test vectors which test the clear, load, and count functions of the counter.

19. In Figure 10-49 we showed you the stepping sequence for a four-phase stepper motor. We also showed you in Figure 10-49 how to interface a stepper motor to a shift register. As the shift register is clocked, the motor will step clockwise or counterclockwise, depending on the direction the register is set to shift. The simple shift register driver, however, cannot easily shift the motor though its half-step sequence or through a combination of half steps and full steps. The purpose of the exercise here is to design a simple state machine which will control the motion of a four-phase stepper motor based on the levels present on three input control signals called CW, CCW, and HS. The three signal names stand for ClockWise, CounterClockWise, and Half Step, respectively. Figure 12-56a shows the response we want the motor to make for each of the combinations of these input signals. Figure 12-56b shows the levels that must be present on the motor controller inputs SW1, SW2, SW3, and SW4, for each of the four full-step positions and each of the eight half-step positions.

CW	CCW	HS	ACTION
0	0	0	HOLD
0	0	1	HOLD
0	1	0	FULL STEP CCW
0	1	1	HALF STEP CCW
1	0	0	FULL STEP CW
1	0	1	HALF STEP CW
1	1	0	ILLEGAL
1	1	1	ILLEGAL

(a)

FULL STEP				HALF STEP						
SW4	SW3	SW2	SW1	SW4	SW3	SW2	SW1	STATE	CW	CCW
0	0	1	1	0	0	1	1	S0		
				0	0	0	1	S1		
1	0	0	1	1	0	0	1	S2		
				1	0	0	0	S3		
1	1	0	0	1	1	0	0	S4		
				0	1	0	0	S5		
0	1	1	0	0	1	1	0	S6		
				0	0	1	1	S7		

(b)

FIGURE 12-56 Stepper controller data for Problem 19.

NOTE: if the motor is in one of the half-step positions, you have to step it to the next full step position before you can step it a full-step!

a. Draw a block diagram for this stepper motor controller showing the basic state machine, the next-state decoder, and the output decoder.

b. Draw a state diagram showing the eight states and the transition arrows and labels for the full steps clockwise and counterclockwise, starting from state S1. (Note: If you put the eight bubbles in a circle, the completed state diagram looks like a flower. Who said engineers can't be artistic?)

c. Add the transition arrows and labels for the half steps clockwise and counterclockwise.

d. Add the transition arrows and labels for the conditions which cause the machine to remain in each state.

e. Write an ABEL or other development tool source program which will program a GAL16V8 to implement the state diagram you have drawn. (Hint: If you use a gray code sequence for the state assignments and reduction level -R2 in your ABEL file, the design will easily fit in a GAL16V8.)

f. To your source file, add the truth table for the output decoder.

g. To your source file, add test vectors which test the basic state machine part of the controller.

h. To your source program, add test vectors which will test the output decoder part of the controller.

20. For the final exercise in this chapter, we want you to design a controller for a coffee, tea, and hot chocolate vending machine. Figure 12-57a shows the basic state diagram for the controller. Figure 12-57b shows a list of the input control signals and the required output signals. Assume that a separate circuit determines if the correct amount of money has been inserted and, if it has, asserts a signal called MONEY.

a. Write the equations for the four output signals and identify which signals are Mealy-type and which signals are Moore-type.

b. Add the output signals to the basic state diagram in the appropriate places.

c. Assuming the binary values for the four states are FAULT = 00, WAIT = 01, PREP = 11, and FILL = 10, write an ABEL or other PAL development tool source program which implements the vending machine controller.

d. To your source file, add test vectors which test both the state machine and output decoder parts of the controller.

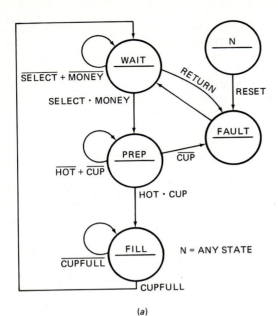

(a)
VENDING MACHINE

INPUT SIGNALS

RESET   —
MONEY   — CORRECT CHANGE INSERTED
HOT   — WATER HOT
CUP   — CUP IN POSITION
CUPFULL   —
SELECT   — USER SELECTS DRINK
RETURN   — USER PRESSES COIN RETURN BUTTON

OUTPUT SIGNALS

DROPCUP   —
HEATERON   —
VALVESON   — WATER, DRINK VALVE ON
DROPMONEY   — RETURN MONEY

(b)

FIGURE 12-57   Problem 20. (a) State diagram.
(b) Control signals.

13 Digital Arithmetic Devices and Microprocessors

In Chapter 7 we introduced you to commonly used numeric codes such as binary, BCD, octal, and hexadecimal. In the first section of this chapter we show you some techniques for performing mathematical operations on numbers which are represented in these codes. For lack of a better term we lump these techniques together under the heading of "Digital Arithmetic" because these are the number systems commonly used with microprocessors and other digital systems. Most scientific calculators now do binary and hexadecimal arithmetic, but it is worth knowing how to do simple calculations so that you are not totally dependent on your calculator. Also, we need to show you the manual techniques so that we can explain how binary arithmetic is done with hardware.

In the second section of the chapter we show you some circuits and devices which perform arithmetic and logic operations on binary numbers. Most of these circuits and devices are quite simple, and we could have discussed them in earlier chapters. However, we left them for this chapter so that we could show you the progression from a device which simply adds two binary numbers to a microprocessor which can be programmed to perform any desired sequence of logical or mathematical operations on binary-coded data.

Finally, and most importantly, in the chapter we give you an overview of the operation and timing of a microcomputer system.

OBJECTIVES

At the conclusion of this chapter you should be able to:

1. Add, subtract, multiply, and divide binary numbers.

2. Convert a negative number to and from a two's complement binary sign-and-magnitude representation.

3. Add and subtract BCD, octal, and hexadecimal numbers.

4. Describe the operation of a full-adder circuit and the operation of a digital-magnitude comparator device.

5. Determine the circuit connections required for programming an arithmetic-logic unit to perform any one of 16 specified logic functions or any one of 16 arithmetic functions.

6. Predict the results produced by ANDing, ORing, and XORing two binary words.

7. Describe how binary multiplication can be implemented with hardware.

8. Describe the function and operation of each part of a simple bit-slice microprocessor system.

9. Describe the sequence of actions that a simple microcomputer will carry out as it fetches and executes a series of instructions.

10. Given the timing diagram for a simple microprocessor system, describe the sequence of signals that occurs on the buses as the microprocessor reads a byte of data from memory or writes a byte of data to memory.

11. Describe how to set up a logic analyzer to produce a trace of the sequence of addresses output by a microcomputer as it executes a program and how to set up an analyzer to produce a trace which shows a sequence of data bytes read in from memory as a microcomputer executes a program.

12. Describe how to set up a logic analyzer to measure the time between when a microprocessor outputs an address and when the valid data comes back to the 8086 from memory on the data bus.

DIGITAL ARITHMETIC OPERATIONS

Binary Addition and Subtraction

DECIMAL-BINARY-DECIMAL CONVERSION REVIEW

To refresh your memory we will show a short review of converting decimal numbers to and from their binary equivalents. To start, Figure 13-1a shows the weights of a few digits in the binary number system. Probably the simplest way of converting a decimal number to its bi-

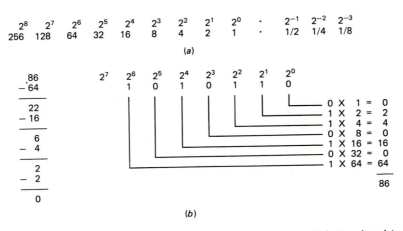

$$
\begin{array}{cccccccccc ccc}
2^8 & 2^7 & 2^6 & 2^5 & 2^4 & 2^3 & 2^2 & 2^1 & 2^0 & \cdot & 2^{-1} & 2^{-2} & 2^{-3} \\
256 & 128 & 64 & 32 & 16 & 8 & 4 & 2 & 1 & \cdot & 1/2 & 1/4 & 1/8
\end{array}
$$

(a)

```
    86                  2⁷   2⁶   2⁵   2⁴   2³   2²   2¹   2⁰
  − 64                       1    0    1    0    1    1    0
  ────
    22                                              0 X  1 =  0
  − 16                                              1 X  2 =  2
  ────                                              1 X  4 =  4
     6                                              0 X  8 =  0
  −  4                                              1 X 16 = 16
  ────                                              0 X 32 =  0
     2                                              1 X 64 = 64
  −  2                                                     ────
  ────                                                       86
     0
```

(b)

FIGURE 13-1 *(a)* Decimal weights of a few powers of 2. *(b)* Review examples of decimal-to-binary and binary-to-decimal conversion.

nary equivalent is to find the largest power of 2 that will fit in the number you are trying to convert, put a 1 in that bit position, and then subtract the decimal weight of that power of 2 from the number. Repeat the process for the remainder from this subtraction and for the remainders from following subtractions. Figure 13-1*b* shows how you use this method to convert the decimal number 86 to binary.

The largest power of 2 that will fit in 86 is 2^6, or 64, so you put a 1 in the 2^6 position and subtract 64 from 86. This gives a remainder of 22. The weight of the next binary digit, 2^5, or 32, is too large to fit in the remainder of 22, so you put a 0 in the 2^5 bit position and try the next lower power of 2. Since 2^4, or 16, fits in the remainder of 22, you put a 1 in that bit position and subtract 16 from 22. This subtraction leaves a remainder of 6. The next power of 2, which is 2^3, or 8, is too large to fit in the remainder of 6, so you write a 0 in that bit position and move on to the next power of 2. Since 2^2, or 4, fits in the remainder of 6, you put a 1 in that bit position and subtract 4 from the 6. The remainder of 2 is exactly equal to 2^1, so you put a 1 in that bit position, a 0 in the 2^0 position, and you are done.

To convert a binary number to its decimal equivalent or to check the results of a decimal-to-binary conversion, you use the reverse of the process we have just described. The right side of Figure 13-1*b* shows how to do this. You just multiply each bit times the decimal weight for that bit and add up all the results. Now that binary numbers are a little fresher in your mind, let's see how you add, subtract, multiply, and divide them.

BINARY ADDITION

Since each binary digit can only have a value of 0 or 1, addition in binary is very simple. Figure 13-2 shows a truth table for the addition of two 1-bit binary numbers A and B. As you can see, the sum S will be a 0 if A = 0 and B = 0. If A = 0 and B = 1 or A = 1 and B = 0, the sum will be a 1. If A = 1 and B = 1, the sum is equal to 2 decimal or 10 binary. This is equal to a sum of 0 and a carry C into the next higher bit position. In other words, since you can't write the 10 in one bit position, you write a 0 in the sum bit position and carry a 1 into

INPUTS		OUTPUTS	
A	B	S	C
0	0	0	0
0	1	1	0
1	0	1	0
1	1	0	1

FIGURE 13-2 Truth table for addition of 2 bits.

the next higher bit position. As a preview of what we will be showing you in the second major section of the chapter, perhaps you can see from this truth table that the sum is $S = A \oplus B$ and the carry is $C = AB$.

The truth table in Figure 13-2 is sometimes called a *half-adder* truth table because it does not show the effect that a carry from a lower bit position, C_{IN}, will have on the sum and the carry. To add binary numbers of more than 1 bit, you need the *full-adder* truth table shown in Figure 13-3*a*. As you can see, a full adder includes the effect of a carry in from a lower digit position. Figure 13-3*b* shows the result produced by using this truth table to add two binary numbers.

As with any addition, you start with the rightmost bits and work your way to the left. To help you work your way across, we have shown the value of C_{OUT} after adding each column in the number. The main columns to note in the example are the case where 1 + 1 + no $C_{IN} = 0$ with a C_{OUT} and the case where 1 + 1 + $C_{IN} = 1$ with a C_{OUT}. Some exercises at the end of the chapter will give you more practice with binary addition.

SIGNED BINARY NUMBERS

Before we show you how to subtract binary numbers, we need to show you how we commonly represent signed, or in other words, positive and negative, binary numbers. Suppose, for example, that you want to store a series of binary values which represent temperatures in some memory locations in your computer. The memory can only store 1's and 0's, not minus signs, so you have to have some other way of showing whether a stored temperature value is negative or positive. The most common method is to use the most significant bit of the binary

INPUTS			OUTPUTS	
A	B	C_{IN}	S	C_{OUT}
0	0	0	0	0
0	0	1	1	0
0	1	0	1	0
0	1	1	0	1
1	0	0	1	0
1	0	1	0	1
1	1	0	0	1
1	1	1	1	1

$$S = A \oplus B \oplus C_{IN}$$
$$C_{OUT} = A \cdot B + C_{IN} \ (A \oplus B)$$

(a)

```
A      0 1 0 1 1 0 1 0
+ B    1 1 0 1 1 1 0 0
      ┌─┬─┬─┬─┬─┬─┬─┬─┐
      │1│1│0│1│1│0│0│0│ ◄ ─ ─ ─ CARRIES
      └─┴─┴─┴─┴─┴─┴─┴─┘
SUM   1 0 0 1 1 0 1 1 0
```

(b)

FIGURE 13-3 *(a)* Full-adder truth table showing effect of carry from a less significant bit position. *(b)* Example of binary addition using full-adder truth table.

word as a *sign bit* and the rest of the bits in the word to represent the magnitude of the stored value. The format usually used to represent signed numbers such as this is called *two's complement binary sign-and-magnitude representation*. The format can be used for binary values with any number of bits, but to make your life easier we will use 8-bit words for our examples. Here's how this scheme works.

When an 8-bit word is used to represent a signed number, the most significant bit is the sign bit and the remaining 7 bits are used to represent the magnitude, or size, of the number, as shown in Figure 13-4a. A 0 in the sign bit position indicates that the number is positive. For a positive number the remaining 7 bits represent the magnitude of the number in standard binary form. As a first example, suppose you want to represent the decimal number +46 as a signed binary number.

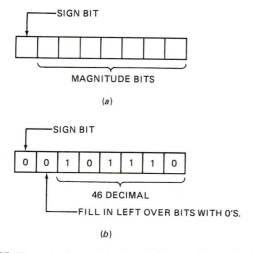

(a)

(b)

FIGURE 13-4 *(a)* Format for 8-bit sign-and-magnitude representation of binary numbers. *(b)* Format for 8-bit sign-and-magnitude binary representation of +46 decimal.

The first step is to convert the decimal number to its binary equivalent 101110. The second step is to construct an 8-bit word with this value, as shown in Figure 13-4b. The number is positive, so write a 0 in the sign bit position. Then write the binary equivalent of 46, 101110, in the lower 6 bits of the word. Finally, fill in any leftover bits between the sign bit and the magnitude bits with copies of the sign bit. The result, as you can see, is 00101110.

Since 7 bits in the 8-bit word are available to represent the magnitude of the number, you can represent positive numbers from 00000000 to 01111111 binary or 0 to 127 decimal with this 8-bit word. With a 16-bit signed number you can represent positive numbers in the range of 0000000000000000 to 0111111111111111 binary or 0 to 32,767 decimal. Now, let's see how you represent negative numbers in this format.

For a negative number, the sign bit is a 1 and the magnitude of the number is represented in *two's complement* form. The simplest and least error prone way to produce the correct 8-bit representation for a negative number is as follows:

1. Write the 8-bit *positive* binary representation of the number, including a 0 sign bit as shown previously.

2. Invert each of the 8 bits, including the sign bit, to form the one's complement.

3. Add 1 to the inverted result to form the two's complement.

Figure 13-5a shows how you use these steps to produce the signed 8-bit two's complement representation of −46 decimal. As we show, the 8-bit binary equivalent of 46 decimal is 00101110. Inverting each bit in this gives a one's complement of 11010001. Adding 1 to the result gives 11010010, which is the correct 8-bit two's complement sign-and-magnitude representation for −46 decimal. As we told you before, the 1 in the sign bit indicates that the number is negative and that the magnitude of the number is expressed in two's complement form. To help fix this conversion process in your mind, Figure 13-5b shows how to produce the 8-bit sign-and-magnitude representation for −1 decimal.

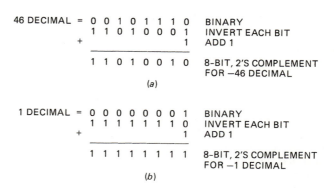

FIGURE 13-5 Conversion of negative decimal numbers to 8-bit sign-and-magnitude representation. *(a)* −46. *(b)* −1.

A 1 in decimal is equal to 0000001 binary, so inverting each bit gives 11111110. Adding 1 to this result gives 11111111 as the 8-bit two's complement representation of −1. Again, the 1 in the sign bit position tells you that the number is negative and that the magnitude is in two's complement form.

Figure 13-6 shows the range of values that can be rep-

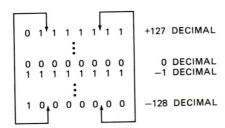

FIGURE 13-6 Range of numbers that can be represented with an 8-bit signed number.

resented with an 8-bit signed number. As you can see, this scheme allows you to represent positive numbers between 0 and +127 and negative numbers from −1 to −128. The total number of values you can represent with the 8 bits is still 2^8, or 256. The range of numbers represented is just shifted down so that half the range is available for negative numbers. Note that the binary values which represent +128 to +255 in positive-number-only representation now represent values from −128 to −1.

This observation brings to mind an important point. If you read an 8-bit number such as 11101101 from a memory location in your computer, how do you know whether this number represents a positive number in the range of +128 to +255 or a negative number in the range of −128 to −1? The answer to this question is that you can't tell just from the number itself. The only way to tell is from the context of how the number is used. In other words, to determine the value of the number, you must know whether the person who wrote the number in memory intended it to be treated as an unsigned number or as a signed number!

The next step is to show you how you determine the magnitude of a binary word that you know represents a signed number. Suppose, for example, that you read the 8-bit signed number 111101101 from the memory in your computer and you want to convert the magnitude of the number to a form more easily understood by humans. The steps to do this conversion are:

1. Invert each bit in the number, including the sign bit.
2. Add 1 to the result.
3. Put a minus sign in front of the final result.

Figure 13-7 shows an example of this conversion process. Inverting each bit in 111101101 gives 000010010, and adding 1 to this result gives 000010011. You put a minus sign in front of the number to remind you that the number represented is negative. To make the result even more understandable, you can convert the binary result to its decimal equivalent of −19.

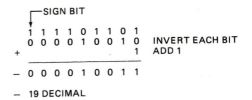

FIGURE 13-7 Determining the absolute value for the magnitude of an 8-bit signed binary number.

As you can see, converting a number from its two's complement representation to an absolute-value representation uses the same "invert each bit and add 1" process as converting from a positive value to the two's complement form. The conversions work this way because the sequence of codes for negative and positive numbers "wrap around," as shown by the arrows in Figure 13-6. Some exercises at the end of the chapter will give you practice converting numbers to and from their signed binary representations.

Now that you know how negative numbers are represented in binary, we can show you two common methods of doing binary subtraction.

BINARY SUBTRACTION: PENCIL METHOD

The first method of binary subtraction we call the *pencil method* because it is essentially the same as the method you probably use to do decimal subtraction by hand. Figure 13-8 shows the truth table or "rules" for subtracting

INPUTS		OUTPUTS	
A	B	D	B_{OUT}
0	0	0	0
0	1	1	1
1	0	1	0
1	1	0	0

DIFFERENCE = $A \oplus B$
BORROW = $\overline{A} \cdot B$

FIGURE 13-8 Truth table for subtracting a 1-bit binary number B from a 1-bit binary number A.

a 1-bit number, B, from another 1-bit binary number, A. The easy cases in this truth table are where A and B are both 0's and where A and B are both 1's. For these cases, the difference D is 0 and no borrow is required to perform the subtraction, so the borrow $B_{OUT} = 0$. For the case where A = 1 and B = 0, D = 1. Since we have defined A to be the "top number" for the subtraction, no borrow is required to perform the subtraction, and $B_{OUT} = 0$. For the final case where the "bottom number" B = 1 and the "top number" A = 0, you have to do a borrow to perform the subtraction, and $B_{OUT} = 1$. After you do the borrow, you are then subtracting 1 from 10 binary. This subtraction then gives a difference D = 1.

The truth table in Figure 13-8 only represents a *half-subtractor* because it does not take into account the effect of a borrow from subtracting a less significant bit. Figure 13-9 shows the truth table for a full-subtractor which takes this input borrow into effect. The values in

INPUTS			OUTPUTS	
A	B	B_{IN}	D	B_{OUT}
0	0	0	0	0
0	0	1	1	1
0	1	0	1	1
0	1	1	0	1
1	0	0	1	0
1	0	1	0	0
1	1	0	0	0
1	1	1	1	1

DIFFERENCE = $A \oplus B \oplus B_{IN}$
BORROW = $\overline{A} \cdot B + \overline{(A \oplus B)} \cdot B_{IN}$

FIGURE 13-9 Full-subtractor truth table showing the effects on a borrow in from a less significant bit.

		[1]				
22 DECIMAL	=	1	0	1	1	0
−12 DECIMAL	=	−0	1	1	0	0
10 DECIMAL	=	0	1	0	1	0

(*a*)

				[1]	[1]	
13 DECIMAL	=	0	0 ~~1~~	0 ~~1~~	0	1
− 6 DECIMAL	=	−0	0	1	1	0
7 DECIMAL	=	0	0	1	1	1

(*b*)

			[1]	[1]		[1]
18 DECIMAL	=	0 ~~1~~	0	0	0 ~~1~~	0
− 5 DECIMAL	=	−0	0	1	0	1
13 DECIMAL	=	0	1	1	0	1

(*c*)

FIGURE 13-10 Binary subtraction examples. *(a)* Using the full-subtractor truth table. *(b)* Using the mental-borrow method. *(c)* Using the full-subtractor truth table.

this truth table are those which will be produced by subtracting B from A. Work your way through the truth table to fix the rules in your mind. Again as a preview of coming attractions, note that the difference can be represented by the simple logic expression $D = A \oplus B \oplus B_{IN}$ and that borrow out to the next bit can be expressed by the simple logic expression $B_{OUT} = \overline{A}B + (\overline{A \oplus B})B_{IN}$. For now, let's see how you use these subtraction rules to do the pencil method of binary subtraction.

The first example in Figure 13-10*a* shows how we subtract the binary equivalent of 12 from the binary equivalent of 22 (22 − 12). Starting from the right, the first 3 bits subtract directly with no borrow required. As shown in line 3 of Figure 13-9, subtracting a 1 from a 0 in the fourth bit position in Figure 13-10*a* produces a difference of 1 and a borrow out to the next bit position. To help remember the borrow, you can write it above the next bit position, as shown in Figure 13-10*a*. For subtracting the fifth bit from the right in Figure 13-10*a*, A is a 1, B is a 0, and B_{IN} from the previous bit is a 1. According to line 6 in the truth table in Figure 13-9, this combination produces a difference of 0 and a B_{OUT} of 0. The final result then is 1010 binary or 10 decimal.

We showed you how to do this first example using the full-subtractor truth table because that is one of the ways it is done in hardware. However, when doing binary subtraction, you can work with the borrows in the same way you do in decimal subtraction. Here's an example of this approach.

Figure 13-10*b* shows an example of subtracting the binary equivalent of 6 from the binary equivalent of 13 (13 − 6). The least significant bit subtracts directly to give a difference of 1. Subtracting the second bit requires a borrow. With a borrow the second bit then gives a difference of 1. Borrowing a 1 from the 1 in the third bit position leaves a 0 in that bit position, as shown. To subtract the bits in the third position, a borrow from the fourth bit position is required. With this borrow the difference in the third bit position then is a 1. The borrow from the third bit position leaves a 0 in the top of the fourth bit position, so the subtraction is done. The result is 111, or 7_{10}.

Figure 13-10c shows a final example of this pencil method of subtraction. The first 2 bits on the right subtract with a simple borrow. Subtracting the third bits requires a borrow, but since the fourth bit in the top number is a 0, it may not be obvious where you go from here. The easiest way is to use the truth table in Figure 13-9. Line 3 in the truth table tells you that the result in the third column will be a difference D = 1 and $B_{OUT} = 1$. B_{OUT} for one bit position is a B_{IN} for the next bit position, so line 2 in the truth table tells you that the result for the fourth bit position is D = 1 and $B_{OUT} = 1$. Finally, line 6 in the truth table tells you that the result for bit five in the example is D = 0 and $B_{OUT} = 0$. This gives a result of 1101, or 13_{10}.

BINARY SUBTRACTION: TWO'S COMPLEMENT ADD METHOD

The pencil method of subtraction works, but it is tedious and prone to errors when you use it on lengthy numbers. A second method of binary subtraction which is simpler and easy to implement in a computer is the two's complement add method. For this method, all you do is add the two's complement representation for the negative of the "bottom" number to the binary representation of the top number.

For a first example, suppose that you want to subtract the binary equivalent of 9 from the binary equivalent of 13. If you remember that the expression (13 − 9) can be written as 13 + (−9), perhaps you can begin to get an intuitive feel for how this method works. As shown in Figure 13-11*a*, the first step in this method is to produce the two's complement sign-and-magnitude representation for the negative of the bottom number. To do this you simply invert each bit of the positive binary representation and add 1 to the result, as we showed you in an earlier section. The second step in this method is to add the two's complement sign-and-magnitude representation of −9 to the binary representation of +13. Note that the sign bits are added just as the other bits are. The result of the addition is 00000100. The final carry produced by the addition is not significant, so you simply ignore it. Since the sign bit of the result is a 0, you know that the result is positive and that its magnitude

$$13 - 9 = 13 + (-9)$$

```
-9 DECIMAL  -    0 0 0 0 1 0 0 1 ⎤ INVERT
                 1 1 1 1 0 1 1 0 ⎦ 
                             + 1    ADD 1
                 ───────────────
                 1 1 1 1 0 1 1 1  =  -9 IN 2'S COMPLEMENT
```

```
+13 DECIMAL    =     0 0 0 0 1 1 0 1
+(-9) DECIMAL  =  +  1 1 1 1 0 1 1 1
               ─────────────────────
+4 DECIMAL     =  1│0 0 0 0 0 1 0 0
                  └─IGNORE THIS CARRY
```

(a)

```
-13 DECIMAL    =  -  0 0 0 0 1 1 0 1 ⎤ INVERT
                    1 1 1 1 0 0 1 0 ⎦
                                + 1   ADD 1
                 ─────────────────
                    1 1 1 1 0 0 1 1   -13 IN 2'S COMPLEMENT
```

```
9 DECIMAL      =     0 0 0 0 1 0 0 1
+(-13) DECIMAL =  +  1 1 1 1 0 0 1 1
               ─────────────────
                    1 1 1 1 1 1 0 0   -4 IN 2'S COMPLEMENT
                    0 0 0 0 0 0 1 1   INVERT
                                + 1   ADD 1
               ─────────────────
-4 DECIMAL     =  -  0 0 0 0 0 1 0 0
```

(b)

FIGURE 13-11 Examples of binary subtraction using signed two's complement method. (a) 13 − 9. (b) 9 − 13.

is in standard binary form. In decimal form the result then is +4, as expected.

Figure 13-11b shows a second example of this two's complement add method of subtracting binary numbers. In this example we subtract the binary equivalent of 13 from the binary equivalent of 9 to produce the signed binary equivalent of −4. As in the previous example, the first step is to produce an 8-bit signed representation for the negative of the bottom number. To do this you invert each bit in it and add 1 to the result. The 8-bit signed representation for −13 then is 11110011. The second step is to add the two's complement representation for −13 to the binary representation of +9. The result is 11111100. The 1 in the sign bit shows that the result is negative as expected and that the magnitude is in two's complement form. To convert the result to a form more recognizable to most people, invert each bit, including the sign bit, add 1 to the result, and put a minus sign out in front to remind you that the number really is negative. The result of this conversion is −00000100, which is equivalent to -4_{10}.

The examples in Figure 13-11a and b not only show you another way of subtracting, they also show you how to add signed binary numbers. In Figure 13-11a we effectively added the signed representations of +13 and −9 to get a result of +4. In Figure 13-11b we effectively added the signed representations for +9 and −13 to get an initial result which was the signed equivalent of −4. The remaining possibilities to explore are the addition of two positive signed numbers and the addition of two negative signed numbers.

Figure 13-12a shows the result produced by adding the binary equivalent of +9 and the binary equivalent of

```
+13     00001101
+ 9     00001001
────    ─────────
+22     00010110
           └─ SIGN BIT IS 0
              SO RESULT IS POSITIVE
```

(a)

```
- 9     11110111 ⎤ 2'S COMPLEMENT,
-13     11110011 ⎦ SIGN-AND-MAGNITUDE FORM
────    ─────────
-22     11101010   SIGN BIT IS 1
        00010101   SO INVERT EACH BIT
      +        1   ADD 1
        ─────────
equals  -00010110  PREFIX WITH MINUS SIGN
```

(b)

FIGURE 13-12 Further examples of adding signed numbers. (a) +9 + (+13). (b) −9 + (−13).

+13. The 0 in the sign bit indicates that the result is positive and that the magnitude is expressed in standard binary. If you convert 0010110 to its decimal equivalent, you should get 22.

Figure 13-12b shows the results produced by adding the signed binary representation for −9 to the signed binary representation of −13. As you can see, the result of the addition is 11101010. The 1 in the sign bit indicates that the result is negative and that its magnitude is expressed in two's complement form. Again you can convert this to a more recognizable form by inverting each bit, including the sign bit, adding 1 to the result, and putting a minus sign out in front. As expected, −00010110 is equivalent to −22 decimal.

A major problem that may occur when adding signed

numbers is *overflow*. If the sum of two signed numbers is too large to fit in the number of bits available for the magnitude, the sum will "overflow" into the sign bit and give an incorrect result. The signed number addition in Figure 13-13 shows an example of this. When the signed

```
 +86        0 1 0 1 0 1 1 0
+106   +    0 1 1 0 1 0 1 0
──────      ───────────────
 -64        1 1 0 0 0 0 0 0
              └── 1 INDICATES A NEGATIVE RESULT,
                  WHICH IS INCORRECT FOR THE SUM
                  OF TWO POSITIVE NUMBERS.
```

FIGURE 13-13 Example of overflow error which can occur when the sum of two signed numbers has more bits than the number of bits allowed for magnitude.

representation of +86 is added to the signed representation of +106, the result is 1100000. The 1 in the sign bit of this result indicates that the result is negative, which is obviously incorrect for the sum of two positive numbers unless it represents the sum of you and your spouse's income after taxes. The problem here is that the sum has overflowed into the sign bit position. Remember that the range of values which can be represented with an 8-bit signed number is −128 to +127. A 16-bit signed number can represent numbers between −32,768 and +32,767. Whenever you are working with signed numbers, you have to watch for this problem.

Binary-Coded Decimal (BCD) Addition and Subtraction

BCD ADDITION

As shown in Table 7-1, BCD is a way of representing a decimal digit with a 4-bit binary code. The BCD codes for the decimal digits 0 to 9 are 0000 to 1001, the standard binary codes for 0 to 9. The decimal number 275, for example, is represented in BCD as 0010 0111 0101. (When writing BCD numbers, we often separate the 4-bit groups to make the individual digits more obvious.)

BCD numbers are made up of 1's and 0's, so you do the basic addition in the same way as you add ordinary binary numbers. When adding BCD numbers, however, you sometimes have to add a correction factor to get the final result in correct BCD form. A few examples are the best way to show you when the correction factor is needed.

The first example in Figure 13-14*a* shows the addition of the BCD representations for 35 decimal and 23 decimal. Following the rules for binary addition gives 0101 1000, which is the correct BCD for 58, the desired result.

The second BCD example in Figure 13-14*b* shows the result of adding the BCD code for 7 to the BCD code for 5. As you can see, the first result is 1100. This is the correct binary result, but it is not the correct BCD result. To get the correct BCD result you add a correction factor of 0110, or 6, to the initial result. This addition produces 0010 in the least significant BCD digit and a carry of 1 into the second BCD digit. The final result

```
            BCD
  35      0011  0101
 +23     +0010  0011
────     ───────────
  58      0101  1000
            (a)

            BCD
   7        0111
 + 5      + 0101
────      ──────
  12        1100        INCORRECT BCD
          + 0110        ADD 6
          ──────
        0001  0010      CORRECT BCD 12
            (b)

            BCD
   9        1001
 + 8      + 1000
────      ──────
  17      0001  0001    INCORRECT BCD
          0000  0110    ADD 6
          ──────────
          0001  0111    CORRECT BCD 17
            (c)
```

FIGURE 13-14 BCD addition examples. *(a)* 35 + 23. *(b)* 7 + 5. *(c)* 9 + 8.

then is 0001 0010, which is the correct BCD for 12, the desired result.

The reason that you have to add the correction factor of 6 is that in BCD addition you want a carry into the next higher 4 bits to be produced when the result is greater than 1001, or 9. In standard binary a carry into the next higher 4 bits is not produced until the result is greater than 1111, or 15. The difference between the carry points in the two systems is 0110, or 6. Any time the result of adding two BCD digits together is greater than 1001, you have to add the correction factor of 0110 to force a carry into the next higher digit position. The result then will be in correct BCD format.

Figure 13-14*c* shows another case where you have to add the correction factor of 0110. Adding the BCD for 9 to the BCD for 8 gives a result of 10001, which is the correct binary result but not the correct BCD result. In this case, the problem is not caused by an illegal BCD number in the lower 4 bits but by the fact that the addition produced a carry into the second 4 bits. This carry means that the result of the initial addition was not only greater than 1001, it was greater than 1111, so the correction factor of 0110 must be added. Adding the correction factor of 0110 then gives the correct BCD result of 0001 0111, or 17.

To summarize, then, you add the correction factor of 0110 if:

1. The result of adding two BCD digits together is greater than 1001.

2. The addition of two BCD digits produces a carry into the next higher digit position.

To show you how this works with multiple-digit BCD numbers, Figure 13-15 illustrates a final example. In this example you have to add 0110 to the least significant

```
    97         1001 0111
   +74         0111 0100
   ___
                     1 0000 [1]1011
             +        0110 0110
   ___
   171         0001 0111 0001
```

FIGURE 13-15   Multiple-digit BCD addition example.

BCD digit because the initial addition produced a result greater than 1001. You have to add 0110 to the second BCD digit because the result of the initial addition produced a carry into the next higher BCD digit. The final result then is 0001 0111 0001, or 171. Note that you do not add another correction factor to the lower 4 bits in response to the carry produced by adding the 0110.

Determining if you have the right final result for BCD addition is easy, because you can simply compare the BCD result with the result you get by adding the decimal numbers directly.

BCD SUBTRACTION

Since BCD numbers are just 1's and 0's, the basic subtraction of BCD numbers follows the same rules as standard binary numbers. Sometimes, however, you have to subtract a correction factor of 0110 from the binary result to get the correct BCD result. As with the correction you have to perform in BCD addition, the reason for this correction is the difference between the carry point in BCD and in standard binary.

Figure 13-16a shows the subtraction of BCD 9 from

```
   17         0001 0111
  - 9         0000 1001
  ___
    8         0000 1110    ILLEGAL BCD
                  -0110    SUBTRACT 6
              _____
              0000 1000    CORRECT BCD

                  (a)

   12         0001 0010
  - 9        -0000 1001
  ___
    3         0000 1001    ILLEGAL BCD
               -    0110   SUBTRACT 6
              _____
              0000 0011    CORRECT BCD

                  (b)
```

FIGURE 13-16   BCD subtraction examples. (a) 17 − 9.
(b) 12 − 9.

BCD 17. As you can see, the initial subtraction leaves 1110 in the lower 4 bits. This number is greater than 1001, so it is not a legitimate BCD number. Subtracting 0110 gives the correct BCD result of 1000, or 8.

Another example in Figure 13-16b shows the subtraction of BCD 9 from BCD 12. As you can see, the initial subtraction gives 1001, which is a legitimate BCD number but not the correct BCD result for the subtraction. The fact that a borrow was required from the second BCD digit to do the subtraction tells you to subtract the

correction factor of 0110. Subtracting 0110 gives the correct BCD result of 0011.

To summarize then, you subtract the correction factor of 0110 if:

1. The result of subtracting two BCD digits gives a result which is greater than 1001.

2. Subtracting two BCD digits requires a borrow from the next higher BCD digit.

Octal Addition and Subtraction

OCTAL ADDITION

The octal, or base 8, number system is not used very much today, but for reference we will briefly show you how to add and subtract octal numbers in case you have to work on an old system which uses octal codes. To refresh your memory of octal, Figure 13-17a shows the

```
   8^4   8^3   8^2   8^1   8^0  .   8^-1  8^-2  8^-3
   4096  512   64    8     1    .   1/8   1/64  1/512

                        (a)

   111   010   101  Binary
    7     2     5   Octal

                        (b)

    3     2     6   Octal
   011   010   110  Binary

                        (c)
```

FIGURE 13-17   Octal numbers. (a) Decimal weights of a few octal digits. (b) Conversion of binary numbers to octal. (c) Conversion of octal numbers to binary.

decimal weights of a few digits in the octal number system. The eight values that an octal digit can have are 0, 1, 2, 3, 4, 5, 6, and 7. Review the appropriate section of Chapter 7 if you need help in remembering how to convert decimal numbers to and from octal.

As we told you in Chapter 7, the main use of octal is to help people remember sequences of binary numbers. Since one octal digit can represent the same number of values as three binary digits, you only need one-third as many digits to represent a given number. To convert a binary number to its octal equivalent, you first mark off groups of 3 bits in the binary number, starting from the right, as shown in Figure 13-17b. You then replace each group of 3 bits with its octal equivalent value. To convert an octal number to its binary equivalent, you just replace each octal digit with its 3-bit binary equivalent, as shown in Figure 13-17c.

Adding octal numbers is similar to adding decimal numbers except that the carry comes at a different point. Whenever the sum of two octal digits is 8 or greater, you subtract 8 from the sum, write down the difference, and add a carry to the next higher digit. Figure 13-18a shows an example of adding two octal numbers.

Adding 2 and 3 in the least significant digit position

```
  3 6 2  Octal        4 +8
+ 4 5 3  Octal        5 2 7  Octal
                    - 2 4 3  Octal
  ┌─┐
1 0 3 5  Octal        2 6 4  Octal
  (a)                   (b)
```

FIGURE 13-18 Octal addition and subtraction.
(a) Addition. (b) Subtraction.

gives a sum of 5. Since the sum is less than 8, you just write down the 5. When you add the 5 and 6 in the next digit position, the result is decimal 11, which is greater than 8. To get the correct value for this digit, you subtract 8 from the 11 and write down the difference of 3. The 8 you subtracted represents a carry into the next digit position, so you include a carry when you add that column. Adding 3 and 4 and a carry in the third column gives a result of 8. This result is equal to 8, so you subtract 8 and write down the difference of 0. Again, the 8 you subtracted from the sum represents a carry into the next higher digit. In this case there are no more digits to add, so you just write this carry as a 1 in the fourth digit position.

The key point to remember about octal addition is that the carry comes at 8 rather than at 10.

OCTAL SUBTRACTION

Octal subtraction likewise is very similar to decimal except that the value of a borrow is 8 instead of 10. Figure 13-18b shows an example of octal subtraction.

Subtracting 3 from 7 in the least significant digit position gives a difference of 4, just as in decimal. To do the subtraction of 4 from 2 in the second column, you need to borrow from the next higher digit. As we said above, the value of a borrow in octal is 8. If you add this borrowed 8 to the 2 already in the top position, you get a result of 10 decimal. Subtracting the 4 on the bottom from 10 gives a result of 6, as shown. Borrowing 1 from the 5 in the third digit position left 4 there. Subtracting the bottom 2 from this 4 gives a result of 2, as shown in Figure 13-18b.

The key point to remember in octal subtraction is that the value of a borrow is 8.

Hexadecimal Addition and Subtraction

HEXADECIMAL REVIEW

For review, Figure 13-19a shows the decimal weights of a few digits in the hexadecimal, or base 16, number system. The 16 symbols used in this number system are the decimal numbers 0 to 9 and the letters A to F, as shown in Figure 13-19b. When you count up in hexadecimal, you go 0, 1, 2, 3, 4, 5, 6, 7, 8, 9, A, B, C, D, E, F, 10, 11, 12, 13, etc. The important point here is that the value of a carry is 16 in hexadecimal instead of 10 as in decimal.

The main use of hexadecimal is to make it easier to work with long binary numbers used in microcomputer and memory systems. Since $16 = 2^4$, each hexadecimal digit is equivalent to four binary digits. The hexadecimal equivalent of a binary number then contains only one-fourth as many digits, so it is much easier to work with.

To convert a binary number to its hexadecimal equivalent, you first mark off groups of 4 bits in the binary number, starting from the right, as shown in Figure 13-19c. Then you just replace each group of 4 bits with its hexadecimal equivalent. To convert a hexadecimal number to its binary equivalent, you simply replace each hexadecimal digit with its 4-bit binary equivalent, as shown in Figure 13-19d. Refer to the appropriate section of Chapter 7 if you need a refresher on how to convert decimal numbers to and from hexadecimal.

HEXADECIMAL ADDITION

Most scientific calculators now do hexadecimal arithmetic, but it is worth knowing how to do simple calculations so that you are not totally dependent on your calculator.

One way to add two hexadecimal numbers is to convert each number to its binary equivalent, add the binary numbers, and convert the result back to hexadecimal.

A more direct method is to add the numbers directly. This may seem strange at first because you are not used to the letter symbols, but it is really not difficult. The main point to remember is that whenever the result of adding two hexadecimal digits is 16 decimal or greater,

```
          16³        16²  16¹  16⁰  ·   16⁻¹    16⁻²    16⁻³
          4096       256  16   1    ·   1/16    1/256   1/4096

                              (a)

DECIMAL   0    1    2    3    4    5    6    7    8    9    10   11   12   13   14   15
HEX       0    1    2    3    4    5    6    7    8    9    A    B    C    D    E    F

                              (b)

          1011  0111  0110  1110   BINARY
           B     7     6     E     HEX

                      (c)

           3     A     C     4     HEX
          0011  1010  1100  0100   BINARY

                      (d)
```

FIGURE 13-19 Hexadecimal number review. (a) Decimal weights of a few hexadecimal digits. (b) Hexadecimal digit symbols and decial equivalents. (c) Binary-to-hexadecimal conversion. (d) Hexadecimal-to-binary conversion.

you subtract 16 decimal from the sum, write down the difference, and add a carry to the next higher digit.

In our first example in Figure 13-20a, adding the least significant digits of 7 and 8 gives a result of 15 decimal. If you don't remember that 15 decimal is equal to F in hex, you can look it up in Figure 13-19b. Adding 9 and 9 in the second digit position gives a decimal 18. This is greater than 16, so you subtract 16, write down the difference of 2, and add a carry to the next higher digit. The final result then is 12F hexadecimal, or 12FH.

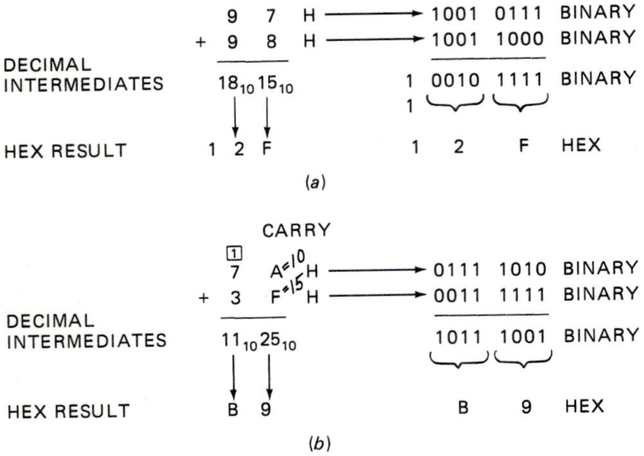

FIGURE 13-20   Hexadecimal addition examples.

The second example in Figure 13-20b shows how you add hex numbers which contain the letter symbols. To add A and F, for example, you mentally convert the A to its decimal equivalent of 10 and mentally convert the F to its decimal equivalent of 15. You then add these two together to get 25 decimal. This 25 is greater than 16, so you subtract 16, write down the remainder of 9, and add a carry to the next digit. In the next digit position a 7 plus a 3 plus a carry gives decimal 11, which is equal to B in hex. The final result then is B9H, as shown. If you feel safer converting the numbers to binary, adding them, and converting back to hex, we have also shown that method for the examples in Figure 13-20.

HEXADECIMAL SUBTRACTION

As with addition, there are two methods of subtracting hexadecimal numbers. One method is to convert the numbers to their binary equivalents, subtract the binary equivalents, and convert the result back to hex. The second method of subtracting is the direct method, similar to the direct method of addition we showed you in the preceding section. For the direct method the main point you have to remember is that the value of a borrow is 16 decimal rather than 10. Figure 13-21 shows a few examples of hexadecimal subtraction.

To subtract 5 from C in the first digit position of the example in Figure 13-21a, you first mentally convert the C to its decimal equivalent of 12. You then subtract 5 from 12 to give a difference of 7 and write this difference down. Likewise, to subtract 2 from F in the second digit

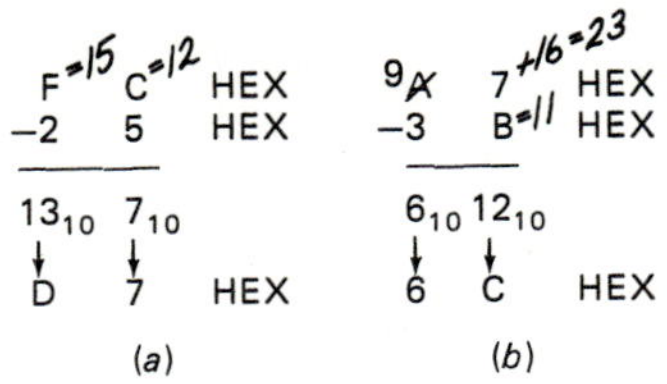

FIGURE 13-21   Hexadecimal subtraction examples.

position, you first mentally convert the F to its decimal equivalent, 15. You then subtract 2 from 15 to get a difference of 13 decimal. Since 13 decimal is equivalent to D in hex, you write the D as the difference for the second digit position. The final result then is D7H.

The second example in Figure 13-21b shows how you deal with borrows in hex subtraction. First you mentally convert the B in the bottom of the first digit position to its decimal equivalent of 11. To do the subtraction you have to do a borrow from the next higher digit. The value of a borrow is 16, so adding the borrowed 16 to the 7 in the top of the first digit position gives 23 decimal. Subtracting the 11 decimal on the bottom from this 23 gives a difference of 12 decimal, or CH. In the second digit position you borrowed 1 from the A which leaves 9. Subtracting 3 from the 9 gives a difference of 6 which you simply write down. The final result of the subtraction then is 6CH.

Binary Multiplication and Division

BINARY MULTIPLICATION

There are several ways of multiplying binary numbers. Figure 13-22 shows an example of what might be called

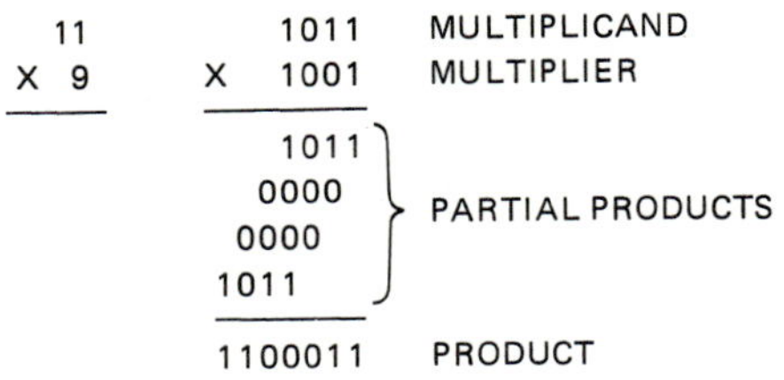

FIGURE 13-22   Pencil method of binary multiplication.

the *pencil method* because it is the way you do it by hand. This method is similar to the way you probably learned to multiply decimal numbers.

In this method you first multiply the top number by the least significant digit of the bottom number. You write the result of this multiplication, called a *partial product*, under the line, as shown. You then multiply the top number by the second digit of the bottom number and write the second partial product under the first, but shifted one bit position to the left. You compute the partial product for each of the remaining bits in the bottom number and write the partial products down, as shown. To get the final result, you just add up all the partial products.

The pencil method is the best one to use if you are multiplying binary numbers by hand, but it is awkward to implement in hardware or with a computer program. Another binary-multiplication method which can easily be implemented with a computer program is the repeated-addition method.

To multiply 7 × 55, for example, the computer can simply add seven 55's. For multiplying large numbers, however, this method is too slow. To multiply 786 × 253 with this method, for example, requires 253 add operations.

A third method which is easier to implement in both hardware and with a computer program uses an *add-and-shift-right* algorithm. An algorithm, remember, is just a formula or set of steps used to solve a problem. To understand this algorithm, it helps to think of the product being developed in a register called an *accumulator* which you can add a number to or you can shift right. The basic algorithm is as follows:

Load all 0's in the accumulator register.
  REPEAT
  IF the multiplier bit is a 0 THEN
    add 0 to the register
  IF the multiplier bit is a 1 THEN
    add the top number (multiplicand) to the register
  Shift the register contents one bit position right
  UNTIL all multiplier bits are used.

For the example in Figure 13-23, the least significant bit of the multiplier is a 1, so after loading all 0's in the accumulator register, you add the top number to the register. You then shift the result of this addition one bit position to the right. This completes one loop through the basic algorithm.

The next significant bit in the multiplier is a 0, so you add 0 to the contents of the accumulator and shift the result right one bit position. Note that the top number is always added to the most significant bits of the accumulator. The easy way to make sure you do this is to draw a vertical line, as shown in Figure 13-23.

The next significant bit in the multiplier is a 0, so you add 0 to the contents of the accumulator and shift the result right one bit position. The most significant bit in the multiplier is a 1, so you add the multiplicand, 1011,

to the accumulator and shift the result right one bit position. Since you have used all the multiplier bits, the multiplication is complete. Work your way through this example to make sure you understand the process.

Note that this add-and-shift method of multiplying takes only one add and one shift operation for each bit in the multiplier. This is much faster than the repeated-addition method.

Incidentally, multiplying a 4-bit number by a 4-bit number can produce a product with as many as 8 bits. Likewise, multiplying an 8-bit number by an 8-bit number can produce a result as large as 16 bits, and multiplying two 16-bit binary numbers can produce a result as large as 32 bits. This relationship tells you how large the register has to be to hold the result of a particular multiplication.

BINARY DIVISION

There are also several ways of doing binary division. Figure 13-24 shows an example of the pencil method which is very similar to the method you probably learned to do decimal long division. You start by trying to divide the divisor into the most significant part of the dividend. For the example in Figure 13-24 you might first try dividing 110 into the first 3 bits of the dividend, 100. Since 110 will not "fit" into 100, you put a 0 in the quotient and try dividing the 110 into a larger part of the dividend. The 110 will fit into 1001, so you put a 1 in the next bit position of the quotient and subtract 110 from 1001. The difference of 11 then becomes the most significant part of the dividend for the next step. The 110 won't fit into 11, so you "bring down" a 0 from the original dividend to produce 110. The divisor of 110 fits exactly into this 110, so you write a 1 in the quotient over the 0 you brought down. Subtracting the divisor of 110 from 110 leaves a difference of 0. Since 110 does not fit into the final 00, you write 00 in the quotient and breathe a sigh of relief.

In the example in Figure 13-24a the divisor divided evenly into the dividend, so there was no remainder. Figure 13-24b shows an example of how you can add 0's to

```
      1011 |        REGISTER B—MULTIPLICAND
     X1001 |        REGISTER C—MULTIPLIER
      0000 |        RESET ACCUMULATOR REGISTER TO 0
      1011 |        SINCE FIRST MULTIPLIER DIGIT IS 1
      1011 |          ADD MULTIPLICAND TO ACCUMULATOR
       101 |1        SHIFT RESULT RIGHT
      0000 |         NEXT MULTIPLIER DIGIT IS 0
      0101 |1          SO ADD 0 TO ACCUMULATOR
      0010 |11       SHIFT RESULT RIGHT
      0000 |         NEXT MULTIPLIER DIGIT IS 0
      0010 |11         SO ADD 0 TO ACCUMULATOR
      0001 |011      SHIFT RESULT RIGHT
      1011 |         NEXT MULTIPLIER DIGIT IS 1 SO
PRODUCT 1100 |011      ADD MULTIPLICAND TO ACCUMULATOR
       110 |0011     SHIFT RESULT RIGHT
```

FIGURE 13-23 Add-and-shift-right method of binary multiplication.

```
               01100   QUOTIENT
DIVISOR   110 ) 1001000   DIVIDEND          12
               −110                       6 ) 72
               ————
                110
               −110
               ————
                  0

                  (a)

               110.01                       6.25
         100 ) 11001.00                   4 ) 25
             −100
             ————
              100
             −100
             ————
              0100

                  (b)
```

FIGURE 13-24 Pencil method of binary division. (a) Example with no remainder. (b) Example showing conversion of remainder to fractional equivalent by continued division.

the right of the binary point and continue the division to convert the remainder to a binary fractional equivalent. The division process can be continued until the remainder is 0 or until you get the desired number of bits to the right of the binary point.

A second method of doing binary division is by repeated subtraction. The principle of this method is to subtract the divisor from the dividend until the difference produced by a subtraction is less than the divisor. The quotient is then simply the number of times you managed to subtract the divisor. This method is simple in concept, but it has the same problem as the repeated-addition method of multiplying—it takes a long time. To divide the binary equivalent of 23,536 by the binary equivalent of 219, for example, would require 107 subtract operations.

A binary-division method which is more easily implemented in hardware and with a computer program is the *subtract-and-shift-left* method. The easiest way to understand this method is to work your way through the example in Figure 13-25*b*.

You start by subtracting the divisor from an equal number of bits on the most significant end of the dividend. If no borrow is necessary to do the subtraction, as in the first operation in Figure 13-25*b*, then the divisor fits into those bits of the dividend. In this case you write a 1 as the most significant bit of the quo-

tient. Then "bring down" the remaining bits of the dividend and join them with the difference produced by the subtraction. Finally, you shift the word formed one bit position to the left to get ready for the next operation.

If a borrow is required to do the subtraction, as shown on line 6 in Figure 13-25*b*, then the divisor is too large to fit into the upper bits of the new dividend. Here you write a 0 for this bit in the quotient, add the divisor back to the new dividend, and shift the new dividend one bit position to the left. You then try the subtraction again, as shown on line 10 in Figure 13-25. This time the subtraction doesn't require a borrow, so you write a 1 for the next bit of the quotient, bring down the remaining bits of the original dividend to join with the difference, and shift the result left one bit position to get ready for the next operation.

You continue the subtract-and-shift process until all the digits of the original dividend are "used up," just as you do in decimal long division. If a difference (remainder) is present, you can add a binary point and bring down 0's to continue the division. You can continue the division until there is no remainder or until you have the desired number of fractional digits.

The main point of this example is to show you that binary division can easily be done in hardware or by a computer program with a sequence of simple subtract-and-shift operations.

OVERFLOW/
CARRY BIT

QUOTIENT REGISTER

Line	Carry	Bits	Bits	Operation	Quotient
1		1010	01000	DIVIDEND	
2		−1000		DIVISOR—SUBTRACT	
3	0	0010	01000	OVERFLOW/CARRY BIT = 0, SO →	1
4	0	0100	1000	SHIFT LEFT	
5		−1000		SUBTRACT	
6	1	1100	1000	OVERFLOW/CARRY BIT = 1, SO →	0
7		+1000		ADD DIVISOR BACK TO RESTORE	
8		0100	1000		
9		1001	000	SHIFT LEFT	
10		−1000		TRY SUBTRACTION AGAIN	
11	0	0001	000	OVERFLOW/CARRY BIT = 0, SO →	1
12		0010	00	SHIFT LEFT	
13		−1000		SUBTRACT	
14	1	1010	00	OVERFLOW/CARRY BIT = 1, SO →	0
15		+1000		ADD TO RESTORE	
16		0010	00		
17		0100	0	SHIFT LEFT	
18		−1000		SUBTRACT	
19	1	1100	0	OVERFLOW/CARRY BIT = 1, SO →	0
20		+1000		ADD TO RESTORE	
21		0100	0		
22		1000		SHIFT LEFT	
23		−1000		SUBTRACT	
24	0	0000		STOP—OVERFLOW/CARRY BIT = 0, SO →	1

$$\frac{41}{8)\,328}$$

QUOTIENT = 101001 = 41_{10}

FIGURE 13-25 Subtract-and-shift-left method of binary division.

Fixed-Point Numbers and Floating-Point Numbers

Most of the numbers we have worked with so far in this chapter have been integers, or, in other words, numbers with no fractional part. Just as you can't store a minus sign in a memory location to show that a number is negative, you can't store a decimal point or binary point in a memory location to indicate which part of a number is the integer part and which part is the fractional part. In this section we show you two methods used to solve this problem.

FIXED-POINT CALCULATIONS

Most simple microprocessor-based systems such as a grocery scale which multiplies weight by price per pound to give total price use fixed-point numbers. This means that all numbers are treated as integers with a binary point or decimal point to the right of the least significant bit. For example, a weight of 9.4 lb would be stored in an 8-bit memory location as 1001 0100 BCD or 01011110 binary. A price of $0.29 per pound would be stored as 0010 1001 BCD or 00011101 binary.

When the binary representation of this weight is multiplied by the binary representation of the price per pound, the result is 101010100110 binary or 2726 decimal. To produce the desired display of $2.73, the programmer must convert the binary number to BCD and determine where to put the decimal point in the result.

For this simple example it is not too difficult to keep track of the position of the binary point or decimal point, but when you are working with very large numbers such as 228×10^8 or very small numbers such as 78×10^{-6}, it is much more difficult. For applications which work with numbers such as these, we usually use a floating-point representation for the numbers.

FLOATING-POINT NUMBERS

Numbers which contain both an integer part and a fractional part are commonly called real numbers. In the fixed-point system we described in the previous section, the hardware or the programmer has to keep track of the position of the binary point in real numbers. In a floating-point system, the position of the binary point is encoded directly in the representation. The techniques used to perform arithmetic operations on these floating-point representations automatically produce a result with the correct representation for the position of the binary point. There are several different floating-point formats. One of the most commonly used formats is the IEEE 754 format shown in Figure 13-26.

In the format shown in Figure 13-26, 32 bits are used to represent each number. Of those 32 bits, 23 bits are used for the significant bits of the number, 8 bits are used for the exponent, and the most significant bit is used as a sign bit for the number. Here's how you convert a real number to this floating-point format.

If you are starting with a decimal number, the first step is to convert the number to its binary equivalent. As a first example we will use the number 105.5 decimal, which is equal to 1101001.1 in binary.

The next step is to normalize the number. Normalizing a binary number is similar to putting a decimal number in "scientific notation." To normalize a binary number you first move the binary point so that there is only one 1 to the left of the binary point. Then you write an exponent to represent the number of bit positions you moved the binary point. For example, to normalize the binary number 1101001.1, you first move the binary point six places to the left to give 1.1010011. You then write $\times 2^6$ after this to show how far the binary point has been moved. The normalized result then is 1.1010011×2^6.

The number part of this normalized result is often called the mantissa, or significand. The fractional part of the mantissa is written in the 23 bits reserved for the mantissa in Figure 13-26. The 1 to the left of the binary point is implied but not actually written as part of the mantissa. You can think of the binary point as being between the bits labeled 23 and 24 in Figure 13-26.

The exponent is stored in the next 8 bits, 24 to 31, in unsigned, biased form. To determine the values for these

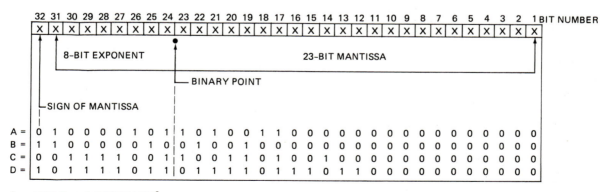

A = +105.5$_{10}$ = 1.1010011 X 2^6
B = −10.25$_{10}$ = 1.01001 X 2^3
C = +0.0000011001101001 = 1.1001101001 X 2^{-6}
D = −0.00010111101110110 = 1.0111101110110 X 2^{-4}

FIGURE 13-26 Floating-point number format with examples. (a) 105.5. (b) −10.25. (c) +0.0000011001101001 binary. (c) −0.00010111101110110 binary.

bits, you simply add 127 decimal or 01111111 binary to the actual exponent. For our example here, the actual exponent is 6. Adding 127 to this gives 133 decimal or 10000101 binary as the value for the exponent bits.

Finally, the sign of the number itself is put in the most significant bit of the first byte. The example number is positive, so the sign bit is a 0. As shown in Figure 13-26, the floating-point representation for 105.5 decimal in this format then is 0 10000101 1010011 00000000 00000000.

The bottom lines of the table in Figure 13-26 show some other examples of real numbers expressed in this binary floating-point format. As you can see from these examples, you can also represent positive numbers with negative exponents, negative numbers with positive exponents, and negative numbers with negative exponents.

The 32-bit floating-point format shown in Figure 13-26 is often called a *single-precision representation*. The 23-bit mantissa in this single-precision representation gives an accuracy of 1 part in 2^{24}, or about 1 part in 7 decimal digits. The 8-bit exponent gives an exponent range of 2^{-128} to 2^{+128}, or about 10^{-38} to 10^{+38}. This range is more than enough for many applications. For applications where this is not enough range, a 64-bit format called *long real*, or *double-precision*, is used. In this format 52 bits are used for the mantissa, 11 bits for the biased exponent, and 1 bit for the sign of the number. This format allows you to represent decimal numbers with an absolute value between 2.3×10^{-308} and $1.7 \times 10^{+307}$.

We will not show you how to do arithmetic operations with numbers in floating-point form because you almost always do these operations with a computer program or with a hardware device specially designed for that purpose. In the sequel to this book, *Microprocessors and Interfacing*, we describe the operation and programming of the Intel 8087/80187/80387 math coprocessor device which performs a wide variety of floating-point computations. In the remaining sections of this chapter we show you some hardware devices and circuits which perform simple binary arithmetic operations and introduce you to a microprocessor.

DIGITAL ARITHMETIC DEVICES AND CIRCUITS

Adder Circuits

In Figure 13-3 we showed you that the sum of two binary digits A and B and a CARRY$_{IN}$ from a previous digit can be expressed as SUM = A $\oplus$ B $\oplus$ CARRY$_{IN}$. The CARRY$_{OUT}$ from the addition of the 2 bits can be expressed as CARRY$_{OUT}$ = AB + CARRY$_{IN}$(A $\oplus$ B). Figure 13-27 shows how these two expressions can be implemented with simple gates. You are not likely to build this circuit because several available MSI devices do the job better. The reason for showing you this circuit is to fix in your mind the point that binary arithmetic functions are just another application of simple logic devices.

If you need to perform simple binary addition for some application, you can use a 4-bit full-adder such as the 74LS83 shown in Figure 13-28. In this diagram A1 rep-

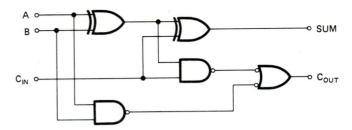

FIGURE 13-27   Simple gate circuit to add two 1-bit binary numbers.

resents the LSB and A4 the MSB of one binary word. B1 represents the LSB and B4 the MSB of the other binary word. The $\Sigma 1$ output represents the sum of the A1 and B1 bits, $\Sigma 2$ represents the sum of the A2 and B2 bits, etc.

The term *full-adder* here means that the device takes into account the effect of a carry in, C0, from the addition of a less significant group of 4 bits. The device also generates a carry output signal C4 which goes high if the addition produces a carry. These devices can be cascaded to add binary words of any length. To do this the C4 output from one device is simply connected to the carry input, C0, of the next higher device. The C0 input of the least significant device is connected to ground.

Note in Figure 13-28 that the carry output signal C4 is produced independently from the sum output signals. This type of carry is called a *look-ahead carry* because it is generated without waiting for the sum to be generated. A 74LS283, for example, generates the C4 signal for two input words in about 16 ns, which is faster than the 24 ns it takes to produce the sum outputs. The advantage of look-ahead carry is that it speeds up the operation of a cascaded group of devices. The more significant devices do not have to wait for the sum to be computed in less significant devices in order to produce their sum outputs. They just have to wait for the look-ahead carries to ripple through the chain of devices.

SUBTRACTION

As we showed you in Figure 13-9, the expression produced by subtracting a 1-bit binary number B from a 1-bit binary number A is DIFFERENCE = A $\oplus$ B $\oplus$ B$_{IN}$, where B$_{IN}$ is a borrow in from a less significant digit. The expression for a borrow out to a more significant bit position is B$_{OUT}$ = $\overline{A}$B + $(\overline{A \oplus B})$B$_{IN}$. Figure 13-29 shows how these full-subtractor expressions can be implemented with simple logic gates.

The main purpose of showing you this circuit is so that you can see how similar it is to the circuitry for adding binary numbers. This observation may lead you to the idea that it is possible to make a circuit which will add or subtract two binary numbers, depending on the logic level on a control input. One example of a device such as this is a 74LS181 Arithmetic Logic Unit which we discuss in a later section of the chapter. This device will add, subtract, or perform one of several other operations on two 4-bit binary numbers.

Another method of doing subtraction in hardware is to produce the two's complement of the "bottom" num-

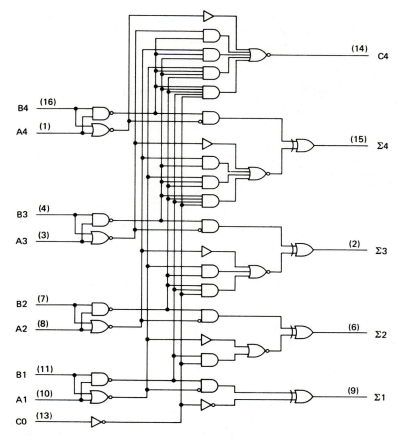

FIGURE 13-28 The 74LS83 4-bit Binary Full-Adder logic diagram. *(Texas Instruments Inc.)*

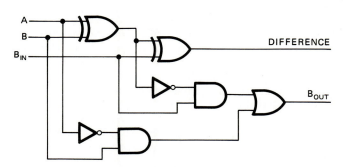

FIGURE 13-29 Simple gate circuit to subtract two 1-bit numbers.

ber and add it to the top number, as we showed you earlier in the chapter. This second method is useful for doing subtraction with a microprocessor.

Magnitude Comparators

In many applications you need to compare two binary words and determine whether one number is larger than the other, smaller than the other, or equal to the other. You might, for example, want to determine whether the actual temperature of a furnace is above, below, or equal to the desired setpoint temperature.

In Problem 19 at the end of Chapter 6, we asked you to design a circuit which determines if two 4-bit binary

numbers are equal. We hope you found out that this circuit can be implemented with XNOR gates and an AND gate. If you use this circuit to compare the numbers and another circuit to subtract the two numbers, the results will show the relative magnitudes of the numbers as follows:

Operation	Results Produced	Conclusion
A minus B	Borrow	A — B
A minus B	No borrow · Not equal	A \ B
A minus B	Equal	A = B

A common MSI device which implements this in hardware is the 74LS85 4-bit Magnitude Comparator shown in Figure 13-30. Several of these devices can be cascaded to compare binary words of any desired length. To cascade these devices you connect the A \ B, A = B, and A — B outputs of the less significant device to the corresponding *inputs* on the more significant device. On the least significant device you simply apply a logic high to the A = B input and logic lows to the A \ B and A — B inputs.

An Arithmetic-Logic Unit

Up to this point we have discussed circuits and devices which perform addition or subtraction on binary num-

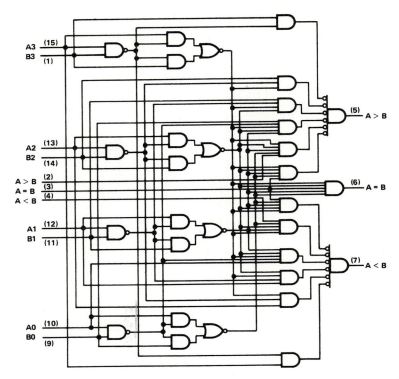

FIGURE 13-30   The 74LS85 4-bit Magnitude Comparator logic diagram. *(Texas Instruments Inc.)*

bers. Figure 13-31*a* shows a logic diagram for a 74LS181 Arithmetic Logic Unit, or ALU, which can perform many different operations on two binary numbers. The operation performed by this ALU is determined by the 4-bit binary code applied to the SELECT inputs S0, S1, S2, and S3 and by the logic level applied to the MODE input M. Figure 13-31*b* shows a table of the arithmetic and logic functions that the device will perform with different combinations of SELECT and MODE input signals.

In this table the letter A represents the 4-bit binary word applied to the A3, A2, A1, and A0 inputs. Likewise, the letter B represents the 4-bit binary word applied to the B3 to B0 inputs, and the letter F represents the 4-bit binary word produced on the F3, F2, F1, and F0 outputs. Also in this table, a plus sign is used to represent the logical OR operation and the word PLUS is used to represent addition. C_N represents the logic level on the carry input. Note that the C_N input is active low.

If the MODE input is high, the device will perform one of the 16 logic functions shown. In logic mode the level on the carry input C_N has no effect on the output word. If the MODE input is low, the device will perform one of the 16 arithmetic functions shown. The specific logic or arithmetic function performed on the two 4-bit input words is determined by the logic levels applied to the SELECT inputs. Let's take a look at some examples of the logic operations this device can perform.

If the MODE input is made high and the S3, S2, S1, and S0 inputs are programmed with HHHL, the 4-bit word applied to the A inputs will be ORed with the 4-bit word applied to the B inputs and the result will be output on the F outputs. Figure 13-32*a* shows the effect of ORing two example words. The words are ORed on a bit-by-bit basis. In other words, the A0 bit is ORed with the

B0 bit and the result is put on the F0 output, the A1 bit is ORed with the B1 bit and the result is output on the F1 bit, etc.

To help you see this, you might think of each bit position being processed by a 2-input OR gate with one input connected to a bit of the A word, the other input connected to a bit of the B word, and the output connected to the corresponding bit of the F word. The output word then will have a 1 in any bit position where either the A word has a 1 or the B word has a 1.

As another example of logic operation on binary words, suppose that the MODE input is high and the select inputs are programmed with HLHH. With this programming, the 74LS181 will AND the word on the A inputs with the word on the B inputs and put the result on the F outputs. Figure 13-32*b* shows the result produced by ANDing two example words. Again, the logic operation is done on a bit-by-bit basis, so a bit in the output word will only be a 1 if the corresponding bit in the A word *and* the corresponding bit in the B word are both 1's.

Another useful logic function that the 74LS181 can perform is to produce the XOR of the words applied to the A and B inputs. To implement this function, you make the MODE input high and program the SELECT inputs with LHHL. Figure 13-32*c* shows the result produced by XORing a couple of example words. As you can see, the output word produced will have a 1 in any bit position where the A bit is different from the corresponding B bit.

Understanding multibit logic operations is a necessary skill when working with microprocessors, so some exercises at the end of the chapter will give you some more practice with this. Now let's look at how the 74LS181 performs arithmetic operations.

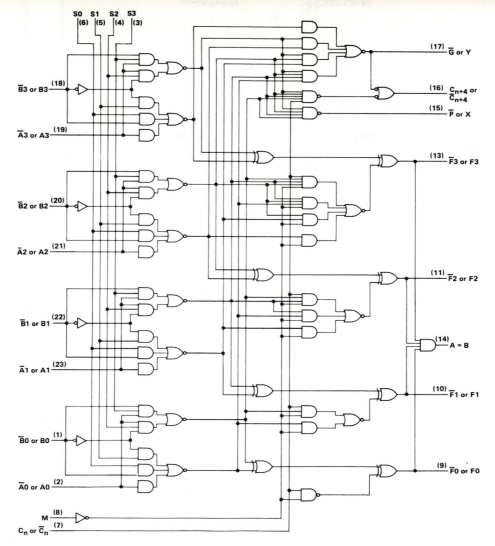

Pin numbers shown on logic notation are for DW, J or N packages.

(a)

SELECTION				ACTIVE–HIGH DATA		
				M = H LOGIC FUNCTIONS	M = L; ARITHMETIC OPERATIONS	
S3	S2	S1	S0		$\overline{C}_N$ = H (NO CARRY)	$\overline{C}_N$ = L (WITH CARRY)
L	L	L	L	F = $\overline{A}$	F = A	F = A PLUS 1
L	L	L	H	F = $\overline{A + B}$	F = A + B	F = (A + B) PLUS 1
L	L	H	L	F = $\overline{A}$B	F = A + $\overline{B}$	F = (A + $\overline{B}$) PLUS 1
L	L	H	H	F = 0	F = MINUS 1 (2's COMPL)	F = 0
L	H	L	L	F = $\overline{AB}$	F = A PLUS A$\overline{B}$	F = A PLUS A$\overline{B}$ PLUS 1
L	H	L	H	F = $\overline{B}$	F = (A + B) PLUS A$\overline{B}$	F = (A + B) PLUS A$\overline{B}$ PLUS 1
L	H	H	L	F = A ⊕ B	F = A MINUS B MINUS 1	F = A MINUS B
L	H	H	H	F = A$\overline{B}$	F = A$\overline{B}$ MINUS 1	F = A$\overline{B}$
H	L	L	L	F = $\overline{A}$ + B	F = A PLUS AB	F = A PLUS AB PLUS 1
H	L	L	H	F = A ⊕ B	F = A PLUS B	F = A PLUS B PLUS 1
H	L	H	L	F = B	F = (A + $\overline{B}$) PLUS AB	F = (A + $\overline{B}$) PLUS AB PLUS 1
H	L	H	H	F = AB	F = AB MINUS 1	F = AB
H	H	L	L	F = 1	F = A PLUS A*	F = A PLUS A PLUS 1
H	H	L	H	F = A + $\overline{B}$	F = (A + B) PLUS A	F = (A + B) PLUS A PLUS 1
H	H	H	L	F = A + B	F = (A + $\overline{B}$) PLUS A	F = (A + $\overline{B}$) PLUS A PLUS 1
H	H	H	H	F = A	F = A MINUS 1	F = A

*EACH BIT IS SHIFTED TO THE NEXT MORE SIGNIFICANT BIT POSITION

(b)

FIGURE 13-31   *(a)* Logic diagram for 74LS181 Arithmetic Logic Unit. *(b)* Truth table for 74LS181 ALU. *(Texas Instruments Inc.)*

$$A = A3 \quad A2 \quad A1 \quad A0$$
$$B = B3 \quad B2 \quad B1 \quad B0$$
$$F = \overline{F3 \quad F2 \quad F1 \quad F0}$$

$$A = 1 \quad 0 \quad 1 \quad 0$$
$$B = 0 \quad 1 \quad 1 \quad 0$$
$$F = A + B = \overline{1 \quad 1 \quad 1 \quad 0}$$

(a)

$$A = 1 \quad 0 \quad 1 \quad 0$$
$$B = 0 \quad 1 \quad 1 \quad 0$$
$$F = A \cdot B = \overline{0 \quad 0 \quad 1 \quad 0}$$

(b)

$$A = 1 \quad 0 \quad 1 \quad 0$$
$$B = 0 \quad 1 \quad 1 \quad 0$$
$$F = A \oplus B = \overline{1 \quad 1 \quad 0 \quad 0}$$

(c)

FIGURE 13-32 Examples of ORing, ANDing, and XORing 4-bit binary words. (a) ORing. (b) ANDing. (c) XORing.

When the MODE input is low and the SELECT inputs are programmed with HLLH, the 74LS181 will add the number on the A inputs to the number on the B inputs and output the sum on the F outputs. If the C_N input is low, indicating a carry in from a lower group of 4 bits, a 1 will be added to this sum. If the addition produces a carry, the C_{N+4} output will go low. The 74LS181 uses a look-ahead-carry scheme so that the sum does not have to be computed before the C_{N+4} can be output.

For low-speed addition of words larger than 4 bits, the C_{N+4} output of one 74LS181 can be connected to the C_N input of the next higher device to cascade the devices. For high-speed addition of larger words, the PROPAGATE carry output $\overline{P}$ and the GENERATE carry output $\overline{G}$ are connected to another device, the 74LS182 Look-Ahead Carry Generator. The 74LS182 uses the and $\overline{P}$ and $\overline{G}$ signals to anticipate and generate carry input signals for up to four 74LS181 ALUs. Using the 74LS182 look-ahead device means that the ALUs adding the more significant bits do not have to wait for a carry to ripple up through the lower devices. Consult the 74LS182 data sheet to see how it is connected with the ALUs.

If the MODE input of a 74LS181 is low, the SELECT inputs are programmed with LLLL, and the C_N input is high, the A word will appear on the outputs. If C_N is made low, the outputs will contain A plus 1. This is an easy way to increment A by 1 which is useful in operations such as forming the two's complement of a number. If the MODE input is low, the SELECT inputs are programmed with HHHH, and the C_N input is high, the word on the output will be equal to A minus 1. This operation is an easy way to decrement the A word by 1.

An output of A plus A, or 2A, can be produced by making the MODE input low and programming the SELECT inputs with HHLL. As shown by the footnote in Figure 13-31, multiplying a binary number by 2 is the same as shifting each bit one bit position to the left. This is useful for some more complex computations.

Still another important operation the 74LS181 can do is subtraction. If the MODE input is low and the SELECT inputs are programmed with LHHL, the device will subtract the word on the B inputs from the word on the A inputs. The 74LS181 does subtraction by generating the one's complement of the B word and adding the result to the A word. We didn't bother to tell you this earlier, but the one's complement of a number is generated by simply inverting each bit in the word. The result of adding the one's complement of B to A gives a result of A minus B minus 1 because the one's complement is 1

less than the two's complement we showed you how to use for subtraction. To get the desired result of A minus B, you simply add 1 to the result by making the C_N input low on the device which is subtracting the least significant 4 bits of a number. Incidentally, for subtract operations the C_{N+4} output represents a borrow.

A 74LS181 can also be used to compare the magnitude of two binary numbers. To do this you make the MODE input low, program the SELECT inputs with LHHL, and make the C_N input low. As we described in the preceding paragraph, this will subtract the B word from the A word. If the two words are the same, the A = B output will go high. If the B word is larger than the A word, the C_{N+4} output will go low. If the A word is larger than the B word, the C_{N+4} output will be high and the A = B output will be low.

The important point about an ALU such as the 74LS181 is that it can perform many different operations on two binary words and that the operation performed is determined by the binary-coded instruction programmed on the MODE and SELECT inputs. After we show you some multiplier devices, we will show you how an ALU is used as the heart of a more powerful programmable device called a microprocessor.

Multiplier Devices

There are several ways of doing binary multiplication in hardware. One way is with a chip set such as the 74284 and 74285 shown in Figure 13-33. Together these chips multiply two 4-bit numbers to produce an 8-bit product. With the help of some adders, several of the basic 4 × 4 multiplier circuits in Figure 13-33 can be cascaded to multiply larger numbers. The 74284/74285 chip set does binary multiplication by simply implementing an 8-input variable truth table with gates. Since the circuit is purely combinational, no clock is required to perform the multiplication. The problem with the basic 74284/74285 approach is that a sizable number of chips are needed to expand the circuit so that it can, for example, multiply two 8-bit numbers.

Fortunately, LSI has made it possible to produce single devices which multiply 8-, 16-, 24-, or 32-bit binary words. One example of this type of device is the Analog Devices ADSP-1010A Multiplier/Accumulator, or MAC, shown in Figure 13-34. The term *accumulator* in the name of this device refers to the large built-in register which holds the partial product and the final product. This device uses clock signals to transfer data words to and from the internal registers, but the basic multiplication is performed by combinational logic circuits, just as it is in the 74284/74285 circuit. The CMOS ADSP-1010A can multiply two 16-bit numbers in 75 ns.

Another single-chip multiplier, the Analog Devices ADSP-3210, can multiply two 32-bit IEEE 754 standard floating-point format numbers in about 240 ns. The speed of a device which works with floating-point numbers is often expressed in *MegaFLOPS*, or just *MFLOPS*, instead of in nanoseconds. The term *MFLOPS*, stands for "million floating-point operations per second." As an example of this type rating, the highest speed version of the ADSP-3210 is rated at 16.7 MFLOPS.

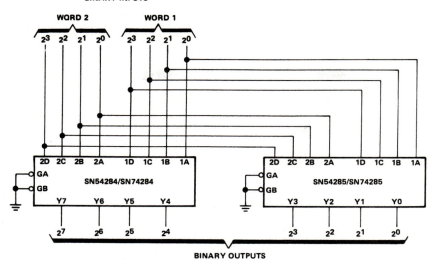

FIGURE 13-33 Chip set 74284 and 74285 used to multiply two 4-bit numbers. *(Texas Instruments Inc.)*

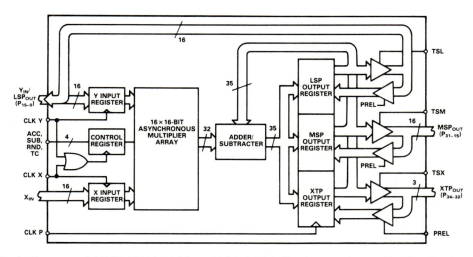

FIGURE 13-34 Block diagram of ADSP-1010A 16-bit × 16-bit Multiplier/Accumulator. *(Analog Devices Inc.)*

INTRODUCTION TO MICROPROCESSORS AND MICROCOMPUTERS

A Microprocessor Bit Slice

In a previous section we described how an ALU can be programmed to perform any one of several logic or arithmetic functions on two binary words. In this section we show you how an ALU is used as the heart of a more general-purpose programmable device called a *microprocessor*.

The two main limitations of a basic ALU such as the 74LS181 we discussed earlier are:

1. The ALU doesn't have registers to hold data words waiting to be operated on or registers to hold words output as the result of an operation.

2. The ALU is not a clocked logic device, so it cannot by itself perform operations which require a sequence of actions.

The Advanced Micro Devices Am2901C shown in Figure 13-35a is a first step in overcoming these limitations. The Am2901C is designed to process 4-bit data words. However, since two or more of the devices can be cascaded to process wider data words, the device is called a *microprocessor bit slice*.

As you can see in the block diagram, the device consists of a basic eight-function ALU, some registers to hold data words before and after operations, and some multiplexers to route data into and out of the ALU.

The operation performed by an Am2901C is determined by a 9-bit instruction word applied to its instruction inputs I0 to I8. Figure 13-35b shows the format for these instruction words.

Three bits in the instruction word are used to specify the desired operation for the ALU. Figure 13-35c shows the codes for each of the eight functions the device can perform on two 4-bit words R and S. These functions are all that is needed because other functions can be

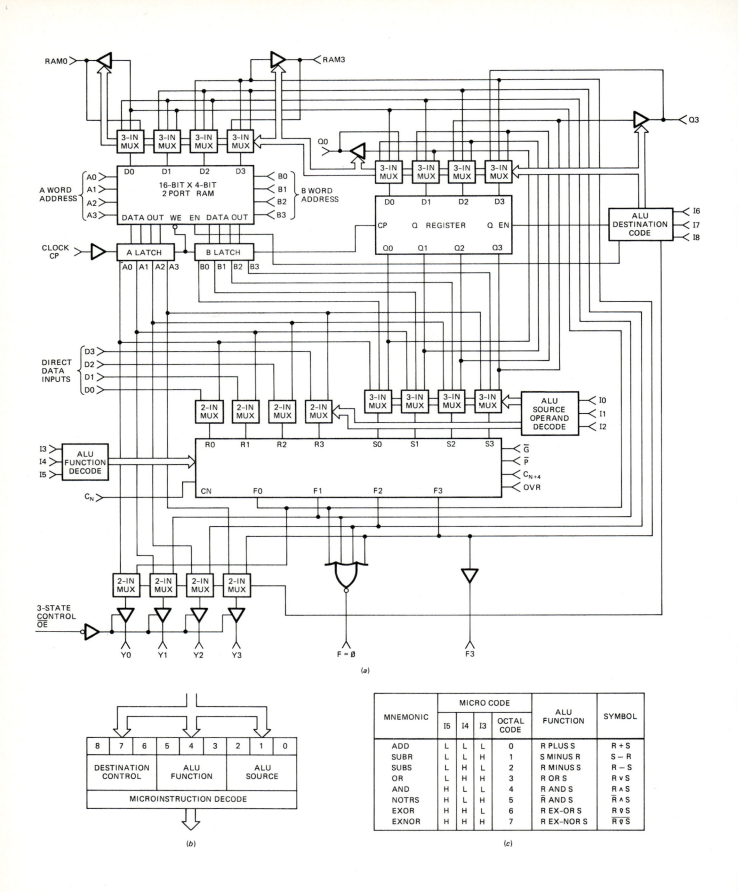

FIGURE 13-35 The Am2901C Microprocess Bit-Slice. *(a)* Internal block diagram. *(b)* Instruction format. *(c)* ALU function codes. *(Copyright © Advanced Micro Devices Inc. July 1988. Reprinted with permission of owner. All rights reserved.)*

implemented with combinations of these eight. The invert function, for example, is not needed because you can invert a binary word by simply XORing the word with a word which is all 1's.

Note in Figure 13-35a that the ALU has a C_{N+4} output which is asserted if an operation produces a carry. It also has an output labeled F3 which is a copy of the MSB of the result. When doing signed arithmetic, this signal is a copy of the sign bit and indicates whether a result is positive or negative. Another output signal is the F signal which will be asserted if the ALU outputs are all 0's after some operation. Still another status-type output signal produced by the 2901C is the overflow signal, OVR, which will be asserted if the result of a signed operation overflows the number of bits allowed for the magnitude. These status signals output by the ALU are often called *flags* because they "flag" some condition.

The source of the data words operated on by the ALU is determined by the three least significant bits in the instruction word. To describe the possible operand sources, we need to first take a closer look at the section of Figure 13-35a labeled 2-PORT RAM. The RAM section of the device consists of sixteen 4-bit registers. When a 4-bit address is applied to the external A0 to A3 inputs, the outputs of the addressed register will be routed to the A LATCH. Likewise, a 4-bit address applied to the external B0 to B3 inputs will route the outputs of one of the 16 registers to the B LATCH.

The operand for the R inputs of the ALU then can come from the A LATCH or from the external data inputs labeled D0 to D3. The operand for the S inputs of the ALU can come from the A LATCH, the B LATCH, or a special register called the Q REGISTER. It doesn't show on the block diagram, but 0000 can also be selected as the operand for either the R or the S inputs. As perhaps you remember from a previous discussion, this is useful when performing binary multiplication with the add-and-shift method. Speaking of multiplication, the main purpose of the Q REGISTER is for holding and shifting the partial product or dividend during multiplication and division operations.

The destination of the data word produced by the ALU is specified by the three most significant bits in the 9-bit instruction word applied to the 2901C. The output word can be directed to the external data outputs Y0 to Y3, to the Q REGISTER multiplexers, or to the RAM multiplexers. Look carefully at how the data lines are connected to the inputs of the Q REGISTER multiplexers. You should see that these multiplexers are connected so that the data word can be sent directly to the Q REGISTER, the data word shifted one bit position to the left can be sent to the Q REGISTER, or the data word shifted one bit position to the right can be sent to the Q REGISTER. Using a multiplexer to shift an operand in this way is a simple example of what is called a *barrel shifter*. The advantage of this method is that a clock signal is not required to do the shift as it is for a shift register. The 3-input multiplexers on the inputs of the RAM likewise allow the data to go directly to the RAM, be shifted one bit position to the left, or be shifted one bit position to the right. The register that a data word will be sent to in the RAM is determined by the external address applied to the A0 to A3 inputs, the external address applied to the B0 to B3 inputs, and the destination bits in the 9-bit instruction word.

To summarize then, the operation to be performed, the source of the operands, and the destination of the result are specified with a 9-bit instruction word and two 4-bit addresses applied to the inputs of the 2901C.

The basic 2901C bit-slice device is more functional than a basic ALU because it has registers to hold source and destination operands. However, it is still limited because it has no mechanism to do a sequence of operations such as an add-and-shift multiplication. Figure 13-36. shows how we connect some other devices with the basic bit slice to form a complete microprocessor which solves this problem.

A Microprocessor Made with Bit Slices

As we told you earlier, a microprocessor is the central part of a microcomputer. In most microcomputers you don't need to know the details of the internal operation of the microprocessor because the microprocessor is a single device which contains all the circuitry in Figure 13-36 and more. A brief run around the operation of the circuit in Figure 13-36, however, should help you appreciate some of what is going on in a single-chip microprocessor such as the Intel 8086 that we discuss in the next major section of the chapter.

First, note the block labeled Am2901C ARRAY in Figure 13-36. This block represents a cascaded group of 2901C bit slices. Four 2901Cs, for example, might be connected in parallel to perform operations on 16-bit words. Second, note that a clock signal is applied to this circuit so that the circuit goes from one state to the next in an orderly manner.

To help explain how the circuit works, let's assume that we want the microprocessor to multiply two numbers using the add-and-shift-right method. From our previous discussions you know that this method of multiplication requires a sequence of operations. To get the array of 2901Cs to perform this sequence of operations, we have to apply a sequence of instructions to their instruction inputs. The sequence of binary-coded instructions for the 2901Cs is stored in the MICRO-PROGRAM PROM. For each operation that this microprocessor can perform there will be a sequence of instructions stored in the MICROPROGRAM PROM. Depending on the design of a particular system, the binary instruction codes stored in the MICROPROGRAM PROM may be 40 to 80 bits wide. The 2901Cs only require a 9-bit instruction, 4 bits for the A REGISTER address inputs, and 4 bits for the B REGISTER address inputs. The other bits in the word on the output of the PIPE-LINE REGISTER are required to generate control signals to the rest of the system and to cause the sequencer to address the desired next word in the MICROPROGRAM PROM. The binary codes stored in the MICROPROGRAM PROM are commonly called *microinstructions*, or *microcodes*. Typically the MICROPROGRAM PROM contains 256 to 4096 microinstructions. Now, let's see how the microprocessor cycles through a sequence of microinstructions to do our multiplication.

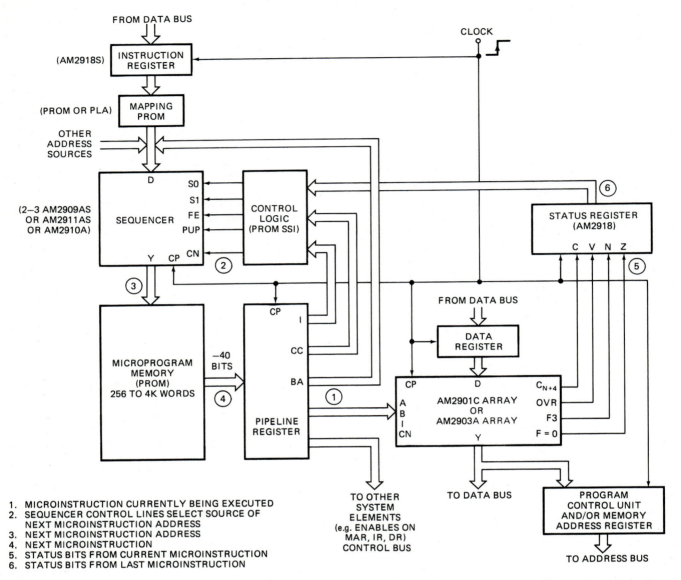

1. MICROINSTRUCTION CURRENTLY BEING EXECUTED
2. SEQUENCER CONTROL LINES SELECT SOURCE OF NEXT MICROINSTRUCTION ADDRESS
3. NEXT MICROINSTRUCTION ADDRESS
4. NEXT MICROINSTRUCTION
5. STATUS BITS FROM CURRENT MICROINSTRUCTION
6. STATUS BITS FROM LAST MICROINSTRUCTION

FIGURE 13-36 Block diagram for a bit-slice microprocessor using Am2901C Microprocessor Bit Slices. *(Copyright © Advanced Micro Devices Inc. July 1988. Reprinted with permission of owner. All rights reserved.)*

For this discussion assume that the binary code for the multiply instruction is applied to the inputs of the INSTRUCTION REGISTER in the upper left corner of Figure 13-36. (Later we will explain how this code arrived there.) On the rising edge of the next clock pulse the basic instruction code will be transferred to the outputs of the INSTRUCTION REGISTER. In response to this code the MAPPING PROM will output the address of the first microinstruction in the MICROINSTRUCTION PROM to the SEQUENCER. The SEQUENCER will pass this address on to the address inputs of the MICROPROGRAM PROM, and the MICROPROGRAM PROM will output the first microinstruction to the inputs of the PIPELINE REGISTER.

On the rising edge of the next clock pulse, the microinstruction will be transferred to the outputs of the PIPELINE REGISTER. Some of the bits on the outputs of the PIPELINE REGISTER will go to the 2901Cs. The 2901Cs

will carry out the first instruction, an add, and output status bits to indicate if, for example, a carry was produced by the addition. Other bits on the outputs of the PIPELINE REGISTER will go to the CONTROL LOGIC PROM where they will be decoded along with the STATUS REGISTER outputs to produce the control signals for the SEQUENCER.

There is no clocked circuitry between the control inputs of the SEQUENCER and its outputs, so the effect of these decoded signals will cause the SEQUENCER to directly output the address of the next desired microinstruction to the MICROPROGRAM PROM. The MICROPROGRAM PROM will then output the next microinstruction to the inputs of the PIPELINE REGISTER. The next rising edge of the clock signal will transfer this next microinstruction to the outputs of the PIPELINE REGISTER.

A major point here is that while the 2901Cs are executing one microinstruction, the CONTROL LOGIC and SEQUENCER are fetching the next microinstruction from the MICROPROGRAM PROM and getting it all set up on the inputs of the PIPELINE REGISTER. Incidentally, the term *pipelining* when used in a computer context means fetching the next instruction while the current instruction is executing.

Reaching back to Chapter 12, perhaps you can see that the outputs of the PIPELINE REGISTER and the STATUS REGISTER during the one state are decoded by the CONTROL LOGIC, SEQUENCER, and MICROPROGRAM PROM to produce the PIPELINE REGISTER inputs required for the PIPELINE REGISTER outputs to transition to the next state when a clock pulse occurs. The whole microprocessor then functions as a complex state machine which steps through a sequence of states determined by the binary instruction applied to the INSTRUCTION REGISTER and by status flags output by the 2901Cs.

In a bit-slice microprocessor the MICROPROGRAM PROM is programmed with a microinstruction sequence for each operation that the designer wants the microprocessor to be able to perform. The designer might, for example, include microinstruction sequences to multiply, divide, calculate square roots, calculate the tangent of an angle, etc. The microinstruction sequence selected for execution at a particular time is determined by the higher-level binary-coded instruction applied to the INSTRUCTION REGISTER in the upper left corner of Figure 13-36. This higher-level binary-coded instruction is often called a *machine-code instruction*.

When you write a microcomputer program in a high-level language such as BASIC, Pascal, or C, your compiler or interpreter converts your high-level language statements to sequences of machine-code instructions. The microprocessor then executes your program by bringing the machine-code instructions into its INSTRUCTION REGISTER one at a time. As we described before, the microprocessor then carries out each machine-code instruction by stepping through a sequence of microinstructions in the MICROCODE PROM. In the next section of the chapter we show you how a sequence of machine-code instructions is fetched from memory by the microprocessor. While the microprocessor circuit is fresh in your mind, however, we wanted to give you an overview of the progression from high-level language to machine codes to microinstructions to a basic ALU made with simple logic gates.

The advantage of building a discrete-component microprocessor such as the one in Figure 13-36 is that the designer can custom program the MICROPROGRAM PROM to implement only the desired instructions. This makes the microprocessor very fast and efficient for those operations. For most applications, however, a microprocessor such as that in Figure 13-36 requires too many devices and too much board real estate. Also, the circuit requires programming at the microcode level as well as at the machine-code level or the high-level language level. For general-purpose applications, many one-chip microprocessors are currently available.

In a later section of the chapter we discuss the operation of a common one-chip microprocessor, the Intel 8086, and show you how it is used in a microcomputer. The 8086 is the first member of a microprocessor family which includes the 80186, the 80286, and the 80386. Microprocessors in this family are used as the central processing unit in many personal computers. We will leave most of our discussion of microprocessors and microcomputers for the sequel to this book, but we want to give you an introduction here so you can see the rest of the progression from a basic ALU to a general-purpose programmable machine. Also, we need you to understand the basic operation of a microcomputer so that you have a framework for the discussion of memory devices and systems in the next chapter.

INTRODUCTION TO MICROCOMPUTERS

In the preceding section we showed you how an ALU can be connected with some register and memory devices to form a microprocessor. In this section we show you how some other components are connected with a microprocessor to make a microcomputer. We also describe in detail how the microcomputer fetches instructions from memory and executes those instructions.

Overview of Microcomputer Structure and Operation

Figure 13-37 shows a block diagram for a simple microcomputer. The major parts are the *central processing unit (CPU)*, *memory*, and the *input/output (I/O)* circuitry. Connecting these parts together are three sets of parallel lines called *buses*. The three buses are the *address bus*, the *data bus*, and the *control bus*. Let's take a brief look at each of these parts.

MEMORY

The memory section usually consists of a mixture of RAM and ROM. It may also have magnetic floppy disks, magnetic hard disks, or optical disks. Memory has two purposes. The first purpose is to store the binary codes for the sequence of instructions you want the computer to carry out. When you write a computer program, what you are really doing is just writing a sequential list of instructions for the computer. The second purpose of the memory is to store the binary-coded data with which the computer is going to be working. This data might be the inventory records of a supermarket, for example.

INPUT/OUTPUT

The input/output, or I/O, section allows the computer to take in data from the outside world or send data to the outside world. Peripherals such as keyboards, video display terminals, printers, and MODEMs are connected to the I/O section. These allow the user and the computer to communicate with each other. The actual physical devices used to interface the computer buses to external systems are often called *ports*. Ports in a computer function just as shipping ports do for a country. An *input port* allows data from a keyboard, an A/D converter, or some other source to be read into the computer under

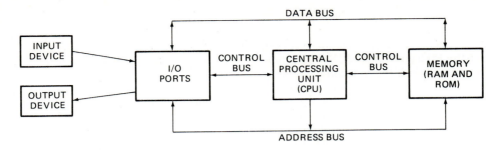

FIGURE 13-37 Block diagram of a simple computer.

control of the CPU. An *output port* is used to send data from the computer to some peripheral such as a video display terminal, a printer, or a D/A converter. Physically, the simplest type of input or output port is just a set of parallel D flip-flops. If they are being used as an input port, the D inputs are connected to the external device and the Q outputs are connected to the data bus which runs to the CPU. Data will then be transferred through the latches when they are enabled by a control signal from the CPU. In a system where they are being used as an output port, the D inputs of the latches are connected to the data bus and the Q outputs are connected to some external device. Data sent out on the data bus by the CPU will be transferred to the external device when the latches are enabled by a control signal from the CPU.

CENTRAL PROCESSING UNIT

The *central processing unit (CPU)* controls the operation of the computer. In a microcomputer the CPU is a microprocessor such as we discussed in an earlier section of the chapter. The CPU fetches binary-coded instructions from memory, decodes the instructions into a series of simple actions, and carries out these actions in a sequence of steps as we described in the previous discussion of a bit-slice microprocessor.

The CPU also contains an *address counter* or *instruction pointer* register which holds the address of the next instruction or data to be fetched from memory, general-purpose registers which are used for temporary storage of binary data, and circuitry which generates the control bus signals.

ADDRESS BUS

The *address bus* consists of 16, 20, 24, or 32 parallel signal lines. On these lines the CPU sends out the address of the memory location that is to be written to or read from. The number of memory locations that the CPU can address is determined by the number of address lines. If the CPU has N address lines, then it can directly address 2^N memory locations. For example, a CPU with 16 address lines can address 2^{16}, or 65,536, memory locations, a CPU with 20 address lines can address 2^{20}, or 1,048,576, locations, and a CPU with 24 address lines can address 2^{24}, or 16,777,216, locations. When the CPU reads data from or writes data to a port, it sends the port address out on the address bus.

DATA BUS

The *data bus* consists of 8, 16, or 32 parallel signal lines. As indicated by the double-ended arrows on the data bus line in Figure 13-37, the data bus lines are *bidirectional*. This means that the CPU can read data in from memory or from a port on these lines or that the CPU can send data out to memory or to a port on these lines. Many devices in a system will have their outputs connected to the data bus, but only one device at a time will have its outputs enabled. Any device connected on the data bus must have three-state outputs so that its outputs can be disabled when that device is not being used to put data on the bus.

CONTROL BUS

The *control bus* consists of 4 to 10 parallel signal lines. The CPU sends out signals on the control bus to enable the outputs of addressed memory devices or port devices. Typical control bus signals are *MEMORY READ, MEMORY WRITE, I/O READ*, and *I/O WRITE*. To read a byte of data from a memory location, for example, the CPU sends out the memory address of the desired byte on the address bus and then sends out a MEMORY READ signal on the control bus. The MEMORY READ signal enables the addressed memory device to output a data word onto the data bus. The data word from memory travels along the data bus to the CPU.

HARDWARE, SOFTWARE, AND FIRMWARE

When working around computers, you hear the terms hardware, software, and firmware almost constantly. *Hardware* is the name given to the physical devices and circuitry of the computer. *Software* refers to the programs written for the computer. *Firmware* is the term given to programs stored in ROMs or in other devices which permanently keep their stored information.

SUMMARY OF IMPORTANT POINTS SO FAR

- A computer or microcomputer consists of memory, a CPU, and some input/output circuitry.

- These three parts are connected by the address bus, the data bus, and the control bus.

- The sequence of instructions, or the program, for a computer is stored as binary numbers in successive memory locations.

The CPU fetches an instruction from memory, decodes the instruction to determine what actions must be done for the instruction, and carries out these actions.

Execution of a Three-Instruction Program

To give you a better idea of how the parts of a microcomputer function together, we will now describe the actions a simple microcomputer might go through to carry out (execute) a simple program. The three instructions of the program are:

1. Input a value from a keyboard connected to the port at address 05H.

2. Add 7 to the value read in.

3. Output the result to a display connected to the port at address 02H.

Figure 13-38 shows in diagram form and sequential list form the actions that the computer will perform to execute these three instructions.

For this example assume that the CPU fetches instruc-

tions and data from memory one byte at a time as is done in the original IBM PC and its clones. Also assume that the binary codes for the instructions are in sequential memory locations starting at address 00100H. Figure 13-38b shows the actual binary codes that would be required in successive memory locations to execute this program on an IBM PC-type microcomputer.

The CPU needs an instruction before it can do anything, so the first action it will do is fetch an instruction byte from memory. To do this the CPU sends out the address of the first instruction byte, in this case 00100H, to memory on the address bus. This action is represented by line 1A in Figure 13-38a. The CPU then sends out a MEMORY READ signal on the control bus (line 1B in the figure). The MEMORY READ signal enables the memory to output the addressed byte on the data bus. This action is represented by line 1C in the figure. The CPU reads in this first instruction byte (E4H) from the data bus and *decodes* it. By decode we mean that the CPU determines from the binary code read in what actions it is supposed to take. If the CPU is a microprocessor, it selects the sequence of microinstructions needed to carry

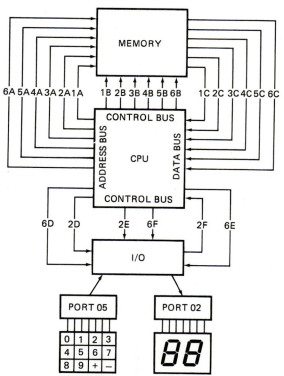

PROGRAM

1. INPUT A VALUE FROM PORT 05.
2. ADD 7 TO THIS VALUE.
3. OUTPUT THE RESULT TO PORT 02.

SEQUENCE

1A CPU SENDS OUT ADDRESS OF FIRST INSTRUCTION TO MEMORY.
1B CPU SENDS OUT MEMORY READ CONTROL SIGNAL TO ENABLE MEMORY.
1C INSTRUCTION BYTE SENT FROM MEMORY TO CPU ON DATA BUS.
2A ADDRESS NEXT MEMORY LOCATION TO GET REST OF INSTRUCTION.
2B SEND MEMORY READ CONTROL SIGNAL TO ENABLE MEMORY.
2C PORT ADDRESS BYTE SENT FROM MEMORY TO CPU ON DATA BUS.
2D CPU SENDS OUT PORT ADDRESS ON ADDRESS BUS.
2E CPU SENDS OUT INPUT READ CONTROL SIGNAL TO ENABLE PORT.
2F DATA FROM PORT SENT TO CPU ON DATA BUS.
3A CPU SENDS ADDRESS OF NEXT INSTRUCTION TO MEMORY.
3B CPU SENDS MEMORY READ CONTROL SIGNAL TO ENABLE MEMORY.
3C INSTRUCTION BYTE FROM MEMORY SENT TO CPU ON DATA BUS.
4A CPU SENDS NEXT ADDRESS TO MEMORY TO GET REST OF INSTRUCTION.
4B CPU SENDS MEMORY READ CONTROL SIGNAL TO ENABLE MEMORY.
4C NUMBER 07H SENT FROM MEMORY TO CPU ON DATA BUS.
5A CPU SENDS ADDRESS OF NEXT INSTRUCTION TO MEMORY.
5B CPU SENDS MEMORY READ CONTROL SIGNAL TO ENABLE MEMORY.
5C INSTRUCTION BYTE FROM MEMORY SENT TO CPU ON DATA BUS.
6A CPU SENDS OUT NEXT ADDRESS TO GET REST OF INSTRUCTION.
6B CPU SENDS OUT MEMORY READ CONTROL SIGNAL TO ENABLE MEMORY.
6C PORT ADDRESS BYTE SENT FROM MEMORY TO CPU ON DATA BUS.
6D CPU SENDS OUT PORT ADDRESS ON ADDRESS BUS.
6E CPU SENDS OUT DATA TO PORT ON DATA BUS.
6F CPU SENDS OUT OUTPUT WRITE SIGNAL TO ENABLE PORT.

(a)

MEMORY ADDRESS	CONTENTS (BINARY)	CONTENTS (HEX)	OPERATION
00100H	11100100	E4	INPUT FROM
00101H	00000101	05	PORT 05H
00102H	00000100	04	ADD
00103H	00000111	07	07H
00104H	11100110	E6	OUTPUT TO
00105H	00000010	02	PORT 02

(b)

FIGURE 13-38 *(a)* Execution steps for a simple three-step program on a microcomputer. *(b)* Memory addresses and memory contents for the three-step program.

out the instruction read from memory. For the example instruction here, the CPU determines that the code read in represents an INPUT instruction. From decoding this instruction byte the CPU also determines that it needs more information before it can carry out the instruction. The additional information the CPU needs is the address of the port that the data is to be input from. This port address part of the instruction is stored in the next memory location after the code for the INPUT instruction.

To fetch this second byte of the instruction, the CPU sends out the next sequential address (00101H) to memory as shown by line 2A in the figure. To enable the addressed memory device, the CPU also sends out another MEMORY READ signal on the control bus (line 2B). The memory then outputs the addressed byte on the data bus (line 2C). When the CPU reads in this second byte, 05H in this case, it has all the information it needs to execute the instruction.

To execute the INPUT instruction, the CPU sends out the port address (05H) on the address bus (line 2D) and sends out an I/O READ signal on the control bus (line 2E). The I/O READ signal enables the addressed port device to put a byte of data on the data bus (line 2F). The CPU reads in the byte of data and stores it in an internal register. This completes the fetching and execution of the first instruction.

Having completed the first instruction, the CPU must now fetch its next instruction from memory. To do this, it sends out the next sequential address (00102H) on the address bus (line 3A) and sends out a MEMORY READ signal on the control bus (line 3B). The MEMORY READ signal enables the memory device to put the addressed byte (04H) on the data bus (line 3C). The CPU reads in this instruction byte from the data bus and decodes it. From this instruction byte, the CPU determines that it is supposed to add some number to the number stored in the internal register. The CPU also determines from decoding this instruction byte that it must go to memory again to get the next byte of the instruction which contains the number that it is supposed to add. To get the required byte, the CPU will send out the next sequential address (00103H) on the address bus (line 4A) and another MEMORY READ signal on the control bus (line 4B). The memory will then output the contents of the addressed byte (the number 07H) on the data bus (line 4C). When the CPU receives this number, it will add the number to the contents of the internal register. The result of the addition will be left in the internal register. This completes the fetching and executing of the second instruction.

The CPU must now fetch the third instruction. To do this the CPU sends out the next sequential address (00104H) on the address bus (line 5A) and sends out a MEMORY READ signal on the control bus (line 5B). The memory then outputs the addressed byte (E6H) on the data bus (line 5C). From decoding this byte the CPU determines that it is now supposed to do an OUTPUT operation to a port. The CPU also determines from decoding this byte that it must go to memory again to get the address of the output port. To do this it sends out the next sequential address (00105H) on the address bus (line 6A), sends out a MEMORY READ signal on the control

bus (line 6B), and reads in the byte (02H) put on the data bus by the memory (line 6C). The CPU now has all the information that it needs to execute the OUTPUT instruction.

To output a data byte to a port, the CPU first sends out the address of the desired port on the address bus (line 6D). Next it outputs the data byte from the internal register on the data bus (line 6E). The CPU then sends out an I/O WRITE signal on the control bus (line 6F). This signal enables the addressed output port device so that the data from the data bus lines can pass through it to the LED displays. When the CPU removes the I/O WRITE signal to proceed with the next instruction, the data will remain latched on the output pins of the port device. The data will remain latched on the port until the power is turned off or until a new data word is output to the port. This is important because it means that the computer does not have to keep outputting a value over and over in order for it to remain on the output.

All the steps described above may seem like a great deal of work just to input a value from a keyboard, add 7 to it, and output the result to a display. Even a simple microcomputer, however, can run through all these steps in a few microseconds.

SUMMARY OF SIMPLE MICROCOMPUTER BUS OPERATION

1. A microcomputer fetches each program instruction in sequence, decodes the instruction, and executes it.

2. The CPU in a microcomputer fetches instructions or reads data from memory by sending out an address on the address bus and a MEMORY READ signal on the control bus. The memory outputs the addressed instruction or data word to the CPU on the data bus.

3. The CPU writes a data word to memory by sending out an address on the address bus, sending out the data word on the data bus, and sending a MEMORY WRITE signal to memory on the control bus.

4. To read data from a port, the CPU sends the port address out on the address bus and sends an I/O READ signal to the port device on the control bus. Data from the port comes into the CPU on the data bus.

5. To write data to a port, the CPU sends out the port address on the address bus, sends the data to be written to the port out on the data bus, and sends an I/O WRITE signal to the port device on the control bus.

Microprocessor Evolution and Types

As we told you in the preceding section, a microprocessor is used as the CPU in a microcomputer. There are now many different microprocessors available, so before we dig into the details of a specific device, we will give you a short microprocessor history lesson and an overview of the different types.

A common way of categorizing microprocessors is by the number of bits that their ALU can work with at a time. In other words, a microprocessor with a 4-bit ALU will be called a 4-bit microprocessor, regardless of the number of address lines or the number of data bus lines that it has. The first microprocessor was the Intel 4004 produced in 1971. It contained 2300 PMOS transistors. The 4004 was a 4-bit device intended to be used with some other devices in making a calculator. Some logic designers, however, saw that this device could be used to replace PC boards full of combinational and sequential logic devices. Also, the ability to change the function of a system by just changing the programming, rather than redesigning the hardware, was very appealing. It was these factors that pushed the evolution of microprocessors.

In 1972 Intel came out with the 8008, which was capable of working with 8-bit words. The 8008, however, required 20 or more additional devices to form a functional CPU. In 1974 Intel announced the 8080 which had a much larger instruction set than the 8008 and only required two additional devices to form a functional CPU. Also, the 8080 used NMOS transistors, so it operated much faster than the 8008. The 8080 is called a *second-generation microprocessor*.

Soon after Intel produced the 8080, Motorola came out with the MC6800, another 8-bit general-purpose CPU. The 6800 had the advantage that it required only a +5-V supply rather than the −5-, +5-, and +12-V supplies required by the 8080. For several years the 8080 and the 6800 were the top-selling 8-bit microprocessors. Some of their competitors were the MOS Technology 6502 used as the CPU in the Apple II microcomputers and the Zilog Z80 used as the CPU in the Radio Shack TRS-80 microcomputers.

After Motorola came out with the MC6800, Intel produced the 8085, an upgrade of the 8080, requiring only a +5-V supply. Motorola then produced the MC6809 which has a few 16-bit instructions but is still an 8-bit processor. In 1978 Intel came out with the 8086 which is a full 16-bit processor. Some 16-bit microprocessors, such as the National PACE and the Texas Instruments 9900 family of devices, were available previously, but the market apparently wasn't ready. Soon after Intel came out with the 8086, Motorola came out with the 16-bit MC68000, and the 16-bit race was off and running.

The 8086 and the 68000 work directly with 16-bit words instead of with 8-bit words, they can address a million or more bytes of memory instead of the 64K bytes addressable by the 8-bit processors, and they execute instructions much faster than the 8-bit processors. Also these 16-bit processors have single instructions for functions that required a lengthy sequence of instructions on the 8-bit processors.

The evolution along this last path has continued to 32-bit processors that work with giga (10^9) bytes or tera (10^{12}) bytes of memory. Examples of these devices are the Intel 80386, the Motorola MC68030, and the National NS32032.

Another direction of microprocessor evolution has been toward devices called *dedicated controllers*. These devices are used to control "smart" machines such as microwave ovens, sewing machines, auto ignition systems, and metal lathes. They are essentially single-chip microcomputers. They have a CPU, ROM, RAM, ports, timers, UARTs, and other functions all in one IC. Some currently available 8-bit devices in this category are the Intel 8051 and the Motorola MC6801. A commonly used 16-bit microcontroller, the Intel 8096, contains a CPU, ROM, RAM, ports, a UART, timers, and a 10-bit A/D converter.

For the examples in the rest of this book we use the Intel 8086 microprocessor. This device was the first member of the 8086, 8088, 80186, 80188, 80286 family. Members of this family are very widely used in personal computers, business computer systems, and industrial control systems. The sequel to this book contains a detailed discussion of the programming and interfacing of the 8086 family devices. Here, however, we only need to look at the basic system connections and operations for an 8086 microcomputer so that the discussion of memory devices and systems in the next chapter will make more sense to you.

A Basic 8086 Microcomputer System

SYSTEM OVERVIEW

Figure 13-39 shows a block diagram of a simple 8086-based microcomputer. This diagram is a closer look at the generalized microcomputer in Figure 13-37. The CPU in the microcomputer in Figure 13-39 is the 8086 microprocessor in the upper left corner. The blocks labeled RAM and PROM represent the memory in this system. The term *RAM* stands for random-access memory. In the next chapter we discuss RAM in detail, but all you need to know about RAM at this point is that you can write data words to it, read the data words from it, and write new data words to it. A RAM then is different from a read-only memory, or ROM, which you can only read data words from. The input and output ports for the system in Figure 13-39 are represented by the block labeled MCS-80 PERIPHERAL. As we discuss in the sequel to this book, there are a very wide variety of port devices available. There are, for example, special port devices which interface to CRTs, port devices which interface with keyboards, and port devices which interface to floppy disks.

Next in Figure 13-39 find the control bus, address bus, and data bus. The basic control bus consists of the signals labeled M/$\overline{IO}$, $\overline{RD}$, and $\overline{WR}$ at the top of the figure. If the 8086 is doing a read from memory or from a port, the $\overline{RD}$ signal will be asserted. If the 8086 is doing a write to memory or to a port, the $\overline{WR}$ signal will be asserted. During a read from memory or a write to memory, the M/$\overline{IO}$ signal will be high, and during port operations, the M/$\overline{IO}$ signal will be low. As we show you in detail later, the $\overline{RD}$, $\overline{WR}$, M/$\overline{IO}$ signals are used to enable addressed devices.

The address bus and the data bus are shown on the right side of Figure 13-39. The 8086 has 20 address lines and 16 data lines. On the right side of Figure 13-39 the address bus and the data bus are shown separately, but where they leave the 8086 the two buses are shown as a single bus labeled ADDR/DATA. The reason for this is that to save pins, the lower 16 bits of an address are

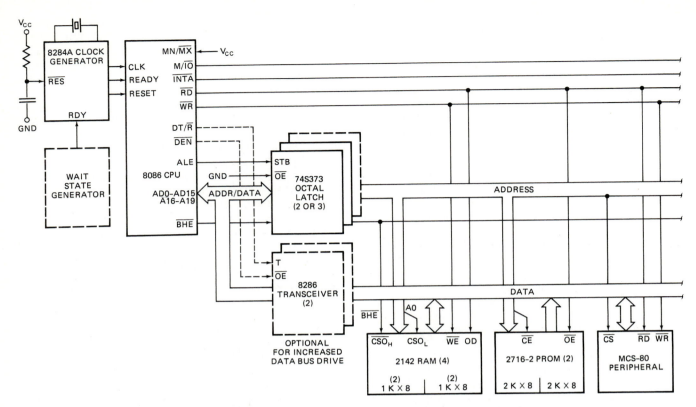

FIGURE 13-39 Block diagram of a simple 8086-based microcomputer. *(Intel Corporation.)*

multiplexed on the data bus. Here's an overview of how this works.

As a first step in any operation where it accesses memory or a port, the 8086 sends out the lower 16 bits of the address on the data bus. External latches such as the 74LS373 octal devices in Figure 13-39 are used to "grab" this address and hold it during the rest of the operation. The 8086 outputs a signal called *ADDRESS LATCH ENABLE*, or *ALE*, to strobe these latches. Once the address is stored on the outputs of the latches, the 8086 removes the address from the ADDRESS/DATA bus and uses the bus for reading or writing data.

Another section of Figure 13-39 to look briefly at is the block labeled 8286 TRANSCEIVER. This block represents bidirectional three-state buffers. For a very small system these buffers are not needed, but as more devices are added to a system they become necessary. Here's why. Most of the devices such as ROMs and RAMs used around microprocessors have MOS inputs, so on a dc basis they don't require much current. However, each input or output added to, for example, the system data bus acts like a capacitor of a few picofarads connected to ground. To change the logic state on these signal lines from low to high, all this added capacitance must be charged. To change the logic state to a low, the capacitance must be discharged. If we connect more than a few devices on the data bus lines, the 8086 outputs cannot supply enough current drive to charge and discharge the circuit capacitance rapidly. Therefore, we add external high-current drive buffers to do the job.

Buffers used on the data bus must be bidirectional because the 8086 sends data out on the data bus and also reads data in on the data bus. The *DATA TRANSMIT/RECEIVE signal, DT/$\overline{R}$*, from the 8086 sets the direction that data will pass through the buffers. When DT/$\overline{R}$ is asserted high, the buffers will be set up to transmit data from the 8086 to ROM, RAM, or ports. When DT/$\overline{R}$ is asserted low, the buffers will be set up to allow data to come in from ROM, RAM, or ports to the 8086.

The buffers used on the data bus must have three-state outputs so that the outputs can be floated when the bus is being used for other operations. For example, we certainly don't want data bus buffer outputs enabled onto the data bus while the 8086 is putting out the lower 16 bits of an address on these lines. The 8086 will enable the three-state outputs on data bus buffers at the appropriate time by asserting $\overline{DEN}$ low.

The final section of Figure 13-39 to look at is the 8284A clock generator in the upper left corner. This device uses a crystal to produce the stable frequency clock signal required to step the 8086 through the execution of its instructions. The 8284 also synchronizes the RESET signal and the READY signal with the clock so that these signals are applied to the 8086 at the proper times. The RESET signal is used to send the 8086 to address FFFF0H to get its next instruction. The first instruction of the system start-up program is usually located at this address, so asserting this signal is a way to "boot," or start, the system. We will discuss the use of the READY input in a later section of the chapter.

Figure 13-40 shows a photo of an Intel SDK-86 microprocessor development board, which is a simple 8086-based microcomputer such as we have described in this section. Perhaps, if you look closely at the photo, you

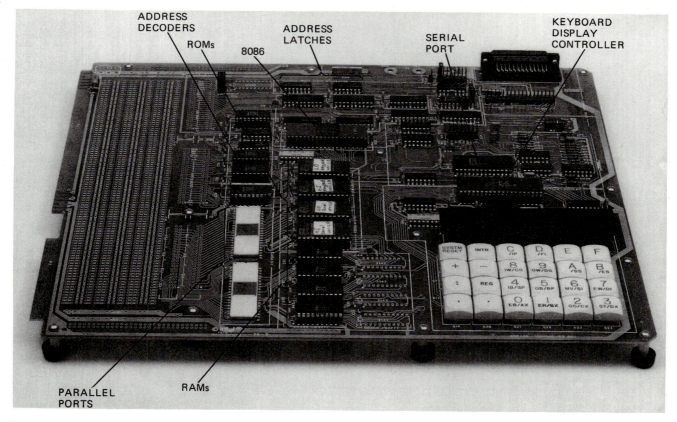

FIGURE 13-40 Intel SDK-86 microprocessor development board. *(Intel Corporation.)*

can identify the CPU, ROM, RAM, ports, and buffer sections of the board. The SDK-86 also has a keypad which allows you to enter simple programs in memory and execute them. A system such as the SDK-86 is used for several exercises in the accompanying laboratory manual.

8086 BUS ACTIVITIES DURING A READ MACHINE CYCLE

Figure 13-41. shows in timing diagram form the activities on the 8086 microcomputer buses during simple read and write operations. Don't be overwhelmed by all the lines on this diagram. Their meaning should become clear to you as we work through the diagram.

The first line to look at in Figure 13-41 is the *clock waveform*, CLK, at the top. This represents the crystal-controlled clock signal sent to the 8086 from an external clock generator device such as the 8284 shown in the top left corner of Figure 13-39. One cycle of this clock is called a *state*. For reference purposes a state is measured from the falling edge of one clock pulse to the falling edge of the next clock pulse. The time interval labeled T_1 in the figure is an example of a state. Different versions of the 8086 have maximum clock frequencies of between 5 and 10 MHz, so the minimum time for one state will be between 100 and 200 ns, depending on the part used and the crystal used.

A basic microprocessor operation such as reading a byte from memory or writing a byte to a port is called a

machine cycle. The times labeled T_{CY} in Figure 13-41 are examples of machine cycles. As you can see in the figure, a machine cycle consists of several states.

The time a microprocessor requires to fetch and execute an entire instruction is referred to as an *instruction cycle*. An instruction cycle consists of one or more machine cycles.

To summarize this then, an instruction cycle is made up of machine cycles, and a machine cycle is made up of states. The time for a state is determined by the frequency of the clock signal. In this section we discuss the activities that occur on the 8086 microcomputer buses during a read machine cycle.

The best way to analyze a timing diagram such as the one in Figure 13-41 is to think of time as a vertical line moving from left to right across the diagram. With this technique you can easily see the sequence of activities on the signal lines as you move your imaginary time line across the waveforms.

During T_1 of a read machine cycle the 8086 first asserts the M/$\overline{\text{IO}}$ signal. It will assert this signal high if it is going to do a read from memory during this cycle, and it will assert M/$\overline{\text{IO}}$ low if it is going to do a read from a port during this cycle. The timing diagram in Figure 13-41 shows two crossed waveforms for the M/$\overline{\text{IO}}$ signal because the signal may be going low or going high for a read cycle. The point where the two waveforms cross indicates the time at which the signal becomes valid for this machine cycle. Likewise, in the rest of the timing diagram crossed lines are used to

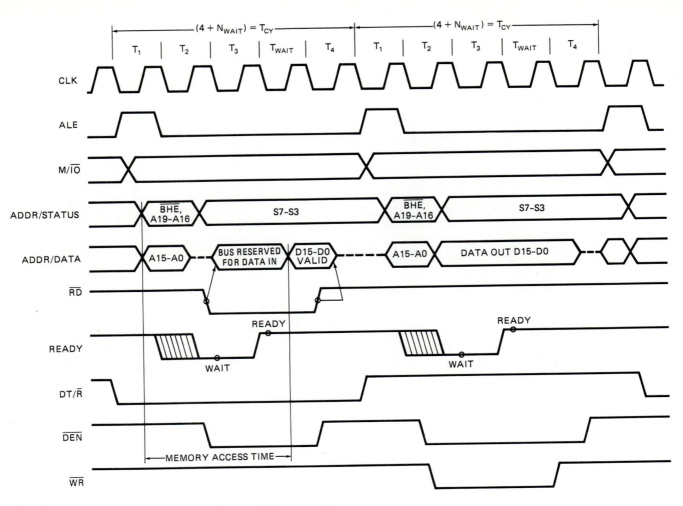

FIGURE 13-41 Basic 8086 system timing. *(Intel Corporation.)*

represent the time when information on a line or group of lines is changed.

After asserting M/IO, the 8086 sends out a high on the ADDRESS LATCH ENABLE signal, ALE. This signal is connected to the enable input (STB) of the 74S373 octal latches, as shown in Figure 13-39. As you can also see in Figure 13-39, the data inputs of these latches are connected to the 8086 AD0 to AD15, A16 to A19, and the *BUS HIGH ENABLE (BHE)* lines. After the 8086 asserts ALE high, it sends out on these lines the address of the memory location that it wants to read. Since the latches are enabled by ALE being high, this address information passes through the latches to their outputs. The 8086 then makes the ALE output low which disables the latches. The address held on the latch outputs travels along the address bus to memory and port devices.

Incidentally, observe in the timing diagram in Figure 13-41 how the activity on the ADDR/DATA lines is represented. The first point at which the two waveforms cross represents the time at which the 8086 has put a valid address on these lines. These two waveforms *do not* indicate that all 16 lines are going high or going low at this point. Again, the crossed lines show the time at which a valid address is on the bus.

After ALE goes low, the address information is held on the latches, so the 8086 no longer needs to send out the addresses. Therefore, as shown by a dashed line on the ADDR/DATA line in Figure 13-41, the 8086 floats the AD0 to AD15 lines so that they can be used to input data from memory or from a port. At about the same time the 8086 also removes the BHE and A16 to A19 information from the upper lines and sends out some status information on those lines.

The 8086 is now ready to read data from the addressed memory location or port, so near the end of state T_2 the 8086 asserts its RD signal low. As you will see in the next chapter, this signal is used to enable the addressed memory device or port device. When enabled, the addressed device will put a byte or word of data on the data bus. In other words, asserting the RD signal low causes the addressed device to put data on the data bus. This cause-and-effect relationship is shown on the timing diagram in Figure 13-41 by an arrow going from the falling edge of RD to the "bus reserved for data in" section of the ADDR/DATA waveforms. The bubble on the tail of the arrow is always put on the signal transition or level that causes some action, and the point of the arrow always indicates the action caused. Arrows of this sort are only used to show the effect a signal from one device will have on another device. They are not usually used to indicate signal cause and effect within a device.

Now, referring to Figure 13-41 again, find the section

of the AD0 to AD15 waveform marked off as memory access time near the bottom of the diagram. This time represents the time it takes for the memory to output valid data after it receives an address and an $\overline{RD}$ signal. If the access time for a memory device is too long, the memory will not have valid data on its outputs soon enough in the machine cycle for the 8086 to receive it correctly. The 8086 will then treat whatever garbage happens to be on the data bus as valid data and go on with the next machine cycle. As long as Murphy's law is still in force, the garbage read in will probably cause the entire program to crash. A section in the next chapter shows you how to calculate whether a particular ROM, RAM, or port device has a short-enough access time to work properly in a given 8086 system. For now, however, we just need you to understand the concept so we can show you one way that an 8086 can accommodate a slow device.

If you look again at the block diagram in Figure 13-39, you should find an input on the 8086 CPU labeled READY. With this pin high, the 8086 is "ready" and operates normally. If the READY input is made low at the right time in a machine cycle, the 8086 will insert one or more *WAIT* states between T_3 and T_4 in that machine cycle. The timing diagram in Figure 13-41 shows an example of this. An external hardware device is set up to pulse READY low before the rising edge of the clock in T_2. After the 8086 finishes T_3 of the machine cycle, it enters a WAIT state. During a WAIT state the signals on the buses remain the same as they were at the start of the WAIT state. The address of the addressed memory location is held on the output of the latches so that it does not change, and as you can see from the timing diagram in Figure 13-41, the control bus signals M/$\overline{IO}$ and $\overline{RD}$ also do not change during the WAIT state, T_{WAIT}. If the READY input is made high again during T_3 or during the WAIT state, as shown in Figure 13-41, then after one WAIT state the 8086 will go on with the regular T_4 of the machine cycle. What we have done by inserting the WAIT state is to freeze the action on the buses for one clock cycle. This gives the addressed device an extra clock cycle time to put out valid data. If, for example, we want to use a slower (cheaper) ROM in a system, we can add a simple circuit which pulses the READY input low each time that ROM is addressed. When reading data from faster devices in the system, no WAIT states will be inserted in the read machine cycle.

Note in Figure 13-39 that a READY input signal is usually passed through the 8284A clock generator IC so that the READY signal actually applied to the 8086 is synchronized with the system clock.

If the 8086 READY input is still low at the end of a WAIT state, then the 8086 will insert another WAIT state. The 8086 will continue inserting WAIT states until the READY input is made high again.

Now let's look back at Figure 13-41 to see how $\overline{DEN}$ and DT/$\overline{R}$ function during a read machine cycle. During T_1 of the machine cycle the 8086 asserts DT/$\overline{R}$ low to put the data buffers in the receive mode. Then, after the 8086 finishes using the data bus to send out the lower 16 address bits, it asserts $\overline{DEN}$ low to enable the data bus buffers. The data put on the data bus by an addressed

port or memory will then be able to come in through the buffers to the 8086 on the data bus.

We can summarize the activities on the buses during an 8086 read machine cycle as follows. The 8086 asserts M/$\overline{IO}$ high if the read is to be from memory and asserts M/$\overline{IO}$ low if the read is going to be from a port. At about the same time, the 8086 asserts ALE high to enable some external latches. It then sends out $\overline{BHE}$ and the desired address on the AD0 to A19 lines. The 8086 then pulls the ALE line low to latch the address information on the outputs of the external latches. After the 8086 is through using lines AD0 to AD15 for an address, it removes the address from these lines and puts the lines in the input mode (floats them). It then asserts its $\overline{RD}$ signal low. The $\overline{RD}$ signal going low turns on the addressed memory or port which then puts the desired data on the data bus. To complete the cycle the 8086 brings the $\overline{RD}$ line high again. This causes the addressed memory or port to float its outputs on the data bus. If the 8086 READY input is made low before or during T_2 of a machine cycle, the 8086 will insert WAIT states as long as the READY input is low. When READY is made high, the 8086 will continue with T_4 of the machine cycle. WAIT states can be used to give slow devices additional time to put out valid data. If a system is large enough to need data bus buffers, then the 8086 DT/$\overline{R}$ signal connected to these buffers will set them for input during a read operation or set them for output during a write operation. The 8086 DEN signal will enable the buffers at the appropriate time in the machine cycle.

8086 BUS ACTIVITIES DURING A WRITE MACHINE CYCLE

Now that we have analyzed the 8086 bus activities for a read machine cycle, let's take a look at the timing diagram for a write machine cycle in the right-hand side of Figure 13-41. Most of this diagram should look very familiar to you because it is very similar to that for a read cycle.

During T_1 of a write machine cycle, the 8086 asserts M/$\overline{IO}$ low if the write is going to be to a port and asserts M/$\overline{IO}$ high if the write is going to be to memory. At about the same time the 8086 raises ALE high to enable the address latches. The 8086 then outputs $\overline{BHE}$ and the address that it will be writing to on AD0 to A19. Incidentally, when writing to a port, lines A16 to A19 will always be low, because the 8086 only sends out 16-bit port addresses. After this address has had time to pass through the latches, the 8086 brings ALE low again to latch the address on the outputs of the latches. Besides holding the address, these latches also function as buffers for the address lines. After the address information is latched, the 8086 removes the address information from AD0 to AD15 and outputs the desired data on the data bus. It then asserts its $\overline{WR}$ signal low. The $\overline{WR}$ signal is used to turn on the memory or port that the data is to be written to. After the addressed memory or port has had time to accept the data from the data bus, the 8086 raises the $\overline{WR}$ signal line high again and floats the data bus.

If the memory or port device cannot absorb the data word within a normal machine cycle, external hardware can be set up to pulse the READY input low each time that device is addressed. If the READY input is pulsed low before or during T_2 of the machine cycle, the 8086 will insert a WAIT state after state T_3. During a WAIT state the signals on the data bus, address bus, and control bus are held constant, so the addressed device has one or more extra clock cycles to absorb the data from the data bus. If the READY input is made high before the end of the WAIT state, the 8086 will go on with state T_4 as soon as it finishes the WAIT state. If the READY input is still low just before the end of the WAIT state, the 8086 will insert another WAIT state. It will continue to insert WAIT states until READY is made high. The point here is that the 8086 can be forced to insert as many WAIT states as are necessary for the addressed device to accept the data.

If the system is large enough to need buffers on the data bus, then DT/$\overline{R}$ will be connected to the direction input on the buffers. During a write cycle the 8086 asserts DT/$\overline{R}$ high to put the buffers in the transmit mode. When the 8086 asserts $\overline{DEN}$ low to enable the buffers, data output from the 8086 will pass through the buffers to the addressed port or memory location.

Work your way across the timing diagrams for the read and write machine cycles in Figure 13-41 until you feel that you understand the sequence of activities that occur. Understanding this well will make the next chapter more easily understood.

8086 MEMORY BANKS AND THE $\overline{BHE}$ SIGNAL

The 8086 has a 20-bit address bus, so it can address 2^{20}, or 1,048,576, addresses. Each address represents a stored byte. The 8086 however has a 16-bit data bus, so to make it possible to read or write a 16-bit word in one bus cycle, the memory for an 8086 is set up as two "banks." Figure 13-42a shows this in diagram form.

Each bank can have up to 524,288 byte storage locations. One memory bank contains all the bytes which have even addresses such as 00000, 00002, and 00004. As shown in Figure 13-42a, the data lines of this bank are connected to the lower eight data lines, D0 to D7, of the 8086. The other memory bank contains all the bytes which have odd addresses such as 00001, 00003, and 00005. The data lines of this bank are connected to the upper eight data lines, D8 to D15, of the 8086.

Address line A0 is used as part of the enabling for memory devices in the lower, or even, bank. The address decoder circuitry in the system is designed so that an even address (A0 = 0) will enable this bank. Address lines A1 to A19 are used to select the desired memory device in the bank and address the desired byte in that device.

The upper, or odd, memory bank is enabled by a separate signal called *BUS HIGH ENABLE*, or $\overline{BHE}$, sent out by the 8086. $\overline{BHE}$ is sent out from the 8086 at the same time as an address is sent out. An external latch, strobed by ALE, grabs the $\overline{BHE}$ signal and holds it stable for the rest of the machine cycle. The $\overline{BHE}$ signal will be asserted low if a *byte* is being accessed at an odd address or if a *word* at an even address is being accessed. Address lines A1 to A19 are used to select the desired memory device in the bank and address the desired byte in that device.

Figure 13-42b shows you the signal levels that will be present on the $\overline{BHE}$ and A0 lines for different types of memory accesses. If the 8086 executes an instruction which reads a byte from or writes a byte to an even address such as 00000H, A0 will be asserted low and $\overline{BHE}$ will be high. The lower bank will be enabled, and the upper bank will be disabled. A byte will be transferred to or from the addressed location in the low bank on D0 to D7.

If the 8086 executes an instruction which reads a word (2 bytes) from memory at an even address such as 00000, both A0 and $\overline{BHE}$ will be asserted low. Therefore, both banks will be enabled. The low byte of the word will be transferred from address 00000H to the 8086 on D0 to D7. The high byte of the word will be transferred from address 00001H to the 8086 on D8 to D15. Note that only one bus cycle is required to transfer the 2 bytes.

When the 8086 executes an instruction which accesses just a byte at an odd address, A0 will be high and $\overline{BHE}$ will be low. Therefore, the low bank will be disabled and the high bank will be enabled. The byte will be transferred from memory address 00001H in the high bank to the 8086 on lines D8 to D15. Note that address 00001H is the first location in the upper bank.

The final case in Figure 13-42b is that where the 8086 reads a word from or writes a word to an odd address. In this case the 8086 requires two machine cycles to copy the 2 bytes from or to memory. During the first machine cycle the 8086 will output address 00001H and assert $\overline{BHE}$ low. A0 will be high. The byte from address 00001H will be read in to the 8086 on lines D8 to D15. During the second machine cycle the 8086 will send out address 00002H. A0 will be low, but $\overline{BHE}$ will be high. The second byte will be read in to the 8086 on lines D0 to D7.

The main reason that the A0 and $\overline{BHE}$ signals function the way they do is to prevent the writing of an unwanted byte into an adjacent memory location when the 8086 writes just a byte. To understand this, think what would happen if both memory banks were turned on for all write operations and you wrote a byte to address 00002. The data on D0 to D7 would be written to address 00002 as desired, but since the upper bank is also enabled, the random data on A8 to A15 would be written into address 00003. The 8086 then is designed so that $\overline{BHE}$ is high during this byte write. This disables the upper bank of memory and prevents the random data on A8 to A15 being written to address 00003.

The main reason we have introduced you to the 8086 memory banks at this point is so that if you use a logic analyzer to look at the bus activities of an 8086 microcomputer, you won't be surprised to see that the 8086 fetches instruction bytes from memory two at a time. In other words, to fetch the instruction bytes for a program which starts at memory address 00100, the 8086 only has to send out addresses 00100, 00102, 00104, etc., because each read automatically brings in both the byte at the even address and the byte at the odd address. Remember that the low, or even, byte will come in on D0 to D7 and the high, or odd, byte will come in on D8 to D15.

A Closer Look at the 8086

Figure 13-43 shows a pin diagram for the 8086. Don't be overwhelmed by all the pins with strange mnemonics

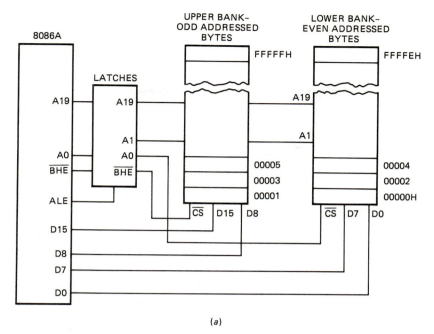

FIGURE 13-42   The 8086 memory banks. (a) Block diagram. (b) Signals for byte and word operations.

ADDRESS	DATA TYPE	$\overline{BHE}$	A0	BUS CYCLES	DATA LINES USED
0000	BYTE	1	0	ONE	D0–D7
0000	WORD	0	0	ONE	D0–D15
0001	BYTE	0	1	ONE	D7–D15
0001	WORD	0	1	FIRST	D0–D7
		1	0	SECOND	D7–D15

(b)

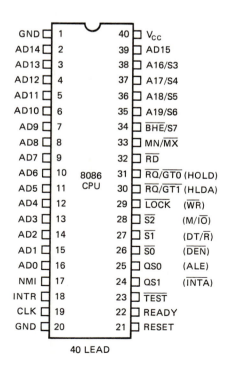

FIGURE 13-43   The 8086 pin diagram. (Intel Corporation.)

next to them. You don't need to learn the detailed functions of all these. The main reason for showing you this is so that if you want to look at some of the 8086 signals with a scope or logic analyzer, you know which pins to connect to. We also want to make a few comments about some of the pins to give you a little more overview of how an 8086-based microcomputer functions.

First in Figure 13-43 find V_{CC} on pin 40 and ground on pins 1 and 20. Next find the clock input labeled CLK on pin 19. As we showed you in the preceding sections, an 8086 requires a clock signal from some external clock generator to synchronize internal operations in the processor. Different versions of the 8086 have maximum clock frequencies ranging from 5 to 10 MHz.

Now look for the ADDRESS/DATA bus lines, AD0 to AD15. Remember from the previous section that the 8086 has a 20-bit address bus and a 16-bit data bus and that the lower 16 address lines are multiplexed out on the data bus to minimize the number of pins needed. The 8086 sends out a signal called ADDRESS LATCH ENABLE, or ALE, on pin 25 to strobe the external address latches. The upper 4 bits of the 20-bit address are sent out on the lines labeled A16/S3 to A19/S6. The double mnemonic on these pins shows that address bits A16 to A19 are sent out on these lines during the first part of a machine cycle, and status information which identifies the type of operation being done in that cycle is sent out on these lines during a later part of the cycle.

Having found the address bus and the data bus, now look for the control bus signal pins. Pin 32 of the 8086 in Figure 13-43 is labeled $\overline{RD}$. This signal will be asserted low when the 8086 is reading data from memory or from a port. Pin 29 has a label $\overline{WR}$ next to it. However, pin 29 also has a label $\overline{LOCK}$ next to it, because this pin has two functions. The function of this pin and the functions of the other pins between 24 and 31 depend on the *mode* in which the 8086 is operating.

The operating mode of the 8086 is determined by the logic level applied to the MN/$\overline{MX}$ input, pin 33. If pin 33 is asserted high, then the 8086 will function in *minimum mode,* and pins 24 to 31 will have the functions shown in parentheses next to the pins in Figure 13-43. For example, pin 29 will function as $\overline{WR}$ which will go low any time the 8086 writes to a port or to a memory location. The $\overline{RD}$, $\overline{WR}$, and M/$\overline{IO}$ signals form the heart of the control bus for a minimum-mode 8086 system. The 8086 is operated in minimum mode in systems such as the SDK-86 where it is the only microprocessor on the system buses.

If the MN/$\overline{MX}$ pin is asserted low, then the 8086 is in *maximum mode.* In this mode pins 24 to 31 will have the functions described by the mnemonics next to the pins in Figure 13-43. In this mode the control bus signals ($\overline{S0}$, $\overline{S1}$, $\overline{S2}$) are sent out in encoded form on pins 26 to 28. An external bus controller device decodes these signals to produce the control bus signals required for a system which has two or more microprocessors sharing the same buses. In the sequel to this book we discuss the operation of multiple microprocessor systems.

Another important pin on the 8086 is pin 21, the RESET input. If this input is asserted and then released, the 8086 will, no matter what it was doing, fetch its next instruction from physical address FFFF0H. At this address, then, you put the first instruction you want the microcomputer to execute after a reset or when the power is first turned on.

Finally, notice that the 8086 has two interrupt inputs, *NONMASKABLE INTERRUPT (NMI)* input on pin 17 and the *INTERRUPT (INTR)* input on 18. A signal can be applied to one of these inputs to cause the 8086 to stop executing its current program and go execute a special program which takes care of the condition that caused the interrupt. You might, for example, connect a temperature sensor from a steam boiler to an interrupt input on an 8086. If the boiler gets too hot, then the temperature sensor will assert the interrupt input. This will cause the 8086 to stop executing its current program and go execute a separate sequence of instructions which turns off the fuel supply to the boiler. At the end of the procedure the 8086 can be returned to execution of the interrupted program.

Looking at Microcomputer Bus Signals with a Logic Analyzer

In Chapter 8 we introduced you to the use of a logic analyzer for making a trace of the sequence of states that a system steps through or a trace which shows the timing relationships between the signals in a system. To refresh your memory, the three major points you have to think about when you connect a logic analyzer up to do a trace are:

1. Connect the data inputs of the analyzer to the system signals you want displayed in the trace.

2. The clock signal specified for the analyzer tells it when to take data samples and store them in its memory. To produce a trace which shows the sequence of states that a system steps through, you usually use an external clock. When you are using an external clock, you specify the clock edge which occurs when valid data is on the data inputs. For making timing measurements, you usually use the crystal-controlled internal clock.

3. The trigger specified for the analyzer tells it when to stop taking samples and display the set of samples stored in its memory. Usually you will use the internal word recognizer to trigger the analyzer when a specified word is present on the data inputs.

MAKING A TRACE OF A SEQUENCE OF ADDRESSES

The first step in using a logic analyzer to look at microcomputer signals is to decide what specific signals you want to look at and then connect the analyzer data inputs to those signals. As a first example, suppose you want to do a trace which shows the sequence of addresses that the 8086 outputs as it executes a test program. In this case you connect the data inputs of the first analyzer pod to the 8086 AD0 to AD15 pins .

The next step is to decide what signal to clock the analyzer on. To make this decision you look carefully at the 8086 timing waveforms in Figure 13-41 to find a signal edge which occurs when valid addresses are on the AD0 to AD15 lines. One possible signal to use for clocking the analyzer is the 8086 CLK signal shown at the top of the waveforms in Figure 13-41. This signal has a falling edge when the address is valid on AD0 to AD15, but it also has falling edges when the lines are floating and when the data from or to memory is on the lines. In other words, if the analyzer is clocked on this signal, the trace will show a mixture of data, addresses, and garbage which you have to sort out.

A better choice for an analyzer clock signal is the 8086 ALE signal, because this signal is only present when addresses are on the AD0 to AD15 lines. To use ALE as a clock signal, connect the EXTERNAL CLOCK input of the analyzer to the 8086 ALE pin. To determine which edge of the ALE signal to clock the analyzer on, look closely at the 8086 timing waveforms in Figure 13-41. At the time when the positive edge of the ALE signal occurs, the 8086 has not yet output the address, so clocking the analyzer on this edge will not grab the addresses. The falling edge of the ALE signal occurs when the address is solidly settled on the AD0 to AD15 lines, so you should set the analyzer to clock on the falling edge of ALE.

The final step is to determine what to trigger the analyzer on. Since you want to make a trace of a sequence of addresses, the logical choice here is to set the internal word recognizer to produce a trigger when the first program address is present on the data inputs. For example, if the first program instruction is in memory at 00100H, you would set the analyzer to trigger when this address is present. When you specify the "trigger position," set the analyzer for "begin" so that the trace listing starts with the specified address.

MAKING A TRACE OF A SEQUENCE OF DATA WORDS

As a second example of using an analyzer to look at microcomputer signals, suppose that you want to do a trace which shows the sequence of data words read in from memory as the 8086 executes a test program. For this trace you connect the analyzer data inputs to the 8086 AD0 to AD15 pins, because the data comes in on these lines.

To determine what signal to clock the analyzer on and which edge of that signal to specify, you again look closely at the 8086 timing waveforms in Figure 13-41. From these waveforms you should see that the 8086 $\overline{RD}$ signal is asserted during a memory read operation, so this is an appropriate signal to connect to the analyzer's EXTERNAL CLOCK input. The rising edge of the $\overline{RD}$ signal occurs when valid data is on the data bus, so set the analyzer to clock on a rising edge.

Since you want a trace of the data words read in from memory by the 8086, you need to look at the test program to determine what to trigger the analyzer on. For this example, assume the simple test program shown here is entered in memory and run:

```
00100  EB  HERE:JMP HERE   ;Endless loop which
00101  FE                  ;does nothing
00102  90        NOP       ;just more words to
00103  90        NOP       ;fetch
00104  04        ADD AL,55H
```

Remember from the discussion of 8086 memory banks and $\overline{BHE}$ in the preceding section that the 8086 can read in a word (2 bytes) at a time if the word starts on an even address. When reading in the code bytes for this program, then, the 8086 will send out address 00100H and assert both A0 and $\overline{BHE}$. Both the low bank and the high bank of memory will be enabled. The byte containing EBH will come into the 8086 from the even bank on AD0 to AD7, and the byte containing FEH will come into the 8086 from the odd bank on AD8 to AD15. The first data word read in from memory then is FEEBH, so this is the word you set the analyzer to trigger on.

When the trace is completed, it will show the sequence of words FEEBH, 9090H, and 0455H over and over. The only part of this program that the 8086 executes is the HERE: JMP HERE instruction represented by the codes EBH and FEH. While the 8086 is decoding the JMP instruction, however, it fetches the codes for the following instructions and stores them in an internal register, ready to be used. This is analogous to the way a helper sets up a stack of bricks for a bricklayer so that the bricklayer does not have to wait for the helper to go to the truck and get each brick as needed. In this program, however, the JMP instruction tells the 8086 to go back and fetch the JMP instruction again. The words 9090H and 0455H then are fetched from memory and stored in the 8086, but they are never used.

USING A CLOCK QUALIFIER IN LOGIC ANALYZER MEASUREMENTS

In the preceding example we showed you how to produce a trace of the data words read in from memory by an 8086. Now suppose that you are executing a program which reads data words from memory and data words from ports. If you simply clock the analyzer on the rising edge of the $\overline{RD}$ signal as you did for the preceding example, the trace will contain both the data words read from memory and the data words read from ports. If you want to, you can use the M/$\overline{IO}$ signal to produce a trace which contains only the words read from memory or only the words read from ports.

Remember from our previous discussions that when the 8086 writes a word to a memory location, or reads a word from a memory location it will assert its M/$\overline{IO}$ signal high. When the 8086 writes a word to a port or reads a word from a port, it will assert the M/$\overline{IO}$ signal low.

To produce a trace of only the data words read from ports, you connect the $\overline{RD}$ signal to the EXTERNAL CLOCK input of the analyzer and connect the M/$\overline{IO}$ signal to an input on the analyzer labeled CLOCK QUALIFIER. The principle here is that if this input is used, the analyzer will only respond to a clock signal if a specified level is present on that input. For this example you want the analyzer to take a sample on the rising edge of the $\overline{RD}$ if M/$\overline{IO}$ is low. Therefore, you will specify a low as the active level for the clock qualifier. Depending on the analyzer, the active level for the clock qualifier input may be set in a menu or by connecting the qualifier signal to one of two inputs on a data pod.

To produce a trace of only the data words read from memory, you can clock on the rising edge of $\overline{RD}$, connect the M/$\overline{IO}$ signal to the CLOCK QUALIFIER input, and specify a high for the CLOCK QUALIFIER. The point here is that by carefully choosing the clock signal and the qualifier signal, you can usually produce a trace of just the data you want.

MEASURING MEMORY ACCESS TIME WITH A LOGIC ANALYZER

As shown in the 8086 timing waveforms in Figure 13-41, one type of memory access time is the time it takes for a memory device to produce valid data on its outputs after an address is applied to its address inputs. With a little thought you can use a logic analyzer to measure the actual memory access time in a system.

The first step in this measurement is to enter and run a test program which reads from the desired memory device over and over. For this example we will use the same program we used in the preceding example. To make it easy to refer to, we repeat it here:

```
00100  EB  HERE:JMP HERE    ;Endless loop which
00101  FE                   ;does nothing
00102  90          NOP      ;just more words to
00103  90          NOP      ;fetch
00104  04          ADD AL,55H
```

The next step is to think about what signals to connect to the analyzer data inputs. To determine the time between a valid address from the 8086 and valid data from the memory, you obviously need to look at the ADDRESS/DATA lines. The number that you can trace and display depends on the particular analyzer you are using. The basic Tektronix 1230 analyzer we discussed in Chapter 8 will sample and display 16 channels in the timing mode which you use for this type of measurement. If you have an analyzer such as this you can connect the analyzer data inputs to the AD0 to AD15 pins on the 8086. The upper four address lines, A16 to A19, do not change during the execution of this example program, so they can be left out.

When making timing measurements with a logic analyzer, you almost always use the crystal-controlled internal clock to tell the analyzer to take samples so that you know the exact time between samples. For an SDK-86 board the memory access time for the RAM that contains the sample program will be around 100 ns. To get the best possible resolution for your timing measurement then, you should set the analyzer clock period for the shortest time possible on your analyzer. The shortest period for the Tektronix 1230 with a 16-channel display is 40 ns per clock, so we will use this setting.

To choose the trigger word for this measurement, look again at the timing waveforms in Figure 13-41. The address goes out on the data bus, and later the data comes back in. Since the address is the first activity, you set the analyzer to trigger on the first address that is sent out.

Once you do a trace, you can determine the memory access time by counting the number of sample points between the address of 0100H appearing on the bus and the data word of FEEBH appearing on the bus. If your analyzer has cursors, you can position one cursor at the time when the address becomes valid, position the other cursor at the time when the data becomes valid, and read the time difference between the two from the on-screen display.

Note that the resolution of this measurement is only 40 ns, because that is the time between samples. In other words, any changes that take place between sample points will not be shown in the display until the next set of samples is taken. On many analyzers you can specify a shorter sampling period if you reduce the number of signal lines being traced. With the Tektronix 1230, for example, you can use a sample clock with a period as short as 10 ns if you can get by with sampling only four signal lines. We usually start by doing a trace of, for example, all 16 lines, and then from the 16 we choose four which show the desired transitions. With just these four lines we can decrease the sample period to 10 ns and thereby increase the resolution of our measurement.

The point of the preceding examples is to show you that a logic analyzer is useful for making a variety of measurements on a digital system such as a microcomputer which has many signal lines.

Where to Next?

This chapter has led you from basic digital arithmetic operations to an overview of the microcomputer operation. In the next chapter we look more closely at the memory devices and systems used in microcomputers and other applications.

CHECKLIST OF IMPORTANT TERMS AND CONCEPTS IN THIS CHAPTER

Binary addition and subtraction

Sign bit

Two's complement binary sign-and-magnitude

Overflow of signed numbers

BCD addition and subtraction

Octal addition and subtraction

Hexadecimal addition and subtraction

Binary multiplication and division
    Partial product
    Add-and-shift-right multiplication
    Subtract-and-shift-left division

Fixed-point and floating-point numbers

Real numbers
    Normalizing a number
    Mantissa, or significand
    Single-precision representation
    Long real, or double-precision, representation

Adder circuits
    Look-ahead carry

Magnitude comparators

Arithmetic Logic Unit (ALU)

Multiplier devices

MegaFLOPS, MFLOPS

Bit-slice microprocessor
    Flags
    Barrel shifter

Microinstructions, microcode

Pipelining

Machine-code instruction

Central processing unit, CPU

Memory

Input and output, I/O

Address bus, data bus, and control bus

Bidirectional bus

Hardware, software, firmware

Input and output ports

Address counter, instruction pointer

Memory read, memory write

I/O read, I/O write

State

Instruction cycle

Machine cycle

$M/\overline{IO}$, ALE, $\overline{BHE}$, $\overline{DEN}$, $\overline{RD}$, $\overline{WR}$, $DT/\overline{R}$ signals

WAIT state

Minimum and maximum mode

Nonmaskable interrupt, NMI

Interrupt, INTR

REVIEW QUESTIONS AND PROBLEMS

1. a.   10110110   b.   11011000
       + 01101110   + 11001111

2. Express the following decimal numbers in 8-bit binary sign-and-magnitude two's complement form:
   a. 26   e. 0
   b. −7   f. −1
   c. −26   g. −97
   d. −125   h. 85

3. Convert the following 8-bit signed numbers to decimal:
   a. 00110011   c. 10001100
   b. 11110110   d. 01011101

4. Perform the following subtractions using the pencil method or the two's complement method:
   a.   01110110   b.   00011100
        − 01001111   − 01100011

5. Perform the indicated operation on the following *signed* numbers:
   a.   01001100   c.   00101010
        + 11110110   − 11110010

   b.   11110000   d.   11100000
        + 11100011   − 00010101

6. Perform the indicated operation on the following BCD numbers.
   a.   0101 0110   c.   0111 1001
        + 0010 0110   − 0110 1000

   b.   0100 0100   d.   0110 0110
        + 0010 1000   − 0011 1001

7. Perform the indicated operation on the following octal numbers:
   a. 37_8 b. 46_8 c. 37_8 d. 52_8
      + 26_8 + 57_8 − 26_8 − 36_8

8. Perform the indicated operation on the following hexadecimal numbers:
   a.   47H   c.   07FFH
      + 59H   − 349H

   b.   3AH   d.   2A92H
      + 9CH   − 03FFH

9. A small microcomputer has 8192_{10} bytes of ROM memory starting at address 2000H. What is the ending address for this memory?

10. Multiply 1001 by 011 using:
    a. The pencil method.
    b. The add-and-shift algorithm.

11. Show the division of 1100100 by 1010 using:
    a. The pencil method.
    b. The subtract-and-shift algorithm.

12. Normalize the following numbers and show the exponent produced:
    a. 0.01101
    b. 110001.0
    c. 0.000101

13. Show how each of the numbers in Problem 12 can be expressed in the single-precision floating-point format shown in Figure 13-26.

14. Using Figure 13-31, show the programming of the SELECT and MODE inputs that the 74181 requires to perform the following arithmetic functions:
    a. A PLUS B
    b. A MINUS B MINUS 1
    c. AB PLUS A

15. Using Figure 13-31, show the programming of the SELECT and MODE inputs of the 74181 required to perform the following positive logic functions:
    a. $\overline{A + B}$
    b. $A \oplus B$
    c. $\overline{AB}$

16. Show the output word produced when the following binary words are ANDed and when they are ORed:
    a. 1010 and 0111
    b. 1011 and 1100
    c. 11010111 and 00111000
    d. ANDing an 8-bit binary number with 11110000 is sometimes called *masking* the lower 4 bits. Why?

17. Describe how high-speed multiplier devices such as the 74284 and 74285 shown in Figure 13-33 actually implement the multiplication in hardware.

18. a. What determines the sources of the operands for the R and S inputs of the Am2901C Microprocessor Bit Slice in Figure 13-35?
    b. Draw a diagram to show how multiplexers can be used to shift a data word left or right one bit position.

19. a. What is meant by the term *microprocessor bit slice?*
    b. What is the purpose of the microcode PROM in a bit-slice microprocessor system?
    c. What is the purpose of the pipeline register in Figure 13-36?
    d. Describe how the bit-slice microprocessor in Figure 13-36 executes an instruction which requires a sequence of operations.

20. a. Draw a labeled block diagram for a simple microcomputer.
    b. Describe the purpose of the address bus, data bus, and control bus in a microcomputer.
    c. Describe the purpose of the ports in a microcomputer.

21. List the sequence of actions that a CPU will take to fetch and execute an instruction which inputs a byte of data from a port.

22. Why are latches required on the AD0 to AD15 bus in an 8086 system?

23. What is the purpose of the ALE signal in an 8086 system?

24. Why are buffers often needed on the address, data, and control buses in a microcomputer system?

25. From what point on the clock waveform is the start of an 8086 state measured?

26. Describe the sequence of events on the 8086 DATA/ADDRESS bus, the ALE line, the M/$\overline{\text{IO}}$ line, and the $\overline{\text{RD}}$ line as the 8086 fetches an instruction word.

27. What logic levels will be on the 8086 $\overline{\text{RD}}$, $\overline{\text{WR}}$, and M/$\overline{\text{IO}}$ lines when the 8086 is doing:
    a. A write to a memory location?
    b. A read from a port?

28. What does an arrow going from a transition on one signal waveform to a transition on another tell you?

29. a. How is an 8086 entered into a WAIT state?
    b. At what point in a machine cycle does an 8086 enter a WAIT state?
    c. What information is on the buses during a WAIT state?
    d. How long is a WAIT state?
    e. How many WAIT states can be inserted?
    f. Why would you want the 8086 to insert a WAIT state?

30. What are the functions of the 8086 DT/$\overline{\text{R}}$ and $\overline{\text{DEN}}$ signals?

31. a. Why is the 8086 memory organized as two banks?
    b. Which half of the data bus is the even bank of 8086 memory connected to?
    c. What logic levels will be on the 8086 A0 line and the 8086 $\overline{\text{BHE}}$ line when the 8086 reads a 16-bit word from address 01240H?
    d. How many bus cycles are required to read a 16-bit word from address 00241H?

32. Describe the response an 8086 will make when its RESET input is asserted high and then returned low.

33. What is the purpose of interrupts in a microcomputer system?

34. Use the timing waveforms in Figure 13-41 to help you answer the following questions. Assuming that the data inputs of a logic analyzer are connected to the AD0 to AD15 outputs of the 8086 in Figure 13-43:
    a. What signal and which edge of that signal would you clock a logic analyzer on to see only the sequence of addresses output by the 8086 as it fetches and executes a sequence of instructions?
    b. If the program starts at address 00240H, how would you set the trigger controls so that the trace starts with this address?

35. a. What signal and which edge of that signal would you clock a logic analyzer on to see the sequence of data words read in from memory or ports?
    b. Describe how you would modify the logic analyzer setup to produce a trace of only the sequence of data words read in from ports. Assume the first data word to be read in is 3A24H.

36. Describe how you would connect and set up an analyzer to measure the time between $\overline{\text{RD}}$ going low and valid data coming back to the 8086 from memory.

CHAPTER
14

Memory Devices and Systems

In the last chapter we wanted to concentrate on the overall operation of a microcomputer, so we did not discuss memory in detail. We treated memory simply as a block which outputs a desired word when you send it an address and a READ signal or a block which stores a data word when you send it an address, the data word, and a WRITE signal. Now that you have an overview of how memories are used, we will discuss the operation of the different types of memory devices currently in use and show you some specifics of how these devices are connected in a system.

Before we start, there is one important point we want to make about the evolution of memory devices. Some years ago Gordon Moore, then president of Intel, stated what has since become known as *Moore's law*. His statement was that "memory density quadruples every two years." Another way of saying this is that the number of memory bits available in a single package increases by four every 2 years. For the examples in this chapter we have chosen devices which best illustrate the principles we are trying to teach. By the time you read this book, larger devices will probably be available unless Moore's law no longer holds true. The discussions here, however, should help you understand the data sheets for new devices as they become available.

OBJECTIVES

At the conclusion of this chapter you should be able to:

1. Describe the mechanism used to store data in several types of semiconductor memory devices.

2. Determine the number of address lines required to select one of a specified number of data words stored in a memory device or system.

3. Define memory timing parameters such as address access time, chip select access time, and data setup time.

4. Describe how EPROM, EEPROM, and flash EPROMs are programmed and erased.

5. Describe how data is stored in a static RAM and how static RAMS can be backed up with a small battery.

6. Describe how dynamic RAMs are written to, read from, and refreshed.

7. Describe how ferroelectric RAMs achieve nonvolatile data storage.

8. Describe how an address decoder selects one of several devices in a system.

9. Design an address decoder which selects a desired device when an address is applied to it.

10. Determine if a given memory device is fast enough to work in a given microprocessor system without wait states.

11. Describe how errors are detected and corrected in data read from an array of dynamic RAMs.

12. Describe how data is stored on magnetic floppy disks, magnetic hard disks, and magnetic tapes.

13. Describe how digital data is stored on and read from optical disks.

14. Describe how a logic analyzer can be connected and set up to measure the actual access times of a memory device in a microcomputer system.

OVERVIEW OF MEMORY TYPES

Memory devices and systems used with microcomputers are often divided into two main categories, primary storage and secondary storage. *Primary storage devices* are connected directly on the microcomputer buses and deliver an addressed word in a few hundred nanoseconds or less. ROMs and RAMs are examples of devices in this category. Most primary memory devices are built with some kind of semiconductor technology.

Secondary storage devices usually hold much more data than primary storage devices, but they typically take a few milliseconds to deliver the first word of a stored series of data words. Examples of secondary storage devices are magnetic floppy disks, magnetic hard disks, magnetic tapes, and optical disks.

Two other terms commonly used to describe memory devices are *volatile* and *nonvolatile*. A nonvolatile device retains the data stored in it when power is removed. A ROM is an example of this type of device. Nonvolatile memory is used to hold the program you want a microcomputer to execute when the power is first turned on. A volatile memory device loses the stored data when the power is removed. Registers and RAMs are examples of volatile memory devices. In a microcomputer you use this type of memory for temporary data storage and to hold programs loaded in from disks.

As a summary of the commonly used memory devices and a preview of the first part of the chapter, we will start with a few words about each of the most commonly used memory types. In the later sections of the chapter we discuss each of these in detail and show how they are connected in a system.

ROM Types

ROM—*Read-only memory*. Programmed during manufacture; cannot be reprogrammed; nonvolatile.

PROM—*Programmable read-only memory*. Programmed by blowing internal fuses; cannot be reprogrammed; nonvolatile.

EPROM—*Erasable programmable read-only memory*. Programmed by storing a charge on insulated gates in the device; erased by shining an ultraviolet light through a quartz window on top of package. After erasing it can be reprogrammed; nonvolatile.

EEPROM—*Electrically erasable programmable read-only memory*. Programmed by storing charges on insulated gates in the device. Can be electrically erased and then reprogrammed; nonvolatile.

FLASH EPROM—Programmed by storing charges on insulated gates in the device. Can be electrically erased and then reprogrammed; nonvolatile.

RAM Types

STATIC RAM OR "S" RAM—*Static random-access memory*. Data stored in an array of flip-flops, so data can be written, read, and rewritten. Volatile unless battery-backed.

DYNAMIC RAM or "D" RAM—*Dynamic random-access memory*. Data stored as charges on tiny capacitors, so stored data must be refreshed every few milliseconds. Volatile unless battery-backed.

PSEUDOSTATIC RAM—Dynamic RAM with on-chip refresh circuitry, so the device appears mostly static to an outside system. Volatile unless battery-backed.

NONVOLATILE RAM, NOVRAM, or NVRAM—Consists of a static RAM and EEPROM in parallel. If there is a power failure, data can be quickly transferred from the RAM section to the EEPROM section so that it will not be lost. When power returns, the data can be transferred back to the SRAM. Nonvolatile.

FERROELECTRIC RAM or "F" RAM—Structure similar to SRAM or DRAM, but the data is held on ferroelectric capacitors when power is removed. Nonvolatile.

MAGNETIC CORE—Data stored by magnetizing small ferrite "doughnuts" in one direction or the other. Data can be written, read, and rewritten. Nonvolatile.

Secondary Storage Devices

FLOPPY DISK—Data stored as magnetized spots in circular tracks on flexible Mylar disk coated with magnetic material. Nonvolatile.

HARD DISK—Data stored as magnetized spots in circular track on rigid aluminum disk coated with magnetic material. Nonvolatile.

OPTICAL DISK—Data stored as pits, bubbles, or magnetized regions on metallized disk. Data read out with laser beam as described in Chapter 11 discussion of digital audio. Nonvolatile.

ROM DEVICES

In the preceding section we gave you a brief overview of the different types of ROM devices. A mask-programmed ROM is programmed during manufacture and cannot be changed later. For high-volume applications where the data does not have to be changed, this type ROM is a good choice because it is cheaper than a programmable device of the same size.

As we described in Chapter 7, a PROM is programmable by blowing fuses, so it is more flexible than a mask-programmed ROM. However, once it is programmed, it cannot be reprogrammed. Also, the fuses take up quite a lot of "real estate" on the chip, so the number of bits that can be stored in a PROM is less than for other type devices. PROMs are useful for small code converter applications such as the example we showed you in Chapter 7.

For prototyping and for applications such as a postal scale where the stored data must be changed when the postal rates change, we use EPROMs, EEPROMs, or flash EPROMs. In the following sections we discuss the details of reading, erasing, and programming these devices.

EPROMs

STRUCTURE AND ADDRESSING

A standard EPROM, such as the one shown in Figure 14-1, is erased by shining short-wavelength ultraviolet light though the quartz window in the top of the package for 15 to 20 min. After erasing, most EPROMs will output all 1's for each data word when read. Once erased, the EPROM can be reprogrammed. A data bit in an EPROM is reprogrammed to a low by forcing electrons onto the normally insulated gate of a MOS transistor with a voltage of 10 to 25 V, depending on the particular device.

One of the most common families of EPROM devices is the 27XXX family developed by Intel and second-sourced by many other companies. Examples of devices in this family are the 2716, 2732, 2764, 27128, 27256, 27512, and 27010. The "7" in these device numbers iden-

FIGURE 14-1 Intel 2716 ultraviolet-erasable EPROM showing quartz window. *(Intel Corporation.)*

tifies them as EPROMs, and the last two or three digits tells you how many kilobits you can store in the device. In a small microcomputer system you might find one or two 2764As which can store 65,536 bits, or 8K bytes, each. In a larger system you might find some 27010s which can store 1,048,576 bits, or 128K bytes, each. To introduce you to the basic characteristics of an EPROM, we will use the 2764A, which is readily available and inexpensive enough for you to experiment with if you want. Incidentally, the IBM PC uses 8K-byte devices such as this to hold its basic start-up and system program.

Figure 14-2 shows the block diagram and pin names for a 2764A EPROM. The basic storage array in the 2764A consists of 512 rows with 128 bits in each row, or a total of 65,536 bits. The device has eight data out-

puts, O0 to O7, so you know that it outputs 8-bit words or bytes. Dividing the number of bits in the basic cell matrix by the number of bits per word tells you that the device stores 65,536 bits/8 bits per byte = 8192 bytes. For this reason the 2764A is often referred to as an $8K \times 8$ device.

As we told you in an earlier chapter, N address lines are required to select one of 2^N words. Since the 2764A contains 8192 words and 8192 is 2^{13}, the 2764A has 13 address inputs labeled A0 to A12. You select the desired byte in a 2764A by applying the address of that byte to the address inputs A0 to A12. Nine of the address bits are used to select one of the 512 rows in the cell matrix, and four of the address bits are used to select one of 16 groups of 8 bits from the selected row.

The CHIP ENABLE ($\overline{CE}$) input on the 2764A is used to switch the device between active mode and standby mode. If the $\overline{CE}$ input is asserted high, the 2764A will be put in a standby mode. As you can see by the I_{SB} line in the dc characteristics table in Figure 14-3, the maximum current in this mode is only 35 mA. Any time the device is not being addressed and read, it can be put in this standby mode to reduce the system power consumption. When the $\overline{CE}$ input is asserted low, the 2764A goes from standby mode to active mode. In active mode all the internal circuitry is powered up and data can be read from the device. Note in Figure 14-3 that the active mode current I_{CC} has a maximum value of 75 mA. The difference between this current and the standby current of 35 mA should show you the advantage of the standby mode. It should also tell you that a 0.1-μF capacitor should be connected between V_{CC} and ground next to each device to prevent switching transients on the V_{CC} line.

The OUTPUT ENABLE ($\overline{OE}$) input is asserted low during a read operation to enable the three-state buffers on the data outputs of the 2764A. EPROMs and most other memory devices are built with three-state output buffers so that the outputs of several devices can be connected on a common data bus. To prevent arguments between the outputs connected on the common bus, a circuit called an *address decoder* makes sure that only one device at a time has its outputs enabled. Incidentally, the three-state outputs of the 2764A are also disabled whenever the device is in its standby mode.

The $\overline{PGM}$ input is pulsed low during programming of a 2764A. For normal read operations the $\overline{PGM}$ input is held at a logic high level. A little later we show you the details of how the device is programmed.

To read a data word from a 2764A EPROM, you apply the address of the desired word to the address inputs, assert the $\overline{CE}$ input low to put the device in the active mode, and assert the $\overline{OE}$ low to enable the output buffers. After a propagation delay, the addressed data word will appear on the outputs. To get a better idea of the relationship of these signals, let's take a look at the timing diagrams for the device.

EPROM TIMING PARAMETERS

Figure 14-4 shows the waveforms and timing parameters for a 2764A read operation. As you can see in Figure 14-4a, it takes some time for valid data to appear on the outputs after an address, $\overline{CE}$, and $\overline{OE}$ are applied to

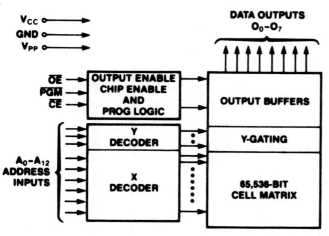

FIGURE 14-2 Block diagram of 2764A EPROM. *(Intel Corporation.)*

Symbol	Parameter	Limits			Conditions
		Min	Max	Unit	
I$_{LI}$	Input Load Current		10	μA	V$_{IN}$ = 5.5V
I$_{LO}$	Output Leakage Current		10	μA	V$_{OUT}$ = 5.5V
I$_{PP}$(2)	V$_{PP}$ Current Read		5	mA	V$_{PP}$ = 5.5V
I$_{SB}$	V$_{CC}$ Current Standby		35	mA	$\overline{CE}$ = V$_{IH}$
I$_{CC}$(2)	V$_{CC}$ Current Active		75	mA	$\overline{CE}$ = $\overline{OE}$ = V$_{IL}$
V$_{IL}$	Input Low Voltage	−0.1	+0.8	V	
V$_{IH}$	Input High Voltage	2.0	V$_{CC}$ + 1	V	
V$_{OL}$	Output Low Voltage		0.45	V	I$_{OL}$ = 2.1 mA
V$_{OH}$	Output High Voltage	2.4		V	I$_{OH}$ = −400 μA
V$_{PP}$(2)	V$_{PP}$ Read Voltage	3.8	V$_{CC}$	V	V$_{CC}$ = 5.0V ± 0.25V

FIGURE 14-3   The dc characteristics of 2764A EPROM. *(Intel Corporation.)*

the device. This time is commonly called the *access time* for the device. In the timing parameters for the 2764A in Figure 14-4*b*, the access time is actually given as three values: t_{ACC}, t_{CE}, and t_{OE}.

The address access time t_{ACC} is the one we usually use to compare the performance of memory devices. This time is the time between a valid address being applied to the address inputs and valid data appearing on the outputs. Although it doesn't look that way in the waveforms, this parameter assumes that the $\overline{CE}$ and the $\overline{OE}$ signals were already asserted when the address was applied. As you can see in Figure 14-4*b*, the maximum, or worst-case, value for this parameter ranges from 180 ns for the 2764A-1 to 450 ns for the 2764A-4.

Devices with longer access times are commonly referred to as *slower*. Usually, slower memory devices are cheaper than faster ones, so a designer will choose the device which has the longest access time that will work in the system. When you replace a defective memory device in a system, it is very important that you use the suffix numbers to get a replacement part which is as fast or faster than the original part. For example, if a system requires a 2764A with a maximum t_{ACC} of 250 ns, it probably will not work correctly with a 2764A-4 which has a maximum t_{ACC} of 450 ns.

Also, remember from a discussion in the last chapter that if a designer wants to use slower, cheaper memory devices with a microprocessor such as the 8086, he or she can add a small circuit which causes the 8086 to insert a wait state every time it accesses those memory devices. The wait state gives the addressed memory device extra time to output valid data.

The t_{CE} shown in Figure 14-4*a* is the time between $\overline{CE}$ being asserted low and valid data appearing on the outputs. Again, this parameter assumes that the address and the $\overline{OE}$ are already present when the $\overline{CE}$ input is asserted. Remember that you must connect a 0.1-μF capacitor between V$_{CC}$ and ground next to each 2764A in a system. The purpose of this capacitor is to prevent V$_{CC}$ voltage spikes which otherwise would be produced by the large change in current when the 2764A is switched from standby mode to active mode or from active mode to standby mode.

The t_{OE} shown in Figure 14-4*a* is the time required for the output buffers to turn on after the $\overline{OE}$ signal is asserted low. This parameter, of course, assumes that

the address and the $\overline{CE}$ signal are already present when $\overline{OE}$ is asserted.

The reason that the three access times are given separately in data sheets is that the parameter which determines the maximum speed a system can operate at varies from system to system depending on the specific circuitry. In other words, t_{ACC} may be the limiting factor in one system, but in another system t_{CE} may be the limiting factor. Later in the chapter we will show you how to determine if a device is fast enough to work in a given microcomputer system.

PROGRAMMING A 2764A

The final point to discuss about a 2764A for now is how it is programmed. To program an EPROM you usually use a computer to develop a *programmer load file* which contains the sequence of data bytes you want to program in the device. You then download the file to a programmer, such as the Data I/O Model 60 shown in Figure 12-6. The programmer must have the correct personality module and be configured for the *exact* device you are programming. This is important because some devices with the same basic number but different suffix letters have different programming voltages. A 2732, for example, has a different programming voltage than a 2732A.

Newer EPROMs such as the 2764A and the 27010 have an Intelligent Identifier Mode which allows a programmer to read out a preprogrammed device identifier code stored in the device. The device identifier code is stored in a part of the device which is not accessible with normal read operations. The programmer accesses this information by raising the A9 input of the device to +12 V and reading it out from the hidden storage locations. A "smart" programmer can read out this device code and use it to automatically select the correct programming algorithm for the device.

Once the programmer is set up for the specific device and the data file has been downloaded to it, the programmer first checks to see that the device is erased and that it is capable of being programmed. If the device is erased, the programmer automatically applies the correct sequence of addresses and pulses to program the device with the data in the downloaded file. Normally you do not have to be concerned with the details of the programming algorithm because the programmer automatically

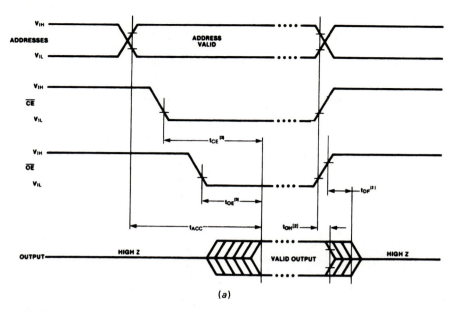

(a)

A.C. CHARACTERISTICS $0°C \leq T_A \leq +70°C$

Versions[4]	$V_{CC} \pm 5\%$	2764A-1		2764A-2 P2764A-2		2764A P2764A		2764A-3 P2764A-3		2764A-4		Unit	Test Conditions
	$V_{CC} \pm 10\%$			2764A-20		2764A-25 P2764A-25		2764A-30 P2764A-30		2764A-45			
Symbol	Parameter	Min	Max	Min	Max	Min	Max	Min	Max	Min	Max		
t_{ACC}	Address to Output Delay		180		200		250		300		450	ns	$\overline{CE} = \overline{OE} = V_{IL}$
t_{CE}	$\overline{CE}$ to Output Delay		180		200		250		300		450	ns	$\overline{OE} = V_{IL}$
t_{OE}	$\overline{OE}$ to Output Delay		65		75		100		120		150	ns	$\overline{CE} = V_{IL}$
t_{DF}[3]	$\overline{OE}$ High to Output Float	0	55	0	55	0	60	0	105	0	130	ns	$\overline{CE} = V_{IL}$
t_{OH}[3]	Output Hold from Address, $\overline{CE}$ or $\overline{OE}$ Whichever Occurred First	0		0		0		0		0		ns	$\overline{CE} = \overline{OE} = V_{IL}$

NOTES:
1. V_{CC} must be applied simultaneously or before V_{PP} and removed simultaneously or after V_{PP}.
2. V_{PP} may be connected directly to V_{CC} except during programming. The supply current would then be the sum of I_{CC} and I_{PP}. The maximum current value is with outputs O_0 to O_7 unloaded.
3. This parameter is only sampled and is not 100% tested. Output Data Float is defined as the point where data is no longer driven—see timing diagram.
4. Model Number Prefixes: No prefix = CERDIP; P = Plastic DIP.

(b)

FIGURE 14-4 Timing waveforms and timing parameters for a 2764A EPROM (a) Waveforms. (b) Timing parameters. *(Intel Corporation.)*

programs and verifies each address. However, to give you an overview of the sequence of operations the programmer does, Figure 14-5 shows a flow chart for the Intel Quick-Pulse Programming algorithm which can be used to program any of the EPROMs from the 2764A to the 27010 and many other EPROMs.

Figure 14-6a and b shows the timing waveforms and parameters for programming and verifying a byte in a 2764A. Basically the programming sequence proceeds as follows.

An address is applied to the address inputs of the 2764A, the data byte to be written at that address is applied to the *outputs* of the device, and the $\overline{CE}$ input is taken to a logic low. Also, the V_{CC} to the device is raised to +6.25 V, and 12.75 V is applied to the V_{PP} input pin. These high voltages are required to force electrons to flow

onto gates which are effectively insulated at lower voltages. After a setup time of 2 μs or more, the $\overline{PGM}$ input is pulsed low for about 100 μs. The data is held on the outputs for a hold time of 2 μs after $\overline{PGM}$ is made high, and then the data is removed. This completes the basic programming cycle.

The next step in the algorithm is to determine if the location was programmed correctly. To do this the $\overline{OE}$ pin is dropped to a logic low. After the t_{OE} access time of the device, the programmed byte will appear on the outputs of the 2764A. The programmer reads this byte and compares it with the original data byte. If the bytes are the same, the programmer goes on to program the next address. If the bytes are not the same, the programmer runs through another programming cycle at that address and checks the programmed byte again. The program-

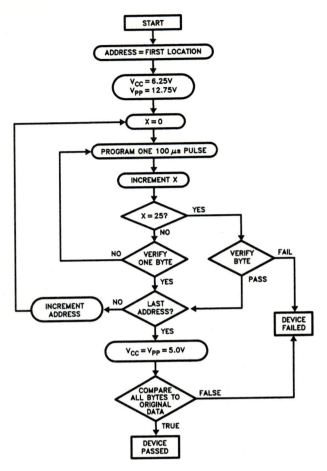

START

ADDRESS = FIRST LOCATION

$V_{CC} = 6.25V$
$V_{PP} = 12.75V$

X = 0

PROGRAM ONE 100 μs PULSE

INCREMENT X

X = 25? — YES

NO

VERIFY ONE BYTE

VERIFY BYTE — FAIL

PASS

NO

YES

INCREMENT ADDRESS — NO — LAST ADDRESS?

YES

DEVICE FAILED

$V_{CC} = V_{PP} = 5.0V$

COMPARE ALL BYTES TO ORIGINAL DATA — FALSE

TRUE

DEVICE PASSED

FIGURE 14-5   Quick-Pulse Programming algorithm for EPROMs. *(Intel Corporation.)*

mer will make up to 25 attempts to program any given byte before it gives up and sends out the "Device Failed" message.

After all locations in the EPROM are programmed and verified, the programmer drops the V_{CC} and the V_{PP} voltages to 5.0 V and checks if all bytes read out correctly at these voltages. If they do, the device passes, and if not, the "Device Failed" message is sent out.

The Quick-Pulse algorithm requires about 1 s to program each 8K bytes in an EPROM such as a 2764A, 27256A, 27010, and other devices which are designed to work with it. This is a major improvement over the previous-generation devices such as the 2716A and the 2732A which require 1.5 to 2 *minutes* to program.

EEPROMs and Flash EPROMs

We have grouped these two devices together because they are similar in structure and can both be electrically erased and reprogrammed. A major use of EEPROMs and flash EPROMs is in remote instruments that must have their operating data changed periodically. An example of this type instrument might be the postal scales in the shipping department of a company. These postal scales automatically determine the weight of a parcel, and, when given the mail zone and class, they compute the

required postage. When the postal rates change, the new rate schedule can be sent to the scale over the phone lines with a modem. The microprocessor which normally controls the scale can use the downloaded data to program the new rate table into an on-board EEPROM or flash EPROM. The main difference between an EEPROM and a flash EPROM is in how the two devices are programmed.

In an EEPROM the oxide insulating layer around each gate used to store bits is very thin. This very thin oxide layer makes it possible to move electrons on or off the gate by a mechanism called *Fowler-Nordheim tunneling*. The tunneling mechanism is used for both programming and erasing an EEPROM. A common EEPROM, the 2816 2K × 8 device, requires about 10 ms to erase and reprogram each byte.

A *flash EPROM* is programmed with the Quick-Pulse method we described for the 2764A in the previous section on EPROMs. The device is erased with the Fowler-Nordheim tunneling mechanism.

Another difference between the two types of device is that in an EEPROM you can erase and reprogram a single byte, but in a flash EPROM the entire device is electrically erased at once, so you must reprogram the entire device. Here's a summary of the characteristics of the two types.

EEPROMs are byte-programmable and require only +5 V to program, but they use more complex storage transistors, so fewer bits can fit on a chip. Another disadvantage is that most EEPROMs require 5 to 10 ms/byte to reprogram. This amounts to about 30 s for a 2K-byte device. Another problem with EEPROMs is that, depending on the particular device, they fatigue and will no longer program after 100 to 10,000 erase/write cycles.

Flash EPROMs use simpler storage transistors, so many more bits can be stored in a single device. Current devices store nearly as many bits as a similar ultraviolet-erasable EPROM. Flash EPROMs can be erased and reprogrammed much faster than most EEPROMs. An entire 32K-byte device such as the Intel 27F256, for example, can be erased in about 1 s and reprogrammed in about 4 s. The disadvantages of flash EPROMs are that they require a programming voltage of 12.00 or 12.75 V and that a single byte cannot be erased and reprogrammed.

The timing waveforms for an EEPROM or a flash EPROM read operation are the same as those shown for a 2732A EPROM in Figure 14-6a.

RAM DEVICES

Static RAMs

STRUCTURE AND ADDRESSING

As we told you before, the term *RAM* stands for random-access memory. This means that you can directly address and read any location in the device, instead of, for example, shifting data bits out in serial to get at a desired bit. The term isn't very definitive because the ROM devices we described in the preceding sections are also

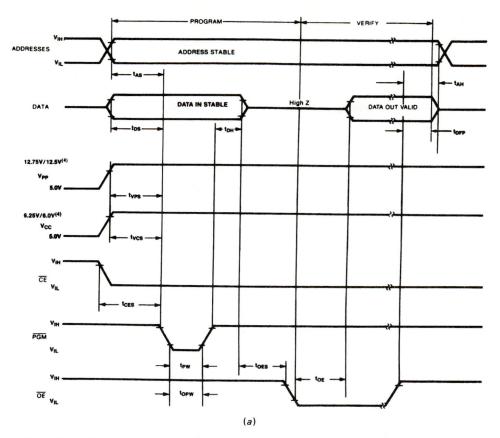

(a)

A.C. PROGRAMMING CHARACTERISTICS

T_A = 25°C ±5°C (see table 2 for V_{CC} and V_{PP} voltages)

Symbol	Parameter	Limits				Test Conditions* (see Note 1)
		Min	Typ	Max	Unit	
t_{AS}	Address Setup Time	2			μs	
t_{OES}	$\overline{OE}$ Setup Time	2			μs	
t_{DS}	Data Setup Time	2			μs	
t_{AH}	Address Hold Time	0			μs	
t_{DH}	Data Hold Time	2			μs	
t_{DFP}	$\overline{OE}$ High to Output Float Delay	0		130	ns	(See Note 3)
t_{VPS}	V_{PP} Setup Time	2			μs	
t_{VCS}	V_{CC} Setup Time	2			μs	
t_{CES}	$\overline{CE}$ Setup Time	2			μs	
t_{PW}	$\overline{PGM}$ Initial Program Pulse Width	0.95	1.0	1.05	ms	int$_e$ligent Programming
		95	100	105	μs	Quick-Pulse Programming
t_{OPW}	$\overline{PGM}$ Overprogram Pulse Width	2.85		78.75	ms	(see Note 2)
t_{OE}	Data Valid from $\overline{OE}$			150	ns	

*A.C. CONDITIONS OF TEST

Input Rise and Fall Times
(10% to 90%)..............................20 ns

Input Pulse Levels ....0.45V to 2.4V

Input Timing Reference Level ......0.8V and 2.0V

Output Timing Reference Level ......0.8V and 2.0V

NOTES:
1. V_{CC} must be applied simultaneously or before V_{PP} and removed simultaneously or after V_{PP}.
2. The length of the overprogram pulse may vary from 2.85 msec to 78.75 msec as a function of the iteration counter value X (int$_e$ligent Programming Algorithm only).
3. This parameter is only sampled and is not 100% tested. Output Float is defined as the point where data is no longer driven—see timing diagram.
4. The maximum current value is with Outputs O_0 to O_7 unloaded.

(b)

FIGURE 14-6   (a) Timing waveforms for programming an Intel 2764A EPROM. (b) Timing parameters for programming an Intel 2764A EPROM. (Intel Corporation.)

random access. A more descriptive term for the devices we discuss here would be *read/write* memories because you can write data to them, read data from them as many times as needed, and then easily write new data to them.

The question that may occur to you at this point is, "How is this different from an EEPROM or a flash EPROM which I can write data to, read data from, and write new data to?" The answer to this question is that RAMs store data bits in what is basically a flip-flop structure or as a charge on a tiny capacitor, so new data bits can be written to a storage location in a few nanoseconds and read out in about the same time. EEPROMs and flash EPROMs can be read out in a few hundred nanoseconds, but it takes between 100 μs and 10 ms to write a new data word to one, depending on the particular device. In short, then, RAMs can be written to and read from with normal microcomputer bus cycles. EEPROMs and flash EPROMs can be written to (programmed) directly in a microcomputer, but they require special timing signals because of the long write times.

RAMs are available in all common semiconductor technologies such as TTL, ECL, NMOS, CMOS, GaAs, etc. Most devices now in use are built with NMOS or CMOS technology, so for those of you who are somewhat familiar with MOS transistors, Figure 14-7 shows the basic

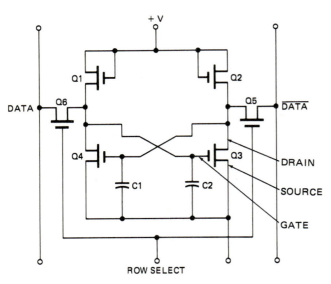

FIGURE 14-7   Circuit for six-transistor static RAM cell.

structure of a static RAM storage cell. Transistors Q1 and Q2 in Figure 14-7 are depletion-mode MOSFETs which function as active loads for the cross-coupled inverters Q3 and Q4. Data sheets commonly use the term *depletion load* to indicate that depletion-mode MOS transistors are being used as loads on the inverters instead of simple resistors. A storage cell which uses transistor loads is often called a *6T* cell, and one which uses resistors as loads is called a *4T* cell.

The cross-coupled inverters Q3 and Q4 in Figure 14-7 function as a very basic flip-flop. These transistors are enhancement-mode MOSFETs which are turned on by a high on their gates. Therefore, if one transistor is on, the other one will be off. For example, if Q4 is on, its

drain will be low. This low connected to the gate of Q3 will cause Q3 to be off, and the drain of Q3 will be high. This high connected back to the gate of Q4 will keep Q4 on, so its drain will be low, etc. Transistors Q5 and Q6 in Figure 14-7 simply function as switches which connect the outputs of the cell to the DATA and $\overline{\text{DATA}}$ lines when the cell is selected.

The basic flip-flop cell in Figure 14-7 stores one data bit. In an actual RAM device the basic storage cells are arranged in a matrix of rows and columns just as they are in ROMs. As an example of a currently available static RAM, Figure 14-8 shows a block diagram and the IEEE symbol for a Micron Technology MT5C2561. This device stores 256K bits in an array of 256 rows and 1024 columns.

When the device is selected by asserting its $\overline{\text{CE}}$ low, part of the address applied to the address inputs is decoded to enable all the cells in a row. A cell is enabled by turning on transistors, such as Q5 and Q6 in Figure 14-7, so that the outputs of the basic flip-flop are connected to the column data lines. The remainder of the applied address bits are decoded by the column decoder to route the output from the desired column to the data output of the device.

The MT5C2561 shown in Figure 14-8 decodes the column data to produce 1-bit output data words, so it is referred to as a 256K × 1 device. One reason the device is decoded this way is to reduce the required number of input and output pins. As you can see in the figure, the MT5C2561 requires only one data input and one data output. Another reason the device is decoded to produce 1-bit output words is so that several devices can easily be connected in parallel to store data words of any number of bits. Nine MT5C2561 devices, for example, can be connected in parallel to store 256K 9-bit words, or 16 devices can be connected in parallel to store 256K 16-bit words. For applications where you don't need to store 256K words, you can use an MT5C2564 which is decoded to output 64K 4-bit words or an MT5C2568 which is decoded to output 32K 8-bit words.

To write a new data word to a cell in a static RAM such as the MT5C2561, the address of the cell is applied to the address inputs, the $\overline{\text{CE}}$ input is asserted low, and the desired data bit is applied to the D_{IN} pin. When the $\overline{\text{WE}}$ input is pulsed low, the write amplifier in the device forces the flip-flop in the addressed cell to the desired state.

From the logic gates in the block diagram of the MT5C2561 in Figure 14-8, you should be able to determine the effects that the $\overline{\text{WE}}$ and $\overline{\text{CE}}$ inputs have on the operation of the device. If the $\overline{\text{CE}}$ input is high, the device will be in a low-power standby mode, and both the input buffer and the output buffer will be disabled. When the $\overline{\text{CE}}$ input is asserted low, the device will be switched to active mode for reading or writing. A high on $\overline{\text{WE}}$ will then enable the output buffer so that the data from an addressed cell will be present on the D_{OUT} pin. A low on $\overline{\text{WE}}$ will enable the input buffer so that a data bit applied to the D_{IN} pin will be written to the addressed cell. To give you a better understanding of the read and write operations of a static RAM, let's take a brief look at the timing waveforms and parameters for one.

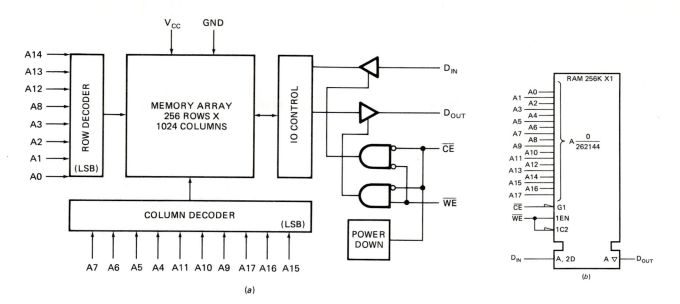

(a)

FIGURE 14-8   The MT5C2561 256K × 1 static RAM. *(a)* Block diagram. *(b)* IEEE symbol. *(Micron Technology, Inc.)*

STATIC RAM TIMING PARAMETERS

Figure 14-9 shows some of the read and write timing waveforms for the MT5C2561. Many of the times in these diagrams should be a review of the discussions for the EPROM timing diagrams in Figure 14-6. The first read waveform in Figure 14-9a assumes that the $\overline{CE}$ input is already asserted low to select the device and that the $\overline{WE}$ input is already high to enable the device for reading. The first parameter to note in Figure 14-9a is the address access time t_{AA} which is the time between a valid address being applied to the address inputs and a valid data bit reaching the data output. At the moment, the

fastest version of the MT5C2561 has a t_{AA} of 25 ns. The second parameter to note in the first read waveforms in Figure 14-9a is t_{RC}. This parameter represents the minimum time that one address must be held on the address inputs before the address of the next desired byte can be applied. This parameter, rather than the access time, is the limiting factor in applications where data bits are being read from memory in rapid succession. The minimum t_{RC} for the fastest version of the MT5C2561 is 25 ns, the same as the t_{AA}, so for this device the t_{RC} is not a problem.

The second read-cycle waveform in Figure 14-9b shows the response that the device will make when $\overline{CE}$ is as-

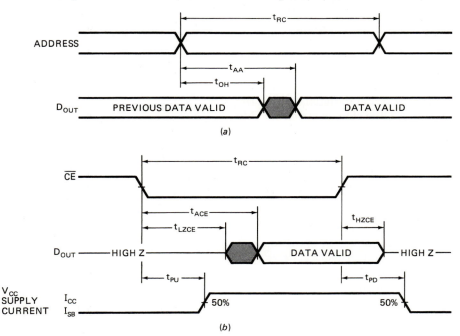

FIGURE 14-9   The MT5C2561 read-cycle waveforms. *(a)* Waveform showing address access time. *(b)* Waveform showing $\overline{CE}$ access time and power-up time. *(Micron Technology Inc.)*

serted low. These waveforms assume that an address was previously applied to the address inputs and that the $\overline{\text{WR}}$ input was asserted high to enable the device for reading. The two parameters to note on this waveform are t_{PU}, the time required for the device to switch from standby mode to active mode, and t_{ACE}, the time between $\overline{\text{CE}}$ going low and valid data appearing on the data output. The data sheet for the MT5C2561 shows the minimum t_{PU} as 0 ns, which means that switching from standby to active mode does not adversely affect the access time. Note, 0.1-μF bypass capacitors are required between V_{CC} and ground next to each device to prevent transients when the device is switched from a standby current of 10 mA to an active current of 100 mA. The t_{CE} for the MT5C2561 is 25 ns, the same as its t_{AA}.

The write-cycle waveforms for the MT5C2561 in Figure 14-10 follow the same basic sequence we described

achieve access times as low as 25 ns. For applications which require shorter access time, ECL RAMs such as the Fujitsu MB7703H-5CF 256 × 16 devices have access times as low as 5 ns. For still higher speed applications, GaAs devices such as the 12G014 256 × 4 devices from GigaBit Logic have access times as low as 1.5 ns.

An important point to make about high-speed RAMs is that although access times are usually used to compare device speeds, the really important parameters are the read- and write-cycle times. These times actually determine how fast successive locations in the device can be read from or written to. As an example of this, the GaAs 12G014 has an access time of 1.5 ns, but its maximum read- and write-cycle times are 2.5 ns. This means that in an actual system application the maximum rate at which successive data words can be written to or read from the device is 1 word every 2.5 ns or 400 MHz.

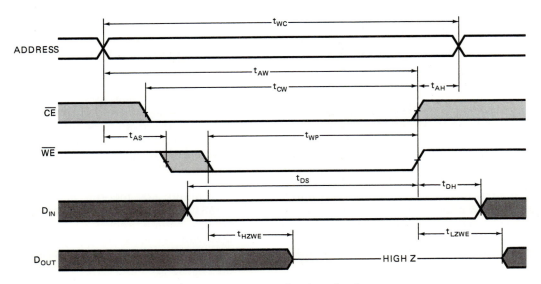

FIGURE 14-10   MT5C2561 write-cycle waveforms. *(Micron Technology Inc.)*

for a microprocessor write cycle at the end of Chapter 13. The microprocessor first outputs the address of the desired data word. As we describe in detail later in the chapter, some of the address line signals are decoded to produce the signal which asserts the $\overline{\text{CE}}$ input of the desired device low to activate it. The microprocessor then sends out the data to be written and pulses the $\overline{\text{WE}}$ input of the memory low to enable it for a write operation. A few nanoseconds after the microprocessor raises the $\overline{\text{WE}}$ input high, it can remove the address and data from the device and go on to its next operation. The write-cycle time t_{WC} for the fastest version of the MT5C2561 is 25 ns. In a later section of the chapter we show you how to determine if the read-cycle and write-cycle times for a given device are fast enough to allow it to be used in a microprocessor system without needing to insert wait states.

HIGH-SPEED STATIC RAMs

The Micron MT5C2561 static RAM we described in the preceding section uses high-speed CMOS technology to

BATTERY BACKUP FOR STATIC RAMs

Static RAMs have the advantage of fast read and write times, but they are volatile. This means that when the power is removed, the stored data is lost. One solution to this problem is to use CMOS RAMs which can easily be backed up with a small battery. An example of this type device is the Motorola MCM60L256 which stores 32K bytes. When the $\overline{\text{E}}$ input of an MCM60L256 is high, the device is switched to a standby mode which requires only a few microamps of V_{CC} supply current. This current is low enough that it can easily be supplied by a small battery or a large electrolytic capacitor. Figure 14-11 shows one simple way this backup battery or capacitor can be connected to take over when the normal power supply fails.

During normal operation the V_{CC} from the power supply provides the current for the memory and a small charging current to the backup battery or capacitor. When the power fails, a special circuit detects the failure and switches the memories to their standby mode. As soon as the V_{CC} voltage from the power supply drops

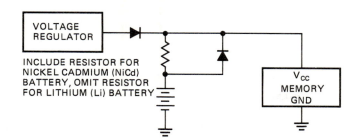

FIGURE 14-11 Battery backup circuit for CMOS static RAMs.

0.7 V below the voltage of the battery, the series diode will turn off and the backup current for the memories will be supplied by the battery.

Some CMOS SRAMs such as the Thompson-Mostek MK48T02 2K × 8 devices contain circuitry which detects when the power drops below 4.5 V and puts the device in standby mode. A tiny lithium backup battery built into the package then retains the stored data. The built-in battery has a life expectancy of about 10 years. Also included in the MK48T02 are a CMOS crystal oscillator and clock/calendar circuit which continue to operate during a power failure. The combination of functions in a device such as this makes it suitable for a small microcomputer system.

NOVRAMs

Still another solution to the volatility problem of SRAMs is a device called a *NVRAM* or *NOVRAM*. An example of this type device is the Xicor X2004M which stores 512 bytes. Each storage location in a NOVRAM consists of a static RAM cell with an EEPROM cell in parallel. During normal operation data is written to and read from the RAM cells just as it is with ordinary static RAM cells. When an external circuit detects a power failure, it asserts a NOVRAM input pin called STORE. Asserting STORE causes the NOVRAM to copy all the data bits from the RAM cells to the EEPROM cells. Since the copy process only takes a few milliseconds, it can easily be done before the power supply drops below the minimum operating voltage. When the power returns, asserting an input called RECALL will cause the NOVRAM to copy all the data bits from EEPROM cells back to the corresponding RAM cells. NOVRAMs have the advantage that they do not require a backup battery, but they have the disadvantage that their storage cells are large, so a relatively small number of bits can be stored in a single device. For applications where sizable amounts of nonvolatile memory are required, battery-backed CMOS static RAMs are usually used.

Dynamic RAMs

STORAGE MECHANISM STRUCTURE AND ADDRESSING

Static RAMs are fast and easy to interface with, but they use much more power and have only about one-fourth as many bits per chip as dynamic RAMs in the same

generation. The reason for this is that the storage cells in static RAMs use four or six transistors to store each bit, but dynamic RAMs, or "dee" RAMs as they are often called, use a single-transistor storage cell, such as that shown in Figure 14-12. The data bit is stored as a charge

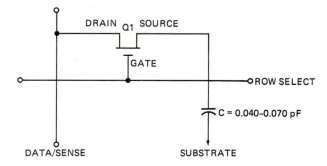

FIGURE 14-12 Dynamic RAM storage cell.

or no charge on a tiny capacitor of perhaps 0.04 to 0.07 pF connected between the source of the transistor and the substrate of the chip. The transistor in the cell functions simply as a switch which connects the storage capacitor to the DATA/SENSE line when the *row* containing the cell is addressed. These storage cells are connected in an array of rows and columns just as the cells for static RAMs and ROMs are. Each column has a sense amplifier to read out the data bit from the capacitor in a selected row and a write amplifier to write a data bit to the capacitor in a selected row.

The major drawback of dynamic RAM cells is that the charges stored on the tiny capacitors drain off after a short time. To keep the stored data valid, each cell must be *refreshed* every 2 to 8 ms, depending on the particular device. A little later we will show you how this refreshing is done. For now, let's take a look at the structure and addressing of a currently available DRAM.

Figure 14-13 shows the block diagram and the IEEE symbol for the Texas Instruments TMS4C1024 which is a 1,048,576 × 1 high-speed CMOS DRAM. The storage array in this device is decoded to produce 1-bit words, so only one data input and one data output are required. Since 1,048,576 = 2^{20}, 20 address bits are required to address one of the words stored in the device. To save pins, however, the 20 address bits are multiplexed into the device on the 10 address inputs A0 to A9. A 10-bit row address is sent into the device, and then a 10-bit column address is sent in. The timing waveforms in Figure 14-14 show the sequence of signals required to address and read a data word from a TMS4C1024.

The 10-bit row address is first applied to the address inputs A0 to A9, and, after a setup time, the ROW ADDRESS STROBE ($\overline{RAS}$) input is asserted low by external hardware. The falling edge of the $\overline{RAS}$ signal latches the row address in the DRAM. Next the 10-bit column address is applied to the address inputs of the DRAM, and, after the appropriate setup time, the COLUMN ADDRESS STROBE ($\overline{CAS}$) input is asserted low by external hardware. The falling edge of $\overline{CAS}$ latches the column address in the device. If the $\overline{W}$ input is high, the addressed data word will become valid on the data output after an ac-

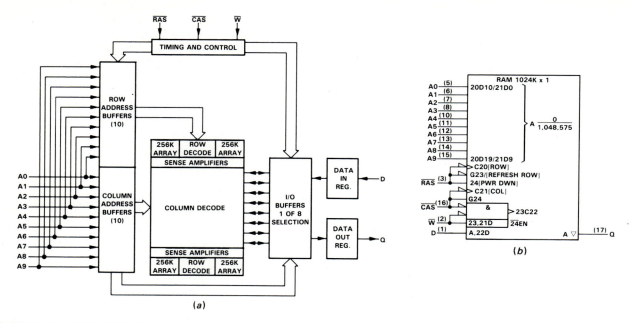

FIGURE 14-13   TMS4C1024 1,048,576-bit dynamic RAM. (a) Block diagram. (b) IEEE symbol. (Texas Instruments Inc.)

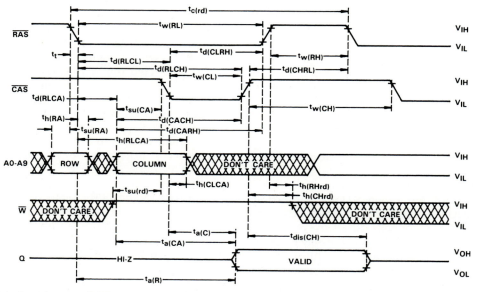

FIGURE 14-14   Read-cycle timing waveforms for TMS4C1024 DRAM. (Texas Instruments Inc.)

cess time of $t_{a(R)}$, $t_{a(CA)}$, or $t_{a(C)}$, depending on which point on the waveforms you measure from. The fastest version of the TMS4C1024 has a maximum $t_{a(R)}$ of 100 ns, which means that valid data will be present on the output 100 ns after $\overline{RAS}$ goes low.

Before another row in the device can be accessed, however, $\overline{RAS}$ has to go back high and stay high for a time labeled $t_{w(RH)}$ in Figure 14-14. This time of about 100 ns is required to *precharge* the DRAM so that it is ready to accept the next row address. The precharge time effectively adds to the access time, so the total time before the next word will be available on the output is considerably longer than the access time. The total time from the start of one read cycle to the start of the next read cycle is identified as $t_{c(rd)}$ in Figure 14-14. The fastest

version of the TMS4C1024 has a maximum $\overline{RAS}$ access time $t_{a(R)}$ of 100 ns, but due to the precharge time, its maximum read cycle time $t_{c(rd)}$ is 190 ns. For applications where data words are being read from random rows, it is this read-cycle time which limits the rate that data words can be read. As we show you later, this is an important factor when the device is connected in a microcomputer which uses a high-frequency clock. For comparison remember that a same-generation static RAM such as the MT5C2561 we discussed earlier has maximum read- and write-cycle times of 25 ns.

Figure 14-15 shows that the sequence of signals for a DRAM write operation is nearly the same as that for a read cycle. The main difference is that after the column address is sent to the address inputs, the data word to

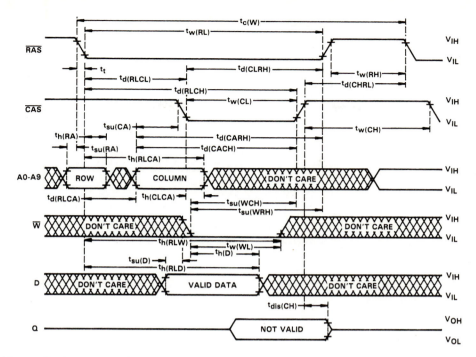

FIGURE 14-15 Write-cycle timing waveforms for TMS4C1024 DRAM. *(Texas Instruments Inc.)*

be written is applied to the data input of the device and the $\overline{W}$ input is pulsed low.

REFRESHING DYNAMIC RAMs

The storage array in the TMS4C1024 is organized as two arrays with 512 rows and 1024 columns in each. For refresh purposes, however, you can think of the device simply as one array with 512 rows and 2048 columns. All the cells in a row are refreshed by addressing that row at least once every 8 ms. Figure 14-16 shows the timing waveforms for the three common ways of doing this.

Probably the most often used refresh method is the $\overline{RAS}$-*only refresh* shown in the set of waveforms in Figure 14-16a. In this method an external counter is used to send a 10-bit row address to the A0 to A9 inputs. The $\overline{CAS}$ input is held high, and the $\overline{RAS}$ input is pulsed low for at least 100 ns. After the required precharge time, the external counter is incremented to send out the next row address and $\overline{RAS}$ is pulsed low again. The process is repeated until all 512 rows have been addressed and strobed with $\overline{RAS}$. The refresh operation on the 512 rows can be done in a *burst* of addresses and $\overline{RAS}$ pulses at the end of every 8-ms period, or it can be *distributed* throughout the 8-ms period.

Assuming a $t_{c(rd)}$ of 190 ns to address and strobe each row, a burst-mode refresh will take at least 190 ns/ row × 512 rows = 97.3 μs every 8 ms. If a distributed-mode refresh is spread evenly over the entire 8 ms, then a row will be refreshed every 8000 μs/512 rows, or every 15.6 μs. Most microcomputer systems use distributed-mode refresh because it interferes less with the operation of the microprocessor.

The second method of refreshing a DRAM such as the TMS4C1024 is the *automatic,* or $\overline{CAS}$-*before-*$\overline{RAS}$,

method shown in the waveforms in Figure 14-16b. Asserting $\overline{CAS}$ low before asserting $\overline{RAS}$ low activates an internal counter which automatically generates the row addresses needed for the refresh operation. External hardware cycles the internal counter through a series of rows by holding $\overline{CAS}$ low and pulsing $\overline{RAS}$ low with timing the same as that for the $\overline{RAS}$-only refresh. The address present on the A0 to A9 inputs is ignored during automatic refresh cycles. According to the timing parameters for the TMS4C1024, the maximum time $\overline{CAS}$ can be low is 10,000 ns, or 10 μs. Since this time is less than the time required for a burst refresh of all the rows, the external hardware must break the refresh into groups of rows. This is not difficult to do, and for small systems the automatic-refresh mode has the advantage that it requires relatively simple external hardware.

The third method of refresh, commonly called *hidden refresh,* is shown in the set of waveforms in Figure 14-16c. Hidden refresh is a variation of the automatic refresh we described in the preceding paragraph. If $\overline{CAS}$ is held low after a normal read operation, the internal row counter will be activated, and succeeding $\overline{RAS}$ pulses will step the counter through a series of rows. The data that was accessed by the initial read operation remains valid on the data output as long as $\overline{CAS}$ is low. During a standard automatic-refresh cycle such as that shown in Figure 14-16b, the data outputs will be floating.

If you need the detailed timing parameters for a specific DRAM, consult the data sheet for it. The reason that we didn't bother to show you all the detailed timing parameters here is that in most systems the timing for DRAM read, write, and refresh operations is taken care of by a device called a *DRAM refresh controller.*

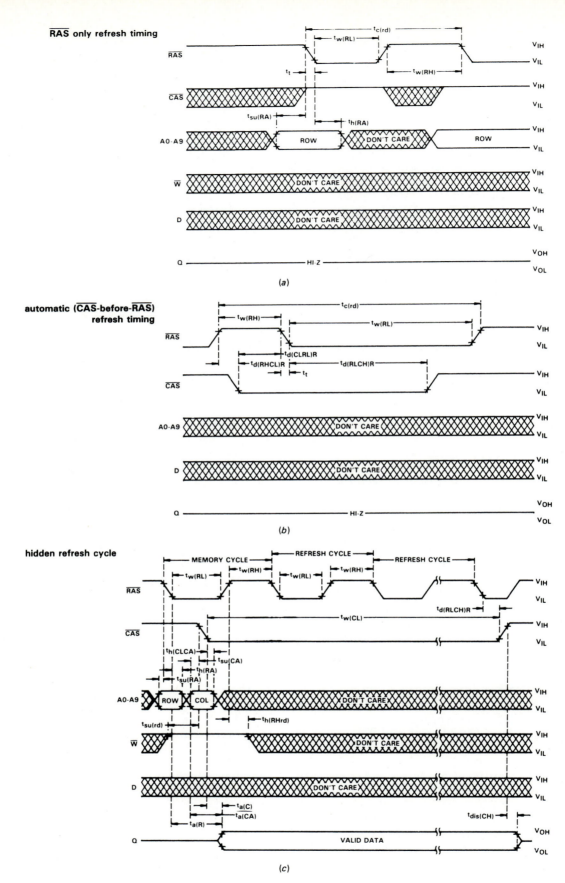

FIGURE 14-16   Refresh-cycle timing waveforms for TMS4C1024 DRAM. *(a)* $\overline{RAS}$-only refresh. *(b)* $\overline{CAS}$-before-$\overline{RAS}$, or automatic, refresh. *(c)* Hidden refresh. *(Texas Instruments Inc.)*

DRAM REFRESH CONTROLLER DEVICES

Interfacing a bank of DRAMs with, for example, a microprocessor involves three major tasks:

1. Multiplexing the two halves of an address into the DRAM devices with the correct $\overline{RAS}$ and $\overline{CAS}$ strobes.

2. Refreshing each row at least once every few milliseconds.

3. Arbitrating or settling the argument that occurs if the microprocessor attempts to access the memory devices at the same time the refresh circuitry is refreshing a row.

The question that may occur to you at this point is, "Why would anyone use dynamic RAMs which all require complicated multiplexing, refreshing, and arbitration, instead of static RAMs which don't?" There are several answers to this question.

First, DRAMs store typically four times as many bits per device as an SRAM in the same generation. Second, the per-bit cost of DRAMs is lower than the per-bit cost of SRAMs. Third, DRAMs require considerably less power per bit than SRAMs. Finally, the multiplexing, refreshing, and arbitration for a whole board of DRAMs can be handled by a single LSI refresh controller IC.

Figure 14-17 shows how an Intel 82C08 DRAM controller IC can be connected to handle the three tasks in an 8086 microcomputer system. The memory devices used in this system are Intel 51C256H 256K $\times$ 1 CMOS DRAMs. As usual for an 8086, the devices are connected as two byte-wide banks, the even bank and the odd bank.

First notice that the 8086 in this microcomputer system is being operated in its *max*imum mode. Instead of sending out control bus signals directly, an 8086 in maximum mode sends out status signals on its $\overline{S0}$ to $\overline{S2}$ lines. These signals are decoded by an external bus controller to produce the control bus signals such as ALE, DEN, DT/$\overline{R}$, $\overline{MRTC}$, $\overline{MWTC}$, $\overline{IORC}$, and $\overline{IOWC}$. The 82C08 also decodes the status signals so that it knows earlier in a machine cycle the type of operation to be carried out. By directly decoding the status signals, the 82C08 decreases the time available for the access time of the DRAMs.

The 82C08 refresh controller automatically refreshes the DRAMs by sending out row addresses with the appropriate $\overline{RAS}$ and $\overline{CAS}$ strobes for $\overline{RAS}$-only refresh. When the 8086 sends out an address to read a byte from the RAM or write a byte to it, the refresh controller multiplexes the address to the DRAMs and supplies the proper $\overline{RAS}$, $\overline{CAS}$, and $\overline{WE}$ strobes. If the controller is doing a refresh cycle when the 8086 attempts to access the RAM, a high on the $\overline{AACK}$ output on the controller will cause the 8086 to insert wait states until the refresh cycle is finished. In other words, when a dispute occurs between the controller and the 8086, the refresh controller simply makes the 8086 wait for its turn. Also, if the controller is involved in a refresh cycle, the 8086 bank select signals A0 and $\overline{BHE}$ will not be clocked through the 74LS74 D flip-flops until $\overline{AACK}$ goes low at the end of the refresh cycle. If the controller is not doing a re-

fresh cycle, the A0 and $\overline{BHE}$ signals will be immediately transferred to the DRAMs by the $\overline{AACK}$ signal.

In summary then, the refresh controller takes care of all the chores required to refresh and interface the dynamic RAM. Except for an occasional wait state, the processor is not aware that the memory is dynamic. Note that in the circuit in Figure 14-17 we added a battery backup and a CMOS oscillator so that the DRAM system will retain its stored data if the power fails.

CURRENT STATUS OF DYNAMIC RAMs

As of this writing, 1M-bit DRAMs are available, and 4M-bit devices have been announced by a few companies. No doubt 16M-bit devices will be available within the half-life of this book. To increase production yields many of these large devices are manufactured with extra or "redundant" rows and columns. As part of the test procedure, a computer-controlled laser beam is used to remove defective rows or columns by "zapping" tiny fuses.

PSEUDOSTATIC DRAMs

Another type of currently available DRAM is the *pseudostatic RAM*. The storage cells in these devices are dynamic, but simple refresh circuitry is included on the chip. The devices then appear static to the user. Examples of this type RAM are the Hitachi HM65256 series devices. These devices are configured as 32,768 $\times$ 8, so one or two devices may be all the RAM that is needed for a small microcomputer system. The HM65256-family devices have nonmultiplexed address inputs, so they do not require a multiplexer to interface them to the address bus as standard DRAMs do.

Ferroelectric RAMs

The static and dynamic RAMs we discussed in the preceding sections are both volatile. In other words, unless they are battery-backed, they lose their stored data when the power goes off. EEPROMs and flash EPROMs are nonvolatile, but it takes a relatively long time to write new data words to them. NOVRAMs are nonvolatile, but the combination RAM and EEPROM cell used to make them nonvolatile takes up a lot of chip real estate. This drastically reduces the number of bits that can be stored in a single device.

The ideal semiconductor memory would have the high density of dynamic RAMs, the fast access time of static RAMs, and the nonvolatility of flash EPROMs. A newly improved device called a *ferroelectric capacitor* may make possible memory devices which are much closer to this ideal. Here's how it works.

To refresh your memory, Figure 14-18 shows a simple capacitor circuit. When you close switch SW1 and "charge" the capacitor, an electric field is set up between the metal plates in the dielectric, as shown. Since charges are not free to move very much in an insulator, the molecules in the dielectric simply become *polarized* in the direction of the applied electric field. This is analogous to the way that a magnetized object such as a compass needle lines up with the flux lines of a magnetic field.

If switch SW1 is opened and switch SW2 is closed, the

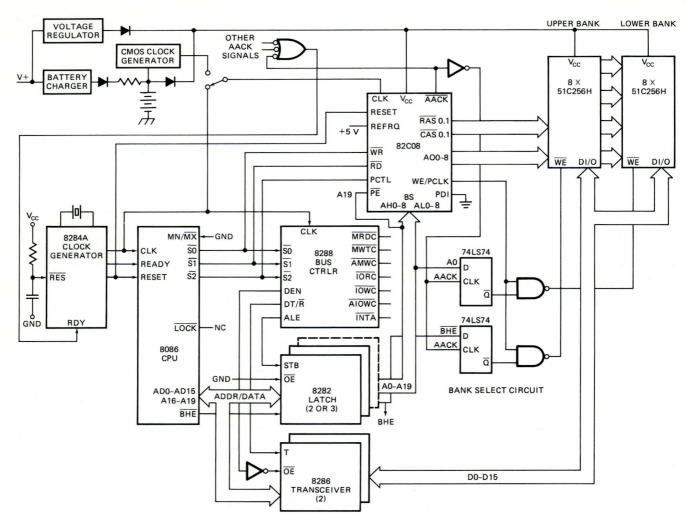

FIGURE 14-17 The 8086 microcomputer system using 82C08 DRAM controller.

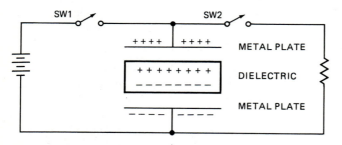

FIGURE 14-18 Simple circuit for charging and discharging a capacitor.

capacitor will discharge through the resistor. When the capacitor is fully discharged, the molecules of a standard dielectric will no longer be polarized. If the power-supply supply terminals are reversed, and switch SW1 is closed again, the molecules of the dielectric will be polarized in the opposite direction. Again, when the capacitor is discharged, the molecules of a standard dielectric will lose their polarization.

However, if the dielectric in the capacitor is a thin film of lead zirconate titanate (PZT for short) compounds, the dielectric will retain some of its polarization when the

capacitor is discharged. A capacitor with a PZT dielectric then can store a logic 0 as polarization in one direction and a logic 1 as polarization in the other direction. Since the polarization remains after the capacitor is discharged, no power is required to retain the data. In other words, the storage is nonvolatile.

A capacitor with a dielectric such as PZT is called a *ferroelectric capacitor*. The prefix *ferro-* in this name simply indicates the analogy between the response of ferromagnetic materials to a magnetic field and the response of PZT-type dielectrics to an electric field. A ferromagnetic material such as iron will become magnetized when it is placed in a magnetic field, and it will retain its magnetism when the field is removed. Likewise, a PZT-type dielectric will become polarized when placed in an electric field, and it will retain its polarization when the electric field is removed.

The polarity of the dielectric in a ferroelectric capacitor can be determined by measuring the amount of current that flows into the capacitor when an external voltage is applied. If the applied voltage is in the same direction as the polarity of the dielectric, only a small current will flow. If the applied voltage is in the opposite direction from the polarity of the dielectric, then a much larger current will flow because the polarity of the dielectric is being reversed.

Figure 14-19a shows how ferroelectric capacitors can be added to a static RAM cell to provide nonvolatile data storage. Ramtron Corporation has used this approach in its FMX801 FRAM. When an external circuit detects a power failure, it asserts the CONTROL input high. This turns on the two switch transistors and stores the logic state of the cell on the ferroelectric capacitors. When the power returns, the CONTROL line is again asserted high to transfer the stored data back to the RAM cell. The operation of this circuit then is similar to the operation of a NOVRAM.

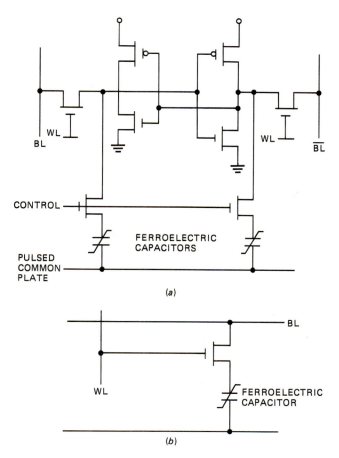

(a)

(b)

FIGURE 14-19   (a) Ramtron static RAM cell using ferroelectric capacitors to give nonvolatile storage. (b) Single-transistor nonvolatile storage cell using ferroelectric capacitor.

The reason that Ramtron used a standard SRAM cell for normal read/write operations and the ferroelectric capacitors simply as backup is that the initial ferroelectric dielectrics fatigued after about 10^{10} read/write cycles. If ferroelectric capacitors are being used only for backup, this means the device will last for several thousand years at a rate of one power failure per hour.

Ramtron Corporation and Krysalis Corporation are now both developing single-transistor ferroelectric capacitor cells, such as the one shown in Figure 14-19b. Newer ferroelectric materials can withstand 10^{12} read/write cy-

cles, so in these cells the data is written directly to and read directly from the ferroelectric capacitor. These cells can be as small as standard DRAM cells, so it should be possible to use these cells to make memories which are as dense as DRAMs, do not need refreshing, and are nonvolatile. Read- and write-cycle times for the evaluation devices now available are 100 to 200 ns, which is in the same range as current DRAMs. It should be interesting to watch the evolution of this technology.

Other Memory Technologies

Starting with stones piled up along a trail, many different technologies have been used to store digital data. In this section we give you a brief overview of a few more of these.

MAGNETIC CORE

Magnetic core is a type of high-speed read/write memory that was used in many early computers before semiconductor memory became widely available. The storage element in core memory is a small ferrite bead or core. The bead is doughnut-shaped and typically about 0.050 in in diameter. If a wire carrying an electric current is passed through the bead, as shown in Figure 14-20, a

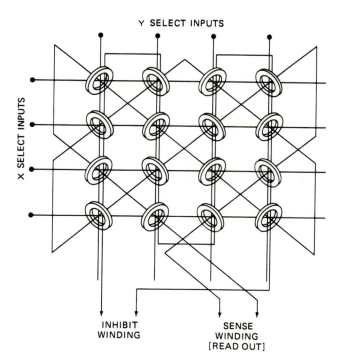

FIGURE 14-20   Magnetic-core memory plane wiring diagram.

magnetic field will be induced in the bead. The direction of the magnetic field depends on the direction of the current. A 1 can be defined as a magnetic flux in one direction and a 0 as a magnetic flux in the opposite direction. Storage is nonvolatile because the core retains its magnetism when the power is turned off.

Data is read from a core in the plane by writing a 0 to

that core. If the core previously contained a 1, the change in flux from a 1 to a 0 will induce a small voltage in the sense wire. If the core previously contained a 0, there will be no change in flux when the 0 is written, so there will be no voltage induced in the sense wire. Reading a core, then, is "destructive," and the data must be rewritten after a read.

Core memory is still used in a few applications because it is physically rugged, nonvolatile, and resistant to radiation. Problems with core memory include its large physical size per bit and the complex interface circuitry it requires.

JOSEPHSON JUNCTIONS

A new technology based on superconducting materials and electron tunneling brings the possibility of having an entire computer and several megabytes of RAM in a single 2-in (5.1-cm) cube. The cycle time of this computer may be 10 times as fast as the fastest 1988 processor, and it will use about 100 times less power. A difficulty with this computer is that the entire unit must be kept at a temperature near absolute zero, which is 273.2°C.

The storage element is a ring of superconducting metal which has no electric resistance at a temperature near absolute zero. Once a current is started flowing around this ring, it will theoretically continue to flow forever. A current flowing one way around the ring can represent a stored 0 or low, and a current flowing the other way can represent a stored 1 or high. A sense line detects which way the current is flowing.

One or two *Josephson junction* switches included in the ring are used to start or change the direction of the circulating current. A Josephson junction switch consists of two strips of superconducting metal separated by an insulator 30 to 50 Å thick. In the absence of an electric field, electrons will "tunnel" through this thin insulator in a way similar to the operation of a tunnel diode. A small electric field applied to the Josephson junction stops the tunneling. Then the junction acts as an open switch. The advantage of Josephson junction switches is that they can be turned on or off in a few picoseconds.

Since logic gates as well as memory cells can be made using this technology, it could be used to build an entire computer. Recently, substances have been found that superconduct at a temperature much nearer room temperature. This has raised hopes that it may be possible to build a Josephson junction computer that does not have to be kept in a flask of liquid helium.

CRYSTALLINE STORAGE

In their never-ending search for ways to store more data bits in smaller spaces, scientists are now researching how data might be stored in and retrieved from crystals. As we told you in Chapter 1, even a small amount of a material contains an immense number of atoms. A crystal of common table salt weighing about 58 g, for example, contains 6.02×10^{23} sodium atoms and 6.02×10^{23} chlorine atoms. If a way could be found to store one data bit on each atom or even on small groups of the atoms in a crystal such as this, it would make possible an im-

mense amount of storage in a small space. The difficulty here is that some way must be found to make individual atoms different from each other to represent 1's and 0's, and some way must be found to write the data bits and to read them out. We're not even close yet, but the possibilities are exciting.

Some of the current experimental devices use a crystalline substance with molecules of an impurity trapped in the crystal lattice. When a particular combination of laser beams is focused on the crystal, the light absorption characteristics of the impurity molecules are changed from their normal state. The difference in absorption can be detected with a low-power laser focused on the crystal. Technologies such as this are a long way from commercial reality, but they are very attractive because they can store gigabits or terabits in a very small volume, they have fast access time, and they are nonvolatile.

ADDRESS DECODING AND PRIMARY MEMORY SYSTEMS

In the preceding section of the chapter we introduced you to some commonly used types of semiconductor memory devices. Most microcomputer systems use a combination of these different devices for their primary memory. ROM devices are usually used to hold the system start-up program, and RAM devices are usually used to hold programs downloaded from secondary storage such as magnetic disks. When the microprocessor outputs a memory address, a device or circuit called an *address decoder* enables the device which contains the data word stored at that address. In this section of the chapter we discuss the operation and design of address decoders.

There are two major reasons why you need to be familiar with address decoding and the system memory map that the decoders produce. The first reason is so that if data does not read out correctly from a particular address, you can easily determine which device contains that address and replace the device. The second reason is so that you can design one if you need to add another set of memory devices to an existing system.

ROM Memory Blocks and Address Decoding

ROM IN AN 8088-BASED SYSTEM

Figure 14-21*a* shows the ROM and microprocessor section of a microcomputer based on the 8088 microprocessor. This circuit is similar to that used in the original IBM PC and its clones. The 8088 used here has the same instruction set and functions almost the same as the 8086 we described in Chapter 13. The only major difference between the 8086 and the 8088 is that the 8086 has a 16-bit data bus and the 8088 has only an 8-bit data bus. In an 8088-based system, then, the memory is configured as one bank with a maximum of 1,048,576 bytes. Remember that, as shown in Figure 13-42, the memory in an 8086 system is set up as two banks with a maximum of 512K bytes in each bank. The reason for the two banks is so that the 8086 can transfer words to and from even addresses with a single bus cycle. The

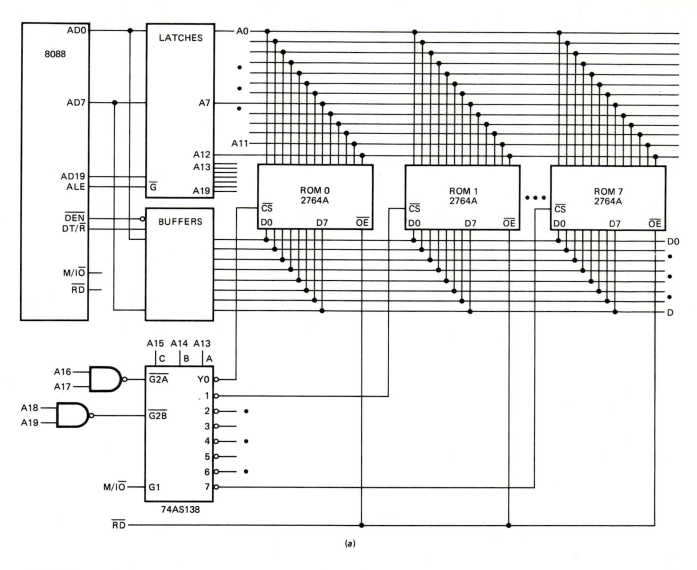

DEVICE	ADDRESS RANGE	A19 A18 A17 A16	A15 A14 A13	A12	A11 A10 A9 A8	A7 A6 A5 A4	A3 A2 A1 A0	HEX EQUIVALENT ADDRESS
ROM 0	START	1 1 1 1	0 0 0	0	0 0 0 0	0 0 0 0	0 0 0 0	F 0 0 0 0
	END	1 1 1 1	0 0 0	1	1 1 1 1	1 1 1 1	1 1 1 1	F 1 F F F
ROM 1	START	1 1 1 1	0 0 1	0	0 0 0 0	0 0 0 0	0 0 0 0	F 2 0 0 0
	END	1 1 1 1	0 0 1	1	1 1 1 1	1 1 1 1	1 1 1 1	F 3 F F F
ROM 2	START	1 1 1 1	0 1 0	0				F 4 0 0 0
	END	1 1 1 1	0 1 0	1				F 5 F F F
ROM 3	START	1 1 1 1	0 1 1	0				F 6 0 0 0
	END	1 1 1 1	0 1 1	1				F 7 F F F
ROM 4	START	1 1 1 1	1 0 0	0				F 8 0 0 0
	END	1 1 1 1	1 0 0	1				F 9 F F F
ROM 5	START	1 1 1 1	1 0 1	0				F A 0 0 0
	END	1 1 1 1	1 0 1	1				F B F F F
ROM 6	START	1 1 1 1	1 1 0	0				F C 0 0 0
	END	1 1 1 1	1 1 0	1				F D F F F
ROM 7	START	1 1 1 1	1 1 1	0	0 0 0 0	0 0 0 0	0 0 0 0	F E 0 0 0
	END	1 1 1 1	1 1 1	1	1 1 1 1	1 1 1 1	1 1 1 1	F F F F F

DECODER ENABLE INPUTS — DECODER SELECT INPUTS — ROM ADDRESS INPUTS

(b)

FIGURE 14-21 ROM section of an 8088-based microcomputer. (a) Schematic. (b) Truth table. (Intel Corporation.)

8-bit data bus of the 8088 means that it can only transfer a single byte at a time to or from memory. However, the 8-bit data bus of the 8088 made it much easier to interface with the 8-bit memory systems and port devices that were available in the early 1980s when the 8088 was designed.

In Figure 14-21a note that the multiplexed addresses from the 8088 are demultiplexed by the external latches strobed by ALE just as in an 8086 system. A 2764A is an 8K × 8 device, so it holds 2^{13}, or 8192, bytes. Since 13 address lines are required to address one of the bytes in each device, address lines A0 to A12 are connected in parallel to the address inputs of the eight 2764A ROMs. When one of the ROMs is selected by asserting its $\overline{CS}$ input low, the 13 address lines will address one of the 2^{13} data words in that ROM.

The eight data outputs from each ROM are connected in parallel on the 8-bit data bus. When the 8088 sends out an appropriate address during a memory read operation, the 74AS138 address decoder will select one of the ROMs by asserting its $\overline{CS}$ input low. The $\overline{RD}$ signal from the 8088 going low during a read cycle will assert the $\overline{OE}$ inputs of the ROMs. As we explained in the earlier discussion of the 2764A, asserting the $\overline{OE}$ input will enable the output buffers if the $\overline{CS}$ input is also asserted. The addressed byte will then be output on the data bus. The data output by a ROM goes to the 8088 through the buffers controlled by the DT/$\overline{R}$ and the $\overline{DEN}$ signals from the 8088 as we described in Chapter 13.

The ROM that will be selected is determined by the address sent out by the 8088 and by the 74AS138 which functions as an address decoder in this circuit. As you may remember from earlier chapters, a 74XX138 is a 1-of-8 low decoder. If the device is enabled by asserting its $\overline{G2A}$, $\overline{G2B}$, and G1 inputs, one of its outputs will be asserted low. The output asserted low is determined by the 3-bit code applied to the A, B, and C select inputs. For example, if 000 is applied to the select inputs, the Y0 output will go low.

In the circuit in Figure 14-21a the $\overline{G2A}$ input of the 74AS138 will be asserted when an address with A16 = 1 and A17 = 1 is output by the 8088. Likewise, the $\overline{G2B}$ input of the 74AS138 will be asserted if the 8088 outputs an address with A18 = 1 and A19 = 1. The G1 input of the 74AS138 is connected to the M/$\overline{IO}$ line from the 8088, so this input will be asserted when the 8088 is reading from memory or writing to memory. To summarize then, the 74AS138 will be enabled if the 8088 is sending out a memory address with A16 to A19 = 1111.

When the 74AS138 is enabled, the output that will be asserted low is determined by the logic levels present on the A15, A14, and A13 address lines. The easiest way to keep track of the range of addresses that will select each ROM is to make a truth table for the circuit, as shown in Figure 14-21b.

To enable the 74AS138 decoder, address lines A16 to A19 must all be 1's, so you write all 1's in these four columns of the truth table. Address lines A0 to A12 are used to address the desired byte in the selected device, so the value for these lines will depend on the specific byte addressed. The range of addresses in each device then will be from all 0's on these lines to all 1's on these lines, as shown in Figure 14-21b. A13, A14, and A15 are connected to the select inputs of the 74AS138, so they determine which ROM is selected. If the 8088 outputs 000 on these three address lines, the Y0 output of the 74AS138 will go low and assert the $\overline{CS}$ input of ROM 0. If the 8088 outputs 001 on these lines, the Y1 output of the 74AS138 will go low and assert the $\overline{CS}$ input of ROM 1. The 3-bit code on these address lines then identifies the ROM that will be selected.

The completed truth table in Figure 14-21b allows you to easily predict the starting and ending addresses for each of the ROMs. The lowest address which will select ROM 0, for example, is 1111 0000 0000 0000 0000 binary or F0000H. The highest address which will select ROM 0 is 1111 0001 1111 1111 1111 binary or F1FFFH. The range of addresses contained in ROM 0 then is F0000H to F1FFFH. Likewise, the range of addresses contained in ROM 1 is F2000H to F3FFFH. One way to describe the operation of the decoder is to say that the address lines connected to the decoder select inputs "count off" 8K blocks of ROM which each have a range of 0000H to 1FFFH. Once you see the pattern here, it is easy to complete the truth table. From the completed truth table you can see that the total address range for the set of eight ROMs is F0000H to FFFFFH or 64K bytes.

ROM is decoded to be at these high addresses in an 8086 or 8088 system because, when the power is turned on or the RESET input of the processor is asserted, the processor automatically goes to address FFFF0H to get its next instruction. The memory at this address must be ROM or some other nonvolatile memory so that the instruction is immediately available for the processor when the power is turned on.

ROM IN AN 8086-BASED SYSTEM

Remember from the discussion in Chapter 13 and Figure 13-42 that the memory in an 8086 system is set up as two banks. The lower bank contains all the bytes with even addresses. The data outputs of devices in this bank are connected on the lower half of the data bus, D0 to D7. A low on system address line A0 is part of the enabling for devices in this bank.

The upper memory bank in an 8086 system contains all the data bytes with odd addresses. The data outputs of devices in this bank are connected to the upper data bus lines D8 to D15. The 8086 $\overline{BHE}$ signal is part of the enabling for devices in this bank.

To give you an idea of how memory banks are implemented in an 8086 system, Figure 14-22a shows a schematic for the ROM section of the SDK-86 microcomputer board shown in Figure 13-40. Just from looking at the schematic symbol for the 2716 ROMs used in this system you can tell what size words and how many words are stored in each. The eight data outputs O0 to O7 tell you that the device stores 8-bit words. The 11 address inputs A0 to A10 tell you that the device stores 2^{11}, or 2048 words. As with the 2764A we discussed earlier, a low on the $\overline{CE}$ input of one of these devices switches it from standby mode to active mode. A low on the $\overline{OE}$ input enables the output buffers. Now let's look at how the 2716s are connected in this system.

System address lines A1 to A11 are connected in par-

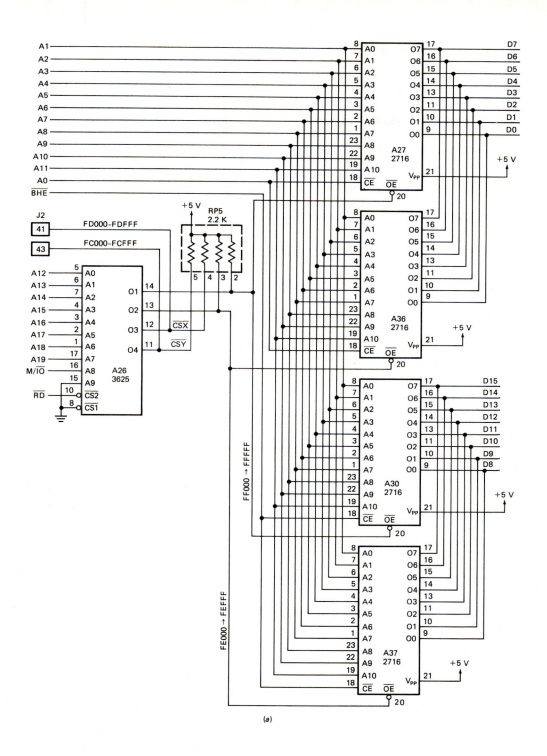

PROM INPUTS				PROM OUTPUTS				PROM ADDRESS BLOCK SELECTED	
M/IO	A14–A19	A13	A12	04	03	02	01		
1	1	1	1	1	1	1	0	FF000H	FFFFFH
1	1	1	0	1	1	0	1	FE000	FEFFF
1	1	0	1	1	0	1	1	FD000H	FDFFFH (CSX)
1	1	0	0	0	1	1	1	FC000H	FCFFFH (CSY)
ALL OTHER STATES				1	1	1	1	NONE	

(b)

FIGURE 14-22   ROM section of Intel SDK-86 microcomputer board. (a) Schematic. (b) Truth table. (Intel Corporation.)

allel to the address inputs of all the ROMs. Note that system address line A0 is not connected to one of the ROM address inputs because it is used as part of the enabling for devices in the lower bank.

The data outputs of the A27 ROM and the A36 ROM are connected in parallel to the lower half of the data bus, D0 to D7. These devices then are in the lower, or even, memory bank. The $\overline{CE}$ inputs of these two devices are connected to system address line A0, so a low on A0 will activate these two devices. The data outputs of devices A30 and A37 are connected in parallel to the upper half of the data bus, D8 to D15, so these two devices are in the odd memory bank. The $\overline{CE}$ inputs of these two devices are connected to the 8086 $\overline{BHE}$ line. As we explained in Chapter 13, the 8086 asserts its $\overline{BHE}$ line when it accesses a byte at an odd address or a word at an even address. Either type of access to the odd bank, then, will enable the A30 and A37 ROMs.

To prevent a fight between device outputs, only one device can be enabled to put data on each half of the data bus at a time. The $\overline{OE}$ inputs of the ROMs in Figure 14-22a are connected so that the O1 line from the address decoder enables a device in the lower bank and a device in the upper bank. The O2 line from the address decoder enables the other device in the lower bank and the other device in the upper bank. The address decoder is designed so that its O1 and O2 outputs are never low at the same time, so only one ROM in each bank will be enabled at a time.

The 3625 address decoder in the circuit in Figure 14-22a is a small bipolar PROM. It has open-collector outputs which require pull-up resistors, as shown. The $\overline{CS2}$ input of the 3625 is connected to the 8086 $\overline{RD}$ signal line, so this decoder will only be enabled by a read operation. This prevents the decoder from enabling the ROMs if an incorrect instruction does a write operation to a ROM address. Writing to an enabled ROM would most likely destroy its outputs.

Unlike the 74AS138 in the preceding example, there is no way you can determine the truth table for the 3625 just by looking at the schematic, so Figure 14-22b shows the truth table for it. According to this truth table, the O1 output of the address decoder will be asserted low if the $M/\overline{IO}$ line is high and system address lines A12 to A19 are all high. The address range that will assert this output then is FF000H to FFFFFH.

When the 8086 reads a byte from an even address in this range, A0 will assert the $\overline{CE}$ input of the A27 ROM and the O1 output of the decoder will assert the $\overline{OE}$ input of the A27 ROM. The addressed byte in A27 will then be output on the lower half of the data bus. If the 8086 reads a word at an even address between FF000H and FFFFFH, A0 and $\overline{BHE}$ will both be asserted, so the A27 ROM and the A30 ROM will both be switched to their active states. The O1 output from the decoder PROM will assert the $\overline{OE}$ inputs of both devices, so A27 will output the even byte on the lower half of the data bus and A30 will assert the odd addressed byte on the upper half of the data bus. When the 8086 is reading just a byte from an odd address in the range FF000H to FFFFFH, A0 will be high, $\overline{BHE}$ will be low, and the O1 output of the decoder will be low. These signals will select and enable only the A30 ROM. It will output the addressed byte on the upper half of the data bus.

A second look at the truth table in Figure 14-22a will show you that the O2 output of the decoder PROM will be asserted by addresses between FE000H and FEFFFH. As determined by the A0 and $\overline{BHE}$ connections, the even addressed bytes in this range will be in the A36 ROM and the odd addressed bytes in this range will be in the A37 ROM.

The truth table in Figure 14-22b also shows that the O3 output of the decoder PROM will be asserted by addresses between FD000H and FDFFFH and the O4 output will be asserted by addresses between FC000H and FCFFFH. These outputs are not used on the SDK-86. They are provided so that additional ROMs can be easily added if desired.

Again, the ROMs are put at this high address because the 8086 goes to address FFFF0H to get its first instruction at power-up or after a RESET. Now that you have an idea of how ROM is set up in an 8086 or 8088 microcomputer system, let's take a look at some of the ways that RAM is organized and decoded.

RAM Memory Blocks and Address Decoding

RAM connections in a microcomputer system are usually quite different from ROM connections for several reasons. First, RAMs must be enabled for both read and write operations. Second, RAM devices have 1-bit or 4-bit output words, so several devices must be connected in parallel to store 8-bit or 16-bit words. Furthermore, for dynamic RAMs, a refresh controller is usually included in the circuitry, and sometimes error-detecting and error-correcting circuitry is included. In this section we discuss some of the considerations involved in connecting and accessing RAM in microcomputer systems. We start with the simple static RAM array on the SDK-86 microcomputer board.

STATIC RAM IN AN 8086-BASED MICROCOMPUTER SYSTEM

Figure 14-23 shows the circuit diagram and truth table for the RAM section of the SDK-86 microcomputer board shown in Figure 13-40. The 2142 RAMs used here are 1K × 4 devices, so system address lines A1 to A10 are connected to the A0 to A9 address inputs of each RAM. Likewise the system $\overline{RD}$, $\overline{WR}$, and $M/\overline{IO}$ signals are connected in parallel to the indicated inputs on each device.

$\overline{RD}$ from the 8086 is connected to the OUTPUT DISABLE (OD) input on each of the 2142s. When the 8086 asserts $\overline{RD}$ low during a read operation, the output buffers on the RAMs selected by the address decoder will be enabled. $\overline{WR}$ from the 8086 is connected to the WRITE ENABLE ($\overline{WE}$) pin on each of the 2142s. Data sent out on the data bus by the 8086 during a write operation will be written to the selected devices when the 8086 asserts $\overline{WR}$ low. The $M/\overline{IO}$ signal from the 8086 is connected to the CS2 chip select input on each of the 2142s to make sure that the RAMs are only enabled for memory operations. This is necessary so that the RAM is not enabled if the 8086 sends out a port address which is the same as a RAM address.

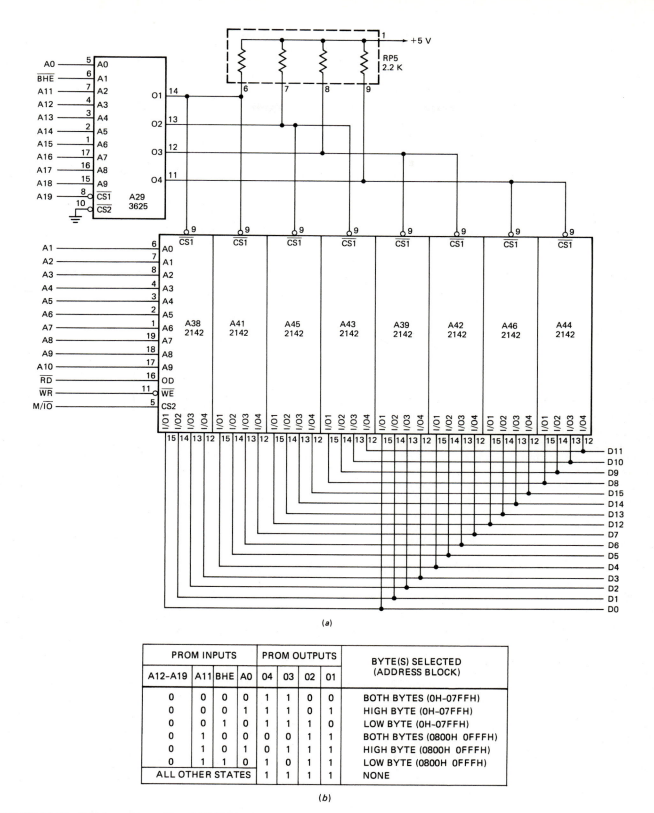

FIGURE 14-23   RAM section of Intel SDK-86 microcomputer board. *(a)* Schematic. *(b)* Truth table. *(Intel Corporation.)*

Since each RAM only stores 4-bit words, two devices must be enabled in parallel to read or write bytes, and four devices must be enabled in parallel to read or write words. Devices A38 and A41, for example, are both enabled by the O1 output of the 3625 address decoder PROM to send out or accept data on the lower half of the data bus. Likewise, devices A45 and A43 are enabled by the O2 output of the address decoder PROM to send out or receive data bytes on the upper half of the data bus. To read or write a word in this block of RAM, all four devices will be enabled. Devices A39 and A42 are enabled by the O3 output of the decoder PROM to give another block of RAM on the lower half of the data bus, and devices A46 and A44 are enabled by the O4 output of the decoder PROM to give another block of RAM on the upper half of the data bus. Now let's look at how the decoder PROM is programmed to enable each of these pairs of devices.

According to the third line of the truth table in Figure 14-23b, the O1 output of the decoder PROM will be asserted if address lines A11 to A19 are low, A0 is low, and $\overline{BHE}$ is high. The address for the first byte in the block of RAM in A38 and A41 then is 00000H. The address of the highest byte in these devices will be that where A1 to A10 are all high. This corresponds to an address of 007FEH.

The second line in the truth table in Figure 14-23b shows that the O2 output of the decoder PROM will be asserted if A11 to A19 are all low, $\overline{BHE}$ is low, and A0 is high. The address for the first odd addressed byte in the block of RAM in the A45 and A43 devices then is 00001H. Again, the address of the highest byte in these devices will be that where A1 to A10 are all high. This corresponds to a 007FFH.

The first line in the truth table in Figure 14-23b shows that if a word at address 00000H is accessed, A0 and $\overline{BHE}$ will both be low and O1 and O2 will both be asserted. The need to assert both O1 and O2 for a word operation is the reason we use a PROM for the address decoder here rather than a simple 1-of-8 low decoder such as a 74AS138. From the first three lines in the truth table, you should now be able to see that devices A38, A41, A45, and A43 provide 2K bytes of RAM in the address range 00000H to 007FFH. Another look at the truth table should show you that devices A39, A42, A46, and A44 provide another 2K-byte block of RAM in the address range 00800H to 00FFFH.

In an 8086 or an 8088 microcomputer system the RAM decoder is designed so that RAM starts at address 00000H. The reason is that when it is executing certain types of programs, the 8086 automatically goes to specific addresses between 00000H and 003FFH to get the data it needs for those programs. We describe this in detail in our earlier book, *Microprocessors and Interfacing, Programming and Harware*, McGraw-Hill, 1986.

Incidentally, whenever the 8086 uses its IN or OUT instruction to access a port device, the port address is sent out on lines AD0 to AD15 and the M/$\overline{IO}$ line is asserted low. The M/$\overline{IO}$ signal is connected to an active low enable input on the address decoder used to select the desired port device. As shown in the previous sections, M/$\overline{IO}$ being high is one of the enabling conditions for the SDK-86 ROM and RAM address decoders.

ACCESSING A DRAM BLOCK IN A MICROCOMPUTER SYSTEM

Remember from the discussion earlier in the chapter that dynamic RAMs need several operations not required by the static RAMs whose connections we showed in the preceding section. First, the two halves of an address must be multiplexed into DRAMs with the appropriate $\overline{RAS}$ and $\overline{CAS}$ strobes. Second, each location in the DRAMs must be refreshed every few milliseconds. Finally, some way must be found to settle the argument that occurs if the microprocessor attempts to access the DRAM when the refresh circuitry is refreshing a row in the DRAM. In most systems a refresh controller device is used to do these tasks. In Figure 14-17 we showed you how an 82C08 DRAM refresh controller can be used to refresh and access two banks of dynamic RAMs. Each bank in the system in Figure 14-17 contains 256K bytes, so the two banks together give a total of 512K bytes. As in the other systems we have looked at, an address decoder is used to determine where this 512K bytes is located in the 1M-byte address space of the 8086.

If you look closely at the 82C08 in Figure 14-17, you should find the PORT ENABLE ($\overline{PE}$) input. This input is asserted low to request access to the DRAM. If the 82C08 is not involved in a refresh operation when $\overline{PE}$ is asserted low, the 82C08 will multiplex the address on the address bus into the DRAMs with the appropriate $\overline{RAS}$ and $\overline{CAS}$ strobes. The 82C08 will also send out an $\overline{AACK}$ signal which clocks the 74LS74 flip-flops to transfer the A0 and $\overline{BHE}$ signals to the two memory banks. For a read operation the addressed byte or word will then be output on the data bus to the 8086. For a write operation the byte or word on the data bus will be written to the addressed locations in the DRAMs.

The output of an address decoder is connected to the $\overline{PE}$ input to assert it for the desired range of addresses. Because the DRAM banks in the circuit in Figure 14-17 are so large, the address decoding is very simple. Each bank in the circuit contains 256K bytes. Since $256K = 2^{18}$, 18 address lines are required to address one of the bytes in a bank. In most systems we connect system address lines A1 to A18 to the 82C08 address inputs and the 82C08 multiplexes these signals into the DRAMs, nine at a time. Address line A0 is used along with the $\overline{BHE}$ signal to select one or both banks. This leaves only the A19 system address line unaccounted for. If we connect the A19 address line directly to the $\overline{PE}$ input of the 82C08, then $\overline{PE}$ will be asserted whenever the 8086 outputs an address with A19 low. In other words, the $\overline{PE}$ input will be asserted when the 8086 outputs any address between 00000H and 7FFFFH. The decoder here is simply a piece of wire or circuit trace which connects A19 to the $\overline{PE}$ input. This connection puts the RAM in the lower half of the 8086 address range, which is appropriate because, as we explained before, we want ROMs containing the start-up program to be at the top of the address range.

Note that in the circuit in Figure 14-17 the status signals from the 8086 go directly to the 82C08 DRAM controller. The 82C08 decodes the status signals directly so that it does not have to wait for the 8288 bus controller to decode them. With this advance decoding, a word can be read from or written to the DRAM without needing a wait state, unless the 82C08 is doing a refresh operation. This decoding inside the 82C08 also means that we do not have to use $\overline{MRDC}$ and $\overline{MWTC}$ as part of the decoding for the $\overline{PE}$ signal to prevent a port address sent out by the 8086 from also enabling memory.

ERROR DETECTING AND CORRECTING IN RAM ARRAYS

Data words read from DRAMs are subject to two types of errors, *hard errors* and *soft errors*. Hard errors are caused by permanent device failure. This may be produced by a manufacturing defect or simply a random breakdown in the chip. Soft errors are one-time errors produced by a noise pulse in the system or by an alpha particle causing the charge to change on the tiny capacitor on which a data bit is stored. As we add larger and larger arrays of DRAMs to a system, the chance of a hard or a soft error increases sharply. This then increases the chance that the entire system will fail. It seems unreasonable that one fleeting alpha particle could possibly cause an entire system to fail. To prevent or at least reduce the chances of this kind of failure, we often add circuitry which detects and in some cases corrects errors in the data read out from DRAMs. There are several ways to do this, depending on the amount of detection and correction needed.

The simplest method for detecting a data error is with a parity-bit approach such as we described in Figure 7-21. A parity-detector device such as a 74LS280 is used to determine the parity of an 8-bit data word as it is being written to a memory location and to generate a parity bit such that the overall parity of the 8 data bits plus the parity bit is, for example, always odd. The generated parity bit for each byte is written into a separate memory device in parallel with the data byte, as shown in Figure 14-24. When the data byte and the parity bit are read from memory, the 74LS280 is used to check the overall parity of the 8 data bits and the parity bit read from memory. If the overall parity of these 9 bits is not odd as it should be, then somewhere in the read/write process an error was introduced. To let you know that an error has occurred, one of the parity outputs of the 74LS280 is connected to an interrupt input on the microprocessor. In response to an interrupt signal, the processor stops execution of the program so that it does not continue with incorrect data and sends an error message to the CRT.

This parity approach is the error-detecting method used by the microcomputers such as the IBM PC and its clones. One difficulty with the simple parity method is that it does not indicate which bit in a word is wrong so that the erroneous bit can be corrected. When a PC detects a parity error during a memory read, for example, it simple displays an error message on the CRT and stops executing whatever program it is executing. The system must be reset and the program reloaded to get it running again. Another difficulty with the simple parity method is that two or more errors in a data word may keep the correct parity and thereby not be detected.

Both the problems of the parity method are to some extent solved by complex error-detecting/correcting codes (ECCs), called *Hamming codes* after the man who did some of the original work in this area. These codes permit you to detect multiple-bit errors in a word and to correct at least 1-bit errors. Figure 14-25a shows how a device called an *error-detecting and correcting* (EDAC) unit is connected in a microcomputer system to do this.

As data words are written to memory, the EDAC generates several check bits for each word. These check bits are stored in memory along with the corresponding data word. Figure 14-25b shows the format for the data words

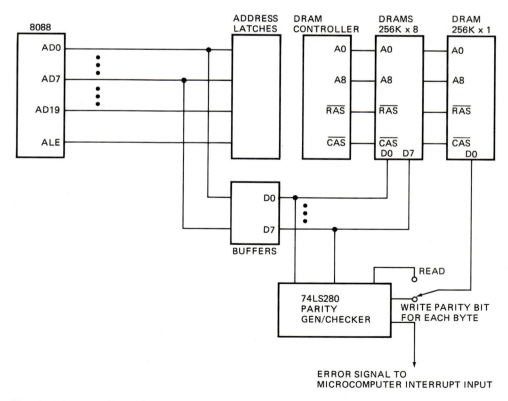

FIGURE 14-24 Circuit using parity method to detect errors in data read from dynamic RAMs.

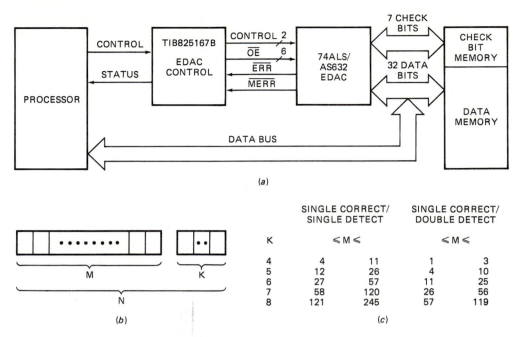

(a)

		SINGLE CORRECT/ SINGLE DETECT		SINGLE CORRECT/ DOUBLE DETECT	
	K	≤ M ≤		≤ M ≤	
	4	4	11	1	3
	5	12	26	4	10
	6	27	57	11	25
	7	58	120	26	56
	8	121	245	57	119

(b)   (c)

FIGURE 14-25   Hamming code error detecting and correcting in DRAM arrays. *(a)* Schematic. *(Texas Instruments Inc.)* *(b)* Format of data bits and check bits. *(c)* Number of check bits required for specific level of detection and correction.

and check bits. Figure 14-25c shows the number of check bits (K) needed for different-size data words (M) and different degrees of detection/correction. According to these values, 5 check bits are required to detect and correct a single-bit error in a 16-bit data word, so the total number of bits that have to be stored for each word in this case is 21. To give you enough information to correct a single error and detect 2-bit errors in a 16-bit word requires 6 encoding bits. Most current systems use the single-bit correct, double-bit detect, so 6 additional bits must be stored for each 16-bit data word. To correct a single-bit error and detect a double-bit error in a 32-bit word requires 7 check bits to be stored for each data word. To store the required check bits, more DRAMs are simply connected in parallel with those holding the data words, as shown in Figure 14-25a.

When a data word is read from memory, the EDAC recalculates the check bits for the data word read. The EDAC then Exclusive NORs these encoding bits with the encoding bits that were stored in memory for that data word. The word produced by this operation is known as a *syndrome word*. The check bits are generated in such a way that the value of this syndrome word indicates which bit, if any, is wrong in the overall data word plus encoding bits. If only a single-bit error is found, the EDAC simply corrects the bit by inverting it before passing the data word on to the microprocessor. If the EDAC finds more than one bit error in the overall data word plus check bits, it sends an error signal to an interrupt input of the microprocessor. In response to the interrupt signal the microprocessor can execute a program which makes the appropriate response to the data error.

For more information on DRAM error detecting/ correcting, consult the data sheets for error-detecting/

correcting devices such as the Intel 8206, the Texas Instruments 74AS632, and the National DP8402A. Incidentally, some of the newer DRAMs such as the Micron Technology MT4C001 1M × 1 device have built-in circuitry which will detect and correct a single-bit error without any access time penalty. If single-bit correction is not sufficient for a given system, then an external unit can be used to gain double-bit error detection.

Memory System Timing Calculations

Earlier in the chapter we promised that we would show how you determine if a given memory device is fast enough to work in a given microcomputer system without requiring the processor to insert wait states. As an example for this section, we will use the ROM section of the SDK-86 system shown in Figure 14-22.

To help you visualize what is involved in determining the timing here, Figure 14-26 shows a block diagram of the SDK-86 circuitry involved in reading a byte of data from ROM. To read a byte from a ROM, the 8086 first sends out an address. Part of the address is decoded to produce the $\overline{CS}$ signal for the ROM. The 8086 then asserts its $\overline{RD}$ signal which enables the output buffers on the ROM so that the addressed byte(s) are output on the data bus. The data byte(s) then come back to the 8086 on the data bus. Depending on the size of the system, the data may pass through buffers on its way to the 8086. The point here is that the 8086 outputs an address at a certain point in its machine cycle, and the data byte(s) from the ROM must get back to the 8086 in time to be accepted. The process is essentially a loop, and your job is to determine if the worst-case path around the loop is fast enough to meet the specified requirements of the microprocessor. Now, suppose that you want to deter-

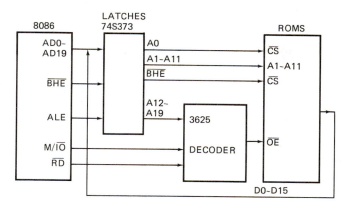

```
        LATCHES
8086    74S373              ROMS
AD0-            A0
AD19           A1-A11          CS
                              A1-A11
BHE            BHE             CS

                A12-
ALE             A19    3625

M/IO                   DECODER   OE

RD

                              D0-D15
```

FIGURE 14-26   Block diagram of SDK-86 CPU, address latches, decoder, and ROM.

mine if the 2716 EPROMs on the SDK-86 board will work correctly with no wait states if you run the 8086 with the 4.9-MHz clock.

First, you look up the access times for the 2716 EPROM in the appropriate data book. According to an Intel data book, the 2716 has a maximum address to output access time t_{ACC} of 450 ns. This means that if the 2716 is already enabled and its output buffers turned on, it will put valid data on its outputs no more than 450 ns after an address is applied to the address inputs. The 2716 data sheet also gives a chip enable to output access time t_{CE} of 450 ns. This means that if an address is already present on the address inputs of the 2716 and the output buffers are already enabled, the 2716 will put valid data on its outputs no later than 450 ns after the $\overline{CE}$ input is asserted low. A third parameter given for the 2716 in the data book is an output enable to output time t_{OE} of 120 ns maximum. This means that if the device already has an address on its address inputs and its $\overline{CE}$ input is already asserted, valid data will appear on the output pins at most 120 ns after the $\overline{OE}$ pin is asserted low.

Now that you have these three parameters for the 2716, the next step is to check that each of these three paths through the overall read loop is short enough for the device to work with the 8086 operating at 4.9 MHz. To determine this you need to look at both the block diagram in Figure 14-26 and the 8086 read-cycle timing waveforms shown in Figure 14-27.

The timing diagram in Figure 14-27 shows that the 8086 sends out $\overline{BHE}$, M/$\overline{IO}$, and an address during T_1 of the machine cycle. Note on the AD15 to AD0 line of the timing diagram that the 8086 outputs this information within a time TCLAV after the falling edge of the clock at the start of T_1. TCLAV stands for *time from clock low to address valid*. According to the 8086 data sheet, the maximum value of this time is 110 ns.

Now look farther to the right on the AD15 to AD0 lines. You should see that valid data must arrive at the 8086 from memory a time TDVCL before the falling edge of the clock at the end of T_3. TDVCL stands for *time data must be valid before clock goes low*. The data sheet gives a value of 30 ns for this parameter.

The time between the end of the TCLAV interval (time from clock low to address valid) and the start of the

TDVCL interval is the time available for getting the address to the memory, the t_{ACC} of the memory device, and the propagation delay time of the data bus buffers if any. If you look at the block diagram in Figure 14-26, you should see that the $\overline{BHE}$ signal and the A0 to A11 address information goes from the 8086 through the 74S373 latches to get to the 2716s. The 12-ns maximum propagation delay of the 74S373s then must be subtracted from the time available for the t_{ACC} of the 2716. For this example we will assume no data bus buffers because the SDK-86 doesn't have buffers in the basic ROM loop. Since the TCLAV is measured from a clock edge in T_1 and TDVCL is measured from the corresponding clock edge in T_3, you can determine the time available for t_{ACC} by subtracting TCLAV, t_{PD}, and TDVCL from the time for three clock cycles. To help you visualize these times, Figure 14-28a shows this operation in simplified form. With a 4.9-MHz clock each clock cycle will be 204 ns. Three clock cycles then total 612 ns. Subtracting a TCLAV of 110 ns, a t_{PD} of 12 ns, and a TDVCL of 30 ns leaves 460 ns available for t_{ACC} of the ROMs.

As we told you in a previous paragraph, the 2716 has a maximum t_{ACC} of 450 ns. Since this 450 ns is less than the 460 ns available, you know that the t_{ACC} of the 2716 is acceptable for the SDK-86 operating with a 4.9-MHz clock. You still, however, must check if the values of t_{CE} and t_{OE} for the 2716 are acceptable for 4.9-MHz operation.

If you look at Figure 14-22, you should see that the $\overline{CE}$ inputs of the 2716s are connected to either A0 or $\overline{BHE}$. The timing for these signals is the same as that for the addresses in the preceding section. As shown in Figure 14-28a, then, the time available for t_{CE} of the 2716 will also be 460 ns. Since the maximum t_{CE} of the 2716 is 450 ns, you know that this parameter is also acceptable for an SDK-86 operating with a 4.9-MHz clock.

The final parameter to check is t_{OE} of the 2716. According to Figure 14-22, the $\overline{OE}$ signals for the 2716s are produced by the 3625 decoder. The signals coming to this decoder are A12 to A19, M/$\overline{IO}$, and $\overline{RD}$. Look at the 8086 timing diagram in Figure 14-27 to see if you can determine which of these signals arrives last at the 3625. You should find that addresses and M/$\overline{IO}$ are sent out during T_1, but $\overline{RD}$ is not sent out until T_2. As shown by the arrow from the falling edge of the $\overline{RD}$ signal, $\overline{RD}$ going low causes the address decoder to send an $\overline{OE}$ signal to the 2716 EPROMs. Since $\overline{RD}$ is sent out so much later than addresses, it will be the limiting factor for this timing calculation. $\overline{RD}$ going low and the EPROM returning valid data must occur within the time of states T_2 and T_3. Now, according to the timing diagram, $\overline{RD}$ is sent out from the 8086 within a time TCLRL after the falling edge of the clock at the start of T_2. From the data sheet the maximum value of TCLRL is 165 ns. As we discussed before, the 8086 requires that valid data arrive on AD0 to AD15 from memory a time TDVCL before the falling edge of the clock at the end of T_3. The minimum value of TDVCL from the data sheet is 30 ns. The time between the end of the TCLRL interval and the start of the TDVCL interval is the time available for the $\overline{OE}$ signal to get produced and for the $\overline{OE}$ signal to turn on the output buffers. To determine the actual time avail-

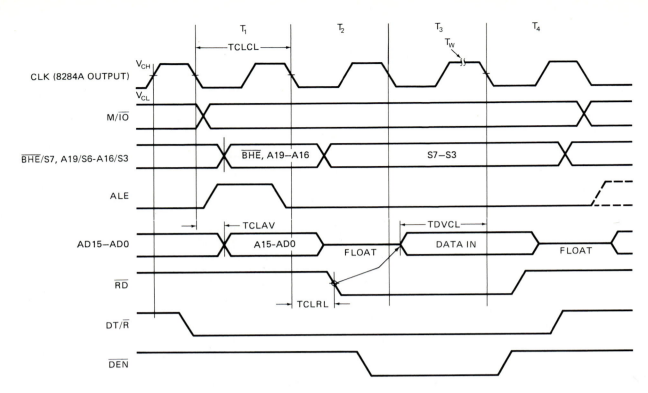

FIGURE 14-27 The 8086 read-cycle timing waveforms. *(Intel Corporation.)*

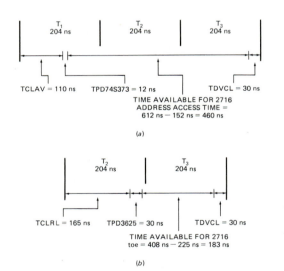

FIGURE 14-28 *(a)* Calculation of maximum allowable access times for 4.9-MHz 8086. *(a)* Time for t_{ACC} and t_{RD}. *(b)* Time for t_{OE}.

able for these operations, first compute the time for states T_2 and T_3. For a 4.9-MHz clock each clock cycle or state will be 204 ns, so the two together total 408 ns. Then subtract the TCLRL of 165 ns and the TDVCL of 30 ns. As shown by the simple diagram in Figure 14-31*b*, this leaves 213 ns available for the decoder delay and the t_{OE} of the 2716. Checking a data sheet for the 3625 would show you that it has a maximum $\overline{CS2}$ to output delay of 30 ns. Subtract this from the available 213 ns to see

how much time is left for the t_{OE} of the 2716. The result of this subtraction leaves 183 ns. As we indicated in a previous paragraph, the 2716 has a maximum t_{OE} of 120 ns. Since this time is considerably less than the 183 ns available, the 2716 has an acceptable t_{OE} value for operating on the SDK-86 board with a 4.9-MHz clock.

You have now checked all three 2716 parameters and found that all three are acceptable for an SDK-86 operating on a 4.9-MHz clock, so no wait states need to be inserted when these devices are accessed.

Here's a final point about calculating the time available for t_{ACC}, t_{CE}, and t_{OE} of some device in a system. Suppose that you want to add another pair of 2716 EPROMs in the prototyping area of the SDK-86 board, and you want to enable the outputs of these added devices with the O3 output of the 3625 ROM decoder in Figure 14-22. The timing for these added devices will be the same as for the previously discussed 2716s except that the data from the added devices must come back through some 8286 buffers. According to an 8286 data sheet, these buffers have a maximum delay of 30 ns. This 30 ns must be subtracted from the times available for t_{ACC}, t_{CE}, and t_{OE}. If you look back at our calculations of the time available for t_{ACC} in Figure 14-28*a*, you will see that we ended up with 460 ns available for t_{ACC}. Subtracting the 30 ns of buffer delay from this leaves only 430 ns, which is considerably less than the maximum t_{ACC} of 450 ns for the 2716. This tells you that, because of the buffer delay, the added 2716s are not fast enough to operate on an SDK-86 board with a 4.9-MHz clock unless wait states are inserted. To take care of this problem, the SDK-86 is designed so that any access to a memory or IO device "off board" will cause a jumper selected number of wait states to be inserted in the machine cycle. For our example here, selecting one wait state will give another 204 ns for the data to get from the 2716s to the 8086. This is more than enough time to compensate for the buffer delay, so the added 2716s will work correctly.

FLOPPY-DISK, HARD-DISK, OPTICAL-DISK, AND TAPE DATA STORAGE

In the previous sections of the chapter we have shown you how primary memory devices such as ROMs and RAMs are used to store data in microcomputer systems. Most high-performance microcomputer systems use a three- or four-level memory structure, as shown in Figure 14-29.

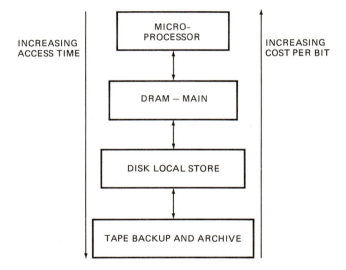

FIGURE 14-29   Three-level memory structure commonly used in microcomputer systems.

The second level in this hierarchy consists of magnetic or optical disks which store megabytes or gigabytes of data and have a much lower cost per bit than primary memory devices such as ROM or RAM. The problem with disks, however, is that their access time is typically tens to hundreds of *milliseconds*. This problem is overcome by downloading code and data from disks to main memory as needed by the microprocessor.

The lowest level in the memory hierarchy is magnetic tape storage. Magnetic tape can store massive amounts of data at a very low cost per bit, but it has a long access time. To better understand this, think of the time it takes to load a cassette audio tape in a player and scan through the tape to find a desired song. Because of their massive storage capacity and their low cost per bit, magnetic tapes are used to back up data stored on hard and floppy disks and to archive data that is seldom needed. Programs and data stored on tape can be transferred to disks or loaded directly into main memory as needed.

The point here is that a combination of memory types is used to produce a cost-effective system. Data is transferred from the slower but cheaper storage to main memory as needed by the microprocessor. In this section of the chapter we introduce you to the operation of some of the devices in the lower two levels of the memory hierarchy.

Floppy-Disk Data Storage

FLOPPY-DISK OVERVIEW

Common sizes for floppy disks are 8-, 5¼-, and 3½-in. Figure 14-30a shows the flexible protective envelope used for 8- and 5¼-in disks, and Figure 14-30b shows the rigid plastic package used for the 3½-in disks.

The disk itself is made of Mylar and coated with a magnetic material such as iron oxide or barium ferrite. The Mylar disk is only a few thousandths of an inch thick, thus the name *floppy*. When the disk is inserted in a drive unit, a spindle clamps in the large center hole or in the center hub and spins the disk at a constant speed of perhaps 300 or 360 rpm.

Data is stored on the disk in concentric, circular tracks. There is no standard number of tracks for any size disk. Older 8-in disks have about 77 tracks per side, common 5¼-in disks about 40 tracks per side, and the new 3½-in disks about 80 tracks per side. Early single-sided drives recorded data tracks on only one side of the disk. Current double-sided disk drives store data on both sides of the disk.

Data is written to or read from a track with a read/write head, such as that shown diagrammatically in Figure 14-31. During read and write operations the head is pressed against the disk through a slot in the envelope.

To write data on a track a current is passed through the coil in the head. This creates a magnetic flux in the iron core of the head. A gap in the iron core allows the magnetic flux to spill out and magnetize a section of the magnetic material along the track. Once a region on the track is magnetized in a particular direction, it retains that magnetism. The polarity of the magnetized region is determined by the direction of the current through the coil. We will say more about this later.

Data can be read from the disk with the same head. Whenever the polarity of the magnetism along the track changes as the track passes over the gap in the read/write head, a small pulse of typically a few millivolts is induced in the coil. An amplifier and comparator convert this small signal to a standard logic level pulse.

The write-protect notch in a floppy-disk envelope can be used to protect stored data from being written over, as do the knockout plastic tabs on audio tape cassettes. An LED and a phototransistor in the drive unit determine if the notch is present and enable the write circuits if it is.

On 8- and 5¼-in disks an index hole punched in the disk indicates the start of the recorded tracks. An LED and a phototransistor are used to detect when the index hole passes as the disk rotates. On 3½-in disks the start of a track is indicated by the position of the hub in the center of the disk.

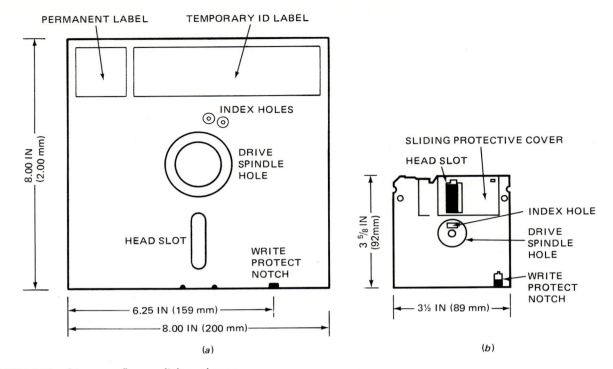

FIGURE 14-30 Common floppy disk packages.
(a) Package used for 8- and 5¼-in disks. *(b)* Hard plastic package used for 3½-in disks.

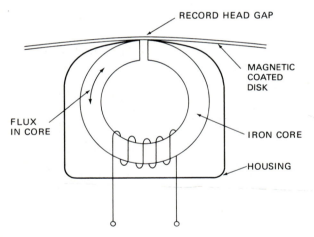

FIGURE 14-31 Diagram of read/write head used for magnetic-disk recording.

DISK-DRIVE AND HEAD-POSITIONING MECHANISMS

The motor used to spin the floppy disk is usually a dc motor whose speed is precisely controlled electronically. It takes about 250 ms for the motor to start up after a start motor command.

One common method of positioning the read/write head over a desired track on the disk is with a stepper motor. A lead screw, or a let-out–take-in steel band such as that shown in Figure 14-32, converts the rotary motion of the stepper motor to the linear motion needed to position the head over the desired track on the disk. As the stepper motor in Figure 14-32 rotates, the steel band

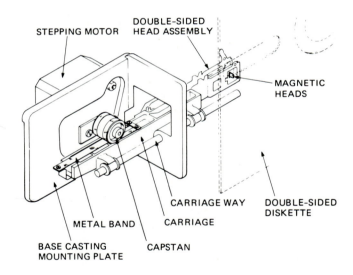

FIGURE 14-32 Common head-positioning mechanism for floppy-disk drive units.

is let out on one side of the motor pulley and pulled in on the other side. This slides the head along its carriage.

To find a given track, the motor is usually stepped to move the head to track zero near the outer edge of the disk. Then the motor is stepped the number of steps required to move the head to the desired track. It usually takes a few hundred milliseconds to position the head over a desired track.

Once the desired track is found, the head must be pressed against the disk, or *loaded*. It takes about 50 ms to load the head and allow it time to settle against the disk.

If you add the times to start the motor, position the head over the desired track, and load the head, you can see that these operations take 100 to 500 ms, depending on the particular drive. When referring to disk drives, two different access times are usually given. One access time is the time required to get the head to the required track, often called the *seek time*. The other access time is the time required to get to the first byte of a desired block of data on a track, commonly called the *latency time*. For comparing the performance of drives, the average seek time is added to the average latency time to give an *average access time*. Average access times for currently available floppy-disk drives range from 100 to 500 ms.

MAGNETIC-DISK DATA-BIT CODING: FM, MFM, AND RLL

A 1 bit is represented on magnetic floppy and hard disks as a change in the polarity of the magnetism on the track. A 0 bit is represented as no change in the polarity of the magnetism. This form of recording is often called *non-return-to-zero*, or *NRZ*, recording because the magnetic field is never zero on a recorded track. Each point on the track is always magnetized in one direction or the other. The read head produces a signal when a region where the magnetic field changes passes over it. As you read through the next section, keep in mind that a transition on the waveforms represents a 1 data bit and no transition represents a 0 data bit.

Figure 14-33 shows the three common methods used

the read head, a pulse is produced. The clock pulses read from the track are used to synchronize a phase-locked loop (PLL) circuit, such as those we described in Chapter 10. The output of the PLL is used to clock a D flip-flop at the center of the bit time where the data bits are written. A transition at the center of the bit time will produce a 1 on the input of the flip-flop, and no transition will produce a 0 on the input of the flip-flop.

The phase-locked loop is required to synchronize the readout circuits because the actual distance, and therefore time, between data bits read from an outer track is longer than it is for data bits read from an inner track. The PLL adjusts its frequency to that of the clock transitions and produces a signal which clocks the D flip-flop at the center of each bit time, regardless of the data rate. Recording clock information along with data information not only makes it possible to accurately read data from different tracks, but it also reduces the chances of a read error caused by small changes in disk speed.

The factor which determines how many data bits can be stored on a track is how close flux changes can be without interfering with each other. A disadvantage of standard F2F recording is that two transitions may be required to represent each data bit. A format which uses only half as many transitions to represent a given set of data bits is the *modified frequency-modulation (MFM)* or *double-density* recording format shown as the second waveform in Figure 14-33. The basic principle of this format is that both clock transitions and 1 data transitions are used to keep the PLL and read circuitry syn-

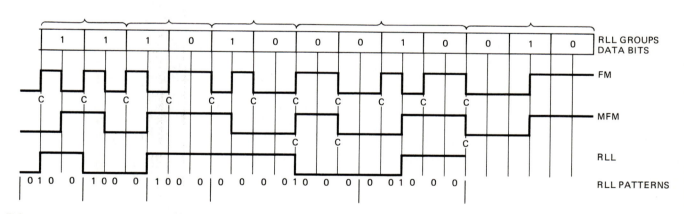

FIGURE 14-33 Comparison of FM, MFM, and RLL coding used for magnetic recording of digital data.

to code data bits on magnetic disks. The top waveform in the figure shows how the example data bits are represented in a format called *frequency-modulation (FM, F2F)* or *single-density* recording. Note the clock transition labeled C at the start of each bit cell in this format. These transitions represent the basic frequency. If the data bit in a cell time is a 1, the magnetic flux is changed again at the center of that bit time. If the data bit in a cell time is a 0, the magnetic flux is left the same at the center of that bit time. Putting in the 1 data transitions modifies the frequency, thus the name *frequency modulation*, or *F2F*.

Remember that whenever a flux transition passes over

chronized. Clock transitions are not put in unless 1 transitions do not happen to come often enough in the data to keep the PLL synchronized. A clock transition will only be put in at the start of the bit cell time if the current data bit is a 0 and the previous data bit was a 0. If you work your way across the second waveform in Figure 14-33, you should see that where two 0's occur in the data sequence, a clock transition is written at the start of the bit cell for the second 0. The MFM waveform in Figure 14-33 is shown on the same scale as the F2F waveform, but since it contains only half as many transitions as the F2F waveform for the same data string, it can be written in half as much distance on a track. This means that twice as much data can be written on a track and explains why this coding is often called *double-density.*

A still more efficient coding for recording data on floppy disks is the *RLL (run-length-limited)* 2, 7 format shown as the third waveform in Figure 14-33. In this format groups of data bits are represented by specific patterns of recorded transitions, as shown in Figure 14-34. The two 1's at the start of our data string,

DATA BIT GROUP	RLL 2, 7 CODE
1 0	1 0 0 0
1 1	0 1 0 0
0 0 0	1 0 0 1 0 0
0 1 0	0 0 1 0 0 0
0 1 1	0 0 0 1 0 0
0 0 1 0	0 0 0 0 1 0 0 0
0 0 1 1	0 0 1 0 0 1 0 0

FIGURE 14-34   RLL 2, 7 codes for data bit groups.

for example, are represented by the pattern 0100. To convert a data string to this coding, the data string is separated into groups chosen from the possibilities in the data column of Figure 14-34. The transition pattern which corresponds to that data bit combination is then written on the track. In MFM format there is at most one 0 between transitions, but in RLL 2, 7 there are between 2 and 7 zeros between transitions, depending on the sequence of data bits. As you can see in Figure 14-33, RLL requires considerably fewer transitions than MFM to represent the example data string. Fewer transitions means that the data string can be written in a shorter section of the track, or, in other words, more data can be stored on a given track. RLL coding typically increases the storage capacity about 40 percent over MFM coding.

Most of the magnetic floppy and hard disks of the last 15 years have used longitudinal recording. This means that the magnetic regions are oriented parallel to the disk surface along the track. Advances in read/write head design have made it possible to orient the magnetized regions vertically along the track. Vertical recording makes it possible to store several times as much data per track as can be stored with longitudinal recording. Toshiba and several other companies now market a disk drive which uses vertical recording to store 4M bytes of data on a single 3½-in floppy disk, such as that in Figure 14-30*b*.

Maximum data transfer rates for currently available floppy disks range from 250K to 1M *bits*/s.

FLOPPY-DISK DATA FORMATS AND ERROR DETECTION

In the preceding section we described the coding schemes commonly used to record data bits on floppy-disk or hard-disk tracks. The next level up from this is to show you the format in which blocks of data bytes are recorded along a track.

There are many slightly different formats commonly used to organize the data on a track, so we can't begin to show you all of them. However, to give you a general idea, Figure 14-35 shows an old standard, the IBM 3740 format, which is the basis of most current formats.

The start of all the data tracks is indicated by an index hole in the disk. Each track on the disk is divided into sectors. In the 3740 format a track has three types of fields. An *index field* identifies the start of the track. *ID fields* contain the track and sector identification numbers for each of the 26 data sectors on the track. Each of the 26 sectors also contains a *data field* which consists of 128 bytes of data plus 2 bytes for a CRC error checking code. As you can see, besides the bytes used to store data, many bytes are used for track and sector identification, synchronization, error checking, and buffering between sectors. *Address marks* shown at several places in this format, for example, are used to identify the start of a field. Address marks, incidentally, have an extra clock pulse recorded with their D2 data bit so that they can be distinguished from data bytes.

Two bytes at the end of each ID field and two bytes at the end of each data field are used to store *checksums* or *cyclic redundancy characters*. These are used to check for errors when the ID and the data are read out. Most disk systems use a cyclic redundancy character (CRC) method.

To produce the two CRC bytes the 128 data bytes are treated as a single large binary number. This number is divided by a constant. The 16-bit remainder from this division is written in after the data bytes as the CRC bytes. When the data bytes and the CRC bytes are read out, the CRC bytes are subtracted from the data string. The result is divided by the original constant. Since the original remainder has already been subtracted, the result of the division should be zero if the data was read out correctly. Higher-quality systems usually write data to a disk and immediately read it back to see if it was written correctly. If an error is detected, then another attempt to write can be made. If 10 write attempts are unsuccessful, then an error message can be sent to the CRT or the write can be directed to another sector on the disk.

The IBM 3740 format shown in Figure 14-38 is set up for single-density recording. An 8-in disk in this format has one index track and 76 data tracks. Since each track has 26 sectors with 128 data bytes in each sector, the total is about 250K bytes. If double-density recording is used, the capacity increases to about 500K bytes. Using both sides of the disk increases the storage to about 1M byte per disk.

The format in Figure 14-35 shows blocks of data stored in sectors on a disk track, but if you have worked with floppy disks at all, you know that your programs, letters, data, etc., are stored on the disk as files. The directory stored on track 0 of your disk contains a list of the names of all your files and some information about each file. A *file allocation table (FAT)* which is also on track 0 remembers which sectors are being used to store the blocks of each file. This two-layer scheme is very helpful because it means that you don't have to know about the tracks and sectors where your file is stored in order to access it.

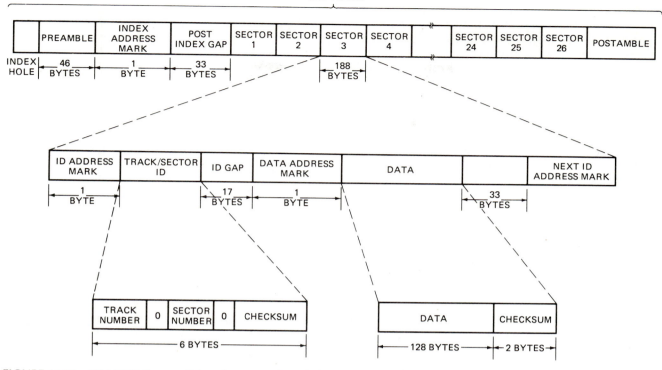

FIGURE 14-35   IBM 3740 floppy-disk soft-sectored track format (single density).

Magnetic Hard-Disk Data Storage

OVERVIEW

The floppy disks that we discussed in the previous section have the advantage that they are inexpensive and removable. However, because the disks are flexible, the data tracks cannot be put too close together. Also, the rate at which data can be read off a disk is limited because a floppy disk can only be rotated at 300 or 360 rpm. To solve these problems, we use a hard-disk system, such as that shown in Figure 14-36.

Common hard-disk sizes are 3½, 5¼, 8, 10½, 14, and 20 in. Most hard disks are permanently fastened in the

FIGURE 14-36   Cutaway photo of Conner Peripherals' CP3100 3-in, 100M-byte hard-disk drive.

drive mechanism and sealed in a dust-free package, but some systems do have removable disk packages. To increase the amount of storage per drive, several disks may be stacked with spacers between them, as shown in the Conner Peripherals' drive in Figure 14-36. A separate read/write head is used for each disk surface.

In some hard-disk drives the read/write heads are positioned over the desired track by a stepper motor and a band actuator, as shown in Figure 14-32. Most current hard-disk drives, however, use a *linear voice coil* mechanism or the *rotary voice coil* mechanism such as that shown in Figure 14-36 to position the read/write heads. This mechanism is essentially a linear motor. A feedback system adjusts the position of the head over the desired track until the strength of the signal read from the track is at its maximum.

On disk drives with more than one recording surface, the tracks are often called *cylinders* because if you mentally connect same numbered tracks on the two sides of a disk or on different disks, the result is a cylinder. The cylinder number then is the same as the track number.

The disks in a hard-disk system are made of a metal alloy coated on both sides with a magnetic material. Hard disks are more dimensionally stable than floppies, so they can be spun faster. Large hard disks are rotated at about 1000 rpm, and smaller hard disks are rotated at about 3600 rpm. Because the rotational speed is about 10 times that of a floppy disk, data is read out 10 times as fast, or about 1M to 10M *bits*/s.

The dimensional stability of hard disks also means that tracks and the bits on the tracks can be put closer together. Also, the high rotational speed and the closely spaced tracks produce a much faster access time than that for floppy disks. Currently available hard disks have average access times of under 20 ms to about 100 ms.

The high rotational speed of hard disks not only makes it possible to read and write data faster, it creates a thin cushion of air that floats the read/write head 10 microinches or so off the disk. Unless the head *crashes,* it never touches the recorded area of the disk, so disk and head wear are minimized. Hard disks must be kept in a dust-free environment because the diameter of dust and smoke particles may be 10 times the distance the head floats off the disk. If dust does get into a hard-disk system, the result will be the same as that which occurs when a plane does not fly high enough to get over some mountains. The head will crash and often destroy the data stored on the disk. When power to the drive is turned off, most hard-disk drives retract the head to a *parking zone,* where no data is recorded, and lock it in that position until power is restored.

Incidentally, hard-disk drives are sometimes called *Winchesters.* Legend has it that the name came from an early IBM dual-drive unit with a planned storage of 30M bytes per drive. The 30-30 configuration apparently reminded someone of the famous rifle, and the name stuck.

Most hard-disk drives record data bits on a disk track using the MFM coding or the RLL coding we described in the floppy-disk section of this chapter. As with floppy disks, there is no real standard for the track and sector format used. Most systems, however, format a track in a manner similar to that shown for floppy disks in Figure 14-35.

As an example of hard-disk sectoring, the hard-disk-drive unit used in the IBM PC/XT uses two double-sided hard disks with 306 tracks on each disk surface. Each track has 17 sectors with 512 bytes per sector. This adds up to about 10M bytes of data storage. Current technology allows as much as 100M bytes of data to be stored in a single 3½-in multidisk drive such as the Conners unit shown in Figure 14-36 or about 700M bytes to be stored on 5½-in drives such as the Control Data Wren unit.

MAGNETIC TAPE BACKUP SYSTEMS

To prevent data loss if a head crashes, hard-disk files are backed up on some other medium such as floppy disks or magnetic tape. The difficulty with using floppy disks for backup is the number of disks required. Backing up an 80M-byte hard disk with 1M-byte floppies would require 80 disks and considerable time shoving disks in and out. Many systems now use a high-speed magnetic tape system for backup. A typical *streaming tape system,* as these high-speed systems are often called, can dump or load the entire contents of an 80M-byte hard disk to a single tape in a few minutes.

Tape systems using small cassettes such as those used for digital audio can store up to 1G byte of data on a single tape. The difficulty with using these tapes for general storage, however, is that the serial access produces a very long latency time to get to the first byte of a stored block of data. The optical disks we discuss in the next section can store a gigabyte or so of data, but individual tracks and sectors can be accessed as they are on magnetic hard disks.

Optical-Disk Data Storage

In Chapter 11 we introduced you to the use of optical disks for storing digital audio data. The same disk technology can be used to store very large quantities of digital data for computers. One unit now available, the Maxtor Tahiti I, for example, stores up to a total of 1 Gbyte (1000M bytes) of data on a single 5½-in disk. This amount of storage corresponds to about 400,000 pages of text. Besides their ability to store large amounts of data, optical disks have the advantages that they are relatively inexpensive, immune to dust, and mostly removable. Also, since data is written on the disk and read off the disk with the light from a tiny laser diode, the read/write head does not have to touch the disk. The laser head is held in position about 0.1 in away from the disk, so there is no disk wear. Also, the increased head spacing means that the head will not crash on small dust particles and destroy the recorded data as it can with magnetic hard disks.

The disk sizes now available in different systems are 3.5, 4.72 (the compact audio disk size), 5.25, 12, and 14 in. The actual drive and head-positioning mechanisms for optical disk drives are very similar to those for magnetic hard-disk drives. A feedback system is used to precisely control the speed of the motor which rotates the disk. Some units spin the disk at a constant speed of 700 to 1200 rpm. Other systems such as those based on the compact audio (CD) format adjust the rotational speed of the disk so that the track passes under the head with a constant linear velocity. In this case the disk is rotated more slowly when reading outer tracks.

Some optical-disk systems record data in concentric tracks as magnetic disks do. Other systems, such as the CD disk systems, record data on a single spiral track in the same way a phonograph record does. A linear voice coil mechanism with feedback control is used to precisely position the read head over a desired track. The head positioning must be very precise, because the tracks on an optical disk are so narrow and so close together. The tracks are typically about 20 microinches wide and about 70 microinches between centers. This spacing allows tens of thousands of tracks to be put on a disk.

Currently the average access time for different units ranges from about 30 to 200 ms, and the data transfer rates range from 2M to 10M bits/s. Early optical disks had slower access times than comparable hard disks, but as these numbers show you, the newer units are in the same range as most current hard disks. .

Optical-disk systems are available in three basic types: read-only, write-once/read, and read/write.

Read-only systems allow only prerecorded disks to be read out. A disk which can only be read from is often called an *optical ROM,* or *OROM.* Examples of this type are the 4.7-in compact audio disks and the optical-disk encyclopedias.

Write-once-read-many, or *WORM,* systems allow you to write data to a disk, but once the data is written, it cannot be erased or changed. The stored data can be read out as many times as desired.

Read/write optical-disk systems, as the name implies, allow you to erase recorded data and write new data on a disk. The recording materials and the recording methods are different for these different types of systems.

Disks used for read-only and write-once/read systems are coated with a substance which will be altered when a high-intensity laser beam is focused on it with a lens. The principle here is similar to using a magnifying glass to burn holes in paper as you may have done in your earlier days. In some systems the focused laser light produces tiny pits along a track to represent 1's. In other systems a special metal coating is applied to the disk over a plastic polymer layer. When the laser beam is focused on a spot on the metal, heat is transferred to the polymer, causing it to give off a gas. The gas given off produces a microscopic bubble at that spot on the thin metal coating to represent a stored 1. Both these recording mechanisms are irreversible, so once written, the data can only be read. Data can be read from this type of disk using the same laser diode used for recording, but at reduced power. A system might, for example, use 25 mW for writing but only 5 mW for reading.

To read the data from the disk, the laser beam is focused on the track and a photodiode used to detect the beam reflected from the data track. A pit or bubble on the track will spread out the laser beam light so that very little of it reaches the photodiode. A spot on the track with no pit or bubble will reflect light to the photodiode. Read-only and write-once systems are less expensive than read/write systems, and for many data storage applications the inability to erase and rerecord is not a major disadvantage. One example of a WORM optical drive is the Control Data Corporation LaserDrive 510 which stores 654M bytes on a removable ANSI/ISO standard 5¼-in disk cartridge.

Most current read/write optical-disk systems use disks coated with an exotic metal alloy which has the required magnetic properties. The read/write head in this type of system has a laser diode and a coil of wire. A current is passed through the coil to produce a magnetic field perpendicular to the disk. At room temperature the applied vertical magnetic field is not strong enough to change the horizontal magnetization present on the disk. To record a 1 at a spot in a data track, a pulse of light from the laser diode is used to heat up that spot. Heating the spot makes it possible for the applied magnetic field to flip the magnetic domains around at that spot and create a tiny vertical magnet. This is called *magneto-optical recording.*

To read data from this magneto-optical-type disk, polarized laser light is focused on the track. When the polarized light reflects from one of the tiny vertical magnets representing a 1, its plane of polarization is rotated a few degrees. Special optical circuitry can detect this shift and convert the reflections from a data track to a data stream of 1's and 0's. A bit is erased by turning off the vertical magnetic field and heating the spot corresponding to that bit with the laser. When heated with no field present, the magnetism of the spot will flip around in line with the horizontal field on the disk.

An example of a currently available read/write optical drive is the Maxtor Corporation Tahiti I which stores about 600M bytes on an ANSI/ISO standard 5¼-in disk cartridge or 1G byte on a special 5¼-in cartridge.

Other techniques for producing read/write disks are now being researched intensely because of the promise this form of data storage has. A read/write disk material recently developed by Tandy, for example, does not require the magnetic field to write bits.

OPTICAL-DISK DATA FORMATS

Data is stored on optical disks in many different formats, and currently there are almost no standards. To give you an idea of the scale of the data storage, Figure 14-37a shows one format used to store digital data on the 4.7-in compact audio disks. In this format data is stored serially in one long spiral track, starting near the center of the disk. The track is divided into blocks, each containing 2K bytes of actual data. Figure 14-37b shows the format for each block. Note that a considerable number of bytes in each block are used for header, synchronization, and error detecting/correcting codes. Extensive error detection/correction is necessary to bring the error

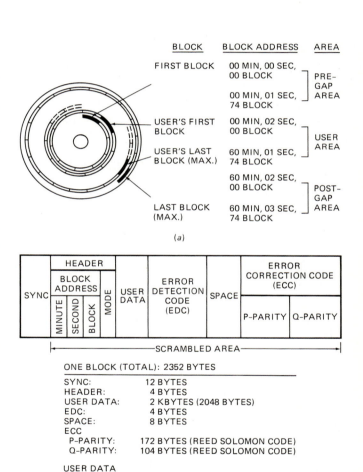

(a)

(b)

FIGURE 14-37   One format used to store digital data on compact audio disks (CDs). *(a)* Disk format. *(b)* Track format. *(Electronic Engineering Times, March 25, 1985.)*

rate down to that of magnetic disks. The position of each block on the track is identified with coordinates of minutes, seconds, and block number. As shown in Figure 14-37a, a second represents 75 blocks numbered 0 to 74. A minute represents 60 s, or a total of 4500 blocks. The entire disk represents 1 hour, or 270K blocks. Note that although data can be read out from the disk at 150K *bytes*/s (about three times the rate for floppy disks), the disk contains so much data that it takes an hour to read out all the data on the disk. Also note that a large area at the start of the track and a large area at the end of the track are used as gaps. In all, about half the total area on an optical disk is used for synchronization, identification, and error correction. This is not a big drawback because of the immense amount of data that can be stored on the disk.

THE FUTURE OF OPTICAL DISKS

For several reasons optical disks have been slow to realize the potential originally predicted for them. First, most of the optical disks that have been available for the last few years have been read-only or WORM types, so they could not easily be used as general-purpose secondary storage for microcomputers. Second, the access times and data transfer rates of the earlier units were considerably lower than those of magnetic hard disks.

Read/write optical disks are now becoming more available, and the access times of the leading read/write optical drives are very close to those of the leading magnetic hard drives. The Maxtor Tahiti I we mentioned earlier, for example, has an average access time of 30 ms, which compares favorably with the 18- to 25-ms access times of the fastest current hard-disk drives. The maximum data transfer rate for the Tahiti I is 10M bits/s, which is the same order of magnitude as the transfer rate for the leading hard-disk units. Also, optical disks have a great deal of data on each track, so often a large block of data can be read off the disk without having to pay the access time penalty of stepping to another track.

Optical disks have the further advantage that the disk cartridge is easily removable and can be locked away for safety and security purposes. There are now available several "jukebox" optical-disk systems which can hold up to 256 removable disks. Typically it takes only a few seconds to load a desired disk into the actual drive so that it can be accessed.

The potentially low cost of a few cents per megabyte and the hundreds of gigabytes of data storage possible for optical-disk systems may change the whole way our society transfers and processes information. The contents of a sizable library, for example, can be stored on a few disks. Likewise, the entire financial records of a large company can be kept on a single disk. "Expert" systems for medical diagnosis or legal defense can use a massive database stored on disk to do a more thorough analysis. Engineering workstations can use optical disks to store drawings, graphics, or IC-mask layouts. The point here is that optical disks bring directly to your desktop computer a massive database that previously was only available through a link to large mainframe computers, or in many cases was not available at all.

CHECKLIST OF IMPORTANT TERMS AND CONCEPTS IN THIS CHAPTER

Primary and secondary storage devices
Volatile and nonvolatile memory
ROM, PROM
EPROM
   Address decoder
   Access time: t_{ACC}, t_{CE}, t_{OE}
   Programmer load file
   Fowler-Nordheim tunneling
SRAM (static RAM)
   Timing parameters: t_{RC}, t_{AA}, t_{PU}, t_{ACE}, t_{WC}
   Battery backup
NOVRAM
DRAM
   Precharge
   $\overline{RAS}$-only refresh
   Automatic, $\overline{CAS}$-before-$\overline{RAS}$ refresh
   Hidden refresh
DRAM controller refresh devices
Pseudostatic RAM
Ferroelectric RAM (FRAM)
Ferroelectric capacitor
Magnetic core
Josephson junction memory
Crystalline storage
Address decoding
Hard and soft errors
Hamming codes
Syndrome word
Memory system timing calculation
Floppy disk
   Latency time
   Seek time
   Average access time
   Frequency modulation (FM, F2F)
   Single-density recording
   Modified frequency modulation (MFM)
   Double-density recording
   RLL 2, 7
   Index, ID, and data fields
   Address marks
   Checksums, cyclic redundancy characters (CRCs)
   File allocation table
Hard disk, Winchester
   Linear and rotary voice coils
   Disk cylinder
   Head crash
   Head parking zone
Streaming tape system
OROM, WORM optical disks
Read/write optical disks
Magneto-optical recording

REVIEW QUESTIONS AND PROBLEMS

1. State Moore's law of semiconductor memory.

2. Define the terms *primary storage* and *secondary storage* and give an example of each type.

3. Define the terms *volatile memory* and *nonvolatile memory*.

4. Describe how each of the following types of memory is programmed and if erasable, how it is erased:
   a. ROM
   b. PROM
   c. EPROM
   d. EEPROM
   e. Flash EPROM

5. Figure 14-38 shows a pin diagram for a common EPROM.

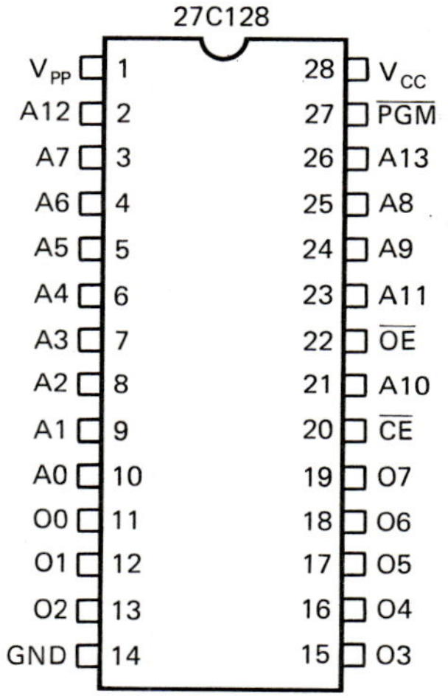

```
                27C128
        V_PP ▢ 1        28 ▢ V_CC
        A12  ▢ 2        27 ▢ PGM
        A7   ▢ 3        26 ▢ A13
        A6   ▢ 4        25 ▢ A8
        A5   ▢ 5        24 ▢ A9
        A4   ▢ 6        23 ▢ A11
        A3   ▢ 7        22 ▢ OE
        A2   ▢ 8        21 ▢ A10
        A1   ▢ 9        20 ▢ CE
        A0   ▢ 10       19 ▢ O7
        O0   ▢ 11       18 ▢ O6
        O1   ▢ 12       17 ▢ O5
        O2   ▢ 13       16 ▢ O4
        GND  ▢ 14       15 ▢ O3
```

FIGURE 14-38   Pin diagram of 27128A EPROM for Problem 5.

   a. What size data words are stored in the device?
   b. How many data words are stored in the device?
   c. How would the data organization of this device be described on a data sheet?
   d. What is the purpose of the $\overline{CE}$ input on the device?
   e. Why should a bypass capacitor be connected between V_{CC} and ground next to this device?
   f. Why do memory devices such as this usually have three-state outputs?
   g. What is the purpose of the $\overline{OE}$ input on this device?

6. The data sheet for a 27128A EPROM indicates that it has a t_{ACC} of 200 ns, a t_{CE} of 200 ns, and a t_{OE} of 75 ns.
   a. Describe what is meant by each of these times.
   b. If the access time of a memory device is not fast enough for a given microcomputer system, how can the microprocessor be "slowed down" so that the device will work in the system without changing the microprocessor clock frequency?

7. What is the major difference between an EEPROM and a flash EPROM?

8. a. Describe the circuit used to store a data bit in a static RAM cell.
   b. Describe the circuit used to store a data bit in a dynamic RAM cell.

9. The Micron Technology MT5C2561 shown in Figure 14-8 is a 256K × 1 static RAM. Draw a circuit diagram showing how eight of these devices can be connected to form a 256K × 8 memory bank for a microcomputer.

10. Figure 14-39 shows the write timing diagram and parameters for a high-speed 2147H static RAM.
    a. How long after an address is applied to the device will valid data appear on the output?
    b. What is the minimum time $\overline{CS}$ must be low before $\overline{WR}$ goes high?
    c. How long after one address is applied must you wait before you apply the next address?
    d. How long must valid data be present on the data input before the $\overline{WR}$ input goes high?
    e. Why are wait states seldom needed when reading data from or writing data to static RAMs in a microcomputer system?

11. The Micron Technology MT1259 shown in Figure 14-40 is a 256K × 1 DRAM.
    a. How many address bits are required to address one of the 256K words in the device?
    b. Draw a simple timing waveform which shows the sequence of signals used to get all these address bits into the device on just the address inputs shown on the pin diagram.
    c. About how often must each location in this device be refreshed?
    d. Describe how a $\overline{RAS}$-only refresh operation can be used to refresh this device.
    e. Define the terms *burst-mode refresh* and *distributed-mode refresh*.

12. a. Describe the three major tasks that a DRAM controller device handles in a microcomputer system.
    b. Describe how the controller arbitrates the dispute that occurs when the microprocessor attempts to read from or write to a bank of DRAMs while the controller is doing a refresh cycle.

13. a. Describe how a NOVRAM achieves nonvolatile data storage.
    b. Describe how a ferroelectric RAM, or FRAM, achieves nonvolatile data storage.

14. Figure 14-41 shows the ROM decoder for an 8088-based microcomputer system.
    a. Construct a truth table for this decoder.
    b. Show the address range enabled by each output on the decoder.
    c. What size ROMs is this decoder designed to enable?

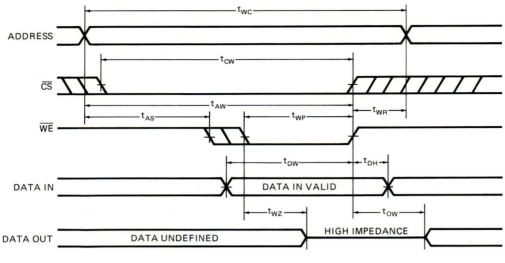

SYMBOL	PARAMETER	2147H-2		UNIT
		MIN	MAX	
t_{WC} (2)	WRITE CYCLE TIME	45		ns
t_{CW}	CHIP SELECTION TO END OF WRITE	45		ns
t_{AW}	ADDRESS VALID TO END OF WRITE	45		ns
t_{AS}	ADDRESS SETUP TIME	0		ns
t_{WP}	WRITE PULSE WIDTH	25		ns
t_{WR}	WRITE RECOVERY TIME	0		ns
t_{DW}	DATA VALID TO END OF WRITE	25		ns
t_{DH}	DATA HOLD TIME	0		ns
t_{WZ} (3)	WRITE ENABLED TO OUTPUT IN HIGH Z	0	25	ns
t_{OW} (3)	OUTPUT ACTIVE FROM END OF WRITE	0		ns

FIGURE 14-39 Diagrams for Problem 10. *(a)* Write timing diagram for 2147H. *(b)* Parameters for 2147H high-speed static RAM. *(Intel Corporation.)*

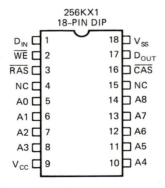

FIGURE 14-40   Pin diagram for Micron Technology MT1259 256K × 1 DRAM for Problem 11.

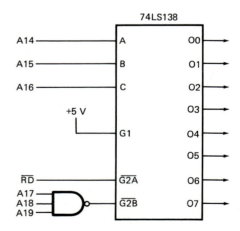

FIGURE 14-41   ROM decoder for Problem 14.

d. Explain why the $\overline{RD}$ signal from the microprocessor is connected to one of the enable inputs on the decoder.

e. Explain why the ROM decoder is designed to put the ROMs in the system address space at the locations you determined in part *b.*

15. Using a 74LS138, design a RAM address decoder which enables one of eight 4K × 8 static RAMs in an 8088 system. The starting address for the first RAM should be 00000H.

16. Figure 14-22*a* shows the ROM section of the SDK-86 microcomputer board.

a. Explain why system address line A0 is connected to two devices and the 8086 $\overline{BHE}$ signal is connected to the other two devices.

b. With the help of the truth table in Figure 14-22*b*, determine which device will be enabled when the 8086 reads a byte from address FFF41H.

c. Why can't a device such as the 74LS138 be used as an address decoder in an 8086 system such as this?

17. Figure 14-23 shows the RAM section of the SDK-86 microcomputer board.
   a. Why is the 8086 M/$\overline{\text{IO}}$ signal connected to one of the enable inputs on each of the RAMs in this system?
   b. Why are two RAM devices connected to each of the decoder outputs?
   c. When the 8086 reads out a word from address 00102H, it reads out 557FH instead of the correct word of 55FFH. Which device in the system does this point to as being bad?

18. Define the terms *hard error* and *soft error* with reference to DRAMs.

19. *a.* Describe how parity is used to check for RAM data errors in microcomputers such as the IBM PC.
   b. What are the major problems with this method of error detection?

20. *a.* Describe how Hamming codes are used to detect and correct errors in digital data words.
   b. How many check bits must be stored with a 32-bit data word to detect/correct a single-bit error and to detect a double-bit error?

21. The 27128-25 is a 16K × 8 EPROM with a t_{ACC} of 250 ns maximum, a t_{CE} of 250 ns maximum, and a t_{OE} of 100 ns maximum. Will this device work correctly without wait states in an 8-MHz 8086-2 system with circuit connections, such as those in the SDK-86 schematics shown in Figure 14-22? Assume the address latches have a propagation delay of 12 ns and the address decoder has a delay of 30 ns.

22. *a.* Describe how a read/write head is used to record and read back data from floppy and hard disks.
   b. Describe how the read/write head is positioned over the desired track on a disk.
   c. What is the advantage of MFM and RLL coding over standard F2F coding for recording data bits on magnetic disks?
   d. Why is a phase-locked loop circuit required to generate a data window signal for reading data from a magnetic-disk track?

23. Explain why magnetic hard disks can store more data than floppy disks and why the data can be read from a hard disk at a higher rate than from a floppy disk.

24. *a.* Describe how data is stored on WORM-type optical disks.
   b. Describe how data is stored on magneto-optical read/write optical disks.
   c. Describe the main advantages of optical disks over magnetic hard disks.
   d. A human brain can store about 10^{10} bits of data, and on good days it has an access time of about a second. Compare these parameters to those of an optical disk, such as the Maxtor Tahiti I described in the chapter.

25. Figure 14-42 shows the pin diagram for a 2732A EPROM.
   a. Draw a circuit showing how the outputs of three cascaded 74LS193s can be connected to the address inputs of the 2732A to cycle the device through all its possible addresses. Also make the connections for the $\overline{\text{CE}}$ input and the $\overline{\text{OE}}$ input so that the device is in its active mode and its outputs are enabled.

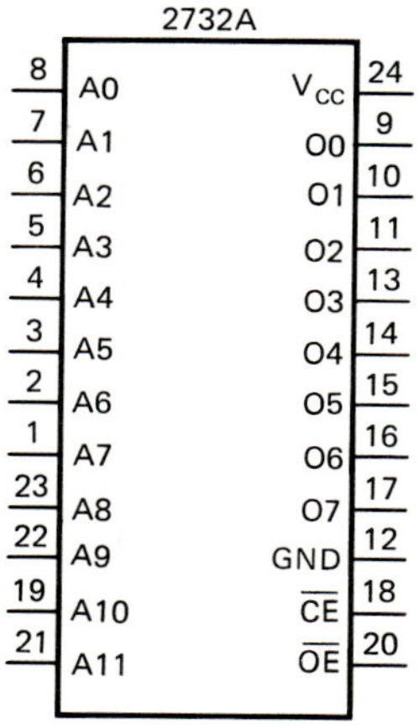

FIGURE 14-42   Pin diagram of 2732A EPROM for Problem 25.

   b. Describe how you would connect a 24-input logic analyzer to this circuit and how you would set up the analyzer to get a trace of the sequence of data words stored in the device.
   c. Describe how you would clock and trigger the analyzer to produce a trace which would allow you to determine the access time for the device.

26. In Chapter 10 we showed you how a phase-locked loop circuit can be used to produce a signal with a frequency N times the frequency of some reference frequency. In Chapter 11 we showed you how a D/A converter such as that used in a digital audio system can generate any waveform by applying a sequence of data words to its inputs. To directly synthesize any waveform, the data values for one cycle of a desired waveform are stored in sequential addresses in a ROM. A counter is connected to the address inputs of the ROM, and the data outputs of the ROM are connected to the inputs of a D/A converter. As the counter is clocked, the ROM outputs cycle through the data points for the waveform over and over. The output of the D/A converter then generates the desired waveform on a point-by-point basis. Different waveforms can be generated by storing different data sequences in different sections of the ROM. Switches connected to higher address lines can be used to select the desired waveform points.
   a. Draw a circuit which shows how an 8-bit synchronous counter can be connected to the lower eight address lines of the 2732A shown in

Figure 14-42. Connect the higher address bits and the enable inputs to ground. Also in the diagram show how an 8-bit D/A converter, such as the MC1408 shown in Figure 11-29, can be connected to the outputs of the ROM.

b. Use a calculator to determine the two-digit hexadecimal equivalent for each of the 256 data points for one cycle of a sine wave. (A programmable calculator and a simple BASIC program does this easily.) Remember that since the range of the D/A converter in Figure 11-29 is centered on 0 V, a value of 00H produces the most negative point on the waveform, a value of 80H produces the zero crossing point, and a value of FFH produces the most positive point.

c. If you have an EPROM programmer, program a 2732A, build the circuit, and experiment with it. As suggested above, you might want to program different sections of the EPROM for different output waveforms and select the desired waveform with switches or a decoded keypad connected to the higher address lines.

d. What determines the frequency of the output sine wave?

e. Assuming the D/A is fast enough, what is the maximum frequency that a direct digital synthesizer with 256 data points such as this could produce if a GaAs RAM with a cycle time of 2.5 ns is used to hold the data values for the waveform?

15 Computer-Aided Engineering of Digital Systems

The largest commonly available ICs now contain 3 to 4 million transistors. Moore's law and many electronics industry analysts predict that by the year 2000, ICs with a billion or more transistors will be common. This number of transistors will make it possible to put an almost unbelievable amount of circuitry in one IC. To design even the million-transistor ICs of today, however, would require several thousand worker-years of effort using manual techniques. Furthermore, it is a major problem to get complex ICs such as these to work together correctly in a circuit. To solve these problems we use computer programs for a large part of the design and testing of ICs and systems.

In Chapter 5 we showed you how a computer program can be used to draw schematics for digital circuits. Then in Chapter 12 we showed you how a computer program such as ABEL can be used to produce fusemaps which implement combinational and sequential circuits in PALs. These two examples are just a small part of how computers are now used in the design and testing of digital systems. In more and more companies the entire design, prototyping, manufacturing, and testing process for a digital instrument is being done with the help of a series of computer programs. The term *computer-aided engineering* (*CAE*) has been used in the past to describe the use of these tools, but now we commonly use the term *electronic design automation*, or *EDA*, instead of the more general term CAE. The EDA term gives a better indication of the extent to which the currently available tools automate parts of the design process.

The first section of this chapter is written as a standalone overview of the entire EDA process and as an introduction to how computer-based "tools" are used in the design of circuits, systems, ICs, printed-circuit boards, and instrument packages. This section also includes some predictions on the effects that these tools will have on the roles of engineers and engineering technicians in the 1990s and beyond. The remaining sections of the chapter describe the use of some of the major types of EDA tools in greater detail.

The second major section of the chapter shows you how an IBM PC-based simulator program can be used to software-breadboard combinational and sequential circuits. The third section of the chapter introduces you to the use of a program called a fault simulator and describes the techniques used to make circuits and systems testable. In the fourth section of the chapter we describe some of the tools and techniques used to design custom application-specific integrated circuits, or ASICs. Finally, at the end of the chapter we show you how a computer program can be used to do most of the work in laying out a PC board for a circuit.

OBJECTIVES

At the conclusion of this chapter you should be able to:

1. Describe the major types of engineering tools now available to design, prototype, and test digital ICs and systems.

2. Describe a typical sequence of steps in the design, prototyping, and testing of a complex digital system or IC using the design automation approach.

3. Describe how design automation tools are changing the roles of engineers and engineering technicians.

4. Describe how a program such as the OrCAD/VST simulator or Mentor Graphics QuickSim is used to simulate the operation of a simple digital circuit.

5. Describe how a fault simulation is carried out on a digital circuit.

6. Describe some simple methods that are used to make digital circuits more easily testable.

7. Describe the tools and techniques currently used to design custom application-specific integrated circuits (ASICs) with logic cell arrays, gate arrays, standard cells, compiled cells, and full-custom design.

8. Describe how a PC board is developed using a PC layout program.

9. Describe how surface-mount components are attached to PC boards and how they are removed when necessary.

OVERVIEW OF DESIGN AUTOMATION APPROACH TO DIGITAL DESIGN

This chapter is written in a top-down manner so that you get an overview before we dig into some specific design automation tools and examples of their use. To help with this overview and introduce you to the commonly used terms, Figure 15-1 summarizes the entire automated design process in diagram form. In this section of the chapter we will show you how a product is moved through this process, and in later sections of the chapter we will show you detailed examples of some of the steps. As we work our way through this process, keep in mind that the entire design process is a sequence of REPEAT-UNTIL loops. Each part of the design process is repeated until the design passes the exit tests for that part.

The Design Review Committee and Design Overview

As we emphasized in Chapter 12, the first and most important step in the design of any digital system is to think very carefully about what you want the instrument to do. In most companies a new product is now defined by a team consisting of design engineers, marketing/sales representatives, mechanical engineers, and production engineers. This team approach is necessary so that the product can be designed using current technology, man-

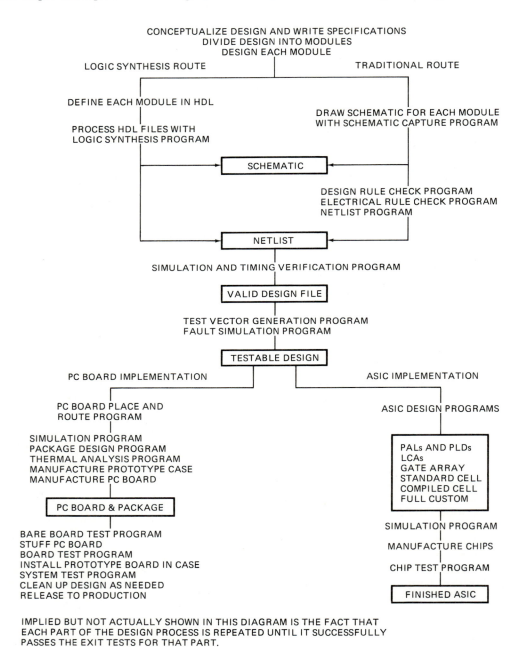

FIGURE 15-1 Summary of automated design process.

ufactured and tested with minimal problems, and marketed successfully.

For an example of this process, suppose that by talking to customers, the marketing representatives in a company determine that there is a need for an inexpensive frequency counter suitable for use in high school and college electronics laboratories. After some discussion the design review committee might arrive at an initial specification list as follows:

- Frequency measurement range of 0 to 100 MHz (optional prescaler to extend input range to 2.0 GHz)
- Period measurement range 0 to 10,000 s
- Sine wave or logic signal input
- Adjustable trigger input level
- Seven display digits

Once the specifications for the new instrument are agreed upon, the design engineers then think about how the circuit for it can be implemented.

Dividing the System into Modules

The next step in the design process is to partition the overall instrument design into major functional blocks, or modules, such as display, counters, timebase, input-signal conditioning circuitry, control logic, and power supply. Each module can then be individually designed or even assigned to different designers.

Initial Design of the Circuit Modules

The next step in the design process is to analyze each functional block to determine how it can best be implemented. To minimize the size and therefore the cost of the power supply, the decision would probably be made to use CMOS devices for as much of the circuitry as possible. After working out the basic design of each module, the design engineers have two choices of how to develop a schematic for each.

THE TRADITIONAL APPROACH

The traditional approach is to design each module at the specific IC level with the help of a 74HC data book, Karnaugh maps, and timing diagrams. When the design is roughed out, the designer uses a schematic capture program such as the OrCAD/SDT III program we discussed in Chapter 5 to draw a schematic for it. For this frequency counter the schematic would probably consist of the timebase and control circuitry from Figure 9-17 and the counter chain and multiplexed display from the circuit in Figure 9-22.

When the schematic design file is completed, it is processed by a program called a *design rule checker,* or *DRC,* which checks that there are no duplicate symbols, overlapped lines, or dangling lines. This step is similar to checking a text file with a spelling checker program.

After the schematic design file passes the DRC check, it is processed by a program called an *electrical rule checker,* or *ERC,* which checks for wiring errors such as two outputs connected together, an output connected to V_{CC}, etc.

When the schematic design file for a module passes the ERC test, a *netlist* program produces a netlist or wiring list for the design. A netlist is a file which lists all devices in the design and all the connections between devices.

THE LOGIC SYNTHESIS APPROACH

A newer approach to producing the schematic for a circuit is to define the function to be performed by the module in a customized high-level language and then use a computer program called a *logic synthesis program* to convert the high-level language circuit description to a schematic. A simple example of a logic synthesis program is the ABEL language processor we use to develop the fusemaps for PALs. As we showed you in Chapter 12, you can design a PAL by simply describing the design for a desired combinational logic circuit or state machine in a Pascal-like high-level language. The ABEL language processor then automatically produces the simplified expressions for the combinational logic or the state machine and generates the fusemap required to implement the design in a PAL or PLD.

The source files for a logic synthesis program are written in languages called *hardware definition languages,* or *HDLs.* One example of an HDL is Verilog-XL from Gateway Design Automation, which uses a syntax similar to the C programming language. Another example is the VHDL (Very High Speed Integrated Circuit Hardware Definition Language) developed by Intermetrics Inc. for the U.S. Department of Defense. VHDL uses a syntax similar to the Ada programming language. Still another HDL is Silc Technologies' language which has its own unique syntax. The logic synthesis programs, or *logic synthesis tools* as we prefer to call them, automatically simplify the logic of a design, check the design for errors, and produce both a schematic and a netlist for the design.

Prototyping the Circuit—Simulation

After the design is polished, the next step is to prototype, or breadboard, the circuit design to make sure it works as expected. In the past this prototyping was usually done by soldering or wire wrapping the circuit on a board, such as that in Figure 3-7c. More and more we now use software breadboarding to test the operation of circuits or ICs. To do this we use a program called a *simulator.* In a later section of the chapter we show you several examples of using a simulator, but basically here's how one works.

The simulator program uses the netlist file which contains software descriptions, or *models,* of the devices in the design and a list of all the connections between these devices. The first step in running a simulation on the circuit is to specify sequences of signals, or *stimuli,* that you want applied to the inputs of the circuit. The second step is to specify the signals you want traced out on the screen during simulation. The final step is to run the simulation.

Running a simulation will produce a logic analyzer type

of trace or a listing of the input signals you specified and the output signals produced by those inputs. By carefully checking the traces of the input and output signals, you can determine if the circuit worked correctly. Besides the logic analyzer type traces, most simulators also give warning messages if a signal line has glitches or if timing parameters such as setup and hold times are violated for a device in the circuit.

The point here is that the simulator uses software *models* of the devices in the design to determine the response that the circuit will make to the specified input signals. As we hope to convince you later, the big advantage of simulation or software breadboarding is that you can change the design and resimulate the circuit in a matter of minutes to hours instead of waiting days for new parts to come in so you can modify a physical prototype and test it. Apollo Computer Corporation reportedly used simulation to cut several months from the prototype debug time for an engineering workstation such as the one shown in Figure 5-22b.

Another advantage of simulation over traditional breadboarding it that simulation often points out marginal timing problems. These timing problems might not show up in a physical prototype because of the particular parts used, but they might cause a failure rate of 20 percent in a production run of the instrument.

Design for Test and Fault Simulation

Once the circuit has passed simulation, the next step is to design in some circuitry which allows the frequency counter to test itself or at least allows it to be easily tested by an external tester. If this were a deluxe frequency counter, we would probably add built-in self-test (BIST) circuitry so that the unit does a complete internal test each time the power is turned on and sends a PASS/FAIL signal to the displays. Since we want to keep the cost of this unit down, however, we will probably just include some built-in-test (BIT) circuitry, which makes the device more easily testable by a production tester.

After the test circuitry is added, the circuit is simulated again to make sure the added test circuitry has not adversely affected the operation of the circuit. If the new circuit passes simulation, the designer writes a set of test vectors for the circuit.

A program called a *fault simulator* is then used to determine the percentage of internal faults that can be detected in the circuit when the specified set of test vectors is applied to it. If the test vectors cannot detect an acceptable percentage of faults in the circuit, then more test vectors must be written or the circuit must be redesigned so that faults are more detectable. In a later section of the chapter we describe how a fault simulator works and how circuitry can be added to make a system more testable.

Second Design Review

As soon as the design passes simulation, the design committee meets to review the design before any more work is invested. Some of the reasons for this meeting are to determine if the design meets specifications and if the unit can be built at a reasonable cost.

The response of the mechanical engineer to a design produced by combining parts of the circuits in Figures 9-17 and 9-22 would most likely be that the design requires too many ICs, and therefore it would require a large PC board or several PC boards to hold all the devices. This would make it difficult to fit the unit in a reasonable-size package.

The response of the production/test engineer to this initial design would surely be that the large number of ICs in the design requires an unreasonably large inventory of parts, excessive labor costs for producing the PC board(s), and excessive labor costs for mounting all the components. The large number of ICs would also make the unit more difficult to troubleshoot if it didn't work when built and more likely to fail after delivery to a customer.

Based on this feedback the design review committee has to decide whether to build the frequency counter with discrete ICs as originally planned or to design a custom application-specific integrated circuit (ASIC) to implement most of the circuit. A key point in making this decision is the number of units that the company expects to sell. The total engineering and production cost for a custom ASIC such as this might be as high as $100,000, depending on the approach used and the number of loops through the design cycle required to get a working part. If the expected sales volume is low, the cheaper way might be the discrete-IC approach. For a medium- or high-volume product, the ASIC approach very rapidly pays for itself by reducing the PC board size, package size, parts inventory, production costs, and testing costs.

For this example let's assume the decision is made to implement as much of the frequency counter circuitry as possible in a single CMOS IC. Once this decision is made, the next step is to refine the circuit design for each module to make sure it implements all the desired functions and that it uses components that can be manufactured in an IC.

One example of the need for this refining in our frequency counter is the control circuitry. The frequency counter circuit in Figure 9-17 uses a 74LS121 monostable to produce the clock pulse for the 74HC175 flip-flops and the reset pulse for the 74HC160 counters. This 74121 circuit works well, but it requires an external capacitor which is too large to easily implement in an IC. Later in the chapter we show you how the 74121 circuit can be replaced with a simple two-flip-flop state machine which can easily be implemented in an IC.

After the circuit design is refined so that it can be implemented in an ASIC, the circuit is simulated again. When the circuit passes simulation, the test vectors are updated and another fault simulation is run to make sure the circuit is still testable. The next step then is to decide how to go about designing the actual ASIC.

Custom Application-Specific Integrated Circuits (ASICs)

Currently there are several different ways of implementing a custom ASIC such as we want to design for our frequency counter. The first thought that might occur to you is to use one or more PLDs such as those we used to implement the elevator controller in Chapter 12. The amount of circuitry in even the largest current PALs, however, is not nearly enough to implement all the needed counters, flip-flops, decoders, and glue logic with one device. Therefore, we need to turn to LSI approaches.

The five major LSI approaches now used to implement custom ASIC designs are logic cell arrays, gate arrays, standard cells, compiled cells, and full-custom design.

For custom ASICs of up to a few thousand equivalent gates, logic cell arrays (LCAs) have the advantage that they can be developed in-house. LCAs consist of an array of functional blocks, I/O blocks, and interconnects. The function of each block and the connections between blocks is determined by the logic levels stored in static memory cells associated with each block. An IBM PC type development system is used to convert a design to the program data for the static memory cells. The process is similar to the way a PLD fusemap is developed with ABEL.

The configuration file or program for the memory cells is usually stored in an external PROM. The LCA is designed so that the configuration file can be automatically transferred from an external PROM to the LCA when the power is turned on. The circuit implemented by an LCA can be changed by simply developing a new configuration file and loading it into the device.

The gate-array, standard-cell, compiled-cell, and full-custom approaches to producing a custom ASIC are used for larger and more complex designs. It would require a complete book to discuss each of these approaches in detail, but a later section of the chapter gives you an introduction to each and compares their tradeoffs. The design tools for all four of these approaches develop a gate- or transistor-level layout for the IC. Therefore, after the layout is developed, an IC manufacturer, or "silicon foundry," must convert the layout to production masks and manufacture the actual prototype ICs. There are several different ways to get from a schematic or netlist to the layout used for making masks.

If the circuit designers in a company have some IC design expertise, they may decide to develop the layout for the ASIC in-house. If the circuit designers in a company do not have IC design expertise, the layout task will be turned over to the silicon foundry or to a company that specializes in IC layout.

In either case, computer-based tools do much of the work in developing the layout. A place-and-route program, for example, is commonly used to produce the layout for a gate-array circuit.

Once the ASIC is laid out, the design is simulated to make sure that placement and routing have not introduced propagation delays or other problems which cause the circuit to fail. If problems are found, the designer can interactively try a different placement or routing to solve them and simulate the result again. When the design passes simulation, the layout is used to generate the production masks. The masks are then used to manufacture the actual IC.

After manufacture, the chips are tested using the test vectors supplied by the designers. The chips which pass are mounted in an appropriate package, tested again, and shipped to the user to be tried in the prototype system.

An ASIC developed with the gate-array, standard-cell, compiled-cell, or full-custom approach has the advantage of being able to implement circuits of up to several hundred thousand equivalent gates in a single package. However, an ASIC developed with one of these approaches also has some disadvantages that you should be aware of before you attempt to develop one.

First, you or the IC manufacturer has to determine the layout for your design. The computer-based tools required to do this are at present quite expensive. Fortunately, we are rapidly moving toward integrated sets of development tools which allow a designer to easily go from a schematic or high-level language description of a design to an optimized and simulated ASIC layout. This leads to their use in more designs, and the cost per design goes down.

Second, there is a foundry start-up charge of several tens of thousands of dollars for the first batch of ICs. This charge is called a *nonrecurring engineering* (NRE) charge because if for some reason your design does not work, you have to cycle through all the development steps and pay this charge again to get a new prototype batch of ICs. If the device is to be used in a high-volume product, the NRE development costs will be spread over many products, so the final cost per device may be very acceptable. However, to minimize the number of times you have to pay the NRE costs, it is important to simulate the design at each development stage.

Third, it usually takes several weeks to get a completed batch of prototype ASIC ICs from the foundry.

Printed-Circuit Board Design

While the ASIC is being designed and manufactured, another part of the design team will design a PC board for the instrument. In the old days we used a light table and large plastic sheets to develop the layout for a PC board. To produce "pads" for IC pins, transistor leads, resistor leads, etc., we stuck opaque "doughnuts" on the sheets. To produce traces between pads, we used opaque tape.

Now we use automatic place-and-route programs such as Board Station from Mentor Graphics, Allegro from Valid Logic Systems, or OrCAD/PCB from OrCAD Systems to lay out PC boards. These programs use the netlist file to determine the best placement of components and the most efficient route for traces between components. The programs allow user interaction so that specific paths can be optimized if needed. For example, in designing a PC board for a very high-speed system, you might determine the actual signal delays from an initial layout attempt and then resimulate the system with these delays. If the resimulation shows a problem, you can manually alter the layout to solve the problem before going on.

The file produced by the PC board layout program is sent to a plotter or laser printer to produce a hard copy. The finished sheet is photographed, and the resultant film is used to photographically produce the desired pattern on a copper-plated PC board. A chemical solution then etches copper from all the areas of the board except those where component pads, traces, ground planes, and power planes are desired. For a multilayer board, several individual boards are produced and then epoxied together under pressure to form a single board.

The board is then drilled under computer control. Finally the plated-through holes and other *vias* which connect traces on different layers are electrochemically added to the board.

After manufacture, the "bare" PC boards are tested with a computer-based tester to check for shorts and opens. On a prototype PC board, minor problems can often be solved by, for example, drilling out a plated-through hole which accidentally got shorted to a power plane. A jumper wire can be added to make a missed connection.

The computerized PC board layout method is much faster than the manual method, and it makes modification easier. Also, since these programs use a set of rules gathered from "expert" PC layout people, they remove most of the human error and usually generate more efficient layouts than manual methods. In a later section of the chapter we describe the use of a PC board layout program in greater detail.

Case Design

Once the PC board, power supply, and display have been designed for an instrument, the mechanical engineer can design the package or case for our frequency counter. A program such as the Mentor Graphics Package Station can be used to do much of this design. This program allows the designer to draw a three-dimensional view of a case and the placement of components in the case. The Package Station program also allows a designer to determine the temperature that will be present at each location in the prototype case for a specified ambient temperature and airflow. This feature allows the designer to determine if the airflow is great enough, if the placement of the PC board(s) in the case is reasonable, and perhaps if devices which produce a large amount of heat are placed too close together on a PC board. Here is another example of software breadboarding which saves much work and materials because, if a problem is found, you can simply go back to the computer screen and try a new design instead of producing a new physical box and trying it.

Production and Testing

When the prototype ASIC, PC board, and case are ready, the ASIC and other components can be inserted in the PC board and tested. If the prototype board works on the bench, it can then be plugged into the prototype case and tested. It is hoped that if all the steps in the design process have been done carefully, the prototype instrument will work the first time. If it does, the instrument can be released to the production department, which will then set up to produce a prototype production run of the instrument. To aid the production department there are several programs available which use the schematic or netlist file to automatically generate a complete list of parts for the instrument.

If the instrument does not work correctly the first time and the reason is not immediately obvious, the trace produced by the simulator program can be compared with a trace produced by connecting a logic analyzer to the corresponding points on the prototype instrument to help find the problem. Once the problem is found, the design team can cycle back through the parts of the design process needed to correct the problem(s) and produce a tested and working prototype.

Production and Test

Once the production prototype run of an instrument is debugged and the appropriate engineering change orders (ECOs) submitted, the design is finalized and released to final production. Many parts of the production, testing, and troubleshooting of the instrument are done with the aid of computer programs.

As we mentioned before, programs are available to generate a parts list from the netlist for a design. Other available programs direct a robot to collect the needed parts from the warehouse for the production run. A computer program running on an automatic tester tests the bare PC boards for shorts and opens before parts are inserted. Another program controls the machine that automatically places the components on the printed-circuit board. The machine which solders all the components on the board is most likely controlled by a microcomputer program. Still another computer program controls the machine which automatically tests the finished PC boards. The program for this automatic test system uses test vectors which were developed as part of the design process. If the product does not have a complete built-in self-test, the finished product is also tested with an automatic test system. The linking together of all the computer-based tools used in the production of a product is called *computer-integrated manufacturing*, or *CIM*.

Summary

As you can see from Figure 15-1, the automated design process involves defining the design at a high level and then using a sequence of computer programs to do the actual circuit design, prototyping, and testing. At the present the path through this sequence of programs is a somewhat bumpy one because there is no one set of tools which smoothly implements all the parts of the design cycle. Also, tools from one manufacturer are often not compatible with tools from other manufacturers.

To solve the first problem, many design automation companies are merging or signing technology trade agreements. VLSI Technology, for example, signed a technology sharing agreement with Mentor Graphics to make VLSI's advanced-cell and gate-array compilers mutually compatible with Mentor's schematic capture and simulation programs. Alliances such as this should soon give a nearly complete EDA path from concept to silicon or system.

To solve the second problem, several groups are now working on standards which will make it easier to transfer a design from one manufacturer's tools to another's. One of these standards that you may hear more about in the future is the electronic design interchange format, or EDIF.

Effects of Design Automation on Roles of Engineers and Technicians

The computer-based tools we described in the preceding sections will have a drastic effect on the roles of design engineers and engineering technicians in the 1990s but will not eliminate the need for either. Here are some of the effects that we feel these tools will have.

- As we move toward billion-transistor ICs, the complexity of digital systems will greatly increase. More and more designs will be implemented with one or a few custom ASICs. As a current example of this trend, the two large Chips and Technologies devices on the PC board in Figure 3-2 implement most of the circuitry required to interface an IBM PC-type microcomputer to an EGA color monitor. Engineers and technicians will use computers to design and prototype these complex devices and systems.

- The increased complexity of ICs will require that design engineers include built-in self-test and redundancy in every design. As an example of this direction, the credit card-sized TRW "superchip" has circuitry that tests all the modules in the device and automatically replaces a malfunctioning module with a working one. An output signal is asserted to indicate the device is operating on a backup module. Incidentally, this device contains 34,700,000 transistors.

- Design engineers will develop a system design in a top-down manner using a high-level language. The high-level language description will be processed by a logic synthesis program to produce schematics, netlists, etc. It will then be processed by a variety of programs to produce a custom ASIC and/or be processed by a PC layout program to produce the PC board. Even the mechanical package will be designed with the aid of a computer program.

- The time between design concept and marketed product will continue to decrease because tools such as logic synthesis programs, simulation programs, silicon compilers, and PC board place-and-route programs can increase the efficiency of engineers and technicians. These tools make it possible to do in days or hours what often took months using manual methods.

- Design teams will be smaller than they have typically been in the past because, once engineers and technicians master the design automation tools, they can accomplish much more in a given time than they could with older manual methods. Much of the "expertise" is built into the tools, so an engineer can, for example, design an ASIC without having to be an expert on IC layout. Even though the design teams will be smaller, the number of new products will continue to increase, so the total employment for electronic engineers and technicians should remain the same or increase. However, these technicians will need knowledge of software engineering concepts and hardware engineering concepts.

- Initially the design automation tools will be used mostly by design engineers, but as we go further up the learning curve, more of the actual use of the tools will be turned over to engineering technicians. In other words, engineering technicians will be breadboarding and testing prototypes with software tools instead of wire-wrap boards and dual-trace scopes. This will enable the design engineers to focus on the higher-level concepts of a design. During the past 15 years we have seen a similar evolution in the tools and methods used to develop software for microcomputers.

The rest of this chapter gives you a more detailed introduction to how some of the design automation tools are used. To start, the next section of the chapter shows you some examples of how circuits are simulated. Later sections discuss some of the basics of designing for test and fault simulation, ASIC design techniques, and PC board layout.

SIMULATION OF DIGITAL CIRCUITS

Overview of Digital Simulation and Models

As we explained briefly in the first section of this chapter, simulation is a way of software-breadboarding a circuit to determine if it functions as desired. The simulator program uses *models* of the devices in the circuit to determine the effect that specified input signals will have on the outputs of the circuit.

Most models of digital devices are just software descriptions of the devices in terms of their basic gate equivalent circuits. These descriptions are usually written in a high-level programming language such as Pascal or C. As a simple example, the model for a basic 3-input AND gate might look something like the following:

```
PROCEDURE ANDGATE ;
    CONST TPLH = 15;
          TPHL = 10;
    VAR IN1, IN2, IN3 : INTEGER ;
        DELAY, OUT : INTEGER ;
    BEGIN
      IF (IN1 = 1) AND (IN2 = 1) AND (IN3 = 1)
         THEN BEGIN
                 DELAY : = TPLH;
                 OUT : =  1;
                 END
         ELSE BEGIN
                 DELAY : = TPHL;
                 OUT : = 0
                 END
END;
```

This model is very primitive, but it should give you the idea. The constants represent the characteristics of

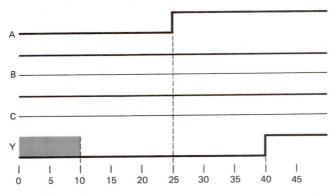

FIGURE 15-2   Simulator trace showing time steps.

the specific device being simulated (TPLH and TPHL). The variables represent the input logic levels (IN1 to IN3), the output logic level (OUT), and the time between a change on the input and the corresponding change on the output (DELAY). Some simulators refer to these characteristics as *properties*. A more complete model would show other properties such as pin name, pin number, part identifier, etc.

When you set up the simulator to do a simulation run, you specify the signals you want applied to the inputs at a particular time, just as you connect signal generators to the inputs of a physical circuit. The simulator uses the model to determine the effects that the specified input signals will have on the output and schedules the output to change appropriately after the delay time for that device. As you can see, the model for the 3-input AND gate device tells the simulator program that if the input signals are all 1's, the output should be scheduled to change to a 1 after 15 ns. If the inputs are not all 1's, the output should be scheduled to change to a 0 after 10 ns. We will use this model to introduce you to how a simulator works and then discuss some common types of models.

The smallest increment of time used by a simulator is called its *time step*. You can think of the time step as the time resolution of the simulator. For simulating TTL and CMOS circuits, simulators usually use a time step of 1 ns because the delay times for these devices are a few nanoseconds. For ECL simulation a time step of 0.1 ns is commonly used. An important point here is that the 1-ns time step is simulator time, not real time. The simulator may take 20 min to determine the effects that some input signal changes produce on the outputs of a complex circuit. The physical circuit would respond to the same input changes in a real time of just a few nanoseconds. The simulator essentially exercises the circuit in slow motion and generates an output which *represents*, or simulates, the real-time operation of the circuit. The simulator output can be used to produce a trace which displays the input stimulus signals and resultant output signals with a desired time scale.

Some simulators evaluate the effect of the specified input signals after each time step. These simulators are called *clock-* or *time-step-driven simulators*. The difficulty with this approach is that a large amount of simulator time is spent evaluating the circuit at time steps where neither the inputs nor the outputs change. A much more efficient and more commonly used type of

simulator is an *event-driven simulator*. This simulator only does a new evaluation of the circuit when a change occurs on some input, intermediate, or output signal. An event-driven simulator is much more efficient because it simply increments the time-step counter over those steps where no changes occur in the signals. To see how this works, let's take a look at Figure 15-2, which shows a simulator trace for our 3-input AND gate.

For a start, suppose that we told the simulator to put a 0 on the A input, a 1 on the B input, and a 1 on the C input at time 0. Then at time step 25 ns we told the simulator to change the signal on the A input to a 1. The output will initially be in an unknown state.

When the simulator runs this simulation, it will first evaluate the effect of the signals specified at time step 0. The specified combination of input signals will cause the output to go from unknown to low after the 10-ns t_{PHL} delay specified for this device, so the simulator will schedule an output change for time step 10 ns. Since there are no events on the input signals between 0 and 10 ns, the simulator will increment the time-step counter to 10 ns and evaluate the circuit again at that point. The only change at this point is to make the Y output a 0.

The next event on the waveforms is the 0-to-1 transition of the A signal at time step 25 ns, so the simulator will increment the time-step counter to this time and evaluate the circuit again. The input signal combination present at this time will cause the output to go to a 1 after a t_{PLH} delay time of 15 ns, so the simulator schedules an output change for 25 ns + 15 ns, or a time step of 40 ns. The simulator will then increment the time-step counter to time step 40 ns and evaluate the circuit again at that time. The only change at this time, of course, is to make the output high.

Now that you have an overview of how a simulator uses models, we need to talk about some of the commonly used types of models. Three of these types are:

Gate-level models

Behavioral models

Hardware models

As we have shown you throughout the book, any digital circuit can be constructed using just basic gates and inverters. Figure 8-11, for example, shows how a D latch is constructed from basic gates, and Figure 13-31*a* shows how an ALU can be constructed from basic gates. We didn't bother to show you, but even a complex device such as an 8086 or 80386 microprocessor can be modeled at the basic gate level for simulation. The difficulty with using gate-level models for complex devices is that simulation using these models requires a very long time. The reason for this is that the simulator must evaluate the effects of each signal change on all the intermediate circuit points (*nodes*) in the device.

If the complex device is a standard part, we usually know that all the internal circuitry works correctly, so we don't need to resimulate at the gate level of detail. To speed up the simulation of circuits containing complex devices, we often use *behavioral models*. Behavioral models simply describe the effects that input signals will have on the output signals and the signal delays between inputs and outputs. A behavioral model of a D flip-flop, for example, will indicate that 20 ns after a positive clock edge, the logic level on the D input will be transferred to the Q output if neither the PRESET nor the CLEAR input is asserted. Behavioral models also contain properties such as setup times, hold times, and minimum pulse widths so that the simulator can check for violations of these times by the signals propagating through the circuit. Sophisticated behavioral models such as the "SmartModels" from Logic Automation Inc. give detailed error messages to pinpoint a timing problem instead of making you work your way through a logic-analyzer type display to find the problem.

For simulating microprocessors there are two types of behavioral models available. One type is called a *hardware verification model*. This model is essentially a black box which will, for example, produce the correctly timed address and control bus signals for a memory-read cycle when given the proper *processor control language (PCL)* file. Hardware verification models are easy to use for checking system timing because all they need is a simple PCL file as a stimulus. However, hardware verification models do not allow simulation of actual microprocessor instructions. If we need this level of simulation, we use *full-functional models*, which do allow the execution of instructions. The disadvantages of full-functional models are that they operate more slowly than hardware verification models and you have to develop the actual object codes for the microprocessor instructions.

The advantage of all the behavioral models is that, except for very simple devices, they execute much faster than the equivalent gate models. Therefore, we use mostly behavioral models. If a needed behavioral model is not available from a model library, most simulators allow you to write a behavioral model in Pascal, C, or a special language such as VHDL.

In cases where a behavioral model of a device is not available and it is not practical to write a model, or in cases where the simulation must interface with external circuitry at real-time speeds, we use *hardware modeling*. In this approach the devices to be simulated are plugged into an external unit such as the Mentor Graphics Hardware Modeling System (HML) shown in Figure 15-3. When using a unit such as this, the simulator program sends stimulus signals to the external devices and evaluates their responses.

FIGURE 15-3   Mentor Graphics HML box.

To develop complex systems such as the engineering workstation shown in Figure 5-22b, we use *multilevel simulators*. An example of a multilevel simulator is the Mentor Graphics QuickSim which can simulate combinations of gate-, behavioral-, and hardware-level models. Multilevel simulators such as this even allow the JEDEC files for PALs to be included in the simulation. To simulate analog circuits, a variety of analog circuit simulators such as PSPICE and LSIM have been available for a number of years. *Mixed-mode simulators* which can simulate circuits such as A/D converters that have both analog and digital parts are currently being developed.

Now that you have an overview of the operation of simulators and the models they use, we will show you how to use a commonly available simulator.

Simulation of a Combinational Logic Circuit

PREPARING A DESIGN FOR SIMULATION

In an early part of Chapter 8 we showed you how propagation delay can cause glitches, hazards, or spikes in combinational logic circuits. As our first example of simulation we will illustrate how a simulator can show you the glitch produced by the circuit in Figure 8-2c.

The first step in using a simulator is to draw a schematic for the circuit using a schematic capture tool, such as the OrCAD/SDT III we showed you how to use in Chapter 5. To refresh your memory of the circuit we want to simulate, Figure 15-4 shows a schematic for it drawn with the OrCAD/SDT III program.

The next step after drawing the schematic is to process the schematic design file with a series of utility programs which prepare it for simulation. Here is a list of these programs and a brief explanation of the function of each.

ANNOTATE—Makes sure the part numbers and pin numbers are accurate and that there are no duplications.

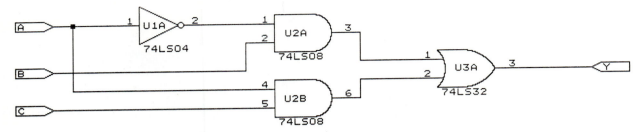

FIGURE 15-4 Schematic of Figure 8-2c drawn using OrCAD/SDT III.

CLEANUP

A design rule check (DRC) program which makes sure the schematic contains no overlapping parts, wires, or junctions.

ERC

Electrical rule checker (ERC) checks for shorts to V_{CC} or ground, inputs with no connection, outputs connected, etc.

NETLIST

Generates the netlist or wiring list for the circuit.

It is usually a very quick process to run the schematic design file through these utilities. However, one common problem you may encounter is that a line (net) which looks just fine on the drawing may not be connected to the desired points. Here, the ERC will generate an error message. We have found that the best way to get rid of this type of error message is to delete the problem net and carefully draw it again. After the design passes through these utility programs with no errors, you are ready for the simulation program.

We chose to use the OrCAD/VST (Verification and Simulation Tools) simulator for the examples in this section because the program is easy to use, readily available, and runs on relatively inexpensive hardware such as IBM PC- and PC/AT-type computers. The general approach and terminology used by the different simulators is similar, so if you use some other one, you should have little difficulty adapting the following discussions to it.

SETTING UP THE STIMULUS FILE

As we said before, a simulator allows you to specify a sequence of signals that you want applied to the inputs of the circuit. As part of setting up the simulator, you create a *stimulus file* which specifies the sequences of signals that you want applied to the inputs of the circuit. The simulator then determines the effects that this sequence of signals will have when propagated through the circuit and generates a logic-analyzer-type trace showing the input and output signals you specify.

The first step in setting up a stimulus file is to think very carefully about the signals that you want to apply to the inputs of the circuit. For this first example, the timing waveforms in Figure 8-3 do most of the thinking for you. From these waveforms you can see that we initially want the A signal to be low, the B signal to be high, and the C signals to be high. After some time interval, such as 100 ns, we want the A signal to go high, and then after perhaps another 100 ns, we want the A signal to return low. Once you have determined the sequence of

stimuli you want to apply to the circuit, you can use OrCAD/VST to create the stimulus file. Here's the process involved in doing this.

When you run the OrCAD/VST program by typing SIMULATE and pressing the return key, it first prompts you for the name of the netlist file you want to simulate. After you enter the netlist file name, OrCAD/VST prompts you for the name of the stimulus file you want it to use. If you have a previously created stimulus file, you enter the name of the file. If you do not already have a stimulus file, you just hit return. OrCAD/VST then tells you the file was not found. After pressing any key, the STIMULUS EDITOR display shown in Figure 15-5a appears. The line across the top of the display shows the commands available in this menu. If you press the left mouse key or the return key, the menu shown in Figure 15-5b will appear on the screen.

You want to add a stimulus, so you move the highlighted box to the ADD line and press the left mouse key again. The display shown in Figure 15-5c will appear.

The first input you want to stimulate in the circuit in Figure 15-5 is the A signal, so move the highlighted box to the SIGNAL NAME line and press the E key to enter the edit mode for that entry. Enter an A for the signal name and press the return key. Move the cursor to the line labeled INITIAL VALUE. This line tells the simulator the logic level you want on the A signal line at time 0 in a trace. Enter a 0 on this line.

Move the highlighted box to the END STIMULUS line in the display in Figure 15-5c and press the return key. Select A for the ADD option, and OrCAD/VST will prompt you with TIME OF FUNCTION. What this prompt is asking you for is the time you want to apply a new logic level to the A signal. The time step can be changed, but the default step for OrCAD/VST is 1 ns. Therefore, the value you enter here will be some number of nanoseconds. To simulate the circuit in Figure 15-4, we decided that a reasonable signal for the A input would be to start the signal low, make it go high at time 100 ns, and then go low at 200 ns. In response to the TIME OF FUNCTION prompt, then, enter 100 and press return. Another menu which shows the signal-level choices will appear. The choices here are 0, 1, TOGGLE, UNKNOWN, and HIGH Z. Move the highlighted box to the 1 line and press return. The display will show that at time 100 the A signal should have a function of 1.

To add the return-to-0 stimulus at time 200 ns, press the A key for ADD, enter 200 for the TIME OF FUNCTION, and enter 0 for the VALUE OF FUNCTION. The

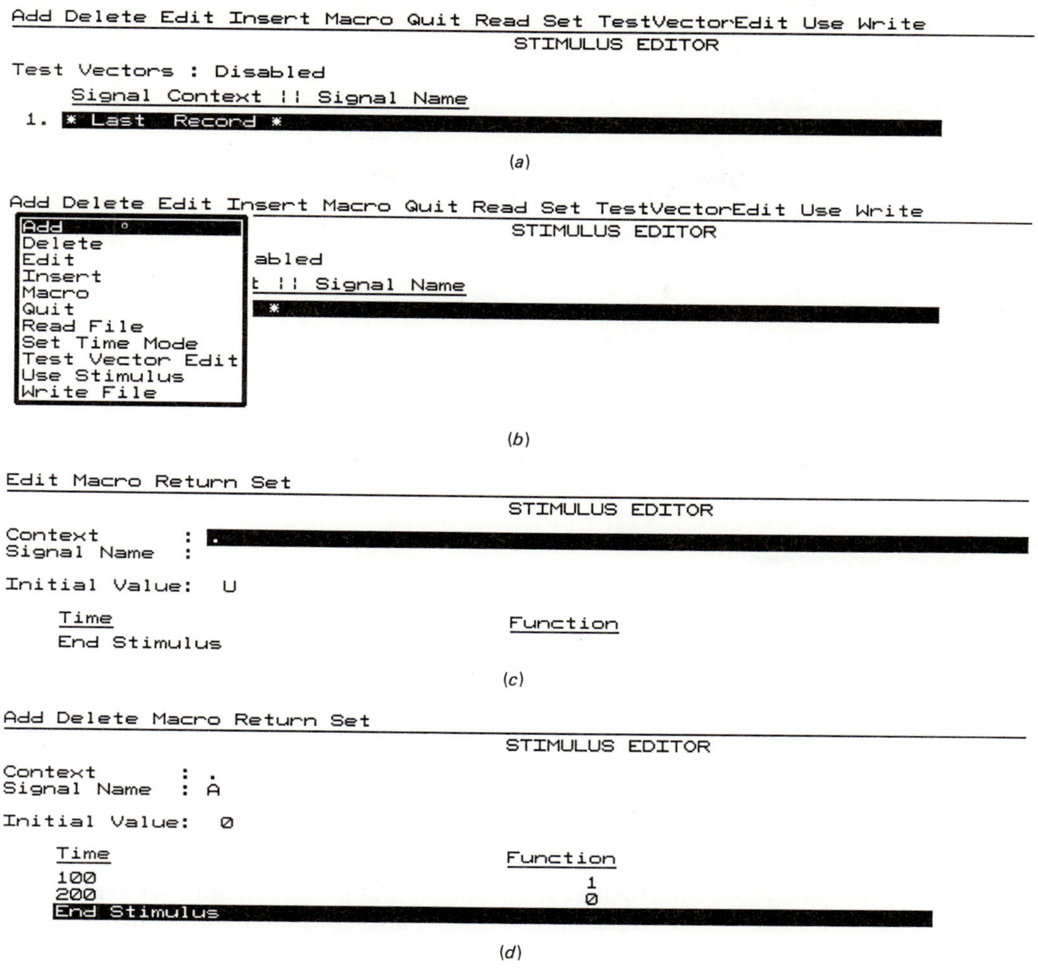

```
                              STIMULUS EDITOR

Test Vectors : Disabled
    Signal Context || Signal Name
  1. █ Last  Record █████████████████████████████████████
```

(a)

```
Add Delete Edit Insert Macro Quit Read Set TestVectorEdit Use Write
┌─────────────────┐
│Add         °    │        STIMULUS EDITOR
│Delete           │
│Edit             │ abled
│Insert           │ t || Signal Name
│Macro            │ █ ████████████████████████████████████
│Quit             │
│Read File        │
│Set Time Mode    │
│Test Vector Edit │
│Use Stimulus     │
│Write File       │
└─────────────────┘
```

(b)

```
Edit Macro Return Set

                              STIMULUS EDITOR

Context      : █ █████████████████████████████████████████
Signal Name  :

Initial Value:  U

     Time                          Function
     End Stimulus
```

(c)

```
Add Delete Macro Return Set

                              STIMULUS EDITOR

Context      : .
Signal Name  : A

Initial Value:  0

     Time                          Function
     100                               1
     200                               0
     End Stimulus ████████████████████████████████████████
```

(d)

FIGURE 15-5   OrCAD/VST. *(a)* Stimulus editor overview. *(b)* Stimulus editor menu. *(c)* Stimulus vector menu. *(d)* Stimulus vector for A input completed.

display should now look like that in Figure 15-5*d*. Press the R key to return to the main STIMULUS EDITOR menu shown in Figure 15-5*a*.

The process we have just described can now be used to enter the desired stimulus signals for the B and C signals. For the transitions we want to look at in this example, all you have to do is give the B signal an initial value of 1 and the C signal an initial value of 1. When you have completed the stimulus file, press the W key to WRITE the stimulus editor information to a disk file.

Unless you are actually using OrCAD/VST, you probably won't remember the details of generating a stimulus file, but from the preceding discussion you should see that it is easily done, once you decide the sequence of signals that you want to apply to the inputs of a circuit.

To tell OrCAD/VST to use the file you generated for the simulation, you press the U key for USE. OrCAD/VST will then display a blank logic-analyzer-type screen. Press the left mouse key or the return key to get a display of the main menu so that you can finish the setup and run the actual simulation.

SETTING UP THE SIMULATOR TO PRODUCE THE DESIRED TRACES

For the next part of the setup, you have to tell OrCAD/VST which signals to trace out on the screen as it does the simulation and which time scale to use for the traces. This is very similar to setting up the trace specifications on a logic analyzer.

From the main menu press T to get to the TRACE submenu, select the TRACE EDIT line, and press the return key. This sequence of actions should produce the TRACE EDITOR display shown in Figure 15-6*a*. The list of signals to be traced is empty, so you need to add the signals you want displayed on the trace.

Pressing the A key for ADD should produce the display shown in Figure 15-6*b*. To modify the fields in this display, you move the highlighted box to that field, press the E key for EDIT, and then enter the value you want in that field. For example, to enter the name you want associated with a trace on the display, you select the DISPLAY NAME line, press the E key, type in the desired name, and press the return key. The logical choice for the first trace on the example circuit is the A signal, so

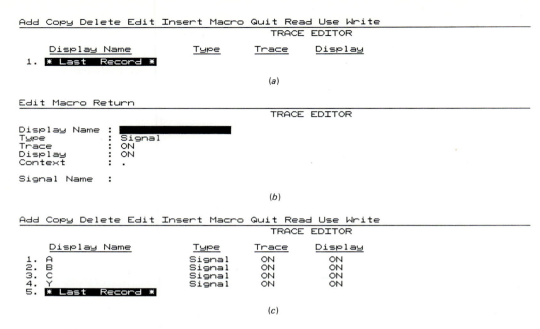

FIGURE 15-6 OrCAD/VST. (a) Trace menu. (b) Trace editor. (c) Completed trace list.

you would enter an A here. As we show you in a later example, this name does not have to be the same as the name of the signal on the schematic.

The other field you need to edit in this menu is the SIGNAL NAME field. In this field you enter the name that identifies the signal on the schematic. If you want to trace a signal that is not identified by name on the schematic, you can identify the signal with its device identifier and pin number. The SIGNAL NAME U2-3, for example, tells OrCAD/VST to do a trace of the signal on pin 3 of the device identified as U2A on the schematic. For our first example here, you just enter an A and press the return key. To get back to the main TRACE menu, press R for RETURN.

You then use the same process to ADD traces for the B, C, and Y signals. When you finish, the display should look like the one in Figure 15-6c. Press the W key to WRITE this trace file to a disk file for future reference. After the file is saved, press the U key to tell OrCAD/VST to USE this file for the trace. Now you are ready to run the simulation.

RUNNING A SIMULATION AND VIEWING THE RESULTS

To run the simulation, select the RUN SIMULATION line in the main menu and press the return key. OrCAD/VST will prompt you for the desired SIMULATION LENGTH. What you enter here is the number of time increments you want the simulator to execute. A simulation run of 300 ns will show the results of the specified inputs signals for this example, so you enter 300 and press return.

After the simulation runs you should see a display such as that in Figure 15-7a. Note that the A signal is initially low and goes high as specified at time 100. The simulator ran for 300 time units, but because the default time scale for the display is 1 ns, only the first 110 of the 300 ns is displayed on the screen. To scroll the

display so you can see the later data points, use the mouse to move the vertical cursor line to the right on the display. As you move the cursor, note that the time position of the cursor is displayed at the bottom of the screen.

When the cursor hits the right side of the screen, the display will jump to a later section of the 300-ns trace. After banging the cursor against the right side of the display a couple of times, you should see a display such as that in Figure 15-7b. In this display you can see that the A signal goes low at 200 ns as specified and that the Y output has an unwanted pulse. You can use the cursor to determine that this pulse goes low at 233 ns and back high again at 248 ns. The simulator then tells you directly that this circuit generates a 15-ns-wide glitch on its output for the specified transitions. Incidentally, unless you specify otherwise, OrCAD/VST uses the maximum value for properties such as propagation delay, setup time, hold time, etc. In most cases this gives a worst-case view of the timing in the circuit.

If you want to see the entire 300 ns of the trace on the screen at once, you can simply change the trace time scale and run the simulation from time 0 again. To do this, press the left mouse key or the return key to get to the main menu. In the main menu select the INITIALIZE line and press the return key. After the simulator reinitializes, get back to the main menu and select the TRACE line to get to the TRACE submenu. In the TRACE menu select the CHANGE VIEW line and press the return key. OrCAD/VST will prompt you for the TRACE DELTA TIME.

The TRACE DELTA TIME tells OrCAD/VST how often to update the trace display. With the TRACE DELTA TIME set for its default value of 1, the trace will be updated after each 1 ns of simulated time. With the TRACE DELTA TIME set for 2, the trace will be updated after every 2 ns of simulated time. Setting this value is anal-

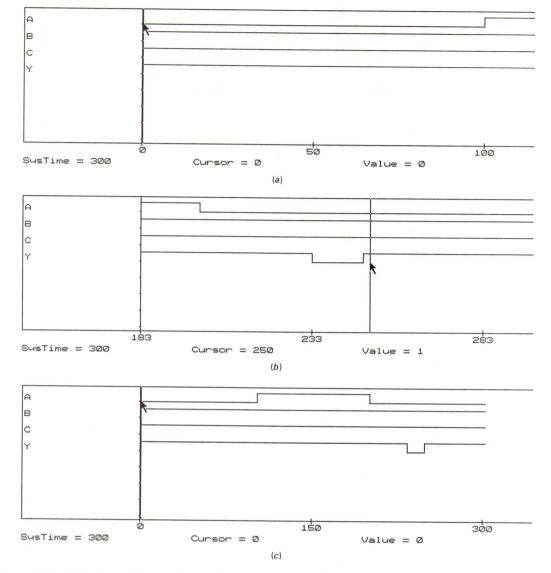

FIGURE 15-7 *(a)* Trace for first 112 ns of simulation. *(b)* Trace scrolled to show glitch. *(c)* Trace showing all 300 ns of simulator time using a TRACE DELTA TIME of 3.

ogous to setting the internal clock rate on a logic analyzer. It does not change the actual signals; it simply changes how often you sample the simulator output to update the display. As with a logic analyzer, the longer the time between updates, the less resolution any time measurements will have. Therefore, we often do one simulation to get an overview and then do another simulation to look closely at critical time periods.

For this example you would like to see a trace of 300 ns of simulated time on the screen. A TRACE DELTA TIME of 3 will produce a display of about 300 ns of simulated time on one screen, as shown in Figure 15-7c. Remember, however, that as you increase the TRACE DELTA TIME to get more displayed on the screen, the resolution of measurements decreases.

Once you become familiar with the OrCAD/SDT III and OrCAD/VST tools, you can almost always draw and simulate a circuit faster than you can find the parts to build the circuit. In the next section we show you how you

can use a simulator to determine if a state-machine design works correctly.

Simulating a State-Machine Design

THE DESIGN PROBLEM AND PROPOSED CIRCUIT SOLUTION

A frequency counter using the basic counter circuit from Figure 9-17 and the multiplexed display from Figure 9-22 works well, but as we told you earlier in the chapter, the control circuit which clocks the flip-flops and resets the counters must be redesigned if the instrument is to be implemented in an ASIC. The 74121 one-shot used in the circuit in Figure 9-17 requires an external capacitor which is too large to be easily built into an ASIC.

As shown in Figure 9-18a, the falling edge of the count WINDOW signal triggers the 74121 to produce a positive pulse which clocks the 74HC175 flip-flops and a negative pulse which resets the main counter chain. What we need to replace the 74121 is a simple state machine which produces these output pulses at the correct times. Figure 15-8 shows the two-flip-flop circuit we designed to do this and the parts of the frequency counter circuitry needed to test it.

will remain in this state until the next falling edge on the count WINDOW signal starts the process over again. The width of the pulses on the Q and $\overline{Q}$ outputs of U2B will be equal to the time between the pulses from QD of U3. As we show you later, the resultant timing diagram for the circuit is similar to the one in Figure 9-18b.

At this point you may wonder how we arrived at the

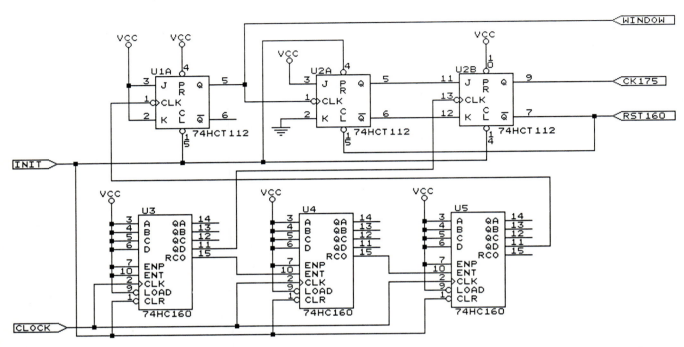

FIGURE 15-8 Schematic for count WINDOW generator, redesigned control circuit, and three counters for timebase counter chain.

The three 74HC160s in Figure 15-8 are simply three of the decade counters in the timebase chain. If a 10-kHz signal is applied to the clock inputs of these counters, U3 will produce a 1-kHz output, U4 will produce a 100-Hz output, and U5 will produce a 10-Hz output. The U1A flip-flop divides the 10-Hz signal by 2 to produce a 5-Hz square-wave signal. The time high for a 5-Hz square wave is 0.1 s, so the output of this flip-flop is a 0.1-s count WINDOW signal. The falling edge of the count WINDOW signal clocks the U2A flip-flop.

To see how the U2A and U2B flip-flop circuit works, assume that initially both of their Q outputs are low. The K input of the U2A flip-flop is tied low and the J input is tied high, so when it is clocked, the Q output will go high and the $\overline{Q}$ output will go low. This will put a high on the J input and a low on the K input of the U2B flip-flop. The falling edge of the next QD pulse from counter U3 will cause the Q output of the U2B flip-flop to go high and the $\overline{Q}$ output to go low. The $\overline{Q}$ output is connected to the CLEAR input, CL, of the U2A flip-flop, so when $\overline{Q}$ of U2B goes low, the Q output of U2A will almost immediately go low and its $\overline{Q}$ output will go high. The falling edge of the next clock pulse from the QD output of U3 will then cause the Q output of U2B to go low again and the $\overline{Q}$ output to go high again. The outputs

design for the U2A and U2B part of the circuit. With some explanation the design should be, for the most part, intuitive to you. As we hinted before, the U2A and U2B flip-flops form a simple state machine. U2B is the register part of the state machine, and U2A is the next-state decoder. Since we want the register to only change state after the falling edge of the count WINDOW, we had to use a negative-edge-activated device in the next-state decoder instead of just combinational logic. Once we put the negative-edge-triggered JK flip-flop in place, all we had to do was connect the J and K inputs of U2A such that when it was clocked by the falling edge of the count WINDOW, a high was transferred to the J input of U2B and a low was transferred to the K input of U2B. The next pulse from the QD output of U3 will step the output register U2B to its next state, which is a high on Q and a low on $\overline{Q}$. When $\overline{Q}$ on U2B goes low, it immediately resets the output of the next-state decoder U2A. The next pulse from the QD output of U3 steps the output register to its other state where Q is low and $\overline{Q}$ is high. The machine will stay in this state until the falling edge on the count WINDOW signal starts the cycle over. Now that you know how this circuit is supposed to work, we will show you how we simulated it to verify that it really does work.

SIMULATION SETUP PROCEDURE

The first step in simulating a design is to draw a schematic for the circuit or part of the circuit you want to simulate. It is best to start with a small block of circuitry and then, when you have verified the operation of that block, add other blocks until you have simulated and verified the entire design. For this example we were initially going to simulate just the state machine made up of U2A and U2B. However, as we started thinking about the required stimulus signals, we decided that it was actually easier to use the three counters and the WINDOW generator flip-flop to produce the stimulus signals for U2A and U2B than to manually figure out the timing for these signals. Figure 15-8 shows the resultant block of circuitry we used for our first simulation.

An important point to notice in the circuit in Figure 15-8 is that we connected the $\overline{CL}$ inputs from the counters and either the $\overline{PR}$ or $\overline{CL}$ input from each flip-flop to a common input line called INIT. In the actual physical circuit these inputs will just be tied high, but when you simulate any circuit which contains flip-flops or devices built with flip-flops, you *must* specify a stimulus which initially puts each device in a known state. Connecting the $\overline{CL}$ and $\overline{PR}$ inputs as shown makes it possible to initialize all the devices with just one stimulus entry. After simulation these connections can be removed for IC or PC board layout.

By including the counters, WINDOW generator, and connections shown in Figure 15-8, the only stimulus signals we have to generate are some initialization signals and a clock signal for the counters.

After you complete the schematic, you cycle it through the annotate, cleanup, ERC, and netlist programs. Then you invoke the simulator on the netlist file. The next step is to generate the stimulus file which specifies the sequences of signals you want applied to the inputs of the circuit by the simulator.

To do this you select EDIT STIMULUS in the main menu and press the return key. When the stimulus editor appears, you ADD the desired stimulus entries as we described for the glitch example. Here is a description of the specific stimuli we needed for this circuit.

First, note that in this circuit some logic device inputs are connected to V_{CC}. The simulator automatically assumes that the power pins on all devices are connected to the power supply V_{CC}, but it does not assume that logic inputs shown as connected to V_{CC} are physically connected. This is done so that logic inputs and power pins can be connected to different V_{CC} voltages. The result is that you have to write a stimulus entry which equates the name V_{CC} on a device input with a logic high. To do this you ADD a stimulus with a signal name of VCC and an initial value of 1.

Next, look at the circuit to see if any logic inputs are connected to ground. The K input on U2A is connected to ground, so you have to write a stimulus which equates this GND symbol with a logic low. To do this you simply ADD a stimulus with a signal name of GND and an initial value of 0. Note that the VCC and GND stimuli do not have any time reference in them because these values will remain the same for the entire simulation run.

Now look at the INIT signal in Figure 15-8. We want this signal to be low for a long-enough time to put all the devices in a known state and then go high so that the devices operate normally. To do this you ADD a stimulus with a signal name INIT and an initial value of 0. To this stimulus you ADD a second line which changes the FUNCTION to a 1 at time 25. The 25 ns is more than enough time for all the flip-flops and counters to get to the state specified by the connection to their $\overline{CL}$ or $\overline{PR}$ inputs. The INIT signal will be left at the 1 level for the rest of the simulation run, so this is all we need for the INIT stimulus.

Finally look at the clock signal in Figure 15-8. For this simulation we want to apply a continuous square-wave signal to the clock input. In the actual frequency counter a 10-kHz signal will be applied to this input, but for the simulation we will increase the frequency to 5 MHz in order to reduce the amount of time we have to run the simulator to see a complete cycle on the CK175 and RST160 outputs. Increasing the clock frequency by a factor of 500 reduces the number of time steps we have to run the simulator by a factor of 500. The relative timing of all the signals will still be the same, but the simulation will run much faster.

The period of a 5-MHz signal is 200 ns, so the signal will be low for 100 ns and high for 100 ns each cycle. As shown in Figure 15-9a the OrCAD/VST simulator provides an easy way to generate repetitive signals such as this. The stimulus is given an initial value of 0, and then at a time equal to half the period it is given a T, or TOGGLE, function. This will cause the signal to go to a 1 at 100 ns. At 200 ns the execution is JMPed back to the TOGGLE instruction. This will cause the clock signal to toggle to a 0 at time 200. The overall effect of this stimulus file is to produce a signal which toggles every 100 ns. Once the complete stimulus file is generated as shown in Figure 15-9b, WRITE it to a file, and press U to USE the stimuli.

The next step is to generate the trace file which tells the simulator the signals you want traced and displayed on the screen. The signals we decided we wanted to look at for this simulation were the CLOCK signal, QD outputs of the three 74HC160 counters (DIV10, DIV100, and DIV1000), the count WINDOW signal, the CK175 signal, the RST160 signal, and the Q output of U2A (Q2). We used the TRACE editor as described in the earlier example to ADD a trace specification for each of these. Figure 15-10a shows the completed trace file. Remember that for a signal which doesn't have a specific name on the schematic, you can give it a display name of your choice and identify it with a device and pin number, as shown in Figure 15-10b. In this example we gave the name Q2 to the Q output signal on pin 5 of the U2A flip-flop.

After the TRACE file is generated, WRITTEN to a disk file, and USED, the next step is to go to the trace menu again and set the TRACE DELTA TIME for 50. The highest frequency signal we want to trace is the clock signal which only changes every 100 ns, so if we tell the simulator to take a sample of the signals every 50 ns, we will be able to see the clock signal and all the other signals displayed. Choosing a TRACE DELTA TIME of 50 ns allows us to see much more of the signals displayed on the screen at once.

```
Edit Macro Return Set
                              STIMULUS EDITOR
Context       : ▐
Signal Name   : CLOCK

Initial Value:  0

        Time                              Function
        100                                  T
        200                               JMP 100
        End Stimulus
```

(a)

```
Add Delete Edit Insert Macro Quit Read Set TestVectorEdit Use Write
                              STIMULUS EDITOR

Test Vectors : Disabled
        Signal Context :: Signal Name
    1.  .VCC
    2.  .GND
    3.  .INIT
    4.  .CLOCK
    5. █ Last  Record █
```

(b)

FIGURE 15-9 *(a)* Stimulus entry to generate 5-MHz square wave on clock input. *(b)* Completed stimulus list.

```
Add Copy Delete Edit Insert Macro Quit Read Use Write
                              TRACE EDITOR

        Display Name      Type     Trace    Display
    1.  CLOCK             Signal    ON        ON
    2.  DIV10             Signal    ON        ON
    3.  DIV100            Signal    ON        ON
    4.  DIV1000           Signal    ON        ON
    5.  WINDOW            Signal    ON        ON
    6.  CK175             Signal    ON        ON
    7.  RST160            Signal    ON        ON
    8.  Q2                Signal    ON        ON
    9. █ Last  Record █
```

(a)

```
Edit Macro Return
                              TRACE EDITOR

Display Name  : Q2
Type          : Signal
Trace         : ON
Display       : ON
Context       : ▐

Signal Name   : .U2-5
```

(b)

FIGURE 15-10 *(a)* Trace specification for frequency counter module simulation. *(b)* Trace editor showing how to enter pin numbers for signal name.

Still another preparatory step is to go back to the main menu and SET the SIMULATOR for SPOOL to DISK so that when the memory buffer is full of data samples, the contents of the buffer will be copied to a disk file to make room for the next set of samples. Finally, we get to run the actual simulation.

RUNNING THE SIMULATION AND VIEWING THE RESULTS

To determine how many system time units of 1 ns you want the simulation to run for, you have to think about how long it will take for the circuit in Figure 15-8 to go through one complete cycle. Each cycle of the input clock signal requires 200 system time units of 1 ns. The QD output of U3 will change after every 10 clock cycles, the QD output of U4 will change after every 100 clock cycles, and the QD output of U5 will change after every 1000 clock cycles. This means that we have to run the simulator for 200×1000, or 200,000, system time units just to see a change on the QD output of U5. The U1A flip-flop divides the U5 QD output signal by 2, so we need to run the simulator for $2 \times 200,000$, or 400,000, system time units to see a complete cycle of the WINDOW signal on its Q output. The CK175 pulse goes high 10 clock cycles after the falling edge of the WINDOW signal and goes back low 20 clock cycles after the falling edge of the WINDOW signal, so we need another 20×200, or 4000, system time units to see both transitions of the strobe signal. To show a little time after the desired pulses, we decided to run the simulation for 408,000 system time units.

Figure 15-11 shows a section of the trace produced by the simulation run. We scrolled the display to show the region of the trace where most of the action occurs. Incidentally, the OrCAD/VST ZOOM menu allows you to position the display anywhere in the trace data and to zoom in on a section of a display so that you can see details more clearly.

As you can see in the trace in Figure 15-11, the falling

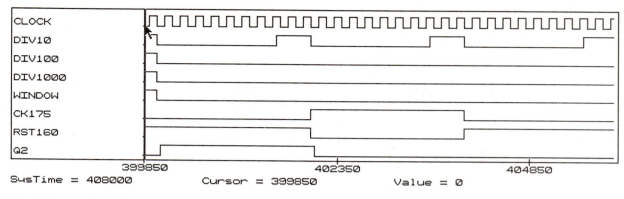

SysTime = 408000   Cursor = 399850   Value = 0

FIGURE 15-11   Section of trace showing generation of control pulses.

edge of the WINDOW signal transfers a high to the Q output of U2A (Q2). Ten CLOCK cycles later this high is transferred to the CK175 output, and the Q2 signal is cleared. After another 10 CLOCK pulses, the cleared state on the Q2 signal is transferred to the CK175 output. This completes the control signal sequence. The cycle will repeat on the next falling edge of the WINDOW signal.

The purpose of the preceding example and discussion is not to make you an expert with OrCAD/VST but to show you how easy it is to simulate the operation of a sequential circuit. We don't have the space to do it here, but with little difficulty we could add the rest of the frequency counter circuitry to the schematic in Figure 15-8 and simulate the operation of the entire circuit in its different modes. As we said before, you can usually draw a schematic and simulate a circuit faster than you can find the parts to build a physical prototype.

We don't have space here to show you the simulation of any more circuits, but another circuit that would be useful and easy to simulate is the microcomputer system in Figure 14-22a. By simulating this circuit with the appropriate models, you could determine if the memory devices are fast enough to operate without wait states when the 8086 is operated with a 4.9-MHz clock. To make it easier to interpret traces with bus signals, OrCAD/VST allows you to display a bus as a single line which shows the binary or hex value on the bus at each point in time. Incidentally, most simulators also allow you to run a simulation to a specified breakpoint condition as well as to a specific time.

In this section of the chapter we have shown you how to simulate circuits to see if they work correctly. It is possible to have a circuit or system which simulates correctly but cannot be fully tested because of the way it is designed. In the next section we show you some design techniques used to make digital circuits and devices fully testable, and we discuss a tool called a fault simulator which is used to determine the testability of a circuit.

FAULT SIMULATION AND DESIGN FOR TEST

Introduction

After they are manufactured, ICs are typically tested with large IC test systems, such as the Hilevel Technology

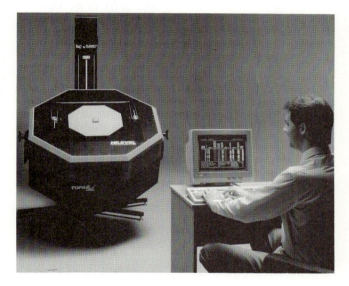

FIGURE 15-12   Hilevel Technology Topaz-V IC ASIC verification system.

Topaz-V shown in Figure 15-12. These systems are essentially a dedicated high-speed computer which applies a large number of test vectors to the device under test (DUT) to determine if it works correctly. In the past a design engineer designed the device, and later a test engineer determined the test vectors needed to test the device. The complexity of digital devices has now reached a point, however, where this divided approach no longer works. For the LSI and VLSI circuits of today, *built-in self-test* (*BIST*) or at least *built-in test* (*BIT*) capability must be part of the initial design effort. In this section of the chapter we describe how a program called a *fault simulator* is used to determine the testability of a circuit, and we show you some of the techniques used to make circuits testable.

TEST VECTORS AND FAULT SIMULATION

As a first example of testing a digital circuit, suppose that you want to test a simple 2-input NAND gate. In Chapter 12 you learned how a PAL programmer uses test vectors to determine if a programmed PAL works correctly, so the first thought that should come to mind is to write a set of test vectors for the NAND gate. Depending on the particular system they are written for, the test vectors might look something like those in Figure 15-13. The test system will then run through the test vectors to determine if the device works correctly for each.

```
([A,B] ————→ Y)
 [0,0] ————→ 1;
 [0,1] ————→ 1;
 [1,0] ————→ 1;
 [1,1] ————→ 0;
```

FIGURE 15-13   Test vectors for 2-input NAND gate.

For a simple circuit such as a 2-input NAND gate, it is quite easy to write a test vector for each of the four possible input signal combinations. Writing a test vector for each of the possible input combinations is sometimes called *exhaustive testing*. For this simple example there are only four test vectors, so the tester will be able to run through them very quickly.

As the complexity of a device or circuit increases, however, the number of test vectors required for exhaustive testing increases very rapidly. To exhaustively test just a single 8-input NAND gate, for example, would require 256 test vectors! To reduce the number of test vectors required for a given circuit, we use two basic approaches, *fault simulation* and *design for test*. The fault-simulation method of testing works as follows.

Fault Simulation

The first step in the fault-simulation approach is to develop a preliminary set of test vectors which represent a variety of but not all possible input conditions. The fault-simulator program is then run with these test vectors, and the results are saved as a reference.

During the next run of the fault simulator, it injects a stuck-at-zero or a stuck-at-one fault at some point in the circuit and again determines the effect of each of the test vectors. If the results are the same as those for the circuit with no injected fault, that fault is said to be an *undetected fault*. If the injected fault causes a result different from that of the circuit with no faults, the fault is said to be a *detected fault*. The fault simulator automatically inserts each of the possible faults in the circuit and determines if that fault is detected or undetected. When the fault-simulation run is complete, the simulator tells you the percentage of possible faults detected by the set of test vectors you supplied. This percentage is usually called the *fault coverage* of the test vectors. For reasonable confidence the fault coverage should be at least 90 percent. Military contracts demand at least 95 percent fault coverage.

If the initial set of test vectors do not produce a high-enough fault coverage, you can add more test vectors and run the fault simulator again. If the percentage of fault detection is still not high enough, you then need to look closely at the design to see how you can make it more testable. To make this more real for you, we will show you a circuit which has an undetectable fault. Then

in a later section we will show you how to make the circuit testable.

As we showed you in Chapter 8 and again in the simulation section of this chapter, the circuit in Figure 15-14a produces a glitch on its output for the 111 to 011 signal transition. The extra, or redundant, AND gate (U2C) in the circuit in Figure 15-14b eliminates the glitch, but it introduces an undetectable fault condition in the circuit. To see this, let's look at how the circuit in Figure 15-14b will respond with a couple of different inserted faults.

When the fault simulator is run with the test vectors shown in Figure 15-14c and no faults inserted, the results will, of course, be those predicted by the test vectors. When the fault simulator is run with the same set of test vectors and a stuck-at-one fault applied to the output of gate U2A, the Y output will be a 1 for all the test vectors. Since these results are different from the no-fault simulation results, this is a detected fault.

If the fault simulator is run with the test vectors shown in Figure 15-14c and a stuck-at-zero fault applied to the output of gate U2C, the results on the Y output will be the same as the no-fault simulation results. This fault then is undetected because it does not cause a change in the output. With this inserted fault the circuit still implements the basic logic, but it is defective because the fault will cause the circuit to produce a glitch for the 111 to 011 transition. In the next section we show you how to make this circuit fully testable and some general techniques used to make circuits testable.

Designing Circuits for Testability

ADDING A TEST INPUT TO A SIMPLE COMBINATIONAL CIRCUIT

The undetectable fault condition in the circuit in Figure 15-14b is caused by the redundant term we introduced by adding the U2C AND gate. To make the circuit testable, we need to temporarily remove the redundancy. The circuit in Figure 15-15a shows one possible way to do this. If the TEST input is high, the circuit will work the same as the circuit in Figure 15-15b and will respond to the test vectors in the same way. If the TEST input is made low, the output of the U2D AND gate will be permanently low, and the circuit will respond to the test vectors as shown in the NO FAULTS Y column of Figure 15-15b. When the fault simulator inserts a stuck-at-zero fault on the output of the U2C AND gate, the results will be those shown in the second Y column of Figure 15-15b. The result for the 011 input combination is now different for the no-fault and fault simulations, so the fault is now detectable. The combination of the vectors in Figure 15-14c with those in Figure 15-15b will completely test the circuit. The TEST input adds another gate to the circuit, but this is a small price to pay for making the circuit fully testable. Incidentally, for this example we wrote test vectors for all the possible input combinations, but the whole point of fault simulation is that this is not necessary to get a high fault coverage.

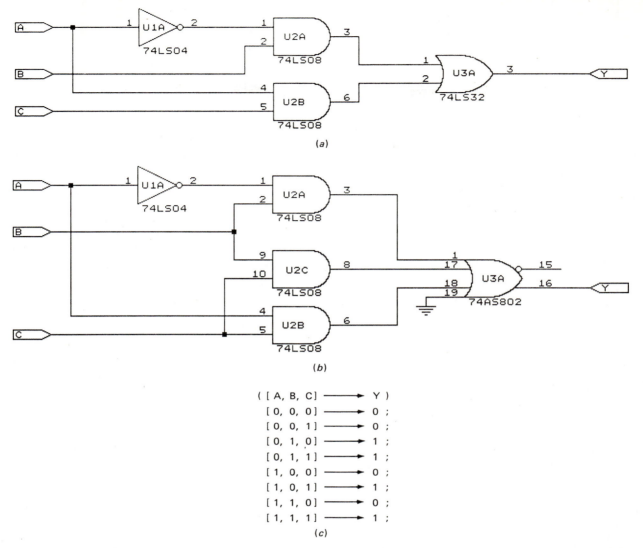

([A, B, C] ⟶ Y)
[0, 0, 0] ⟶ 0 ;
[0, 0, 1] ⟶ 0 ;
[0, 1, 0] ⟶ 1 ;
[0, 1, 1] ⟶ 1 ;
[1, 0, 0] ⟶ 0 ;
[1, 0, 1] ⟶ 1 ;
[1, 1, 0] ⟶ 0 ;
[1, 1, 1] ⟶ 1 ;

(c)

FIGURE 15-14 (a) Circuit which produces a glitch. (b) Circuit with redundant gate to prevent glitch. (c) Test vectors for circuit.

OVERVIEW OF ADDING TESTABILITY TO A CIRCUIT

The fault-simulation and test-input method we described in the preceding section is suitable for combinational circuits, but it is not suitable for state machines and large functional blocks. The three major steps in making a complex design testable are:

1. Partition the circuit into easily testable blocks.

2. Make the inputs of each block accessible so that they can be stimulated from the inputs of the device.

3. Make the outputs of each block accessible so that the results of input stimuli can be observed.

One of the most common methods of accomplishing these three tasks is by adding multiplexers and demultiplexers to the design so that either the normal input signals or the test signals can be routed into or out of the chip. As an example of how this is done, we will show you some test circuitry we might add to the ASIC-based frequency counter that we are developing as the main project of this chapter. As we go through these examples, keep in mind that the purpose of the added circuitry is to make it possible to test the major functional blocks of the chip with the minimum number of test vectors. We are not trying to localize exactly which specific part in the device is not working. If the IC fails any part of the test, we obviously just throw it out and test the next one. Also keep in mind that we only have space to show you how to test the major functions, not all the glue logic.

ADDING TEST CIRCUITRY TO THE FREQUENCY COUNTER

For reference, Figure 15-16 shows the complete schematic for the parts of the frequency counter we want to implement in an ASIC. As a start, suppose that we want to make the block of circuitry which includes the main

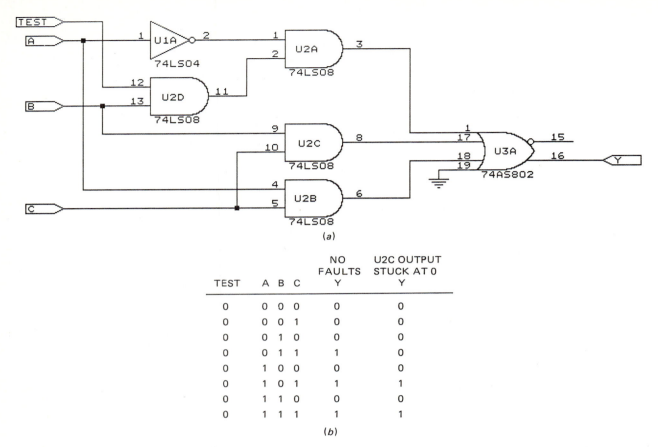

FIGURE 15-15 (a) Circuit from Figure 15-14b modified to make it testable. (b) Fault-simulator results for no-faults and for stuck-at-zero fault on output of U2C.

TEST	A	B	C	NO FAULTS Y	U2C OUTPUT STUCK AT 0 Y
0	0	0	0	0	0
0	0	0	1	0	0
0	0	1	0	0	0
0	0	1	1	1	0
0	1	0	0	0	0
0	1	0	1	1	1
0	1	1	0	0	0
0	1	1	1	1	1

(b)

counter chain, flip-flops, multiplexers, display decoder, and digit decoder.

The first step in developing test circuitry is to partition the circuit into blocks which can be more easily tested. The main problem with the counter-display multiplexer part of the circuit in Figure 15-16 from a testing standpoint is the seven 74HC160 counters. With these counters cascaded as they are in the circuit, it would take 10 million clock pulses to step them through a complete count sequence. Assuming one clock pulse per test vector, this is an unrealistic number of test vectors.

The solution to this problem is to add a 2-input multiplexer on the ENABLE TRICKLE (ENT) input of each of the 74HC160 counters, as shown in Figure 15-17. For normal operation the multiplexers are switched so that the ENT input of each counter is connected to the RIPPLE CARRY OUT (RCO) of the previous counter. For testing, the multiplexers are switched so that the ENT inputs of the counters are all connected directly to +5 V. With these connections the counters will all increment on each clock pulse, so they are connected in parallel rather than in series. Only 10 test vectors then are required to step all the counters through their 10 states.

The second step in making a circuit testable is to make the data and control inputs of each block accessible so that they can be stimulated by test vectors. Let's take a close look at the circuit in Figure 15-16 to see which

circuit points we need to make accessible so that we can properly stimulate the circuit.

To refresh your memory, the counters accumulate a count during the count window time. At the end of the count window, the accumulated count is transferred to the outputs of the 74HC175 flip-flops and the 74HC160 counters are reset to zero. The 74HC151 multiplexers route the BCD code from one 74HC175 at a time to the inputs of the 74LS47. The 74LS47 converts the BCD code to the equivalent seven-segment code for the common-anode LED displays. At the same time the 74HC138 decoder produces a signal which turns on V_{CC} to the anode of the desired LED digit.

The multiplexing scheme used here is ideal from a test standpoint because if we access the proper circuit points, we can route the outputs from one counter at a time through the flip-flops, multiplexers, and decoder to output pins on the device. The circuit points we need to access are:

1. The $\overline{CLR}$ inputs on the 74HC160 counters
2. The CLK inputs on the 74HC160 counters
3. The CLK inputs on the 74HC175 flip-flops
4. The CLKA input on the 74HC93 counter used to step the multiplexers and the display-digit selector
5. The RO1 and RO2 inputs of the 74HC93 counter.

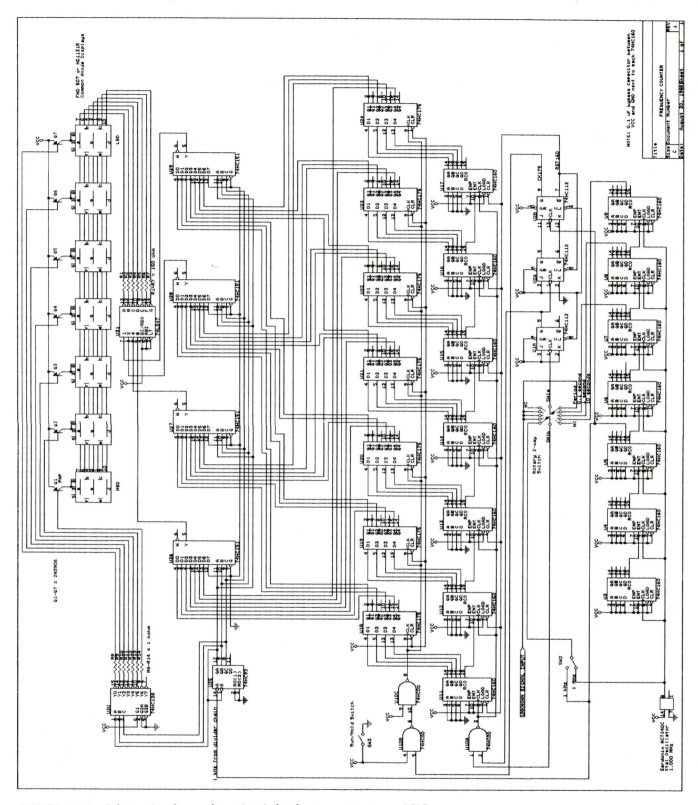

FIGURE 15-16   Schematic of complete circuit for frequency counter ASIC.

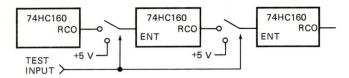

FIGURE 15-17 Adding multiplexers to separate 74HC160s into separate counting units for testing.

For simplicity we initially assume that external pins will be used to get the stimulus signals to the desired points in the circuit. Later we will show you how device pins not needed for testing can be used to get the stimulus signals into the device.

First let's look at the 74HC160 $\overline{CLR}$ inputs. These inputs are all connected because in normal operation they are all asserted by a control signal at the end of a count cycle. As shown in Figure 15-18a, adding a simple 2-input multiplexer will allow us to route either the nor-

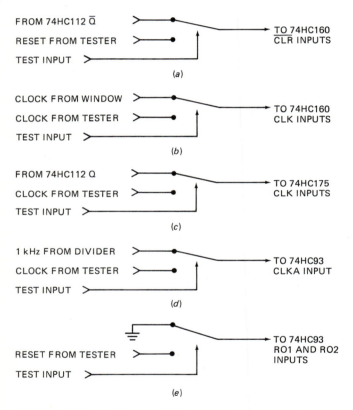

FIGURE 15-18 Adding multiplexers to give access to 74HC160, 74HC175, and 74HC93 control signals. (a) 74HC160 $\overline{CLR}$ inputs. (b) 74HC160 CLK. (c) 74HC175 CLK. (d) 74HC93 CKA. (e) 74HC93 RO1 and RO2.

mal clear signal or an external clear signal produced by the tester to the $\overline{CLR}$ inputs of the 74HC160s. The multiplexer control input can be connected to the same signal which controls the multiplexers on the ENT inputs of the 74HC160s.

The second signal we have to access is the CLK signal for the 74HC160 counters. Again, as shown in Figure

15-18b, adding a simple 2-input multiplexer allows us to select either the normal clock signal from the internal control circuitry or an external clock signal produced by the tester. The control input on this multiplexer will be connected to the same signal line as the control input of the 74HC160 $\overline{CLR}$ signal multiplexer.

Adding three more 2-input multiplexers, as shown in Figure 15-18c to e, allows us to access the CLK input of the 74HC175, the CLKA input of the 74HC93 counter which drives the multiplexers, and the reset inputs (RO1 and RO2) of the 74HC93.

Now that we have access to these points in the circuit, here's the sequence of signals we might apply to test this part of the circuit.

1. Assert the TEST input pin to put the multiplexers in the test position.

2. Assert the signal to clear the 74HC160 counters and the 74HC93 counter.

REPEAT
    Send a clock pulse to the 74HC160 counters to increment them all by 1.
    Send a clock pulse to the CLK inputs of the 74HC175 flip-flops to transfer the count to the outputs of the flip-flops.

    REPEAT
        Check the segment outputs on the 74LS47 and the digit strobe output on the 74HC138.
        Send a clock pulse to the CLKA input of the 74HC93 to point the multiplexers at the next set of 74HC175 outputs.

    UNTIL all seven digits checked
    Send one more clock pulse to 74HC93 CLKA input to get outputs back to 0.

UNTIL all 10 states of 74HC160 counters checked

If you think your way through this sequence a couple of times, you should see that it thoroughly tests the main counter and multiplexer circuitry. The next block we want to set up the test circuitry for is the timebase divider chain.

The problem with this chain of counters is that it would take 10 million clock pulses to step it through its complete count sequence. Again, the solution to this problem is to divide the chain into individual counting units by putting a 2-input multiplexer on the ENT input of each of the 74HC160 counters, just as we did for the 74HC160s in the main counter chain. In the same way as shown in Figure 15-18, two more 2-input multiplexers can be used to give us access to the $\overline{CLR}$ inputs and to the CLK inputs of the 74HC160s. To save pins, the CLK signal for the counters can be sent in on the pin normally used to output the 1-MHz signal.

Now that we have access to the circuit points that we need to stimulate, the next step is to set up some way of observing the outputs of the counters. The desired QD output signals are already available on pins of the device because these signals go to range switches, etc. The RIP-

PLE CARRY OUT (RCO) signals, however, are not available. If we connect the inputs of an 8-input multiplexer to the RCO outputs, as shown in Figure 15-19, we can selectively route one RCO output at a time to an output pin on the device. This multiplexer can be enabled by the same signal which enables the $\overline{CLR}$ and CLK mul-

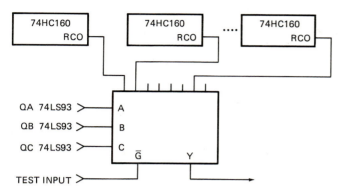

FIGURE 15-19 An 8-input multiplexer used to access RCO outputs of 74HC160s in timebase counter chain.

tiplexers. The multiplexer input routed to its output is determined by the 3-bit code on its select inputs. The select inputs of this multiplexer can be connected to the outputs of the 74HC93 counter. As clock pulses are applied to the 74HC93 CLKA input, the multiplexer will be stepped along the RCO outputs of the 74HC160s.

The test sequence for this part of the circuit then might proceed as follows.

```
Clear 74HC93 and 74HC160 counters.
REPEAT
    REPEAT
        Check QD outputs and RCO output.
        CLKA pulse to 74HC93 to step to RCO output of
          next 74HC160
    UNTIL all RCO signals checked
    Clear 74HC93 counter.
    Clock 74HC160 counters.
UNTIL all states of 74HC160 counters checked
```

As you can see, this test procedure is much simpler than stepping the cascaded counter chain through 10 million or more test vectors. Incidentally, we did not bother to do it, but the same type of multiplexer could be used to look at the RCO outputs of the 74HC160 counters in the main divider chain if desired.

The final section of the circuit in Figure 15-16 to develop a test for is the control circuitry we simulated in the previous section of this chapter. For this block of circuitry we already have access to the window generator flip-flop via the line from SW1a. We can generate the desired stimulus pulse for flip-flop U2B by applying test vectors to the 74HC160 counters as described in the preceding section. Therefore, all we have to do for this circuit block is to connect the CK175 signal and the RST160 signal to external pins so that we can observe them externally.

For this first attempt at developing testability in the circuit in Figure 15-16, we have not worried about adding extra pins to get stimulus signals in and desired signals out of the chip. The cost of packages increases rapidly as the number of pins increases, so to keep package size and cost down, we need to think about ways to reduce the number of pins required. The most common way to do this is to use some more multiplexers.

For an example of this, suppose we add a 2-input multiplexer on each output of the 74HC138 digit decoder, as shown in Figure 15-20. We can use those seven pins

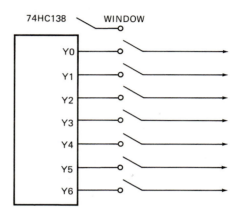

FIGURE 15-20 Adding multiplexers to 74HC138 output signals so that signal pins can be used for control signals during test.

to output signals such as the WINDOW, CLK175, RST160, and the output of the 8-input multiplexer we used to look at the RCO outputs of the 74HC160s in the timebase counter chain. As a second example, if the segment driver outputs are implemented as three-state outputs, these outputs can be floated during some parts of the test procedure. These pins can then be used to input stimulus signals such as the $\overline{CLR}$ signal for the 74HC160s in the timebase divider chain.

We hope the preceding discussions have given you some feeling for one way that test circuitry is added to a design. After the test circuitry is added, the entire circuit is simulated again to make sure that the added test circuitry does not adversely affect the timing or some other part of the operation of the overall circuit. If the circuit passes simulation, then you are ready to design an ASIC for it. In the next major section of the chapter we discuss the tools and techniques used to do this. Before we get into that, however, we want to briefly discuss a few more testing terms you are likely to hear.

SCAN-PATH TESTING

For complex ICs such as microprocessors which have many inaccessible internal registers, state machines, and combinational logic blocks, a technique called *scan-path testing* is used. To use this mode the circuit is designed with a multiplexer on the input of each register. When a TEST signal is asserted, the multiplexers connect all the registers in one long shift register called a *scan path*, as shown in Figure 15-21. The principle here is that the

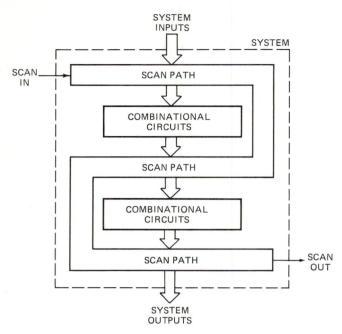

SCAN IN

SYSTEM INPUTS

SYSTEM

SCAN PATH

COMBINATIONAL CIRCUITS

SCAN PATH

COMBINATIONAL CIRCUITS

SCAN PATH

SCAN OUT

SYSTEM OUTPUTS

FIGURE 15-21   Block diagram showing how internal registers are connected for scan-path testing.

stimulus signals for the combinational circuits are shifted in and the responses to the stimuli are shifted out. The advantages of this scheme are that all the registers are tested and that only one input and one output pin are needed.

SIGNATURE ANALYSIS

Signature analysis is a technique used to implement built-in self-test. The circuit to be tested is first divided into functional blocks. A pseudo-random number generator such as those we described in Chapter 9 is then used to internally supply stimulus patterns to each block. Each stimulus pattern produces a unique pattern, or signature, on the outputs of the block being tested. The actual signature is compared with the correct signature to determine if the block of circuitry is working correctly.

FUTURE OF BUILT-IN SELF-TEST

As ICs become more and more complex built-in self-test will become more necessary. One device developed by TRW does a complete internal self-test each time the power is turned on. If the self-test finds a bad block of circuitry, it automatically switches in a reserve block of circuitry to replace the bad block and asserts a signal which lets the user know that the system is operating on the backup block. This is analogous to the way a reserve gas tank system on some cars works. Soon automatic test-and-replace circuitry should be included in most complex ICs, especially those used in military systems, automobile control systems, and medical systems.

ASIC DESIGN TECHNIQUES

As we told you in the first section of this chapter, the common methods of producing a custom ASIC are PLDs,

logic cell arrays (LCAs), gate arrays, standard-cell devices, compiled-cell devices, and full-custom devices. In Chapter 12 we showed you how to work with PLDs, and here we show you how an ASIC is developed with each of these other methods. We will use our frequency counter as an example of a circuit that might be implemented with some of these methods.

Logic Cell Arrays

ARCHITECTURE AND PROGRAMMING

Figure 15-22a shows the basic construction of a logic cell array (LCA) such as those in the Xilinx or AMD XC30XX series devices. The center part of the device consists of a matrix of *configurable logic blocks*, or *CLBs*. As shown in Figure 15-22b, each CLB consists of two D flip-flops and a section of combinational logic. Surrounding the CLBs shown in Figure 15-22a are configurable I/O blocks or *IOBs*. As shown in Figure 15-22c, these blocks connect to the I/O pads on the chip. They can be programmed for combinational input/output or for clocked input/output. Between adjacent CLBs and between IOBs and CLBs the LCA has interconnect lines. The programmable switches in these lines allow the blocks to be connected in an almost infinite number of possible ways.

The functions performed by each block and the interconnections between blocks are determined by how the device is programmed. The programming for each type of element in the LCA is determined by the logic states stored in static memory cells associated with each block. As an example of this, the IOB in Figure 15-22c shows how memory cell bits are used to determine output polarity, combinational or clocked output, and output enable polarity.

Once the configuration program is developed for an LCA, it is programmed into an external EPROM. For parallel transfer of the configuration file to the LCA, the EPROM is connected to the LCA, as shown in Figure 15-23. Each time power is applied to the LCA or a control signal on the LCA is asserted, the configuration program is automatically transferred from the external EPROM to the static memory cells in the LCA. The transfer takes a few milliseconds.

To develop the configuration program for an LCA, you use a series of programs such as those in the Xilinx XACT Development System. The first step in the development process is to draw a schematic for the design using a schematic capture program such as Schema or OrCAD SDT III. You can use standard MSI device symbols or symbols from a library of macro symbols. The advantage of the macros is that they are optimized so they use less CLBs to implement a desired function than a strict implementation of the standard device would.

Once you have entered your design in schematic form and generated a netlist for it, the *automated design implementation* (ADI) programs automatically minimize any combinational logic in the design and optimize state machine designs for implementation in an LCA. The ADI then automatically places the design in the CLBs and IOBs and routes connections between them. If necessary, the designer can interact with the route-and-place op-

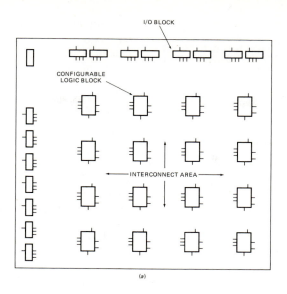

(a)

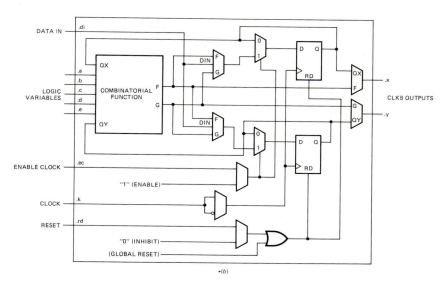

(b)

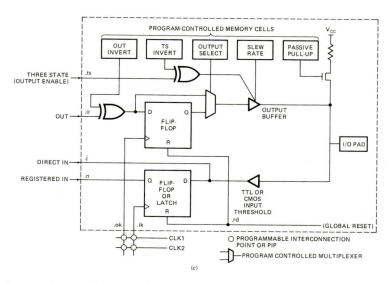

(c)

FIGURE 15-22   *(a)* Internal organization of logic cell array (LCA). *(b)* Circuitry in configurable logic block (CLB) of LCA. *(c)* Circuitry in configurable I/O block (IOB). *(Xilinx Inc.)*

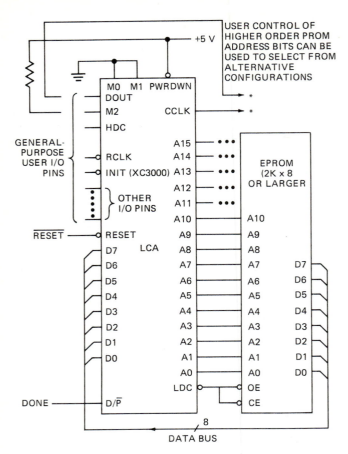

FIGURE 15-23   Connection of external EPROM to LCA for parallel load of configuration file at power-up. *(Xilinx Inc.)*

eration to optimize delay paths in critical parts of the circuit.

After the place and route is completed, a simulator program in the development system can be used to determine if the timing of the design for the LCA is correct. When the design passes simulation, a program called Makebits is used to develop the configuration file for the static memory cells. The configuration file can be tried in one of two different ways.

One approach is to convert the configuration file to a programmer load file and use it to program an EPROM with a standard EPROM programmer. The programmed EPROM then can be plugged into a prototype board and used to test the design. An alternative approach is to use the Xactor Design Verifier shown in Figure 15-24. A plug on the end of the emulator cable plugs into the LCA socket on the prototype board, and the configuration file is downloaded directly into an LCA contained in the "pod" of the emulator. The prototype circuit can then be exercised from the computer.

Logic cell arrays have the advantage that you don't have to wait for a silicon foundry to manufacture the prototype devices for you. With a development system such as XACT, you can develop the configuration file for a device, program the device, and then test the device without leaving your desk, similar to the way you design and test PALs. LCAs are reprogrammable, so to alter a de-

sign all you have to do is develop a new configuration file, load the new configuration file in the device, and test it. The largest currently available LCAs have a capacity of about 9000 equivalent 2-input gates. This is greater than the capacity of the largest currently available PALs.

One disadvantage of LCAs is that they have to be reprogrammed from an external ROM or some other source each time the power is turned on. Also, the chip real estate used by the configuration RAM cells reduces the space available for gates and interconnecting wires. Therefore, LCAs cannot implement as much circuitry per device as the standard gate arrays we discuss in a later section of the chapter. Incidentally, LCAs are often called *programmable gate arrays,* or *PGAs,* but the term logic cell array is really more descriptive because when you design a circuit using one of these devices, you work with logic blocks rather than individual gates.

IMPLEMENTING THE FREQUENCY COUNTER IN AN LCA

Now that you have an overview of LCAs, let's take a look at how we might use one to develop a custom ASIC for our frequency counter. The first step in this process is to determine the available LCAs and the number of CLBs in each. Figure 15-25 shows the capacities of the devices in the Xilinx XC30XX family.

Before we can choose a device from this group, we have to determine how many CLBs and/or IOBs are required to implement each of the devices in the frequency counter. The macro library in the XACT Development System has a list which indicates the number of CLBs required to implement each macro. The macro list contains many common TTL MSI devices and many "stripped-down" devices which require less CLBs than the standard TTL devices. As an example of this, a standard 74XX160 requires six CLBs to implement, but a simple decade counter without the load feature only requires three CLBs. From the list in the XACT manual we determined the number of CLBs required for the frequency counter as follows.

Device	Number	CLBs per Device	Total CLBs
Decade counter	14	3	42
Quad D flip-flops	7	2	14
8-input MUX	4	4	16
3-bit binary counter	1	2	2
3 to 8 line decoder	1	4	4
BCD to 7-segment	1	4	4
JK flip-flops	3	½	2
Glue	2	1	2
Crystal oscillator	1	0	0
	34		86

Looking at Figure 15-25 you can see that the XC3042 is the smallest device which has enough CLBs to implement all our frequency counter circuitry in a single device.

FIGURE 15-24   XC-DS24 Xactor in-circuit emulator. *(Xilinx Inc.)*

BASIC ARRAY	LOGIC CAPACITY (USABLE GATES)	CONFIG-URABLE LOGIC BLOCKS	USER I/O'S	PROGRAM DATA (BITS)
XC3020	2000	64	64	14779
XC3030	3000	100	80	22176
XC3042	4200	144	96	30784
XC3064	6400	224	120	46064
XC3090	9000	320	144	64160

FIGURE 15-25   Comparison of currently available LCAs in the Xilinx and AMD XC3000 family. *(Xilinx Inc.)*

Once the design is partitioned among devices, the next step is to draw a schematic for the LCA portion using the desired macro for each device, check the schematic for rule violations, and develop a netlist. As described in the preceding section, the ADI program can then be used to place the design in the array of CLBs and route connections between CLBs and IOBs. For this design the decade counters in the main divider chain have to operate at the highest frequency, so their placement is most critical. Therefore, these counters would be manually placed first and the rest of the circuit placed automatically by the ADI programs.

Once the place and route is complete, a configuration file is developed. The configuration file can be programmed into an EPROM, or if an in-circuit emulator is available, the file can be downloaded directly into the LCA in the emulator to test the prototype.

Note that in this LCA implementation of the frequency counter, we did not include the internal test circuitry we developed in the previous section on design for test. The reason for this is that the LCA does not require this test circuitry and the associated test vectors. LCA devices are exhaustively tested at the factory for programmability and functionality. Therefore, if the configuration file is thoroughly verified in the prototype, then the pro-

duction devices will all work correctly when programmed with that file.

LCAs have the advantage that they are user programmable, but they are currently limited to about 9000 equivalent gates. To implement larger designs in a single IC device, we turn to techniques which involve the actual manufacture of the IC. The easiest of these is the gate-array approach which we discuss in the next section.

Gate Arrays

OVERVIEW OF GATE-ARRAY STRUCTURE AND DEVELOPMENT

Gate arrays, as the name implies, initially consist of a matrix of unconnected basic gates, usually 2-input NAND gates, on a single silicon die, as shown in Figure 15-26. The wafers containing these dies are fully fabricated ex-

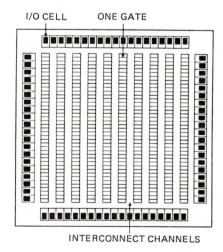

FIGURE 15-26   Internal organization of a basic gate-array device. *(Xilinx Inc.)*

cept for the final metallization layers which make connections between gates and between gates and bonding pads. A custom design is implemented in a gate array by adding the final interconnection layers. When the fabrication is complete, the dies are tested, cut apart, mounted in packages, tested again, and shipped to the user.

To develop the layout for a gate array, the circuit is first designed, simulated, made testable, and fault-simulated as we described in an earlier section of the chapter. Then a *place-and-route* program is used to produce the "floor plan" of the design in the array of gates. Most gate-array development packages have a library of predefined gate structures, or *macros*, which simplify this task. The macros for critical sections of the circuit are usually placed manually first, and then the place-and-route program is allowed to complete the layout automatically. Some manufacturers such as VLSI Technology and LSI Logic now have programs which automatically generate the layout for a gate array directly from a schematic.

Once the gate array is laid out, the design is simulated again to make sure that placement and routing have not introduced propagation delays which cause the circuit to fail. If problems are found, the designer can interactively try a different placement or routing to solve them and simulate the result again. When the design passes simulation, the layout is used to generate the masks which are used to add the interconnection layers to the basic gate-array dies. The completed dies are then tested, and the ones which pass are mounted in an appropriate package. The completed ICs are then tested again and shipped to the user.

One advantage of the gate-array approach is that a very large amount of circuitry can be implemented in a single device. As we said at the beginning of this section, currently available devices such as the LSI Logic LCA100K devices allow circuits of up to 100,000 equivalent gates to be implemented with a single device. Also, the turnaround time for prototype ICs is shorter for gate arrays than for some other ASIC methods because the IC dies are already 85 to 90 percent manufactured. Still another advantage is that for very high volume products, the cost per device is lower for gate arrays than it is for other devices such as LCAs.

The traditional gate-array ASIC approach, however, has several disadvantages that you should be aware of before you attempt to develop one. First, you or the silicon foundry has to place and route the design in the array. Second, developing the prototype ICs involves several actual fabrication steps, so there is a start-up charge of several tens of thousands of dollars for the first batch of ICs. This charge is called a *nonrecurring engineering (NRE)* charge because if for some reason your design does not work, you have to cycle through all the development steps and pay this charge again to get a new prototype batch of ICs. If the device is to be used in a high-volume product, the NRE development costs will be spread over many products, so the final cost per device may be very acceptable. However, to minimize the NRE costs, it is important to simulate the design at each development stage. Still another disadvantage of the gate-array ap-

proach is that it usually takes 4 to 6 weeks to get a completed batch of prototype ICs.

To get faster turnaround time on gate arrays, a company called Laserpath has developed a special gate array which has all the gates interconnected with two layers of metal connections. A computer-controlled laser vaporizes unwanted connections to implement your design.

Another new system developed by Actel Corporation may overcome many of the disadvantages of the standard gate-array development process for gate arrays under 10,000 gates. This system allows you to develop a gate-array prototype in a manner similar to the way prototype PALs and LCAs are produced.

The devices used in the Actel system are specially designed gates arrays which have "antifuses" connecting all the gates in a matrix. A development program converts a schematic to the antifuse pattern needed to implement the design. Then a special programmer called an *activator* blows the required antifuses. As with the LCA approach we described in the preceding section, the Actel approach has the advantage of small NRE cost and rapid turnaround time.

IMPLEMENTING THE FREQUENCY COUNTER IN A GATE ARRAY

In a previous section we described how our frequency counter could be implemented with a PLA and a few MSI CMOS devices. The LCA implementation has the advantage of short turnaround time and small NRE cost, but it has the disadvantage that it requires the external-configuration EPROM. Also, the actual cost per device for an LCA is currently higher than the cost of an equivalent gate-array device. Therefore, if we think that the sales volume of the frequency counter will be high enough to amortize the NRE cost, we will very likely develop a gate array to implement the frequency counter. With current prices the crossover point for this decision would probably be between 20,000 and 50,000 units.

The steps in developing a gate array for the frequency counter are basically the same as those we described in the preceding section. A schematic is generated using library macros for the components in the circuit. The design is then simulated, enhanced with test circuitry, and fault-simulated. A place-and-route program then produces the layout for the design in the "sea of gates." The design is then simulated again, and if it passes simulation, it is sent off to the foundry to create the masks for adding the connection layers. After fabrication the devices must be tested with user-supplied test vectors.

Standard Cells, Compiled Cells, and Full-Custom Design

OVERVIEW

All the custom ASIC development methods we have discussed so far use devices which are already fabricated, or in the case of gate arrays, almost completely fabricated. The custom ASIC techniques we have grouped together in this section all involve fabricating the custom device from scratch. In other words, the layout is used to produce masks for fabricating the entire die, not just the metallization layers as is done with gate arrays. The layout for each of these three methods specifies the placement of individual transistors and the connections between them. Since they require more design steps and more fabrication steps, these methods all cost more and have a longer turnaround time than gate arrays. However, we use these three methods for circuits that contain elements such as RAMs or ROMs which cannot easily be implemented in gate arrays, require higher density than can be achieved in a gate array, or must operate at higher frequencies than possible with a gate-array implementation.

The tradeoffs among these three approaches are speed of development and control of circuit performance. Standard cell is the fastest to design, but it has the least flexibility in design. At the other end of the spectrum, full-custom design has great flexibility but requires a long development time because the design is done at the individual transistor level.

A wide variety of computer-based tools are available to aid in development of a custom ASIC using one of these methods. Any discussion of these ASIC design tools, however, is complicated by the number of different tools available and the rate at which they are evolving. At this time there are probably a hundred ASIC companies, each with its own way of implementing a design, so the development tools are far from standardized. Even the terminology has not yet been standardized. Different companies, for example, use the term *silicon compiler* very differently. Hopefully this will all settle out in the next few years.

In the following discussions we will try to give you an overview of how a custom ASIC is developed using standard cells, compiled cells, and full-custom methods. To get the detailed sequence of steps for a particular set of tools, consult the manuals for those tools.

STANDARD CELLS

Remember that for standard-cell, compiled-cell, and full-custom cell approaches, the layout which will be used to make the fabrication masks specifies the location of individual transistors and the connections between them. Standard cells are predefined transistor layout patterns for commonly used circuit building blocks such as encoders, flip-flops, counters, shift registers, ALUs, etc. These "cells" are contained in a cell library.

To implement your design with standard cells, you draw a schematic which executes your design using these cells. This is somewhat analogous to designing a circuit with a pencil and a TTL data book.

After the appropriate netlist has been generated and simulated, a place-and-route program is used to produce the transistor level layout for the IC. Standard cells are designed so that they physically all have the same height, or "pitch." Therefore, any cell can be placed next to any other cell. The process of placing cells is often called *floor planning.*

When the layout is completed, the design is simulated again to make sure the place-and-route process did not introduce any problems such as excessive delay in a signal line. If the design passes simulation, the layout is used to generate the masks needed to manufacture the IC.

Examples of standard-cell ASICs are the Chips and Technologies' devices on the circuit board in Figure 3-2a. These devices do most of the work of interfacing an IBM PC-type computer to an EGA CRT display. Using standard LSI devices and support chips for this circuit would require a sizable PC board.

The use of standard cells eliminates the need to develop each part of the circuit from individual transistors, and the chip only has to be as large as needed for the specific circuit. However, since the actual IC is manufactured from scratch, the NRE cost is relatively high, turnaround time to prototype ICs is 10 to 12 weeks, and if an error is found, the whole design-fabrication cycle must be repeated.

COMPILED CELLS

Compiled cells are a compromise between standard cells and full-custom design. In the compiled-cells approach many of the cell layouts are individually developed to fit the needs of a particular design. The computer-based tools used to develop these individualized cell layouts are called *silicon compilers.*

One type of silicon compiler called a *data-path compiler* is used to develop the cell layout for circuits elements such as RAMs, ROMs, ALUs, etc., which have parallel data paths. The data-path compiler allows you to specify the width of the data path and then duplicates a basic structure to implement a cell layout for that width element.

Another type of silicon compiler called a *state-machine compiler* is used to develop the cell layout for circuit elements such as state machines, encoders, decoders, and random logic.

Although we refer to it as the compiled-cell approach, standard cells for basic functions are often included in a design. Standard "megacells" for circuit elements such as the basic core of a microprocessor may also be included.

When the cell layouts for all the parts in a design including test circuitry have been generated, a schematic is drawn using representations of these cells. After simulation and fault simulation, a sophisticated place-and-route program such as VLSI Technology's Chip Compiler or Seattle Silicon's APAR+ is used to produce the complete chip layout.

The advantages of the compiled-cell approach are that each cell has only the functions needed in the particular design, the cell can be laid out to optimize specific signal paths, and the physical dimensions of each cell can be adjusted for best packing on the silicon. One disadvantage of the compiled-cell approach is that the development tools are more complex and expensive than those for standard cells. Other disadvantages are the long turnaround time of 12 to 18 weeks for prototype ICs and the relatively high NRE cost.

FULL-CUSTOM DESIGN

The full-custom method of ASIC design is used for applications that require higher speed, more specialized functions, higher density, or other features which cannot be achieved with standard-cell or compiled-cell techniques. In the full-custom approach the IC layout and interconnections are designed almost totally at the individual transistor level. Some computer-based tools such as Caeco's Layout Synthesis program are now available to help with this level of design.

The full-custom method has the advantages that it makes very efficient use of chip real estate, specialized circuits can be implemented, and signal paths can be optimized for minimum delay, etc. However, it has the disadvantages of high cost, long turnaround time, and difficulty in changing a design.

Summary and Comparison of Custom ASIC Development Methods

One difficulty with many of the computer-based tools currently available for custom ASIC development is that they are foundry- and process-specific. A tool might, for example, produce a layout that will only work for VLSI Technology's 1.5-micron CMOS fabrication process. Several companies are now developing chip compilers which produce foundry-independent layouts. Examples are Mentor Graphics AVID and Seattle Silicon's APAR+.

Another difficulty has been that most of the computer-based tools for custom ASIC development do not allow integration of analog and digital circuitry on the same chip. A variety of programs are now being developed to address this need.

Throughout the preceding sections we have tried to give the tradeoffs between the different methods of developing a custom ASIC, but to help you see the whole picture, Figure 15-27 shows a comparison of the methods we have discussed. As technology evolves, you can write the current numbers in the figure for reference.

To further help you see why we go to the extra cost, time, and trouble of designing an ASIC with standard-cell, compiled-cell, or full-custom methods, Figure 15-28 shows a visual comparison of a specialized microprocessor circuit implemented with three different methods. As you can see in the figure, the chip real estate required by the standard cell is less than that required for the gate-array implementation. The compiled-cell implementation is less than one-third the size of the gate-array implementation and has a maximum operating frequency twice that of the standard-cell implementation. We did not have one available for comparison, but a full-custom implementation would most likely require less chip area and have a still higher maximum operating frequency.

PRINTED-CIRCUIT BOARD DEVELOPMENT

Terminology, Manufacture, and Component Mounting

PRINTED-CIRCUIT BOARD TERMINOLOGY

Figure 15-29a shows the two sides of a printed-circuit (PC) board which have traditionally been called the circuit side and the component side. Surface-mount devices have made it possible to mount components on both sides of a board, but usually one side of a board has a higher concentration of components and the other side has a higher concentration of circuit traces.

Most PC boards are made of an epoxy compound mixed with glass fibers for reinforcement. The circuit traces on a PC board are made of copper and are usually plated with tin to keep them from oxidizing. As a carryover from schematic capture terminology, some people refer to the circuit traces as *nets*. Straight sections of a trace or net are sometimes called a *track*. The round or rectangular copper sections where a trace connects to a component lead is called a *pad*. If the component has traditional through-the-board leads, then the pads for that component have plated-through holes which connect to a pad on the other side of the board. Besides supplying an electrical connection to the other side of the board, the plated-through holes help hold the components firmly on the board. For surface-mount devices (SMDs), a pad only needs a plated-through hole if it has to reach a trace on the other side of the board.

Figure 15-29b and c shows some cross-sectional views of PC boards. The board in Figure 15-29b is called a *two-layer* board because it only uses the component side and the circuit side. In this view you can see how the plated-through holes make connections between the two sides of the board. A connection between different layers in a PC board is called a *via*.

Figure 15-29c shows a cross-sectional view of a multilayer PC board. As you can see, this board is like a sandwich with interleaved layers of copper and fiberglass epoxy. The ground and power layers in a board such as this are mostly solid, so they are commonly called *ground planes* and *power planes*. Multilayer boards such as this are used for high-speed applications where a transmission-line environment is needed for signals between devices. A trace separated from the a ground plane by the fiberglass-epoxy dielectric forms a transmission line with a predictable impedance. As we explained in Chapter 10, these transmission lines can be terminated with a resistor to prevent signal reflections. Also, sandwiching the power plane between two ground planes creates a bypass capacitor which helps filter switching transients on V_{CC}.

Multilayer PC boards are also used in applications where there is not enough space on the two outer layers to route all the needed traces between components. For these applications signal layers are included inside the board as well as on the two outer layers. As we said before, the plated-through connections between layers are called vias. A via which makes a connection between one side of the board and the other is called a *through-hole via*. A via which makes a connection from an outside layer to one of the inner layers is called a *blind via*, and a via which makes a connection between two inner layers is called a *buried via*.

Now that you know some of the commonly used PC board terms, we will discuss how the layout for a PC board is developed, how a PC board is manufactured, how components are mounted on a board, and how they are removed.

TYPE	FULL CUSTOM	COMPILED CELLS	STANDARD CELLS	GATE ARRAYS	LCA	FLD
PROGRAMMED BY	FACTORY	FACTORY	FACTORY	FACTORY	CUSTOMER	CUSTOMER
PROGRAMMING MECHANISM	MASK, ALL LAYERS	MASK, ALL LAYERS	MASK, ALL LAYERS	MASK, 4 LAYERS	RAM SWITCHES	FUSE, RAM OR EPROM
CIRCUIT TYPES	DIGITAL AND ANALOG	DIGITAL	DIGITAL AND ANALOG	DIGITAL	DIGITAL	DIGITAL
ARCHITECTURE	UNRESTRICTED	TAILORABLE FUNCTIONS	SSI-VLSI	SSI-LSI	PROGRAMMABLE CELL	PROGRAMMABLE AND-OR
EQUIVALENT GATES, TYP MAX	N/A 300,000	50,000 200,000	10,000 100,000	4,000 100,000	1,000 9,000*	500 2,000*
RELATIVE NONRECURRING ENGINEERING COSTS	HIGH $75K UP	HIGH $75K-100K	MEDIUM $40K-75K	LOW $20K-40K	NONE	NONE
COST AND COMPLEXITY OF DESIGN TOOLS	VERY HIGH	HIGH	HIGH	MEDIUM	LOW	LOW
RELATIVE TURNAROUND	18-24 WEEKS	12-18 WEEKS	10-12 WEEKS	2-6 WEEKS	NONE	NONE

*GATE EQUIVALENCE OF THESE DEVICES IS HIGHLY DESIGN-DEPENDENT.

FIGURE 15-27 Comparison of custom ASIC development methods.

DEVELOPING THE LAYOUT FOR A PC BOARD

As we mentioned in the first section of this chapter, in the traditional method of laying out a PC board we start with several sheets of translucent plastic and a light table. One sheet of plastic is used to develop the copper pattern for each layer of the board. The patterns are laid out on the plastic at a scale of two to four times the actual size they will be on the finished board. When the layouts are photographed, they are reduced to the actual size needed to manufacture the board.

To represent a pad on the layout, an opaque "doughnut" is stuck on a plastic sheet. If the pad is to be part of a plated-through hole, another doughnut is put in the same position on the sheet which represents the other side of the board. For a multilayer board, doughnuts with large holes are placed on the sheets for the power and ground planes so that the plated-through hole will not make contact with these layers unless a connection is desired. Ready-made patterns for standard components such as connectors, IC sockets, and transistors are available to speed up the process somewhat. Traces are produced on a layout by connecting opaque tape between pads.

Manual layout of a PC board is a very slow, tedious, and error-prone process, especially for a multilayer board. It often takes many hours to determine the best placement for the components and the best routing for the signal lines. Furthermore, if an error is found, a whole section of pads and traces may have to be removed and laid down again.

Fortunately, we now have computer programs which can assist greatly in determining the best placement of the components and can often route the entire board automatically. These programs are essentially *expert systems* which use a set of rules derived from experienced PC layout people to determine the best placement and routing. When using many of these programs, you can specify that a given trace is an ECL signal line, and the program will automatically route the line so that it has no stubs that might cause signal reflections. These PC board layout programs also allow you to interact with the routing so that you can equalize propagation delays, etc., if needed.

When completed, the computer-generated layout for each layer of the board is sent to a Gerber plotter or a laser plotter. The output from the plotter can then be photographed or used directly to manufacture the PC board. In a later section of the chapter we describe in detail how you use a computer-based PC board layout program, but here we want to tell you how the layout produced manually or by computer is used to manufacture a board.

MANUFACTURING A PRINTED-CIRCUIT BOARD

After the layout for a PC board is developed manually or with a computer, the layout for each copper layer is photographically reduced to the correct size for the board. The negatives produced are then used as part of the process of producing the board.

When the fiberglass-epoxy boards come from the manufacturer, they are completely coated on one or both sides with a thin copper sheet. The copper is usually covered with an adhesive-backed plastic sheet to keep it from oxidizing. After the protective sheet is removed, the production process proceeds as follows:

1. The bare board is rinsed with a solvent to remove any residues.

2. A thin coating of a material called *photoresist* is applied to the board and allowed to dry. The photoresist undergoes a chemical reaction when exposed to light.

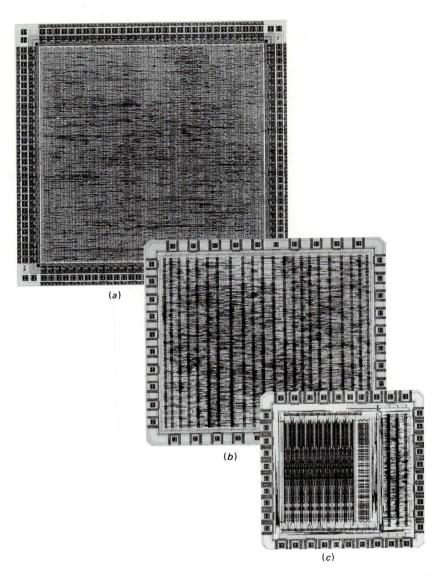

FIGURE 15-28   Special microprocessor implemented with *(a)* gate-array method, *(b)* standard-cell method, and *(c)* compiled-cell method. *(VLSI Technology Inc.)*

3. The negative for that layer is placed on the photosensitized board and ultraviolet light allowed to fall on it. A chemical change takes place in the areas of the photoresist where the light hit. (Some newer systems directly draw the trace pattern on the photoresist with a laser instead of using negatives and a UV light source.)

4. The board is "developed" in a solvent which removes the photoresist from all the areas of the board except those where the UV light "hardened" it. This process is similar to making photographic prints, thus the name *printed*-circuit board.

5. The next step is to etch away the copper from all areas of the board except those protected by the hardened photoresist. A solution of ferric chloride, sulfuric acid, or ammonium hydroxide is used to do this etching.

6. After etching, the remaining photoresist is removed with a solvent.

7. If the board is part of a multilayer board, the next step is to laminate the different layers into one board with heat and pressure.

8. Next, all the holes are drilled in the board. This is done with a *Numerical Control*, or NC, drilling machine. The NC here simply means that the drilling is controlled by a computer program. Most PC board layout programs now have an option which allows you to automatically produce the drilling program for standard NC drilling machines.

9. After the holes are drilled, the holes that are to be plated through are chemically etched to make space for the plating.

10. The holes are then electrochemically plated.

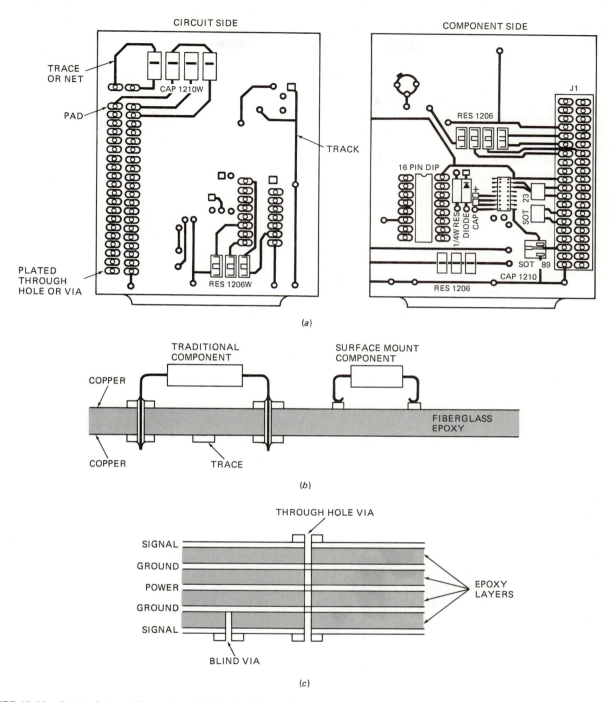

FIGURE 15-29 Printed-circuit boards. *(a)* Circuit-side and component-side views. *(b)* Cross-sectional view of two-layer board. *(c)* Cross-sectional view of multilayer board.

11. Next the entire board may be plated with tin by dipping it in a solution. This prevents oxidation of exposed copper and makes soldering easier.

12. The board is then tested with an automatic tester to make sure there are no shorts or opens in the board.

13. A solder mask/heat shield is applied to the board. This is the green coating you may have seen on PC boards.

14. The component identification labels are then silkscreened or in some other way printed on the board.

15. The boards are delivered to the customer for component insertion and testing.

INSERTING AND REMOVING COMPONENTS FROM PC BOARDS

Once a PC board is delivered, the next step is to "stuff" it with components and solder them in place. If the board is a traditional through-hole mount type, then an automatic insertion machine may be used to put in the components. After most of the components are inserted, this type of board is passed through a flow soldering machine on a conveyer belt. In most of these machines a fountain of molten solder contacts the circuit side of the board as it passes and quickly solders all the components in place. The solder mask on the board prevents solder bridges between adjacent traces, etc. After flow soldering, any specialized components are hand-soldered in place, component leads are trimmed as necessary, and soldering flux is removed from the board with a solvent.

If the board uses surface-mount components, a drop of paste containing solder and flux is first applied to each pad. An automatic place machine is then used to position components on the board. The paste holds the components in place on the board. After the majority of components are placed, the board is passed through a solder machine. For SMD boards a hot vapor is often used to melt the solder contained in the paste and to solder the components to the board. Again, after most of the components are soldered by machine, any specialized components are hand-soldered in and the remaining solder flux removed from the board with a solvent.

The best way to remove a component such as an IC from a printed-circuit board is to use a heat source which heats all the pins at once. Figure 15-30 shows one such

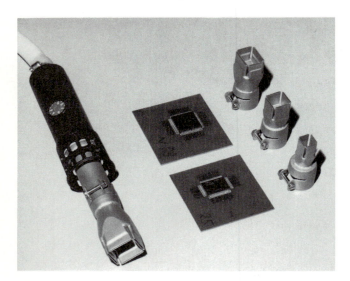

FIGURE 15-30   Leister-Labor S hot-air contactless desoldering and soldering tool for removing and replacing leaded and surface-mount components on PC boards.   *(Courtesy Brian R. White Co., Inc., Ukiah, California.)*

tool. When the solder is melted on all the pins, it can easily be removed and the holes cleaned out as needed to insert a new part. If a heat source such as this is not available, then the best way to remove a part such as an

IC with many pins is to first cut all the pins to free the body of the IC. Then remove the IC pins from the board one at a time.

Removing SMD components from a PC board requires considerable care, because the pads are smaller, closer together, and more fragile than those on through-hole boards. For these boards the only practical way to remove components without damaging the board is with a heater which heats all the leads on the device at once. The tool in Figure 15-30 uses a jet of hot air to heat component leads. It can be used to remove a wide variety of devices with pins and surface-mount devices by simply changing the nozzle used to direct the blast of hot air. If the tool is positioned carefully, the solder on the desired component will be melted before the heat spreads to adjacent components. A scrap piece of PC board or aluminum foil can be used to direct heat away from other components if they are very close to the part you are removing. The main precaution when using a tool such as this is to keep the heat on the component just long enough to melt the solder! This usually takes only a few seconds.

To replace components on a surface-mount board, you put a small dab of solder-flux paste on each pad, press the component in place, and then heat all the leads at once with the tool shown in Figure 15-30.

It is important for you to be familiar with surface-mount techniques, because it is predicted that the use of SMDs will soon far surpass the use of components with leads. The main point to remember is that the small scale of SMD boards means that you have to work on them very carefully.

Now that you have an overview of how PC boards are laid out, manufactured, and populated with components, we will finish the chapter by showing you how a computer program is used to produce a PC board layout from a schematic.

Computer-Aided PC Board Layout

Examples of currently available low-cost PC board layout programs are OrCAD/PCB, Accel Technologies' Tango-PCB, and Douglas Electronics' Professional Layout. These programs allow you to lay out and route the traces for a PC board in a manner very similar to the way a schematic capture program allows you to draw a circuit diagram. For the discussion in this section we chose the OrCAD/PCB program because it is easy to use, and it is fully compatible with the OrCAD Schematic Design Tools we showed you how to use in Chapter 5 and the OrCAD/VST tools we showed you how to use earlier in this chapter. As an example circuit for this section we will use the frequency counter control circuit shown in Figure 15-8.

After considerable thought we decided that the best way to lead you through the computer-aided PC board development process is with a list of numbered steps as follows.

1. *Draw a schematic for the design using the OrCAD Schematic Design Tools.*

2. *Simulate the design with OrCAD/VST to make sure the design works as desired.*

3. *Add the package data to the schematic file.* Each device in a schematic has a group of "parts fields" associated with it. The parts fields are used to hold data such as the part name, part identifier, tolerance, package type, etc. If, for example, the design contains a 74HC160, you have to "stuff" one of the parts fields with a code that tells the layout program whether to use a standard 16-pin DIP or a surface-mount package.

An OrCAD STD utility program called FLDSTUFF is usually used to add the package data to the schematic file. We used the simple command

FLDSTUFF ONESHOT.SCH 8 TTL.STF /O/V

to add a standard DIP code to part field 8 of each of the devices in our frequency counter circuit.

4. *Run NETLIST again to generate a net file suitable for the PCB program.* A command such as the one shown here (but typed on one line) is used to do this:

NETLIST ONESHOT.SCH ONESHOT.NET
ORCADPCB /S

5. *Run the PCB program.* This is done by simply typing:

PCB ‹CR›

6. *Set up the PCB with the desired layout parameters.* The setup submenu allows you to specify track width, via size, number of layers, via type, grid spacing on CRT, etc. For our example here we used the default settings.

7. *Draw the outline of the board on the screen.* This is done using the PLACE command and the EDGE subcommand. To decide how large to make the board, we use a rough rule of 1 square inch for each 14-pin or 16-pin DIP and the associated nets. If this turns out to be too much or not enough, you can always modify the edges later. To draw in the edges you put the cursor where you want one corner of the board and press the B key, move the cursor to the next corner and press the B key again, etc. When you get back to the starting corner, you press the E key to end the sequence.

8. *Define any copper zones, no-via zones, or forbidden zones.* Forbidden zones are areas of the board where you don't want tracks or vias. You use the PLACE command and the ZONE subcommand to draw the boundaries of zones in about the same way you place the edges of the board. Once a zone is defined, you use the TYPE menu to specify whether the zone is copper, no via, or forbidden. You then put the cursor in the zone and SEED the zone by pressing the S key. The zone region of the display will then be filled in with a characteristic pattern which identifies that type zone. The layout in Figure 15-31 shows an example of a forbidden zone in the upper left-hand corner.

9. *Bring the circuit modules (parts) on to the display from disk.* To do this you go from the QUIT com-

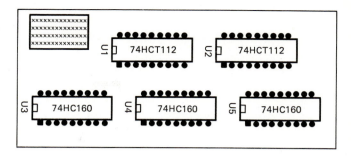

FIGURE 15-31 Plot of basic printed-circuit board showing forbidden zone and placement of five device modules.

mand in the MAIN menu to the INITIALIZE command in the next menu to the USE NETLIST command in the next-level menu. When prompted, you enter the name of the netlist file. A stack containing all the modules in the circuit will then appear at the cursor location on the screen.

10. *Place the circuit modules on the board.* To do this go through the menus PLACE, MODULE, and then PLACE. Move the cursor to the stack of modules, and press the left mouse key to select the top module. A box representing this module will move with the cursor and can be positioned on the board. When you get a module positioned where you want it, press the left mouse key again and the module will stay in position. Repeat the process until all modules are removed from the stack and placed on the board. If you change your mind about the placement of a module, you can select it by putting the cursor on it and pressing the left mouse key, moving the module to where you want it, and pressing the left mouse key again to hold/leave the module in that position. Figure 15-31 shows our basic board with the five modules placed.

11. *Compile the NETLIST for routing.* This is done by selecting ROUTING in the MAIN menu, NETLIST in the next menu, and COMPILE in the next menu.

12. *Check that the modules are placed for minimum trace lengths.* The two ways to do this are the OrCAD VECTOR and RATSNEST commands. To get to these options you select ROUTING in the main menu, NETLIST in the ROUTING menu, and VECTOR or RATSNEST in the NETLIST menu.

When the VECTOR command is executed, you get a display such as that shown in Figure 15-32a. The diamonds indicate the "center of gravity" for each of the modules. Ideally the physical center of a module should be the same as its center of gravity. To improve a layout, you can move modules around and try the vector command again. To help you determine if moving modules helped, VECTOR gives you a total net length in the lower right-hand corner of the display. This *Manhattan length* is computed by routing all the nets with rectangular connections and adding all the X distances and all the Y distances. In other words:

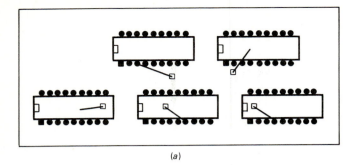

(a)

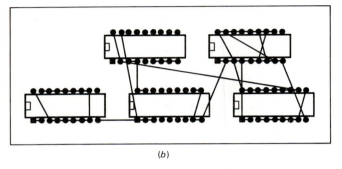

(b)

FIGURE 15-32 Optimizing placement of modules.
(a) Layout showing vectors. (b) Layout showing RATSNEST.

$$\text{Manhattan length} = \Sigma X + \Sigma Y$$

The RATSNEST option produces a display such as that shown in Figure 15-32b. As you can see, RATSNEST makes direct connections between pins. By looking at the ratsnest you can usually see how to move modules to reduce the Manhattan length shown in the lower right-hand corner of the display. When you finish with the VECTORS or the RATSNEST, you ERASE them before routing.

13. *Route critical paths manually.* Select the ROUT-ING option in the main menu, put the cursor on one end of a net you want to route, and press the B key. Move the cursor to where you want the first turn, and press the B key again. When you get to the destination pin, press the N key to draw a NEW net or the E key to END manual routing. If at some point in a net you need to insert a via to go to the other layer, press the O key, continue the net, and press the O key again to get back to the original layer. Incidentally, for a two-layer board, the component side is referred to as layer 2 and the circuit or solder side is referred to as layer 1. Traces are usually run vertically on even-numbered layers and horizontally on odd-numbered layers. This convention makes it easier for signal lines to cross each other by going to the other side of the board. PCB uses an X cursor when you are on layer 2 and a + cursor when you are on layer 1.

14. *Route rest of paths automatically.* Before autorouting a board you may want to specify a rout-ing strategy. To see the options for this select AUTOROUTE in the ROUTING menu and STRAT-

EGY in the AUTOROUTE menu. Normal strategy for-bids 45° connections to traces and discourages 45° connections to pads. Flexible strategy relaxes these rules somewhat, and EXTENSIVE eliminates these rules. Usually you try to autoroute a board with NORMAL, and if the routing is not fully successful, you select FLEXIBLE or EXTENSIVE and try autorouting again.

You can autoroute all or part of a board. To autoroute all the board, select ALL in the AUTOROUTE menu and press the left mouse key. When autorouting is complete, the screen will show a display similar to that in Figure 15-33a for layer

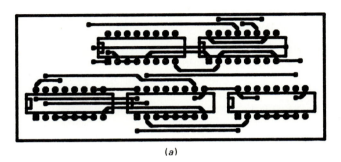

(a)

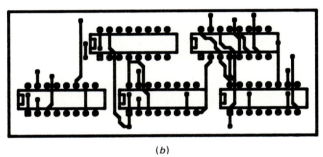

(b)

FIGURE 15-33 Plot after AUTOROUTE ALL commands.
(a) Solder side. (b) Component side.

1 or Figure 15-33b for layer 2. If any nets are left unconnected, you can try a different strategy or try manual routing to connect them.

15. *Plot the layout for each layer.* For inspection pur-poses you can plot each layer on a standard pen plotter. Figure 15-33 shows a plot of both layers of our board. However, for the plots that will be used to make the board, you use a Gerber photoplotter or a laser photoplotter. These plotters have the res-olution required to produce smooth lines in the plot.

As with our previous examples where we used specific programs, we don't expect you to remember all the de-tails of OrCAD PCB unless you are actually working with it. However, the first sentence of each of the preceding steps should give you a good overview of developing a PC board layout with a computer program. From personal experience we can testify that this method is much eas-ier than the traditional manual method. It should be-come even easier as artificial intelligence techniques are used to improve the speed and flexibility of PC board place-and-route programs.

CHECKLIST OF IMPORTANT TERMS AND CONCEPTS

Computer-aided engineering (CAE)
Electronic design automation (EDA)
Design rule checker (DRC)
Electrical rule checker (ERC)
Netlist
Logic synthesis program
Logic synthesis tools
Hardware definition language (HDL)
Custom application-specific integrated circuit (ASIC)
Nonrecurring engineering (NRE) charge
Simulation
   Properties
   Time step
   Clock- or time-step-driven simulator
   Event-driven simulator
   Node
Behavioral model
Hardware verification model
Processor control language (PCL)
Full-functional model
Hardware modeling
Simulators, multilevel and mixed-mode
Stimulus file
Built-in self-test (BIST)

Built-in test (BIT)
Fault simulator
Exhaustive testing
Design for test
Undetected and detected faults
Fault coverage
Scan-path testing
Signature analysis
Logic cell array (LCA)
Configurable logic block (CLB)
Configurable I/O block (IOB)
Automated design inplementation (ADI)
In-circuit emulator
Programmable gate array (PGA)
Gate array
Place-and-route program
Standard-cell, compiled-cell, full-custom design
Floor planning
Silicon compiler
Data-path compiler
State-machine compiler
Printed-circuit board development
   Net, track
   Pad
   Via: through-hole, blind, buried
   Ground and power planes
Expert systems
Manhattan length

REVIEW QUESTIONS AND PROBLEMS

1. One of the most complex microprocessors now available is the Intel 80386 which contains about 285,000 transistors. Assuming 50 percent utilization, how many of these microprocessors could be implemented in a single billion-transistor device?

2. a. Describe how a simulator is used to test the prototype of a circuit.
   b. Describe the advantages of this software-breadboard approach to testing the prototype of a circuit.

3. What information does a fault-simulator program give you about a circuit?

4. List the five major methods used to develop custom ASICs.

5. Describe the effect that electronic design automation will have on each of the following and explain why you think it will have that effect:
   a. The complexity of systems.
   b. Built-in test and redundancy.
   c. The time from design concept to marketed product.
   d. The size of design teams.
   e. The role of engineering technicians.

6. What determines the time resolution of a simulator?

7. a. What is the advantage of a behavioral model over a gate-level model when simulating a device such as a microprocessor?
   b. When are hardware models used in a simulation?

8. a. What information do you put in a stimulus file before you run a simulation?
   b. How do you tell a simulator which signals you want to see on the screen after a simulation is run?

9. When simulating a flip-flop or a counter, why is it necessary to set or reset it before clocking it?

10. Write a list of stimuli that you would use to determine if the circuit in Figure 9-34 steps through the desired count sequence as it is clocked.

11. a. Describe how a fault simulator is used to determine the testability of a circuit.
    b. Approximately what level of fault coverage is considered acceptable for a design?

12. Figure 8-38 shows a diagram for a simple digital clock. As shown, this circuit would require 43,200 or 86,400 clock pulses to step it through an entire count sequence.
    a. Draw a diagram showing how test circuitry can be added so that the circuit can be fully tested

with only 12 or 24 clock pulses when a test input pin is asserted low.

b. Draw a diagram showing circuitry that could be added so that the carryout from each counter could be routed to an output pin one at a time for testing.

13. a. Describe the basic structure of a logic cell array.

b. What is the maximum-size circuit that can be implemented in currently available LCAs?

c. Describe how the program for an LCA is developed.

d. Describe how an LCA is programmed to implement a desired circuit.

e. What are the major advantages of LCAs over other custom ASIC methods such as the standard-cell method?

14. a. Describe the basic structure of a gate array.

b. Describe how the layout for a gate-array design is produced with computer-based tools.

c. Define the term *NRE* and explain its relevance to gate-array ASICs.

15. a. Why are the production costs higher and the turnaround times longer for a cell-based ASIC than for a gate-array ASIC?

b. What is a major advantage of the compiled-cell approach over the standard-cell approach to custom ASIC design?

c. For what type applications do we resort to full-custom layout of an ASIC design?

16. a. Describe the manual method for laying out a PC board.

b. What are the major advantages of the computer-based layout method over the manual method?

17. Briefly describe the steps in the actual manufacture and component mounting of a PC board.

18. Describe how large surface-mount components are removed and replaced on a PC board.

19. Describe the major steps in producing a PC board layout from a schematic using computer-based tools.

BIBLIOGRAPHY

The major sources of further information on the topics discussed in this book are manufacturers' data books, application notes, and articles in current engineering periodicals. With the foundation provided in this text, you should be able to read these materials comfortably.

Listed below, by chapter, are some materials you may find helpful. Following the chapter listings is a list of periodicals which we found to be particularly helpful in keeping up with the latest advances in digital devices and systems. Most of these periodicals have articles which review the basic principles of a particular area of electronics and then discuss the latest developments in that area.

Chapters 1–9

TTL Data Book, Volume 2, Texas Instruments Inc., Dallas, Texas, 1985.

ALS/LS Logic Data Book, Texas Instruments Inc., Dallas, Texas, 1986.

IEEE Standard Graphic Symbols for Logic Functions, ANSI/ IEEE Std. 91-1984, The Institute of Electrical and Electronics Engineers, New York, 1984.

Logic Symbols and Diagrams, The Institute of Electrical and Electronics Engineers, New York, 1987.

Kampel, Ian, *A Practical Introduction to the New Logic Symbols,* Second Edition, Butterworths, Essex, England, 1986.

High-Speed CMOS Logic Data Book, Series A, Motorola Semiconductor Products Inc., Austin, Texas, 1983.

High-Speed CMOS Logic Data Book, Texas Instruments Inc., Dallas, Texas, 1988.

F100K ECL Data Book, Fairchild Semiconductor Corp., Puyallup, Washington, 1986.

Fast Data Manual, Signetics Corp., Sunnyvale, California, 1986.

Motorola MECL Data Book, Series C, Motorola Semiconductor Products Inc., Phoenix, Arizona, 1982.

Chapter 5

Schematic Design Tools Manual, OrCAD Systems Corp., Hillsboro, Oregon, 1988.

Chapter 9

Component Data Catalog, GE-Intersil Inc., Cupertino, California, 1987.

Chapter 10

MECL System Design Handbook, Fourth Edition, Motorola Semiconductor Products Inc., Phoenix, Arizona, 1983.

Telecommunications Device Data Book, Series B, Motorola Inc., Austin, Texas, 1985.

Fike, John L. and George E. Friend, *Understanding Telephone Electronics,* Second Edition, Texas Instruments Inc., Dallas, Texas, 1984.

CMOS/NMOS Special Functions Data Book, Motorola Inc., Austin, Texas, 1986.

Chapter 11

Pohlmann, Ken C., *Principles of Digital Audio,* Howard W. Sams & Co., Indianapolis, Indiana, 1987.

Data Acquisition and Conversion Handbook, GE and Datel, Mansfield, Massachusetts, 1979.

Component Data Catalog, GE–Intersil Inc., Cupertino, California, 1987.

Chapter 12

Roth, Charles H., Jr., *Fundamentals of Logic Design,* Third Edition, West Publishing Co., St. Paul, Minnesota, 1985.

PAL Device Handbook and Data Book, Monolithic Memories–Advanced Micro Devices Inc., Sunnyvale, California, 1988.

GAL Handbook, Lattice Semiconductor Corp., Portland, Oregon, 1986.

System Design Handbook, Second Edition, Monolithic Memories–Advanced Micro Devices Inc., Sunnyvale, California, 1985.

ABELTM 3.0 Manual, Data IO Corp., Redmond, Washington, 1988.

Chapter 13

Hall, Douglas V., *Microprocessors and Interfacing, Programming and Hardware,* McGraw-Hill, New York, 1986.

Mick, John and Jim Brick, *Bit-Slice Microprocessor Design,* McGraw-Hill, New York, 1980.

Microprocessors and Peripherals Handbook, Volumes 1 and 2, Intel Corporation, Santa Clara, California, 1988.

Chapter 14

Hall, Douglas V., *Microprocessors and Interfacing, Programming and Hardware,* McGraw-Hill, New York, 1986.

Memory Management Applications Handbook 1987—DRAM Controllers/Cache Memories/EDAC/FIFO/BiCMOS Drivers, Texas Instruments, Dallas, Texas, 1987.

Chapter 15

The Programmable Gate Array Design Handbook, Xilinx Inc., San Jose, California, 1986.

Verification and Simulation Manual, OrCAD Systems Corp., Hillsboro, Oregon, 1988.

PC Board Layout Tools Manual, OrCAD Systems Corp., Hillsboro, Oregon, 1988.

Periodicals

EDN, ISSN 0012-7515, Cahners Publishing Co., Boston, Massachusetts.

Electronic Design, USPS-172-080, Hayden Publishing Co., Rochelle Park, New Jersey.

Electronics, ISSN 0883-4989, VNU Business Publications, Hasbrouck Heights, New Jersey.

Electronic Engineering Times, ISSN 0192-1541, Electronic Engineering Times, Manhasset, New York.

APPENDIX A
PIN CONNECTION DIAGRAMS FOR SELECTED TTL DEVICES

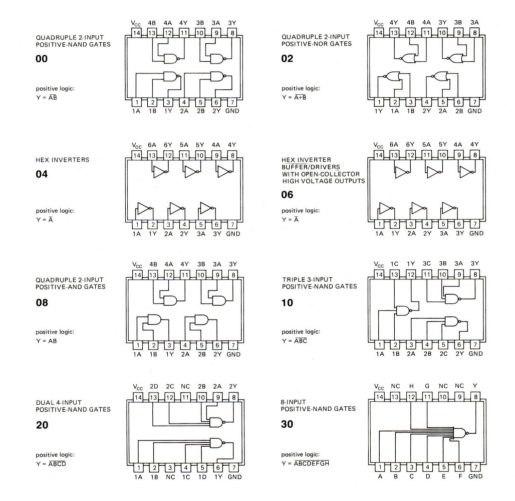

QUADRUPLE 2-INPUT
POSITIVE-NAND GATES

00

positive logic:
$Y = \overline{AB}$

QUADRUPLE 2-INPUT
POSITIVE-NOR GATES

02

positive logic:
$Y = \overline{A+B}$

HEX INVERTERS

04

positive logic:
$Y = \overline{A}$

HEX INVERTER
BUFFER/DRIVERS
WITH OPEN-COLLECTOR
HIGH VOLTAGE OUTPUTS

06

positive logic:
$Y = \overline{A}$

QUADRUPLE 2-INPUT
POSITIVE-AND GATES

08

positive logic:
$Y = AB$

TRIPLE 3-INPUT
POSITIVE-NAND GATES

10

positive logic:
$Y = \overline{ABC}$

DUAL 4-INPUT
POSITIVE-NAND GATES

20

positive logic:
$Y = \overline{ABCD}$

8-INPUT
POSITIVE-NAND GATES

30

positive logic:
$Y = \overline{ABCDEFGH}$

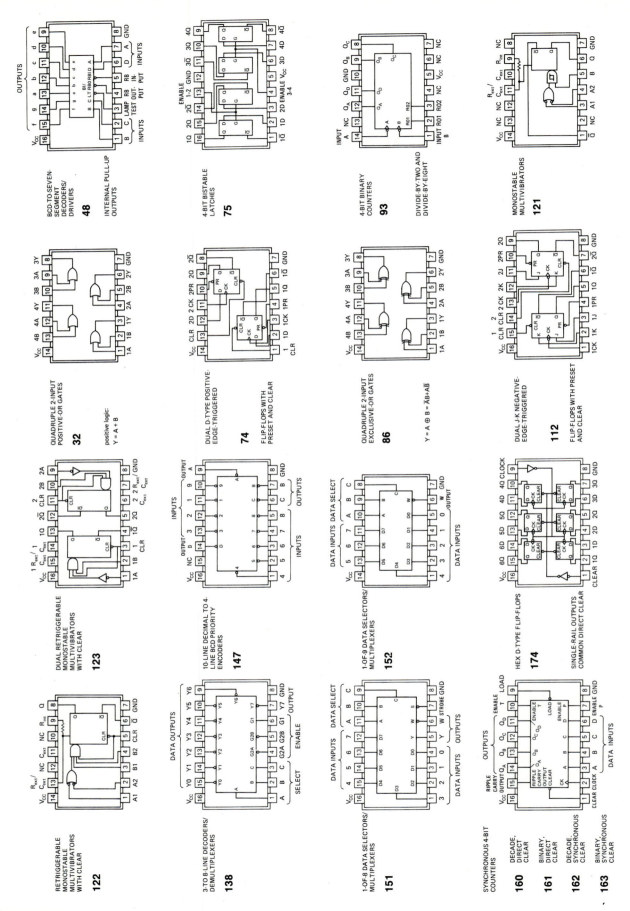

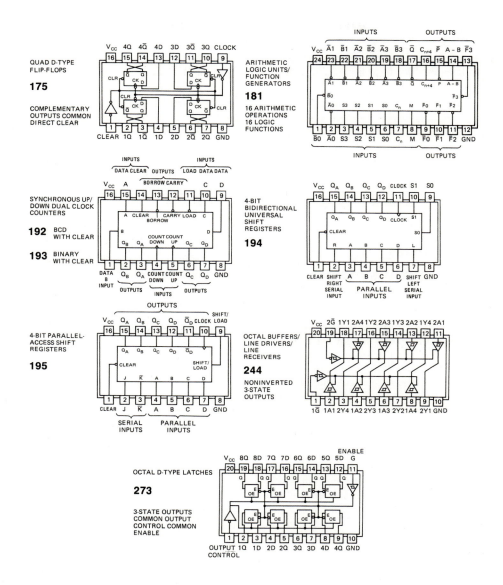

QUAD D-TYPE
FLIP-FLOPS

175

COMPLEMENTARY
OUTPUTS COMMON
DIRECT CLEAR

ARITHMETIC
LOGIC UNITS/
FUNCTION
GENERATORS

181

16 ARITHMETIC
OPERATIONS
16 LOGIC
FUNCTIONS

SYNCHRONOUS UP/
DOWN DUAL CLOCK
COUNTERS

192 BCD
WITH CLEAR

193 BINARY
WITH CLEAR

4-BIT
BIDIRECTIONAL
UNIVERSAL
SHIFT
REGISTERS

194

4-BIT PARALLEL-
ACCESS SHIFT
REGISTERS

195

OCTAL BUFFERS/
LINE DRIVERS/
LINE
RECEIVERS

244

NONINVERTED
3-STATE
OUTPUTS

OCTAL D-TYPE LATCHES

273

3-STATE OUTPUTS
COMMON OUTPUT
CONTROL COMMON
ENABLE

APPENDIX B
INTERNAL STRUCTURE AND OPERATION OF LOGIC GATES

Most of the time you don't need to know about the internal structure of a logic gate because the data sheet parameters give you a "black-box" model which allows you to successfully interface it to other devices. However, to better understand the data sheet parameters, it is sometimes useful to have some knowledge of the internal structure. We therefore present here a brief introduction to the internal structure of basic gates for several common logic families.

BIPOLAR LOGIC FAMILIES

Review of Basic Bipolar Transistor Switch Operation

Logic gates are essentially electric switches. Integrated-circuit logic gates use bipolar or MOS transistors to make these switches. Figure B-1 shows the circuit for an NPN bipolar transistor switch that functions as an inverter. If a logic high voltage of $+5$ V is applied to the input, the base-emitter junction of the transistor will be forward-biased and the transistor turned on. The base will be at about 0.7 V. This leaves 5 V $- 0.7$ V, or 4.3 V, across the 10-kΩ base resistor. Therefore, a current of 4.3 V/10 kΩ, or 0.43 mA, will flow into the base. This base current will cause a much larger current flow from the collector to the emitter of the transistor. The ratio of collector current to base current is called beta (β), and is typically about 100. This means that the collector current will be 100 times the base current. For the circuit shown, the base current of 0.43 mA would theoretically produce a collector current of 43 mA. However, a current of 43 mA through the 1-kΩ collector resistor would produce a voltage drop of 43 V across it (43 mA $\times$ 1 kΩ = 43 V). This is clearly impossible with a 5-V supply. What actually happens is that the collector current increases until the voltage drop across the collector resistor pushes the collector voltage down to within a few tenths of a volt of that on the emitter. For the example shown, the collector will be at about 0.1 V, which is a logic low. In this condition the transistor is said to be *saturated*. This term comes from the fact that the base region of the transistor is

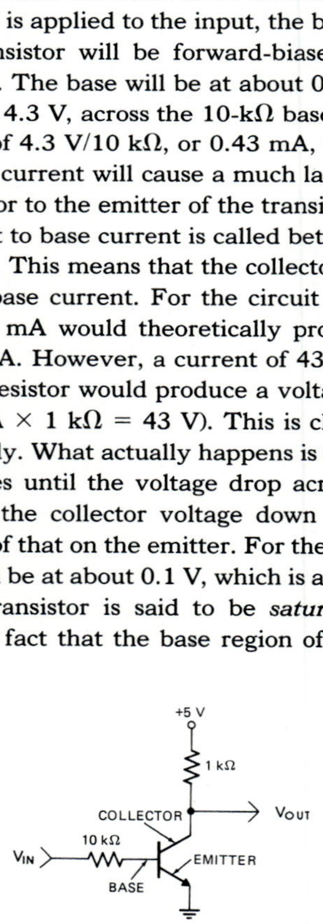

saturated with charges because there is not enough voltage on the collector to pull out all the available charge there. The important point here is that a saturated bipolar transistor has a voltage of a few tenths of a volt between its collector and its emitter. For the circuit in Figure B-1, then, a logic high on the input produces a logic low on the output, so the circuit functions as an inverter. If the input of the circuit in Figure B-1 is tied to ground, a logic low, the base-emitter junction will not have the 0.7 V required for it to be forward-biased. Therefore, the transistor will be off. With no base current there is no collector current. With no collector current there is no voltage drop across the 1-kΩ collector resistor. The collector of an "off" transistor floats to whatever voltage it is pulled to. In this circuit the collector will be pulled to $+5$ V, which is a logic high. A logic low on the input turns off the transistor and produces a logic high on the output. The transistor in this circuit is either off or saturated.

TTL NAND Gate Circuitry

Figure B-2 shows the internal circuitry of a standard TTL NAND gate. To analyze the operation of this circuit, assume the B input is tied high to $+5$ V, so that it has no effect. Assume also that the A input is grounded or at a logic low state, as in Figure B-2b. Grounding the A input gives a path for the base current to flow through Q1 and turn on Q1. The base of Q1 is at V_{BE} above ground, or about $+0.7$ V. This leaves 4.3 V across the 4-kΩ resistor. With Ohm's law you can calculate the base current to be about 1.1 mA. Since Q1 has no source of collector current, 1.1 mA of base current will pull Q1 into saturation, and its collector voltage will be only about 0.1 V above its emitter voltage. Since the emitter is grounded, the collector of Q1 and the base of Q2 are about 0.1 V. This is much less than the 0.7 V required to turn on Q2, so Q2 is off. With no current flowing through Q2, no voltage is developed across the 1-kΩ resistor, and the base of Q3 is at ground. Therefore, Q3 is also off and has no effect on the output, since the collector of an off transistor acts as an open switch.

Now that you have followed the emitter path of Q2 to the output, go back and follow the collector path of Q2 to the output. Since Q2 is off, its collector has no effect on the base of Q4. Q4 is supplied with base current by the 1.6-kΩ resistor connected to $+5$ V. The maximum output load current the Q4 would have to supply in this state is the 0.4 mA, or 400 μA, needed to drive the inputs of up to 10 load gates. The small base current needed to produce this output develops little voltage drop across the 1.6-kΩ resistor. The base of Q4 is near 5 V. The output is a V_{BE} drop of 0.7 V plus a diode drop across D3 of 0.7 V below this, or about 3.6 V. For standard TTL, an

FIGURE B-1 NPN bipolar transistor switch/inverter.

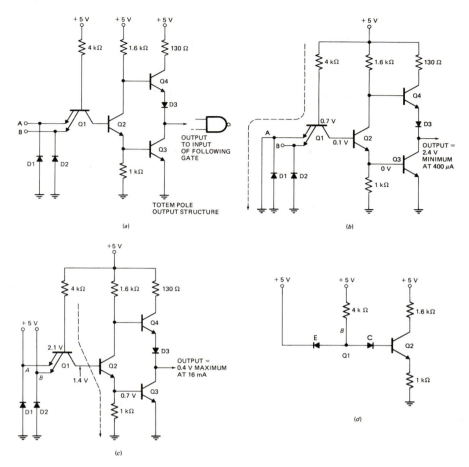

FIGURE B-2   (a) Circuit of standard TTL NAND gate with totem-pole output structure. (b) Voltages with input A grounded. (c) Voltages with input A high. (d) Ohmmeter view of Q1.

output voltage greater than 2.4 V is a legal logic high level. A low or ground on the A input makes the output high. You can trace currents and voltages to show that the B input is equivalent to A in function. From this you can deduce that the output of this TTL NAND gage is high if A is low, if B is low, or if both A and B are low. These cases are represented by the bubbled input OR symbol for a NAND gate. Next consider the effect of making input A high (see Figure B-2c). Again assume the B input is tied high to +5 V. With input A also at a positive logic high, or tied to +5 V, there is not a path for Q1 base-emitter current to flow, so Q1 is off. However, the ohmmeter view of an NPN transistor in Figure B-2d shows an available current path to the base of Q2 through the 4-kΩ resistor and the collector-base diode of Q1. This path supplies sufficient base current to Q2 to saturate it. Q2 supplies enough drive to Q3 to turn it on and to drive it into saturation. The voltage at the output now is the voltage across a saturated transistor—a few tenths of a volt—which is a logic low. Now go back and follow the other path to see how the high on the A input affects Q4. The emitter of Q2 is at 0.7 V as set by V_{BE} of Q3. Since Q2 is saturated, its collector is only about 0.1 V above this, or about 0.8 V with respect to ground. This is considerably less than the 1.4 V necessary to turn on the series combination of the Q4 base-emitter junction and D3. Therefore Q4 is off and has no effect on the low output. The purpose of D3 is to en-

sure, as much as possible, that Q4 is never on at the same time as Q3. If Q3 and Q4 were both on for very long, a large current would flow directly from V_{CC} to ground, heating the chip and wasting power. Even with this totem-pole output structure using D3, there is a brief instant each time the output changes states at which both transistors are on. This allows a short surge of current through the on transistors to ground, which produces voltage spikes or transients on the V_{CC} line. Unless you have sufficient bypass capacitors, these transients may disturb other circuits.

The purpose of the input clamp diodes D1 and D2 in Figure B-2 is to limit the amplitude of negative-going signal reflections on the input lines.

Schottky TTL Internal Gate Structure

In their eternal quest for speed, integrated-circuit designers sought a way of overcoming the storage time problem of the saturated transistors in the standard TTL gate. The solution came in the form of a unique diode made with a metal-semiconductor junction. This Schottky diode, as it is called, has a forward voltage drop of about 0.3 V and turns on or off very fast because it has no storage time problem. Referring to Figure B-3a, a Schottky diode connected between the base and

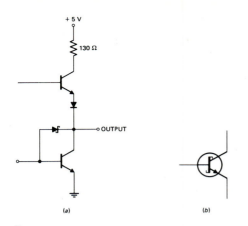

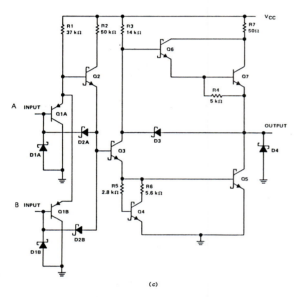

FIGURE B-3 (a) Schottky diode clamp to prevent TTL output transistor collector from reaching saturation. (b) Schottky diode-clamped transistor schematic symbol. (c) Internal structure of a 74ALS00 NAND gate.

collector of a transistor goes into conduction when the collector voltage of the transistor drops to about 0.3 V below the base voltage. This clamps the collector voltage at about 0.4 V and prevents the transistor from going into saturation. Since the transistor is not saturated and the Schottky diode has no storage time, both can be turned off very rapidly.

In logic ICs the Schottky diode is often made as part of the transistor. Figure B-3b shows the schematic symbol for a Schottky diode-clamped transistor. By making most of the transistors in the IC gates Schottky diode-clamped and using slightly different circuits, manufacturers have produced the various Schottky TTL families which are faster and/or use less power than standard TTL. Figure B-3c shows the internal structure of a 74ALS00 NAND gate. To start the analysis of this circuit, assume that the A input is high so it has no effect, and that the B input is low. A low on the B input will turn on Q1B and pull current through R1. This will rob the base cur-

rent from the Q2, Q3, and Q5 path, so they will all be off. With Q3 off, R3 will supply base current to Q6. The emitter current from Q6 will turn on Q7 and pull the output to a logic high. The Darlington output structure consisting of Q6 and Q7 supplies increased output drive for the output high state.

If the A and B inputs are both high, Q1A and Q1B are both off. R1 then supplies base current to Q2 and turns it on. The emitter current from Q2 supplies base current to Q3 and turns it on. The emitter current of Q3 turns on Q4 and Q5. When Q5 turns on, it pulls the output to a logic low. Q3 robs the base current from Q6 so it is off, and consequently Q7 is off. The Q4 transistor in this circuit is called an active turnoff. Its purpose is to help Q5 turn off more rapidly when the output switches from low to high. This reduces the transients on V_{CC} caused by both transistors being on for a short time during the transition.

Emitter-Coupled Logic Gate Structure

Emitter-coupled logic, or ECL, is another group of logic families that utilizes a very different circuit from that of the TTL families to achieve very short propagation delays. Figure B-4a shows the circuit of a typical ECL gate of the MECL 10,000 family. Note the supply voltages of ground and of −5.2 V. The simplest gate in these families is an OR-NOR gate rather than the NAND gate of the TTL families. Refer to Figure B-4a and mentally remove Q2, Q3, and Q4. You can see that Q1 and Q5 have their emitters coupled together to form a simple differential amplifier. The Q7 and Q8 transistors provide low-impedance emitter follower outputs from the collectors of the differential amplifier. Q5 supplies a stable reference V_{BB} of −1.3 V to Q5, one side of the differential input amplifier. If the bases of Q1 and Q5 were at the same voltage, −1.3 V, then the two transistors would share the emitter current equally. When the base voltage on Q1 is pulled low to the −5.2 V V_{EE}, no current will flow through Q1 and its collector voltage will go up to give a high on the NOR output. The increase in current through Q5 will drop its collector voltage to give a low on the OR output. As you can see from Table 3-2 (Chapter 3), a minimum output high level is −0.980 V and a maximum output low level is −1.95 V. Internal resistors RP tie the bases of unused inputs low to −5.2 V V_{EE}, so they do not affect the output. If input A is now switched from a low state to a high state of −0.980 V, current flow will switch from Q5 to Q1, since the base of Q1 is now at a higher voltage than the base of Q5. Q5 turning off will raise the OR output to a high level, and Q1 turning on will drop the NOR output to a low level. Raising any one or all the inputs to a high level will cause the OR output to go high, and the NOR output will go low. Emitter pull-down resistors, typically 510 Ω, are tied from each output to −5.2 V to equalize the output risetimes and falltimes. Output signal lines more than a few inches long are terminated with a 50-Ω resistor to −2 V, as described in Chapter 10.

MOS LOGIC FAMILIES

All the logic families discussed so far use bipolar transistors in their circuitry. The families to be discussed next use MOS (metal-oxide semiconductor) transistors.

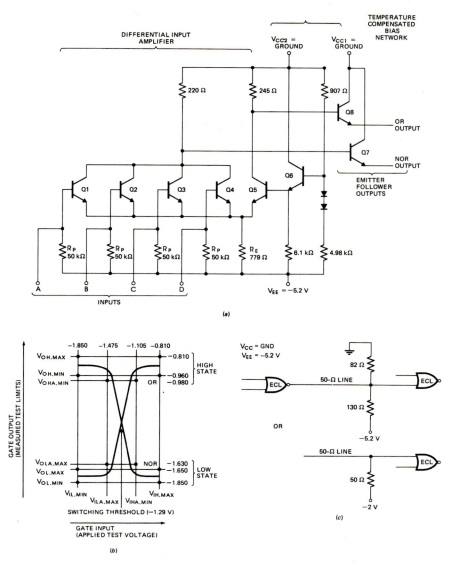

FIGURE B-4 (a) Typical emitter-coupled logic gate of the Motorola MECL 10,000 family. (b) ECL OR-NOR transfer curves *(Courtesy of Motorola)* (c) Terminations for ECL 50-Ω transmission line.

Review of MOS Transistors

To refresh your memory concerning MOS transistors, Figure B-5 shows the structure, conductance curves, and schematic symbols with proper voltage connections for the common types of MOSFET. The structure of the first of these, the enhancement-mode type, is shown in Figure B-5a. The N+ regions connected to the source and drain are just heavily doped N-type silicon in which most carriers are electrons. A P-type substrate is electrically insulated from the control gate by a layer of silicon dioxide and a layer of silicon nitride, which protects the silicon dioxide from contamination. As the device is shown, there is no conduction path from the source to the drain because the substrate, source, and drain form two back-to-back PN junctions. At room temperature the P substrate will contain some free electrons produced by thermal agitation as well as the "holes" produced by doping. If the substrate is tied to the source and a positive voltage is applied to the gate, these free electrons will be attracted to the edge of the substrate near the gate. The attracted electrons form a conducting electron or N channel (labeled "induced channel" on Figure B-5a) between the source and drain. Figure B-5b shows that the more positive the gate is made, the more drain current flows through the device. Negligible drain current flows until a threshold voltage $V_{GS,TH}$ of typically 2 to 3 V between gate and source is reached. Note that the schematic symbol in Figure B-5c shows the channel as a broken line, which reminds you that no channel exists unless the gate is made positive to enhance the conduction; hence the name enhancement-only (E-only) MOSFET.

The structure of the depletion-enhancement MOSFET shown in Figure B-5d is identical to the E-only MOSFET except that an N-type channel is diffused between the N+ source and drain regions. This means that even with zero voltage applied between the gate and source substrate, a conduction channel exists between source and drain, so current will

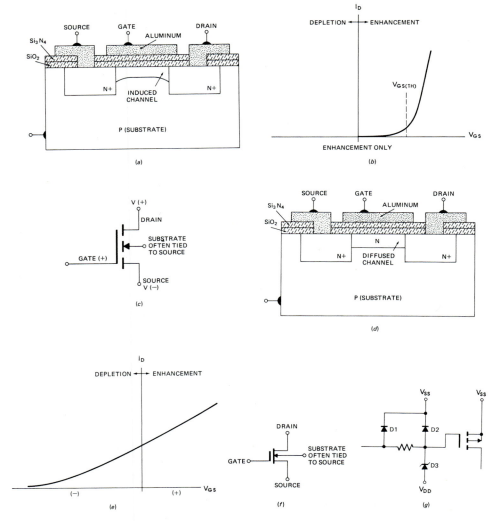

FIGURE B-5 (*a*) E-only MOSFET structure. (*b*) E-only MOSFET transconductance curve. (*c*) E-only MOSFET schematic symbol. (*d*) D-E MOSFET structure. (*e*) D-E MOSFET transconductance curve. (*f*) D-E MOSFET schematic symbol. (*g*) Typical MOS input static protection circuitry.

flow. If the gate is made more positive with respect to the source, free electrons are attracted from the substrate and the channel is "enhanced," so more drain current will flow. If the gate is made more negative, electrons are repelled from the channel region. This "depletes" the channel and raises its resistance, so less current flows. The conductance curve in Figure B-5*e* illustrates the effect of gate source voltage on drain current for the D-E, or depletion-enhancement, MOSFET. The schematic symbol in Figure B-5*f* shows the channel as an unbroken line, reminding you that the channel exists as a ready-made part of the device, before any gate voltage is applied.

E-only MOSFETs and D-E MOSFETs also are easily made as P-channel devices by changing the N regions to P and the P regions to N. The polarity of the applied voltages also is reversed. For a P-channel, E-only MOSFET, the source is connected to a positive supply and the drain to a more negative one. To get a P-channel, E-only device to conduct, the gate is made more negative than the source.

Because the input of a MOS transistor is electrically isolated, almost no input current is required and the input acts as only a capacitive load to the output of a preceding gate.

The extremely high input impedance of MOS-based logic creates severe handling problems because the thin layer of oxide isolating the gate is easily ruptured by static electricity. An attempt to protect the device is made by including protective circuitry on each input such as shown in Figure B-5*g*. Diodes D1 and D2 will conduct if the input voltage goes 0.7 V above V_{SS}. D3 is a sharp-knee zener with a breakdown voltage of about 35 V, and it will conduct if an input voltage goes 35 V positive from V_{DD}. Since the gate oxide breakdown voltage usually is between 100 and 200 V, this circuit should protect the oxide, and it does up to a point. However, the static voltage generated by walking across a carpet on a low-humidity day may be 10 times as much as the protection circuit is capable of shunting. The protection circuitry cannot be improved without degrading the performance of the gate, so care in handling is required.

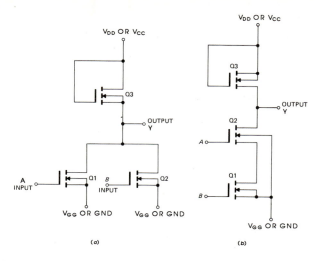

FIGURE B-6 (a) NMOS NOR gate. (b) NMOS NAND gate.

MOS Logic Families

Because gates using MOS transistors are low in power dissipation and require little chip area, MOS logic has been used in many LSI circuits, such as shift registers, memories, and microprocessors. Common variations of MOS logic include PMOS, NMOS, CMOS, DMOS, HMOS, VMOS, and SOSMOS. PMOS is essentially obsolete because of its low speed, so we will start with NMOS.

NMOS

PMOS was the first MOS family because P-channel processing had fewer problems with contamination than N-channel processing did. The higher speed and lower chip area per transistor possible with N-channel devices led manufacturers to develop NMOS devices as soon as processing technology permitted. Most of the present MOS memories and microprocessors employ some variation of N-channel MOS. Figures B-6a and b show the circuit of a basic NMOS NOR gate and NAND gate. For the NOR gate (Figure B-6a), Q3 is an N-channel depletion-mode FET, which functions as a load resistor for Q1 or Q2. If inputs A and B are at a voltage near V_{GG}, the Q1 and Q2 will be off and the output will be pulled high by Q3. If either the A or B input is pulled to a voltage near V_{DD}, Q1 or Q2 will turn on and pull the output low to V_{GG}. This satisfies the expression for a positive logic NOR gate, $Y = \overline{A + B}$. You can analyze the circuit in Figure B-6b to see how it functions as a positive NAND gate.

VMOS, DMOS, and HMOS

VMOS, DMOS, and HMOS are structural variations of NMOS processing that produce circuits with much less propagation delay. Figures B-7a, b, and c compare the structures of standard NMOS, VMOS, and DMOS. VMOS derives its name from its V-shaped structure and the fact that current flows vertically from source to drain, rather than horizontally, as in the standard NMOS. VMOS transistors, because of their

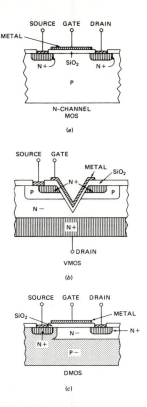

FIGURE B-7 (a) NMOS structure. (b) VMOS structure. (c) DMOS structure. (Electronic Design)

low capacitance and high speed, have shown promise as radio-frequency power amplifiers as well as in LSI logic circuits. DMOS shortens the effective channel length to decrease propagation delay by double-diffused doping in the gate region. In Figure B-7c this is represented by the N-region under the gate, which is not present in the standard NMOS.

HMOS, or high-performance MOS, achieves short propagation delays by scaling down proportionately all the dimensions of the N-channel MOS transistors on the chip. The only difficulty with HMOS seems to be that optimum performance requires a supply voltage less than the de facto standard of +5 V.

CMOS

At about the same time as P-channel technology was being developed for LSI circuits, C, or complementary, MOS was used to produce a logic family with logic functions comparable to those found in TTL, but with the low-power dissipation typical of MOS devices. CMOS circuits use much more chip area than PMOS, but they are faster, have input and output characteristics more compatible with common logic families, and can run on a single supply voltage between 3 and 15 V. Figures B-8a and b show the circuits for basic CMOS NOR and NAND gates. Note the use of complementary N- and P-channel enhancement-mode MOS transistors, which gives the family its name. In Figure B-8a, a basic CMOS NOR gate, if input A is at a high level near V_{DD} (V_{CC}), then the P-channel Q3 will be off and the N-channel Q2 will be on. Q2 then pulls

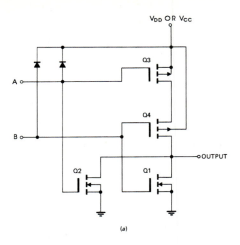

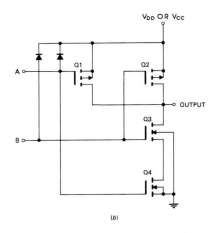

FIGURE B-8 (a) CMOS NOR gate structure. (b) CMOS NAND gate structure.

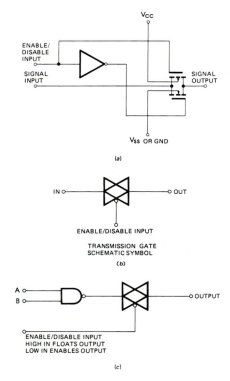

FIGURE B-9 (a) Simple CMOS transmission gate. (b) Schematic symbol. (c) Transmission gate used to produce 3-state output on standard gate.

the output to a low level near ground. A high on the B input will hold off Q4 and turn on Q1, also pulling the output low. If both the A and B inputs are at a low level near ground, Q1 and Q2 will be off and Q3 and Q4 will be on. This pulls the output to a high level near V_{DD}. Since the output is low with A, B, or both A and B high, this circuit performs the NOR function. You can use similar analysis to show how the CMOS circuit in Figure B-8b performs the NAND function.

The high speed of the 74HC series of CMOS devices is achieved by scaling the dimensions of the transistors down. Scaling reduces capacitance and allows faster switching.

CMOS Transmission Gates and Three-State Outputs

Standard CMOS outputs cannot be tied together for the same reason that totem-pole TTL outputs can't be tied together. If one output is trying to go high and the other output low, a direct current path to ground results. This can be fatal to one or both chips. For applications requiring several gate outputs to be tied to a common data bus line, and only one output selected at a time, 3-state output CMOS devices are used. As with TTL, the three output states are high, low, and floating (or "don't care"). Some CMOS devices use a unique circuit called a transmission gate to produce the 3-state output (see Figure B-9a). If both the N- and P-channel transistors in this circuit are turned on, then the output is connected to the input by the low channel resistance of the parallel transistors. The output then follows the input high or low.

If both transistors are off, they appear as open circuits or an open switch. There is then no connection between the input and the output, so the output floats. The inverter in the transmission gate gives the opposite polarity signals necessary to control the N- and P-channel transistors. Figure B-9b shows the schematic symbol for a transmission gate. A low on the enable/disable input turns on the output. Figure B-9c shows the connection of a transmission gate onto a standard CMOS structure to give a 3-state output. Transmission gates are essentially logic-controlled switches. They are useful for multiplexing analog or digital signals onto a common line. An example device is the CD4016 CMOS quad bilateral switch.

APPENDIX C
EXPLANATION OF IEC/IEEE LOGIC SYMBOLS

Written by F.A. Mann, the following explanation of the IEC/IEEE logic symbols is reproduced courtesy of Texas Instruments Inc.

3 Explanation of Logic Symbols†

3.1 Introduction

The International Electrotechnical Commission (IEC) has been developing a very powerful symbolic language that can show the relationship of each input of a digital logic circuit to each output without showing explicitly the internal logic. At the heart of the system is dependency notation, which will be explained in section 3.4.

The system was introduced in the USA in a rudimentary form in IEEE/ANSI Standard Y32.14-1973. Lacking at that time a complete development of dependency notation, it offered little more than a substitution of rectangular shapes for the familiar distinctive shapes for representing the basic functions of AND, OR, negation, etc. This is no longer the case.

Internationally, IEC Technical Committee TC-3 has approved a new document (Publication 617-12) that consolidates the original work started in the mid 1960s and published in 1972 (Publication 117-15) and the amendments and supplements that have followed. Similarly for the USA, IEEE Committee SCC 11.9 has revised the publication IEEE Std 91/ANSI Y32.14. Now numbered simply IEEE Std 91-1984, the IEEE standard contains all of the IEC work that has been approved, and also a small amount of material still under international consideration. Texas Instruments is participating in the work of both organizations, and this document introduces new logic symbols in accordance with the new standards. When changes are made as the standards develop, future editions will take those changes into account.

The following explanation of the new symbolic language is necessarily brief and greatly condensed from what the standards publications now contain. This is not intended to be sufficient for those people who will be developing symbols for new devices. It is primarily intended to make possible the understanding of the symbols used in various data books and the comparison of the symbols with logic diagrams, functional block diagrams, and/or function tables to further help that understanding.

3.2 Symbol Composition

A symbol comprises an outline or a combination of outlines together with one or more qualifying symbols. The shape of the symbol is not significant. As shown in Figure 3-1, general qualifying symbols are used to tell exactly what logical operation is performed by the elements. Table 3-1 shows general qualifying symbols defined in the new standards. Input lines are placed on the left and output lines are placed on the right. When an exception is made to that convention, the direction of signal flow is indicated by an arrow as shown in Figure 3-11.

† Written by F. A. Mann.

3-3

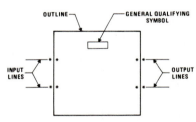

*Possible positions for qualifying symbols relating to inputs and outputs

Figure 3-1. Symbol Composition

All outputs of a single, unsubdivided element always have identical internal logic states determined by the function of the element except when otherwise indicated by an associated qualifying symbol or label inside the element.

The outlines of elements may be abutted or embedded in which case the following conventions apply. There is no logic connection between the elements when the line common to their outlines is in the direction of signal flow. There is at least one logic connection between the elements when the line common to their outlines is perpendicular to the direction of signal flow. The number of logic connections between elements will be clarified by the use of qualifying symbols, and this is discussed further under that topic. If no indications are shown on either side of the common line, it is assumed there is only one connection.

When a circuit has one or more inputs that are common to more than one element of the circuit, the common-control block may be used. This is the only distinctively shaped outline used in the IEC system. Figure 3-2 shows that, unless otherwise qualified by dependency notation, an input to the common-control block is an input to each of the elements below the common-control block.

A common output depending on all elements of the array can be shown as the output of a common-output element. Its distinctive visual feature is the double line at its top. In addition, the common-output element may have other inputs as shown in Figure 3-3. The function of the common-output element must be shown by use of a general qualifying symbol.

539

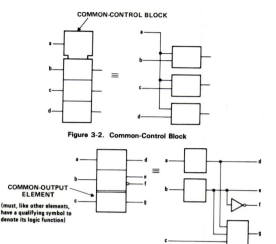

COMMON-CONTROL BLOCK

Figure 3-2. Common-Control Block

COMMON-OUTPUT ELEMENT

(must, like other elements, have a qualifying symbol to denote its logic function)

Figure 3-3. Common-Output Element

3.3 Qualifying Symbols

3.3.1 General Qualifying Symbols

Table 3-1 shows general qualifying symbols defined by IEEE Standard 91. These characters are placed near the top center or the geometric center of a symbol or symbol element to define the basic function of the device represented by the symbol or of the element.

Table 3-1. General Qualifying Symbols

SYMBOL	DESCRIPTION	CMOS EXAMPLE	TTL EXAMPLE
&	AND gate or function.	'HC00	SN7400
≥ 1	OR gate or function. The symbol was chosen to indicate that at least one active input is needed to activate the output.	'HC02	SN7402
= 1	Exclusive OR. One and only one input must be active to activate the output.	'HC86	SN7486
=	Logic identity. All inputs must stand at the same state.	'HC86	SN74180
2k	An even number of inputs must be active.	'HC280	SN74180
2k + 1	An odd number of inputs must be active.	'HC86	SN74ALS86
1	The one input must be active.	'HC04	SN7404
▷ or ◁	A buffer or element with more than usual output capability (symbol is oriented in the direction of signal flow).	'HC240	SN74S436
⎍	Schmitt trigger; element with hysteresis.	'HC132	SN74LS18
X/Y	Coder, code converter (DEC/BCD, BIN/OCT, BIN/7-SEG, etc.).	'HC42	SN74LS347
MUX	Multiplexer/data selector.	'HC151	SN74150
DMUX or DX	Demultiplexer.	'HC138	SN74138
Σ	Adder.	'HC283	SN74LS385
P − Q	Subtracter.	*	SN74LS385
CPG	Look-ahead carry generator.	'HC182	SN74182
π	Multiplier.	*	SN74LS384
COMP	Magnitude comparator.	'HC85	SN74LS682
ALU	Arithmetic logic unit.	'HC181	SN74LS381
⊓	Retriggerable monostable.	'HC123	SN74LS422
1⊓	Nonretriggerable monostable (one-shot).	'HC221	SN74121
G ⊓⊓	Astable element. Showing waveform is optional.	*	SN74LS320
!G ⊓⊓	Synchronously starting astable.	*	SN74LS624
G! ⊓⊓	Astable element that stops with a completed pulse.	*	*
SRGm	Shift register. m = number of bits.	'HC164	SN74LS595
CTRm	Counter. m = number of bits; cycle length = 2^m	'HC590	SN54LS590
CTR DIVm	Counter with cycle length = m.	'HC160	SN74LS668
RCTRm	Asynchronous (ripple-carry) counter; cycle length = 2^m	'HC4020	*

*Not all of the general qualifying symbols have been used in TI's CMOS and TTL data books, but they are included here for the sake of completeness.

Table 3-1. General Qualifying Symbols (Continued)

SYMBOL	DESCRIPTION	CMOS EXAMPLE	TTL EXAMPLE
ROM	Read-only memory.	*	SN74187
RAM	Random-access read/write memory.	'HC189	SN74170
FIFO	First-in, first-out memory.	*	SN74LS222
I = 0	Element powers up cleared to 0 state.	*	SN74AS877
I = 1	Element powers up set to 1 state.	'HC7022	SN74AS877
Φ	Highly complex function; ''gray box'' symbol with limited detail shown under special rules.	*	SM74LS608

*Not all of the general qualifying symbols have been used in TI's CMOS and TTL data books, but they are included here for the sake of completeness.

3.3.2 General Qualifying Symbols for Inputs and Outputs

Qualifying symbols for inputs and outputs are shown in Table 3-2, and many will be familiar to most users, a likely exception being the logic polarity symbol for directly indicating active-low inputs and outputs. The older logic negation indicator means that the external 0 state produces the internal 1 state. The internal 1 state means the active state. Logic negation may be used in pure logic diagrams; in order to tie the external 1 and 0 logic states to the levels H (high) and L (low), a statement of whether positive logic (1 = H, 0 = L) or negative logic (1 = L, 0 = H) is being used is required or must be assumed. Logic polarity indicators eliminate the need for calling out the logic convention and are used in various data books in the symbology for actual devices. The presence of the triangular polarity indicator indicates that the L logic level will produce the internal 1 state (the active state) or that, in the case of an output, the internal 1 state will produce the external L level. Note how the active direction of transition for a dynamic input is indicated in positive logic, negative logic, and with polarity indication.

The internal connections between logic elements abutted together in a symbol may be indicated by the symbols shown in Table 3-2. Each logic connection may be shown by the presence of qualifying symbols at one or both sides of the common line, and, if confusion can arise about the number of connections, use can be made of one of the internal connection symbols.

The internal (virtual) input is an input originating somewhere else in the circuit and is not connected directly to a terminal. The internal (virtual) output is likewise not connected directly to a terminal. The application of internal inputs and outputs requires an understanding of dependency notation, which is explained in section 3.4.

Table 3-2. Qualifying Symbols for Inputs and Outputs

Logic negation at input. External 0 produces internal 1.

Logic negation at output. Internal 1 produces external 0.

Active-low input. Equivalent to ─◁ in positive logic.

Active-low output. Equivalent to ─▷ in positive logic.

Active-low input in the case of right-to-left signal flow.

Active-low output in the case of right-to-left signal flow.

Signal flow from right to left. If not otherwise indicated, signal flow is from left to right.

Bidirectional signal flow.

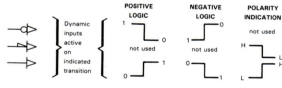

	POSITIVE LOGIC	NEGATIVE LOGIC	POLARITY INDICATION
Dynamic inputs active on indicated transition	not used	not used	not used

Nonlogic connection. A label inside the symbol will usually define the nature of this pin.

Input for analog signals (on a digital symbol) (see Figure 3-14).

Input for digital signals (on an analog symbol) (see Figure 3-14).

Internal connection. 1 state on left produces 1 state on right.

Negated internal connection. 1 state on left produces 0 state on right.

Dynamic internal connection. Transition from 0 to 1 on left produces transitory 1 state on right.

Internal input (virtual input). It always stands at its internal 1 state unless affected by an overriding dependency relationship.

Internal output (virtual output). Its effect on an internal input to which it is connected is indicated by dependency notation.

In an array of elements, if the same general qualifying symbol and the same qualifying symbols associated with inputs and outputs would appear inside each of the elements of the array, then these qualifying symbols are usually shown only in the first element. This is done to reduce clutter and to save time in recognition. Similarly, large identical elements that are subdivided into smaller elements may each be represented by an unsubdivided outline. The SN54HC242 or SN54LS440 symbol illustrates this principle.

3.3.3 Symbols Inside the Outline

Table 3-3 shows some symbols used inside the outline. Note particularly that open-collector (open-drain), open-emitter (open-source), and 3-state outputs have distinctive symbols. An EN input affects all the external outputs of the element in which it is placed, plus the external outputs of any elements shown to be influenced by that element. It has no effect on inputs. When an enable input affects only certain outputs, affects outputs located outside the indicated influence of the element in which the enable input is placed, and/or affects one or more inputs, a form of dependency notation will indicate this (see 3.4.10). The effects of the EN input on the various types of outputs are shown.

It is particularly important to note that a D input is always the data input of a storage element. At its internal 1 state, the D input sets the storage element to its 1 state, and at its internal 0 state, it resets the storage element to its 0 state.

The binary grouping symbol will be explained more fully in section 3.8. Binary-weighted inputs are arranged in order, and the binary weights of the least significant and the most significant lines are indicated by numbers. In this document, weights of input and output lines will usually be represented by powers of two only when the binary grouping symbol is used; otherwise, decimal numbers will be used. The grouped inputs generate an internal number on which a mathematical function can be performed or that can be an identifying number for dependency notation (Figure 3-31). A frequent use is in addresses for memories.

Reversed in direction, the binary grouping symbol can be used with outputs. The concept is analogous to that for the inputs, and the weighted outputs will indicate the internal number assumed to be developed within the circuit.

Other symbols are used inside the outlines in accordance with the IEC/IEEE standards but are not shown here. Generally, these are associated with arithmetic operations and are self-explanatory.

When nonstandardized information is shown inside an outline, it is usually enclosed in square brackets [like these]. The square brackets are omitted when associated with a nonlogic input, which is indicated by an X superimposed on the connection line outside the symbol.

3-9

Explanation of Logic Symbols

Table 3-3. Symbols Inside the Outline

Symbol	Description
	Postponed output (of a pulse-triggered flip-flop). The output changes when input initiating change (e.g., a C input) returns to its initial external state or level. See paragraph 3.5.
	Bi-threshold input (input with hysteresis)
	N-P-N open-collector or similar output that can supply a relatively low-impedance L level when not turned off. Requires external pull-up. Capable of positive-logic wired-AND connection.
	Passive-pull-up output is similar to N-P-N open-collector output but is supplemented with a built-in passive pull-up.
	N-P-N open-emitter or similar output that can supply a relatively low-impedance H level when not turned off. Requires external pull-down. Capable of positive-logic wired-OR connection.
	Passive-pull-down output is similar to N-P-N open-emitter output but is supplemented with a built-in passive pull-down.
	Three-state output.
	Output with more than usual output capability (symbol is oriented in the direction of signal flow).
EN	Enable input When at its internal 1-state, all outputs are enabled. When at its internal 0-state, open-collector and open-emitter outputs are off, three-state outputs are in the high-impedance state, and all other outputs (i.e., totem-poles) are at the internal 0-state.
J, K, R, S	Usual meanings associated with flip-flops (e.g., R = reset to 0, S = reset to 1).
T	Toggle input causes internal state of output to change to its complement.
D	Data input to a storage element equivalent to:
m	Shift right (left) inputs, m = 1, 2, 3, etc. If m = 1, it is usually not shown.
+m	Counting up (down) inputs, m = 1, 2, 3, etc. If m = 1, it is usually not shown.
	Binary grouping. m is highest power of 2.

3-10

Explanation of Logic Symbols

Table 3-3. Symbols Inside the Outline (Continued)

Symbol	Description
CT = 15	The contents-setting input, when active, causes the content of a register to take on the indicated value.
CT = 9	The content output is active if the content of the register is as indicated.
	Input line grouping . . . indicates two or more terminals used to implement a single logic input. e.g., The paired expander inputs of SN7450.
"1"	Fixed-state output always stands at its internal 1 state. For example, see SN74185.

3.4 Dependency Notation

3.4.1 General Explanation

Dependency notation is the powerful tool that sets the IEC symbols apart from previous systems and makes compact, meaningful symbols possible. It provides the means of denoting the relationship between inputs, outputs, or inputs and outputs without actually showing all the elements and interconnections involved. The information provided by dependency notation supplements that provided by the qualifying symbols for an element's function.

In the convention for the dependency notation, use will be made of the terms "affecting" and "affected." In cases where it is not evident which inputs must be considered as being the affecting or the affected ones (e.g., if they stand in an AND relationship), the choice may be made in any convenient way.

So far, eleven types of dependency have been defined, and all of these are used in various TI data books. X dependency is used mainly with CMOS circuits. They are listed below in the order in which they are presented and are summarized in Table 3-4 following 3.4.12.

Section	Dependency Type or Other Subject
3.4.2	G, AND
3.4.3	General Rules for Dependency Notation
3.4.4	V, OR
3.4.5	N, Negate (Exclusive-OR)
3.4.6	Z, Interconnection
3.4.7	X, Transmission
3.4.8	C, Control
3.4.9	S, Set and R, Reset
3.4.10	EN, Enable
3.4.11	M, Mode
3.4.12	A, Address

3-11

Explanation of Logic Symbols

3.4.2 G (AND) Dependency

A common relationship between two signals is to have them ANDed together. This has traditionally been shown by explicitly drawing an AND gate with the signals connected to the inputs of the gate. The 1972 IEC publication and the 1973 IEEE/ANSI standard showed several ways to show this AND relationship using dependency notation. While ten other forms of dependency have since been defined, the ways to invoke AND dependency are now reduced to one.

In Figure 3-4 input b is ANDed with input a, and the complement of b is ANDed with c. The letter G has been chosen to indicate AND relationships and is placed at input b, inside the symbol. A number considered appropriate by the symbol designer (1 has been used here) is placed after the letter G and also at each affected input. Note the bar over the 1 at input c.

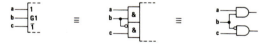

Figure 3-4. G Dependency Between Inputs

In Figure 3-5, output b affects input a with an AND relationship. The lower example shows that it is the internal logic state of b, unaffected by the negation sign, that is ANDed. Figure 3-6 shows input a to be ANDed with a dynamic input b.

Figure 3-5. G Dependency Between Outputs and Inputs

3-12

Explanation of Logic Symbols

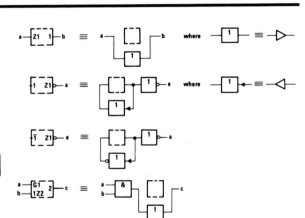

Figure 3-6. G Dependency with a Dynamic Input

The rules for G dependency can be summarized thus:

When a G*m* input or output (*m* is a number) stands at its internal 1 state, all inputs and outputs affected by G*m* stand at their normally defined internal logic states. When the G*m* input or output stands at its 0 state, all inputs and outputs affected by G*m* stand at their internal 0 states.

3.4.3 Conventions for the Application of Dependency Notation in General

The rules for applying dependency relationships in general follow the same pattern as was illustrated for G dependency.

Application of dependency notation is accomplished by:

1) labeling the input (or output) *affecting* other inputs or outputs with the letter symbol indicating the relationship involved (e.g., G for AND) followed by an identifying number, appropriately chosen, and

2) labeling each input or output *affected* by that affecting input (or output) with that same number.

If it is the complement of the internal logic state of the affecting input or output that does the affecting, then a bar is placed over the identifying numbers at the affected inputs or outputs (Figure 3-4).

If two affecting inputs or outputs have the same letter and the same identifying number, they stand in an OR relationship to each other (Figure 3-7).

Figure 3-7. ORed Affecting Inputs

If the affected input or output requires a label to denote its function (e.g., "D"), this label will be *prefixed* by the identifying number of the affecting input (Figure 3-15).

If an input or output is affected by more than one affecting input, the identifying numbers of each of the affecting inputs will appear in the label

3-13

---

of the affected one, separated by commas. The normal reading order of these numbers is the same as the sequence of the affecting relationships (Figure 3-15).

If the labels denoting the functions of affected inputs or outputs must be numbers (e.g., outputs of a coder), the identifying numbers to be associated with both affecting inputs and affected inputs or outputs may be replaced by another character selected to avoid ambiguity, e.g., Greek letters (Figure 3-8).

Figure 3-8. Substitution for Numbers

3.4.4 V (OR) Dependency

The symbol denoting OR dependency is the letter V (Figure 3-9).

When a V*m* input or output stands at its internal 1 state, all inputs and outputs affected by V*m* stand at their internal 1 states. When the V*m* input or output stands at its internal 0 state, all inputs and outputs affected by V*m* stand at their normally defined internal logic states.

Figure 3-9. V (OR) Dependency

3.4.5 N (Negate) (Exclusive-OR) Dependency

The symbol denoting negate dependency is the letter N (Figure 3-10). Each input or output affected by an N*m* input or output stands in an Exclusive-OR relationship with the N*m* input or output.

When an N*m* input or output stands at its internal 1 state, the internal logic state of each input and each output affected by N*m* is the complement of

3-14

---

what it would otherwise be. When an N*m* input or output stands at its internal 0 state, all inputs and outputs affected by N*m* stand at their normally defined internal logic states.

If a = 0, then c = b
If a = 1, then c = $\bar{b}$

Figure 3-10. N (Negate) (Exclusive-OR) Dependency

3.4.6 Z (Interconnection) Dependency

The symbol denoting interconnection dependency is the letter Z.

Interconnection dependency is used to indicate the existence of internal logic connections between inputs, outputs, internal inputs, and/or internal outputs.

The internal logic state of an input or output affected by a Z*m* input or output will be the same as the internal logic state of the Z*m* input or output, unless modified by additional dependency notation (Figure 3-11).

3.4.7 X (Transmission) Dependency

The symbol denoting transmission dependency is the letter X.

Transmission dependency is used to indicate controlled bidirectional connections between affected input/output ports (Figure 3-12).

When an X*m* input or output stands at its internal 1 state, all input-output ports affected by this X*m* input or output are bidirectionally connected together and stand at the same internal logic state or analog signal level. When an X*m* input or output stands at its internal 0 state, the connection associated with this set of dependency notation does not exist.

Although the transmission paths represented by X dependency are inherently bidirectional, use is not always made of this property. This is analogous to a piece of wire, which may be constrained to carry current in only one direction. If this is the case in a particular application, then the directional arrows shown in Figures 3-12, 3-13, and 3-14 are omitted.

3-15

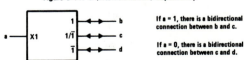

Figure 3-11. Z (Interconnection) Dependency

If a = 1, there is a bidirectional connection between b and c.

If a = 0, there is a bidirectional connection between c and d.

Figure 3-12. X (Transmission) Dependency

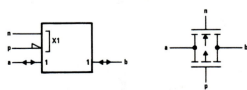

Figure 3-13. CMOS Transmission Gate Symbol and Schematic

3-16

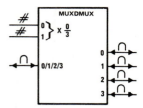

Figure 3-14. Analog Data Selector (Multiplexer/Demultiplexer)

3.4.8 C (Control) Dependency

The symbol denoting control dependency is the letter C.

Control inputs are usually used to enable or disable the data (D, J, K, R, or S) inputs of storage elements. They may take on their internal 1 states (be active) either statically or dynamically. In the latter case, the dynamic input symbol is used as shown in the third example of Figure 3-15.

When a Cm input or output stands at its internal 1 state, the inputs affected by Cm have their normally defined effect on the function of the element; i.e., these inputs are enabled. When a Cm input or output stands at its internal 0 state, the inputs affected by Cm are disabled and have no effect on the function of the element.

Explanation of Logic Symbols 3

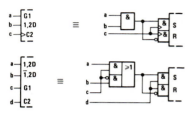

Note AND relationship of a and b

Input c selects which of a or b is stored when d goes low.

Figure 3-15. C (Control) Dependency

3.4.9 S (Set) and R (Reset) Dependencies

The symbol denoting set dependency is the letter S. The symbol denoting reset dependency is the letter R.

Set and reset dependencies are used if it is necessary to specify the effect of the combination R = S = 1 on a bistable element. Case 1 in Figure 3-16 does not use S or R dependency.

When an Sm input is at its internal 1 state, outputs affected by the Sm input will react, regardless of the state of an R input, as they normally would react to the combination S = 1, R = 0. See cases 2, 4, and 5 in Figure 3-16.

When an Rm input is at its internal 1 state, outputs affected by the Rm input will react, regardless of the state of an S input, as they normally would react to the combination S = 0, R = 1. See cases 3, 4, and 5 in Figure 3-16.

When an Sm or Rm input is at its internal 0 state, it has no effect.

Note that the noncomplementary output patterns in cases 4 and 5 are only pseudo stable. The simultaneous return of the inputs to S = R = 0 produces an unforeseeable stable and complementary output pattern.

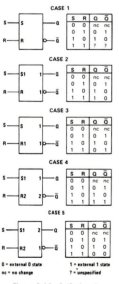

0 = external 0 state   1 = external 1 state
nc = no change   ? = unspecified

Figure 3-16. S (Set) and R (Reset) Dependencies

3.4.10 EN (Enable) Dependency

The symbol denoting enable dependency is the combination of letters EN.

An ENm input has the same effect on outputs as an EN input, see 3.3.3, but it affects only those outputs labeled with the identifying number m. It also affects those inputs labeled with the identifying number m. By contrast, an EN input affects all outputs and no inputs. The effect of an ENm input on an affected input is identical to that of a Cm input (Figure 3-17).

Explanation of Logic Symbols 3

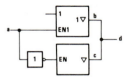

If a = 0, b is disabled and d = c
If a = 1, c is disabled and d = b

Figure 3-17. EN (Enable) Dependency

When an ENm input stands at its internal 1 state, the inputs affected by ENm have their normally defined effect on the function of the element, and the outputs affected by this input stand at their normally defined internal logic states; i.e., these inputs and outputs are enabled.

When an ENm input stands at its internal 0 state, the inputs affected by ENm are disabled and have no effect on the function of the element, and the outputs affected by ENm are also disabled. Open-collector outputs are turned off, three-state outputs stand at their normally defined internal logic states but externally exhibit high impedance, and all other outputs (e.g., totem-pole outputs) stand at their internal 0 states.

3.4.11 M (MODE) Dependency

The symbol denoting mode dependency is the letter M.

Mode dependency is used to indicate that the effects of particular inputs and outputs of an element depend on the mode in which the element is operating.

If an input or output has the same effect in different modes of operation, the identifying numbers of the relevant affecting Mm inputs will appear in the label of that affected input or output between parentheses and separated by solidi (Figure 3-22).

3.4.11.1 M Dependency Affecting Inputs

M dependency affects inputs the same as C dependency. When an Mm input or Mm output stands at its internal 1 state, the inputs affected by this Mm input or Mm output have their normally defined effect on the function of the element; i.e., the inputs are enabled.

When an Mm input or Mm output stands at its internal 0 state, the inputs affected by this Mm input or Mm output have no effect on the function of the element. When an affected input has several sets of labels separated by solidi (e.g., C4/2→/3 +), any set in which the identifying number of the Mm input or Mm output appears has no effect and is to be ignored. This represents disabling of some of the functions of a multifunction input.

The circuit in Figure 3-18 has two inputs, **b** and **c**, that control which one of four modes (0, 1, 2, or 3) will exist at any time. Inputs **d**, **e**, and **f** are D inputs subject to dynamic control (clocking) by the **a** input. The numbers 1 and 2 are in the series chosen to indicate the modes so inputs **e** and **f** are only enabled in mode 1 (for parallel loading), and input **d** is only enabled in mode 2 (for serial loading). Note that input **a** has three functions. It is the clock for entering data. In mode 2, it causes right shifting of data, which means a shift away from the control block. In mode 3, it causes the contents of the register to be incremented by one count.

Note that all operations are synchronous.

In MODE 0 (b = 0, c = 0), the outputs remain at their existing states as none of the inputs has an effect.

In MODE 1 (b = 1, c = 0), parallel loading takes place thru inputs **e** and **f**.

In MODE 2 (b = 0, c = 1), shifting down and serial loading thru input **d** take place.

In MODE 3 (b = c = 1), counting up by increment of 1 per clock pulse takes place.

Figure 3-18. M (Mode) Dependency Affecting Inputs

3.4.11.2 M Dependency Affecting Outputs

When an M*m* input or M*m* output stands at its internal 1 state, the affected outputs stand at their normally defined internal logic states; i.e., the outputs are enabled.

When an M*m* input or M*m* output stands at its internal 0 state, at each affected output any set of labels containing the identifying number of that M*m* input or M*m* output has no effect and is to be ignored. When an output has several different sets of labels separated by solidi (e.g., 2,4/3,5), only those sets in which the identifying number of this M*m* input or M*m* output appears are to be ignored.

Figure 3-19 shows a symbol for a device whose output can behave as either a 3-state output or an open-collector output depending on the signal applied to input **a**. Mode 1 exists when input **a** stands at its internal 1 state, and, in that case, the three-state symbol applies, and the open-element symbol has no effect. When **a** = 0, mode 1 does not exist so the three-state symbol has no effect, and the open-element symbol applies.

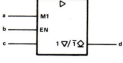

Figure 3-19. Type of Output Determined by Mode

In Figure 3-20, if input **a** stands at its internal 1 state establishing mode 1, output **b** will stand at its internal 1 state only when the content of the register equals 9. Since output **b** is located in the common-control block with no defined function outside of mode 1, the state of this output outside of mode 1 is not defined by the symbol.

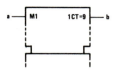

Figure 3-20. An Output of the Common-Control Block

In Figure 3-21, if input **a** stands at its internal 1 state establishing mode 1, output **b** will stand at its internal 1 state only when the content of the register equals 15. If input **a** stands at its internal 0 state, output **b** will stand at its internal 1 state only when the content of the register equals 0.

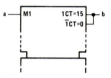

Figure 3-21. Determining an Output's Function

In Figure 3-22, inputs **a** and **b** are binary weighted to generate the numbers 0, 1, 2, or 3. This determines which one of the four modes exists.

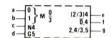

Figure 3-22. Dependent Relationships Affected by Mode

At output **e**, the label set causing negation (if **c** = 1) is effective only in modes 2 and 3. In modes 0 and 1, this output stands at its normally defined state as if it had no labels. At output **f**, the label set has effect when the mode is not 0 so output **e** is negated (if **c** = 1) in modes 1, 2, and 3. In mode 0, the label set has no effect so the output stands at its normally defined state. In this example, $\bar{0}$,4 is equivalent to (1/2/3)4. At output **g**, there are two label sets: the first set, causing negation (if **c** = 1), is effective only in mode 2; the second set, subjecting **g** to AND dependency on **d**, has effect only in mode 3.

Note that in mode 0 none of the dependency relationships has any effect on the outputs, so **e**, **f**, and **g** will all stand at the same state.

3.4.12 A (Address) Dependency

The symbol denoting address dependency is the letter A.

Address dependency provides a clear representation of those elements, particularly memories, that use address control inputs to select specified sections of a multidimensional array. Such a section of a memory array is usually called a word. The purpose of address dependency is to allow a symbolic presentation of the entire array. An input of the array shown at a particular element of this general section is common to the corresponding elements of all selected sections of the array. An output of the array shown at a particular element of this general section is the result of the OR function of the outputs of the corresponding elements of selected sections.

Inputs that are not affected by any affecting address input have their normally defined effect on all sections of the array, whereas inputs affected by an address input have their normally defined effect only on the section selected by that address input.

An affecting address input is labeled with the letter A followed by an identifying number that corresponds with the address of the particular section of the array selected by this input. Within the general section presented by the symbol, inputs and outputs affected by an A*m* input are labeled with the letter A, which stands for the identifying numbers, i.e., the addresses, of the particular sections.

Figure 3-23 shows a 3-word by 2-bit memory having a separate address line for each word and uses EN dependency to explain the operation. To select word 1, input **a** is taken to its 1 state, which establishes mode 1. Data can now be clocked into the inputs marked "1,4D." Unless words 2 and 3 are also selected, data cannot be clocked in at the inputs marked "2,4D" and "3,4D." The outputs will be the OR functions of the selected outputs; i.e., only those enabled by the active EN functions.

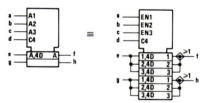

Figure 3-23. A (Address) Dependency

The identifying numbers of affecting address inputs correspond with the addresses of the sections selected by these inputs. They need not necessarily differ from those of other affecting dependency inputs (e.g., G, V, N, . . .), because, in the general section presented by the symbol, they are replaced by the letter A.

If there are several sets of affecting A*m* inputs for the purpose of independent and possibly simultaneous access to sections of the array, then the letter A is modified to 1A, 2A, ... Since they have access to the same sections of the array, these sets of A inputs may have the same identifying numbers. The symbols for 'HC170 or SN74LS170 make use of this.

Figure 3-24 is another illustration of the concept.

3.5 Bistable Elements

The dynamic input symbol, the postponed output symbol, and dependency notation provide the tools to differentiate four main types of bistable elements and make synchronous and asynchronous inputs easily recognizable (Figure 3-5). The first column shows the essential distinguishing features; the other columns show examples.

544 APPENDIX C

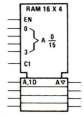

Figure 3-24. Array of 16 Sections of Four Transparent Latches with 3-State Outputs Comprising a 16-Word × 4-Bit Random-Access Memory

Table 3-4. Summary of Dependency Notation

TYPE OF DEPENDENCY	LETTER SYMBOL*	AFFECTING INPUT AT ITS 1-STATE	AFFECTING INPUT AT ITS 0-STATE
Address	A	Permits action (address selected)	Prevents action (address not selected)
Control	C	Permits action	Prevents action
Enable	EN	Permits action	Prevents action of inputs ◊outputs off ▽outputs at external high impedance, no change in internal logic state Other outputs at internal 0 state
AND	G	Permits action	Imposes 0 state
Mode	M	Permits action (mode selected)	Prevents action (mode not selected)
Negate (Ex-OR)	N	Complements state	No effect
Reset	R	Affected output reacts as it would to S = 0, R = 1	No effect
Set	S	Affected output reacts as it would to S = 1, R = 0	No effect
OR	V	Imposes 1 state	Permits action
Transmission	X	Bidirectional connection exists	Bidirectional connection does not exist
Interconnection	Z	Imposes 1 state	Imposes 0 state

* These letter symbols appear at the AFFECTING input (or output) and are followed by a number. Each input (or output) AFFECTED by that input is labeled with that same number. When the labels EN, R, and S appear at inputs without the following numbers, the descriptions above do not apply. The action of these inputs is described under "Symbols Inside the Outline," see 3.3.3.

Figure 3-25. Four Types of Bistable Circuits

TRANSPARENT LATCHES   1/2 SN74HC75

EDGE-TRIGGERED   1/2 SN74HC74   1/2 SN74HC107

PULSE-TRIGGERED   SN74L71   1/2 SN74107

DATA-LOCKOUT   SN74110   1/2 SN74111

Transparent latches have a level-operated control input. The D input is active as long as the C input is at its internal 1 state. The outputs respond immediately. Edge-triggered elements accept data from D, J, K, R, or S inputs on the active transition of C. Pulse-triggered elements require the setup of data before the start of the control pulse; the C input is considered static since the data must be maintained as long as C is at its 1 state. The output is postponed until C returns to its 0 state. The data-lockout element is similar to the pulse-triggered version except that the C input is considered dynamic in that, shortly after C goes through its active transition, the data inputs are disabled, and data does not have to be held. However, the output is still postponed until the C input returns to its initial external level.

Notice that synchronous inputs can be readily recognized by their dependency labels (1D, 1J, 1K, 1S, 1R) compared to the asynchronous inputs (S, R), which are not dependent on the C inputs.

3.6 Coders

The general symbol for a coder or code converter is shown in Figure 3-26. X and Y may be replaced by appropriate indications of the code used to represent the information at the inputs and at the outputs, respectively.

Figure 3-26. Coder General Symbol

Indication of code conversion is based on the following rule:

Depending on the input code, the internal logic states of the inputs determine an internal value. This value is reproduced by the internal logic states of the outputs, depending on the output code.

The indication of the relationships between the internal logic states of the inputs and the internal value is accomplished by:

1) labeling the inputs with numbers. In this case, the internal value equals the sum of the weights associated with those inputs that stand at their internal 1-state, or by

2) replacing X by an appropriate indication of the input code and labeling the inputs with characters that refer to this code.

The relationships between the internal value and the internal logic states of the outputs are indicated by:

1) labeling each output with a list of numbers representing those internal values that lead to the internal 1-state of that output. These numbers shall be separated by solidi as in Figure 3-27. This labeling may also be applied when Y is replaced by a letter denoting a type of dependency

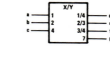

TRUTH TABLE

INPUTS			OUTPUTS			
c	b	a	g	f	e	d
0	0	0	0	0	0	0
0	0	1	0	0	0	1
0	1	0	0	0	1	0
0	1	1	0	1	1	0
1	0	0	0	1	0	1
1	0	1	0	0	0	0
1	1	0	0	0	0	0
1	1	1	1	0	0	0

Figure 3-27. An X/Y Code Converter

(see section 3.7). If a continuous range of internal values produces the internal 1 state of an output, this can be indicated by two numbers that are inclusively the beginning and the end of the range, with these two numbers separated by three dots (e.g., 4 . . . 9 = 4/5/6/7/8/9) or by

2) replacing Y by an appropriate indiction of the output code and labeling the outputs with characters that refer to this code as in Figure 3-28.

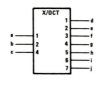

TRUTH TABLE

INPUTS			OUTPUTS						
c	b	a	j	i	h	g	f	e	d
0	0	0	0	0	0	0	0	0	0
0	0	1	0	0	0	0	0	0	1
0	1	0	0	0	0	0	0	1	0
0	1	1	0	0	0	0	1	0	0
1	0	0	0	0	0	1	0	0	0
1	0	1	0	0	1	0	0	0	0
1	1	0	0	1	0	0	0	0	0
1	1	1	1	0	0	0	0	0	0

Figure 3-28. An X/Octal Code Converter

Alternatively, the general symbol may be used together with an appropriate reference to a table in which the relationship between the inputs and outputs is indicated. This is a recommended way to symbolize a PROM after it has been programmed.

3.7 Use of a Coder to Produce Affecting Inputs

It often occurs that a set of affecting inputs for dependency notation is produced by decoding the signals on certain inputs to an element. In such a case, use can be made of the symbol for a coder as an embedded symbol (Figure 3-29).

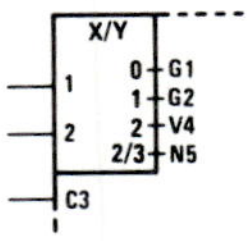

Figure 3-29. Producing Various Types of Dependencies

If all affecting inputs produced by a coder are of the same type as their identifying numbers shown at the outputs of the coder, Y (in the qualifying symbol X/Y) may be replaced by the letter denoting the type of dependency. The indications of the affecting inputs should then be omitted (Figure 3-30).

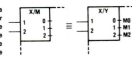

Figure 3-30. Producing One Type of Dependency

3.8 Use of Binary Grouping to Produce Affecting Inputs

If all affecting inputs produced by a coder are of the same type and have consecutive identifying numbers not necessarily corresponding with the numbers that would have been shown at the outputs of the coder, use can be made of the binary grouping symbol. k external lines effectively generate 2^k internal inputs. The bracket is followed by the letter denoting the type of dependency followed by m1/m2. The m1 is to be replaced by the smallest identifying number and the m2 by the largest one, as shown in Figure 3-31.

3.9 Sequence of Input Labels

If an input having a single functional effect is affected by other inputs, the qualifying symbol (if there is any) for that functional effect is preceded by the labels corresponding to the affecting inputs. The left-to-right order of these preceding labels is the order in which the effects or modifications must be applied. The affected input has no functional effect on the element if the logic state of any one of the affecting inputs, considered separately, would cause the affected input to have no effect, regardless of the logic states of other affecting inputs.

If an input has several different functional effects or has several different sets of affecting inputs, depending on the mode of action, the input may be shown as often as required. However, there are cases in which this method of

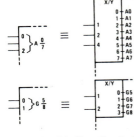

Figure 3-31. Use of the Binary Grouping Symbol

presentation is not advantageous. In those cases, the input may be shown once with the different sets of labels separated by solidi (Figure 3-32). No meaning is attached to the order of these sets of labels. If one of the functional effects of an input is that of an unlabeled input to the element, a solidus will precede the first set of labels shown.

If all inputs of a combinational element are disabled (caused to have no effect on the function of the element), the internal logic states of the outputs of the element are not specified by the symbol. If all inputs of a sequential element are disabled, the content of this element is not changed, and the outputs remain at their existing internal logic states.

Labels may be factored using algebraic techniques (Figure 3-33).

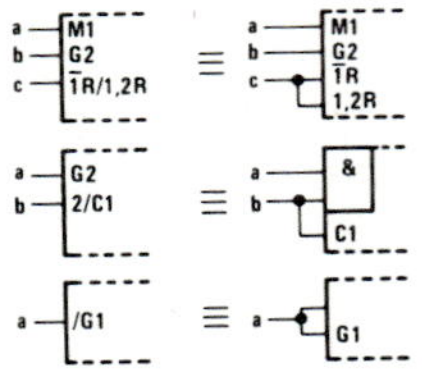

Figure 3-32. Input Labels

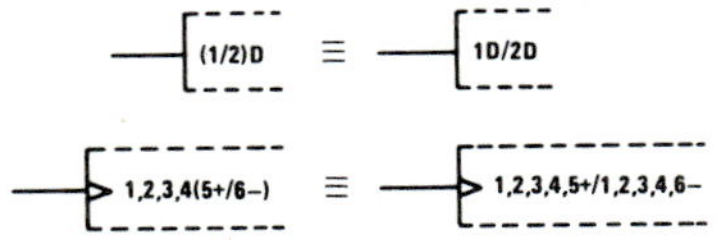

Figure 3-33. Factoring Input Labels

3.10 Sequence of Output Labels

If an output has a number of different labels, regardless of whether they are identifying numbers of affecting inputs or outputs or not, these labels are shown in the following order:

1) If the postponed output symbol has to be shown, this comes first, if necessary preceded by the indications of the inputs to which it must be applied
2) Followed by the labels indicating modifications of the internal logic state of the output, such that the left-to-right order of these labels corresponds with the order in which their effects must be applied
3) Followed by the label indicating the effect of the output on inputs and other outputs of the element.

Symbols for open-circuit or 3-state outputs, where applicable, are placed just inside the outside boundary of the symbol adjacent to the output line (Figure 3-34).

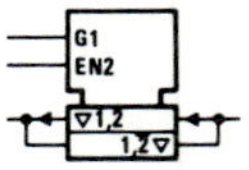

Figure 3-34. Placement of 3-State Symbols

If an output needs several different sets of labels that represent alternative functions (e.g., depending on the mode of action), these sets may be shown on different output lines that must be connected outside the outline. However, there are cases in which this method of presentation is not advantageous. In those cases, the output may be shown once with the different sets of labels separated by solidi (Figure 3-35).

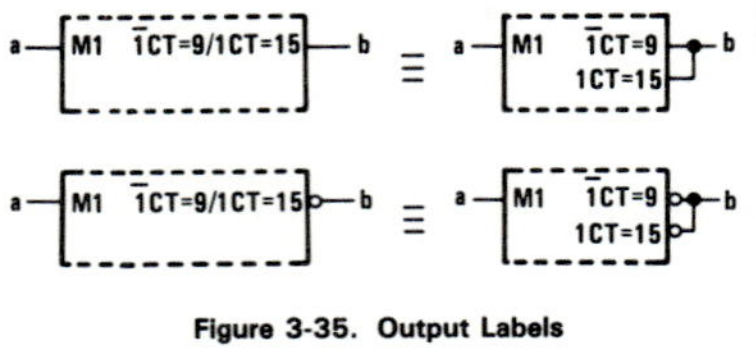

Figure 3-35. Output Labels

Two adjacent identifying numbers of affecting inputs in a set of labels that are not already separated by a nonnumeric character should be separated by a comma.

If a set of labels of an output not containing a solidus contains the identifying number of an affecting Mm input standing at its internal 0 state, this set of labels has no effect on that output.

Labels may be factored using algebraic techniques (Figure 3-36).

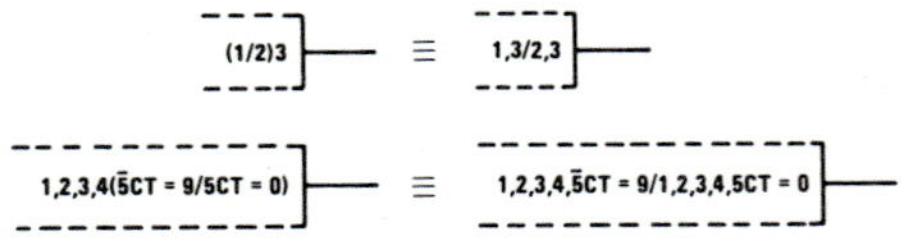

Figure 3-36. Factoring Output Labels

If you have questions on this Explanation of Logic Symbols, please contact:

Texas Instruments Incorporated
F.A. Mann, MS 49
P.O. Box 655012
Dallas, Texas 75265

Telephone (214) 995-2659

IEEE Standards may be purchased from:

Institute of Electrical and Electronics Engineers, Inc.
IEEE Standards Office
345 East 47th Street
New York, N.Y. 10017

International Electrotechnical Commission (IEC) publications may be purchased from:

American National Standards Institute, Inc.
1430 Broadway
New York, N.Y. 10018

INDEX

NUMBERED ENTRIES